Handbuch der mikroskopischen Anatomie des Menschen

Begründet von Wilhelm von Möllendorff
Fortgeführt von Wolfgang Bargmann
Herausgegeben von A. Oksche und L. Vollrath

VI/7

Lutz Vollrath

The Pineal Organ

With 190 Figures

Springer-Verlag Berlin Heidelberg New York 1981

Professor Dr. L. Vollrath
Anatomisches Institut der Universität, Saarstraße 19–21, D-6500 Mainz

Professor Dr. A. Oksche
Zentrum für Anatomie und Cytobiologie der Justus Liebig-Universität, Aulweg 123, D-6300 Giessen

Handbuch der mikroskopischen Anatomie des Menschen
Band VI: Blutgefäß- und Lymphgefäßapparat · Innersekretorische Drüsen
7. Teil: The Pineal Organ

ISBN-13: 978-3-642-81527-0 e-ISBN-13: 978-3-642-81525-6
DOI: 10.1007/978-3-642-81525-6

Library of Congress Cataloging in Publication Data Main entry under title: Handbuch der mikroskopischen Anatomie des Menschen, bearb. Later volumes: Begründet von Wilhelm v. Möllendorff; fortgeführt von Wolfgang Bargmann. Includes bibliographies. CONTENTS. – 1. Bd. Die lebendige Masse. T. 1- – 2. Bd. Die Gewebe. T. 1- – 3. Bd. Haut und Sinnesorgane. T. 1- – (etc.) 1. Histology. I. Möllendorff, Wilhelm Hermann Wichard von, 1887– ed. II. Bargmann, Wolfgang, 1906– ed. QM551.H15 611'.018 55-37658
ISBN-13: 978-3-642-81527-0

Softcover reprint of the hardcover 1st edition 1981

2122/3120-543210

To the memory of

Professor
WOLFGANG BARGMANN

and

Professor
Sir FRANCIS KNOWLES, Bart.

Preface

When I was asked by the late Professor Wolfgang Bargmann in 1973 to write the continuation of his review article on the epiphysis cerebri published in the present Handbuch Series in 1943 I felt greatly flattered and readily agreed, not knowing that it is virtually impossible for a single person to cover comprehensively the vast field of pineal research carried out over the last 35 years. Although I have done my best to include as much of the relevant work as possible (up to July 1979), I am constantly haunted by the references I have missed. I sincerely apologize for these omissions: first, to the authors of these papers, who, in addition to working on one of the most stony grounds of biological research, now may experience the additional frustration of not being quoted; secondly, to my fellow pinealogists or pinealogists-to-be, who may carry out painstaking experiments only to find that they have been done before, or who may be led astray because of lack of information.

Nevertheless, I do hope that this monograph will fulfil its aim, namely to show how pineal research has progressed over the last 35 years and where it stands today. It is hoped that it will aid in planning relevant experiments and that it shows which approaches to the pineal problem would be better avoided. Those who expect to find the final answer to the pineal problem in this monograph will be disappointed like so many others in the past who have attempted to unravel the mystery of the pineal body.

Finally, I wish to express my sincere thanks to those who have helped me: Foremost, to Frau Iris Liebthal, my secretary, who for years and with never-ending patience has deciphered my cryptic, handwritten notes and brought them into a legible form so that they could be worked on again. In the final stages of preparing this monograph, Frau Kristina Bartelt shared a fair deal of the secretarial work involved. Many colleagues have contributed by putting illustrations at my disposal. Their names are mentioned in the figure legends. Frau Sabina Soda is thanked for photographic work and Herr Gerhard Schlich for contributing a number of line drawings. I am also greatly indebted to the publishers for the expert processing of this monograph. Lastly, I apologize to all my younger colleagues and to students in my laboratory to whom, because of the writing of this review, I was unable to give as much attention and support as they deserved.

Mainz, January 1981 LUTZ VOLLRATH

Contents

A. General Introduction

The epiphysis cerebri, or pineal organ, which has been described as "perhaps the last great mystery in the physiology of mammalian organs" (WURTMAN and AXELROD, 1965), has ceased to be an enigma. This is not to say that its precise role in the economy of the body has been clearly established. It would appear, however, that over the last 10–20 years sufficient structural and functional data have been obtained to provide a firm basis from which it should not be too difficult to unravel the precise functional significance of this epithalamic appendage.

Two findings in particular have proved to be most powerful incentives for further research into the epiphysis cerbri. The first was the discovery that the organ synthesizes a hormone-like indoleamine termed melatonin. An endocrine nature of the epiphysis cerebri had long been suspected, however. The actual proof for this assumption was difficult to obtain; LERNER et al. (1958, 1960) required 100 kg or 200,000 cattle pineals to isolate the active principle. The subsequent demonstration that melatonin exerted hormonal effects on a number of organs (WURTMAN et al., 1965a) firmly established the endocrine nature of the pineal. The pineal body soon took its place in the rapidly growing field of endocrinology – especially neuroendocrinology – as an important organ for consideration: many review articles (see Appendix) have been written and a number of conferences and workshops held (SIMONNET and THIÉBLOT, 1963; KAPPERS and SCHADE, 1965; REITER, 1970a; WOLSTENHOLME and KNIGHT, 1971; ALTSCHULE, 1975a; NIR et al., 1978a; KAPPERS and PEVET, 1979), illustrating the ever-increasing interest in this organ.

The second finding, perhaps of even greater biological significance than the discovery of yet another endocrine gland, was the observation that the pineal, unlike other endocrine organs, is more or less directly influenced by environmental factors, among which photic stimuli appear to play a leading role (MILINE, 1957a; FISKE et al., 1960, 1962; WURTMAN et al., 1961; QUAY, 1961a; AXELROD and WURTMAN, 1968). This observation beautifully corroborated Bargmann's prophetic statement, published in the present series (BARGMANN, 1943), that the "pineal cells of the mammalian epiphysis cerebri, no longer under the direct influence of light rays, might receive impulses for their metabolism via tracts of the central nervous system, linked to the optic system", which Bargmann modestly called a "working hypothesis which, for the understanding of epiphyseal function, might perhaps not be unfruitful".

At the time of writing it is probably still best to characterize the epiphysis cerebri as "a neuroendocrine transducer" (WURTMAN and AXELROD, 1965; WURTMAN and ANTON-TAY, 1969; COLLU and FRASCHINI, 1972) similar to the anterior pituitary, the posterior pituitary and the adrenal medulla; in all these organs a neuronal signal is transformed into a hormonal message. The usage of the concise and very instructive term transducer should not blind us to the fact that the epiphysis cerebri is a very complex organ both structurally and functionally.

Despite the fact that in the mammalian epiphysis cerebri both the formation of melatonin and its regulation have been elucidated to an extent (p. 5) that can be termed admirable, it is probably wise not to overemphasize and to generalize the results obtained in a relatively small number of animal species. Admittedly, the melatonin story is fairly coherent and seemingly convincing. Yet it may represent just one facet of pineal function. In the melatonin concept of pineal function merely two of the different structural components of the epiphysis cerebri are required, namely the noradrenergic sympathetic nerve fibres, which provide the neuronal input, and the intrinsic cells of the pineal gland, the pinealocytes, which function as effectors. Glial elements, which in the pineal may be involved in the formation of substances other than melatonin, have received relatively little attention. The same applies to central nerve fibres which by way of the habenular and posterior commissures may provide, at least quantitatively, another major nervous input to the epiphysis cerebri. Even the main structural component of the epiphysis cerebri, the pinealocyte, is far from clear. Is it a typical endocrine cell, or a special type of glial or nerve cell? What are the morphological correlates of its various alleged secretory products? Why does the pinealocyte possess a long cytoplasmic process, which an ordinary endocrine cell does not require?

The reasons for not being in a position to propose a unifying concept of pineal function are manifold. To begin with, the epiphysis cerebri shows great structural diversity, even at a gross anatomical level, and it would appear, especially in the most widely studied rodents, that parts of the epiphyseal complex are prone to regression. Other findings point to the possibility that certain parts of the pineal complex may differ in function. There is clear evidence that the epiphysis cerebri in lower vertebrates differs from that in higher vertebrates: while in petromyzontids, fish, amphibians and lacertilians the epiphyseal complex contains photoreceptor cells similar to those in the lateral eyes and is capable of responding to light directly, pineals in birds and mammals lack typical photoreceptors and are only indirectly light sensitive.

With respect to the input to the mammalian pineal, the effects of light and darkness mediated by the sympathetic nerve fibres have been most thoroughly investigated. However, as will be shown below, other factors that have received very little attention may be equally or more important than photic stimuli. Recently it has been shown that the mammalian pineal organ possesses receptors for sex steroids (p. 283) and that these (p. 284) and other hormones, e.g. those secreted by the adrenal medulla during stress (p. 308), influence pineal function. These observations led Ellis (1976a, b) to propose that the epiphysis cerebri is an endocrine–endocrine transducer.

Further, it has been shown that administration of melatonin may exert different effects, depending on the amount administered, the time of the day and the mode of administration (p. 283). Also melatonin is capable of directly affecting metabolic processes in the pineal (p. 264). Often the question has been posed whether melatonin represents a true hormone that exerts its effects outside the pineal or whether it is merely a special type of transmitter present in the pineal and in other organs, e.g. the retina, Harderian glands and enterochromaffin cells (p. 237). So far this question has remained unanswered, although perhaps the majority of investigators tend to regard melatonin as a pineal hormone, or at least as one of the different active substances in the pineal that may possess hormonal properties. At the International Pineal Symposium in Jerusalem (1977), 5-methoxytryptophol was strongly advocated as representing another pineal indoleamine that exerts hormonal effects (WURTMAN and OZAKI, 1978).

Because of the presence of melatonin and its precursor, serotonin (GIARMAN and DAY, 1958/59), the indoleamines have long been the main topic of pineal research, and the epiphysis has even been termed the "centre for the regulation of biogenic amines" (ARVY, 1966). Yet, there are other substances in the pineal, such as arginine vasotocin (PAVEL, 1963–1975; CHEESMAN, 1970), as well as substances showing antidiuretic activity (KÖNIG and MEYER, 1967, 1971; KÖNIG et al., 1970), which bring the epiphysis cerebri close to the classic neurosecretory system, hypothalamic releasing hormones (WHITE et al., 1974), renin-like activity (HAULICA et al., 1975) and a large number of only partly defined peptides (p. 242). The presence of all these substances indicates that a substantial amount of research remains to be done before one can expect to achieve a clear picture of the pineal.

Although a large amount of data have accumulated with respect to possible target organs of the pineal, we are still far from knowing what it performs under physiological conditions, except in a few species. One of the reasons for this difficulty is that in commonly used experimental animals the pineal is not essential for survival, at least under laboratory conditions. In most species pinealectomy is not followed by any dramatic changes. Hence, it would appear that the influences which the pineal exerts on other organs are either relatively subtle or not easily detectable. For a relatively long period of time, peripherally located endocrine organs have been regarded as primary sites of pineal action, although the results were not very striking. More recently, the advent of new and sophisticated techniques has made possible the study of possible functional relationships between the pineal and the brain. These studies have made clear that of the various cerebral structures the hypothalamus may be the most important target organ of the pineal; this would explain why studies aimed at demonstrating a peripheral action of the epiphysis cerebri have failed to produce the expected results. When judged by the amount of work done and the striking results obtained in hamsters in which the epiphysis cerebri is responsible for gonadal atrophy in the winter (REITER, 1973c, d, 1973/74), it would appear that it is mainly the hypothalamus–anterior pituitary–gonadal axis that is influenced by the pineal. However, in Sect. B. XI of this monograph it will be shown that the epiphysis cerebri appears to influence so many different biological parameters that one wonders in which field its true significance is to be sought.

Finally, a remark by RALPH (1975) should be heeded: "It would be unfortunate if endocrinologists were to accept prematurely the conclusion that the pineals of all vertebrates work in the way we now presume they do in the very small sample of mammals that have been only moderately well studied".

The present survey is meant as a continuation of the monograph on the epiphysis cerebri by BARGMANN (1943) published in the present series. Its aim is to show how pineal research has progressed over the last 35 years and where it stands today. It is hoped that it will aid in planning relevant experiments and that it shows which approaches to the pineal problem would be better avoided. The main theme is the morphology of the epiphysis cerebri. Since morphological studies alone give only limited insight into the complexity of the epiphysis cerebri, biochemical, pharmacological and physiological data have been incorporated into the monograph. It is hoped that by doing so both the morphological and the non-morphological results will be seen in the right perspective. As very little is known about the pineal organ in humans, great emphasis has to be placed on results obtained in animal species.

For a general survey, the excellent book on the pineal by WURTMAN, AXELROD and KELLY (WURTMAN et al., 1968a) should be consulted. Chemical aspects have been most thoroughly discussed in the book by QUAY (1974a). Recently, a series of reviews on the pineal has been published, the first two volumes being entitled *The Pineal 1976* (RELKIN, 1976) and *The Pineal 1977* (REITER, 1977). In addition, there is a recent book on the pineal by BARDASANO RUBIO (1978). The interrelationships between the pineal and the gonads are dealt with in the books by REITER (1978) and MESS et al. (1978). Readers familiar with Polish may consult the booklet by KARASEK (1976a).

Regarding the older literature, the reader is referred to the publications by GLADSTONE and WAKELEY (1940), BARGMANN (1943), KITAY and ALTSCHULE (1954a) and THIÉBLOT and LEBARS (1955).

B. The Mammalian Epiphysis Cerebri

I. Basic Biochemistry, Pharmacology and Physiology

It is evident from perusal of the vast amount of literature concerning the epiphysis published over the last 30 years that mainly biochemical and pharmacological studies have been responsible for putting pineal research on a firm basis within a relatively short time. Hence, it is appropriate to mention some biochemical and pharmacological aspects before embarking on a detailed description of pineal morphology.

That the mammalian epiphysis cerebri may occupy a unique position among cerebral structures was first hinted at by observations showing that the uptake of radioactive phosphorus by the epiphysis cerebri was considerably higher than in any other part of the brain, including the pituitary (BORELL and ÖRSTRÖM, 1945, 1947; REISS et al., 1949a, b). Corroborative evidence was obtained by WISLOCKI and DEMPSEY (1948), who found that parenchymal cells of the pineal were actively engaged in nucleoprotein synthesis. The breakthrough in pineal research came in the late 1950s, when LERNER et al. (1958) isolated, from bovine pineal glands, the active principle that causes a lightening of amphibian skin described by MCCORD and ALLEN (1917) 40 years previously. The active principle was termed melatonin because its effect was opposite to the action of the melanocyte stimulating hormone (MSH) produced by the pars intermedia of the pituitary. Whereas MSH leads to a dispersion of melanin granules in the melanophores, melatonin causes pigment aggregation (LERNER and CASE, 1959). However, mammalian melanocytes do not appear to respond to melatonin (SNELL, 1965).

Chemically melatonin is a relatively simple substance, i.e. 5-methoxy-*N*-acetyltryptamine (Fig. 1) (LERNER et al., 1959a). Serotonin (5-hydroxytryptamine) is one of its precursors. According to WURTMAN and ANTON-TAY (1969), the biosynthesis of all pineal indoles (Fig. 2) is initiated by the uptake of circulating tryptophan into pinealocytes, the uptake apparently being unaffected by a high serotonin content of the pineal body (WURTMAN and SHEIN, 1975).

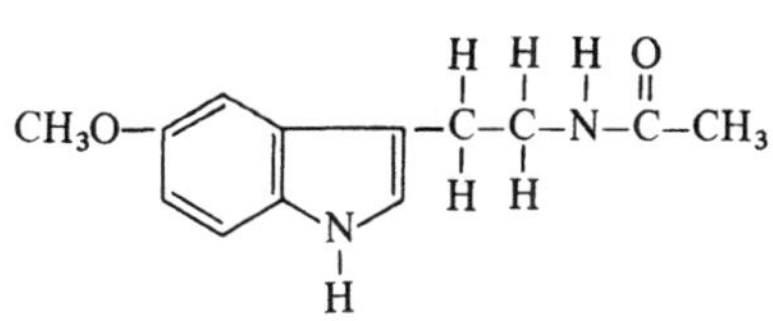

Fig. 1. Chemical structure of melatonin, the most widely studied pineal substance

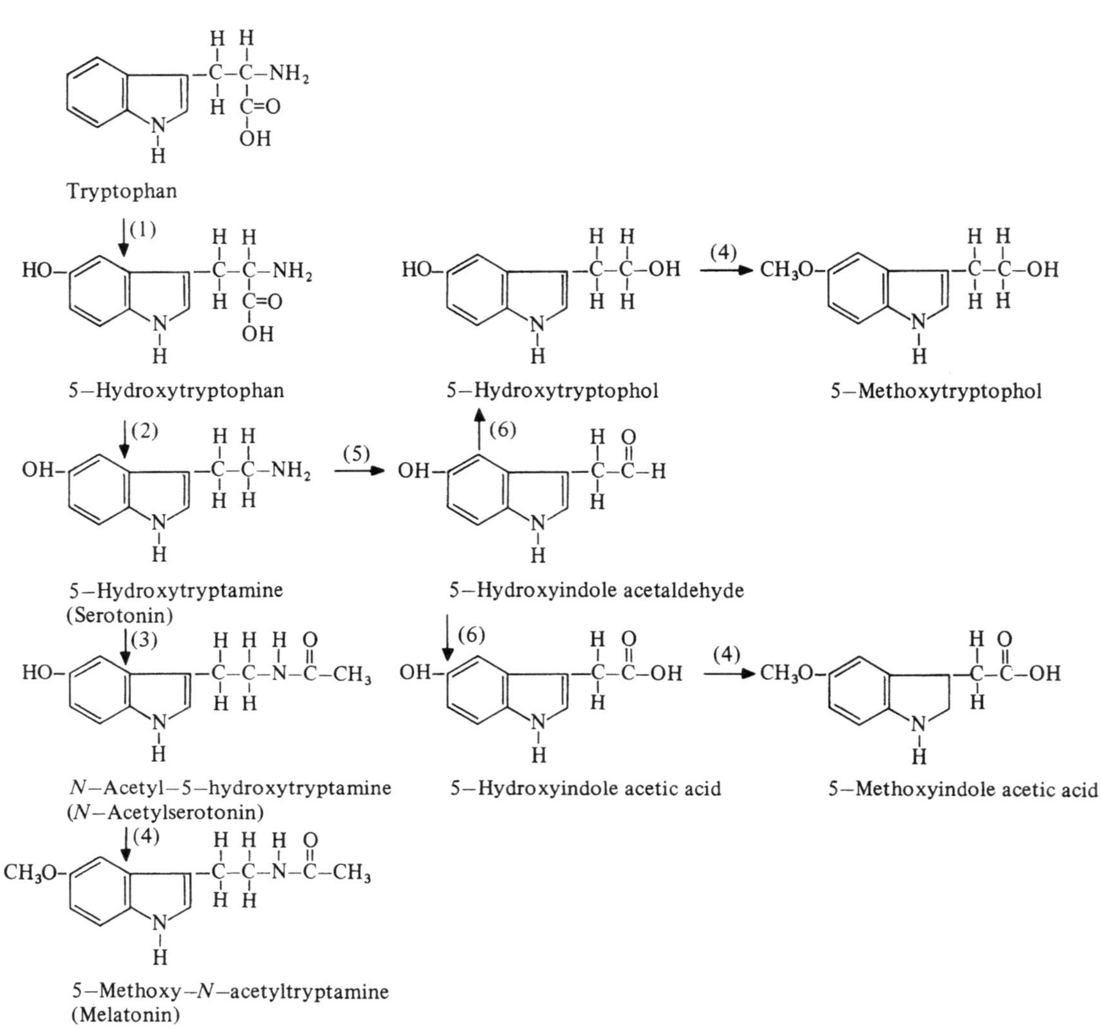

Fig. 2. Biosynthesis of melatonin and other pineal indoleamines. The numbers in brackets refer to enzymes which catalyse the reactions indicated: (1) tryptophan hydroxylase, (2) aromatic L-amino acid decarboxylase, (3) *N*-acetyltransferase (NAT), (4) hydroxyindole-*O*-methyltransferase (HIOMT), (5) monoamine oxidase (MAO), (6) aldehyde dehydrogenase

A considerable fraction of the amino acid is then utilized for the synthesis of epiphyseal proteins (WURTMAN et al., 1969). That part which is destined to become serotonin, or one of its derivatives, is initially hydroxylated at the 5-position through the action of tryptophan hydroxylase (LOVENBERG et al., 1967; JEQUIER et al., 1969; HORI, 1975), that in bovine pineals has been shown to reside in the mitochondrial fraction (HORI et al., 1976). The resulting 5-hydroxytryptophan is rapidly decarboxylated by aromatic L-amino acid decarboxylase to form the biogenic amine 5-hydroxytryptamine (SNYDER et al., 1965a). Under the influence of serotonin-*N*-acetyltransferase (NAT) (WEISSBACH et al., 1960, 1961), 5-hydroxytryptamine is converted to *N*-acetylserotonin. This com-

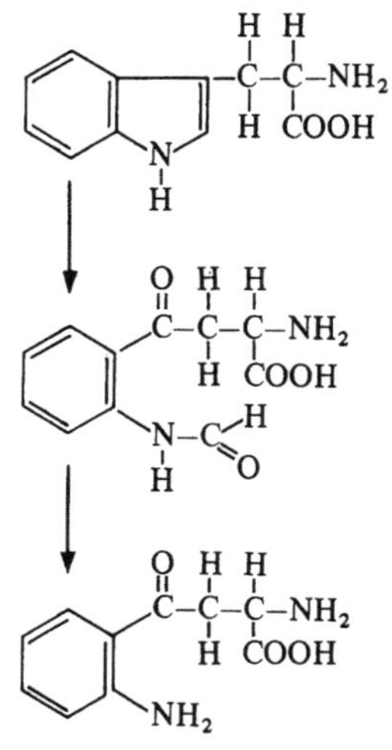

Fig. 3. Alternative metabolic pathway of tryptophan, leading to the formation of kynurenine

pound is then O-methylated through the action of hydroxyindole-*O*-methyltransferase (HIOMT) to form melatonin (AXELROD and WEISSBACH, 1960, 1961), the required metabolic intermediate *S*-adenosylmethionine being present in the pineal in high levels (BALDESSARINI and KOPIN, 1966). The enzyme involved in the formation of this compound has been demonstrated by GUCHHAIT and GRAU (1978). NAT and HIOMT are particularly important components since their activities represent some of the most frequently studied parameters of pineal function. An alternative metabolic pathway of tryptophan in the pineal is its conversion into kynurenine, due to the action of indoleamine 2,3-dioxygenase (FUJIWARA et al., 1978) (Fig. 3).

It should be noted that the presence of NAT and HIOMT is not restricted to the pineal (see p. 237). NAT, a sensitive assay of which has been described by DEGUCHI and AXELROD (1972c), is present also in the brain and liver. While PAUL et al. (1974) are of opinion that the NAT of the brain may represent a different enzyme, DEGUCHI (1975b) claimed that the enzyme has similar properties in the pineal and liver, but that its regulation differs. Bovine pineal HIOMT has been purified and characterized by means of cell fractionation, ammonium sulphate treatment, Sephadex G-200 gel filtration and anion exchange chromatography and shown to be a polymer consisting of units with molecular weights of 21,800; 83,000; 100,000; 125,000; and 150,000 (KARAHASANOĞLU and ÖZAND, 1972). HIOMT has been purified also by immunoadsorption chromatography (KUWANO et al., 1978), and this purified substance has been used to raise antibodies (KUWANO and TAKAHASHI, 1978). An HIOMT technique using endogenous indoleamines as substrates has been described by BALEMANS et al. (1978b, c). An inhibition of HIOMT by pyridoxal 5′-phosphate has been mentioned by BORCHARDT et al. (1978). According to OTANI et al. (1968a) rabbit pineals may be left in situ for up to 8 h post mortem at room temperature without any distinct decrease in the activities of HIOMT, 5-hydroxytryptophan decarboxylase and monoamine oxidase. QUAY and SMART (1967) carried out similar studies in the rat pineal and noted that the post-mortem decline of HIOMT activity was 5% after 1 h, 35% after 3 h, 40% after 6 h and 57% after 12 h. Recently it has been stressed again that in the epiphysis cerebri the enzyme HIOMT is not only involved in the formation of melatonin, but participates also in the synthesis of 5-methoxytryptophol (Fig. 2), which, like melatonin, may be of hormonal nature (WURTMAN and OZAKI, 1978). Activating substances of L-tryptophan-5-monooxygenase (E.C. 1.14.16.4) in bovine pineal glands have been described by HORI and OHOTANI (1978).

As pointed out in the Introduction, epiphyseal function is greatly influenced by environmental lighting conditions. Already prior to the isolation of melatonin, there were some indications that pineal morphology was not independent of light and darkness. QUAY (1956a) noted that in white-footed mice (*Peromyscus leucopus*) maintained in darkness the nuclei of pinealocytes in the posterior part of the pineal contained significantly larger amounts of phloxine-staining granules than those from animals exposed to increasing day-length. Direct evidence for an influence of light on the epiphysis cerebri was obtained by MILINE (1957a), who showed that in rats and rabbits exposed to continuous illumination for several weeks the parenchymal cells atrophied, whereas under constant darkness they hypertrophied. The paper by FISKE et al. (1960) had greater impact on pineal research. These authors exposed young rats for 6–25 weeks to different lighting patterns and noted that constant light, in contrast to constant darkness, led to an approximately 25% decrease in pineal weight. The results of this crucial experiment were soon confirmed. WURTMAN et al. (1961) found that rats exposed to constant light for 56 days had lighter pineals and larger ovaries than animals kept in darkness. Since the increase in ovarian weight was prevented by the administration of bovine pineal extracts, it was concluded that the effect of light upon the weight of the ovaries was mediated via the pineal. In the same year QUAY (1961a, b) showed that the depressive effect of continuous illumination was not restricted to rats but was also present in hamsters. Moreover, it was demonstrated that the pineal response to light was blocked by bilateral intra-orbital transection of the optic tracts.

Greater insight into specific aspects of pineal function was obtained by biochemical investigations dealing with enzymes engaged in melatonin formation. Pineal function could now be assessed with greater precision than had hitherto been possible. In contrast, direct measurement of melatonin was rarely carried out.

According to WURTMAN et al. (1968a), chemical determinations of melatonin by means of fluorometry in 3 N HCl after extraction into organic solvents such as chloroform (AXELROD and WEISSBACH, 1961) or *p*-cymene (QUAY, 1963a) are not sufficiently sensitive or specific to measure the small amounts of melatonin normally present in the pineal glands of most species. A fluorometric method using *o*-phthalaldehyde has been described by MILLER and MAICKEL (1970). See also MAICKEL and MILLER (1968).

Bio-assay methods, on the other hand, are very sensitive and can measure as little as 0.001 μg of the indoleamine. They depend on the ability of melatonin to blanch isolated skin of *Rana pipiens* (LERNER and WRIGHT, 1960) or intact *Xenopus laevis* larvae (QUAY and BAGNARA, 1964; QUAY, 1968a; RALPH and LYNCH, 1970; LYNCH et al., 1971). In the first case, skin samples are placed in Ringer's solution and darkened with either caffeine or MSH. The concentration of melatonin is then measured by its ability to increase the transmission or decrease the reflection on skin of an incident light beam (MORI and LERNER, 1960). When *Xenopus laevis* larvae are used, the extent of blanching is measured using the melanocyte index (QUAY and BAGNARA, 1964). Melatonin assays of this type have been carried out, e.g. with bovine (LERNER et al., 1960), rat (PROP and KAPPERS, 1961) and kangaroo pineals (QUAY and BAKER, 1965). A separation of melatonin by means of Sephadex G25 followed by elution with distilled water has been described by VAN DE VEERDONK (1967a). Information on chromatographic and gas-chromatographic techniques for the determination of melatonin may be obtained from the papers by GREER and WILLIAMS (1967), PENNINGTON (1968), COLE and CRANK (1971) and CATTABENI et al. (1972).

Recently immunoassays and radio-immunoassays for melatonin have been developed (GROTA and BROWN, 1974; BUBENIK et al., 1974; ARENDT et al., 1975;

LEVINE and RICEBERG, 1975; WURZBURGER et al., 1976; WETTERBERG et al., 1978; THORESEN, 1978), which will undoubtedly greatly contribute to elucidating the role of the pineal in humans. The synthetic strategy required for the development of a radio-immunoassay for melatonin has been discussed by BLAIR and SEABORN (1979). Regarding the use of gas chromatography–mass spectrometry for the determination of 5-methoxytryptophol, see LEONE et al. (1979).

In the 1960s, HIOMT activity was a widely used parameter for estimation of epiphyseal function. Following the initial observation by AXELROD and WEISSBACH (1961) that HIOMT activity was high in the pineal bodies of the monkey, cow and cat, WURTMAN et al. (1963a) studied HIOMT activity under experimental conditions and noted that in rats and guinea-pigs the activity of this enzyme was low under constant light and high under constant darkness. It was therefore felt that melatonin synthesis in the pineal organ was controlled by light. Further studies (AXELROD et al., 1965a) showed that HIOMT activity responds relatively quickly to changes in environmental lighting conditions and that the enzyme exhibits a characteristic 24-h cycle, with low enzyme levels at the end of the light period and a striking increase after the onset of darkness.

Once the control of pineal function by light had been established, attempts were made to examine the nervous pathways and the chemical processes involved in the regulation of pineal function. Following the important finding by KAPPERS (1960a) that the rat epiphysis cerebri is innervated by sympathetic nerve fibres originating in the superior cervical ganglia, it was demonstrated that information concerning the environmental lighting conditions was conveyed to the epiphysis cerebri via these fibres (WURTMAN et al., 1964a, b). The suggestion by DE ROBERTIS and PELLEGRINO DE IRALDI (1961a, b) that the granular vesicles present in pineal sympathetic nerve fibres contained noradrenaline was subsequently confirmed by electron microscopic autoradiographic studies (WOLFE et al., 1962). It soon became clear that noradrenaline is of crucial importance for the regulation of pineal function. Since the noradrenaline content of the pineal increased after the onset of darkness and decreased during the day (WURTMAN and AXELROD, 1966a), showing a behaviour similar to HIOMT, it was thought that noradrenaline acted as chemical mediator. Since rat pineal tissue cultured in vitro was shown to be capable of forming melatonin (SHEIN et al., 1967; WURTMAN et al., 1968b), this system became a widely used model to study the effects of different drugs on melatonin formation. In view of the sympathetic innervation of the epiphysis cerebri it was important to demonstrate that noradrenaline stimulates melatonin formation (KLEIN, 1969; KLEIN et al., 1971a; DEGUCHI, 1973; KLEIN and WELLNER, 1973) and that isolated pineals have the ability to concentrate noradrenaline (DENGLER et al., 1962; cat, calf). As might be expected, enhancement of melatonin synthesis requires the formation of RNA and protein (AXELROD et al., 1969; WURTMAN et al., 1969; DEGUCHI, 1973; KLEIN and WELLER, 1973; CARDINALI et al., 1973a; MORISSEY and LOVENBERG, 1978a, b, c).

Protein synthesis stimulated by noradrenaline involves several steps. Already in the late 1960s, WEISS (1968a, 1971) and WEISS and COSTA (1967a, b, 1968a, b) had shown that adenyl cyclase was present in the pineal gland (see also EBADI et al., 1970a) and that this enzyme activity was enhanced by noradrenaline, although more recently it has been shown that stimulation of the β-adrenergic receptor of the pineal body may cause a rapid *decrease* in hormone-sensitive adenyl cyclase activity (KEBABIAN et al., 1975). In view of the concept that the adenyl cyclase–cyclic adenosine monophosphate (cAMP) system acts as second messenger in a number of target cells, it was not surprising to find that the administration of cAMP or dibutyryl cAMP stimulates melatonin and serotonin synthesis in cultured rat pineals (SHEIN and WURTMAN, 1969, 1971; BERG and KLEIN, 1970, 1971; KLEIN and BERG, 1970; KLEIN and WELLER, 1970a, b, 1973; KLEIN et al., 1970a, b, 1973a, b; DEGUCHI and AXELROD, 1973a). When pineals were exposed to β-adrenergic agonists, either in vivo or in vitro, their cAMP content increased rapidly (STRADA et al., 1972; DEGUCHI, 1973). That adenyl cyclase and cAMP play an important role in pineal function was also shown by the not yet fully understood observations that adenyl cyclase activity increased during light and decreased during darkness (WEISS, 1970) and that the level of cAMP was about six times higher at the end of the light period than at the end of the dark period (EBADI et al., 1970b).

The effect of noradrenaline on the synthesis of serotonin and melatonin in the epiphysis cerebri is mediated by β-adrenergic receptors (KLEIN and WELLER, 1970b; WURTMAN et al., 1971; DEGUCHI and AXELROD, 1972a; BROWNSTEIN et al., 1973a,b,c) as indicated by the fact that after administration of β-receptor blockers (e.g. propranolol) the response of the epiphysis cerebri to a number of stimuli is blocked (DEGUCHI, 1973; STRADA et al., 1972; LYNCH et al., 1973a; HOLZ et al., 1974; PARFITT et al., 1976).

The involvement of β-adrenergic receptors has also been demonstrated by use of the sympathomimetic drug, terbutaline, which enhances NAT activity in vitro (BÄCKSTRÖM and WETTERBERG, 1973a,c). Similarly, NAT activity is stimulated by amphetamine (BÄCKSTRÖM and WETTERBERG, 1973b) and isoprenaline (BÄCKSTRÖM and WETTERBERG, 1973c). Therefore, the sequence of events in the epiphysis cerebri appears to be as follows: After the onset of darkness, the terminals of the sympathetic nerve fibres release large quantities of noradrenaline that interact with, and bind to, the β-adrenergic receptor on the postsynaptic (pinealocyte!) cell membrane (ZATZ et al., 1976b). The adenyl cyclase system is then stimulated, causing a rapid increase of cAMP, which, via a cAMP-dependent protein kinase (FONTANA and LOVENBERG, 1971, 1972, 1973), enhances the metabolism in the end organ involving both transcription (ROMERO et al., 1975a, b; ZATZ et al., 1976a) and translation (AXELROD et al., 1969; DEGUCHI and AXELROD, 1972a, 1973a).

The enzyme responsible for the hydrolysis of cAMP, a phosphodiesterase, has been characterized by SANKARAN et al. (1978). A nuclear protein appears to be phosphorylated in early stages of NAT induction, but not in later stages (WINTERS et al., 1977; MORISSEY and LOVENBERG, 1977). The possible role of cyclic nucleotides in the regulation of noradrenaline *release* from the rat pineal has been investigated by PELAYO et al. (1978).

The intensity of the effect in the end organ is related not only to the amount of neurotransmitter released but also to the responsiveness of the receptor to its agonist. Interesting in view of the regulation of melatonin synthesis is the fact that the pineal β-receptors exhibit differences in their sensitivity to noradrenaline and related compounds. It was shown, for example, that injection of

L-dopa, a precursor of catecholamines, led in denervated rat pineals to a much higher increase in *N*-acetyltransferase than in normal controls, indicating super-induction due to supersensitivity of β-receptors (DEGUCHI and AXELROD, 1972b). Supersensitivity was likewise noted following noradrenaline depletion, reserpine administration, exposure to continuous illumination, denervation or bilateral superior cervical ganglionectomy (WEISS and COSTA, 1967a; WEISS, 1969; WEISS and STRADA, 1972; DEGUCHI and AXELROD, 1973a). Subsensitivity, on the other hand, was observed following repeated administration of catecholamines (DEGUCHI and AXELROD, 1973a).

Interestingly, the responsiveness of the β-adrenergic receptor exhibits a characteristic daily cycle, the responsiveness at the end of the light period being nearly ten times greater than at the end of the dark period (ROMERO and AXELROD, 1974). According to KEBABIAN et al. (1975) stimulation of the β-adrenergic receptors causes a rapid decrease in the number of available receptors, while lack of stimulation causes an increase. WILKINSON (1978) demonstrated that the sensitivity of the β-receptors depends on the presence of calcium ions (see also ZATZ and ROMERO, 1978a). Supersensitivity and subsensitivity under the influence of cholera toxin have been studied by ZATZ (1977).

In contrast to β-receptors, an action of α-adrenergic mechanisms has been denied in the mammalian pinealocyte (WURTMAN et al., 1971; WEISS and STRADA, 1972; HOLZ et al., 1974). However, HAUSER et al. (1973, 1974) have presented evidence that the stimulatory effects of noradrenaline on pineal phospholipid metabolism are mediated by α-adrenergic receptors. An additional factor in favour of the involvement of α-receptors is the observation that an α-adrenergic blockade with phenoxybenzamine potentiates the stress-response of pineal NAT (LYNCH et al., 1977). α-Receptors have also been implied in the regulation of noradrenaline release in the rat pineal body (PELAYO et al., 1977).

From what has been said so far it can be seen that biogenic amines are of special significance in the mammalian pineal organ. In fact, serotonin, one of the precursors of melatonin, is present in the pineal in higher concentrations than in any other mammalian organ (GIARMAN and DAY, 1958/59; GIARMAN et al., 1959, 1960; QUAY, 1963c, 1968b; MILINE et al., 1958; QUAY and HALEVY, 1962).

In addition to melatonin (EBELS et al., 1972a) and serotonin, a number of related compounds are present in the epiphysis cerebri (e.g. MCISAAC et al., 1965; THIÉBLOT et al., 1967) (Fig. 2). The enzyme monoamine oxidase converts serotonin to the unstable 5-hydroxyindole acetaldehyde, which is either oxidized to 5-hydroxyindole-3-acetic acid (LERNER et al., 1960; QUAY, 1964a; QUAY and BAKER, 1965; FELDSTEIN and WILLIAMSON, 1968; ZURBURG and EBELS, 1975) or reduced to 5-hydroxytryptophol (MCISAAC et al., 1965; KLEIN and NOTIDES, 1969; OTANI et al., 1969). Further derivatives are 5-methoxyindole-3-acetic acid (LERNER et al., 1959c, 1960; EBELS et al., 1972b; ZURBURG and EBELS, 1975; QUAY, 1964a) and 5-methoxytryptophol (MCISAAC et al., 1964; OTANI et al., 1968b, 1969; KLEIN and NOTIDES, 1969; QUAY and SMART, 1967; EBELS and HORWITZ-BRESSER, 1976).

Moreover, the mammalian epiphysis cerebri may synthesize small amounts of β-carbolines (WURTMAN et al., 1968a; QUAY, 1974a), which are of interest in view of their neuropharmacological properties. In addition to noradrenaline, which appears to be present mainly in sympathetic nerve fibres (PELLEGRINO DE IRALDI and ZIEHER, 1966), dopamine has also been demonstrated in the pineal (rat, CSILLIK, 1963; ZIEHER and PELLEGRINO DE IRALDI, 1966; HYYPPÄ et al., 1971; BROWNSTEIN and AXELROD, 1974), but it has received little attention. Since the dopamine content is not affected by denervation it appears to have an extraneuronal localization (PELLEGRINO DE IRALDI and ZIEHER, 1966). For discussion of histamine, see p. 90.

The finding that biogenic amines play a leading role in pineal function should not blind us to the fact that other substances, e.g. peptides, may be equally or more important (see p. 240). Little is currently known about the functional significance of guanosine 3′:5′-cyclic monophosphate (O'DEA and ZATZ, 1976), GABA (KANAZAWA et al., 1976) and biopterin (EBELS et al., 1979; VAN DER HAVE-KIRCHBERG et al., 1977; ZURBURG and EBELS, 1974).

Finally, it should be mentioned that the epiphysis cerebri exhibits very pronounced circadian changes. This can be expected when one considers that light and darkness play important roles in regulating pineal function. It is relevant to note, however, that the different parameters investigated may show different rhythmic patterns (for details, see below). Here it suffices to mention that melatonin formation is usually low during the day and high during the night.

Methods for assessing the biological activity of the mammalian pineal organ have been reviewed by CARDINALI and WURTMAN (1975), see also WARTENBERG (1973).

II. Gross Anatomy

1. General Aspects

Although the pineal bodies of different mammals possess a common developmental origin from the posterodorsal region of the diencephalic roof (see p. 44), there are considerable variations with respect to pineal form and position. OKSCHE (1965a) distinguished elongate, conical and pear-shaped pineals. A classification based on the shape of the organ alone does not appear to be entirely satisfactory, because in some mammalian species the epiphysis is not a single compact structure, but consists of more or less isolated cell clusters, which may differ in function.

For a classification based on the location of the epiphysis cerebri the hind part (the splenium) of the corpus callosum has been used as a point of reference (BLIN and MAURIN, 1956) and pineals with subcallosal, postcallosal and supracallosal positions can be distinguished. According to this classification, a subcallosal position is found in primates and carnivores, whereas a postcallosal position is supposed to be typical of the cow, horse, ass, mule, sheep and goat (BLIN and MAURIN, 1956; FRAUCHIGER, 1963). The situation in rodents, to which a subcallosal position of the epiphysis has been ascribed, indicates a shortcoming of the above classification. In this group of mammals, the epiphysis shows great variation in both shape and position, so that they may occupy either subcallosal or supracallosal positions, or both. Moreover, this classification does not adequately cover the topographical relationships between the epiphysis and the third ventricle of the brain, which may be of importance (see p. 266).

In the present survey, a tentative classification of the positions of the mammalian pineals is given, which is independent of the corpus callosum, but takes into account the relationship of the pineal to the third ventricle as well as the form and arrangement of pineal tissue. According to this classification, pineals belongs to the proximal type (or type A) if the bulk of the pineal tissue lies closely related to the third ventricle. If the pineal is elongate, the length of the organ being about or slightly more than twice the largest width, it is allocated to the proximo-intermediate type, designated as type AB. If the pineal is very long and a substantial amount of pineal tissue lies either superficially or closely related to the cerebellum, it belongs to the proximo-intermedio-distal type, designated as type ABC. If any of these parts are substantially reduced in amount the capital letters are replaced by small letters of the Greek alphabet. If an area is lacking, the corresponding letter is omitted. Type $\alpha\beta$C (e.g. in the rat) is an elongated structure in which the bulk of the pineal tissue lies superficially and in which the proximal and intermediate parts are greatly reduced. The pineal complex of the hamster, consisting of a superficial and a deep pineal, belongs to type αC (for details, see below).

2. Form, Size and Location

Details on shapes and dimensions of pineal organs are given in Tables 1 and 2. Pineal weights and volumes are dealt with in Sect. B.II.3. The sequence in which the various orders of mammals are mentioned follows that used by REMANE et al. (1976), although in a reverse manner (cf. BARGMANN, 1943).

Table 1. Shape of mammalian pineal organs

Order, Species	Shape	Author(s)
Primates		
Man	Like a pine cone	BARGMANN (1943)
Rhesus monkey	Spindle-shaped organ with a fairly long, tapering tip	HÜLSEMANN (1967), ECKHARDT (1972)
Artiodactyla		
Cow	Peanut-like	see BARGMANN (1943)
Sheep	Spherical or pea-like	see BARGMANN (1943)
Goat	Spherical or pea-like	see BARGMANN (1943)
Pig	Rye grain-like	see BARGMANN (1943)
Hyracoidea		
Hyrax	Principal pineal tissue mass bounding the pineal recess antero-dorsally and accessory pineal tissue on the dorsal side of habenular commissure	QUAY and MILLAR (1971)
Proboscidea		
Elephant	Diffusely arranged pineal tissue bordering the pineal and supra-pineal recesses	HAUG (1972)

Table 1 (continued)

Order, Species	Shape	Author(s)
Perissodactyla		
Horse	Round to pear-shaped or oval	FASSBENDER (1962)
Ass	Olive-shaped	BLIN and MAURIN (1956)
Mule	Conical	BLIN and MAURIN (1956)
Lagomorpha		
Rabbit	Elongate, flattened organ with a club-shaped thickening distally	HÜLSEMANN (1967), ROMIJN (1973a)
Carnivora		
Fissipedia		
Dog	Cone-shaped; lancet-like	ZACH (1960), HÜLSEMANN (1967)
Cat	Pear-shaped	ZACH (1960)
Pinnipedia		
Northern fur seal	Heart-shaped; flattened dorso-ventrally	ELDEN et al. (1971)
Weddell seal	Flattened body of rhomboidal shape	CUELLO and TRAMEZZANI (1969a, b)
Rodentia		
Sciuromorpha		
Geomys bursarius	Pineal body with 3–4 mm long stalk	SHERIDAN and REITER (1973)
Dipodomys sp.	Elongate organ with club-shaped ending	GREGOREK et al. (1977)
Gliromorpha		
Glis glis	Round, compact organ	HÜLSEMANN (1967)
Myomorpha		
Cricetidae		
Peromyscus leucopus	Basically an elongate, dumbbell-shaped organ, showing signs of regression in the thin intermediate part	QUAY (1956a)
Meriones unguiculatus	Elongate, dumbbell-shaped organ	GREGOREK et al. (1977)
Cricetinae		
Mesocricetus auratus	Two separated parts; a large superficial pineal and a small deep pineal	SHERIDAN and REITER (1970a)
Cricetulus griseus	Elongate organ with club-shaped distal part	GREGOREK et al. (1977)
Microtinae		
Dicrostonyx groenlandicus	Elongate, dumbbell-shaped organ	QUAY (1978)

Table 1 (continued)

Order, Species	Shape	Author(s)
Muridae		
Rat	Basically an elongate organ with club-shaped distal thickening, the intermediate and proximal parts sometimes showing signs of regression	KAPPERS (1960a), BJÖRKLUND et al. (1972), GREGOREK et al. (1977), VOLLRATH and BOECKMANN (1978)
Mouse	Elongate organ of apparently equal width throughout or showing a thickened intermediate part or a club-shaped distal thickening	COLLIER (1943), VON BARTHELD and MOLL (1954), HÜLSEMANN (1967), UPSON et al. (1976), GREGOREK et al. (1977)
Dipodidae		
Jaculus jaculus	Elongate organ with distal club-shaped thickening	HÜLSEMANN (1967)
Caviomorpha		
Guinea-pig	Elongate, dumbbell-shaped organ	HÜLSEMANN (1967)
Xenarthra		
Dasypus novemcinctus	Pineal tissue diffusely arranged around intercommissural recess	HOFER et al. (1976)
Chiroptera		
Myotis sodalis	Compact organ broadly attached to roof of diencephalon	QUAY (1965a, 1970c)
Myotis lucifugus	See *M. sodalis*	QUAY (1965a, 1970c)
Tupaioidea		
Tupaia glis	Small organ triangular in shape and projecting into third ventricle	HÜLSEMANN (1967)
Insectivora		
Blarina brevicauda	Compact organ broadly attached to roof of diencephalon	QUAY (1965a)
Erinaceus europaea	Rod-like organ lying inside third ventricle	HÜLSEMANN (1967)
Talpa europaea	Compact, oval organ lying in front, below and behind habenular commissure	HOFER et al. (1976)
Marsupialia		
Petaurus breviceps	Lacking	J. JOHNSON (1977)
Petaurus norfolcensis	Lacking	J. JOHNSON (1977)
Pseudocheirus	Diffusely arranged around pineal recess	J. JOHNSON (1977)
Trichosurus	Diffusely arranged around pineal recess	J. JOHNSON (1977)

Table 1 (continued)

Order, Species	Shape	Author(s)
Caenolestes obscurus	Diffuse to solid structure around pineal recess	J. Johnson (1977)
Thylogale billarderi	Solid organ in vicinity of pineal recess	J. Johnson (1977)
Macropus fuliginosus	Solid organ in vicinity of pineal recess	J. Johnson (1977)
Didelphys virginiana	Hollow organ surrounding pineal recess	Hofer et al. (1976)
Megaleia rufa	Symmetrical bilobate structure	Quay and Baker (1965)
Macropus canguru	Uniformly rounded shape	Quay and Baker (1965)

Table 2. Dimensions and classification of mammalian pineal organs

Order, Species	Length (mm)	Breadth (mm)	Height (mm)	Author(s)	Classification type[a]
Primates					
Man	12	8	4	Kopsch (1953)	A
Orangutan	4.59 1.7	3.3	1.35 1.1	See Bargmann (1943) Quay (1970a, b)	A
Gibbon	2.1	2.4	1.4	See Bargmann (1943)	A
Chimpanzee	2	3		See Bargmann (1943)	A
Rhesus monkey	4	1	1	Wartenberg (1973)	A$\beta\gamma$
Artiodactyla					
Cow	16 15	8 7	5.5 7	Lang (1959) Blin and Maurin (1956)	AB
Sheep	7.5 7	5.5 5.5	4.5 4.5	Lang (1959) Blin and Maurin (1956)	A
Goat	7	5.5	4.5	Blin and Maurin (1956)	A
Pig	10	5	4	Blin and Maurin (1956)	AB
Hyracoidea					
Hyrax				Quay and Millar (1971)	$A_{diff.}$
Proboscidea					
Elephant				Haug (1972)	$A_{diff.}$
Perissodactyla					
Horse	10	7	6	Blin and Maurin (1956)	A
Ass	15	6	6	Blin and Maurin (1956)	AB
Mule	15	6	6	Blin and Maurin (1956)	AB

Table 2 (continued)

Order, Species	Length (mm)	Breadth (mm)	Height (mm)	Author(s)	Classification type[a]
Lagomorpha					
Rabbit	8	1.5		ROMIJN (1973a)	αβC
Carnivora					
Fissipedia					
Dog	3.1	1.2		ZACH (1960)	AB
	4	3	1.5	BLIN and MAURIN (1956)	A
Cat	3.2	1.4		ZACH (1960)	
Pinnipedia					
Weddell seal	20–30	10		CUELLO and TRAMEZZANI (1969a, b)	ABC
	21–27	11–17	3–5	TEDMAN (1977)	AB
Northern fur seal	7–14			ELDEN et al. (1971)	ABC
Mirounga leonina	25–32	5–9	5–7	TEDMAN (1977)	ABC
Rodentia					
Sciuromorpha					
Pocket gopher				SHERIDAN and REITER (1973)	αβC
Dipodomys sp.	1.2–2.2			GREGOREK et al. (1977)	αβC
Gliromorpha					
Glis glis				HÜLSEMANN (1967)	A
Myomorpha					
Cricetidae					
Peromyscus leucopus				QUAY (1956a)	αβC; αC
Meriones unguiculatus	2.4–2.8	The enlarged proximal part measures 200–230 μm, the parenchymal stalk 50–100 μm and the distal part 280–350 μm in thickness		GREGOREK et al. (1977)	AβC
Cricetinae					
Mesocricetus auratus	Deep pineal averages 80–280 μm in thickness; superficial pineal varies in length from 600 to 820 μm; the non-parenchymal stalk measures 2,700 to 3,200 μm in length			GREGOREK et al. (1977)	αC
Cricetulus griseus	Superficial portion ranges between 800 and 1,300 μm in length and 350 and 735 μm in diameter			GREGOREK et al. (1977)	αβC

Table 2 (continued)

Order, Species	Length (mm)	Breadth (mm)	Height (mm)	Author(s)	Classification type[a]
Microtinae					
Lemming	3.5	0.15 (Calculated from Fig. 1)		QUAY (1978)	AβC
Muridae					
Rat	Superficial pineal measures between 1,600 and 1,950 μm in length with a maximal average diameter of 820 μm; further literature on which the classification is based is given in Table 1			GREGOREK et al. (1977)	Usually C or βC; occasionally C
Mouse	Length varies from 850 to 1,200 μm; the distal portion is enlarged and measures 350 μm in diameter			GREGOREK et al. (1977)	ABC or αCγ
	0.8	0.4–0.5		COLLIER (1943)	
	1.0	0.3		UPSON et al. (1976)	
Dipodidae					
Jaculus jaculus				HÜLSEMANN (1967)	$\alpha\beta$C
Caviomorpha					
Guinea-pig	Dimensions see Table 3			HÜLSEMANN (1967)	ABC or AβC
Xenarthra					
Dasypus novemcinctus				HOFER et al. (1976)	$A_{diff.}$
Chiroptera					
Myotis sodalis				QUAY (1965a, 1970c)	A
Myotis lucifugus				QUAY (1965a, 1970c)	A
Rhinolophus ferrum-equinum				STAMMER (1972)	"Superficial position"
Tupaioidea					
Tupaia glis				HÜLSEMANN (1967)	$A_{intraventr.}$
Insectivora					
Blarina brevicauda				QUAY (1965a)	A
Erinaceus europaea				HÜLSEMANN (1967)	$A_{intraventr.}$
Talpa europaea				HOFER et al. (1976)	A
Marsupialia					
Petaurus breviceps				J. JOHNSON (1977) KENNY and SCHEELINGS (1979)	Lacking A_{hollow}

Table 2 (continued)

Order, Species	Length (mm)	Breadth (mm)	Height (mm)	Author(s)	Classification type[a]
Petaurus norfolcensis				J. JOHNSON (1977)	Lacking
Pseudocheirus				J. JOHNSON (1977)	$A_{diff.}$
Trichosurus				J. JOHNSON (1977)	$A_{diff.}$
Caenolestes obscurus				J. JOHNSON (1977)	$A_{diff.-sol.}$
Thylogale billarderi				J. JOHNSON (1977)	A
Macropus fuliginosus				J. JOHNSON (1977)	A
Didelphys virginiana				HOFER et al. (1976)	A_{hollow}
Megaleia rufa				QUAY and BAKER (1965)	A
Macropus canguru				QUAY and BAKER (1965)	A
Dasyuroides byrnei				KENNY and SCHEELINGS (1979)	A
Sminthopsis crassicaudata					A A
Antechinomys laniger					A
Pseudocheirus peregrinus					A
Schoinobates volans					A
Trichosurus vulpecula					A
Phascolarctos cinereus					A
Petrogale					A
Wallabia bicolor					A
Monotremata					
Tachyglossus aculeatus				KENNY and SCHEELINGS (1979)	A
Ornithorhynchus anatinus					A

[a] For details see text. *diff.* diffusely arranged parenchyma; *diff.-sol.* diffuse to solid parenchyma; *intraventr.* intraventricular location.

a) Man

In man the pineal body is a small, piriform organ, the colour of which can be grey to white, yellow (BARGMANN, 1943) or red-grey (WARWICK and WILLIAMS, 1973). It belongs to the proximal type, or to type A of the above-mentioned classification. It lies near the posterior margin of the diencephalic roof and occupies the depression between the superior colliculi of the mesencephalon (Fig. 4). Proximally, the organ is intimately related to the third cerebral

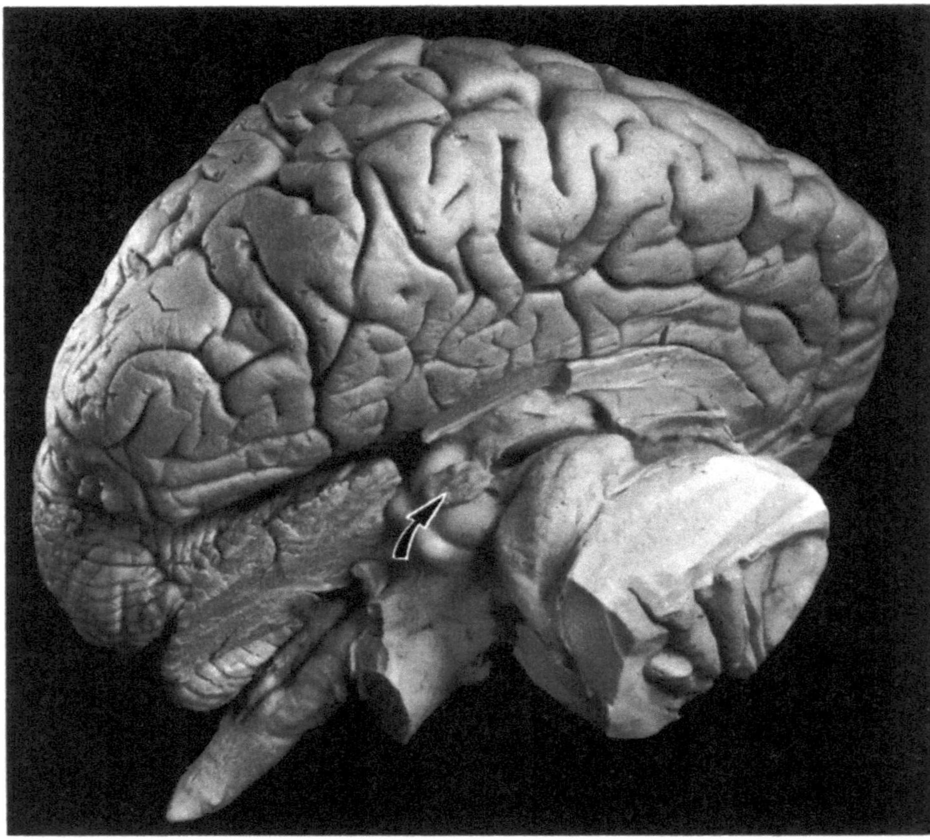

Fig. 4. Dissected human brain showing the epiphysis cerebri in situ (*arrow*). Dissection and photography by Dr. S. HEIL, Department of Anatomy, University of Mainz

ventricle, an evagination of which penetrates into the organ as the pineal recess (Fig. 5). The epiphysis is attached to the brain by means of a stalk, or peduncle, which divides anteriorly into two laminae a superior and an inferior one. The inferior lamina comprises the posterior commissure and the superior lamina houses the habenular commissure. The two laminae are separated from each other by the pineal recess.

The major part of the epiphysis is covered by tissue of the tela choroidea, which reflects over the tectum. In view of the concept that the pineal may affect the hypothalamus via the cerebrospinal fluid (CSF) (p. 266), it is interesting that, apart from the close relationship to the pineal recess of the third ventricle, the pineal comes into close contact with the suprapineal recess of the third cerebral ventricle and lies within the subarachnoid space (Fig. 5), and hence is bathed by subarachnoid CSF at its outer aspect (COOPER, 1960). Although perhaps not directly applicable to the adult human pineal, it may be of significance that in human fetuses ruthenium red has been shown to penetrate from

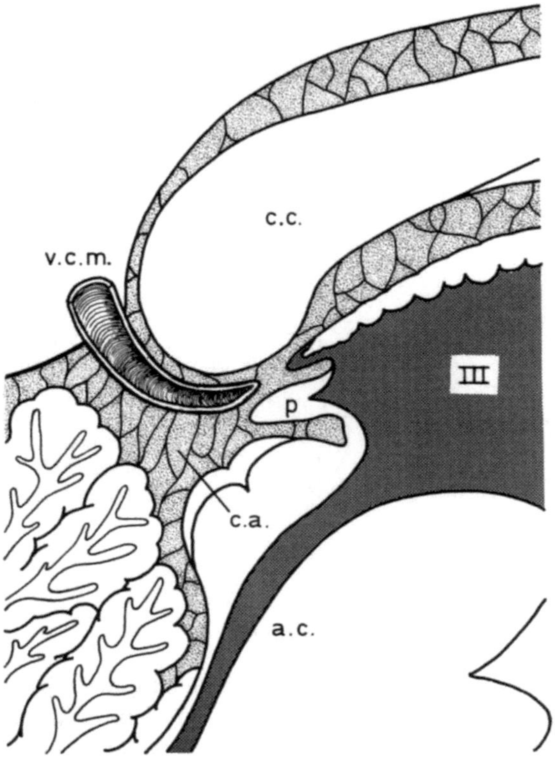

Fig. 5. Relationships of the human pineal gland to the CSF. *a.c.*, Aqueductus cerebri; *c.a.*, cisterna ambiens; *c.c.*, corpus callosum; *p*, pineal gland; *v.c.m.*, vena cerebri magna; *III*, third ventricle. Design by Dr. S. HEIL, Department of Anatomy, University of Mainz

the subarachnoid space into the extracellular spaces of the pineal parenchyma, which superficially lacks tight junctional elements (MØLLER, 1976).

For radiological aspects of the pineal body, see MOREAU (1966).

b) Other Primates

As pointed out by QUAY (1970b), the pineal structure is known in detail for relatively few of the some 500 species of living, non-human primates. From the published results it appears (Table 2) that in the majority of cases the epiphysis cerebri lies near to the third ventricle; the pineal of the orangutan, *Pongo pygmaeus* (QUAY, 1970b), having an especially extensive exposure to the ventricle along the pineal recess and the suprapineal recess or dorsal sac. In view of the presence in some species of elongated pineals extending in a posterior direction, it is interesting to note that a comparable situation is found in the rhesus monkey, *Macaca mulatta* (HÜLSEMANN, 1967; WARTENBERG, 1973) (Fig. 6), in which the pineal can be tentatively classified as type Aβ or ABγ. Which type of pineal organ is most typical of primates remains to be established.

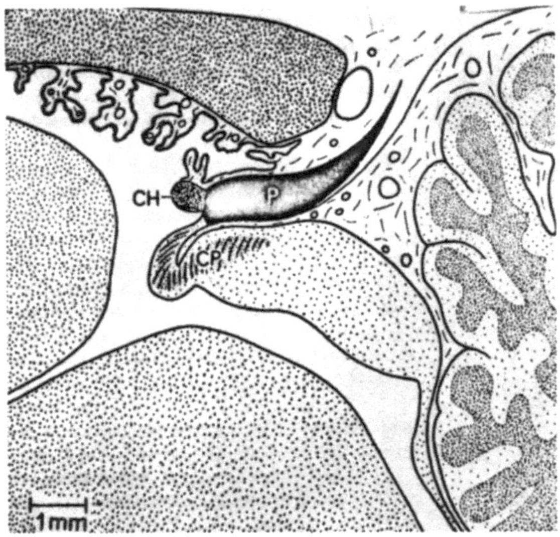

Fig. 6. Schematic drawing of midsagittal section of the epithalamic region of a rhesus monkey, *Macaca mulatta*. *CH*, habenular commissure; *CP*, posterior commissure; *P*, pineal gland. WARTENBERG (1973) after HÜLSEMANN (1967)

c) Cetacea

According to BARGMANN (1943) a pineal is lacking in *Delphinus longirostris, Balaenoptera borealis* and *Phocaena communis*. In view of the situation in the elephant (HAUG, 1972) and in *Dasypus* (p. 33), in which a pineal was thought lacking until pineal tissue was found to be inconspicuously arranged in the immediate vicinity of the intercommisural recess, a statement concerning the absence of an epiphysis cerebri in whales should be viewed with caution.

d) Artiodactyla

In contrast to the afore-mentioned order, distinct pineal organs are present in the artiodactyla. From the studies by LANG (1959) and BLIN and MAURIN (1956) it can be seen that while the pineals of sheep and goat represent proximal types, those of the cow and pig can be classified as proximo-intermediate types. From Fig. 7 it can be seen that in the sheep the pineal recess tapers to become slit-like and that the pineal comes into close contact with the suprapineal recess.

e) Hyracoidea

According to QUAY and MILLAR (1971), adult male *Procavia capensis* do not possess a compact pineal body, but rather a relatively thin sheath of parenchymal tissue covering the dorsal and dorsolateral parts of what could be regarded as an unusually large and complex pineal recess. A small amount of pineal tissue is also found on the dorsal side of the thickest part of the habenular commissure (dorsal or accessory pineal). As the organ is not compact, but rather diffuse, it may be classified as Type $A_{diff.}$.

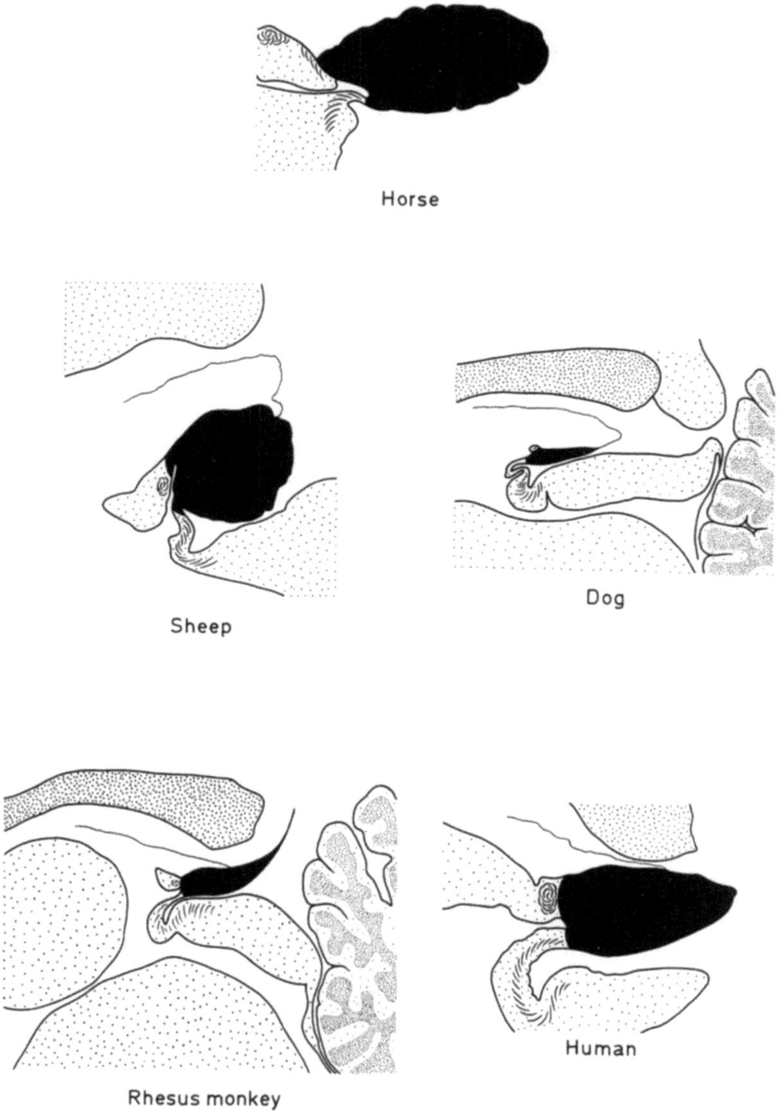

Fig. 7. Schematic representation of the pineal glands in a number of mammalian species. HÜLSEMANN, 1967

f) Proboscidea

A similar type of pineal organ to that seen in *Procavia* has been described in a 14-year-old elephant, *Loxodonta africana* (HAUG, 1972). Again, two completely separated parts of diffusely arranged pineal tissue were observed, one part bordering the pineal recess, and the other lying dorsal to the habenular commissure and abutting on the suprapineal recess.

No recent data appear to be available on the pineal organ in Sirenia.

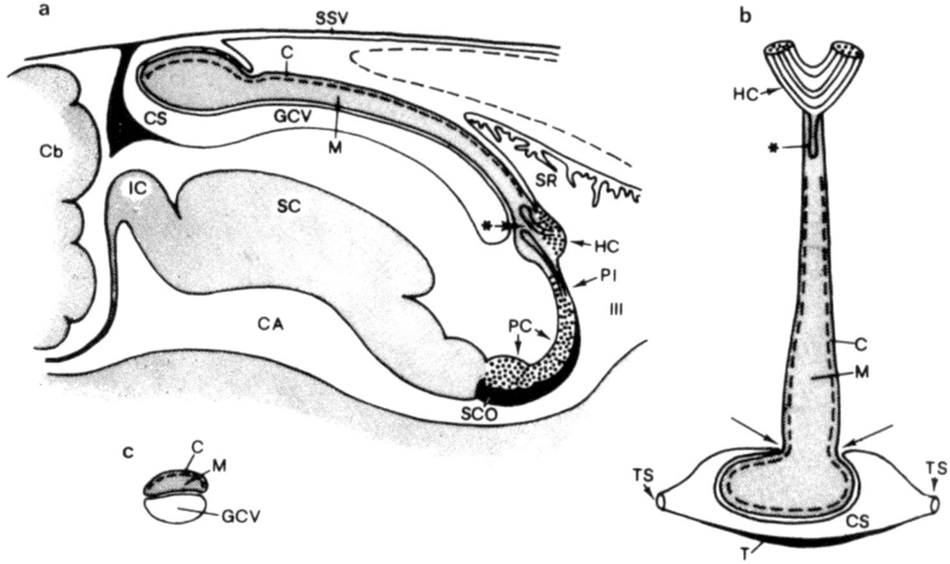

Fig. 8a–c. Schematic representation of the rabbit pineal organ. **a** sagittal section, **b** horizontal section, **c** transverse section. *CA*, cerebral aqueduct; *C*, cortex of pineal organ; *Cb*, cerebellum; *CS*, confluens sinuum; *GCV*, great cerebral vein; *HC*, habenular commissure; *IC*, inferior colliculus of the quadrigeminal bodies; *M*, medulla of pineal organ; *PC*, posterior commissure; *PI*, pars intercalaris; *SC*, superior colliculus of the quadrigeminal bodies; *SCO*, subcommissural organ; *SR*, suprapineal recess; *SSV*, superior sagittal vein; *T*, tentorium cerebelli; *TS*, transverse sinus; *III*, third ventricle; →, entrance of both nervi conarii; *, aberrant commissural fibres. ROMIJN, 1973a

g) Perissodactyla

The data given by BLIN and MAURIN (1956) (Table 2) and Fig. 7 clearly indicate that in the horse, ass and mule the pineals are all of the proximo-intermediate type. Figure 7 shows that in the horse the actual pineal recess is relatively small and is in continuity with the slit-like intercommissural recess.

No recent data on the pineal organ in Tubulidentata appear to be available.

h) Lagomorpha

The pineal organ in the most widely studied lagomorph, the rabbit, *Oryctolagus cuniculus*, differs from the afore-mentioned species in that it is a long organ (Fig. 8), increasing in size in a proximodistal direction (HÜLSEMANN, 1967; ROMIJN, 1973a). An interesting feature is that a distal, deflected thickening exists that is invaginated into the lumen of the confluence of sinuses and partly also into the right or left transverse sinuses (A. SMITH, 1971; ROMIJN, 1973a). Since a distinctly thickened part of the pineal organ lies superficially and since the elongate organ is continuous with the commissural region, it is classified as type $\alpha\beta C$.

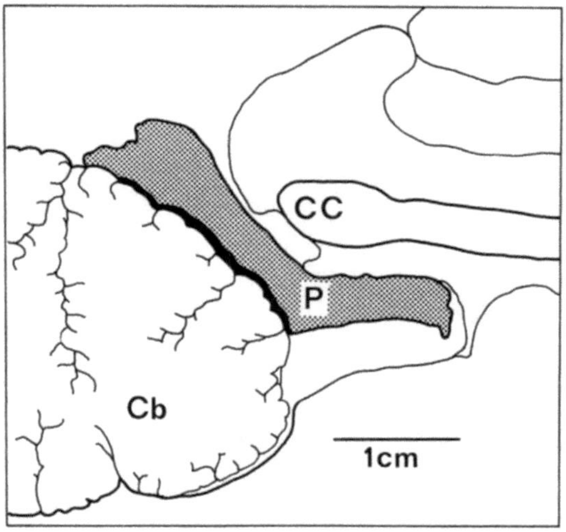

Fig. 9. Schematic drawing of the pineal region of a Weddell seal. Note the large-sized pineal organ (*P*). *Cb*, cerebellum; *CC*, corpus callosum. After CUELLO and TRAMEZZANI, 1969a

For the topographical relationship between the epiphysis cerebri and the subcommissural organ in *Lepus cuniculus*, see CANEPA (1949).

i) Carnivora

From the few species studied it would appear that in the two suborders of Carnivora, the fissiped and pinniped carnivores, the pineal organs shows distinct differences. In the fissiped carnivores, dog (BLIN and MAURIN, 1956; ZACH, 1960), cat (ZACH, 1960) and ferret, *Mustela vulgaris* (TRUEMAN and HERBERT, 1970b), the pineal body tends to become elongate, belonging to type AB. That the width of the dog pineal is more likely to be 1.2 mm than 3 mm (Table 2) is suggested by the schematic drawing in Fig. 7.

In the pinniped carnivores, *Leptonichotes weddeli*, the Weddell seal (CUELLO and TRAMEZZANI, 1969a, b), and *Callorhinus ursinus*, the northern fur seal (ELDEN et al., 1971), the pineal organs are relatively large and appear to belong to type ABC. In the Weddell seal the body of the organ characteristically makes an angle at the level of the superior colliculi (Fig. 9).

The topographical relationships between the pineal and the subcommisural organ in the dog and cat have been studied by CANEPA (1949).

j) Rodentia

In rodents the epiphysis cerebri is very variable in both shape and location. When studying the shape of rodent pineal one is inclined to agree with HÜLSEMANN's (1967) statement that in rodents the pineal is basically a long, rod-like structure extending from the third ventricle to immediately beneath the skull.

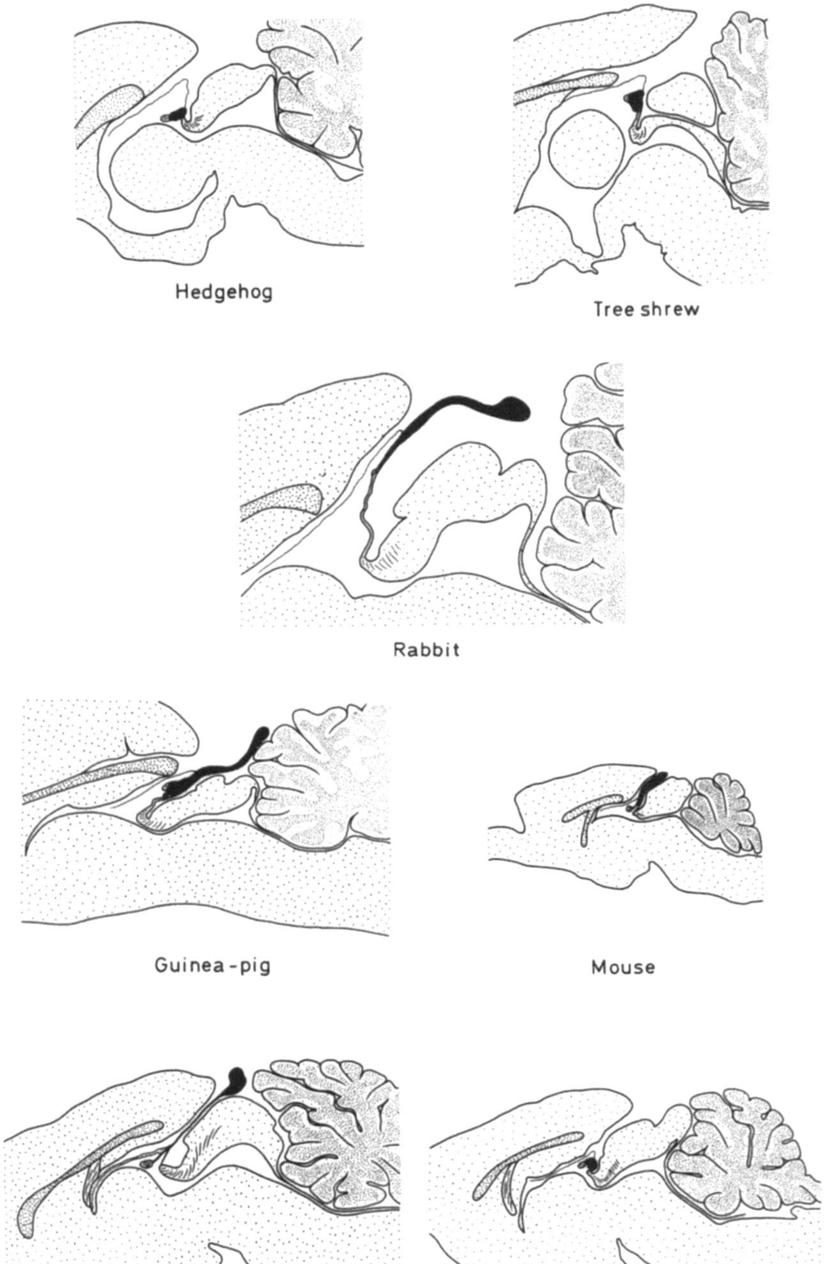

Fig. 10. Schematic representation of the pineal glands in a number of mammalian species. HÜLSEMANN, 1967

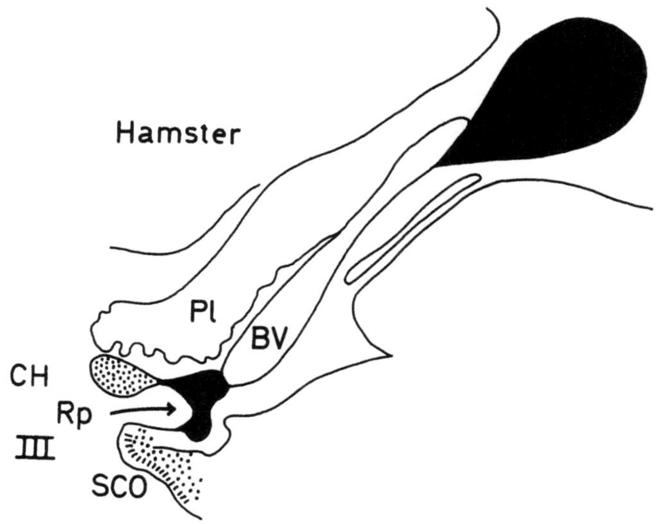

Fig. 11. Schematic drawing of midsagittal section of a hamster pineal complex (*depicted in black*). Note the separation of the pineal complex into a large superficial and a small deep pineal. *BV*, blood vessel; *CH*, habenular commissure; *Pl*, choroid plexus; *Rp* pineal recess of the third ventricle (*III*); *SCO*, subcommissural organ

In some species it consists of two or more clearly separated parts, giving the impression that certain areas of the pineal complex are prone to regression. According to QUAY (1965a) the gross shapes of the pineal generally seem to reflect the available space between the adjacent areas of the brain, particularly cerebellum, cerebral hemispheres and mid-brain. As to the significance of the more dorsal location of the pineal in some rodents, QUAY offered three explanations which he termed quite speculative: (1) early meningeal attachments of the pineal lead to its dorsal position, (2) a more superficial position may have had adaptive significance in reception of light, and (3) a dorsoposteriad migration of the pineal may have occurred along its path of primary innervation, the nervi conarii, which enter the rat pineal via the tentorium cerebelli. An important field of research will be to elucidate whether the various parts of the pineal complex are identical in function; some indications suggest that they may differ.

Unfortunately, relatively few detailed quantitative and qualitative studies on pineal shape in various rodents are currently available. Hence, it is premature to decide whether a particular type of pineal shape is typical of a given suborder of rodents.

Among the Caviomorpha, only guinea-pigs have been studied in detail. In this species the pineal is an elongate organ, possessing two prominent thickenings, proximal and distal (Fig. 10) (HÜLSEMANN, 1967). It may be classified as type ABC or AβC depending on the size of the intermediate part.

Recent planimetric studies carried out in our laboratory (JUNG, unpublished observations) have shown that in histological sections the length of the entire pineal organ amounted to 6.140 mm and that the different parts of the pineal

Table 3. Dimensions and volumes of male guinea-pig pineal organs[a]

	Proximal part	Intermediate part	Distal part	
Length	1.432 ±0.185[b]	3.592 ±0.542	1.116 ±0.050	Total length 6.140 mm
Breadth	0.466 ±0.068	0.214 ±0.041[c] 0.267 ±0.049[d]	0.503 ±0.062	
Height	0.432 ±0.046	0.136 ±0.019[c] 0.275 ±0.034[d]	0.496 ±0.058	
Volume	0.150 ±0.034[e]	ca. 0.140	0.148 ±0.036	Total volume 0.438 mm^3

[a] Unpublished results by D. JUNG, Dept. of Anatomy, University of Mainz.
[b] Mean ±SD in mm.
[c,d] As the intermediate part varies considerably in its dimensions the thinnest (c) and thickest (d) parts were separately assessed.
[e] Mean ±SD in mm^3.

complex are virtually identical in volume despite their differing dimensions (Table 3).

In the Myomorpha investigated to date the pineal organs show great interspecies and intraspecies variation. In *Mus musculus* (VON BARTHELD and MOLL, 1954; HÜLSEMANN, 1967) the pineal organ is elongated and is apparently of equal width and height along its entire length, belonging to type ABC.

An interesting gross anatomical structure of the pineal organ is found in the hamster, *Mesocricetus auratus*. While it was previously thought that all of the pineal tissue was superficially located, near the confluence of sinuses ("superficial pineal"), it later became apparent that pineal tissue was also present near the habenular commissure where it lay closely related to the pineal recess of the third ventricle ("deep pineal") (SHERIDAN et al., 1969; SHERIDAN and REITER, 1970a, b) (Fig. 11). The tissue connecting the deep and the superficial pineal is apparently made up of numerous myelinated and unmyelinated axons and blood vessels (SHERIDAN, 1969). Occasionally, clusters of cells closely resembling pinealocytes were seen between the two main parts of the pineal complex (SHERIDAN and REITER, 1970a). The volume of the deep pineal amounts to between 3% and 12% that of the superficial pineal (VOLLRATH and BOECKMANN, 1978).

Great intraspecies variation is typical of the rat pineal complex. Basically, it consists of three parts, namely (a) the deep pineal, also termed lamina intercalaris, which extends into the (b) pineal stalk and (c) the pineal organ proper or superficial pineal (KAPPERS, 1960a; BJÖRKLUND et al., 1972). It may be classified as type $\alpha\beta C$ (see below). According to KAPPERS (1960a) the pineal stalk is made up of connective tissue, blood vessels, nerve fibres and parenchymal cells, the latter is thought to be more or less well differentiated pinealocytes.

Recent quantitative studies have shown that the stalk region of the rat pineal complex shows great intraspecies variation (BOECKMANN, 1980). Depending on the amount and location of parenchymal tissue in the stalk region, three basic types of pineal complexes can be distinguished. In 96 male Sprague-Dawley rats investigated, only 18% showed a pineal complex as described above, namely with a continuous parenchymal stalk linking the deep and superficial pineals (Fig. 12). In 6% of the rats investigated parenchymal cells were not present in the pineal stalk. Here the pineal complex consisted of only two parenchymal parts, namely the superficial and the deep pineals (Fig. 13). In 76% of the animals investigated a slender parenchymal stalk extended from the deep pineal towards the superficial pineal without reaching it (Fig. 14). Detailed measurements revealed that the lengths of the parenchymal stalks not reaching the superficial pineal showed a normal distribution curve, roughly 50% of the parenchymal stalks extending halfway along the length between the deep and the superficial pineals.

Quantitative studies have shown that deep tissue, i.e., deep pineal + parenchymal stalk not reaching the superficial pineal, amounted to between 0.5% and 2.5% that of the superficial pineal.

There are no statistically significant differences in the size of the deep pineal in male and female Wistar rats (BOECKMANN, 1980). Moreover, great intraspecies variation is found in the topographical relationships between the deeply located pineal tissue and the third ventricle, depending on the width of the intercommissural region on the one hand, and the dorsal extension of both the habenular commissure and the pineal stalk on the other. As shown in Figs. 15 and 16, the intercommissural contact area varies from relatively narrow to extremely wide, and the parenchymal pineal stalk may lie broadly exposed to the CSF of the suprahabenular recess or be covered, to varying degrees, by tissue seemingly belonging to the habenular commissure. It should be noted that the fairly consistent suprahabenular recess was never found to reach the superficial pineal.

While in commonly used laboratory rats the superficial pineal via the stalk and the deep pineal is structurally linked to the diencephalon, the pineal of the hooded rat (GARDNER, 1953) has been claimed not to be in contact with the epithalamus. Instead it was thought to be supported by connective tissue strands that connect the pineal with the tela choroidea rostrally and with the inferior surface of the transverse sinus caudally.

In *Jaculus jaculus* (HÜLSEMANN, 1967) (Fig. 10) the bulk of pineal tissue lies superficially, as in rats. It is tentatively classified as type $\alpha\beta$C.

As in rats, great intraspecies variation of the pineal is found in the white-footed mouse, *Peromyscus leucopus* (QUAY, 1956a). It fundamentally consists of two parts, the more anterior and basal of these surrounding a dorsal evagination of the third ventricle and the more posterior and dorsal part lying near the surface, between the cerebral hemispheres and over the anterior part of the cerebellum. Between the two parts of the pineal there was a continuous strand of pineal tissue in 31% of the animals, an interrupted strand of pineal tissue in 44%, scattered islets of pineal tissue in 22% and no intermediate pineal tissue in 3%. There was no correlation between sex and abundance or arrangement of intermediate pineal tissue.

In the collared lemming, *Dicrostonyx groenlandicus*, QUAY (1978) found relatively large pineal organs which can be classified as type ABC.

In the glirimorph rodent, *Glis glis* (HÜLSEMANN, 1967), the pineal differs from all the other rodents studied by reason of the fact that pineal tissue

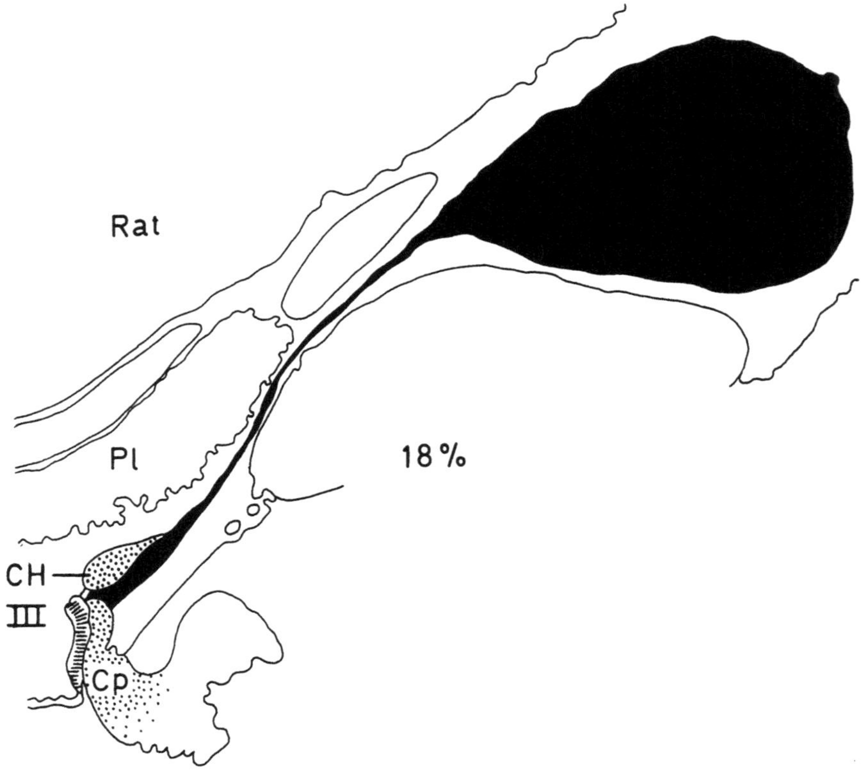

Fig. 12

Figs. 12–14. Schematic representation of the rat pineal complex, as derived from serial sagittal sections. Drawn from specimens provided by Dr. D. BOECKMANN (see BOECKMANN, 1980). **Fig. 12.** Eighteen percent of the rats investigated showed a pineal complex consisting of deep and superficial pineals linked by a continuous parenchymal stalk. **Fig. 13.** In 6% of the rats investigated the parenchymal stalk linking the deep and the superficial pineals was lacking. **Fig. 14.** In 76% of the specimens the parenchymal stalk extended from the deep pineal in a posterior direction and terminated somewhere along its course to the superficial pineal. *CH*, habenular commissure; *Cp*, posterior commissure; *Pl*, choroid plexus; *SCO*, subcommissural organ; *III*, third ventricle

is restricted to the commissural region. It is a round, compact structure belonging to type A (Fig. 10).

A representative of the Sciuromorpha, the pocket gopher, *Geomys bursarius*, has a pineal complex where the pineal organ proper lies superficially and possesses a 3–4 mm long pineal stalk (SHERIDAN and REITER, 1973). Whether it possesses a deep pineal has not been established. It is therefore tentatively classified as type βC.

k) Pholidota

According to BARGMANN (1943), *Manis javanensis* lacks a pineal organ.

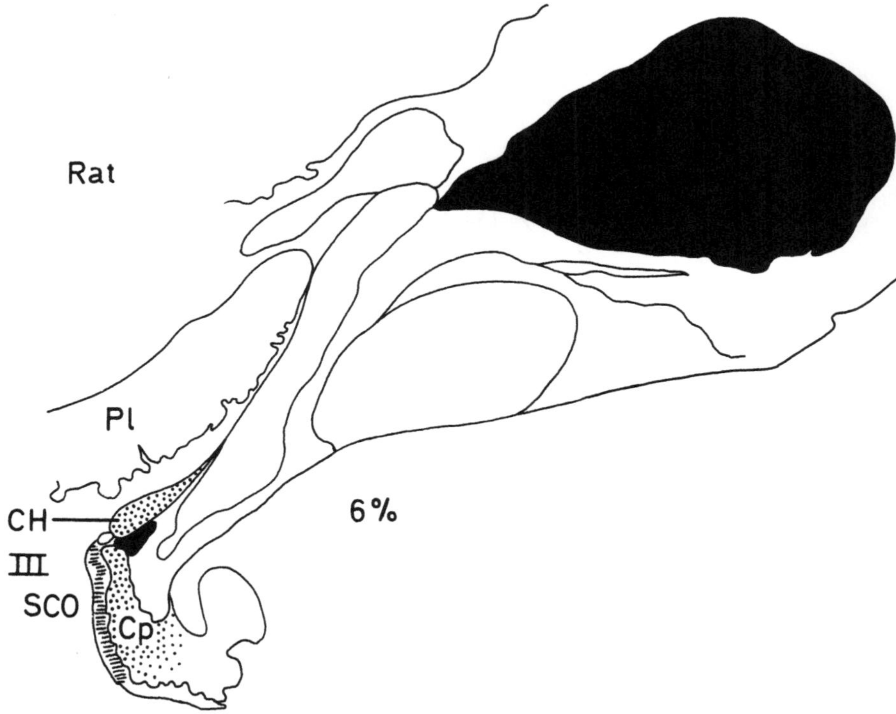

Fig. 13

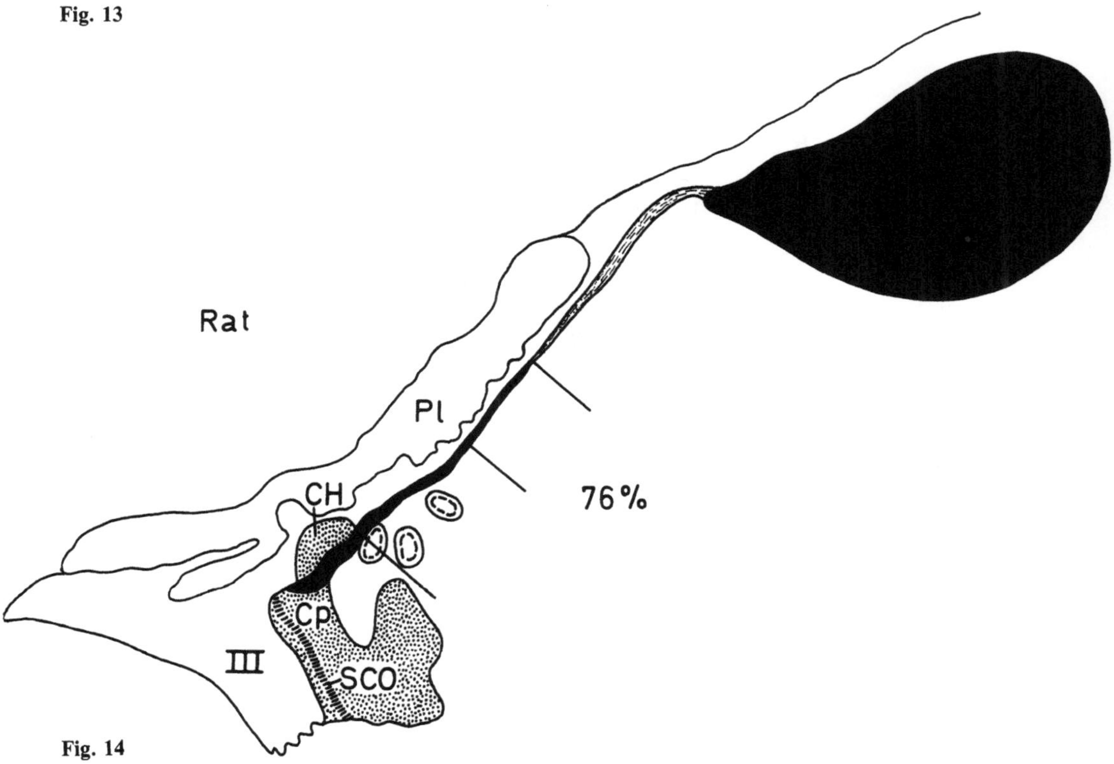

Fig. 14

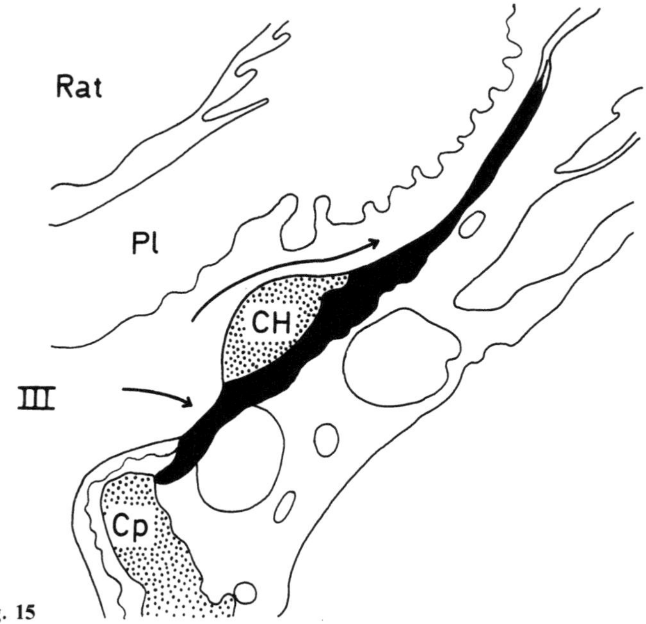

Fig. 15

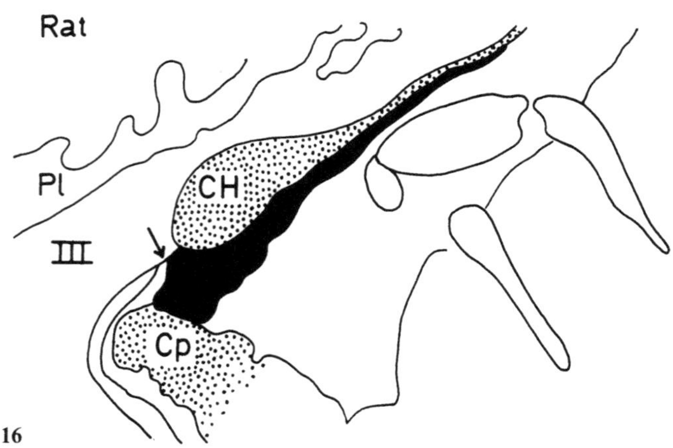

Fig. 16

Figs. 15, 16. Topographical relationships of the deep pineal to the CSF. **Fig. 15.** In some specimens the deep pineal is broadly exposed to the CSF either at the intercommissural recess (*short arrow*) or at the suprahabenular recess (*long arrow*). **Fig. 16.** In other specimens there is only a narrow or no direct zone of contact with the CSF at the intercommissural area. The direct contact with the CSF of the suprahabenular recess may also be lacking, either because the parenchymal stalk is very short or because it is covered by tissue from the habenular commissure extending in a dorsal direction. Abbreviations and material as in Figs. 12–14

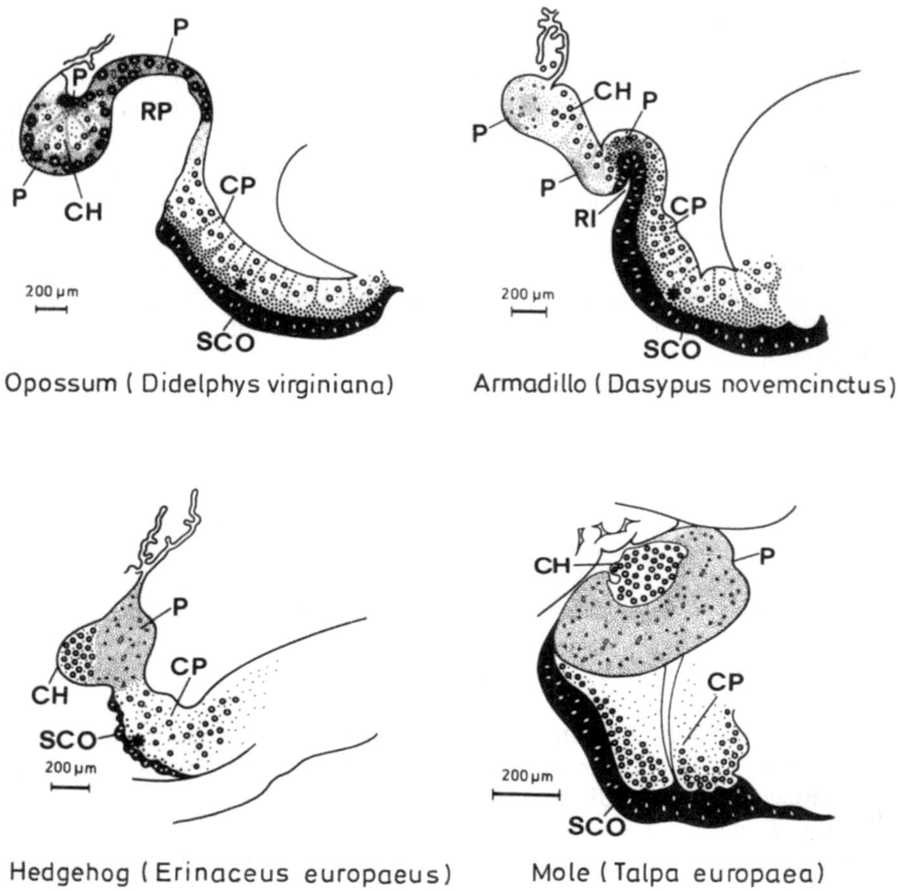

Fig. 17. Schematic representation of atypical pineal organs in four mammalian species. *CH*, habenular commissure; *CP*, posterior commissure; *P*, pineal organ; *Ri*, intercommissural recess; *RP*, pineal recess; *SCO*, subcommissural organ. HOFER et al., 1976

l) Xenarthra

Recent studies by HOFER et al. (1976) have shown that *Dasypus novemcinctus*, contrary to previous belief (BARGMANN, 1943), clearly possesses pineal tissue, which is inconspicuously arranged in the vicinity of the intercommissural recess and does not form a distinct compact or vesicular organ (Fig. 17). In view of this finding, the reported absence of a pineal in *Dasypus villosus* and *Bradypus* (BARGMANN, 1943) should be viewed with some caution.

There is no recent information available on the pineal organ in Dermoptera.

m) Chiroptera

As pointed out by QUAY (1970c) the pineal organ of bats has never been the primary subject of any published investigation. Hence, very little relevant information is currently available on this topic. QUAY (1970c) has reviewed the scanty information on the embryology, anatomy and physiology of the bat pineal that was available at that time and added a few of his own observations. In the bats, *Myotis sodalis* and *M. lucifugus* (QUAY, 1965a, 1970c) the pineal was found to be a compact organ broadly attached to the roof of the diencephalon, probably belonging to type A. A notable feature is the very insignificant pineal recess and the fact that there is little, if any, development of a suprapineal recess or dorsal sac (QUAY, 1970c).

In contrast to *Myotis*, the pineal organ of the great horseshoe bat, *Rhinolophus ferrum equinum*, has a superficial position (STAMMER, 1972), but it is not possible to classify it due to the little information available. In both young and adult *Rhinolophus* the pineal was surprisingly large when compared with brain size (STAMMER, 1972).

n) Tupaioidea

In *Tupaia glis* (HÜLSEMANN, 1967) the pineal body is triangular in shape, protruding deeply into the third ventricle (Fig. 10). It is classified as a type $A_{intraventricular}$ pineal.

No information on the pineal organ appears to be available for Macroscelidae.

o) Insectivora

The pineal organs of the Insectivora studied so far all belong to the proximal type or type A. In the shrew, *Blarina brevicauda* (QUAY, 1965a), the pineal is a compact organ broadly attached to the roof of the diencephalon. In the hedgehog, *Erinaceus europaeus* (HÜLSEMANN, 1967), the pineal protrudes deeply into the third ventricle (Fig. 10), but in the diagram published by HOFER et al. (1976) this is not so conspicuous. Shape and position of the pineal organ in the mole, *Talpa europaea* (HOFER et al., 1976), are shown in Fig. 17. The pineal organs of the African insectivores *Setifer setosus, Potamogale velox, Chlorotalpa stuhlmanni, Elephantulus fuscipes* and *Rhynchocyon stuhlmanni* have been diagrammatically depicted by STEPHAN and SPATZ (1962); they appear to be solid organs lying in the immediate vicinity of the commissures.

p) Marsupialia

In the Marsupialia studied to date the epiphysis cerebri is rather variable. According to J.I. JOHNSON (1977), *Petaurus breviceps* and *P. norfolcensis* appear to have no pineal tissue. A modest amount of pineal tissue confined to the ventricular surface between the habenular and posterior commissures was found in *Pseudocheirus* and *Trichosurus*. In *Caenolestes obscurus* pineal tissue invested the habenular commissure in a doughnut-shape configuration, a more typical mammalian pineal protuberance being present in *Thylogale billarderi* and *Macropus fuliginosus*. In the opossum, *Didelphys virginiana* (HOFER et al., 1976), the pineal was found to be a hollow structure surrounding the pineal recess and covering the anterior and basal parts of the habenular commissure (Fig. 17).

Solid pineals broadly attached to the posterior roof of the diencephalon are present in kangaroos. In the red kangaroo, *Megaleia rufa*, the pineal was found to be a symmetrical bilobate structure, and in the grey kangaroo, *Macropus canguru*, a uniformly rounded organ (QUAY and BAKER, 1965). The pineal recess is especially deep and appears to be incompletely lined by ependymal cells, which become extremely attenuated in this region (QUAY, 1966c). It thus appears that in marsupials the pineal organ, if at all present, can be classified as type A. This conclusion has been substantiated in a recent study in which the pineal organs of 11 marsupial species belonging to five families were studied (KENNY and SCHEELINGS, 1979).

q) Monotremata

In the Monotremata, *Ornithorhynchus anatinus* and *Tachyglossus aculeatus*, the pineal is a well-developed solid organ situated between the habenular and posterior commissures (KENNY and SCHEELINGS, 1979).

3. Weight and Volume

a) Weight

Data on pineal weight are available for a great number of mammalian species (Table 4) and in experimental studies pineal weight has often been used as an indicator of the functional activity of the organ. The data published

Table 4. Pineal weight

Order, Species	Weight (mg)	Author(s)
Primates		
Man	138.6[a]	LEGAIT and LEGAIT (1977)
		LEGAIT and OBOUSSIER (1977)
	129.06	VON EYL and GUSEK (1975)
	137.5	ARIETI (1954)
	117	BASINSKA et al. (1969a)
Macaca irus	5.3–10.0	KENNY et al. (1973)
Artiodactyla		
Nesotragus	8.9	LEGAIT and OBOUSSIER (1977), LEGAIT et al. (1977)
Rhynchotragus	7.6	
Cephalophus	14.8	
Raphicerus	19.4	
Oreotragus	26.3	
Hyemoschus aquaticus	9.9	
Ouerebi	13.7	
Pelea	32.5	

Table 4 (continued)

Order, Species	Weight (mg)	Author(s)
Gazella	36.8	
Antidorcas	29.4	
Redunca	38.9	
Ovis aries	47.7	
Sheep	69.3	KENNAWAY and SEAMARK (1975)
	182	BLIN and MAURIN (1956)
	101	BASINSKA et al. (1969a)
	81.34	SANTAMARINA and VENZKE (1953)
Aepyceros melampus	9.7	LEGAIT and OBOUSSIER (1977), LEGAIT et al. (1977)
Capra aegagrus	26.7	
Goat	75	BLIN and MAURIN (1956)
Lama guanicoe	87.4	LEGAIT and OBOUSSIER (1977), LEGAIT et al. (1977)
Phacochoerus	46.5	LEGAIT et al. (1977)
Cervus	244.5	
Damaliscus	92.8	
Oryx gazella	135.0	
Tragelaphus	89.3	
Alcelaphus	132.3	
Connochaetes	425.4	
Kobus	48.8	
Hippotragus	81.3	
Alces alces	148.4	
Syncerus	334.5	
Bos banteng	157.8	LEGAIT and OBOUSSIER (1977), LEGAIT et al. (1977)
Bos taurus domesticus	157.8	
Cattle (see also Table 7)	324	SANTAMARINA (1958)
Ox	300	BLIN and MAURIN (1956)
Okapia	59.9	LEGAIT and OBOUSSIER (1977), LEGAIT et al. (1977)
Taurotragus	155	
Giraffa camelopardalis	360.9	
Pig	40–50	BLIN and MAURIN (1956)
	66	HEINECKE (1959)
	34	BABINSKA et al. (1969a)
	40–50	SANTAMARINA and VENZKE (1953)
Perissodactyla		
Dolichohippus grevyi	378.2	LEGAIT and OBOUSSIER (1977)
Equus onager	163.3	
Equus caballus	166.4	

Table 4 (continued)

Order, Species	Weight (mg)	Author(s)
Horse	150–200	BLIN and MAURIN (1956)
Ass	520	
Mule	860	
Lagomorpha		
Oryctolagus cuniculus	12.4	LEGAIT and OBOUSSIER (1977)
Carnivora		
Fissipedia		
Felis domesticus	10.4	LEGAIT and OBOUSSIER (1977)
Felis sylvestris	6.5	
Ailurus fulgens	1.6	
Vulpes vulpes	7.4	
Canis aureus	1.6	
Lutra lutra	5.9	
Canis (vulgaris) familiaris	8.8	
Scotch terrier	6.8	
Canis lupus	31.0	
Hyaena crocuta	27.2	
Panthera tigris	11.1	
Ursus arctos	15.6	
Pinnipedia		
Phoca vitulina	787.9	
Arctocephalus pusillus	145.5	
Zalophus californiensis	70.5	
Otaria byronia	100.6	
Thalarctos maritimus	24.6	
Odobenus rosmarus	3,439.6	
Leptonychotes weddelli	3,500–3,600 (newborn) 400–600 (adult)	TEDMAN (1977)
	1,000	CUELLO and TRAMEZZANI (1969a, b)
Callorhinus ursinus	260	ELDEN et al. (1971)
Rodentia		
Mus musculus	0.2	LEGAIT and OBOUSSIER (1977)
Eliomys quercinus	2.8	
Rattus norvegicus	1.5	
Sprague-Dawley rats	1.56	REITER (1973e)

Table 4 (continued)

Order, Species	Weight (mg)	Author(s)
Pigmented Long-Evans rats	0.90	REITER (1973e)
Long-Evans rats males	1.269	QUAY (1961a)
Mesocricetus auratus males	0.202	QUAY (1961a)
Cavia aperea	2.5	LEGAIT and OBOUSSIER (1977)
Pedestes angola	2.5	
Dolichotes patagonia	4.4	
Marsupialia		
Macropus benetti	11.0	LEGAIT and OBOUSSIER (1977)
Macropus rufus	35.6	
Macropus giganteus	25.1	
Macropus canguru	10.6–14.61	QUAY and BAKER (1965)
Megaleia rufa	6.36–40.12	QUAY and BAKER (1965)

[a] Usually given as mean.

and the conclusions drawn should be viewed with some caution for a number of reasons. To begin with, pineal gross anatomy may be fairly complex so that it may be virtually impossible to remove the pineal in toto and to weigh it. Moreover, the often tiny organs dry out quickly, which again leads to changes in weight. Comparison of the data obtained is seriously hampered by the fact that the organs were weighed either unfixed or fixed. Moreover, there are indications that pineal weight varies depending on the time of day, season and a great number of environmental factors that are difficult to control. Finally, it should be noted that according to MEIJERS and ZWEENS (1962) weight is not a good indicator of the functional activity of the pineal organ because the weight greatly depends on the amount of intercellular body fluid, which is reduced by oestradiol and increased by progesterone.

On the whole the data on pineal weight published so far do not lend themselves to a meaningful interpretation. Hence, only some points will be mentioned. As can be expected, pineal weight has been thoroughly studied in humans and varies considerably. In 747 cases studied, overall mean pineal weight was 138.6 mg; interestingly, mean pineal weight differed strikingly in different months (March, 135 mg; April, 172 mg; August, 116 mg; September, 181 mg) (LEGAIT and LEGAIT, 1977). Mean pineal weight was 129.06 mg when calculated from the data published by VON EYL and GUSEK (1975). According to RODIN and OVERALL (1967), no correlation exists between pineal weight on the one hand and body weight, brain weight, sex or colour, on the other. The largest human pineal ever reported weighed 940 mg and came from a 25-year-old male patient suffering from "polyglandular syndrome" (BARTLETT, 1913).

There are clear indications that in human adults pineal weight changes with age. Studying 585 pineals, VON EYL and GUSEK (1975) observed an increase

Table 5. Average human pineal weight (mg) according to age (VON EYL and GUSEK, 1975)[a]

	0–10	11–20	21–30	31–40	41–50	51–60	61–70	71–80	81–90 years
♂	77	113	127	127	151	132	152	148	121
♀	70	83	106	148	160	133	164	155	156

[a] No. of specimens studied 585.

Table 6. Ratios of pineal weight to brain weight in humans according to age (VON EYL and GUSEK, 1975)

	0–10	11–20	21–30	31–40	41–50	51–60	61–70	71–80	81–90 years
♂	1:9250	1:11750	1:10520	1:10540	1:8710	1:10120	1:8570	1:8640	1:10570
♀	1:8920	1:14410	1:11970	1:8620	1:7770	1:9300	1:7490	1:7890	1:7620

in pineal weight from birth onwards, reaching a maximum between 40 and 50 years of age, followed by a decline in the 6th decade and a renewed increase in the 7th decade. In the 8th and 9th decades pineal weight showed a slight decrease (Table 5). By contrast, RODIN and OVERALL (1967) noted a slight decrease in pineal weight at about 35 years of age, followed by an increase between 50 and 65 years and a decrease thereafter. That a decrease in pineal weight seems to occur regularly between 30 and 50 years of age is also indicated by data from BERBLINGER and UEMURA (see BARGMANN, 1943): BERBLINGER found a decrease between 41 and 50 years and UEMURA between 31 and 40 years.

Although the presence of sex differences in pineal weight has been denied (RODIN and OVERALL, 1967), it is interesting to note that in VON EYL and GUSEK's (1975) material females had lighter pineals up to the age of 30, while thereafter the pineals were always heavier than in males. The ratios of pineal weight to brain weight according to age are given in Table 6.

In *animal* species pineal weight also varies considerably. In rats it ranges from 0.4 to 1.8 mg (AXELROD et al., 1965a) and, as shown in Fig. 18, undergoes diurnal changes, exhibiting lowest values at 18.00 h. Diurnal changes are not always readily detected, as emphasized by MERRITT and SULKOWSKI (1969), who were unable to observe clear-cut differences in rats.

The problem of age-dependent changes of pineal weight in adult animals has been studied by SANTAMARINA and VENZKE (1953). The results obtained are shown in Table 7. It can be seen that, apart from great intraspecies variation, there is a tendency, as in humans, toward a decrease in weight of the pineal in certain phases of adult life. Not shown in Table 7 is the finding in dairy cows that the pineals were larger than in beef cows.

The problem of sex differences in pineal weight has received little attention. SANTAMARINA and VENZKE (1953), studying 425 adult male rats and 250 adult female rats, found average weights of 0.9 mg (range 0.2–1.9 mg) in males and 0.7 mg (0.3–1.4 mg) in females.

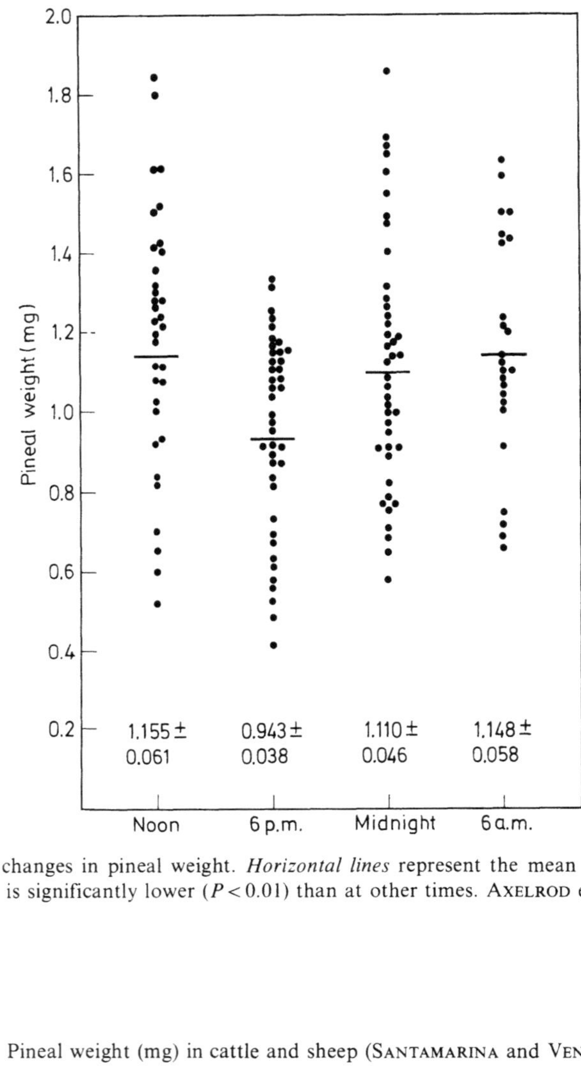

Fig. 18. Diurnal changes in pineal weight. *Horizontal lines* represent the mean values. The weight at 6 p.m. is significantly lower ($P < 0.01$) than at other times. Axelrod et al., 1965a

Table 7. Pineal weight (mg) in cattle and sheep (Santamarina and Venzke, 1953)

Age	Cows	Bulls	Ewes	Rams
12–24 months	133.80 ± 24.60	160 ± 33.90[a]	56.50 ± 12.30[c]	
2 yr	161.69 ± 39.58		95.40 ± 27.32	96.64 ± 27.07
3 yr	274.50 ± 78.27		78.33 ± 44.27	102.96 ± 40.28
4 yr	261.37 ± 95.77		86.00 ± 18.38	87.37 ± 45.78
5 yr	262.00 ± 88.63			
6 yr	193.66 ± 62.53	174.00 ± 53.71[b]		
7 yr	178.00 ± 46.69			
8 yr	229.27 ± 73.92			
10/11 yr	238.00 ± 72.83			

[a] 6–18 months.
[b] 2–6 years.
[c] 1 year.

Seasonal changes in pineal size have been demonstrated by volumetric studies. This aspect will be dealt with in the following section.

b) Volume

Pineal volume has been studied in a great number of species with the aim of relating it to body weight and brain weight (LEGAIT et al., 1975a, 1976a, b). The results obtained are shown in Table 8. Without going into great detail it suffices to mention that in all orders examined pineal volume showed great variation, as indicated by the absolute volumes of the pineal as well as the epiphyseal index (for details, see LEGAIT et al., 1976a), rendering correlation studies difficult. Nevertheless, the large amount of data obtained shows fairly clearly that pineal volume and body weight (correlation coefficients between 0.6503 and 0.8601) as well as pineal volume and brain weight (0.6702–0.8076) are positively correlated. With respect to pineal function the positive correlation with body weight has been interpreted to be due to the fact that the pineal is an endocrine organ and that its size can be expected to be correlated with the amount of circulating blood and hence body weight.

Table 8. Pineal volumes (LEGAIT et al., 1976a, b)

Species	Volume (mm^3)	Species	Volume (mm^3)
Homo and **Simians**		*Cheirogaleus medius*	0.231
Homo sapiens	46.12	*Lemur fulvus*	0.430
Ateles paniscus	1.64	*Lemur variegatus*	1.729
Colobus badius	4.48	*Hapalemur simus*	0.316
Pan troglodytes	20.5	*Lepilemur ruficaudatus*	0.316
Hylobates lar	4.66	*Microcebus murinus*	0.075
Gorilla gorilla	3.28	*Avahi laniger*	0.286
Nasalis larvatus	10.04	*Indri indri*	2.203
Pygathrix nemaeus	1.89	*Propithecus verreauxi*	1.226
Papio doguera	22.75	*Daubentonia madagascariensis*	1.421
Erythrocebus patas	7.28	*Loris loris*	0.196
Cercopithecus ascanius	3.20	*Nycticebus cougang*	0.529
C. mitis	8.21	*Perodicticus potto*	0.707
C. talapoin	1.60	*Galago crassicaudatus*	0.580
Macaca mulatta	2.21	*Galago demidovi*	0.197
Cercocebus albigena	13.24	*Tarsius syrichta*	0.996
Saimiri sciurea	1.29	*Tarsius bancanus*	0.367
Lagothrix lagotricha	3.09	*Urogale everetti*	0.630
Cebus albifrons	3.02		
Callicebus moloch	0.80	**Rodents** and **lagomorphs**	
Aotes trivirgatus	0.06	*Sciurus vulgaris*	0.5331
Tamarin tamario	0.09	*Marmotta marmotta*	2.849
Callithrix jacchus	0.51	*Citellus citellus*	0.2753
		Glis glis	0.0653
Prosimians		*Muscardinus avellanarius*	0.0226
Tupaia glis	0.258	*Eliomys quercinus*	0.1592
Cheirogaleus major	0.424	*Graphiurus graphiurus*	0.0206

Table 8 (continued)

Species	Volume (mm^3)	Species	Volume (mm^3)
Sigmodon hispidus	0.3179	*Saccopteryx canescens*	0.0131
Cricetus cricetus	0.7296	*S. bilineata*	0.0634
Mesocricetus auratus	0.1926	*S. leptura*	0.0498
Microtus orcadensis	0.4050	*Rhinilophus hipposideros*	0.0343
M. senegalensis	0.0120	*R. ferrum equinum*	0.2069
Meriones libycus	0.0466	*R. euryale*	0.0533
M. shawi	0.1429	*Asellia tridens*	0.0176
M. crassus	0.1006	*Eptesicus brasilensis*	0.0148
M. persicus	0.0331	*M. myotis*	0.0476
Dasymys incomtus	0.0157	*M. bechsteinii*	0.0177
Lemniscomys striatus	0.0137	*Myotis cappacini*	0.0105
Rattus norvegicus	0.4237	*M. oxygnathus*	0.0192
Rattus (albinos)	0.6149	*M. natteri*	0.0103
Praomys morio	0.0183	*M. nigricaus*	0.0055
Cricetomys gambianus	0.0726	*Rhogeessa parvula*	0.0150
Myocastor coypus	3.892	*Pipistrellus pipistrellus*	0.0211
Mus musculus	0.0503	*Lasiurus borealis*	0.0288
Cavia aperea	0.9172	*Chaerophon leucostigma*	0.0116
Oryctolagus cuniculus	6.0284	*Molossus ater*	0.0413
		M. molossus	0.0505
Chiroptera		*Molossops planirostris*	0.0191
Chilonycteris personata	0.0079	*M. temmincki*	0.0109
C. rubiginosa	0.0860		
Mormoops megalophylla	0.0174	**Insectivora**	
Phyllostomus discolor	0.0463	*Sorex minutus*	0.00241
P. hastatus	0.8588	*S. araneus*	0.00443
Choeroniscus godmani	0.0002	*Crocidura russula*	0.00726
Anoura geoffroyi	0.0237	*C. occidentalis*	0.0161
Glossophaga longirostris	0.0492	*Suncus murinus*	0.0312
G. soricina	0.0306	*Echinops telfairi*	0.1052
Lonchophylla thomasi	0.0001	*Hemicentetes semispinosus*	0.092
Lonchoglossa caudifer	0.0211	*Setifer setosus*	0.2587
Leptonycteris curasoae	0.0312	*Tenrec ecaudatus*	0.2343
Carollia castenea	0.0031	*Erinaceus europaeus*	0.247
C. perspicillata	0.0001	*Aetechinus algirus*	0.1533
Sturnira lilium	0.0544	*Hemiechinus auritus*	0.0681
S. tildae	0.0244	*Neomys fodiens*	0.0114
Artibeus cinereus	0.0212	*Chlorotalpa stuhlmanni*	0.0307
A. jamaicensis	0.0677	*Nesogale talazaci*	0.0439
A. lituratus	0.2033	*N. dobsoni*	0.0223
Mesophylla macconelli	0.0105	*Elephantulus fuscipes*	0.160
Uroderma bilobatum	0.0505	*Galemys pyrenaicus*	0.0391
Vampyrops helleri	0.0251	*Talpa europaea*	0.212
Vampyressa pusilla	0.0324	*Limnogale mergulus*	0.0450
Vampyrops vittatus	0.0236	*Desmana moschata*	0.2037
Enchisthenes hasti	0.0182	*Rhynchocyon stuhlmanni*	0.705
Sphaeronycteris toxophyllium	0.0163	*Potomogale velox*	0.0016
Desmodus rotundus	0.0984	*Solenodon paradoxus*	0.735
Diphylla ecaudata	0.1189	*Crizoryctes talpoides*	0.0216
Pteropus lylei	2.0157	*Microgale cowani*	0.0068
Peropteryx macrotis	0.1299	*Chrysochloris asiatica*	0.0146
Rhynchiscus naso	0.0150	*C. leucorhina*	0.0533

Interesting is the observation that the mean pineal index was found to increase from insectivores (103.2), over simians (132.2), prosimians (208.9), and rodents (220.1) to Chiroptera (462.5) (LEGAIT et al., 1976a), but interpretation of these figures is difficult because the number of animals studied for each species was usually very small and because the pineals may have been in different and hence not comparable states of functional activity. In female Wistar rats mean pineal volume was larger than in males; however, the difference was not statistically significant (LEGAIT et al., 1976c). In Wistar rats weighing 6 to 411 g the correlation coefficient between pineal volume and body weight was found to be 0.8731 and the allometry coefficient, 0.60 (LEGAIT et al., 1976c).

That the state of the functional activity of the pineal gland may be important for pineal volume is shown by the fact that in the dormouse, *Glis glis*, pineal volume is small in July and large in October and January, when the animals hibernate (LEGAIT et al., 1973). In view of the influence of light and darkness on pineal function, it is interesting to note that in 32 rodent species pineal volume is small in rodents from Africa and large in rodents from countries with less daily exposure to light (LEGAIT et al., 1973, 1975a).

Changes in pineal weight or volume under experimental conditions will be dealt with in Sect. B. XI.

4. Epiphysectomy

Epiphysectomy, a widely used technique to study epiphyseal function, can be easily accomplished in some of the common laboratory rodents, e.g., the rat, in which the epiphysis lies superficially. Epiphysectomy in hamsters was also regarded as a simple procedure until it was recognized that in this species the pineal complex consisted of both a superficial and a deep pineal (SHERIDAN and REITER, 1970a). Complete pinealectomy also appears to be difficult in the rabbit, in which a major part of the distal end of the pineal is invaginated into the transverse sinus (A. SMITH, 1971). References on pinealectomy in various animals have been given by KITAY and ALTSCHULE (1954a).

Technical details on the procedure may be found in the publications by KITAY and ALTSCHULE (1954a), THIÉBLOT and LEBARS (1955), QUAY (1965c), SCHERZER (1964), HOFFMAN and REITER (1965a) and HSIEH and OTA (1969) to mention just a few papers. It should be noted that whereas KITAY and ALTSCHULE (1954a) described the procedure as difficult, with an operative mortality of up to 98%, in recent years epiphysectomy has turned out to be a highly successful procedure with nearly 100% survival (HOFFMAN and REITER, 1965a). The standard procedure used in our laboratory on rats follows that described by HOFFMAN and REITER (1965a) and BLISS and BATES (1973). Following anaesthesia (5 mg Nembutal/100 g body weight, IP) the scalp is cut anteroposteriorly along the midline from between the eyes to beyond the base of the skull. The skin flaps are lifted and the head is tightly held by an assistant, or preferably fixed by means of a special device, comparable to a stereotactic instrument. After carefully scraping the underlying tissues in an area approximately 1 cm in diameter and overlying the epiphysis, a cylindrical drill with a diameter of 5 mm is centred on the confluence of the superior sagittal and transverse sinuses and a ring drilled, using only slight pressure. To avoid jamming, the drill is lifted every 5 s. The ring is drilled to the desired depth, reducing the speed when approaching the dura. After removal of the bone disc, the area is carefully swabbed to avoid excess bleeding and the clearly visible epiphysis cerebri is removed by means of a pair of watchmaker's or iridectomy forceps.

After applying a small amount of antiseptic powder, the bone disc is rapidly returned to avoid excessive haemorrhage, and the skin flaps are pulled together and stapled with a suitably sized standard desk stapler from which the anterior end of the base has been ground off up to the front margin of the stapling plate. For a few days after the operation the animals are kept singly to avoid scratching the wounds. Unlike HOFFMAN and REITER (1965a), we do not attach an adjustable collar to the drill to preset the required depth. The important points for a successful operation are to use adequate pressure during drilling and to avoid excessive haemorrhage by quick and careful swabbing.

A method for pinealectomizing rat fetuses in utero, carried out as early as 15 days 16.5 h post coitum and with a survival rate of 60%, has been described by OWMAN (1963a). KINCL and BENAGIANO (1967) used hypothermia instead of ether anaesthesia for pinealectomies in neonatal rats.

Further papers on technical aspects of pinealectomy include ROCHE and DZIUK (1969), BRAUN et al. (1977), KUSZAK and RODIN (1977) and LETELLIER et al. (1978).

III. Development

The mammalian epiphysis cerebri develops from a diencephalic ependymal area, usually lying between the habenular and posterior commissures (see BARGMANN, 1943). A notable exception is found in the opossum, in which the pineal anlage is subcommissural rather than intercommissural (A. MACHADO, 1965). The concept that the pineal anlage is a uniform one has been challenged by FRAUCHIGER (1963), who claimed that the epiphysis cerebri is composed of two genetically different parts, namely a glandular one and a neuronal (cerebral) one. Diencephalic evaginations, which FRAUCHIGER compared with the neurohypophysis, were thought to act as inducers influencing dorsally lying mesenchymal or ectodermal cells to become pineal parenchymal cells which in turn would mingle with cells of cerebral origin (ependymal cells, astrocytes) to form the definitive epiphysis. It would appear, however, that there is insufficient evidence as yet for this new concept (see p. 64).

1. Man

Recent studies do not give a complete picture of all the developmental stages of the human epiphysis cerebri; rather they deal with specific aspects, such as its innervation and its fine structure. Hence, for obtaining the basic

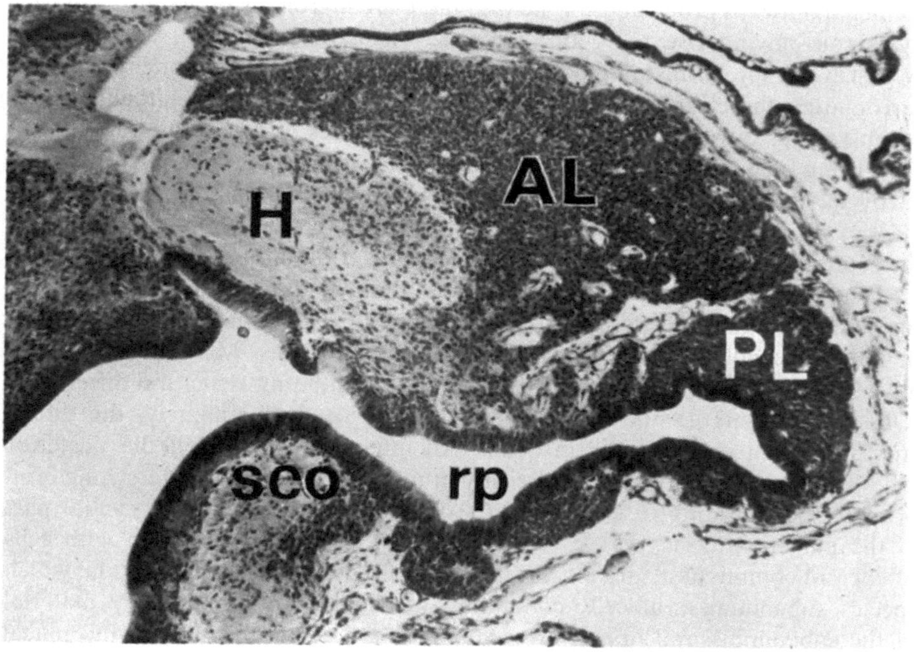

Fig. 19. Sagittal section through the pineal gland of a human fetus, CRL = 100 mm. Note the separation into an anterior lobe (*AL*) and a posterior lobe (*PL*). *H*, habenular area; *rp*, pineal recess; *SCO*, subcommissural organ. MØLLER, 1974; ×85

information the reader is referred to early publications (GLADSTONE and WAKELEY, 1935, 1940; BARGMANN, 1943). Here it suffices to give a brief summary. According to BARGMANN (1943), referring to papers by HOCHSTETTER, KRABBE and TURKEWITSCH, the human epiphysis first appears in embryos of 6–8 mm total length, corresponding to the beginning of the 2nd month of gestation. O'RAHILLY (1968) dated the appearance of the pineal body to the 33rd postovulatory day. At this stage, the pineal primordium lies in front of the posterior commissure and consists of two components, namely a band of thickened ependyma that is slightly raised above the general surface and a shallow evagination lying underneath it. The further development is characterized by cell proliferation, accompanied by cell migration, at the anterior and anterolateral regions of the thickened band of ependyma, giving rise to a solid cell mass, the so-called anterior lobe of the epiphysis cerebri (Fig. 19). As a result of the forward growth of the thickened ependyma, the groove deepens, assuming the typical form of the primary pineal diverticulum in lower classes of vertebrates. The posterior lobe of the epiphysis is first represented by the ependymal lining of the diverticulum; later a posterior lobe proper is formed by cell proliferation in the basal region of the diverticulum (Fig. 19). While the posterior lobe is being formed, the uniform pineal diverticulum disappears. Prior to the disappearance of the diverticulum, the wall becomes folded, and between the hollow epithelial outward projections of the wall there are ingrowths of vascular connective tissue, giving rise to a lobulation of the parenchyma (GLADSTONE and WAKELEY, 1940). In histological sections the deeply stained parenchymal cells surrounding the clear vascular areas have the appearance of epithelial cords cut in various directions and appear either as rosettes or cell cords.

During further development, the anterior and posterior lobes, partly separated by a septum of vascular connective tissue (Fig. 19), fuse to form a solid mass of tissue. At 150 days of development [197 mm crown-rump length (CRL)], the septum has almost completely disappeared (HÜLSEMANN, 1971).

Whereas according to BARGMANN (1943) the two epiphyseal lobes are structurally indistinguishable, GLADSTONE and WAKELEY (1940) noted that the anterior lobe, especially in its centre, was more homogeneous. More recent studies have confirmed the presence of two pineal lobes exhibiting structural differences (OLSSON, 1961; HÜLSEMANN, 1971; MØLLER, 1974). Unfortunately, the differences have not been studied in detail. An interesting feature of the posterior lobe is the occasional presence of richly granulated cells with distinct tinctorial properties. According to OLSSON (1961) these cells stem from the caudal part of the pineal ventricle, whose cells have many features in common with cells of the subcommissural organ. Similar results were obtained in *Bos*, in which species subcommissural cells containing the characteristic secretory material of the subcommissural organ were present in the caudal portion of the pineal organ (OLSSON, 1958). The cellular arrangement differs as well. According to HÜLSEMANN (1971), the cells of the anterior lobe show a net-like arrangement whereas those of the posterior lobe have a palisade-like arrangement. Moreover, during the later phases of development cavities occur in the posterior lobe, giving rise to a spongy appearance (HÜLSEMANN, 1971).

Different developmental stages of the human epiphysis are depicted in Fig. 20. Note the shift of the longitudinal axis of the pineal anlage from a more or less vertical position to a horizontal position. This is brought about by intensive growth of the anterior part, especially in a caudal direction, aided by developmental changes of the adjacent brain tissue (HÜLSEMANN, 1971).

The postnatal development of the pineal recess has been studied by PALKOVITS et al. (1962c). Up to 1 year of age, the pineal recess measured 1,000–2,200 μm in depth and 900–1700 μm in width. Subsequently it increased to 1,600–3,000 μm in depth and 1,600–3,500 μm in width, but at 18 years it was shallow, measuring 350–600 μm in depth.

A characteristic feature of the adult epiphysis cerebri is that the organ is richly innervated by sympathetic and commissural fibres (p. 186). According to HÜLSEMANN's (1971) electron microscopic studies, commissural fibres are probably first to arrive in the epiphysis cerebri. They reach the epiphysis before the 60th day of development (47 mm CRL), and MØLLER (1976) noted them in fetuses of 30 mm CRL. Sympathetic nerve fibres, forming the nervus conarii, penetrate the distal pole of the epiphysis between the 60th (47 mm CRL) and 82nd (83 mm CRL) day of development in HÜLSEMANN's (1971) material, while MØLLER (1976) detected them at as early as 30 mm CRL. In the vicinity of blood vessels sympathetic nerve fibres were not noted before the 120th day (HÜLSEMANN, 1971). Varicosities of sympathetic nerve fibres were frequently

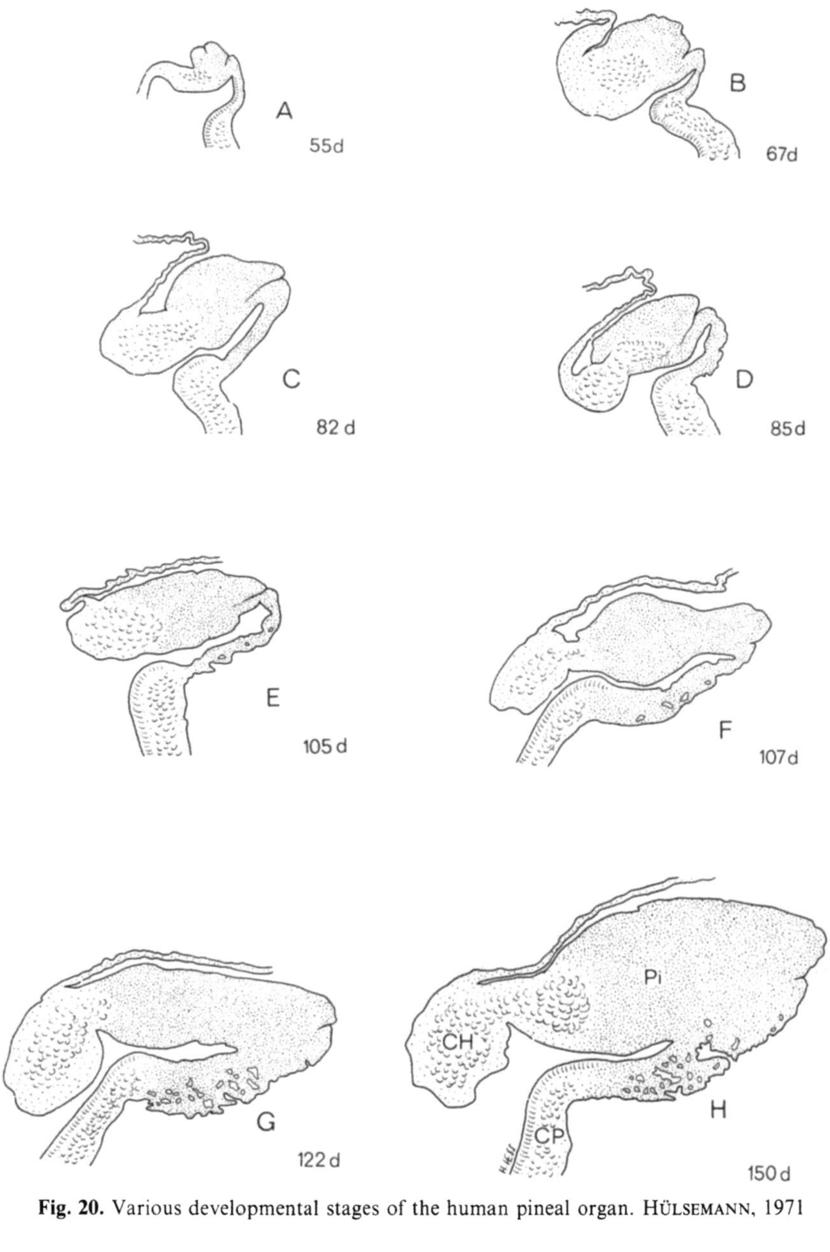

Fig. 20. Various developmental stages of the human pineal organ. HÜLSEMANN, 1971

found at 150 days of gestation. They contained both clear vesicles and dense-core vesicles typical of sympathetic fibres in general (p. 195). However, the cytochemical technique by WOOD revealed that the vesicles were still immature at this stage (MØLLER, 1976).

In human fetuses a supposedly newly discovered nerve linking the posterior part of the epiphysis with the subcommissural organ has been described

(Møllgaard and Møller, 1973). Closely related to these nerve fibres was an accumulation of ganglion cells, lying at the boundary of the external limiting membrane at the dorsal aspect of the brain stem and partly inside the brain tissue, closely associated with the hypendyma of the rostral-most part of the subcommissural organ. Recently Møller (1978) corrected this point and mentioned that in the material studied nerve cells were located all along the nerve from the penetration of the external limiting membrane to the apex of the pineal gland. An equivalent nerve has been found in sheep and rabbit fetuses, though an actual ganglion was lacking (Møller et al., 1975). This nerve may be homologous to the pineal nerve of anurans, which extends from the frontal organ via the posterior wall of the pineal body to the subcommissural organ (cf. Oksche, 1971; Wake et al., 1974). In view of the current concept that the mammalian pineal organ contains only efferent and no afferent nerve fibres (p. 186) it is interesting to note that Møller et al. (1975) claim that the fetal pineal nerve contains afferent fibres as indicated by branching of these fibres in an afferent direction and the presence of axodendritic synapses. The precise destination of this nerve is not known (Møller, 1978).

The *vascular network* of the developing epiphysis cerebri is prominent and possesses perivascular spaces of varying widths. With increasing age, the spaces decrease in size, first in the anterior lobe and later in the posterior lobe. The endothelial cells of the capillaries lack fenestrations and possess tight junctions (Møller, 1974).

Pinealocyte differentiation has also been studied. Opinions vary as to the exact stage at which pinealocytes can be identified. Møller (1974) recognized pinealocytes at as early as 30 mm CRL; Hülsemann (1971) noted some at 83–92 mm CRL, but found it difficult to identify them even at 197 mm CRL. The most detailed account of the prenatal differentiation of the human pinealocyte is that by Møller (1974), who studied 18 specimens ranging in size between 30 and 178 mm CRL. Additional information may be obtained from the paper by Hülsemann (1971). During the period studied, human pinealocytes exhibited relatively few changes. The main difference was that pinealocytes of older fetuses showed a larger amount of cell organelles than those of younger ones. The fetal pinealocytes (Fig. 21) measured 8–12 μm in diameter and possessed several processes often terminating in club-shaped endings. The nuclei were predominantly ovoid in shape and showed a slight condensation of chromatin near the nuclear envelope, which possesses pores with a diameter of between 50 and 80 nm. Mitochondria, most of them elongated, were randomly located in the cytoplasm. Mitochondria with tubular cristae typical of the fully differentiated mammalian pinealocyte could not be detected. The amount of randomly distributed granular endoplasmic reticulum was moderate. In addition to free ribosomes, groups of up to 15 polyribosomes were often seen. Smooth endoplasmic reticulum was rare and lay mainly in terminal clubs of pinealocyte processes. The Golgi complex consisted of a stack of 2–5 parallel membrane-bounded sacs with the typical concave–convex appearance and was actively engaged in the production of smooth and coated vesicles. Granular bodies which may represent secretory products were most often spherical or slightly elongated. They measured 100–600 nm in diameter and were located both in the cell body

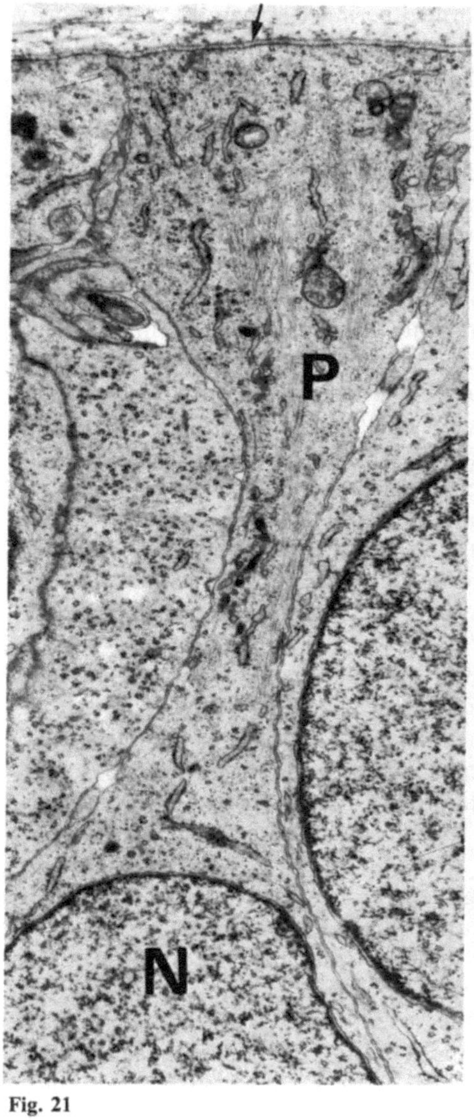

Fig. 21

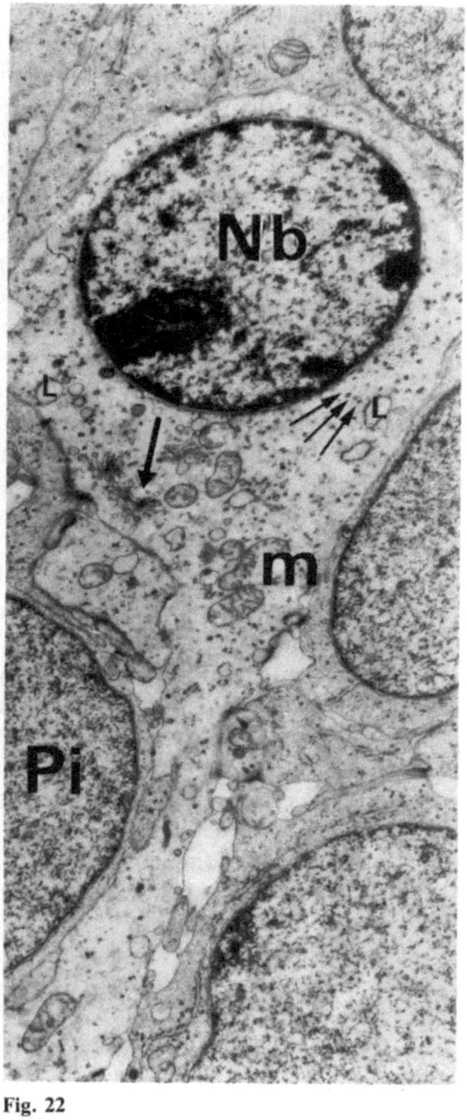

Fig. 22

Fig. 21. Electron micrograph of a pinealocyte at the periphery of the pineal gland of a human fetus. The cytoplasm around the nucleus (*N*) is free of filaments, in contrast to the numerous filaments in the process (*P*). The surface of the gland facing the subarachnoid space is covered by a basal lamina (*arrow*). MØLLER, 1974; × 12,000

Fig. 22. Electron micrograph of a neuroblast (*Nb*) among pinealocytes (*Pi*). The nucleus of the neuroblast is characterized by a coarse chromatin pattern, and the cytoplasm contains many ribosomes (*arrows*) and single cisternae of the granular endoplasmic reticulum (*L.*). *m*, mitochondria; *large arrow*, Golgi apparatus. MØLLER, 1974; × 9,000 and the cell processes. Lysosomes were rare. Microfilaments of 5–8 nm in diameter were abundant. While some of them formed perinuclear bundles, the majority were seen in cell processes, especially in the centre. Microtubules with a diameter of between 23 and 27 nm and a width of the wall of 8 nm were rare and were randomly distributed in the cytoplasm. Centrioles and cilia were present. Whereas lipid droplets were rare, glycogen was abundant, especially in the cell processes. Coated vesicles were common and omega figures were often seen on the cell membrane, which was found to be equipped with gap junctions, intermediate type junctions and desmosomes.

The comparison with the fully differentiated pinealocyte (p. 110) shows that the human fetal pinealocyte is ultrastructurally basically similar to that of adult mammals. Without detailed ultrastructural knowledge of the *adult* human pinealocyte, the functional activity of the *fetal* pinealocyte is difficult to assess, but the assumption by MØLLER (1974) that the latter possess all the organelles necessary for hormone synthesis seems reasonable. Yet it has to be pointed out that the fetal pinealocytes do not possess synaptic ribbons (MØLLER, 1976). This principal difference may be of crucial functional importance (p. 146). The presence of numerous gap junctions points to possible functional interrelationships between fetal pinealocytes (MØLLER, 1976). The capability of the pinealocytes to synthesize indoleamines is indicated by the observation that as early as 155 mm CRL some pinealocytes show a yellow serotonin fluorescence (Falck-Hillarp technique) (OLSON et al., 1973).

In addition to developing pinealocytes, the fetal epiphysis cerebri contains neuroblasts (Fig. 22) (MØLLER, 1974) and filamentous astrocytes (HÜLSEMANN, 1971). The latter cells were first noted at 83–93 mm CRL. They were particularly conspicuous in the periphery of the organ, where they were attached to one another by desmosomes. Moreover, they were found around blood vessels. In specimens with a CRL of 197 mm, melanosomes were found in glial cells lying near the tip of the epiphysis.

Postnatally, major histological changes occur. They are difficult to interpret functionally since electron microscopic studies are lacking. Whereas immediately after birth the pineal cells are relatively homogeneous in appearance, a mosaic pattern soon becomes apparent. These changes, already hinted at by KRABBE as well as by GLOBUS and SILBERT (see GLADSTONE and WAKELEY, 1940), have been thoroughly described by KERÉNYI and SARKAR (1968). The mosaic pattern is due to the fact that within a period of 2–3 weeks after birth large pineal parenchymal cells with large and pale nuclei occur, forming aggregations which are separated from one another by anastomosing narrow bands of densely packed smaller cells with chromatin-rich nuclei, resembling lymphocytes in appearance. The mosaic pattern was found to be well preserved until approximately 6 months of age. Different reasons have been given for the disappearance of the mosaic pattern, which is almost complete by about 9 months of age. KERÉNYI and SARKAR (1968) suggested that the large, pale cells became smaller and more chromophilic and that the small cells developed into elongate, fibroblast-like cells. A differentiation of the small, dark cells into fibroblasts has also been described by GLOBUS and SILBERT (see GLADSTONE and WAKELEY, 1940). KRABBE, on the other hand (see GLADSTONE and WAKELEY, 1940), claimed that the small cells develop into parenchymal cells with large, pale nuclei, and hence represent "proparenchymal" cells. In view of the relative scarcity of fibroblasts in the mammalian pineal organ, KRABBE's opinion appears to be the most likely one.

In the 1st year of postnatal life rosette-like structures consisting of ependymal cells of the subcommissural organ are regularly present in the pineal (PALKOVITS et al., 1962c).

2. Rat

The course of pineal development is best known in rats. QUAY (1974a), who has thoroughly reviewed this topic, distinguished three major developmental and maturational phases, which overlap to some extent. First there is a morphogenetic phase that begins on the 12th day of embryonic life and extends until birth. A phase of cellular proliferation starts on the 16th day of embryonic life and terminates a few days after birth. A phase characterized by cellular hypertrophy and differentiation is the longest one, beginning at birth and ending 9–12 weeks later.

As in man, the first indication of the pineal anlage is the occurrence of a shallow evagination of the diencephalic roof in the region of the habenular and posterior commissures between 12 and 14.5 days of intra-uterine life (GARDNER, 1949; KAPPERS, 1960a; OWMAN, 1960/61, 1964a; CLABOUGH, 1973). Between 14.5 and 16 days the shallow epiphyseal evagination becomes elongate, forming a sac-like structure (Fig. 23), its longitudinal axis pointing in a dorsocaudal direction. Moreover, the rostrodorsal and caudoventral walls of the epiphyseal sac thicken. Proliferation of the ependymal cells, especially in the rostrodorsal area, is very pronounced between 16 and 17 days, giving rise to folds and follicles. At the same time, in most of the epiphyseal cells a secretory (?) material intensely stainable with paraldehyde-fuchsin was noted, especially in cell pro-

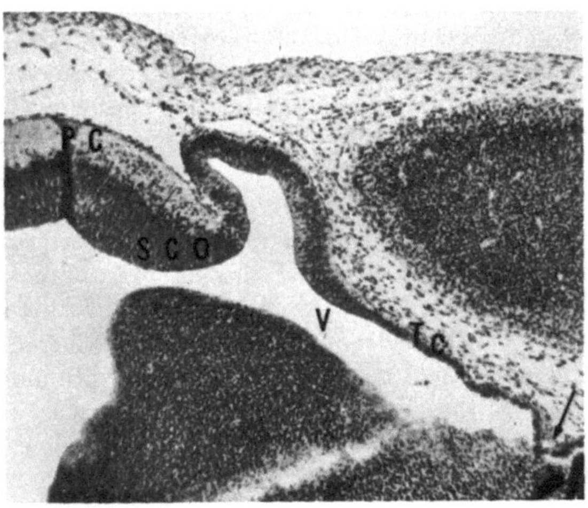

Fig. 23. Pineal anlage and adjacent structures in a 16-day-old fetal rat. Midsagittal section. Posterior to the aulage are the posterior commissure (*PC*) and the developing subcommissural organ (*SCO*); rostrally is the tela chorioidea (*TC*), which terminates as a downward projection (*arrow*) into the third ventricle (*V*). CLABOUGH, 1973; ×50

cesses extending to the epiphyseal lumen (OWMAN, 1960/61, 1964a). Mesenchyme and blood vessels penetrated the epiphysis between the 17th and 18th day, though in one study capillaries were noted already at 16 days 20 h (BOURGES and BOUCHER, 1975). They form connective tissue strands and an extensive vascular network, lying close to parenchymal follicles. Moreover, a rearrangement of the epiphyseal cells occurs, as a result of which the cells form perivascular palisades.

The follicular appearance of the fetal epiphysis cerebri is a relatively short-lived one. As early as day 19 the epiphysis has lost its follicular structure and consists of a very dense, compact mass of cells. KERÉNYI and VON WESTARP (1968, 1971), by contrast, noted rosette-like structures even after birth and found that when animals were kept in constant darkness the parenchymal rosettes persisted. Due to cell proliferation between day 19 and birth the epiphysis grows rapidly and increases in compactness. Simultaneously, perivascular connective tissue forms septa which separate groups of cells, giving rise to islands which are clearly visible at birth (GARDNER, 1949). As soon as islands are present, regional differences become apparent: while in marginally located islands the nuclei are pale, those of the centrally located islands exhibit distinctly more chromatin. By 150 days of age the islands appear uniform (GARDNER, 1949).

Regarding the development of innervation, nerve fibres from at least two sources have to be considered. *Commissural fibres* were found at 18 days of gestation, penetrating three-quarters of the length of the epiphysis; in the vicinity of blood vessels no nerves were present at this stage (GARDNER, 1953). *Sympathetic nerve fibres* (GARDNER, 1953; HÅKANSON et al., 1967; A. MACHADO et al., 1968; C. MACHADO et al., 1968; A. MACHADO, 1971; A. MACHADO and LEMOS, 1971) reach the epiphysis only postnatally. Immediately after birth they were found mainly in the capsule of the organ, as assessed by formaldehyde-induced fluorescence, but under the electron microscope some rare nerve terminals were occasionally observed in contact with pineal cells. A few isolated nerve bundles penetrated the organ 24 h after birth and in 2-day-old animals nerve bundles and varicose terminals were abundant in the dorsal and apical regions of the epiphysis, but were sparse ventrally and anteriorly. One day later varicose terminals were present throughout the gland, the nerve fibres showing signs of branching. During the first days after birth the nerve bundles became ensheathed by Schwann cells and a basal lamina. Right from the beginning, the nerve fibres exhibited a positive reaction for acetylcholinesterase. At days 6–7 the green fluorescence of the nerve fibres changed to yellow, coinciding with the occurrence of a yellow serotonin fluorescence in pinealocytes. The adult pattern of fluorescence intensity was reached at 10–11 days and that of the number of varicose terminals by 21 days. The ultrastructure of the developing sympathetic nerve fibres is described on p. 201. The effects of administration of nerve growth factor antiserum have been studied by SCHOTT et al. (1970). They found that when the antiserum was given to 19-day-old rat embryos there was a permanent reduction of the capacity of the pineal to take up noradrenaline. Administration to neonates was followed by a 25% reduction in noradrenaline uptake.

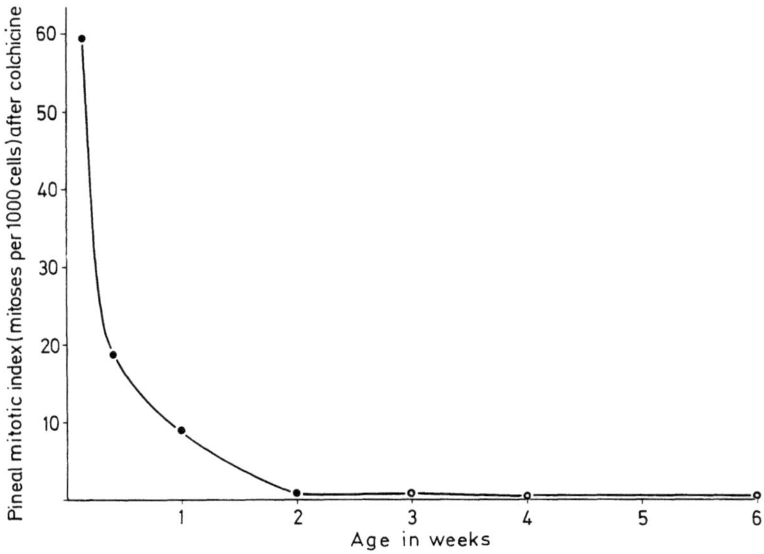

Fig. 24. Mitotic index in the rat pineal gland after colchicine treatment. Note the rapid decrease of the mitotic index during the first 2 weeks of postnatal development. QUAY and LEVINE, 1957a

After the 4th postnatal day, the epiphysis is gradually displaced in a caudal and dorsal direction along the quadrigeminal lamina of the mesencephalon, but it retains its connexion with the commissural region by means of a slender stalk. The stalk is formed by fusion of the two epiphyseal peduncles, which originate from the rostrodorsal part (anterior peduncle) and the caudoventral part (posterior peduncle) of the epiphyseal sac (KAPPERS, 1960a), but according to MEINEL (1967) fusioning does not occur proximally, but only in the intermediate and distal areas of the epiphyseal sac. Concurrently with the formation of the epiphyseal stalk, the epiphyseal recess disappears (KAPPERS, 1960a).

A characteristic feature of the postnatal development of the rat pineal is a pronounced increase in size. Since immediately after birth the mitotic activity of the epiphysis, as assessed by means of colchicine treatment (QUAY and LEVINE, 1957a, b; QUAY, 1974a) and ^{3}H-thymidine autoradiography (WALLACE et al., 1969; TRAKULRUNGSI and YEAGER, 1977), is high and decreases precipitously during the 1st and 2nd weeks (Figs. 24, 25), epiphyseal growth in this developmental phase is due mainly to cellular hyperplasia (WALLACE et al., 1969). Thereafter, epiphyseal growth is brought about by cellular hypertrophy, as illustrated by a decreasing cell density (QUAY and LEVINE, 1957a, b; WALLACE et al., 1969) (Fig. 26) and a rapid increase of the mean nuclear diameter (BLUMFIELD and TAPP, 1970). The decrease in cell density and the increase in nuclear size were particularly pronounced during the first 8 weeks of life (BLUMFELD and TAPP, 1970). Whereas the cell density remained fairly constant between 8 and 30 weeks, nuclear size reached a maximum at about 12 weeks. Thereafter, there was a slight drop, but subsequently nuclear size remained at a fairly constant level. Moreover, the average diameters of the nuclei in the central area of the organ appeared to be slightly greater than those in the periphery.

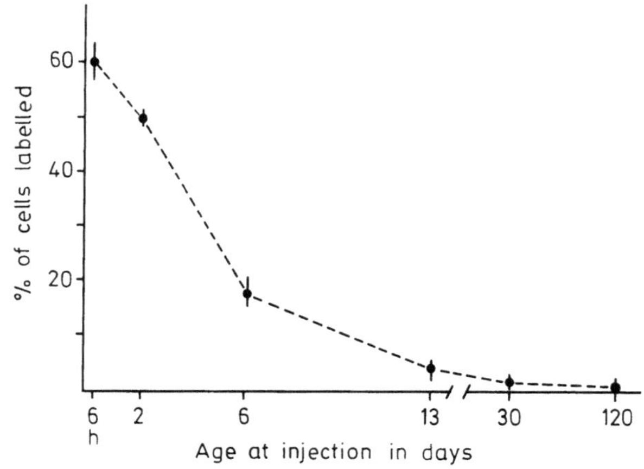

Fig. 25. Cell proliferation in the rat pineal body as studied by ^{3}H-thymidine autoradiography. The percentage of labelled cells in the pineal body in animals injected at different ages and surviving for 120 days is shown. Redrawn from WALLACE et al., 1969

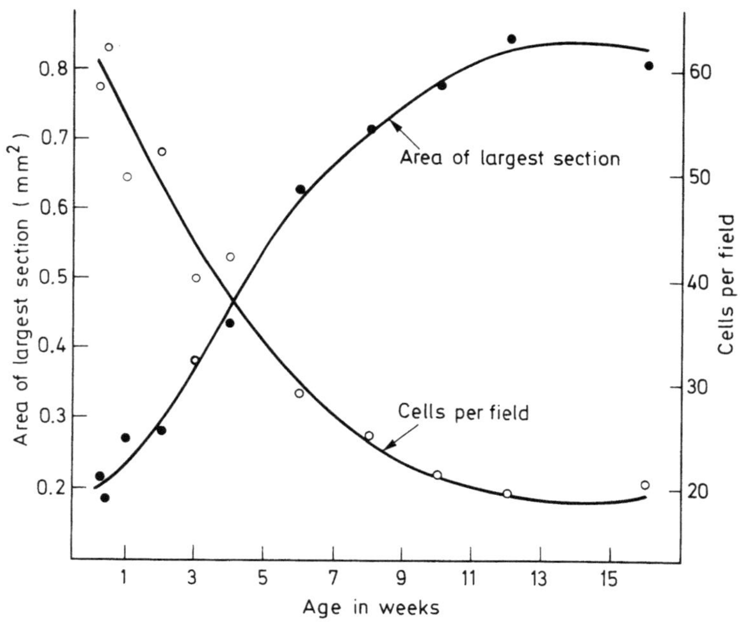

Fig. 26. Pineal growth from birth to 16 weeks of age in the rat. Note the increase in size, as indicated by an increase in the area of largest section, i.e. mean area of the largest section in the pineals of a particular age measured from camera lucida outline drawings by planimetry. Cells per field = mean number of nuclei of all cells counted in microscopic fields selected at set intervals (1 field area = 0.00168 mm^2) in the largest section in the pineals. QUAY and LEVINE, 1957a

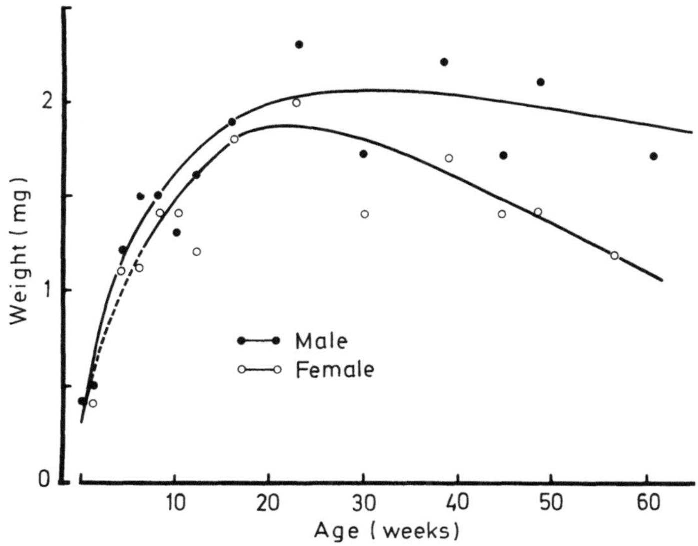

Fig. 27. Age-dependent changes in the weight of the rat pineal gland. BLUMFIELD and TAPP, 1970

It should be noted, however, that results of other studies favour the notion that the nuclei in the centre are smaller than those in the periphery (p. 69).

Volumetric data show that epiphyseal growth continues up to 8.5 weeks of age in the hooded rat (WALLACE et al., 1969) and up to 12 weeks in Sprague-Dawley rats (QUAY and LEVINE, 1957a, b). Gravimetric data confirm the rapid postnatal growth of the rat epiphysis, but authors disagree regarding the growth pattern. QUAY (1974a) concluded that the available data do not show a linear increase in epiphyseal weight but rather fluctuations of unknown and perhaps doubtful character. An almost linear increase in epiphyseal weight from birth to 84 days, with a slight decrease between 20 and 24 days, was described by TAIT et al. (1969). BLUMFIELD and TAPP (1970) noted a rapid increase during the first 8 weeks, followed by a slower increase, and reaching a maximum at 23 weeks in females and at 30 weeks in males (Fig. 27). LOMBARD DES GOUTTES (1966, 1967a) distinguished two growth phases and described a sinusoidal weight curve. The first phase, from 1 to 6.5 days, was characterized by a 200% increase in weight and the second phase, from 6.5 days to puberty, by a 50% increase. During the first phase, a particularly sudden increase occurred between the 6th and 7th days postnatally, when pituitary and thymus weights showed prominent peaks as well. The sinusoidal curve, mentioned above, was characterized by an amplitude of 0.1 mg. Peaks were noted at 6, 10, 13, 16 and 23 days of age and troughs at 2, 7, 11, 14 and 19 days. No changes in weight occurred between 25 and 35 days. Judging from the graph published, mean pineal weight increases from 0.45 to 0.85 mg between 2 and 25 days of age.

The development of pineal weight under the influence of pineal grafts has been studied by DAS GUPTA (1969). Following the implantation of one to eight pineals into the thighs, pineal weight decreased. Changes were most noticeable

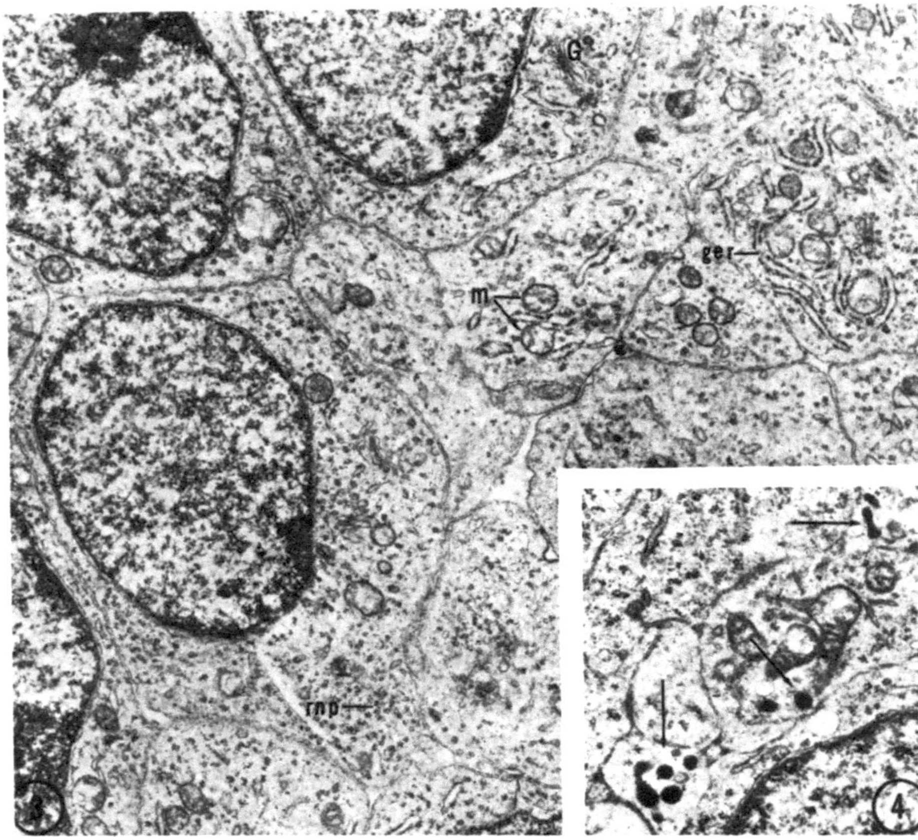

Fig. 28 **Fig. 29**

Fig. 28. A cluster of pineal cells typical of the 18-day-old fetal rat pineal organ, showing granular endoplasmic reticulum (*ger*), mitochondria (*m*), Golgi apparatus (*G*), and polysomes (*rnp*). CLABOUGH, 1973

Fig. 29. Eighteen day-old fetal rat. Various types of dense bodies or granules (*arrows*) are found in pineal cells. CLABOUGH, 1973

when the recipients were 1–2 weeks old and when the grafts were left in place for 7–8 weeks. Although a decrease in weight of the host pineal was already noted with one pineal transplanted, it was even more pronounced when three to five or five to eight transplants had been grafted. *Structural* changes were noted only when five to eight pineals had been used. The pinealocytes were then much smaller in size than in control animals and the contents of total lipids and phospholipids as well as the activity of oxidative and hydrolytic enzymes had decreased. Ultrastructurally the perivascular space appeared extremely narrow.

Differentiation at the *cellular level* has been studied both pre- and postnatally. According to electron microscopic studies (CLABOUGH, 1973), epiphyseal cells at 16 days of embryonic life possess large oval nuclei, numerous polyribosomes,

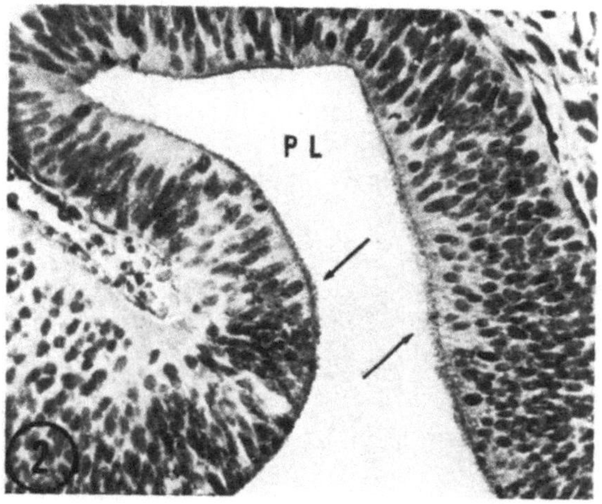

Fig. 30. Pineal anlage in a 16-day-old fetal rat showing apical cytoplasmic knobs (*arrows*) which project from the circumluminal cells into the pineal lumen (*PL*). Midsagittal section. CLABOUGH, 1973

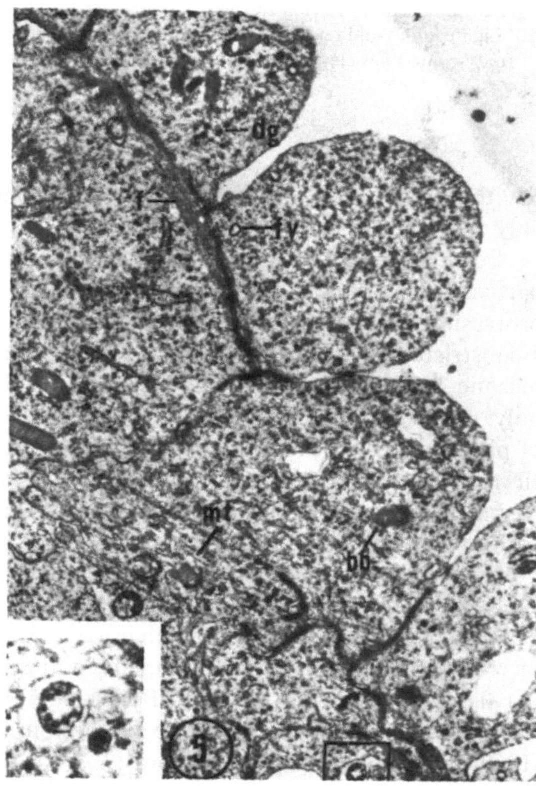

Fig. 31. Apical knobs of circumluminal cells in an 18-day-old fetal rat pineal organ showing microtubules and filaments, dense granules, fuzzy-coated vesicles and a basal body. A cross-section of a ciliary shaft (*enlarged at left*) reveals nine pairs of peripheral tubules. *bb*, basal body; *dg*, dense granule; *f*, filament; *fv*, fuzzy-coated vesicle; *mt*, microtubule. CLABOUGH, 1973

a few vesicles and some small mitochondria. At 18 days (Figs. 28, 29) distinct nucleoli were noted and, in addition to polyribosomes, granular endoplasmic reticulum began to appear. A small juxtanuclear Golgi apparatus was frequently present and dense bodies measuring 0.05–0.3 μm occurred occasionally through-

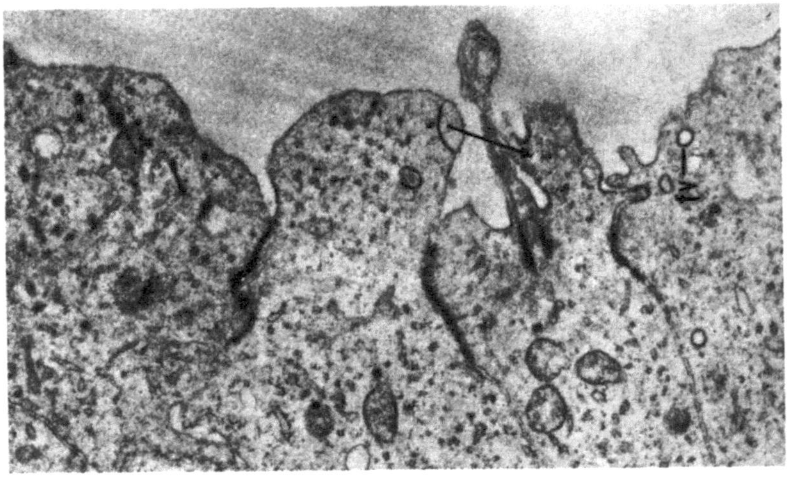

Fig. 32. Longitudinal section through a ciliary derivative which arises from a pineal circumluminal cell. Eighteen-day-old fetal rat. The distal portion is irregularly enlarged and contains a few vesicles. *fv*, fuzzy-coated vesicle; *arrow*, possible cross-section through the med-region of a ciliary shaft. CLABOUGH, 1973

out the cytoplasm. The majority of mitochondria were small and contained only a few cristae.

For phylogenetic considerations the observation is of interest that, particularly at 18 days, the circumluminal cells of the epiphysis possessed apical protrusions (Figs. 30–32), measuring 5 μm in diameter, which sometimes showed a constricted neck (OWMAN, 1960/61; CLABOUGH, 1973) and cilia-bearing cytoplasmic bulges (CLABOUGH, 1970). Whereas OWMAN, using light microscopy only, noted a certain resemblance between these protrusions and outer segments of pineal photoreceptors of lower vertebrates, CLABOUGH, using both light and electron microscopy, concluded that the apical knobs resembled "ellipsoids" of inner segments of photoreceptors and that the cilia-bearing bulges were an abortive attempt to produce surface membrane specializations, possibly of a photoreceptive nature. B. ZIMMERMAN and TSO (1975) observed comparable signs of photoreceptor differentiation between 4 and 12 days after birth.

The further development is characterized, at 20 days of gestation, by an increase in the amount of cell organelles and the occurrence of glycogen (Fig. 33) (CLABOUGH, 1973; BOURGES and BOUCHER, 1975), the latter authors confirming the presence of what looked like the secretory material described by OWMAN (1960/61) using the light microscope (p. 257). Lipid droplets, abundantly present in fully differentiated rat pinealocytes (p. 79), were already noted in newborn rats (MACHADO, 1966).

For the *postnatal* differentiation of the rat pinealocyte *quantitative* data are available. From the results of BAYEROVÁ and MALINSKÝ (1972) it can be seen (Fig. 34) that the various compartments of the rat pinealocyte change differently. While Golgi apparatus and agranular vesicles show little change from birth

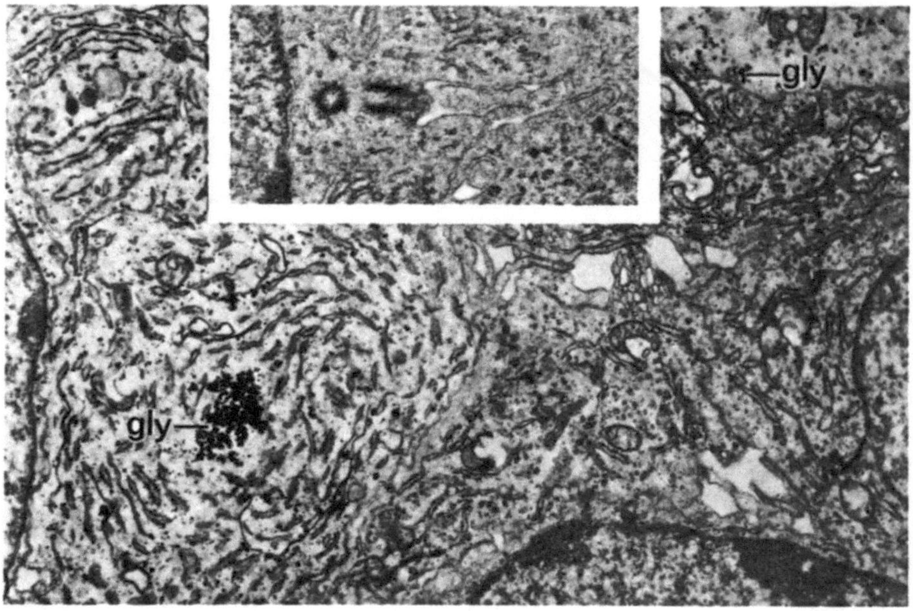

Fig. 33. Pineal cells in a 20-day-old fetal rat showing glycogen (*gly*). CLABOUGH, 1973

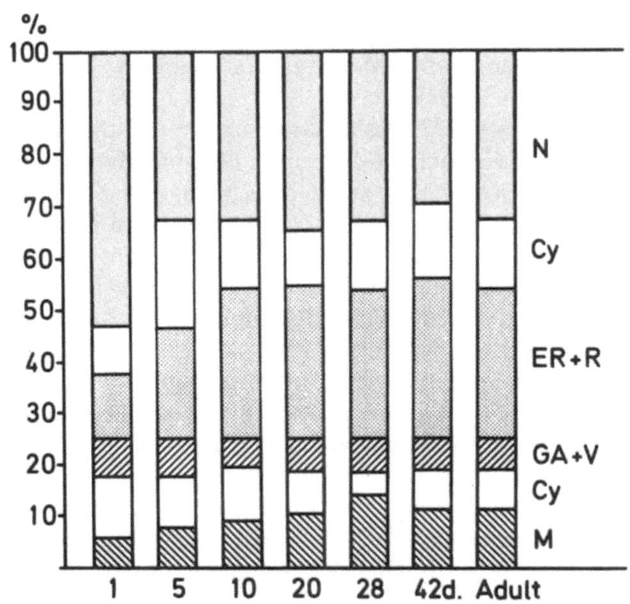

Fig. 34. Development of various cellular organelles in rat pinealocytes, expressed as percentage of total cell colume. *Cy*, cytosol; *ER+R*, endoplasmic reticulum and zones comprising ribosomes; *GA+V*, Golgi apparatus and agranular vesicles; *M*, mitochondria; *N*, nucleus. Adapted from BAYEROVÁ and MALINSKÝ, 1972

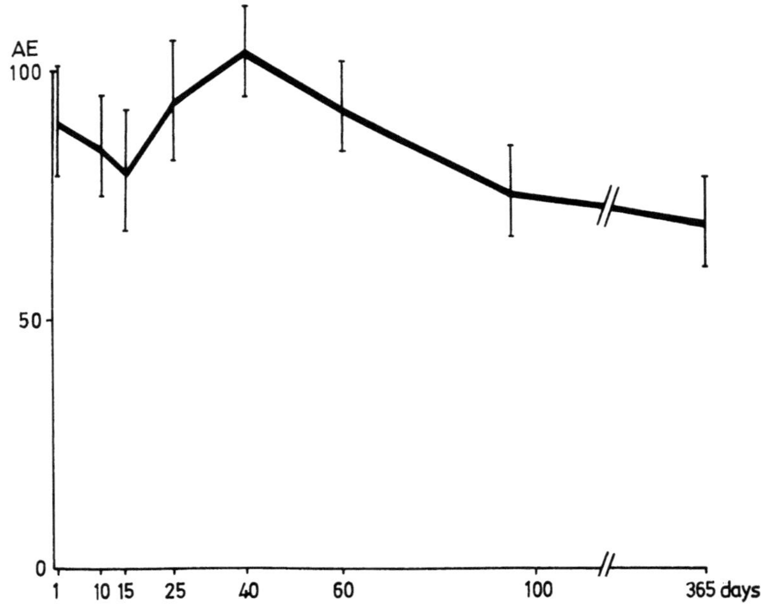

Fig. 35. RNA content of pineal parenchymal cells (rat) during postnatal development, as expressed in working units (*AE*). Cytophotometry after gallocyanin-chrome alum staining. NOVÁKOVÁ et al., 1969

to adulthood, endoplasmic reticulum and ribosomes increase strikingly between birth and 10 days, thereafter remaining fairly constant. An active RNA synthesis soon after birth has been confirmed by autoradiography using ^{14}C-uridine (MACKINNON and SIMPSON, 1968).

BAYEROVÁ and MALINSKÝ (1972) have further shown that the relative amounts of mitochondria increased up to 28 days of age, subsequently showing distinctly lower, but virtually identical values at 42 days of age and in adulthood. In view of the increase of ribosomes and endoplasmic reticulum it is not surprising that the nucleic acids undergo demonstrable changes during development. As shown in Fig. 35, the RNA content was found to decrease between birth and 15 days of age when studied by means of cytophotometry (NOVÁKOVÁ et al., 1969). Both cytophotometric (NOVÁKOVÁ et al., 1969) and chemical (Table 9) (NIR et al., 1969a) determinations agree that in the subsequent developmental phase (ca. 15 to 40 days of age) an increase of the pineal RNA content occurs. In the later phases, the results obtained are at variance: while NIR et al. (1969a) observed a further increase up to the age of 51 days (Table 9), NOVÁKOVÁ et al. (1969) noted a decrease from 40 days of age to adulthood (Fig. 35).

The results of *DNA* determinations are also at variance. While NOVÁKOVÁ et al. (1969) could not detect any change throughout the 1st year of life, NIR et al. (1969a) reported a steady increase from 21 days to 51 days of age (Table 9).

Pineal protein content, protein composition and free amino acids showed changes as well. As can be seen in Table 9, the protein content increases (NIR et al., 1969a). Fast-moving proteins have been shown to decrease, while slow-moving proteins increase (NIR et al., 1974a). The same authors found that

Table 9. RNA, DNA and protein contents of the rat pineal during development (NIR et al., 1969a).[a]

Age	RNA	DNA	Protein
21 days	6.01 ± 0.27	4.86 ± 0.19	68.4 ± 1.5
31 days	9.27 ± 0.29	5.76 ± 0.22	88.0 ± 2.3
41 days	10.93 ± 0.34	6.66 ± 0.22	112.7 ± 5.3
51 days	11.93 ± 0.29	7.01 ± 0.29	115.4 ± 3.2

[a] μg/pineal gland ± SEM.

glutamic acid, tryptophan, leucine, lysine and phenylalanine showed a continuous increase with age, while valine and 5-*OH*-indole, a possible derivative of serotonin but not identified as a metabolite in any tissue, exhibited a striking increase only between 10 weeks and 1 year. By contrast, glycine, ornithine, histidine, iso-leucine, alanine, serine, glutamine, threonine and asparagine decreased between 31 days and 10 weeks of age.

Experimentally it was shown that when developing rats were exposed to continuous light, the RNA and protein contents increased more slowly than in control animals kept under alternating light and darkness. DNA levels were not markedly affected. Exposure to continuous darkness had virtually no effect (NIR et al., 1969a).

While the majority of parameters investigated show progressive changes during development, regressive changes occur as well. These are particularly prominent between 3 days and 3 weeks after birth and include cell death followed by resorption and phagocytosis (QUAY and LEVINE, 1957a). For the development of melatonin formation see p. 258.

3. Mouse

Relatively little is known about pineal development and differentiation in mice. As in rats, pineal cell proliferation declines rapidly after birth, a striking decrease occurring between 2 and 9 days of age. At 21 days DNA synthesis is found in less than 1% of the cells, and by 1 month it had essentially ceased (DILL and WALKER, 1966). Epiphyseal growth is particularly pronounced between birth and 40 days of age and between 60 and 90 days, when adult weight values were attained (ITO and MATSUSHIMA, 1967). Between birth and 40 days, the amount of parenchymal nuclei per unit area decreased strikingly, indicating cellular hypertrophy. Moreover, the nuclei become pleomorphic (ITO and MATSUSHIMA, 1967, 1968a, b). Histochemical studies have shown that the glycogen content of the pineal decreases after the 5th postnatal day and reaches the 5-day levels again at 60 days of age; day–night differences in glycogen content become apparent at 22 days of age (KACHI et al., 1975).

4. Hamster

Pineal development in hamsters is of interest in view of the subdivision of the epiphyseal complex into superficial and deep pineals (p. 28). The prenatal development has been studied by CLABOUGH (1970, 1973) and CLABOUGH and SEIBEL (1968). It is basically similar to that in rats, though starting distinctly earlier, namely at 12 days of gestation, with the occurrence of a discrete epiphyseal evagination, preceded by a thickening of the diencephalic roof due to an increased mitotic activity (Fig. 36a, b). The further stages of pineal development are depicted in Fig. 36c–f. For further details see the legend to Fig. 36.

The first signs of a subdivision of the hamster pineal complex into superficial and deep pineals are visible by day 6, when the pineal organ, after a phase of rapid growth followed by elongation, shows a distinct, attenuated parenchymal stalk linking what is later to become the deep and the superficial pineals (HEWING, 1976). By 8 days the stalk has doubled in length and the deep pineal has increased in size. The next 7 days of development are characterized by a progressive thinning of the pineal stalk, which at 22 days of age consists of a few parenchymal cells, blood vessels and a bundle of non-myelinated axons (HEWING, 1976). Since in adult hamsters the pineal stalk lacks parenchymal cells, it would be of interest to examine at what stage of development these cells disappear.

Differentiation at the cellular level has also been studied. According to HEWING (1976) pinealocyte differentiation is more rapid in the cortical regions than in the central regions of the superficial pineal. While light microscopically the pinealocytes of the cortical zones of 16-day-old animals were indistinguishable from those in adults (HEWING, 1976), HATTORI (1971) using the electron microscope concluded that the pinealocytes were mature at 70 days of age, although lysosomes occurred only between 80 and 90 days of age. Dense-core vesicles measuring 70–130 nm in diameter appeared in the pinealocytes at 50 days. At the same time oligodendrocytes and astrocytes could be distinguished, after a phase of rapid proliferation of glial cells between 30 and 40 days of age. Synaptic ribbons appear in the pinealocytes at 4 days after birth, reaching a maximum number at 9 days followed by a decrease to adult numbers at 16 days (HEWING, 1979).

The postnatal development of the sympathetic innervation and the serotonin fluorescence in the *Djungarian hamster, Phodopus sungorus,* has been studied by VAN VEEN et al. (1978). The *nervi conarii* reach the pineal capsule at one day after birth and a few fibres begin to invade the parenchyma. By day 5 a rich green-fluorescing fibre network has spread throughout the organ, including the stalk and deep pineal. Serotonin fluorescence in the parenchyma is first noted at 10 days. At 15 days the adult state is reached, the intense serotonin fluorescence masking the green catecholamine fluorescence of the nerve fibres. At this latter stage, pineals of *Phodopus* and *Mesocricetus* showed the same degree of differentiation.

5. Rabbit

Little recent information is available on the development of the rabbit epiphysis cerebri. HOLMES (1957) noted that in the epiphyses of 20–26 day old fetuses many of the parenchymal cells formed follicles, or rosettes, and that the follicular

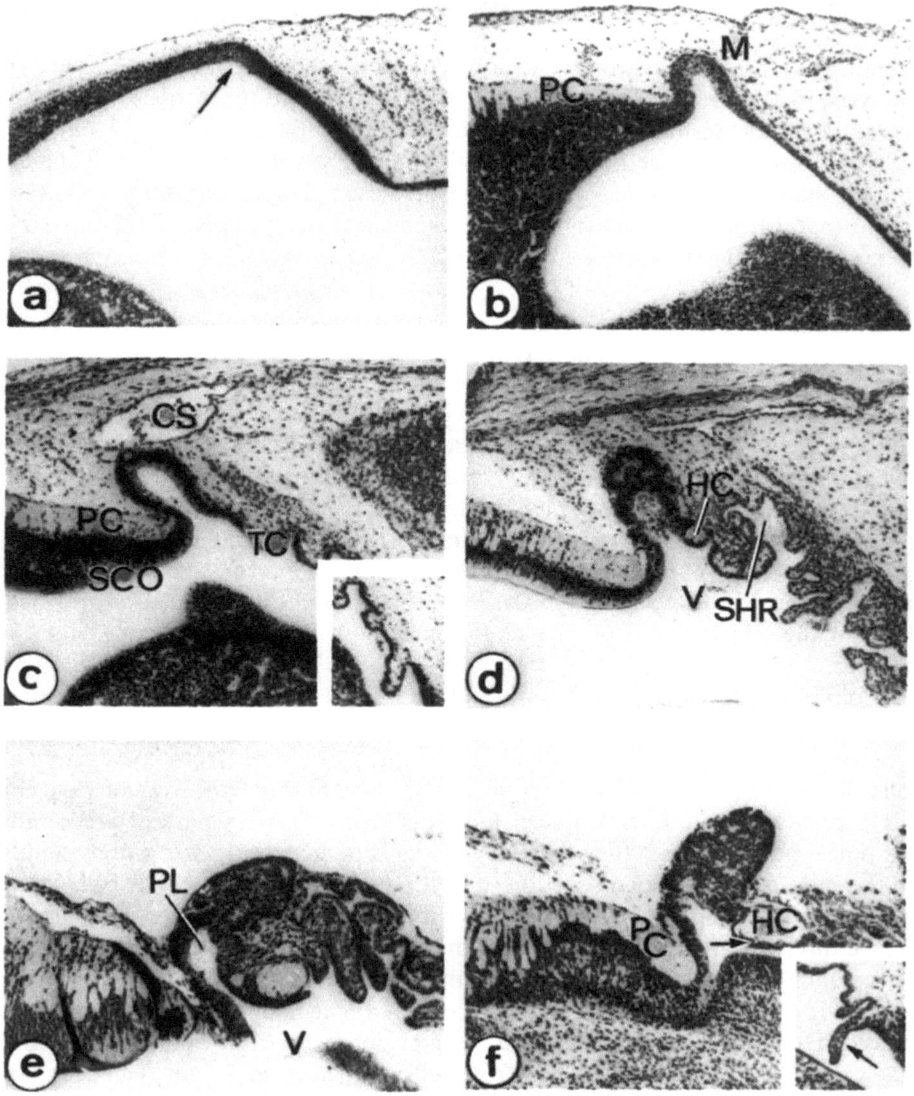

Fig. 36a–f. Development of the hamster pineal organ, from day 11 to 16 (term). Posterior aspect is to the left. **a** Day 11: a thickening of the diencephalic roof (*arrow*) and high mitotic activity precede epiphyseal evagination in this region. **b** Day 12: the pineal anlage lies just rostral to the differentiating posterior commissure. The area between the diencephalic roof and the integument is filled with mesenchyme. **c** Day 13: the pineal evagination has elongated in a dorsocaudal direction and overlies the developing subcommissural organ and posterior commissure. The apex is associated with the developing confluens sinuum, and the anterior wall is continuous with the folded tela choroidea, which terminates rostrally in a deep fold. **d** Day 14: cellular proliferation and secondary evaginations of the pineal wall have resulted in the formation of follicles in the distal end of the pineal organ. The differentiating habenular commissure, suprahabenular recess and choroid plexus are evident at this stage. **e** Day 15: a lumen persists in only the proximal one-third of the pineal organ. The distal portion consists of cells which are arranged randomly or in rosettes. **f** Day 16 (term): the pineal is nearly devoid of follicles and a lumen. Some cells which at high magnification resemble the cells of the pineal are associated with the inferior border of the habenular commissure (*arrow*). A prominent velum transversum is present at this stage (*inset*, *arrow*). *CS*, confluens sinuum; *HC*, habenular commissure; *M*, mesenchyme; *PC*, posterior commissure; *PL*, pineal lumen; *SHR*, suprahabenular recess; *TC*, tela choroidea; *V*, third ventricle. CLABOUGH, 1973

structure became less obvious as development proceeded. According the KERENYI and VON WESTARP (1971) most of the follicles had disappeared by 11 days of age. Pinealocyte nuclei became spindle-shaped between 20 and 25 days of age and decreased in size between 35 and 40 days; simultaneously, the distance between neighbouring pinealocyte nuclei increased. Under constant darkness, the postnatal transformation of the rabbit pineal is delayed (KERENYI and VON WESTARP, 1968, 1971; KERENYI et al., 1972).

6. Cow and Sheep

The development of the epiphysis cerebri of cow and sheep is basically similar to that of other mammals. A detailed account has been given by BRACK (1962), who studied the bovine epiphysis in 67 specimens between 1.5 cm CRL and 1 year of age. ANDERSON (1965) has given a brief summary of epiphyseal development in cow and sheep and HOLLMANN (1963) studied the origin of glial cells. Data on pineal weight in developing and adult sheep have been published by KENNAWAY and SEAMARK (1975) (Table 10), who also provided data on the development of one of the melatonin-forming enzymes, HIOMT (p. 260).

Epiphyseal development in cows is of interest because it has been suggested by FRAUCHIGER (1963) and MEYBURG (1965) that in this species pinealocytes are not of neuro-ectodermal origin, but rather originate from mesenchymal cells. This conclusion was based on the observation that in fetal cows of 7–30 cm CRL, cells with small, dark nuclei migrated from distal meningeal areas into the pineal anlage, where they formed sheaths around blood vessels and islands in the centre of the organ. As the development proceeded, these cells were thought to develop into pinealocytes, while the intrinsic cells of the pineal anlage differentiated into glial cells. Although there can be no doubt that cells of mesenchymal origin penetrate into the pineal anlage, there is as yet little support for the notion that these cells give rise to pinealocytes (KAPPERS, 1965a). Indeed, HOLLMANN (1963) presented clear evidence that both pinealocytes and glial cells in the bovine pineal arise from cells of ependymal origin. The presence of synaptic ribbons argues against a mesenchymal origin of pinealocytes (p. 147), which appear to be restricted to cells of neuro-ectodermal origin.

Table 10. Development of pineal weight in sheep (KENNAWAY and SEAMARK, 1975)

Gestational age (days)	Pineal wt (mg)[a]	Gestational age (days)	Pineal wt (mg)[a]	Gestational age (days)	Pineal wt (mg)[a]
130	7.2 ± 0.9	140	12.3 ± 2.0	150	15.0 ± 2.0
132	6.1 ± 0.4	142	11.1 ± 0.5	1–3 day old lambs	13.7 ± 1.5
134	8.8 ± 2.4	144	10.4 ± 1.1		
136	8.6 ± 1.0	146	12.1 ± 1.6	Adult sheep	69.3 ± 7.6
138	9.6 ± 1.2	148	15.4 ± 1.2		

[a] Mean ± SEM

According to BRACK (1962) the bovine epiphysis develops as follows: At 4 cm CRL, the epiphyseal anlage is a small caudally directed diverticulum which subsequently develops lateral extensions. Cell proliferation starts to be prominent in the cranial part of the epiphyseal anlage at 9 cm CRL. Capillaries and connective tissue were noticeable at 10 cm CRL and primitive astrocytes at 12 cm CRL.

At 16 cm CRL the epiphysis is a uniform, oval and compact body as a result of cell proliferation in the caudal wall of the epiphyseal diverticulum. Non-myelinated nerve fibres of central origin penetrate the organ at 23.5 cm CRL and connective tissue trabeculae are present at 25 cm CRL. A rosette-like arrangement of pineal cells, due to partly severed tubular extensions of the epiphyseal lumen, is typical at 26.5 cm CRL. Myelinated nerve fibres passing through the habenular commissure are noted at 34 cm CRL and glial patches consisting of a feltwork of glial fibres surrounded by densely packed pinealocytes with dark nuclei occur at 40 cm CRL. The final shape of the epiphysis is acquired at 60 cm CRL, simultaneously with an increase in the amount of fibrous astrocytes, which form a dense feltwork just beneath the surface of the organ.

IV. General Histology

1. Survey of the Basic Structural Components

The mammalian epiphysis cerebri is a mainly cellular organ. The major cellular component, the pinealocyte, also termed pineocyte, parenchymal cell, or chief cell, amounts to 82% of the cells scanned in the rat epiphysis cerebri (WALLACE et al., 1969). Characteristically, a pinealocyte consists of a cell body with a varying number of processes. The processes are of undefined length and possess club-shaped endings which terminate either between adjacent pinealocytes or in the immediate vicinity of blood vessels. It is unclear as yet whether different types of pinealocytes exist or whether alleged differences are merely the expression of different functional stages of one and the same cell type (p. 167). In addition to pinealocytes, glial cells are present in varying amounts. In the rat pineal they amount to 12% (WALLACE et al., 1969). Although it is difficult to classify some of the glial cells observed, astrocytes have been clearly identified in some species. The cytoplasmic processes of astrocytes, if present in sufficient amounts and arranged in bundles, may appear as glial fibres (p. 178). Nerve *cells* are not a regular and consistent feature of the mammalian epiphysis cerebri. Because of the close topographical relationship of the epiphysis cerebri to the third ventricle in some species, ependymal cells may be attached to, or form part of, the pineal. Besides cells, nerve *fibres* are an important component of the pineal (p. 186). All fibres appear to be efferent. They stem from three sources. Sympathetic fibres, coming from the superior cervical ganglia and reaching the pineal as nervus or nervi conarii, regulate melatonin synthesis. Parasympathetic fibres are infrequently present and nothing precise can be said about their function. A regular feature is the presence of commissural fibres, which are mostly confined to the proximal regions of the pineal. They are still regarded as being without significance for pineal function, though there is increasing evidence that this may not be the case (see p. 208).

The blood supply of the mammalian pineal organ is affluent (p. 216). Connective tissue is usually scarce (p. 223). Ground substance is likewise rare. An interesting morphological feature is the presence of a system of intercellular canaliculi of varying widths, apparently linking pinealocytes and blood vessels (QUAY, 1974b) (see p. 173).

2. General Histological Pattern

a) Man

In the epiphysis cerebri of adult humans the pinealocytes are not randomly oriented (see BARGMANN, 1943). Varying amounts of connective tissue, in continuity with the capsule of the epiphysis, penetrate the parenchyma, forming septa, and surround groups of pinealocytes, giving rise to cell cords or lobules (QUAY, 1965a; SCHARENBERG and LISS, 1965; HÜLSEMANN, 1967). Occasionally the lobules have the appearance of follicles. Often groups of pinealocytes are incompletely surrounded by stromal elements.

In addition to follicles or rosettes consisting of pinealocytes, rosettes of ependymal cells stemming from the subcommissural organ may be present within the pineal parenchyma (PALKOVITS et al., 1962a, b). Altogether the histological structure of the human pineal is very variable. WILDI and FRAUCHIGER (1965) studying lobulation of human pineals quantitatively noted that follicles were present in only 52% of the specimens investigated (50% in males, 54% in females) and that the number of follicles increased with age (Table 11). The fact that in males the first lobules were observed at 8 years of age and in females at 18 years seems to suggest sex differences.

In old age the histological appearance of the pineal appears to be especially variable. ARIETI (1954), studying pineals from subjects aged 68 to 96, noted that no two glands resembled each other. Nevertheless, the attempt was made to classify the pineal into four major types. Glands of the first type assumed a pseudo-alveolar appearance, showing lobules separated by thin connective

Table 11. Lobulation of the human pineal according to age. (Adapted from WILDI and FRAUCHIGER, 1965)

Age (yr)	Frequency of follicles
0–20	18.5%
21–40	37.5%
41–60	45.5%
61–90	58.0%

tissue trabeculae. In glands of the second type the connective tissue seemed to have expanded between the lobules. In the third type the connective tissue invaded the parenchyma and separated the cells from one another in such a way that no alveolar structure was recognizable. The fourth type consisted of a mixture of the first three types. Despite the great morphological variation, ARIETI concluded that the pineal, even in old age, is functionally active. For further details on pineal function and ageing , see p. 261.

The structural complexity of the human epiphysis cerebri is also illustrated by the fact that even individual parenchymal lobules have a distinct architecture. SCHARENBERG and LISS (1965) noted that after silver staining the peripheral layers of a lobule, which they termed the marginal plexuses, consisted of a ring of widely differing pinealocytes (p. 71) with numerous processes, whereas the central areas contained fewer cells which did not show a distinct architectural arrangement.

b) Other Mammals

According to BARGMANN (1943), the distinct segregation of the epiphyseal parenchyma into lobules or follicles by means of prominent connective tissue strands, as observed in humans, is not found in other mammals, though he pointed out that comparative studies in primates were lacking. Systematic studies on the distribution of connective tissue within the epiphyseal parenchyma are still not available. Since such information is not very likely to contribute greatly to a better understanding of pineal function, no attempt was made in the present survey to collect the data from the various original publications. Here it suffices to mention that in small laboratory rodents connective tissue is usually scarce (see p. 223). A notable exception is *Jaculus jaculus*, whose epiphysis cerebri is strongly pervaded by connective tissue, giving rise to distinct lobules (HÜLSEMANN, 1967). A lobulation due to the presence of connective tissue has also been reported in horse and in sheep (HÜLSEMANN, 1967). In *Macaca mulatta* (NOVOTNÁ et al., 1966) variously wide and irregular connective tissue septa were noted, but a subdivision of the parenchyma into lobules was not conspicuous. In pig pineals (HEINIGER, 1965a), a lobulation was absent in 50% of the specimens studied; in 32.7% it was moderately developed and in 17.3% strongly developed. Altogether it would appear that lobulation is a highly variable feature of the mammalian epiphysis cerebri.

The arrangement of the parenchymal cells to one another varies as well (see Table 12). According to QUAY (1965a), the parenchymal cells are generally arranged in solid clusters, cords, pseudo-alveoli or incompletely separated lobules, depending on species and age. In some species, especially at certain developmental stages, they form tubules, follicles or rosettes.

3. Regional Differences

There are several indications that the histological structure of the epiphysis cerebri may not be uniform throughout. Although the presence of regional

Table 12. Arrangement of pinealocytes in different mammals

Species	Arrangement of pinealocytes	Author(s)
Man	Cords, clusters	HÜLSEMANN (1967)
	Evenly distributed	HÜLSEMANN (1967)
Rhesus monkey	No special arrangement	ECKHARDT (1972)
Orangutan	Partially lobules	QUAY (1970a, b)
Sheep	Connective tissue septa only peripherally; pinealocytes loosely arranged; perivascular accumulations	HÜLSEMANN (1967)
	Follicles restricted to tip of epiphysis; sometimes arranged in cords	LANG (1959)
Horse	Connective tissue septa; lobules with evenly distributed pinealocytes	FASSBENDER (1962), HÜLSEMANN (1967)
Rabbit	Often pseudo-acinar or pseudofollicular arrangement; cords are rare	HÜLSEMANN (1967), ROMIJN (1973a)
Dog	Irregularly scattered nuclei; indistinct trabeculae	VENZKE and GILMORE (1941)
	Pseudo-acini, mainly near base of organ	HÜLSEMANN (1967)
	Cell cords	ZACH (1960)
Cat	Lobules or follicles surrounded by connective tissue in the periphery of the organ	ZACH (1960)
Ferret	Uneven distribution; no follicles	TRUEMAN and HERBERT (1970a)
Seal	Islets surrounded by little connective tissue	CUELLO and TRAMEZZANI (1969a, b), CUELLO (1973)
Pocket gopher	Clusters or possibly acini	SHERIDAN and REITER (1973)
Mongolian gerbil	Chromophilic pinealocytes arranged in lobules separated by interstitial cells	NIELSEN and MØLLER (1978)
Rat	Follicles or rosettes	KAPPERS (1960a)
	Cords or rosettes	BUSS (1965)
Mouse	Often rosettes with central lumen	HÜLSEMANN (1967)
Jaculus jaculus	Abundant connective tissue delimiting lobules with densely packed cords of parenchymal cells	HÜLSEMANN (1967)
Bat	Cords and clusters	QUAY (1965a)
Shrew	Cords, clusters	QUAY (1965a)
Hedgehog	Multilayered, tortuous cords forming convolutions or pseudo-acini, no connective tissue septa	HÜLSEMANN (1967)
Golden mole	Homogeneously dispersed	PEVET and KUYPER (1978)
Opossum	Tightly packed or forming rosettes	HOFER et al. (1976)

differences has long been advocated, especially by QUAY (1965a), this concept is still in its infancy, partly because the differences are not easily detectable and systematic studies are lacking. Since the functional significance of most of the pertinent findings is obscure, no attempt will be made to deal with this aspect exhaustively.

Regional differences may be present in a proximodistal direction and peripherally. A good example for regional differences in a proximodistal direction is the North American rodent, *Peromyscus,* in which the epiphyseal complex consists of proximal, intermediate and distal regions (p. 29). Here the distal portion contains fewer phloxinophilic fibres and pinealocytes of greater size and mean nuclear diameter than the proximal portion. Moreover, the proximal region shows a greater response to environmental lighting than the distal zone (QUAY, 1956a, 1965a). In the rat, the nuclei of the cells in the deep pineal are distinctly smaller than those in the superficial pineal (VOLLRATH and BOECKMANN, 1978). In hamsters the deep pineal contains more glial cells than the superficial pineal (SHERIDAN and REITER, 1970b). A prominent basal to distal zonation of the pineal parenchyma has been noted in kangaroos (QUAY, 1966c). In mice, glycogen levels and nuclear densities show regional differences that change with age (KACHI et al., 1975).

Structural differences between central and peripheral regions of the epiphysis cerebri have been relatively frequently reported. In the rat pineal the cortical regions have more lipid (QUAY, 1957a), greater blood content (QUAY, 1958a) and show cytologically less response to continuous light than the medulla (QUAY, 1963b). Nuclear size measurements yielded different results: in one group of studies the pinealocyte nuclei were found to be larger in the centre of the organ than in the periphery (QUAY and RENZONI, 1966a; BLUMFIELD and TAPP, 1970), in other studies the reverse was found (QUAY, 1963b; MILINE et al., 1969a; DIEHL, unpublished results). Differences in the packing density of cell nuclei have been reported as well. A greater packing density in the periphery as against the centre was found in cat (ZACH, 1960) and dog (ZACH, 1960; WELSER, 1968), the reverse being the case in horse (FASSBENDER, 1962).

Differences between centre and periphery may also be due to differences in cell shape and cytoplasmic appearance. Differences in cell shape have been noted in cow, sheep (LANG, 1959), dog, cat (ZACH, 1960) and horse (FASSBENDER, 1962), all of which possessed monopolar pinealocytes triangular in shape in the periphery and pinealocytes with 2–4 distinct processes in the centre. The horse differed slightly from the other species investigated by also having bipolar pinealocytes in the periphery.

Peculiarities in cytoplasmic appearance are responsible for regional differences in the rabbit. As shown by light and electron microscopy (ROMIJN, 1973a, b), the cortical tissue contains "light" and "dark" pinealocytes in a ratio of 3:1, a transitional area being characterized by equal amounts of these types. According to ROMIJN (1973a), the cortex is represented by a relatively thin layer of parenchyma which covers the medulla at the dorsal and lateral sides of the gland and is absent in the ventral regions where the organ is closely related to the great cerebral vein.

What looks like an unusual reason for a subdivision into cortical and medullary regions is present in the Weddell seal. Here, pinealocytes are restricted to the cortical region, the medulla being made up by connective tissue, blood vessels, nerve fibres and glial cells (CUELLO and TRAMEZZANI, 1969a, b). Other species with a clear zonation into cortex and medulla are sheep (NESIĆ, 1962) and red-cheeked sousliks, *Citellus erythrogenys major* (POPOVA et al., 1975).

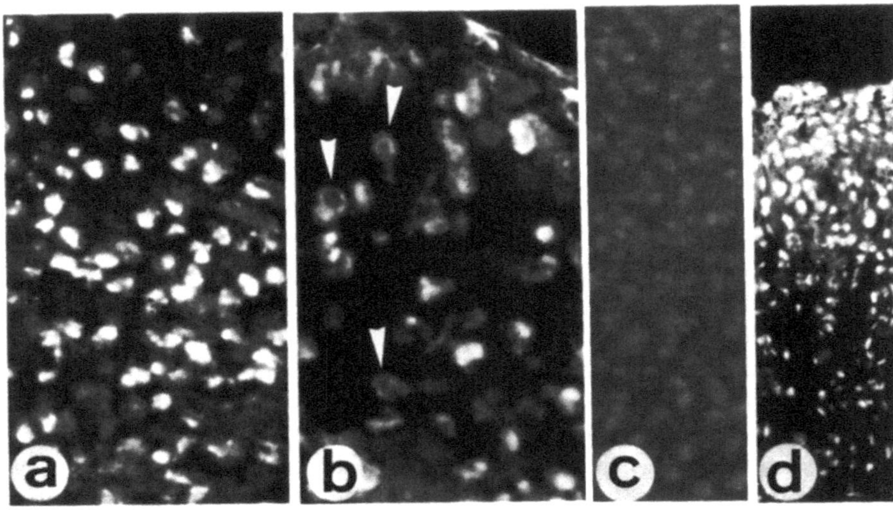

Fig. 37a–d. Immunohistochemical demonstration of melatonin in the rat pineal gland. **a** Rat killed at 24.00 h. Note the intense fluorescence in a number of clearly outlined structures. ×128; **b** Rat killed at 22.30 h. The *arrows* point to parenchymal cells in which the sites of the nuclei are non-fluorescent. ×200; **c** Control experiment in which antiserum saturated with melatonin was used. Note the lack of specific fluorescence. Rat killed at 24.00 h. ×128; **d** Survey picture of a rat pineal gland showing strong fluorescence in the marginal zone of the pineal and lack of specific fluorescence in the centre of the organ. The fluorescence in the lower part of the micrograph is due to the presence of non-specific precipitates. Rat killed at 24.00 h. ×128. FREUND et al., 1977

What the subdivision into cortical and medullary regions means in terms of function is not yet known, but it would appear that this aspect of pineal morphology deserves special attention. Frequently it has been reported that under certain experimental conditions changes were particularly clear in the periphery of the pineal organ. Support for the concept of functional differences between cortex and medulla has recently come from studies in rats. DAFNY et al. (1975) found that when monopolar recordings following amygdaloid, optic tract or acoustic click stimulations were made from the pineal at different depths, the response was similar from 0.0–0.8 mm, but changed at 1.2 mm below the surface. Moreover, by means of single unit recording, only within the first 200 μm below the surface was it possible to detect spontaneous spike activity (MCCLUNG and DAFNY, 1975; DAFNY and MCCLUNG, 1975). Unfortunately these results are difficult to interpret since little is known about the electrophysiological properties of mammalian pinealocytes. The fact that in immunohistochemical studies melatonin or melatonin-like substances were demonstrable in the cortex only (FREUND et al., 1977) (Fig. 37d) points to differences in melatonin formation. Differences in the packing density of pinealocyte nuclei in the centre of the pineal have been related to running activity: packing was dense in slow-running rats, while wide spacing was observed in fast runners (REISS et al., 1967). Fluorescence histochemical studies revealed that in rats the yellow serotonin fluorescence increased in intensity from the periphery to the centre and that after superior cervical ganglionectomy some residual fluorescence was apparent deep within the gland (MORGAN and HANSEN, 1978).

Regional differences other than between cortex and medulla and between the proximal and distal portions have also been noted. In the rat, paraldehyde fuchsin-stainable cells, which were not affected by cutting the pineal stalk, were restricted to the basal tip of the superficial pineal (QUAY, 1957c). In the ferret, an antero-inferior region differed by having less densely packed pinealocytes than the rest of the organ (TRUEMAN and HERBERT, 1970b).

V. The Pinealocyte

1. Light Microscopy

a) Shape

The main cellular constituent of the mammalian epiphysis cerebri, the pinealocyte, is a light microscopically clearly outlined, polymorphous element with varying numbers of processes of variable widths and lengths (BARGMANN, 1943). Routinely used light microscopic staining procedures do not give a clear picture of the complexity of pinealocyte shape. Particularly favourable for the demonstration of the complete outline of pinealocytes is the silver carbonate method of DEL RIO-HORTEGA (see BARGMANN, 1943, for details), which has been relatively little used in pineal research. In man the pinealocytes appear as mono-, bi- or multipolar elements, some of the processes bearing club-shaped endings (BARGMANN, 1943). These different types are apparently not evenly distributed in the pineal parenchyma. SCHARENBERG and LISS (1965) noted that unipolar cells with a round cell body and one wide and frequently arched process were exclusively present in the marginal areas of the organ. In peripheral regions of parenchymal lobules (see p. 67) five different cell types were distinguished, based on the number, length and width of processes, the presence or absence of club-shaped endings and their topographical relationships. By contrast, in the central parts of the lobules a single cell type characterized by an oval cell body and a maze of intertwined processes was present. Though the polymorphism of human pinealocytes is impressive, its functional significance is obscure. Since systematic light microscopic studies on pinealocyte shape are lacking, the reader is referred to the section dealing with the ultrastructure of the pinealocyte (p. 111) for further details.

b) Staining Properties

Pinealocyte cytoplasm has relatively little affinity for dyes (ZACH, 1960), rendering detailed histological examination difficult. A notable difference between the pineal and some other endocrine organs is that the pinealocytes cannot be subdivided into structurally and functionally distinct groups of cell.

Nevertheless there are slight tinctorial differences, and a distinction has been made between chromophobic pinealocytes and chromophilic pinealocytes (QUAY, 1956b; HEINECKE, 1959; TAPP and BLUMFIELD, 1970a).

The most detailed study on the chromophilia of pinealocytes has been carried out by QUAY (1956b) applying orange G-aniline blue and chrome alum haematoxylin-phloxine to Sprague-Dawley rat pineals. This author estimated that 41% to 90% of the pinealocytes were chromophobic, roughly 1%–12% presecretory chromophils, roughly 1%–14% secretory chromophils and 6%–41% postsecretory chromophils. Chromophobic cells varied in appearance, ranging from small cells with little apparent cytoplasm and small, darkly stained nuclei to large cells with large, pale nuclei and prominent nucleoli, the latter cells showing transitional forms to chromophils. TAPP and BLUMFIELD (1970a) noted that the cytoplasm of the chromophobes was faintly granular and contained clear vacuoles, and that in many cases the acidophilic nucleolus had a dense chromatin mass adjoining it. Some of the chromophobic cells contained glycogen, as assessed by PAS staining combined with diastase digestion. Chromophilic cells (QUAY, 1956b) showed variable amounts of material stainable with orange G or phloxine. The material occurred first in the immediate vicinity of the nucleus and subsequently in cell processes, which were supposed to be particularly rich in the specific material. Postsecretory cells were characterized by little cytoplasm and a small granular nucleus.

Chromophobic and chromophilic pinealocytes may have a preferential localization. In the pig pineal, chromophilic cells were abundant in the marginal areas and rare in the central parts (HEINECKE, 1959).

c) Application of Special Optical Methods

Special optical methods have been little used in studying pinealocyte structure. BAYEROVÁ et al. (1962) and OBRUCNIK and BAYEROVÁ (1962), investigating the human pineal by means of polarizing and fluorescence microscopy, demonstrated the presence of birefringent (cholesterol, concretions) and autofluorescent (neutral fats, lipoproteins and pigments) inclusions.

Autofluorescence, probably due to the presence of a tryptophan-rich protein (A. SMITH et al., 1972a; KAPPERS et al., 1974), has been found to be a characteristic feature of some pinealocytes in rabbit (A. SMITH et al., 1972a, b), rat (SMITH and KAPPERS, 1974; KAPPERS et al., 1974; PEVET et al., 1975a), hedgehog and mole (PEVET et al., 1975a, 1976a). In the rabbit pineal, the autofluorescent cells showed an even distribution. The fluorescent material is granular in the cell bodies and homogeneous in the cell processes. Following formaldehyde vapour treatment the autofluorescent cells also showed the typical serotonin fluorescence generally present in pinealocytes (SMITH et al., 1972a). After gonadectomy the autofluorescent pinealocytes increase in number (KAPPERS et al., 1974; SMITH and KAPPERS, 1974; PEVET et al., 1975a), probably at the expense of non-autofluorescent serotonin-containing pinealocytes (SMITH et al., 1972b). Administration of PMS leads also to an increase of autofluorescent pinealocytes, but after 7 and 14 days the changes reverse to normal (SMITH et al., 1975).

d) Nucleus and Nucleolus

Mammalian pinealocytes may contain more than one nucleus with one or two nucleoli (BARGMANN, 1943). The nuclei are often polymorphic and may exhibit *Kernkugeln* which in all likelihood are cytoplasmic invaginations into the highly lobulated nuclei (KEVORKIAN and WESSEL, 1959; for further details see p. 115). The nucleoli are usually very prominent (OKSCHE, 1956), but they appear to contain little RNA (p. 93). They possess a pronounced affinity for acidic dyes, such as orange G (HEINIGER, 1965b).

Size measurements of nuclei and nucleoli have infrequently been carried out, although the results obtained are promising. Thus it was found that differences in nuclear size exist between proximal and distal regions of the pineal complex, as shown in the white footed mouse, in which the mean nuclear diameter was 6.13 ± 0.06 µm in the proximal portion and 6.49 ± 0.08 µm in the distal part (QUAY, 1956a). Nuclear size differences between cortical and medullary regions have been mentioned above (p. 69). BLUMFIELD and TAPP (1970), studying the rat, distinguished three types of pinealocytes with different nuclear sizes, namely type I cells with nuclei measuring 8.3 µm in diameter, type II cells with nuclei measuring 6.7 µm and type III cells with nuclei measuring 5.4 µm in diameter. The usefulness of measurements of nuclear and nucleolar size for functional studies is illustrated by the fact that nuclear and nucleolar sizes undergo circadian changes (Fig. 38), showing peaks at midday, followed by a gradual decrease toward the end of the light period and a slight increase during the night (RENZONI and QUAY, 1964; QUAY and RENZONI, 1966a).

Seasonal changes in nuclear size have been reported for hamster (PFLUGFELDER, 1957; MOGLER, 1958), sheep (NESIĆ, 1962), *Glis glis* and *Eliomys quercinus* (LEGAIT et al., 1975b) and bats (QUAY, 1976).

That nuclear size may depend on the age of the individuals investigated has been demonstrated in both pigs (HEINIGER, 1965a) and rats (CASSANO et al., 1961). In pigs, pinealocyte nuclei measured 7.98 µm in diameter in old female specimens and 5.8 µm in young females, and in 1,000-day-old rats pinealocyte nuclei were distinctly larger than in 190-day-old animals.

A relationship of nuclear size to the state of the gonads is suggested by the observation that in pigs the largest nuclei were found in hermaphroditic animals (8.72 µm in diameter) and that castration in males was followed by a slight increase in nuclear size from 5.4 µm in the controls to 5.8 µm in the experimental group (HEINIGER, 1965a).

Altogether it would appear that nuclear size changes are relatively small and that sensitive statistical tests have to be applied to demonstrate statistically significant differences. Even under experimental conditions which lead to pronounced changes in pineal function nuclear size reacts relatively little. Exposure of rats to continuous illumination for several weeks, which strongly inhibits melatonin formation (p. 244), led to an only slight decrease of nuclear size (QUAY, 1963b). Virtually no changes could be observed by ROTH et al. (1962) following exposure to either continuous illumination or constant darkness. In guinea-pigs, constant light led to a 15% increase in nuclear size while constant dark resulted in a 7% decrease (VOLLRATH, 1975). In hamsters pinealocyte

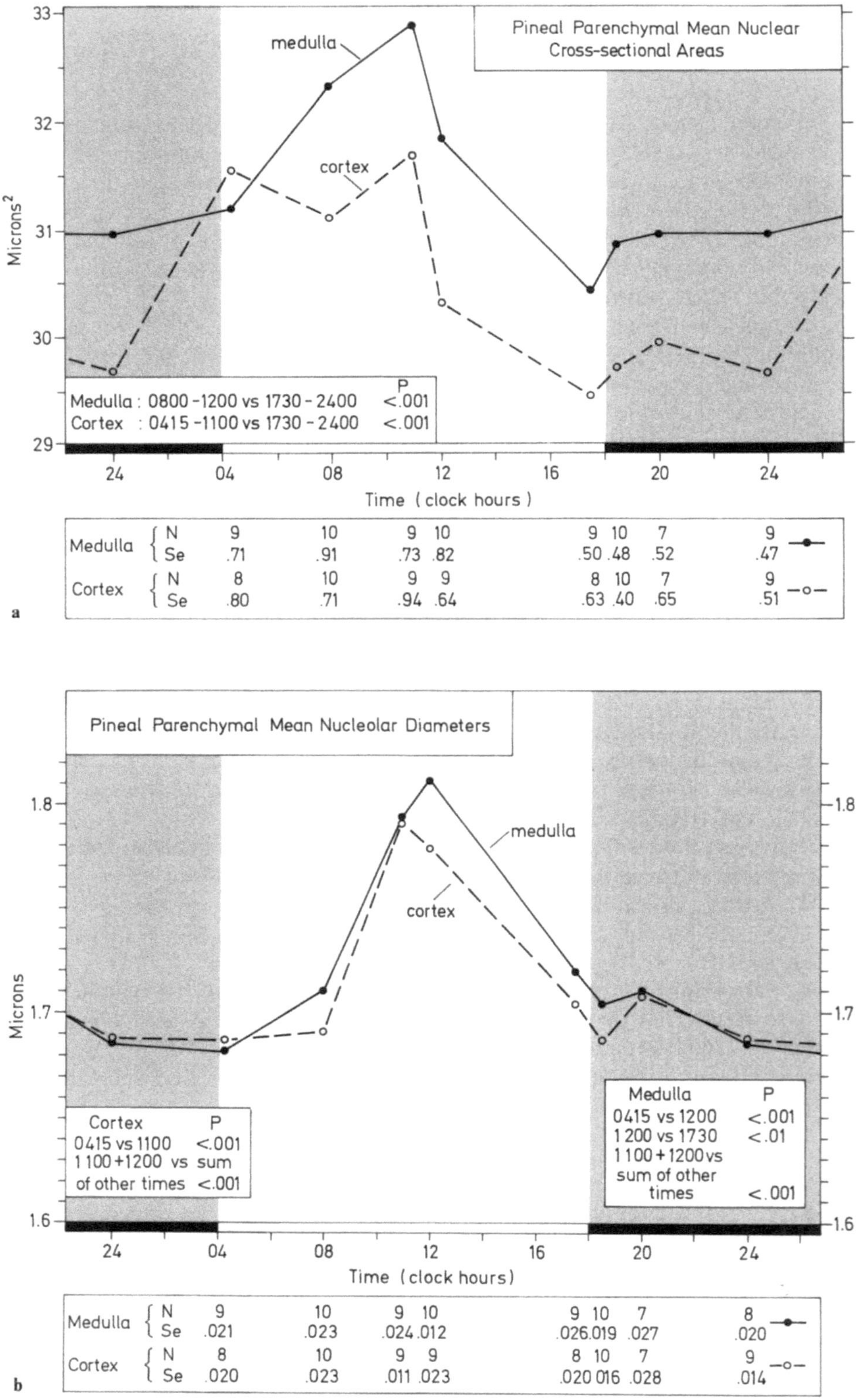

Fig. 38a, b. Circadian rhythm of **a** nuclear and **b** nucleolar diameters in the cortex and medulla of the rat pineal gland. QUAY and RENZONI, 1966a nuclei measured 5.90±0.16 μm under short photoperiods and 5.62±0.13 μm under long photoperiods (HOFFMAN and REITER, 1965b).

e) Mitotic Activity

Fully differentiated pinealocytes show little cell proliferation, as shown by a scarcity of mitotic figures (BARGMANN, 1943; QUAY and LEVINE, 1957a, b; HEINIGER, 1965a) and extremely low ^{3}H-thymidine incorporation (WALLACE et al., 1969). Quantitative studies carried out in 16-week-old rats revealed a mitotic index (metaphases and anaphases/1,000 cells) of 0.008, which after colchicine treatment increased to 0.098 (QUAY and LEVINE, 1957a, b). Despite the low number of mitotic figures present, the mean number of mitoses shows a circadian rhythm, with a peak of nine mitoses per pineal at midday and less than one at night (RENZONI and QUAY, 1964; QUAY and RENZONI, 1966a). Administration of testosterone or 17β-oestradiol did not lead to significant differences in pineal mitotic activity (QUAY and LEVINE, 1957a, b).

f) Circadian Changes

As can be expected from the pronounced circadian changes of pineal activity and the results mentioned above, the pinealocytes as a whole also show circadian changes. As an indirect measure of pinealocyte size the nuclear density, i.e. the number of pinealocyte nuclei per unit area of pineal tissue, has been used. This extrapolation appears reasonable since there is hardly any extracellular matrix present in the mammalian epiphysis cerebri. With this technique it has been shown (Table 13) that in mice pinealocytes are largest early in the morning, followed by a decrease up to 21.00 h and a subsequent increase (KACHI et al., 1971b, 1973). Whether or not the changes in pinealocyte size are accompanied

Table 13. Circadian changes of nuclear density[a] in the mouse pineal. (Adapted from KACHI et al., 1971b, 1973)

Time	Male mice	Female mice	
		Pro-oestrus/oestrus	Metoestrus/dioestrus
6.00	27.0±2.3	24.7±0.5	22.4±1.9
9.00	24.1±3.0[b]	24.5±1.6	22.0±2.7
12.00	26.2±2.2		
13.00		24.4±1.7	23.4±1.0
15.00	27.8±2.0		
18.00	28.5±2.2	27.5±1.5	26.4±4.2
21.00	33.6±3.1	28.6±3.4	29.5±2.5
24.00	29.4±5.0		
1.00		27.4±2.2	28.5±2.6

[a] Number of pinealocyte nuclei per unit area.
[b] Value given differs from that in the original publication but was corrected by the authors in the reprints distributed.

by corresponding changes in organ size has not been studied. Corresponding changes cannot necessarily be expected since, as shown by QUAY (1973a, 1974b) in rats, there is a system of 0.2–0.4 μm wide intercellular canaliculi which change in size, so that there is the possibility that pinealocytes expand at the expense of these intercellular canaliculi (for further details, see p. 174).

Other morphological parameters exhibiting circadian changes are nuclear and nucleolar size (p. 73), mitotic index (p. 75), amount of glycogen (p. 78) and, as shown by electron microscopy, number of synaptic ribbons (p. 150) and amount of dense-core vesicles (p. 165) of pinealocytes.

g) Pigments

Pineal pigments have been thoroughly reviewed by QUAY (1974a). Hence it suffices to give a brief survey, all the more so since pigments show great intra- and interspecies variations and since there is little if any reason to believe that these variations have any pineal-specific functional significance (QUAY, 1974a).

Two types of pigments have been noted in pinealocytes: melanin and lipofuscin. Melanin is not very abundant. It has been clearly identified in cattle (SANTAMARINA and MEYER-ARENDT, 1956; SANTAMARINA, 1958) and tentatively identified in pig, horse and goat (MIKAMI, 1951).

In humans, lipofuscin pigment appears around puberty and increases with age, showing great interindividual variation (QUAY, 1957a, 1974a; S. MILCU and VREJOIU, 1959; HORANYI, 1960; BAYEROVÁ et al., 1962, 1965). It has also been found in horse, pig and goat (MIKAMI, 1951) and is particularly abundant in old sows (HEINIGER, 1965a). In rat pinealocytes, lipofuscin was detected by electron microscopy (BOSTELMANN, 1965; GUSEK and SANTORO, 1960, 1961).

A third type of pigment, hemosiderin, was not present in pinealocytes, but was found in epiphyseal macrophages (QUAY, 1974a).

2. Histochemistry

With the advent of highly sophisticated histochemical techniques attempts were made to elucidate pineal structure and function. As will be seen, these studies have filled many gaps, but apart from the formaldehyde-induced fluorescence method for the demonstration of monoamines they have yielded relatively little insight into the functional significance of the epiphysis cerebri.

a) Inorganic Constituents

The presence of inorganic constituents in the epiphysis cerebri has recently been reviewed (QUAY, 1974a). Apart from earlier studies (see BARGMANN, 1943) there do not appear to be any recent topochemical investigations which would indicate the precise localization of these components within the epiphysis. Hence, it can only be surmised that they are present within pinealocytes.

Table 14. Electrolyte composition of pineal and anterior and posterior pituitaries (PIECHOWIAK and SCHNIZER, 1976)[a]

		Pineal	Posterior pituitary	Anterior pituitary
Na^+	Light:	212.2 ± 16.6	219.3 ± 3.7	221.2 ± 6.5
	Dark:	193.2 ± 4.2	204.5 ± 1.8	272.8 ± 5.8
K^+	Light:	611.7 ± 16.3	465.9 ± 17.1	478.0 ± 6.2
	Dark:	519.9 ± 27.1	440.4 ± 9.7	524.1 ± 7.5
Ca^{2+}	Light:	10.8 ± 1.2	11.1 ± 0.9	10.4 ± 3.0
	Dark:	10.9 ± 1.7	10.7 ± 0.9	18.5 ± 3.1
Mg^{2+}	Light:	37.4 ± 1.8	31.6 ± 1.2	38.9 ± 1.6
	Dark:	33.0 ± 1.4	30.1 ± 0.7	44.8 ± 0.8

[a] mmol/kg dry weight ± SEM.

Although few studies dealing with inorganic constituents of the mammalian epiphysis cerebri are currently available, they clearly indicate the unique position of the epiphysis among cerebral structures. Neutron activation analysis in different regions of the CNS including hippocampus, cerebral cortex, cerebellum, thalamus, mid-brain, pons, medulla and spinal cord in calves, cows and pigs revealed that the highest contents of copper, manganese and zinc were found in the epiphysis cerebri (WONG and FRITZE, 1969). With the same technique DEMMEL et al. (1979) confirmed the higher levels of zinc in the human pineal when compared to the brain; cobalt and selenium are likewise higher in the pineal, whereas rubidium exhibits lower concentrations than in the brain. The iron content of the pineal did not differ from that of the brain. A high concentration of manganese, exceeded only by that in the pituitary, was also noted in the rabbit pineal (FORE and MORTON, 1952). Atomic absorption spectroscopy (PIECHOWIAK and SCHNIZER, 1976) showed that in rats the Na^+, Ca^{2+} and Mg^{2+} contents of the pineal were roughly similar to those of the pituitary, but that the K^+ content was distinctly higher (Table 14). With the exception of Ca^{2+}, all the electrolytes studied showed lower values during the night than during the day. In the case of K^+ and Na^+ a striking decrease occurred immediately after the onset of darkness.

Further data on inorganic constituents will be given in the section dealing with pineal concretions (p. 228).

b) Carbohydrates

In the mammalian epiphysis cerebri three carbohydrates in particular have been investigated, namely glycogen, sialic acid and ascorbic acid.

α) Glycogen

The presence of glycogen is not a regular feature of mature pinealocytes. In the human epiphysis cerebri very few pinealocytes possessed demonstrable amounts of glycogen, which were extremely small (BAYEROVÁ and BAYER, 1960).

Table 15. Circadian changes of glycogen[a] in the mouse pineal. (Adapted from KACHI et al., 1971b, 1973)

Time	Male mice	Female mice	
		Pro-oestrus/oestrus	Metoestrus/dioestrus
6.00	7.2 ± 4.1	9.3 ± 2.0	9.6 ± 7.8
9.00	9.3 ± 2.8	11.4 ± 4.3	10.1 ± 3.0
12.00	10.2 ± 3.1		
13.00		10.2 ± 2.5	13.0 ± 5.0
15.00	14.4 ± 5.0		
18.00	24.9 ± 4.1	25.1 ± 6.7	25.7 ± 7.0
21.00	24.7 ± 6.6	14.2 ± 5.3	27.7 ± 11.1
24.00	12.4 ± 6.3		
1.00		7.3 ± 2.0	11.6 ± 3.4

[a] Expressed as "glycogen score", i.e. the number of intersections (out of 441) coinciding with PAS-positive glycogen granules, at a magnification of × 1,000.

Slightly more glycogen appears to be present in rhesus monkeys, in which the pinealocytes were classified according to the presence or absence of glycogen (WISLOCKI and DEMPSEY, 1948). According to LUKASZYK and REITER (1975) glycogen is present in pinealocytes lying peripherally in the organ.

In non-primate mammals glycogen has also been found. In pig, goat and horse glycogen was attributed to a relatively rare second type of parenchymal cell (MIKAMI, 1951), but QUAY (1974a) raised doubts as to whether these cells represent pinealocytes.

In rodents, the glycogen content is rather variable. In rats, glycogen could not be demonstrated electron microscopically (BUSS, 1965), but a chemical analysis revealed 1.30 μg/mg pineal tissue (QUAY, 1963b). Glycogen, as assessed by electron microscopy, was abundant in some pinealocytes of gerbil (GREGOREK, 1973), mole-rat (PEVET et al., 1976b) and golden mole, *Amblysomus hottentotus* (PEVET and KUYPER, 1978), and was clearly demonstrable in hamsters (SHERIDAN, 1967). Apparently enough glycogen was present in mice to enable one to study possible diurnal changes (see below). Moreover, glycogen was found in pinealocytes of rabbits (LEONHARDT, 1966), in which it has been thought to be restricted to the small, dark pinealocytes (ROMIJN et al., 1977a) and noctule bats (PEVET et al., 1977a, b).

The glycogen content has also been studied under various physiological and experimental conditions. Circadian changes have been found in mice, the highest content being present between 18.00 and 21.00 h (Table 15) (KACHI et al., 1971b, 1973, 1974). Blinding lowered the glycogen content (GREGOREK, 1973) and abolished its diurnal rhythm (KACHI et al., 1971b). Short-term exposure to continuous illumination resulted in an increase of the glycogen content, followed by a decrease after 3 days and a return to normal values after 3 weeks (KACHI et al., 1971a). Long-term exposure (11 weeks) led in rats to a drop of the glycogen content from 1.30 μg/mg pineal tissue to 0.55 μg (QUAY, 1963b). In continuous *darkness* the diurnal rhythm of the glycogen content persists for the first 7 days (KACHI et al., 1974). Reversal of environmental lighting conditions causes a complete reversal of the glycogen rhythm as early as the 5th day (KACHI et al., 1974). Recent studies by KACHI and ITO (1977) suggest a neural control of the pineal glycogen content and diurnal rhythm, as shown by the fact that the nocturnal decrease of glycogen is abolished by reserpin, superior cervical ganglionectomy, decentralization and continuous light. The development of the glycogen rhythm during *ontogenesis* has been studied by KACHI et al. (1975) (see p. 61).

β) Sialic Acid

Little information is available regarding the sialic (neuraminic) acid content of the epiphysis cerebri. In the gangliosides of the human pineal the sialic acid content was considerably lower than in those of the brain (LANDOLT et al., 1966). However, in the bovine epiphysis the sialic acid content (5.85 µmol/g fresh tissue) exceeded that of the brain (J. GREEN et al., 1962).

γ) Ascorbic Acid

According to histochemical studies, the ascorbic acid content of mammalian pinealocytes is both small and variable. In the human pineal ascorbic acid was noted in some pinealocytes only (LEOPOLD, 1941; CLARA, 1942).

In the rat pineal, the content of ascorbic acid appears to be very variable. Whereas COLLIER (1943) failed to demonstrate it in rat (and mouse) pineals, PROP and KAPPERS (1961), using the method of BACCHUS, noted a positive reaction resulting in an uneven distribution of granular material. Electron microscopically electron dense precipitates were regularly present in pinealocytes (BUSS, 1965; GUSEK et al., 1965), but due to poor tissue preservation precise localization was difficult. The presence of ascorbic acid within lipid droplets as described by PROP and KAPPERS (1961) could not be confirmed by BUSS (1965).

δ) Mucopolysaccharides

Pineal mucopolysaccharides have been only to a limited extent investigated. In the human epiphysis cerebri, acid mucopolysaccharides were particularly noticeable in granulated cells resembling mast cells in form and position (BAYEROVÁ and BAYER, 1960). In the bovine epiphysis sulphomucopolysaccharides could not be detected chemically (J. GREEN et al., 1962).

c) Lipids

Pineal lipids have attracted special attention, particularly at a time when relatively little was known about pineal function and when it was hoped that chemical and histochemical analyses might provide some clues. A major train of thought was that pineal lipids represented, or were somehow related to, secretory products (see, e.g., FARRELL, 1960a). With the benefit of hindsight, it would appear that this is not so. QUAY (1974a), who carefully reviewed the chemistry and histochemistry of pineal lipids, concluded that no lipid compounds have been isolated, or identified, which are peculiar to pineal tissue. His interpretation was that pineal lipids are probably involved primarily in general metabolism and in membrane structure and function within the organ, rather than as having roles as hormones or precursors to hormones. Moreover, the lipid content has been used to determine whether pinealocytes resemble neurons or glial cells or whether they have a distinct individuality.

Chemical analysis revealed (BASINSKA et al., 1969a, b) that the pineal resembles liver and kidney in its total lipid, cholesterol and phospholipid contents. The total lipid content, related to wet weight, was 2.9% in human, 3.7% in bovine and 4.0% in ovine pineals. According to CZARNOCKI et al. (1969) it amounted to 9.2%, 5% and 10.8%, respectively. Cholesterol, over 90% being in its free form, amounted to 14% of the total lipid content (BASINSKA et al., 1969a, b). CZARNOCKI et al. (1969) who detected cholesterol in its free form only, found that it amounted to 3.64 mg/g fresh tissue in the human, 6.09 mg in the bovine and 5.65 mg in the ovine pineals. Phospholipids amounted to 58% in human, 71% in bovine, 60% in ovine (BASINSKA et al., 1969a, b) and 32% in porcine pineals (JOUAN et al., 1964). Further analysis showed that the major phospholipids were phosphatidyl choline (45%), phosphatidyl ethanolamine (22%), sphingomyelin (14%), phosphatidyl serine (9%), monophosphoinositide (8%) and diphosphatidyl glycerol (2%) (BASINSKA et al., 1969a, b). The presence of phosphatidyl serine has been confirmed by BERG et al. (1971). Moreover, phosphatidyl inositol has been identified (see also BERG et al., 1971), which together with phosphatidyl glycerol showed an enhanced ^{32}P incorporation after administration of noradrenaline, dopamine, tyramine, octopamine or propranolol (EICHBERG et al., 1972), the drug-induced stimulation of the phospholipid metabolism apparently involving α-adrenergic receptors (HAUSER et al., 1973, 1974). In *young* animals the phospholipids did not differ from those in adults, apart from an additional presence of cardiolipin (3%) (BASINSKA et al., 1973).

Polyphosphoinositides have not been identified in the epiphysis cerebri (BASINSKA et al., 1969a, b). Variable results have been obtained with respect to gangliosides and cerebrosides. While gangliosides, as assessed by their sialic acid content, have been found by LANDOLT et al. (1966), BASINSKA et al. (1969a, b) were unable to detect them. Cerebrosides are present in minute amounts (7.2 pg/cell; 1.1% of dry weight) requiring ultramicrochemical methods for their detection (LANDOLT and HESS, 1966), which would explain why other authors (J. GREEN et al., 1962; FRAUCHIGER and SELLEI, 1968) failed to demonstrate them.

Paper chromatographic studies carried out on rat pineals revealed the presence of the phospholipids lecithine, phosphatidyl ethanolamine, phosphatidyl serine and sphingomyelin as well as triglycerides and fatty acids (PROP, 1963; KAPPERS et al., 1964). Thin layer chromatography was used by FRAUCHIGER and SELLEI (1968).

In view of the problem whether the pinealocytes resemble neurons or glial cells, QUAY (1974a), using data given in other literature, compared the pineal lipid content with that of neuronal perikarya and astrocytes of young rat brains possessing little myelin. This comparison revealed that the pineal differs from neurons in that it has more cholesterol and less phosphatidyl serine and differs from astrocytes in that it has a larger amount of phosphoinositides. It differs from both cell types in having larger amounts of sphingomyelin and galactolipid. It was therefore concluded that in its lipid composition the pinealocyte has a distinctive individuality and shows nearly equivalent degrees of differentiation from astrocytes as from neurons.

For the *histochemical* demonstration of lipids a number of different methods have been used. The most detailed studies are:

a) By QUAY (1957a) using Sudan black, oil red O, oil blue N, SELIGMAN and ASHBEL's method for carbonyl lipids, Nile blue for neutral and acidic lipids after CAIN, WINDAUS' digitonin reaction, BAKER's acid hematein test for phospholipid and the SCHULTZ method for cholesterol;

b) by KAPPERS et al. (1964) employing the acid hematein test, ELFTMAN's technique for the demonstration of phospholipids, KLÜVER-BARRERA's technique for phospholipids with controlled chromation after ELFTMAN, McMANUS' technique for the demonstration of phospholipids and possibly lipoproteins, BERENBAUM's technique for bound or masked lipids, staining with Fettrot for neutral lipids, LANDING's method for choline-containing lipids, the nile blue-sulphate method, the MOLISCH reaction modified by DIETZEL for sugar-containing lipids and CAIN's reaction for acetal lipids.

For the demonstration of fat in the rat pineal, PROP (1961) found the osmic acid-sodium iodide reagent of CHAMPY-COUJARD especially useful.

Histochemical, histological and ultrastructural studies have shown that the lipid content of the epiphysis cerebri is both very variable and subject to artefactual modification (KAPPERS et al., 1964). First, there is characteristic interspecies variation of the lipid content. In human epiphyses between 3% and 58% of the pineal cells contained small lipid droplets either alone or in association with pigment (QUAY, 1957a). A species with abundant lipid droplets is the rat, as assessed by light (e.g. QUAY, 1957a; PROP and KAPPERS, 1961) and electron microscopy (see p. 158). Abundant lipid was seen also in cattle (GUTTE and GRÜTZE, 1977; OKSCHE, personal communication; see, however, ARVY, 1961). In pigs, GUTTE and GRÜTZE (1977) observed large amounts of lipid, whereas ARVY (1961) classified the lipid content as small. Small amounts of lipid in the pineal were described in sheep (ARVY, 1961) and, based on electron microscopy (p. 158), in guinea-pig, mouse, gopher, dog, mole and seal. No lipid could be detected in monkey pinealocytes by light microscopy (WISLOCKI and DEMPSEY, 1948).

In addition to interspecies variation, the lipid content shows changes under various physiological conditions. A highly consistent daily rhythm was found in rats as assessed by extracting and measuring the ethanol-soluble lipid fraction (QUAY, 1965b, 1974a) and by counting the lipid-containing pinealocytes (QUAY, 1974a), the lipid content being highest at midday and lowest at midnight. The lipid content changes also in relation to the oestrous cycle. In the rat pineal, it was lowest just before ovulation; it increased after ovulation, reaching a maximum at dioestrus (ZWEENS, 1963, 1964, 1965). The lipid content is apparently also *age*-dependent. In the human pineal it increases with age (QUAY, 1957a). In the pineal of female rats a decrease was noted between 8 and 15.5 weeks of age (ZWEENS, 1965).

The location of the lipids also varies. In pineals of young humans the lipid-containing cells were few and lay peripherally, whereas in older specimens they were found throughout most of the organ (QUAY, 1957a). In the rat epiphysis cerebri regional differences are also conspicuous. According to QUAY (1957a) and MILINE et al. (1969a), the greatest amount of lipid is usually found in cells towards the peripheral and posterior (basal: MARKINA, 1970) parts of the organ, though it should be noted that in the medulla a greater *proportion* of pinealocytes contained lipid droplets than in the cortex (QUAY, 1974a). Moreover, special types of lipids are confined to particular types of pinealocytes (see below).

Histochemical analysis of pineal lipids revealed (QUAY, 1957a) that human pinealocytes contain a greater amount of neutral lipids and a smaller amount of carbonyl lipid than rat pinealocytes. In the rat pineal (QUAY, 1957a; KAPPERS et al., 1964; PROP, 1963, 1965) neutral lipids and phospholipids are closely associated, the phospholipids surrounding the neutral lipids in a ring-like fashion when stained with Nile blue sulphate. Cerebrosides could not be detected histochemically (KAPPERS et al., 1964).

According to QUAY (1957a) the lipid-containing pineal cells are not a homogeneous population, but comprise two major types. Whereas type I cells contained primarily ethanol-soluble carbonyl lipids coated with phospholipids, type II cells were characterized by a heavy staining of the cytoplasmic matrix with the acid hematein technique and a scarcity of microscopically visible, cytoplasmic droplets or vacuoles. In a subsequent investigation QUAY (1962a), using morphological and histochemical criteria, subdivided the acid hematein-positive type II cells into orthocytes, spirocytes, polygonocytes and contocytes (see also QUAY, 1974a). Massive subcutaneous injections of noradrenaline or the related catecholamine, amphetamine, yielded a strong increase in the amount of type II cells within 3 to 6 h (QUAY, 1962a). The effects of various treatments on pineal lipid content are listed in Table 16. Pinealocytes in vitro show prominent lipid droplets (Fig. 39), which increase in amount when the cells are cultured in the presence of thyroxine (Fig. 40) (S. MILCU et al., 1968).

Table 16. Change of lipid content in mammalian pineal glands as assessed by light or electron microscopy

Treatment	Author(s)
A. Increase	
Ovariectomy	ZWEENS (1965)
Gonadectomy	DAS GUPTA (1968a)
HCG	ZWEENS (1965)
Testosterone	S. MILCU et al. (1962)
Immobilization	MILINE et al. (1968)

Table 16. (continued)

Treatment	Author(s)
Bilateral superior cervical ganglionectomy	Kappers et al. (1964)
Induced fibrosarcomas	Huxley and Tapp (1972)
DOCA	Panagiotis and Hungerford (1962)
Oxypertine	Duffy and Markesbery (1970)
Desmethylimipramine	Duffy and Markesbery (1970)
p-Chlorophenylalanine	Duffy and Markesbery (1970), Sheridan and Keppel (1971)
p-Chlorophenylalanine + pargyline	Duffy and Markesbery (1971)
Nialamide	Buss (1965), Buss and Gusek (1965)
6-Hydroxydopamine	Sheridan and Keppel (1971)
Noradrenaline for 7 days	Buss and Gusek (1965)
B. Decrease	
Continuous light	Quay (1961 a, b), Zboray (1965), Prop and Ebels (1968)
Castration	S. Milcu et al. (1962)
Castration + adrenalectomy	Hungerford and Panagiotis (1962)
Ovariectomy + oestradiol	Zweens (1965)
Adrenalectomy	Markina (1970), Panagiotis and Hungerford (1961 a, b)
Hypophysectomy	Zweens (1965), Hungerford and Panagiotis (1962)
Pineal extract	Buss (1965)
Noradrenaline	Gusek et al. (1965), Buss and Gusek (1965)
Aludrin	Buss (1965)
Chlorpromazine	Horita et al. (1978)
Propylthiouracil	Panagiotis and Hungerford (1962), Hungerford and Panagiotis (1962)
Sodium administration	Prop (1965)
Sodium deficiency	Panagiotis and Hungerford (1961 a)
Angiotensin II	Panagiotis and Hungerford (1962), Hungerford and Panagiotis (1962)
C. Without visible effect	
Thyroxine	Hungerford and Panagiotis (1962)
Testosterone	
Hydrocortisone	
Adrenaline	
Serotonin	
ACTH	
β-MSH	
Alloxan	
Castration	

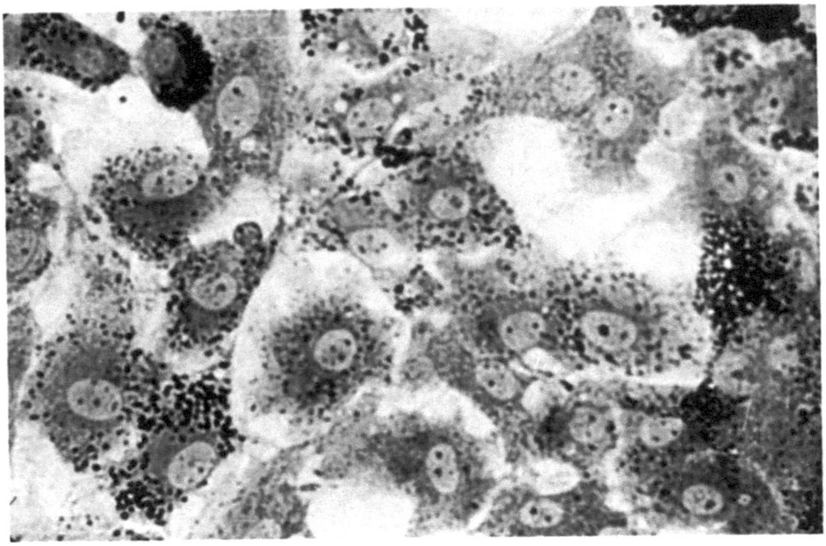

Fig. 39. Cultured pinealocytes showing prominent intracytoplasmic lipid droplets. Nile-blue stain. S. MILCU et al., 1968; ×200

d) Amino Acids

Histochemical studies dealing with pineal amino acids are scarce. In view of the key-role of tryptophan, this amino acid attracted special attention, but the results obtained are not very revealing. Traces of tryptophan were detected in the rat pineal, using the histochemical technique after ADAMS (GUSEK and BUSS, 1966a). With the same method tryptophan could not be visualized in the guinea-pig pineal (LUES, 1971). In cattle and pig moderate amounts were demonstrated by GUTTE and GRÜTZE (1977). In the rabbit, the use of the apparently more sensitive method by BRUEMMER et al. revealed tryptophan in one type of pinealocyte which is characterized by autofluorescence and is evenly distributed throughout the organ, but lying predominantly near blood vessels (A. SMITH et al., 1972b).

Biochemical studies of the free amino acid contents revealed (JOUAN et al., 1968; LABELLA et al., 1968; VELLAN et al., 1970; KRASS et al., 1971; QUAY, 1974a) that as far as the aliphatic amino acids are concerned, the pineal resembles astrocytoma more than it does brain tissue. Pineal and astrocytoma concentrations of leucine, valine, lysine, arginine and isoleucine are greater, and those of glutamic acid and γ-aminobutyric acid are less than in brain. Sulphur-containing amino acids, such as cystathionine and taurine, show remarkably high concentrations in the epiphysis cerebri. The amino acid derivatives ethanolamine and phosphoethanolamine exhibit lower concentrations in the pineal than in the pituitary, and pineal concentrations are higher in females than in males.

Recent studies (NIR et al., 1973a) dealing with the rat pineal have shown that the free amino acids undergo a diurnal rhythm and that they can be

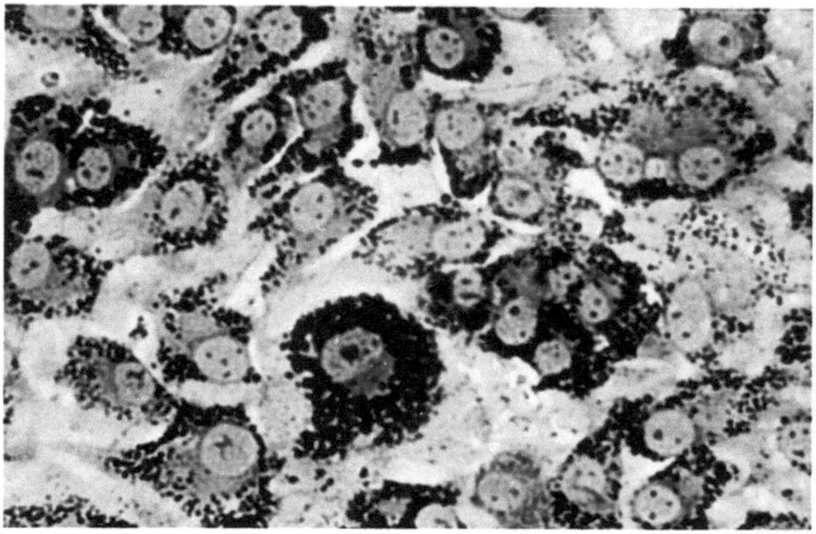

Fig. 40. Pinealocytes cultured *in vitro* under the influence of 50 µg/ml thyroxine. Note the increased amount of lipid droplets. Nile-blue stain. S. MILCU et al., 1968; ×200

influenced by environmental lighting conditions. At midday, lysine, arginine, histidine and taurine are at their height, whereas between 1.00 and 2.00 in the morning tryptophan and methionine are present in higher percentages. Exposure to continuous illumination brings about a decrease in pineal amino acid levels, except of tryptophan, tyrosine and glycine, whereas constant darkness causes an increase, except of taurine. Glutamine, by contrast, increases after both continuous darkness and light.

e) Biogenic Amines

As mentioned above (p. 11), the mammalian pineal is rich in biogenic amines. These amines play a role (a) as secretory product (melatonin) and its precursor (serotonin) and (b) as neurotransmitter (noradrenaline). Moreover, dopamine and histamine are present. The pinealocytes contain only some of these substances, namely serotonin, melatonin and dopamine, the latter having been little investigated. Histamine appears to be present mainly in mast cells, though some authors have postulated its presence in pinealocytes also. Noradrenaline is primarily present intraneuronally, and will be dealt with in Sect. B. VII. 2. It should be noted that serotonin is not exclusively present in pinealocytes. It may be taken up by sympathetic nerve fibres.

α) Serotonin

Thanks to the availability of the formaldehyde-induced fluorescence technique of FALCK-HILLARP for the visualization of monoamines, serotonin has

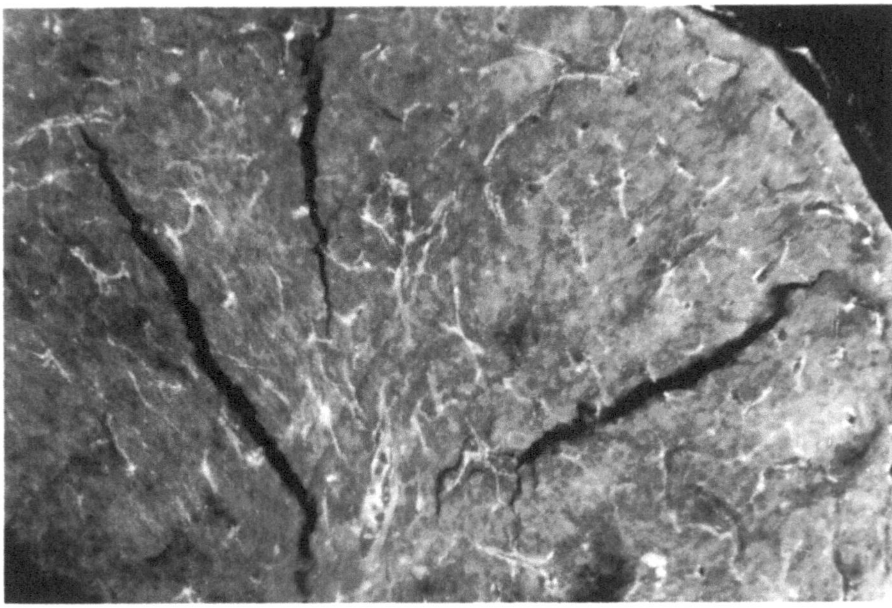

Fig. 41. Demonstration of serotonin in both parenchymal cells and nerve fibres of the rat pineal gland by means of the Falck-Hillarp technique. Falck et al., 1966; × 100

become the histochemically most widely studied pineal indoleamine. Following the reports by a number of authors (Miline et al., 1958; Giarman and Day, 1958/59; Giarman et al., 1959, 1960; Quay and Halevy, 1962; Jouan et al., 1966; Quay, 1963c, 1968b; Sellei and Frauchiger, 1968) that the mammalian gland contains large amounts of serotonin and the human pineal very variable (0.36 to 22.82 μg/g, Giarman et al., 1960) amounts, Bertler et al. (1963, 1964) and Csillik (1963) demonstrated serotonin *histochemically* and noted that a yellow (= serotonin) fluorescence is present in both pinealocytes (Fig. 41) and sympathetic nerve fibres, the latter containing ca. 30% of the total pineal serotonin (Neff et al., 1968, 1969). The presence of serotonin was subsequently amply confirmed (e.g., Owman, 1964a, b, 1965, 1968; Falck et al., 1966; Wood and Duncan, 1968; Wood, 1972; C. Machado et al., 1968; Björklund et al., 1972) and corroborated by autoradiographic studies (Taxi and Droz, 1966; Bak et al., 1970; Louis et al., 1970; Romijn et al., 1977b). It soon became apparent that interspecies differences exist. By means of fluorescence microscopy, serotonin was demonstrable in rat, mouse, hamster, guinea-pig, pig (Bertler et al., 1963; Owman, 1964a, 1965, 1968; Håkanson and Owman, 1966; Sheridan and Sladek, 1975), rabbit (A. Smith et al., 1972a, b) and ferret (Trueman and Herbert, 1970a, b), but not in man (Håkanson and Owman, 1966; Owman, 1968), dog, cat and cattle (Bertler et al., 1963; Owman, 1964a, 1965; Håkanson and Owman, 1966). Very intense fluorescence was noted in the rhesus pineal (Fig. 42) (Sheridan and Sladek, 1975). These histochemical observations appear to demonstrate fairly accurately the varying amounts of serotonin present

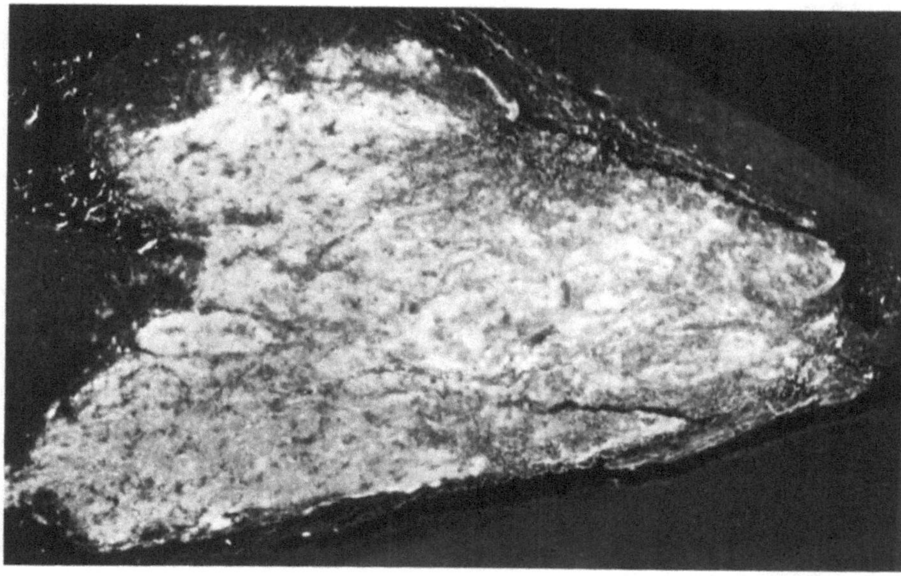

Fig. 42. Serotonin fluorescence in the pineal of a rhesus monkey. The fluorescence is so extensive that the entire gland appears to fluoresce. SHERIDAN and SLADEK, 1975; ×80

in pinealocytes, since, in contrast to other histochemical techniques, the results show a close correspondence with data obtained by chemical or biological assays of serotonin (WURTMAN et al., 1968a; QUAY, 1974a), as well as other parameters, such as Dopa decarboxylase and monoamine oxidase activities (HÅKANSON and OWMAN, 1966). Moreover, quantitative microfluorimetric studies applied to histological sections of rat pineals following the administration of a number of drugs known to influence serotonin concentration have shown that a linear relationship exists between fluorescence intensities and chemically detectable serotonin, and that the variations in fluorescence intensities are small in adjacent sections from the same pineal and in pineals from different animals (TILDERS et al., 1974).

The localization of serotonin-containing pinealocytes shows species differences as well. In the rat, yellow fluorescing cells are not only found in the pineal proper, but also in the stalk and the lamina intercalaris or deep pineal (BJÖRKLUND et al., 1972) (Fig. 43). In the pineal proper greenish-yellow fluorescing pinealocytes amount to 3,000–5,000 cells per mm^3 (CSILLIK, 1963). In the rabbit (A. SMITH et al., 1972a) they are particularly abundant (11–18 cells/0.05 mm^2) in a narrow zone lying just below the dorsal surface of the organ, whereas in the remainder of the organ they amount to 0–10 cells/0.05 mm^2.

In the rabbit pineal, the situation of the fluorescent cells is fairly complex. Altogether it would appear that three categories of pinealocytes are present: (a) pinealocytes showing formaldehyde-induced serotonin fluorescence only (type I), (b) autofluorescent pinealocytes (type II) which also exhibit formaldehyde-induced fluorescence and (c) pinealocytes which cannot be demonstrated by means of fluorescence (type III). The situation is complicated by reason of the fact that these

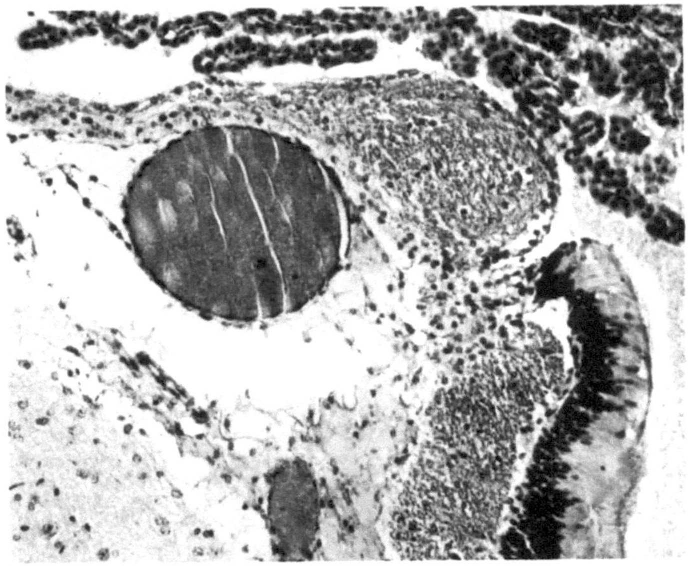

a

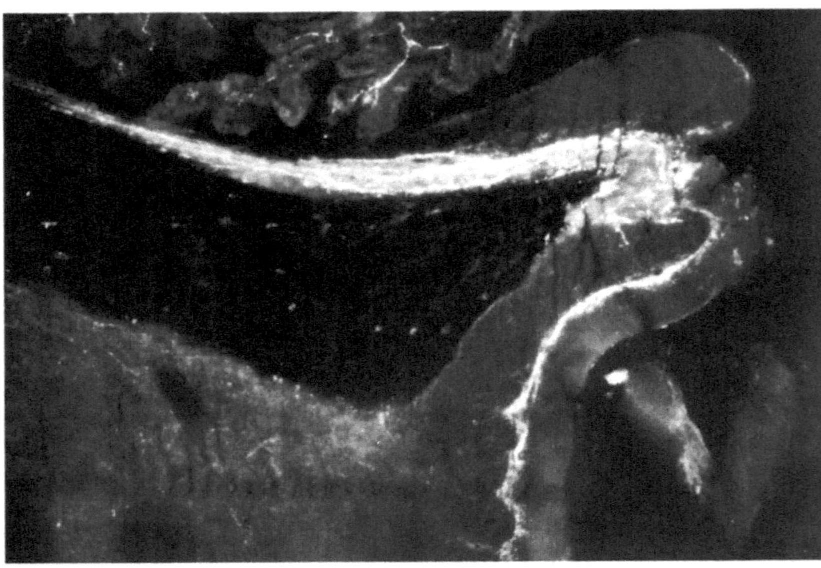

b

Fig. 43a, b. Deep pineal gland of a rat. **a** Light micrograph. Compare with Figs. 15 and 16. **b** Note the intense serotonin fluorescence in this region. BJÖRKLUND et al., 1972; ×190

categories have a different distribution within the gland and that they behave differently under experimental conditions. Autofluorescent cells are evenly distributed throughout the epiphysis, whereas the pinealocytes showing formaldehyde-induced fluorescence are particularly prominent only in the "cortex", which is a thin strip along the dorsal aspect of the epiphysis cerebri. After administration of *p*-chlorophenylalanine, type I cells decrease in number whereas type II cells show a numerical increase, but only in the cortical region. The unique situation of the cortex is also illustrated by the observation that circadian changes in the numbers of type I and type II cells appear to be restricted to this region. Since during the night type I cells decrease and type II cells increase and since elements of type II contain a tryptophan-rich protein, it has been suggested that during the day tryptophan is metabolized into serotonin and during the night is incorporated into protein (A. SMITH et al., 1972a, b).

In the ferret, the intense parenchymal fluorescence is reduced in the antero-inferior part of the epiphysis, where the pinealocytes are scanty (TRUEMAN and HERBERT, 1970a, b). In the pig pineal, which shows a very pronounced yellow fluorescence, the fluorescence intensities varied considerably from cell to cell (OWMAN, 1965).

The precise intracellular localization of serotonin in the pinealocytes is not always easy to determine, apart from the fact that the nuclei are non-fluorescent. There can be no doubt that the pinealocyte perikarya contain serotonin. Moreover, pinealocyte processes may comprise the indoleamine. In the ferret serotonin was only occasionally demonstrable in the proximal parts of the processes (TRUEMAN and HERBERT, 1970a, b). In the pig, some pinealocytes were completely loaded with small, intensely yellow-fluorescing granules (OWMAN, 1965). Autoradiographic studies at the electron microscopic level did not reveal many more details, due to the fact that it is difficult to accurately identify the position of the label in relation to small tissue components. However, it was clearly shown that following injection of ^{3}H-5-hydroxytryptophan, a precursor of serotonin, both pinealocytes and nerve fibres were labelled (TAXI and DROZ, 1966). Quantitative studies revealed that 53% of the silver grains counted lay over pinealocytes, 43% over sympathetic nerve fibres and 4% over perivascular spaces (BAK et al., 1970), which agrees well with the findings that serotonin is equally distributed between parenchymal cells and sympathetic nerve fibres (BERTLER et al., 1963, 1964) and that following denervation the serotonin content of the pineal decreases by 50%–73% (PELLEGRINO DE IRALDI et al., 1963).

In view of the fact that serotonin shows prominent day–night changes with high levels during the day and low levels during the night (QUAY, 1963c, 1964b; BERTLER et al., 1964; SNYDER et al., 1964a, 1965b, 1967a), it was only logical to find a decrease of serotonin-containing cells in the night (A. SMITH et al., 1972b; rabbit).

Experimental studies in a number of species have shown that the fluorescence intensity of the pinealocytes can be altered. In rats the fluorescence is abolished after administration of the decarboxylase inhibitors Ro 4-4602 or NSD 1015 (FALCK et al., 1966). It decreases slightly after bilateral cervical sympathectomy (BERTLER et al., 1963; OWMAN, 1964a; HÅKANSON and OWMAN, 1966), although negative findings have also been reported (CSILLIK, 1963), after reserpine treatment (BERTLER et al., 1963; CSILLIK, 1963) or after administration of nialamide combined with decarboxylase inhibitors (BERTLER et al., 1963). An *increase* of parenchymal serotonin fluorescence was detected after administration of the monoamine oxidase inhibitor nialamide combined with 5-hydroxytryptophan (BERTLER et al., 1963), which is consistent with the observation that injection of 5-hydroxytryptophan alone leads to a 400% increase in pineal serotonin, as assessed by bio-assay (PELLEGRINO DE IRALDI et al., 1963). Microspectrofluorimetric studies confirmed the increase of the fluorescence intensities following administration of nialamide, tryptophan and 5-hydroxytryptophan and demonstrated a decrease after *p*-chlorophenylalanine, while reserpine had no effect (TILDERS et al., 1974). Visual assessment revealed no change in parenchymal fluorescence intensity after chemical sympathectomy by means of 6-hydroxydopamine (ERÄNKÖ and ERÄNKÖ, 1971). In the ferret, reserpine, nialamide, L-dopa, 5-hydroxytryptophan, decentralization, unilateral superior cervical ganglionectomy or sham operations did not strikingly affect pinealocyte fluorescence, apart from the fact that after 11–12 days the fluorescence changed from yellow to yellow-green (TRUEMAN and HERBERT, 1970a, b). In the hamster, the yellow serotonin fluorescence of the pinealocytes decreased after reserpine treatment and increased after administration of nialamide (OWMAN, 1965). In guinea-pig pinealocytes, the weak yellow fluorescence was unchanged after nialamide and after nialamide in combination with serotonin, reserpine or α-methyl-*m*-tyrosine; but after metaraminol it changed to a diffuse greenish colour of somewhat lower intensity than normal (OWMAN, 1965). The slightly positive reaction in rabbit pinealocytes was unaffected by nialamide, bilateral removal of the superior cervical ganglia, or α-methyl-*m*-tyrosine. An intense green fluorescence was noted after *m*-tyrosine (OWMAN, 1965). The effect of *p*-chlorophenylalanine on rabbit pinealocytes has been mentioned on p. 88.

Other histochemical tests for serotonin in the pineal have been little used. DUCHESNE (1964), using a method by GEREBTZOFF and DRESSE, described a positive reaction in the rat pineal.

β) Melatonin

Suitable histochemical techniques for the demonstration of melatonin have become available only recently. BUBENIK et al. (1974) demonstrated melatonin and its precursor *N*-acetylserotonin by means of a double antibody technique using fluorescein or peroxidase labelling and showed that in rat pineals almost all parenchymal cells exhibit a specific fluorescence, though differing intensities were recorded in individual cells. Relatively strong fluorescence was found in the capsular membrane and in ependymal cells overlying the capsule and the adjacent colliculi. FREUND et al. (1977), on the other hand, found specific fluorescence only in the cortical regions of the organ (Fig. 37a–d).

γ) Dopamine

Relatively little is known about the catecholamine dopamine in the epiphysis cerebri, apart from the fact that its presence is firmly established (CSILLIK, 1963; HYYPPÄ, 1971; HYYPPÄ et al., 1971; BROWNSTEIN and AXELROD, 1974). It appears to be present in pinealocytes, since it is not affected by denervation (PELLEGRINO DE IRALDI and ZIEHER, 1966).

δ) Histamine

It is still an open question whether the histamine present in the mammalian pineal gland (GIARMAN and DAY, 1958/59) is exclusively bound to mast cells or whether some of the amine is present in pinealocytes. That histamine may reside in pinealocytes has recently been suggested by the finding that the histamine content of the rat pineal exhibited diurnal fluctuations in phase with those reported for melatonin, showing an approximately twofold increase at the beginning of the dark period (GARBARG et al., 1974). However, there is as yet no convincing histochemical evidence for its presence within pinealocytes. Although rat pinealocytes develop a distinct blue-white fluorescence upon treatment with *O*-phthaldialdehyde (OPT) (BERNÁD and CSABA, 1970; HÅKANSON et al., 1974), which is used for the fluorimetric assay of histamine, this cannot be taken as proof for the presence of histamine, since cytospectrofluorimetric studies combined with ion-exchange chromatography have shown that in rat pinealocytes the OPT-reactive compounds are different from histamine and may represent histidine or peptides possessing histidine in the NH_2-terminal position (HÅKANSON et al., 1974). For further details on the presence of histamine, see p. 224.

f) Peptides

In view of the fact that the mammalian pineal gland contains peptides and polypeptides (e.g. JOUAN et al., 1965, 1968) which may play an important role (see p. 240), staining reactions for peptides have been applied in a variety of species. Altogether it would appear that the presence of stainable peptide material is not a regular feature of fully differentiated mammalian pinealocytes (e.g. DAS GUPTA, 1968b, hamster). In the rat, a few granulated cells stainable with paraldehyde fuchsin were found in the basal tip of the epiphysis cerebri (QUAY, 1957c). Some peptide material of unspecified nature was present in pinealocytes of 16-day-old fetuses (OWMAN, 1960/61; and p. 51). Recently, arginine vasotocin has been demonstrated by immunocytochemical techniques in cells of the adult rat pineal, which were thought to represent a second line of pinealocytes or glial cells (BOWIE and HERBERT, 1976). In the pineal of the gerbil (*Meriones unguiculatus*) abundant paraldehyde fuchsin-positive material was attributed to tentatively identified glial cells (JAPHA et al., 1977). In view of possible functional differences between the superficial and the deep portions of the pineal it is interesting to note that in this species the stainable material of the deep pineal, in contrast to that of the superficial pineal, was not noticeably affected by trypsin (JAPHA et al., 1977). Presumptive peptidic secretory material has also been found in some monkey and bovine pinealocytes (LUKASZYK and REITER, 1975). In the monkey a few pinealocytes lying in the centre of the organ contained small droplets of a colloid-like substance which was also demonstrable between pinealocytes. In the bovine pineal, "secretory" material stainable with PAS and chrome alum haematoxylin or aldehyde fuchsin was restricted to pinealocyte processes. In addition, the presumptive secretory product was found in the walls of blood vessels, glial cells and associated with corpora arenacea.

g) Proteins and Glycoproteins

Histochemical studies on pineal glycoproteins and proteins other than enzymes are rare. Chemical studies have revealed that pineal proteins can be subdivided into a great number of different fractions. In the rat pineal, 17 fractions were distinguished by means of polyacrylamide gel microelectrophoresis (PUN and LOMBROZO, 1964). In the same species, the water-soluble proteins, when investigated by means of microdisc electrophoresis, showed 22 different peaks, which could be separated into 12 major fractions, of which 11 protein peaks and 6–7 major fractions were in the fast-migrating pre-albumin region and the remainder, in the slow post-albumin region (NIR et al., 1973b). By contrast only four major fractions of water-soluble proteins were described in bovine pineal extracts (FAJER et al., 1962).

In isolated sectioned human pinealocytes studied microchemically, the proteolipid protein amounted to 1.57% of dry weight or 10.2 pg/cell, and residue protein, which is total protein minus proteolipid protein, totalled 53.9% of dry weight or 351 pg/cell (LANDOLT and HESS, 1966).

Like most other biological parameters of the pineal, the proteins also show typical changes under various physiological conditions. Diurnal changes have been described by two groups of authors, though the results obtained show slight differences. While NIR et al. (1971a, 1973b) found peaks of protein levels at 16.00 h in males and at 20.00 h in females and lowest levels at 24.00 h and 4.00 h respectively, MERRITT and SULKOWSKI (1969) observed peak levels at 18.00 h and lowest levels at 12.00 h. According to NIR et al. (1973b) the fast-migrating proteins showed highest levels during daytime, while the slow-moving fractions were particularly prominent at night. During the oestrous cycle there was a clear and consistent tendency to high protein levels at dioestrus, intermediate levels at pro-oestrus and metoestrus, and low levels at oestrus (NIR et al., 1972a).

In view of the fact that many proteins with diverse functions such as hormones, antibodies, membrane components, serum proteins, and enzymes are glycoproteins, it has been regarded as important that the rat pineal gland actively synthesizes and releases glycoproteins, as demonstrated by studies involving L-[^{3}H] fucose (LOTT et al., 1971, 1972). Whether or not the released substances represent secretory products remains to be established, but it is interesting to note that the application of cytochemical staining methods for the visualization of glycoproteins at the electron microscopic level showed that presumptive secretory granules gave a positive reaction. Lysosomes, Golgi vesicles, microtubules, microtubular sheaves (p. 139), zonulae adhaerentes and synaptic ribbons (p. 146) also reacted (KRSTIĆ, 1975a, 1976a).

h) Nucleic Acids

Relatively few studies are available concerning epiphyseal nucleic acids and the results obtained are not always consistent.

α) Deoxyribonucleic Acid (DNA)

The DNA content of the human epiphysis cerebri, as assessed by ultramicrochemical analysis, amounted to 10.13 μg/mg dry weight, greatly exceeding that of other brain regions (LANDOLT et al., 1966). In rats there were no statistically significant diurnal changes of the DNA content, though there was a trend to higher values at midday and lower levels at midnight (NIR et al., 1971a). During the oestrous cycle there were likewise no significant changes (NIR et al., 1972a), but in immature female rats a single injection of 10 µg oestradiol led to a brief increase of DNA in the pineal (NIR et al., 1970a). Developmental studies have shown that the DNA content of individual pinealocytes remains constant throughout the 1st year of life (NOVÁKOVÁ et al., 1969).

β) Ribonucleic Acid (RNA)

RNA is present in both nucleoli and cytoplasm. Despite the prominence of the nucleoli (p. 116), they appear to contain little RNA, as indicated by a slight affinity to pyronin Y and a strong staining reaction with acidic dyes such as orange G (HEINIGER, 1965b; pig). The fact that in the monkey nucleolar staining was not diminished after ribonuclease treatment (WISLOCKI and DEMPSEY, 1948) points in the same direction.

Cytoplasmic RNA, as assessed by cytochemical staining methods, has been found to be present in relatively small amounts in pinealocytes of man (BAYEROVÁ and BAYER, 1960; HEINIGER, 1965b), monkey (WISLOCKI and DEMPSEY, 1948), pig (HEINIGER, 1965b; MIKAMI, 1951), horse and goat (MIKAMI, 1951); yet studies by GUTTE and GRÜTZE (1977) point to large amounts of RNA in cattle and pig. Individual pinealocytes contained 11.0 pg RNA, as shown by ultramicrochemical studies carried out on freeze-dried sections of 40 µm thickness (LANDOLT et al., 1966). That the total RNA content of the pineal (15.89 µg/mg dry weight) exceeded that in a number of different CNS regions (LANDOLT et al., 1966) is understandable when one realizes that the pinealocytes are more densely packed than nerve cells.

The RNA content of the epiphysis cerebri changes under various physiological and experimental conditions. Cytophotometric studies (Table 17) have shown that the RNA content is higher during the day than during the night (NOVÁKOVÁ et al., 1971). In a more detailed study NIR et al. (1971a) found that the RNA content increased slightly in the morning, reached a maximum at 12.00 h, decreased strikingly until midnight, when the levels were lowest, and increased prominently between midnight and 8.00 h. The increase of cytoplasmic RNA between midnight and 8.00 h has been confirmed by MERRITT and SULKOWSKI (1969), but they found lowest values (ca. 6.3 µg/mg pineal tissue) at 12.00 h followed by an increase between 12.00 h and 18.00 h. ZATZ et al. (1976a) have postulated that RNA, presumably messenger RNA, which is required for the

Table 17. RNA content[a] of rat pinealocytes (NOVÁKOVÁ et al., 1971)

Time	Normal lighting conditions	Continuous light	Continuous darkness
11.00	106.0 ± 4.5	139.0 ± 3.9	98.0 ± 17.7
23.00	96.7 ± 4.2	119.2 ± 4.7	91.6 ± 15.4

[a] Expressed as arbitrary cytophotometric units; means ± SD.

induction of NAT, is synthesized and accumulated during the first half of the night. Studies in relation to the oestrous cycle have shown that the RNA content was high at dioestrus, intermediate at pro-oestrus and metoestrus and low at oestrus (NIR et al., 1972a).

Experimentally, it has been shown that oestradiol treatment briefly raised the pineal RNA content (NIR et al., 1970a). Exposure to continuous illumination or darkness led to equivocal results. While NIR et al. (1969a) found that continuous illumination resulted in a decrease of the RNA content, NOVÁKOVÁ et al. (1971) described an increase (Table 17). In continuous darkness, NIR et al. (1969a) were unable to detect any changes while NOVÁKOVÁ et al. (1971) observed a decrease (Table 17). Administration of the anticholinergic psychotogenic drug IB 336 was followed by a decrease of the RNA content (MERRITT and SULKOWSKI, 1969). Changes in the amounts of ribosomes will be dealt with in the appropriate sections.

i) Enzymes

Histochemical enzyme studies on mammalian pinealocytes have been carried out in relatively few species, and it would appear that they have contributed virtually nothing to elucidate pineal function. A disadvantage is that a fair number of these studies stem from the early days of enzyme histochemistry when the techniques employed were not sophisticated enough to yield meaningful results. Another drawback is that the results obtained cannot be directly compared because often one and the same enzyme has been demonstrated by means of different techniques. Moreover, the enzyme activities have not been objectively assessed and only some of the histochemically demonstrable enzymes have been studied by biochemical techniques so as to allow direct correlation. The sequence in which the various enzymes are being dealt with follows that given by the Enzyme Commission. It should be noted that histochemically demonstrable enzymes only have been mentioned. A detailed account of all the enzymes present in the epiphysis cerebri has been given by QUAY (1974a).

α) Oxidoreductases

Alcohol dehydrogenase (EC 1.1.1.1) has been demonstrated histochemically in rat pinealocytes cultured in vitro, but no details concerning its distribution have been given (TAŞCĂ et al., 1968). Biochemically the enzyme could not be demonstrated in bovine pineals (THIÉBLOT et al., 1966c).

Lactate dehydrogenase (EC 1.1.1.27), which, according to biochemical studies in cattle (THIÉBLOT et al., 1965, 1966c), exhibits a higher activity in the pineal than in brain tissue, shows histochemically a relatively high activity in pinealocytes of squirrel monkeys (MANOCHA and BOURNE, 1966) and rats (BARA et al., 1971; GEREBTZOFF and DEVEČERSKI, 1972; KRSTIĆ and TARSOLY, 1972; BOTTICELLI et al., 1972), especially on the surface of lipid droplets (TAPP et al., 1973). A positive enzyme reaction was also obtained in rat pinealocytes cultured in vitro (TAŞCĂ et al., 1968). Moderate activity was noted in human pineals (BAYEROVÁ and BAYER, 1967) and human pinealoma (KAWASHIMA et al., 1965). In the

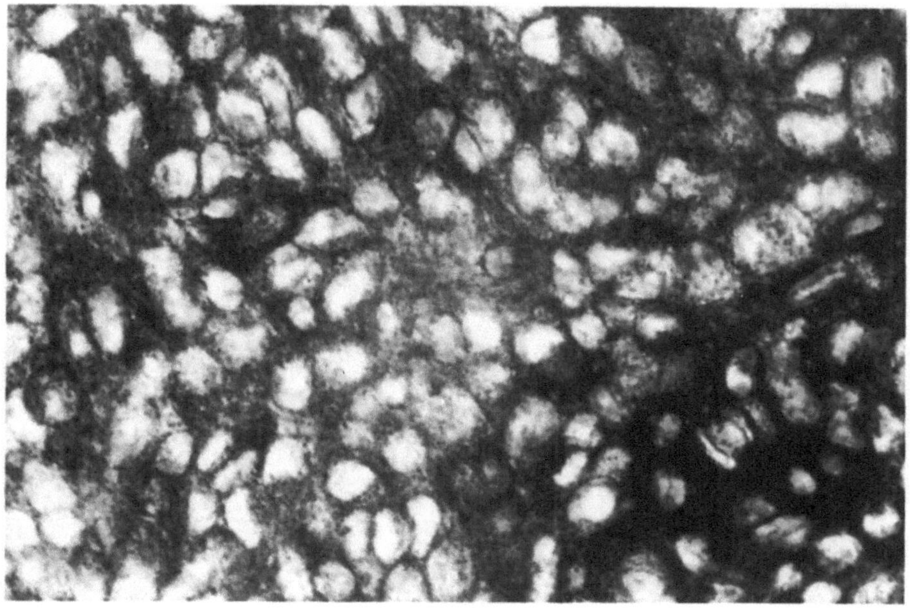

Fig. 44

Figs. 44–46. Histochemical demonstration of lactate dehydrogenase activity in the guinea-pig pineal gland. **Fig. 44** Lactate dehydrogenase; **Fig. 45** Lactate dehydrogenase "H fraction"; **Fig. 46** Lactate dehydrogenase "M fraction". Unpublished photomicrographs from the late Dr. W. Köhl, Department of Anatomy, University of Mainz

guinea-pig pineal, the enzyme reaction was negative when a conventional incubation medium was used (VOLLRATH and SCHMIDT, 1969). However, when gel media which prevent diffusion of water soluble enzymes into the incubation medium were used, lactate dehydrogenase activity was clearly demonstrable (KÖHL, unpublished results) (Fig. 44). As can be seen from Figs. 45 and 46, both H- and M-type isoenzymes are present.

Changes in enzyme activity have also been demonstrated. During the oestrous cycle, maximal enzyme activity was present at pro-oestrus and minimum levels at late dioestrus (FASANO et al., 1971a). A decrease in enzyme activity was observed in 7-day pregnant rats (DEVEČERSKI, 1972a) and an increase following exposure to cold (DEVEČERSKI, 1972b).

Hydroxybutyrate dehydrogenase (EC 1.1.1.30) has been demonstrated in the rat pineal both in vivo (BOSTELMANN, 1968a) and in vitro (TAŞCĂ et al., 1968). In the guinea-pig pineal, the enzyme activity was low (KÖHL, unpublished results). In the rat, exposure to continuous illumination was followed by a decrease in enzyme activity, whereas after constant darkness an increase was noted (BOSTELMANN, 1968a). Moderate enzyme activity was typical of a human pinealoma (KAWASHIMA et al., 1965).

Malate dehydrogenase (EC 1.1.1.37) revealed low to high activities in rat pineals both in vivo (BARA et al., 1971; KRSTIĆ and TARSOLY, 1972) and in vitro (S. MILCU et al., 1968; TAŞCĂ et al., 1968). No activity could be demonstrated

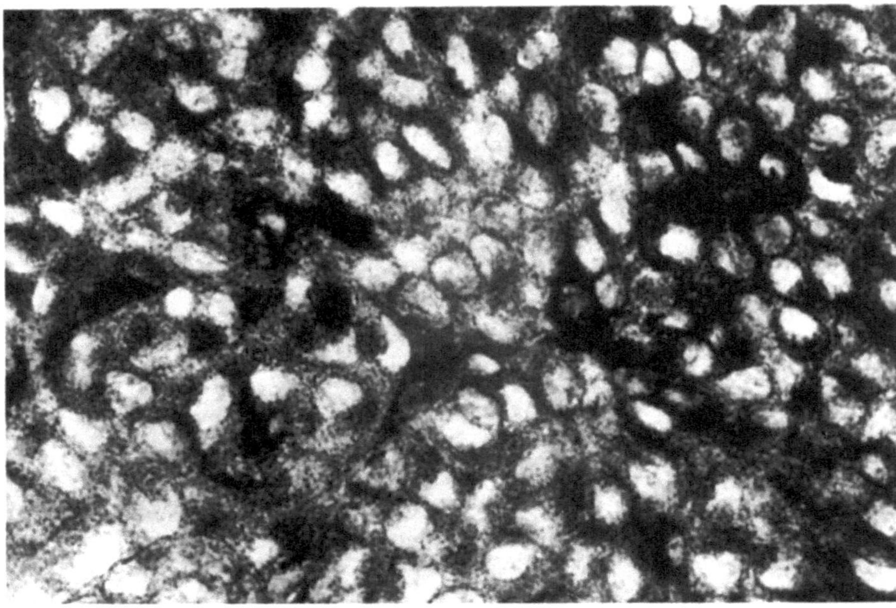

Fig. 45

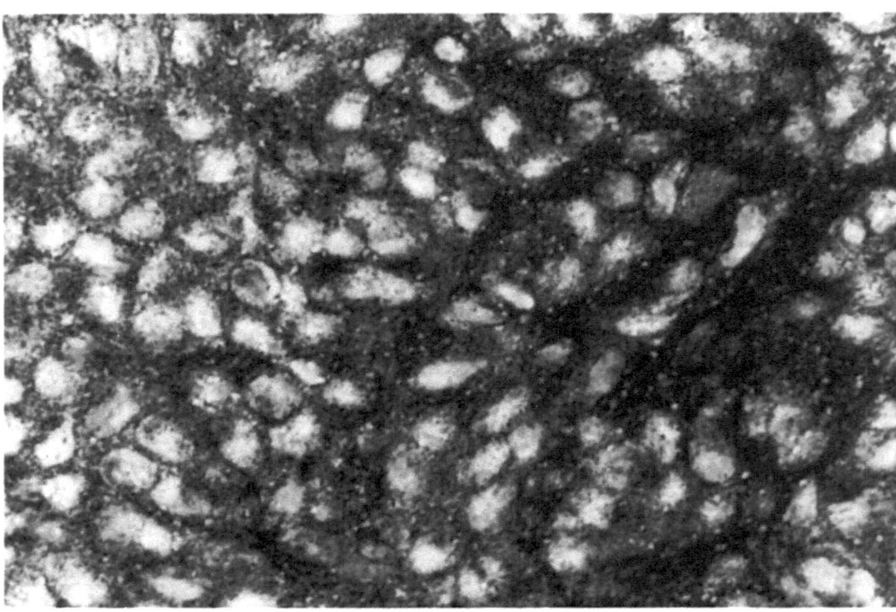

Fig. 46

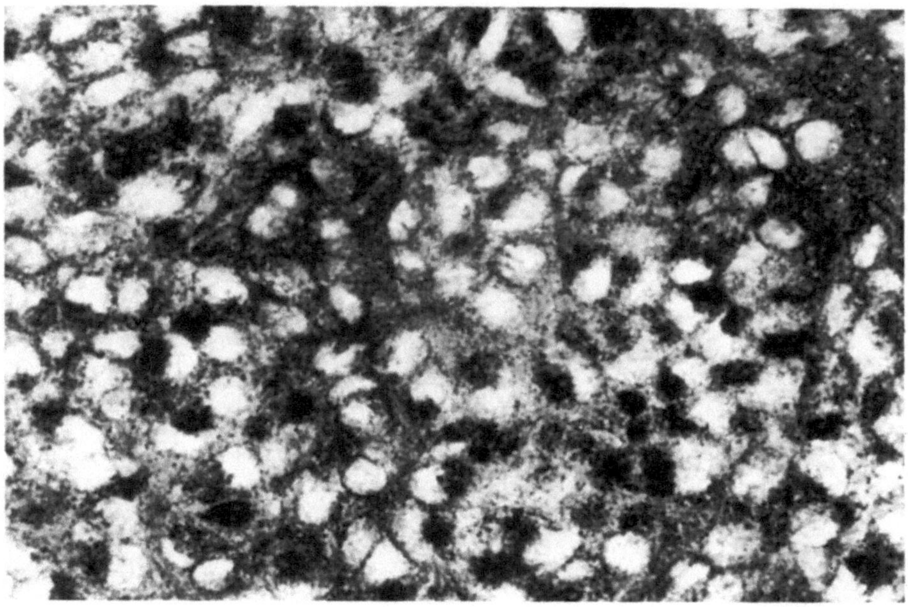

Fig. 47. Histochemical demonstration of the NAD-dependent malate dehydrogenase. Guinea-pig pineal organ. Unpublished photomicrograph from the late Dr. W. Köhl, Department of Anatomy, University of Mainz

in the guinea-pig pineal when conventional incubation media were used (VOLLRATH and SCHMIDT, 1969). When gel media were used (KÖHL, unpublished results), the NAD-linked enzyme was clearly demonstrable (Fig. 47) in circumscribed, round to slightly oval areas lying closely related to the cell nucleus, which appear to correspond to the so-called active sites of pinealocytes, as revealed by electron microscopy (p. 136). Moderate activity has been found in human pinealoma (KAWASHIMA et al., 1965).

Isocitrate dehydrogenases (EC 1.1.1.41 and 1.1.1.42). In view of the fact that these enzymes are substantially soluble, it is not surprising that the results obtained are fairly variable. While KRSTIĆ and TARSOLY (1972) described a moderate enzyme activity in the rat pineal and BOTTICELLI et al. (1972), a fairly strong one, GEREBTZOFF and DEVEČERSKI (1972) found a weak reaction when the organ was incubated for the NAD-linked enzyme and an even weaker outcome of the reaction for the NADP-linked enzyme. The use of gel media in the guinea-pig (KÖHL, unpublished results) revealed (Fig. 48) that the NAD-linked enzyme is not evenly distributed in the pinealocyte cytoplasm, but is restricted to sites where, according to electron microscopic studies, mitochondria are abundant. The reaction for the NADP-linked enzyme was only weakly positive.

In tissue culture, a prominent reaction was achieved, which was distinctly enhanced by administering thyroxine (S. MILCU et al., 1968). TAŞCĂ et al. (1968) noted that a thin perinuclear area was devoid of NAD-linked enzyme activity.

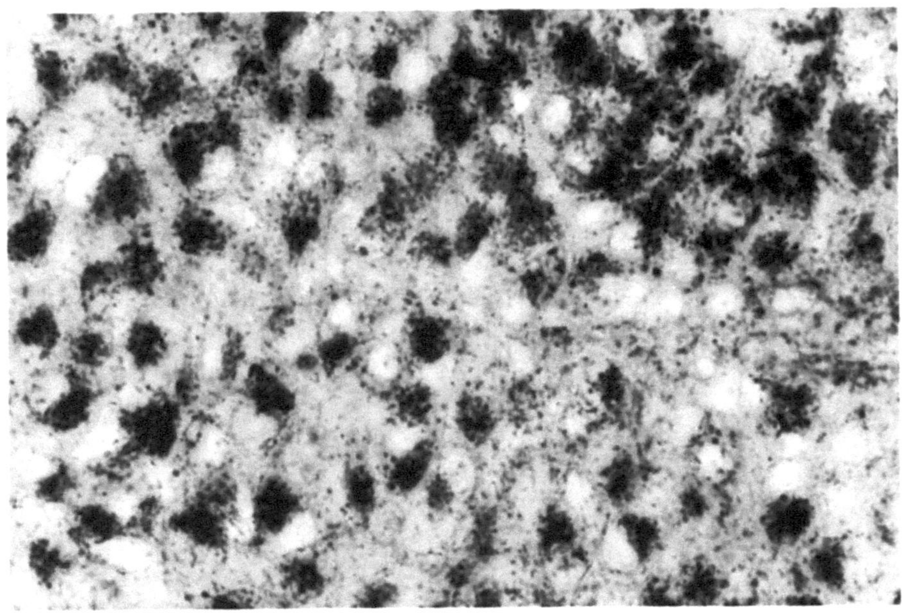

Fig. 48. Histochemical demonstration of the NAD-linked isocitrate dehydrogenase in the guinea-pig pineal organ. Note that the reaction product is confined to round or oval areas. Unpublished photomicrograph by the late Dr. W. KÖHL, Department of Anatomy, University of Mainz

In the pineal of *pregnant* rats, the NADP-linked isocitrate dehydrogenase showed a slight reduction at 7 days of gestation and a strong decrease at 15 days (DEVEČERSKI, 1972a). Exposure to cold revealed a different behaviour of the two enzymes. After 7 days of exposure, the NAD-linked enzyme showed an increase and the NADP-linked enzyme a decrease, but after 18 days both reverted to normal (DEVEČERSKI, 1972b). Moderate activity has been found in human pinealoma (KAWASHIMA et al., 1965).

6-Phosphogluconate dehydrogenase (EC 1.1.1.44) has been demonstrated in the guinea-pig pineal, where it showed a distinct distribution pattern (KÖHL, unpublished results) (Fig. 49). The enzyme was also demonstrable in rat pinealocytes in vitro (S. MILCU et al., 1968) and in vivo (BOTTICELLI et al., 1972).

Glucose-6-phosphate dehydrogenase (EC 1.1.1.49), an enzyme of the pentose phosphate pathway, is of special interest, since the epiphysis cerebri has an active pentose phosphate cycle in common with other endocrine organs (KRASS and LABELLA, 1966, 1967; LABELLA and KRASS, 1966). However, *histochemically* the activity was not always particularly high. Moderate activity was noted in human pineal (BAYEROVÁ and BAYER, 1967) and a moderate to very strong activity in the rat pineal (ABE et al., 1963; GUSEK and BUSS, 1966a; TAPP et al., 1973; BARA et al., 1971; GEREBTZOFF and DEVEČERSKI, 1972; KRSTIĆ and TARSOLY, 1972; BOTTICELLI et al., 1972). In tissue culture, rat pinealocytes exhibited low glucose-6-phosphate dehydrogenase activity (TAŞCĂ et al., 1968). Typical of the guinea-pig pineal is an uneven distribution of the enzyme. While

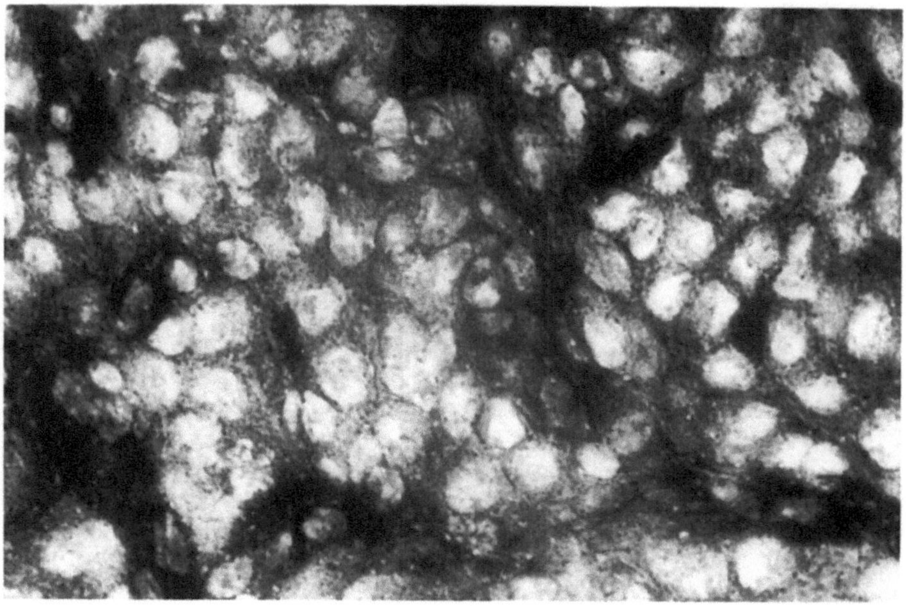

Fig. 49. Histochemical demonstration of 6-phosphogluconate dehydrogenase activity in the guinea-pig pineal organ. Unpublished photomicrograph by the late Dr. W. Köhl, Department of Anatomy, University of Mainz

the majority of pinealocytes revealed relatively little activity, there are groups of pinealocytes which show an intense reaction in the perikarya and sometimes also in the cell processes (Vollrath and Schmidt, 1969) (Fig. 50). Fetal guinea-pigs show a slightly lower activity than adults (Vollrath and Schmidt, 1969). An increase in enzyme activity was found in rat pinealocytes after administration of gonadotropic hormone for 7 days (Gusek and Buss, 1966a) or exposure to cold (Devečerski, 1972b). In pregnant rats (Devečerski, 1972a) and guinea-pigs (Vollrath and Schmidt, 1969) the enzyme activity was unchanged when compared to non-pregnant controls. Moderate activity was seen in human pinealoma (Kawashima et al., 1965).

Steroid dehydrogenase (EC 1.1.1.50 and 1.1.1.51) reactions were weakly positive in the human (Bayerová and Bayer, 1967) and negative in the guinea-pig pineal (Vollrath and Schmidt, 1969).

α-Glycerophosphate dehydrogenase (EC 1.1.2.1) activity has been demonstrated in rat pinealocytes both in vivo (Bostelmann, 1963, 1968a) and in vitro (Taşcă et al., 1968) as well as in ungulate pinealocytes (Arvy, 1961). In rat and rabbit (Bostelmann, 1963), the enzyme activity was variable; typically, groups of pinealocytes stood out clearly by reason of an intense reaction, whereas the remainder revealed a much lower activity. Exposure of rats to continuous illumination was followed by a decrease of enzyme activity, whereas continuous darkness stimulated the enzyme (Bostelmann, 1968a). Strong activity was found in a human pinealoma (Kawashima et al., 1965).

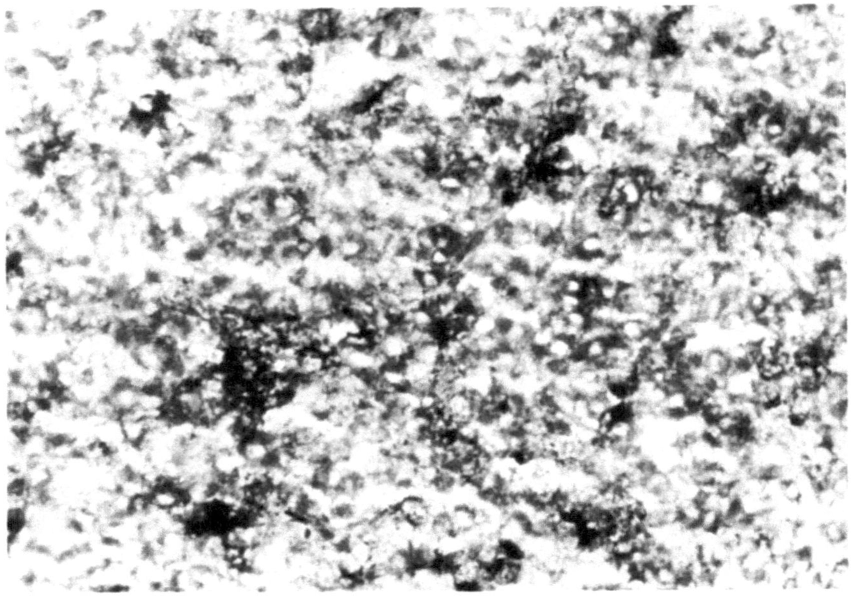

Fig. 50. Histochemical demonstration of glucose-6-phosphate dehydrogenase activity in the guinea-pig pineal organ. Note that the reaction product is mainly confined to groups of strongly reacting cells. VOLLRATH and SCHMIDT, 1969; ×315

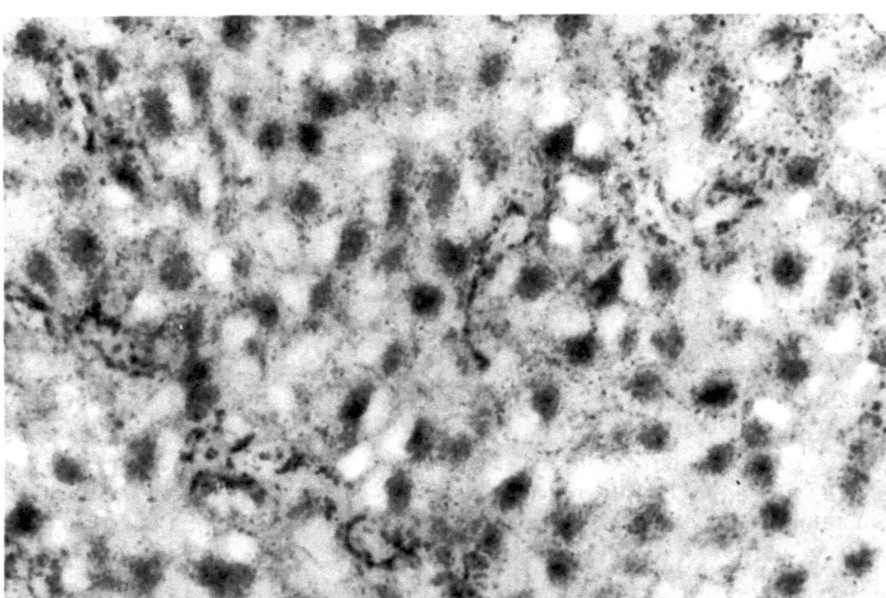

Fig. 51. Histochemical demonstration of glyceraldehyde phosphate dehydrogenase activity. Note that the reaction product occurs in two forms: fine granular in round to oval areas and coarse granular in cell processes (?). Unpublished photomicrograph by the late Dr. W. KÖHL, Department of Anatomy, University of Mainz

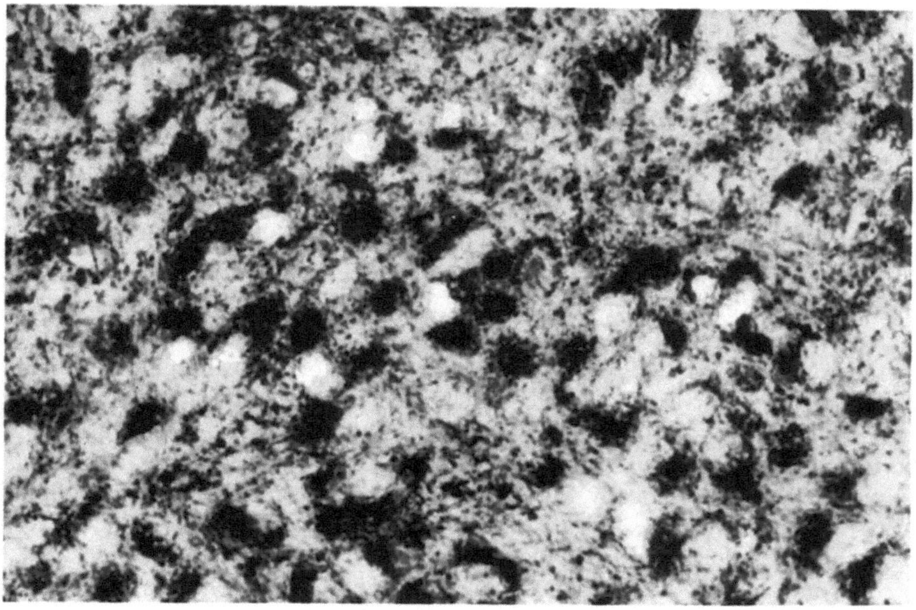

Fig. 52. Histochemical demonstration of succinate dehydrogenase activity in the guinea-pig pineal organ. The most intense reaction is seen in round to oval areas corresponding to mitochondria-rich sites. Unpublished photomicrograph by the late Dr. W. KÖHL, Department of Anatomy, University of Mainz

Glyceraldehyde phosphate dehydrogenase (EC 1.2.1.12) has been demonstrated in the guinea-pig pineal by KÖHL (unpublished results). As can be seen in Fig. 51, its distribution differs from all the other enzymes mentioned so far in that in some cells the granular reaction product is evenly distributed in the perikarya and is present also in cytoplasmic processes. For further details see the legend of Fig. 51.

Succinate dehydrogenase (EC 1.3.99.1) activity has been demonstrated in all species investigated, though in varying intensities. Low activities have been reported for humans (BAYEROVÁ and BAYER, 1967), rabbit (BOSTELMANN and BIENENGRÄBER, 1963), guinea-pig (VOLLRATH and SCHMIDT, 1969), mouse (SHIMIZU and MORIKAWA, 1957) and rat (LEDUC and WISLOCKI, 1952; BARA et al., 1971; KRSTIĆ and TARSOLY, 1972). Other authors have attributed moderate (TAPP et al., 1973) or strong activities (GEREBTZOFF and DEVEČERSKI, 1972) to at least some of the rat pinealocytes. Extremely high enzyme activities have been described in cattle, sheep and pig (ARVY, 1961). Strong activity has been noted in human pinealoma (KAWASHIMA et al., 1965).

Relatively little detailed knowledge regarding the precise intracellular localization of the enzyme is currently available. In the guinea-pig pineal the use of gel media demonstrated that the reaction product accumulated in clearly outlined juxtanuclear sites (Fig. 52).

While histochemical studies point to considerable interspecies differences in enzyme activity, biochemical studies do not support this assumption (QUAY,

1960). When compared with other tissues, pineal activity equals 50% of that of liver, 75% of choroid plexus, 90% of cerebral cortex (area 18) and 160% of the neurohypophysis (QUAY, 1958b, 1974a).

Succinate dehydrogenase activity has been studied under various physiological and experimental conditions. Biochemical studies by QUAY (1974a) have shown a clear circadian rhythm with maximum levels at 20.00 h and lowest levels at 12.00 h (lights on from 4.00 h to 18.00 h). However, histochemical results point to a circadian pattern in which the lowest enzyme activity was present at 20.00 h and the highest at 4.00 h (MØLLER and HØYER, 1979). Histochemically detectable changes have been claimed to occur also in relation to the oestrous cycle, with maximum activity during late oestrus, intermediate activity during dioestrus and pro-oestrus and minimum activity during metoestrus (FASANO et al., 1971). According to histochemical studies succinate dehydrogenase activity is virtually unchanged in pregnant rats (DEVEČERSKI, 1972a) and pregnant guinea-pigs (VOLLRATH and SCHMIDT, 1969). During puberty, the enzyme activity doubles when compared with 4-week-old rats (BOSTELMANN, 1969b). Biochemically, the enzyme activity was found to more than double in both sexes during the first 6 weeks postnatally, followed after about a year by a gradual decrease (QUAY, 1958b, 1959a, 1974a).

Enzyme activity was enhanced by exposing rats to continuous darkness (BOSTELMANN, 1968a, 1969a) or cold (QUAY, 1958b; DEVEČERSKI, 1972b). It was reduced by continuous light (BOSTELMANN, 1968a, 1969a), the latter finding being in agreement with biochemically assessed levels of enzyme activity (QUAY, 1963b). Gonadectomy resulted in an increase of enzyme activity (DAS GUPTA, 1968a), but biochemically no changes could be detected after interfering with gonadal, thyroid or adrenal cortical function (QUAY, 1958b, 1959a).

Glutamate dehydrogenase (EC 1.4.1.2 and 1.4.1.3) activity has been demonstrated in a number of species (man: BAYEROVÁ and BAYER, 1967; rat: BOSTELMANN, 1963, 1968a; GEREBTZOFF and DEVEČERSKI, 1972; TAŞCĂ et al., 1968; rabbit: BOSTELMANN, 1963; guinea-pig: KÖHL, unpublished results). However, the activities were extremely low, except in the guinea-pig. Under suitable conditions, fibrous astrocytes could be selectively stained in the rat pineal (GEREBTZOFF and DEVEČERSKI, 1972). Exposure of rats to continuous darkness or constant illumination apparently has no effect on pineal glutamate dehydrogenase activity (BOSTELMANN, 1968a). Enzyme activity increases strikingly during puberty (BOSTELMANN, 1969a).

Monoamine oxidase (EC 1.4.3.4) activity is present in both pinealocytes and sympathetic nerve fibres (rats, rabbit: HÅKANSON and OWMAN, 1965, 1966; man: GORIDIS and NEFF, 1972), about 85% of the enzyme activity being present in pinealocytes and 15% in nerve fibres (GORIDIS and NEFF, 1971). It would appear that the enzyme is not identical in the two sites, as shown by the fact that the enzyme present in the nerves is readily affected by the monoamine oxidase inhibitor clorgyline, whereas that of the pinealocytes is not (GORIDIS and NEFF, 1972; YANG et al., 1972).

The enzyme reveals striking species differences. HÅKANSON and OWMAN (1965, 1966), who combined biochemical determinations with cytochemical visual assessments of enzyme activity and noted a close correspondence of the results, found highest enzyme levels in cows, intermediate levels in humans, rats and rabbits and low values in pigs and cats. In humans, the enzyme activity ranged between 0.48 and 2.72 µmol/mg per hour and was in the same order of magnitude as in the brain (VOGEL et al., 1969).

Histochemical studies have yielded variable results. Little if any, staining for monoamine oxidase was obtained in pinealocytes of mouse, guinea-pig and rabbit by SHIMIZU et al. (1959). In squirrel monkey (MANOCHA and BOURNE, 1966) and guinea-pig (VOLLRATH and SCHMIDT, 1969) the habenular fibres invading the epiphysis cerebri show a positive reaction. In the vervet monkey, two main tracts of monoaminergic fibres enter the pineal body from the posterior and habenular commissures (HAYES et al., 1970). In the rat, the majority of authors have not reported positive staining reactions (B. SMITH, 1963; BOSTELMANN, 1968a; TRENTINI et al., 1969; TAPP et al., 1973), in contrast to SHIMIZU et al. (1959) and BOTTICELLI et al. (1972). In vitro studies were likewise variable. TAŞCĂ et al. (1968) failed to get a positive reaction. However, the enzyme could be demonstrated when the pineal was cultured in a diffusion chamber inserted into the peritoneal cavity. The formazan deposits were then found lying scattered throughout the perinuclear cytoplasm and extending into the cell processes (O'STEEN and DILL, 1964).

Monoamine oxidase activity has been investigated also under different physiological and experimental conditions. In humans, the enzyme showed no change with increasing age (WURTMAN et al., 1964c). In rats, adult enzyme levels were found at 10 days of age (WURTMAN et al., 1964c), but BOSTELMANN (1969a), using histochemistry, noted a further striking increase of enzyme activity during puberty. Histochemically, no changes in enzyme activity were noted after exposure to continuous illumination, darkness (BOSTELMANN, 1968a) and sympathectomy (HÅKANSON and OWMAN, 1965), though the latter procedure led to a biochemically detectable decrease (HÅKANSON and OWMAN, 1965). Interference with the pituitary–gonadal axis did affect the histochemically detectable monoamine oxidase activity (TRENTINI et al., 1969, 1973).

NADPH diaphorase (EC 1.6.99.1) activity has been demonstrated in rat pinealocytes cultured in vitro (TAŞCĂ et al., 1968; S. MILCU et al., 1968) and in guinea-pig pinealocytes in vivo (VOLLRATH and SCHMIDT, 1969). In the guinea-pig, the enzyme was among those which showed the highest activities, and two types of pinealocytes could be distinguished. One type showed a very intense reaction so that the delicate pinealocyte processes were clearly shown. They lay grouped together, forming either clusters or chain-like structures. The remainder of the pinealocytes showed a considerably weaker activity. In pregnant guinea-pigs, the enzyme activity was greatly enhanced in the second half of gestation, giving rise to a strikingly even distribution of the reaction product throughout the parenchyma (Fig. 53). Between 2 and 4 days after delivery the enzyme activity reverted to normal. In pregnant animals, the increased enzyme levels could be depressed by oophorectomy or hysterectomy, but not by fetectomy. No changes were noted following 2 weeks treatment with chorionic gonadotrophin, oestradiol, progesterone, or placental homogenates or after gonadec-

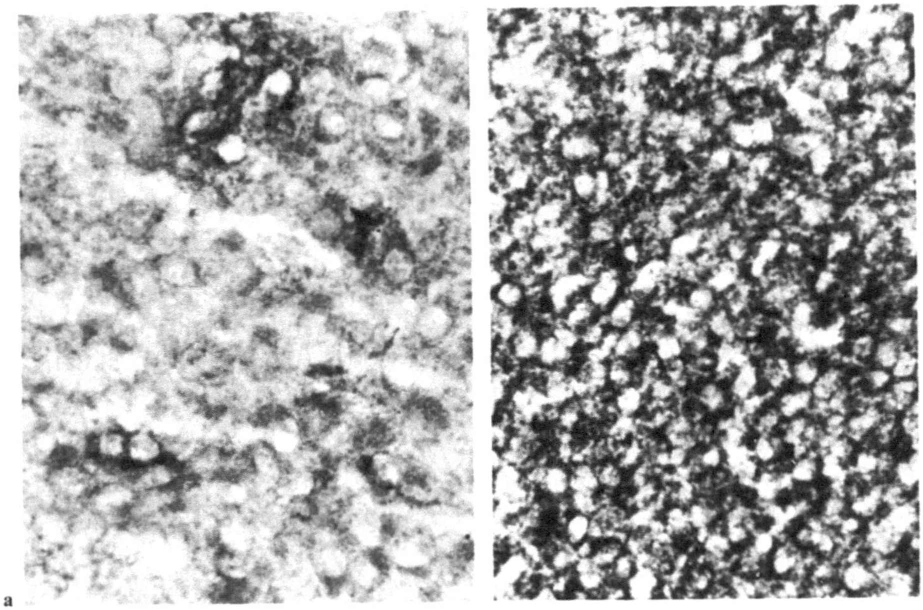

Fig. 53a, b. Histochemical demonstration of NADPH-diaphorase activity in pineal organs of **a** non-pregnant and **b** pregnant (second half of gestation) guinea-pigs. Note the striking increase in reaction product in the pregnant animal. VOLLRATH and SCHMIDT, 1969; ×540

tomy. In pineals of fetal guinea-pigs, the NADPH diaphorase activity was slightly lower than in adults (VOLLRATH and SCHMIDT, 1969).

NADH diaphorase (EC 1.6.99.3) activity has been shown to be strong in human pinealocytes (BAYEROVÁ and BAYER, 1967) and strong to moderate in rat pinealocytes in vivo (BOSTELMANN, 1963, 1968a; TAPP et al., 1973) and in vitro (TAŞCĂ et al., 1968; S. MILCU et al., 1968), in rabbit pinealocytes (BOSTELMANN, 1963) and in guinea-pig pinealocytes (VOLLRATH and SCHMIDT, 1969). In the guinea-pig, the enzyme shows an identical behaviour to that of NADPH diaphorase (Fig. 53). During puberty, the enzyme activity doubles (BOSTELMANN, 1969a; rat). Exposure to constant light lowers the enzyme activity, whereas constant darkness stimulates the enzyme (BOSTELMANN, 1968a). In tissue culture, administration of thyroxine led to a slight increase in enzyme activity (S. MILCU et al., 1968).

Cytochrome oxidase (EC 1.9.3.1) has been found to be evenly distributed in rat and rabbit pinealocytes (BOSTELMANN, 1963, 1968a). In the rat, the enzyme activity has been claimed to be weak (KRSTIĆ and TARSOLY, 1972) to moderate (TAPP et al., 1973). In the guinea-pig no activity could be detected histochemically (VOLLRATH and SCHMIDT, 1969). During puberty the enzyme activity doubled (BOSTELMANN, 1969a). It was lowered by exposure to continuous illumination and raised by constant darkness (BOSTELMANN, 1968a).

Peroxidase (EC 1.11.1.7) activity has been reported to be weak in the rat pineal (GUSEK and BUSS, 1966a).

β) Transferases

For transferases, some of which play a most important role in melatonin formation, e.g. hydroxyindole-*O*-methyltransferase (EC 2.1.1.4) or *N*-acetyltransferase (EC 2.3.1.5), little histochemical information is currently available. *N*-acetyltransferase activity, studied at the electron microscopic level, was demonstrable mainly in mitochondria where the reaction product lay between the inner and outer mitochondrial membranes and inside the cristae (KERENYI et al., 1975a). Activity which may be due to an acetyltransferase (EC 2.3.1.?) has been reported to be high in hamster pinealocytes, especially on mitochondrial membranes, as shown by electron microscopic cytochemistry (PENNEY and REITER, 1967).

Phosphorylase (EC 2.4.1.1), a glycosyltransferase, has been demonstrated in pineals of 1-day-old rats. In older animals a decrease in enzyme activity was noted and in adult animals negative results were obtained (SHIMIZU and OKADA, 1957). Recently the enzyme has also been found in adult rat pinealocytes, where it exhibited low activity (TAPP et al., 1973). Results obtained from cultured rat pinealocytes are contradictory. A positive reaction has been reported by S. MILCU et al. (1968) and a negative one by TAŞCĂ et al. (1968).

γ) Hydrolases

Non-specific esterases (EC 3.1.1.1; 3.1.1.2; 3.1.1.6; for histochemical and biochemical differences, see PEARSE, 1972) have been demonstrated in the pineals of different species. Moderate activities were usually found in pinealocytes of man (BAYEROVÁ and BAYER, 1967), squirrel monkey (MANOCHA, 1970b), rat (LEDUC and WISLOCKI, 1952; BOSTELMANN, 1963; ARSTILA, 1967; TAŞCĂ et al., 1968; MILINE et al., 1969a; TAPP et al., 1973), rabbit (BOSTELMANN, 1963) and gerbil (JAPHA et al., 1977). In the guinea-pig, some parenchymal cells showed a very intense reaction (VOLLRATH and SCHMIDT, 1969).

The localization of the reaction product showed clear regional differences in the rat pineal, with low amounts in the central regions and large amounts peripherally, especially posteroventrally (MILINE et al., 1969a). The intracellular distribution also varied. In guinea-pig pinealocytes the reaction product was confined to the cell bodies. In rat and rabbit the cell processes were also stained (BOSTELMANN, 1963). Electron microscopically, the enzyme was demonstrated in lysosomes (ARSTILA, 1967). In male Mongolian gerbils (*Meriones unguiculatus*) (JAPHA et al., 1977) the esterase reaction gave varying results depending on the substrates used. With α-naphthyl acetate, the pinealocytes showed moderate esterase activity. Two types of pinealocytes could be distinguished, namely cells irregular in shape, the reaction product being observed in both cytoplasmic granules and cytosol, and a second type more regular in outline, with the reaction product confined to cytoplasmic granules. In addition, "high-esterase cells" were seen lying along the course of blood vessels and in the interstitial tissue, probably corresponding to glial cells. When α-naphthyl butyrate was used as substrate instead of α-naphthyl acetate only the "high-esterase cells" were distinctly stained.

In fetal guinea-pigs the enzyme activity was considerably lower than in adults (VOLLRATH and SCHMIDT, 1969). In 1-week-old rats, the enzyme was demonstrated by PROP (1965).

Under experimental conditions, the enzyme activity was enhanced after administration of Prolan for 21 days (GUSEK and BUSS, 1966a, b), upon exposure to either continuous illumination (BOSTELMANN, 1968a) or cold (MILINE et al., 1970) and after castration (BOSTELMANN, 1969b).

Lipase (EC 3.1.1.3). Biochemical studies have shown that lipase activity is lower in the pineal than in the brain (THIÉBLOT et al., 1965, 1966c), but *histochemically* more reaction product was found in the pineal than in brain tissue (BOSTELMANN, 1963; BOSTELMANN and BIENENGRÄBER, 1963). PROP (1965) noted a close association of lipase activity and fat droplets in rat pinealocytes. In vitro the enzyme was apparently also detectable (TAŞCĂ et al., 1968). During puberty, lipase activity doubles (BOSTELMANN, 1969a). Under experimental conditions, lipase activity has been shown to increase following constant darkness (BOSTELMANN, 1968a) or adrenalectomy (MARKINA, 1970). A decrease was found after continuous illumination (BOSTELMANN, 1968a).

Acetylcholinesterase (EC 3.1.1.7) activity is clearly demonstrable in the mammalian pineal organ. In the bovine pineal the enzyme shows low activity (23 μM/g per hour) (LABELLA and SHIN, 1968). It is distinctly higher before sexual maturation (ARO et al., 1973). The precise localization of the enzyme is not known. RODRIGUEZ DE LORES ARNAIZ and PELLEGRINO DE IRALDI (1972) estimated that the enzyme is equally distributed between pinealocytes and nerve fibres, since superior cervical ganglionectomy is followed by a 50% decrease of enzyme activity. Later the enzyme activity reverts to normal (PELLEGRINO DE IRALDI and RODRIGUEZ DE LORES ARNAIZ, 1976). Histochemically the enzyme can be regularly demonstrated in intrapineal sympathetic nerve fibres (ARVY, 1961; ERÄNKÖ et al., 1970a, b; ERÄNKÖ and ERÄNKÖ, 1971; A. MACHADO and LEMOS, 1971; TRUEMAN and HERBERT, 1970a, b) (Fig. 54). The pinealocytes showed a mild to moderate reaction in the squirrel monkey (MANOCHA, 1970b), while in rat (ARSTILA, 1967), ferret (TRUEMAN and HERBERT, 1970a, b) and guinea-pig (Fig. 54) they are negative.

Non-specific cholinesterase (EC 3.1.1.8) activity has been demonstrated biochemically in bovine pineals, where the enzyme showed similar activities as in brain or pituitary tissue (LABELLA and SHIN, 1968). Histochemically the enzyme has been detected in porcine, ovine and bovine pinealocytes (ARVY, 1961, 1963, 1965). In rat pinealocytes, it was present immediately after birth and disappeared within 2 weeks (A. MACHADO and LEMOS, 1971). In adult rat pinealocytes it may also be present, since some biochemically detectable activity was noted after denervation (RODRIGUEZ DE LORES ARNAIZ and PELLEGRINO DE IRALDI, 1972).

Alkaline phosphatase (EC 3.1.3.1) exhibits the highest activity in the walls of blood vessels (see p. 218). In pinealocytes reaction product was not regularly found. If present, it may not necessarily be due to the presence of non-specific alkaline phosphatase (in brief, alkaline phosphatase) but to other phosphatases.

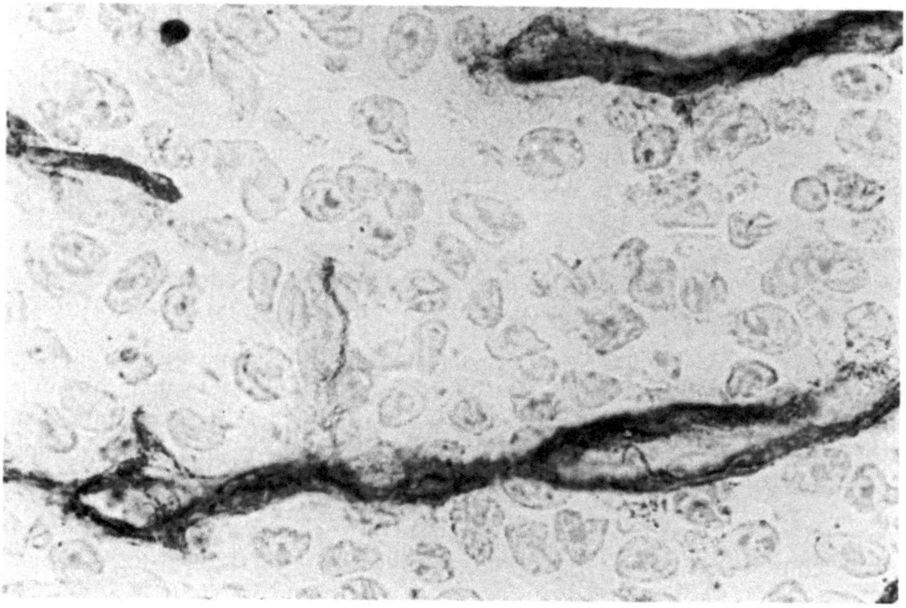

Fig. 54. Histochemical demonstration of acetylcholinesterase activity in the guinea-pig pineal organ. Note the strongly staining nerve fibre bundles. Unpublished photomicrograph by the late Dr. W. KÖHL, Department of Anatomy, University of Mainz

In human pinealocytes, either no (A. MACHADO and MACHADO, 1965) or little activity (BAYEROVÁ and BAYER, 1967) has been detected. Activity also appears to be low in monkeys (WISLOCKI and LEDUC, 1948). In the squirrel monkey activity was restricted to the peripheral parts of the parenchymal cells (MANOCHA, 1970a). Low activity was noted in rat pinealocytes by TAPP et al. (1973), but other authors have reported negative results (BOSTELMANN, 1963, 1968a; ARSTILA, 1967). According to GUSEK and BUSS (1966a), slight activity was occasionally seen along the plasmalemma of rat pinealocytes. Delicate granules throughout the cells were described by KAPPERS et al. (1962). In the guinea-pig small amounts of reaction product were found in the immediate vicinity of pinealocyte nuclei (VOLLRATH and SCHMIDT, 1969). In *Bos*, the pinealocyte nucleoli were positive (STEFANESCU-GAVAT, 1970), and a positive reaction in both nuclei and cytoplasm was obtained in pinealocytes of pig, horse and goat (MIKAMI, 1951). Rabbit pinealocytes were negative (BOSTELMANN and BIENENGRÄBER, 1963). Biochemically, pineal alkaline phosphatase activity exceeded that of brain in sheep (THIÉBLOT et al., 1965, 1966c) and pig (JOUAN and ROCABOY, 1965).

Acid phosphatase (EC 3.1.3.2), a lysosomal enzyme, has been found in the pinealocytes of all species investigated. Biochemically the enzyme was found to possess higher levels of activity in ovine pineals than in ovine brain tissue (THIÉBLOT et al., 1965, 1966c). Histochemical techniques usually revealed a moderate enzyme activity in pinealocytes (man: BAYEROVÁ and BAYER, 1967; squirrel monkey: MANOCHA, 1970a; rat: TAPP et al., 1973; guinea-pig: VOLLRATH and

SCHMIDT, 1969; rabbit: BOSTELMANN and BIENENGRÄBER, 1963; Mongolian gerbil: JAPHA et al., 1977; pig, sheep and cattle: ARVY, 1961, 1965; STEFANESCU-GAVAT, 1970). Rat (ARSTILA, 1967) and guinea-pig pinealocytes (VOLLRATH and SCHMIDT, 1969) showed great intraspecies variation in enzyme activity. Regional differences in enzyme activity were not noted in the rat (MILINE et al., 1969a). Electron microscopically, the enzyme was demonstrated in lysosomes and in vesicles or cisternae of the Golgi complex (ARSTILA, 1967). Acid phosphatase activity was enhanced during early oestrus and low during late oestrus and metoestrus (FASANO et al., 1971). An increase was also noted after castration (DAS GUPTA, 1968a; BOSTELMANN, 1969b) and exposure to continuous illumination, constant darkness (BOSTELMANN, 1968a) or cold (MILINE et al., 1970).

5-Nucleotidase (EC 3.1.3.5) activity was strong in bovine pinealocytes and glial cells (STEFANESCU-GAVAT, 1970). This is in accordance with biochemical studies in sheep, which revealed higher enzyme activities in pineal than in brain tissue (THIÉBLOT et al., 1965, 1966c). Negative histochemical results were obtained in guinea-pigs (VOLLRATH and SCHMIDT, 1969).

Glucose-6-phosphatase (EC 3.1.3.9) has been histochemically demonstrated in bovine pinealocytes, where cytoplasm and nucleoli were stained (STEFANESCU-GAVAT, 1970). Electron microscopically it was found in smooth and rough endoplasmic reticulum and perinuclear space (KRSTIĆ, 1975c, 1976a).

Arylsulphatase (EC 3.1.6.1) activity in rat pinealocytes was either not demonstrable (GUSEK and BUSS, 1966a) or weak (ARSTILA, 1967). In a few cells, granular reaction product was abundant. Electron microscopically it was found to be closely related to lysosomes and the two outermost cisternae of the Golgi apparatus. Occasionally it was closely associated with the nuclear membrane and cisternae of the agranular endoplasmic reticulum (ARSTILA, 1967; KERÉNYI et al., 1975b). Arylsulphatase activity was not affected by administration of Prolan over long periods (GUSEK and BUSS, 1966a). In rat pinealocyte cultures in vitro thyroxine apparently stimulated the enzyme (S. MILCU et al., 1968).

β-Glucuronidase (EC 3.2.1.31), another enzyme often associated with lysosomes, was histochemically found to exhibit medium activity in porcine, ovine and bovine pinealocytes (ARVY, 1961, 1965). Biochemically, the enzyme activity in lamb pineals was twice that of brain tissue (THIÉBLOT et al., 1965, 1966c). A positive histochemical reaction was obtained in rat pinealocytes, both in vivo (WALTIMO and TALANTI, 1965) and in vitro (S. MILCU et al., 1968). In a human pinealoma, moderate activity was found (KAWASHIMA et al., 1965).

Leucine aminopeptidase (EC 3.4.1.1.1) activity has been reported to be extremely high in the pineals of pig, sheep, cattle (ARVY, 1961, 1963, 1965) and rat (NIEMI and IKONEN, 1960; TRENTINI et al., 1969; MILINE et al., 1969a), which is in agreement with biochemical findings by THIÉBLOT et al. (1965, 1966c) in lambs and by JOUAN and ROCABOY (1966) in pigs. Histochemically, the enzyme could also be demonstrated in rat pinealocytes cultured in vitro (TAŞCĂ et al., 1968; S. Milcu et al., 1968). In the guinea-pig pineal, the enzyme reaction was negative (VOLLRATH and SCHMIDT, 1969). Long-term administration of human chorionic gonadotrophin (HCG) has been claimed to enhance the histochemically detectable leucine aminopeptidase activity (TRENTINI et al., 1969, 1973).

ATPase (EC 3.6.1.3) has been found to be moderately active in the pinealocytes of man (BAYEROVÁ and BAYER, 1967) and squirrel monkey (MANOCHA, 1970a). In the rat, the enzyme has been demonstrated light microscopically in pineal tissue cultured in vitro (TAŞCĂ et al., 1968; S. MILCU et al., 1968) and electron microscopically in vivo (TORACK and BARRNETT, 1964). Electron microscopically, the enzyme activity was demonstrated in the cell membrane of pinealocytes where they were in close contact with what looked like glial cells (TORACK and BARRNETT, 1964). Biochemically, ATPase activity of the lamb pineal was lower than that of the brain (THIÉBLOT et al., 1965, 1966c).

Nucleosidediphosphatase (EC 3.6.1.6) could apparently not be demonstrated in pinealocytes. It was found, however, in the walls of blood vessels (TORACK and BARNETT, 1964, rat).

Thiamine pyrophosphatase (not listed by EC), a marker enzyme of the Golgi apparatus of nerve cells, was demonstrated by means of electron microscopy in the cisternae of both the Golgi apparatus and the smooth and rough endoplasmic reticulum of rat pinealocytes (KRSTIĆ, 1976a). In guinea-pig the enzyme reaction was negative (VOLLRATH and SCHMIDT, 1969).

δ) Lyases

Aldolase (EC 4.1.2.13) was biochemically demonstrable in lamb pineals (THIÉBLOT et al., 1965, 1966c). Histochemically, it showed a weak activity in the rat pineal (MAEDA et al., 1968).

Carbonic anhydrase (EC 4.2.1.1) was demonstrated in the rat pineal and showed a diurnal rhythm (QUAY, 1972a). The enzyme activity measured is probably entirely bound to red blood cells.

Adenyl cyclase (EC 4.6.1.1) plays an important role in mediating the effect of noradrenaline on melatonin formation (p. 10). Histochemical studies at the electron microscopic level have shown reaction product on the surface of pinealocytes in gerbils (Fig. 55) but not in rats. In both species the main activity was noted in capillary endothelial cells, their basal lamina, pericapillary amorphous material and nerve endings (KRSTIĆ, 1977a).

ε) Conclusion

In view of the numerous histochemical findings it would be highly desirable to give a general interpretation of the enzymatic equipment of the mammalian pinealocyte. Points of interest are how the enzymatic equipment of the pinealocyte relates to pineal function as a whole, and to what extent it resembles or differs from that of other endocrine glands, nervous tissue or, more specifically, other aggregates of neuro-endocrine cells. As most of the histochemical findings have not been corroborated by biochemical studies and as no systematic

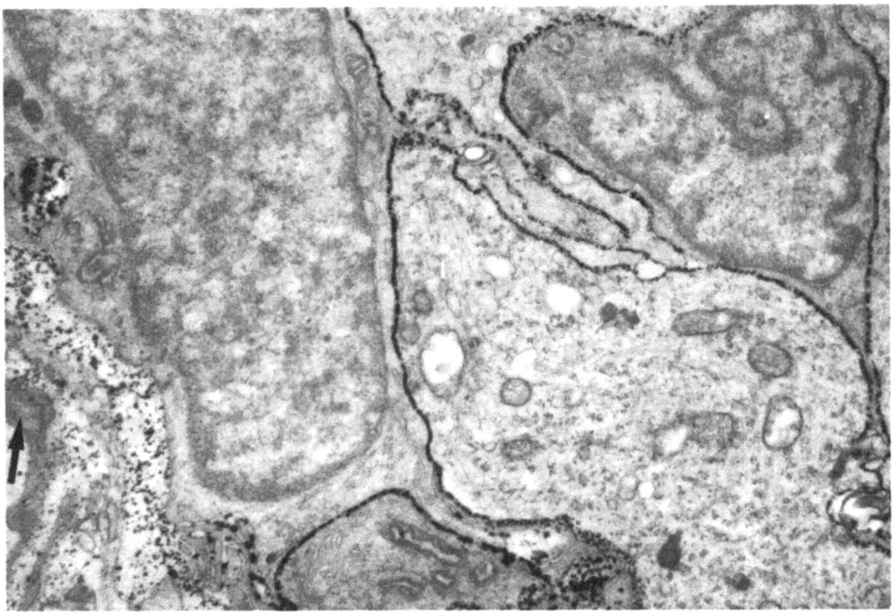

Fig. 55. Ultracytochemical demonstration of adenyl cyclase activity in pinealocytes of a Mongolian gerbil. Reaction product fills the intercellular clefts and the pericapillary space (*arrow*). KRSTIĆ, 1977a; ×14,000

studies have been carried out simultaneously in different tissues, it is premature to comment on the aspects mentioned above. QUAY (1974a) concluded that "this organ has biochemical individuality ... and that mammalian pinealocytes in their enzyme activities, particularly those active in monoamine biosynthesis, resemble neuronal and retinal cells more than neuroglia, ependyma or various cells of the mesodermal series".

3. Ultrastructure

a) General Aspects

Electron microscopic studies, in addition to clearly establishing the basic ultrastructural features of the mammalian epiphysis cerebri, have revealed that the mammalian pinealocyte is a unique type of cell: it is a modified photoreceptor cell (p. 111) which despite its endocrine nature does not contain a sufficiently large number of secretory granules to identify it prima facie as a secretory cell. Although the pineal contains different types of what looks like secretory products, there is insufficient evidence as yet for the presence of functionally distinct pinealocytes. On morphological grounds different types of pinealocytes have been distinguished (for a thorough review of this problem, see PEVET, 1977a). Frequently a distinction has been made between "light" and "dark" pinealocytes. While some authors claim that light and dark pinealocytes are merely different functional stages of one and the same cell type, others regard them as separate cell types. It should be noted that dark pinealocytes are not to be confused with glial cells, which often look darker than light pinealocytes. Recently, different types of pinealocytes have been distinguished on the grounds of differences in the secretory processes; while pinealocytes of population I contain granular vesicles, pinealocytes of population II comprise flocculent material in the cisternae of the granular endoplasmic reticulum (PEVET, 1977a). Since light pinealocytes and pinealocytes of population I predominate, the following description refers to these cells unless otherwise stated.

b) Pinealocyte Shape

Electron microscopic studies have clearly demonstrated that the perikarya of mammalian pinealocytes possess typical cytoplasmic processes. This is so even in rats, in which the presence of such structures had been denied by light microscopists (cf. BARGMANN, 1943).

Since the number and form of the processes have a bearing on pinealocyte shape, it is appropriate to deal briefly with pinealocyte processes. In most studies no details as to the exact number of processes that emerge from a perikaryon have been given. A notable exception is the paper by WARTENBERG (1968), in which an attempt was made to answer this question by cutting serial sections. With this technique it was shown that a single cat pinealocyte gave rise to one or perhaps two processes. Other species with few processes appear to be monkey (WARTENBERG, 1968), rat (WOLFE, 1965), rabbit (WARTENBERG and GUSEK, 1965; LEONHARDT, 1967) and dog (SANO and MASHIMO, 1966). In the seal the pinealocytes have been considered as monopolar cells (CUELLO, 1973). Larger numbers of processes appear to be present in guinea-pig (LUES, 1971), cow, sheep (ANDERSON, 1965), mole (PEVET, 1974), hedgehog (PEVET and SABOUREAU, 1973) and ground squirrel (POVLISHOCK et al., 1975), just to mention a few. An interesting problem is whether the pinealocytes of a given species possess a more or less constant number of processes or not. This question has been most thoroughly discussed for the pinealocytes of moles, in which the attempt was made to investigate in what way the mammalian pinealocyte differs from or resembles its phylogenetic ancestor, the pineal photoreceptor cell. Such a study showed (PEVET and COLLIN, 1976) that in the mole basically three types of pinealocytes can be distinguished according to the number and form of the processes and other features.

Very briefly, pineal photoreceptors of lower vertebrates show a distinct polarity, with outer and inner segments at one pole and a basal process or a pedicle at the other. As shown in Fig. 56 and described by PEVET and COLLIN (1976), the relatively thin pedicle shows little if any variation, so that in mammalian pinealocytes this process is always retained. According to PEVET and COLLIN (1976) the rare type I pinealocytes possess, in addition to the slender basal process, a relatively short bulbous process of large calibre with a neck-like constriction, the whole supranuclear cytoplasmic projection being regarded as a remnant of the inner segment of the pineal photoreceptor. Using the terminolo-

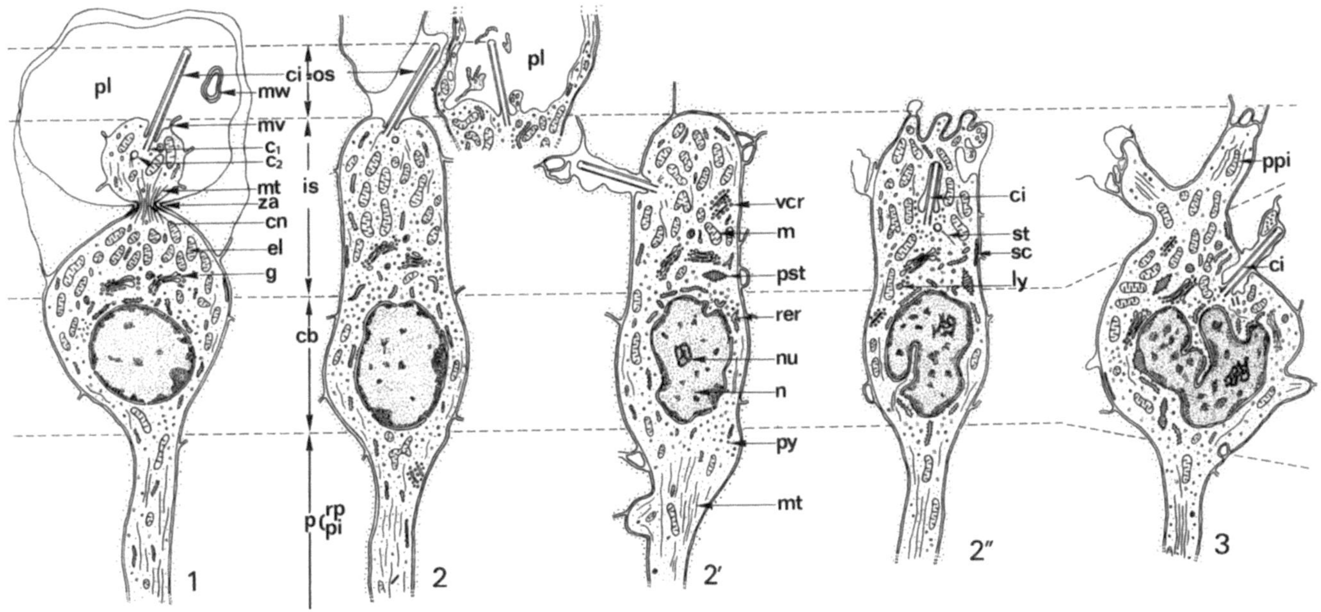

Fig. 56. Plurality of the derivatives of the photoreceptor cell line in a mammalian pineal organ (adult mole, *Talpa europaea*) according to Pevet and Collin (1976). *1*, rudimentary photoreceptor cell; *2*, *2′*, *2″*, cells intermediate between photoreceptor cells and pinealocytes; *3*, pinealocyte *sensu stricto*. *Abbreviations*: c_1, distal dentriole; c_2, proximal centriole; *ci*, cilium; *cb*, cell body; *cn*, cellular neck; *el*, mitochondria of "ellipsoid"; *g*, Golgi complex; *is*, inner segment; *ly*, lysosome; *m*, mitochondria; *mt*, microtubules; *mv*, microvilli; *mw*, membranous whorl; *n*, nucleus; *nu*, nucleolus; *os*, outer segment; *pl*, pineal lumen; *p* (*rp*, *pi*), process of a rudimentary photoreceptor cell or pinealocyte or an intermediate cell; *ppi*, process of a pinealocyte; *pst*, paracrystalline structure; *py*, polysomes; *rer*, rough endoplasmic reticulum; *sc*, subsurface cisternae; *st*, striated ciliary rootlet; *vcr*, vesicle-crowned ribbon; *za*, zonula adherens. Pevet and Collin, 1976

gy employed with regard to nerve cells, this type of pinealocyte would correspond to a bipolar cell. In type II pinealocytes, by contrast, no neck-like constriction is seen, so that the apical part of the cell is not a real process. In view of the presence of the pedicle these cells can be classified as unipolar elements. Type III pinealocytes, the most frequently observed cell type, have been characterized as neuronal, rather than photoreceptor-like cells. Here the apical process branches near the cell body (Fig. 56), so that these cells can be classified as multipolar cells.

How these findings relate to other species is not known, but it is clear that they will provide an incentive for studying pinealocyte shape more closely than has hitherto been done. Due to the length and the tortuosity of pinealocyte processes it is difficult to follow their course and hence to obtain a complete picture of pinealocyte shape. Here it suffices to mention that pinealocyte processes may be long or short and that they may terminate in a bulbous swelling.

Since the perikarya and the processes of pinealocytes differ in ultrastructure, they will be considered separately.

c) Perikaryon

α) Nucleus

The nuclei of mammalian pinealocytes display the general fine structural features of metabolically highly active interphase nuclei (WOLFE, 1965; ARSTILA, 1967; BUCANA et al., 1974). They are characterized by a high electron-lucency of their nucleoplasm (rat: WOLFE, 1965; rabbit: WARTENBERG and GUSEK, 1965; mouse: PELLEGRINO DE IRALDI, 1969), corroborating the numerous reports by light microscopists that pinealocyte nuclei are "vesicular" or "chromatin-poor" (WOLFE, 1965). It should be noted that dark pinealocytes possess a somewhat more electron-dense nucleoplasm than the light cells and also more condensed chromatin. As can be implied from the pale appearance of the pinealocyte nuclei, the amount of condensed chromatin is usually small, but it varies from pinealocyte to pinealocyte (ARSTILA, 1967; WELSER et al., 1968) and from species to species (LUES, 1971). In many species it predominates in an uneven marginal zone of the nucleus, closely related to the nuclear envelope (rat: WOLFE, 1965; cow, sheep: ANDERSON, 1965; hamster: SHERIDAN and REITER, 1968; guinea-pig: LUES, 1971; pocket gopher: SHERIDAN and REITER, 1973; rabbit: ROMIJN, 1973b; dog: WELSER et al., 1968; golden mole: PEVET and KUYPER, 1978). Widely dispersed chromatin, on the other hand, was described in young Long-Evans rats (ARSTILA, 1967), seal (CUELLO, 1973), hedgehog (PEVET and SABOUREAU, 1973), mole (PEVET, 1974), mole-rat (PEVET et al., 1976b), noctule bat, *Nyctalus noctula* (PEVET et al., 1977a), gerbil (WELSH and REITER, 1978) and others. Occasionally, a confluence of marginally and centrally located chromatin has been described (WOLFE, 1965). Perinucleolar chromatin has also been found (WOLFE, 1965; LUES, 1971).

Electron microscopically, pinealocyte nuclei show a pronounced polymorphism in some species. In humans, the round to slightly oval nuclei possess shallow indentations and deep channels (MALINSKÝ et al., 1963). The presence

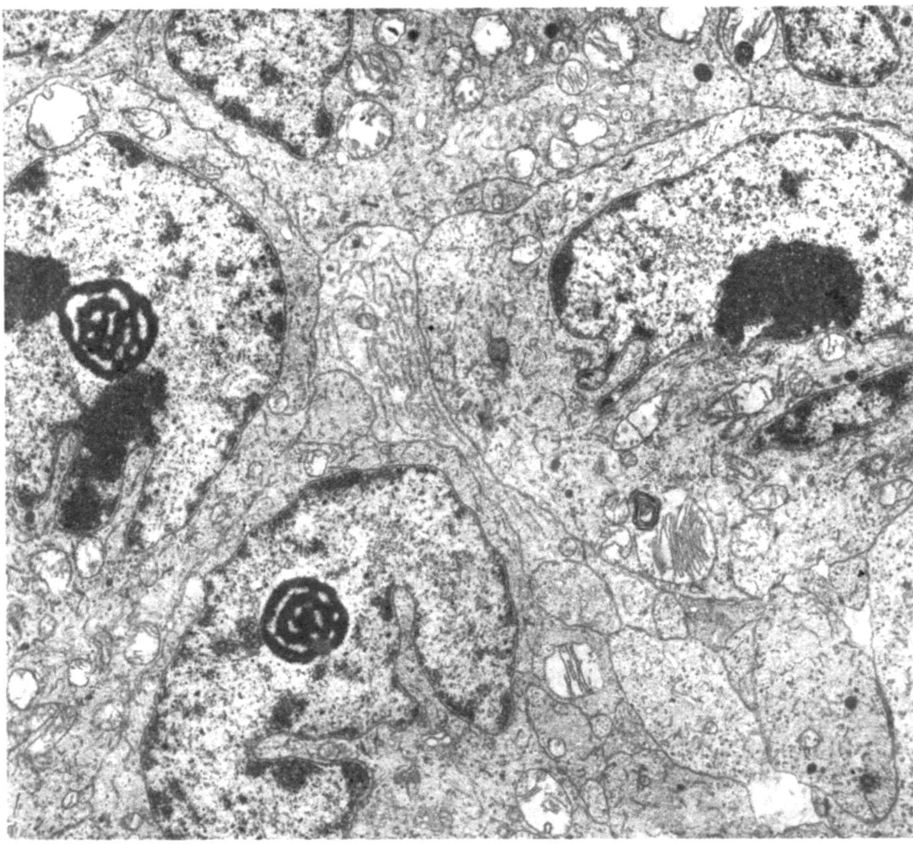

Fig. 57. Survey electron micrograph of guinea-pig pinealocytes showing perikarya with indented nuclei and prominent nucleoli. Between the perikarya, pinealocyte processes are seen. VOLLRATH, 1975

of bizarre-shaped, lobulated nuclear profiles has been described in rat (ARSTILA and HOPSU, 1964; OROFINO, 1964; WOLFE, 1965; BOSTELMANN, 1965; ARSTILA, 1967; KRSTIĆ, 1974: freeze-etching electron microscopy), pocket gopher (SHERIDAN and REITER, 1973), mole (PEVET, 1974) and ground squirrel (POVLISHOCK et al., 1975). In rats, in which irregularly shaped nuclei were a characteristic feature of older animals (BOSTELMANN, 1965), nuclear profiles sometimes exhibited the appearance of amitotic figures (WOLFE, 1965). In other species the nuclear profiles are similar to those in humans. In the cat, usually one flat impression at one nuclear pole was observed (WARTENBERG, 1968). In the guinea-pig a few such shallow indentations on one side of the nucleus were described (LUES, 1971), but channel-like indentations can also be observed (Fig. 57). One or a number of deep invaginations can be seen in pinealocyte nuclei of monkey (WARTENBERG, 1968; BERERHI and ABBAS-TERKI, 1970), rabbit (WARTENBERG and GUSEK, 1965), hamster (SHERIDAN and REITER, 1970b, deep pineal; BUCANA et al., 1974), mouse (PELLEGRINO DE IRALDI, 1969) and also in the various species exhibiting bizarre-shaped nuclei. In sheep and cow, the most conspicuous invaginations were observed in nuclei located in the anterior regions of the organ, especially in the vicinity of the pineal recess (ANDERSON, 1965). In the dog, nuclear invaginations are not present (SANO and MASHIMO, 1966; WELSER et al., 1968).

As can be expected, the nuclear invaginations comprise cytoplasm which, probably depending on the width of the invaginations, may or may not contain cytoplasmic inclusions. Cell organelles have been found in nuclear invaginations of humans (MALINSKÝ et al., 1963) and rats (ARSTILA and HOPSU, 1964; WOLFE, 1965); in the rabbit cell organelles were seldom present (WARTENBERG and GUSEK, 1965). Occasionally, round lipid inclusions were encountered (rat: ARSTILA and HOPSU, 1964; cow: ANDERSON, 1960). Altogether there are no indications that the cytoplasm invaginated into the nucleus shows any basic differences to that of the rest of the perikaryon.

The deep nuclear invaginations are particularly relevant with respect to the occurrence of the characteristic nuclear bodies (*Kernkugeln*, "nuclear pellets") described in pinealocytes in earlier (cf., BARGMANN, 1943) or more recent (HEINECKE, 1959; HEINIGER, 1965a) studies using light microscopy. KEVORKIAN and WESSEL (1959) have thoroughly reviewed the older literature and, by means of light microscopy, confirmed the presence of what looked like nuclear pellets. In the same material, examined electron microscopically, they did not find special intranuclear bodies. Instead they noted that many nuclei were excessively folded and irregular in outline and that multiple parallel finger-like projections of cytoplasm reached almost across the nucleus. They therefore concluded that nuclear pellets peculiar to pinealocytes do not in fact exist as such and that what looks like a *Kernkugel* is in all likelihood simply a misconstrued artefact in relatively thick light microscopic sections due to overlapping and unusual views of nuclear folds, lobulations and invaginations (see also KARASEK, 1967). In view of the presence of prominent nucleoli, they suggested that the only real pellets can be nothing other than enlarged nucleoli which are undergoing either a normal physiological process of intranuclear secretion or a retrogressive phase of necrosis. In this context it is interesting to note that HEINECKE (1959) also found a close association of nucleoli and inclusion bodies and that he likewise concluded that the inclusions have supposedly developed from nucleoli.

Rare, but regularly occurring bundles of 10 to 30 microfilaments have been described in rat pinealocyte nuclei (ARSTILA and HOPSU-HAVU, 1967). The bundles, measuring 0.5–4 µm in diameter, were seen both in the centre and in the periphery of the nucleus, without being in contact with the nucleolus or the nuclear envelope. Individual microfilaments measured 7–8 nm in diameter and up to 3 µm in length and appeared to consist of two electron-dense lines each measuring 2–3 nm in thickness and a lighter line in between of equal width. Adjacent microfilaments were separated from one another by a clear space of 10 nm. The function of the filaments is unknown. Since similar bundles were found in the cytoplasm it has been suggested that they may be somehow involved in nucleocytoplasmic interaction.

The nuclear envelope of the mammalian pinealocyte consists of the usual two nuclear membranes, i.e. inner and outer, which embrace the perinuclear space (WOLFE, 1965; ARSTILA, 1967; SHERIDAN and REITER, 1968; WARTENBERG, 1968; LUES, 1971; PEVET, 1974), measuring ca. 20 nm in width (BOSTELMANN, 1965). The nuclear envelope does not appear to possess any unusual features, apart from the fact that the two nuclear membranes have been reported to be sharply staining (cow, sheep: ANDERSON, 1965) or to stand out sharply from both cytoplasm and nucleoplasm (rat: BOSTELMANN, 1965). The outer nuclear membrane undulated in monkeys (WARTENBERG, 1968), bulged outwards as if to form cisternae of the surrounding endoplasmic reticulum in the rat (ARSTILA, 1967; POLETTI and CASTELLANO, 1967) and had a crinkled appearance as a result of blebbing in cow and sheep, particularly in cells from juvenile animals (ANDERSON, 1965). The outer membrane is usually a smooth-surfaced one, but in certain areas it may possess ribosomes (cow, sheep: ANDERSON, 1965; rat: POLETTI and CASTELLANO, 1967; dog: WELSER et al., 1968; mole: PEVET, 1974). In the mole-rat proteinaceous material has been found in the perinuclear space (PEVET et al., 1976b).

Nuclear pores are a regular feature of the nuclear envelope and have been mentioned in most species investigated. Diaphragms bridging the pores have been described in rat (POLETTI and CASTELLANO, 1967), hamster (SHERIDAN and REITER, 1968), cow and sheep (ANDERSON, 1965). Precise quantitative data on nuclear pores are not available. Whereas in the hamster (SHERIDAN and REITER, 1968) an abundance of these structures has been noted, they were few in number in the monkey (WARTENBERG, 1968). Numerous pores have been described in rat pinealocytes examined after freeze-etching (Fig. 58) (KRSTIĆ, 1974). The size of the pores has apparently not been measured.

β) Nucleolus

Relatively little detailed information is currently available on pinealocyte nucleoli, probably because they do not possess any unusual features.

In some species the nucleoli have been described as prominent, conspicuously large or massive (cow, sheep: ANDERSON, 1965; rat: ARSTILA, 1967; hamster: SHERIDAN and REITER, 1968; BUCANA et al., 1973; guinea-pig: LUES, 1971; mouse: PELLEGRINO DE IRALDI, 1969; mole: PEVET, 1974; ground squirrel: POVLISHOCK et al., 1975) though it would appear that their prominence is partly due to the fact that the nucleoli are strikingly electron-dense, contrasting with the relatively uniform, mottled appearance of the nucleoplasm and cytoplasm (WOLFE, 1965, rat). The nucleoli have been reported to be relatively inconspicuous or small in rabbit (WARTENBERG and GUSEK, 1965), dog (SANO and MASHIMO, 1966) and monkey (WARTENBERG, 1968).

Some of the details of nucleolar structure mentioned include the presence of a reticular (WARTENBERG and GUSEK, 1965, rabbit) or irregular (ARSTILA, 1967) nucleolonema, embracing amorphous areas in the rat (WOLFE, 1965), guinea-pig (LUES, 1971) and seal (CUELLO, 1973). Typical guinea-pig pinealocyte nucleoli are shown in Fig. 57. In cow and sheep (ANDERSON, 1965), on the other hand, the nucleolus consisted of dense particles which were tightly packed in the central region and loosely packed in the periphery. The presence of

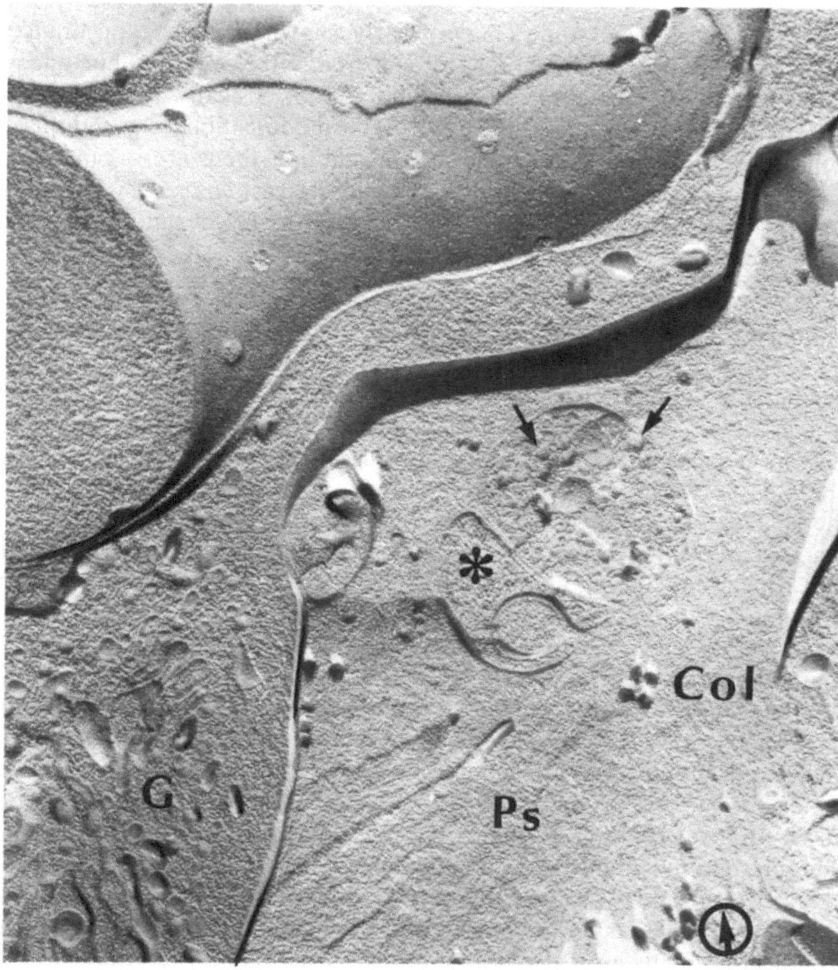

Fig. 58. Freeze-fractured pineal cell with pericapillary space (*Ps*) under low magnification (× 27,500). The nucleus shows numerous pores in the nuclear envelope. The Golgi complex (*G*) appears as an aggregation of concave and convex vesicles. A transversely fractured nerve terminal with part of a Schwann cell and two axons is discernible in the pericapillary space (*). *Arrows* indicate synaptic vesicles. *Col*, collagen fibrils. The *encircled arrow* indicates the direction of shadowing.
KRSTIĆ, 1974

perinucleolar chromatin has been described in rat (WOLFE, 1965), guinea-pig (LUES, 1965) (Fig. 57) and dog (WELSER et al., 1968).

The functional significance of the findings obtained is difficult to evaluate since quantitative data are lacking. WOLFE (1965) was of the opinion that the nucleoli seemed to be disproportionately large in size and complexity, considering the paucity of cytoplasmic ribosomes. It should be recalled that the nucleoli appear to comprise relatively little RNA as assessed by histochemistry (p. 93).

γ) *Cytoplasm*

Electron microscopic studies have clearly shown that the appearance of the pinealocyte cytoplasm is such that the function of the cell cannot be unequivocally deduced. While most of the organelles do not basically differ from those in other mammalian cells, there are some notable modifications (e.g. endoplasmic reticulum, centrioles and mitochondria). In addition there are organelles, the so-called synaptic ribbons, which are not regularly encountered in mammalian cells. A striking difference to most other endocrine organs is that morphological correlates of secretory products are not conspicuous.

Endoplasmic Reticulum. That the mammalian pinealocyte may possess unique functional properties is reflected inter alia by the endoplasmic reticulum. In a very thorough electron microscopic study of the rat pinealocyte WOLFE (1965) noted that the major part of the endoplasmic reticulum resembled neither the typical granular endoplasmic reticulum nor the typical smooth endoplasmic reticulum. It lacked the crisp tubular profiles, the pattern of branching and the peculiar tortuosity of the smooth endoplasmic reticulum and was devoid of the ribosomes typical of the rough endoplasmic reticulum, the occasional presence of ribosomes being regarded as merely topographical. To distinguish it from rough and smooth endoplasmic reticulum, WOLFE (1965) proposed the term "intergrade reticulum". However, most authors do not make this distinction and just distinguish between smooth and rough endoplasmic reticulum.

Without relevant morphometric data it is impossible to evaluate and compare the amounts of endoplasmic reticulum present and especially the relative proportions of its two types. Regarding the total amount of endoplasmic reticulum, it would appear that it is altogether rare and definitely much less abundant than in endocrine cells producing either protein or steroid hormones. A randomly taken electron micrograph of guinea-pig pinealocytes (Fig. 59) shows virtually no endoplasmic reticulum. Occasionally substantial amounts have been noted (e.g. cat: RODIN and TURNER, 1965a; DUNCAN and MICHELETTI, 1966; deep pineal of hamster: SHERIDAN and REITER, 1970b; horseshoe bat: STAMMER, 1972). Large amounts have also been reported in the rabbit (LEONHARDT, 1967), but the description by WARTENBERG and GUSEK (1965) that the endoplasmic reticulum was concentrated in certain peripheral, often evaginated parts of the pinealocytes suggests that the amounts present were relatively small. Endoplasmic reticulum is explicitly scarce in guinea-pig (Fig. 59) (LUES, 1971) and hedgehog pinealocytes (PEVET and SABOUREAU, 1973). Recent morphometric data of rat pinealocytes show that $5.37\% \pm 0.07\%$ of the cell volume is occupied by rough and smooth endoplasmic reticulum (KRSTIĆ, 1977b).

In the majority of species, smooth endoplasmic reticulum exceeds the rough type in amount (rat: WOLFE, 1965, ARSTILA, 1967; mouse: PELLEGRINO DE IRALDI, 1969; hamster: SHERIDAN, 1967, SHERIDAN and REITER, 1968, 1970; LIN et al., 1975; rabbit: WARTENBERG and GUSEK, 1965, LEONHARDT, 1967; guinea-pig: LUES, 1971; cow and sheep: ANDERSON, 1962, 1965; hedgehog: PEVET, 1972; gerbil: WELSH and REITER, 1978). Occasionally, the rough endoplasmic reticulum was more abundant than the smooth type (mole: PEVET, 1974; mole-rat: PEVET, 1976; PEVET et al., 1976b; dormouse: ROUX et al., 1974).

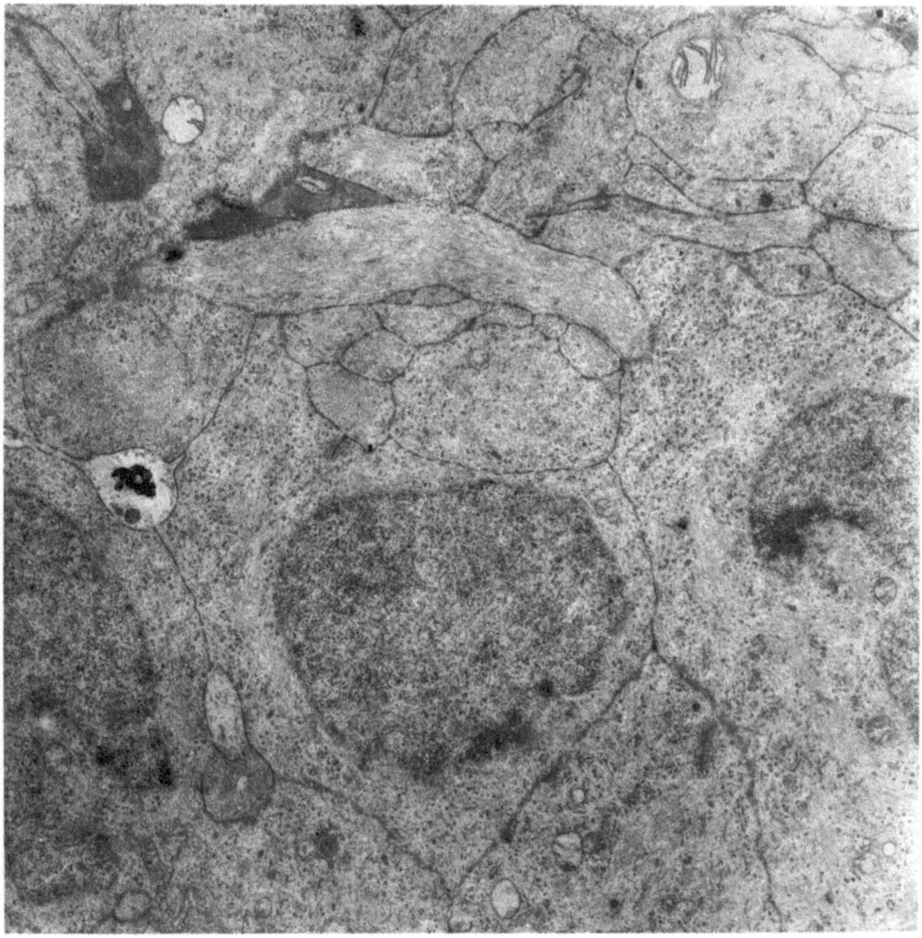

Fig. 59. Randomly taken electron micrograph of adult guinea-pig pinealocytes and processes showing virtually no rough endoplasmic reticulum. VOLLRATH, previously unpublished picture

In the majority of papers, no details regarding the intracellular localization of endoplasmic reticulum have been given. A peripheral localization of rough endoplasmic reticulum has been described in rat (ARSTILA, 1967) and guinea-pig (LUES, 1971). An even distribution was noted in rabbit (LEONHARDT, 1967), cat, monkey (WARTENBERG, 1968), mole-rat (PEVET et al., 1976b) and noctule bat (PEVET et al., 1977a). *Smooth* endoplasmic reticulum has been reported to be plentiful near the nucleus (rat: ARSTILA, 1967), but in other species it was evenly dispersed in the cytoplasm (cat, monkey: WARTENBERG, 1968; guinea-pig: LUES, 1971).

The structural appearance of the endoplasmic reticulum is variable. *Rough* endoplasmic reticulum has been reported to consist of isolated vesicles 30 nm to 0.3 μm in size (BOSTELMANN, 1965, rat), fragments with slight dilatations

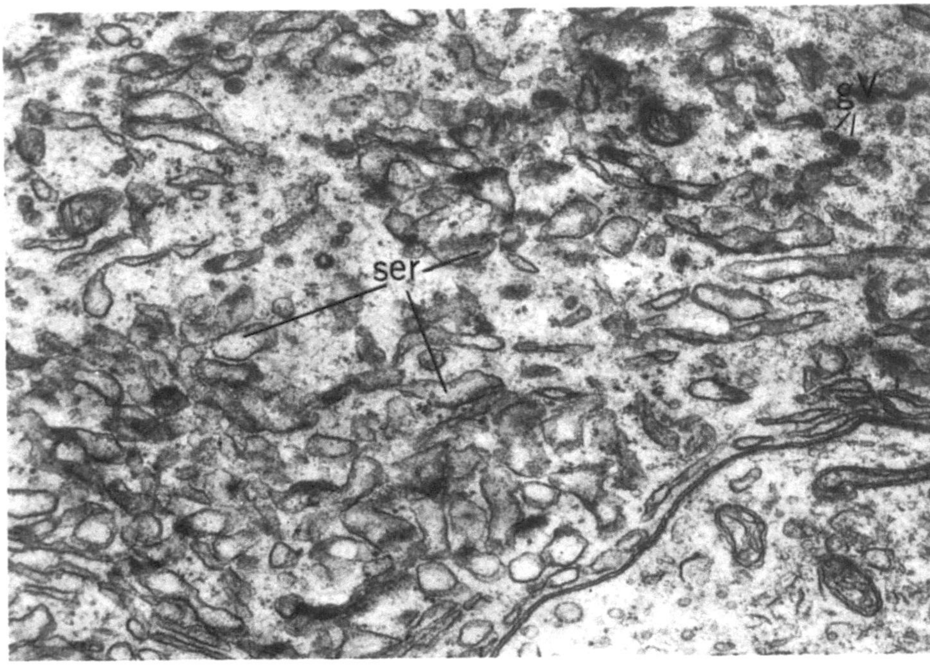

Fig. 60. Electron micrograph of accumulation of smooth endoplasmic reticulum (*ser*) in hamster pinealocyte (*gV*) granulated vesicle. SHERIDAN and REITER, 1968

measuring 60 nm in width (LEONHARDT, 1967, rabbit), isolated flattened sacs (SHERIDAN and REITER, 1968, 1970b, hamster) or isolated cisternae (ARSTILA, 1967, rat). A vacuolated appearance has been described in the dog (SANO and MASHIMO, 1966). Occasionally stacks of rough endoplasmic reticulum are present. In the rat, they consisted of between six and eight parallel cisternae measuring 2–5 µm in length and 30 nm in width (BOSTELMANN, 1965) and resembling Nissl bodies (WOLFE, 1965). On freeze-etching replicas, pores of the endoplasmic reticulum could not be demonstrated (KRSTIĆ, 1974).

Smooth endoplasmic reticulum consists of randomly oriented vesicular and elongated components which apparently do not form a continuous interconnecting network (cow, sheep: ANDERSON, 1965; guinea-pig: LUES, 1971). In the rabbit, short single or branched fragments have been described (LEONHARDT, 1967), as have small vesicles and tubules (WARTENBERG and GUSEK, 1965). Tortuous tubules were present in the pocket gopher (SHERIDAN and REITER, 1973). Especially concentrated structures of smooth endoplasmic reticulum have been described in the hamster (SHERIDAN and REITER, 1968; Fig. 60). In the rat, the membrane profiles were likewise "... sometimes so closely knit that other cytoplasmic organelles seem to be crowded aside" (WOLFE, 1965). In the mole-rat, stacks of smooth endoplasmic reticulum have been found (PEVET et al., 1976b).

The *contents* of the cisternae of the endoplasmic reticulum are inconspicuous in most species investigated. Notable exceptions are the pinealocyte of popula-

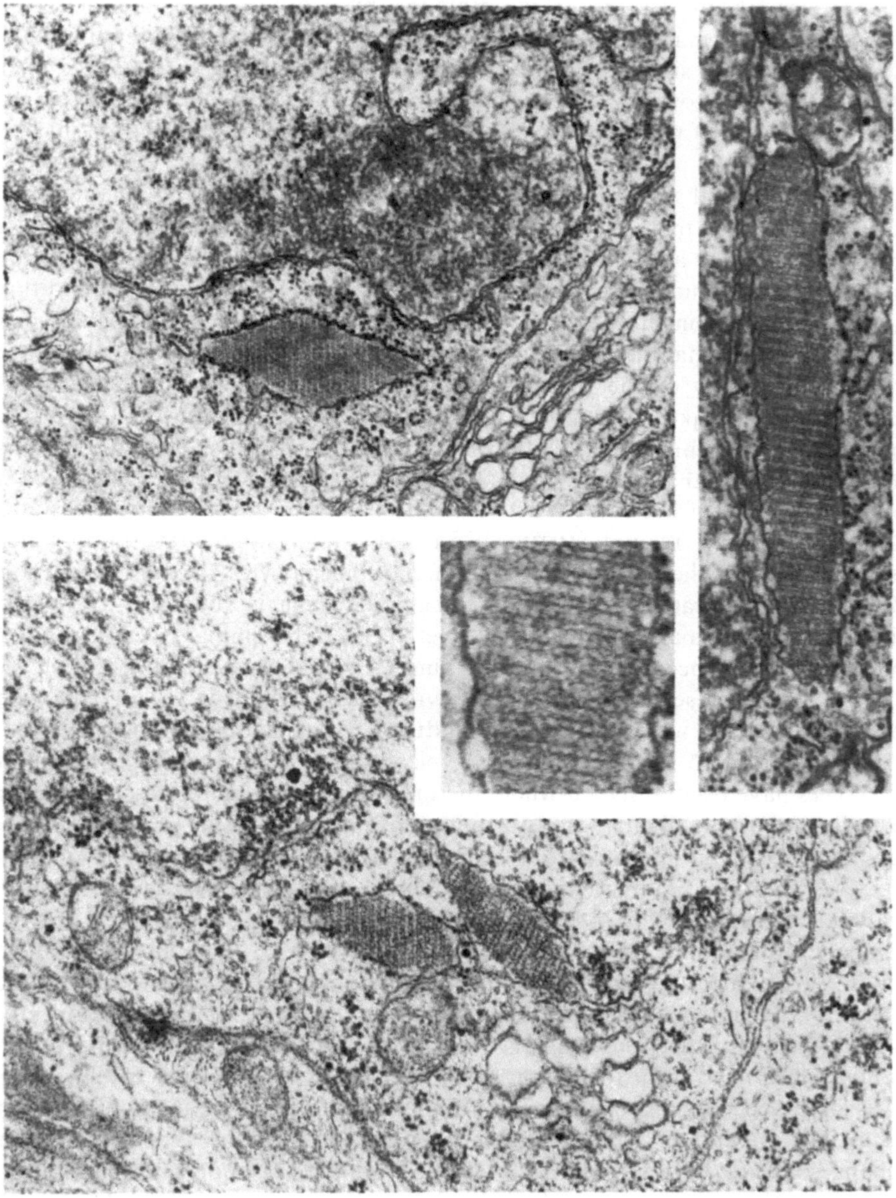

Fig. 61. Paracrystalline structures in cisternae of the rough endoplasmic reticulum (mole, *Talpa europaea*). PEVET, 1974

tion II (for further details see p. 168) of mole and mole-rat, and some unspecified pinealocytes of the dormouse. In the mole, rough endoplasmic reticulum comprised lozenge-shaped paracrystalline structures (Fig. 61) measuring 0.5–1.5 μm in length and 0.15–0.3 μm in width (PEVET, 1974, 1976, 1977a; PEVET and SMITH,

1975b). As shown by pronase treatment, these structures apparently consist of proteins (PEVET, 1977b). In the mole-rat, by contrast, moderately electron-dense flocculent material was present (PEVET, 1976, 1977a; PEVET et al., 1975b, 1976b). Roughly similar material was noted in the dormouse (ROUX et al., 1974, 1977). In cow and sheep, an homogeneous substance was noted, the electron density of which approximated that of the surrounding cytoplasm (ANDERSON, 1965).

The functional significance and the precise chemical nature of the materials contained within the endoplasmic reticulum has yet to be established. PEVET (1977a) and PEVET and KAPPERS (1977a, b) regard it as a special type of secretory material that occurs under certain physiological and experimental conditions only and that is somehow related to the antigonadotropic activity of the pineal. The fact that the material is especially abundant in animals in which the antigonadotropic activity of the pineal is enhanced due to the subterrestrial living conditions was taken as support for this hypothesis (PEVET et al., 1975b). A possible relationship to the state of the sexual organs is also suggested by the observation that in the mole the paracrystalline structures increased in males in phases of high sexual activity and in females during oestrus, gestation and lactation (PEVET and SMITH, 1975a). Also interesting is the finding that, in castrated rats, the endoplasmic reticulum of pinealocytes gives rise to vacuoles filled with moderately electron-dense material which is subsequently released into the intercellular spaces (KARASEK et al., 1976a). Similar vacuoles, but of unknown origin, occurred in hedgehog pinealocytes at the beginning of a physiological phase of sexual quiescence (PEVET and SABOUREAU, 1973) and when sexual quiescence was experimentally induced (PEVET and SABOUREAU, 1974).

In addition to the more or less typical forms of endoplasmic reticulum, instances have been observed where the membranes arrange themselves concentrically and eventually may transform into myelin bodies. This process was especially noticeable in pinealocytes of hamsters exposed to cold (Figs. 62–65) (BUCANA et al., 1973). Similar, though less conspicuous, structures were found in rabbit (LEONHARDT, 1966, 1967), rat (LIN, 1967), guinea-pig (LUES, 1971), infertile diabetic mutant mice (MCNEILL, 1978), mole (PEVET, 1974), mole-rat (PEVET et al., 1976b), noctule bat (PEVET et al., 1977a) and hedgehog (PEVET and SABOUREAU, 1973). In the latter species the concentric lamellae developed in November and December when the animals were sexually quiescent. In the dormouse, *Eliomys quercinus*, transformation of endoplasmic reticulum into myelin bodies was observed after exposure to continuous illumination for 6 months (ROUX et al., 1977), and in hamsters after blinding (CLABOUGH, 1971). How these changes are to be interpreted in terms of function is not known, although it is clear that the concentric lamellae become more numerous under extreme experimental conditions. BUCANA et al. (1973) suggested that they "may be focal sites of degenerative activity prior to the formation of new membrane systems".

Subsurface Cisterns. Cisterns of the endoplasmic reticulum closely related to the cell membrane ("subsurface cisterns"), which are common features of excitable cells (ROSENBLUTH, 1962), have also been noted in pinealocytes, where

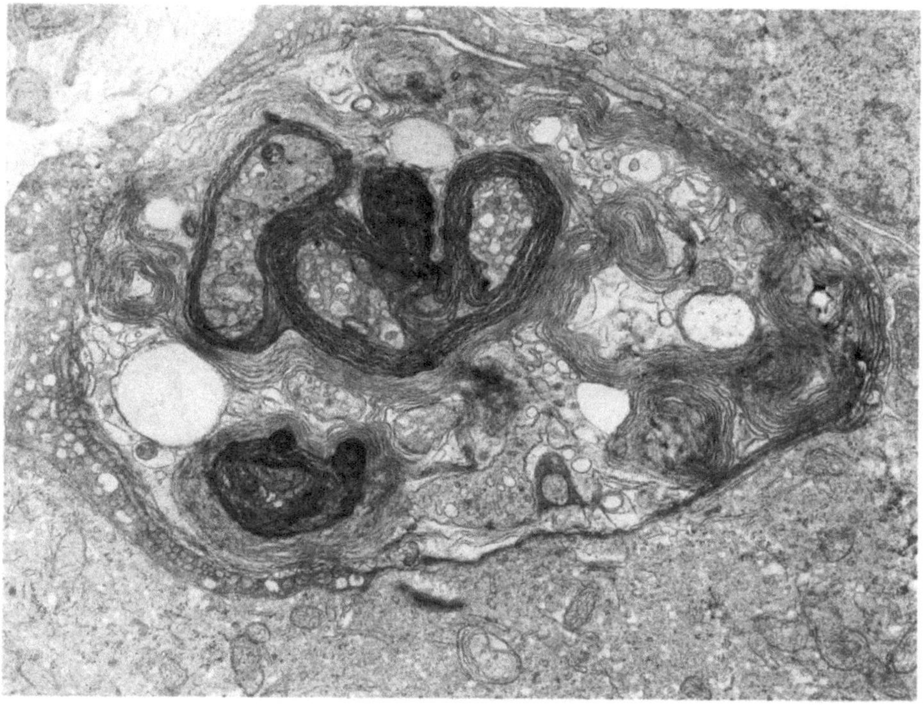

Fig. 62

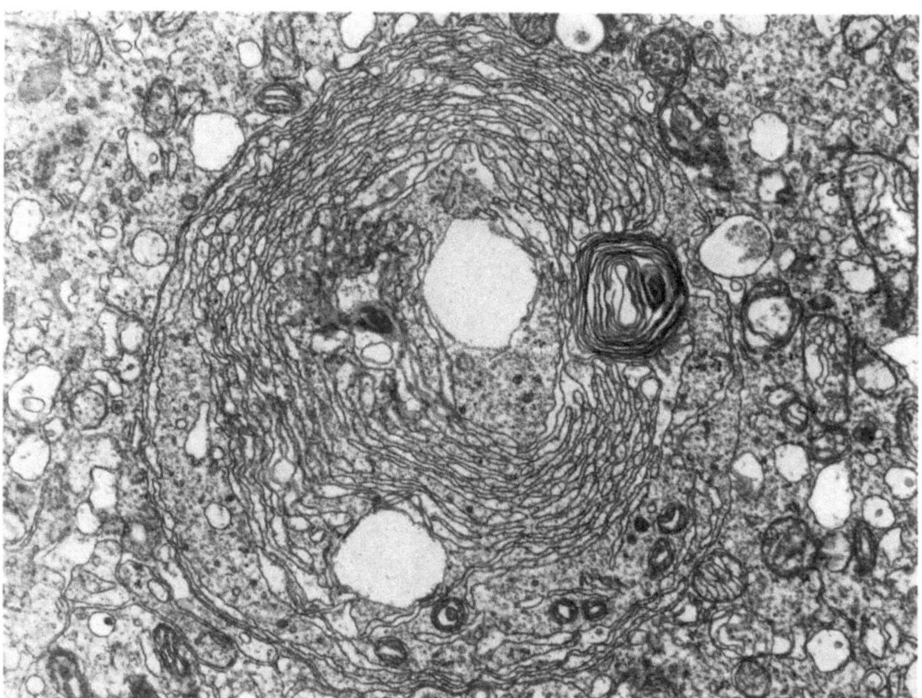

Fig. 63

Figs. 62–65. Various types of lamellar whorls in pinealocytes of hamsters exposed to cold. BUCANA et al., 1973

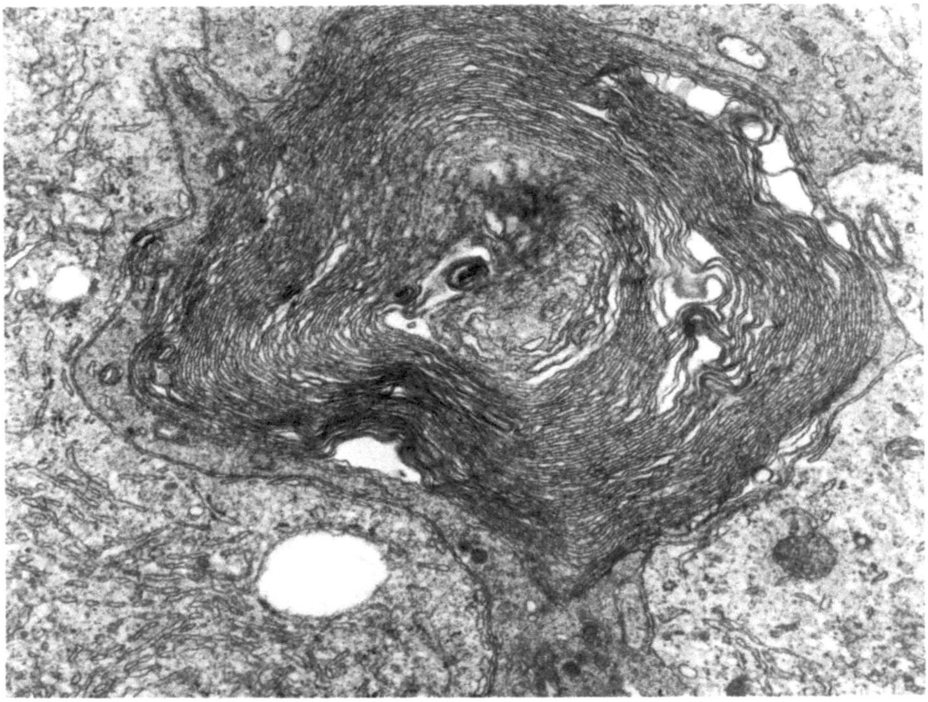

Fig. 64

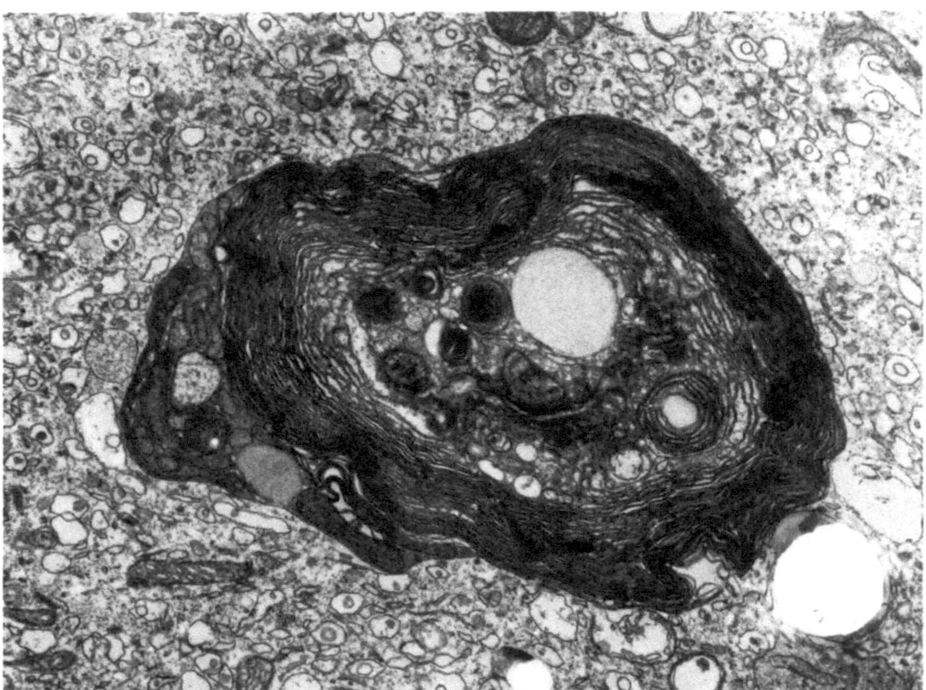

Fig. 65

they may be involved in a possible transport of substances to adjacent structures (ANDERSON, 1965; WARTENBERG, 1965; PEVET, 1974).

The most detailed account of the pinealocyte subsurface cisterns is that by WOLFE (1965). In the rat, the cisterns lacked ribosomes either entirely or wherever they confronted the plasmalemma. They remained 15–20 nm apart from the plasmalemma and underlay a significant fraction of the pinealocyte surface, apparently in the form of a broadly fenestrated cisterna. Often they occurred in register with corresponding cisternae in adjacent pinealocytes, without visible modification of the interjacent membranes or cleft. Occasionally the cisterns were found subjacent to autonomic nerve terminals. According to KRSTIĆ (1974), subsurface cisterns were rarely visible on freeze-etched replicas; when present, they were much smaller than on thin sections.

The relationships of the subsurface cisterns to neighbouring structures have been studied by WARTENBERG (1965) in the cat. Here they were often observed at sites where pinealocyte perikarya faced other pinealocyte perikarya, pinealocyte processes or glial cells. They were rarely present in two adjacent pinealocyte processes. In the seal (CUELLO, 1973), the cisterns were frequently noted in places where the plasmalemma lay closely related to a pinealocyte nucleus.

The subsurface cisterns described in other species generally conform to the above pattern (monkey: WARTENBERG, 1968; guinea-pig: LUES, 1971; dog: WELSER et al., 1968; mole: PEVET, 1974, 1976; hedgehog: PEVET, 1976; noctule bat: PEVET et al., 1977a; gerbil: WELSH and REITER, 1978). In the mole-rat, they were thought to be especially abundant (PEVET et al., 1976b).

Annulate Lamellae (Fig. 66). The functional significance of annulate lamellae is unknown; they have also been found in mammalian pinealocytes, but in only a few species. The most detailed description of pinealocyte annulate lamellae is that by BUCANA et al. (1971), who found them in light, but not dark, pinealocytes of the superficial hamster pineal. As can be expected in view of the nuclear origin of these structures, they were found mainly in the vicinity of the nucleus. Occasionally they lay near the Golgi apparatus. They consisted of between 2 and 15 double membranes (Fig. 66), the lateral margins of which appeared dilated with a number of irregularly shaped vesicles scattered along them. Moreover, a close association with microtubules was a regular feature.

Similar annulate lamellae were present in blinded hamsters (LIN et al., 1975), and probably also in mole (PEVET, 1974), mole-rat (PEVET et al., 1976b) and noctule bat (PEVET et al., 1977a).

In addition to the annulate lamellae shown in Fig. 66, so-called canaliculate lamellar bodies have been found (Fig. 67), which are structurally related to annulate lamellae. The first detailed description of these structures was that by LIN (1967), who published a picture of an unusually large body (Fig. 67) found in the rat. Distinctly smaller structures have recently been described in untreated rats (MCNEILL, 1977) and in rats given melatonin (FREIRE and CARDINALI, 1975) or cyproterone acetate, an anti-androgenic substance (GUSEK, 1976). Whether or not these structures are a peculiar feature of the rat pinealocyte remains to be established. Since the structures are altogether rare [MCNEILL

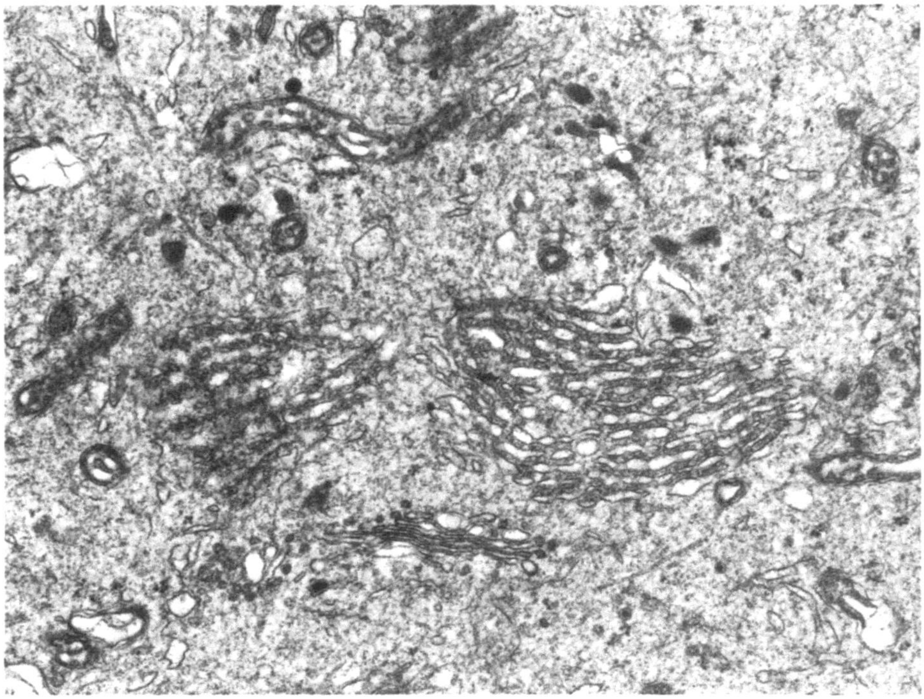

Fig. 66. Annulate lamellae in hamster pinealocyte. BUCANA et al., 1971

(1977) found them in 4 out of 22 glands surveyed], it seems premature to ascribe to them a specific function.

Ribosomes. Relatively little detailed information on pinealocyte ribosomes is currently available, probably because they do not present any unusual features. As can be expected from the small amounts of rough endoplasmic reticulum present in mammalian pinealocytes, the majority of ribosomes lie freely in the cytoplasm, either singly or in groups forming polysomes. According to WARTENBERG and GUSEK (1965, rabbit) and WOLFE (1965, rat), free ribosomes are altogether rare. In the guinea-pig (Fig. 59), they may be fairly abundant. In the dormouse (ROUX et al., 1974), they were more abundant in dark pinealocytes than in light ones. Free ribosomes have been reported to lie randomly scattered in pinealocytes of cow, sheep (ANDERSON, 1965), cat, monkey (WARTENBERG, 1968), hamster (LIN et al., 1975), dog (WELSER et al., 1968) and mole (PEVET, 1974). However, regional differences in the amounts of free ribosomes have also been noted. In the guinea-pig (Fig. 59), they were lacking in regions where microfilaments were abundant. The same applies to regions especially rich in mitochondria. In the rat pinealocyte, there were few around the nucleus but they increased in number in the cell periphery (ARSTILA, 1967).

Frequently, the amount of ribosomes has been investigated under various physiological and experimental conditions. Ribosomes increased in amount dur-

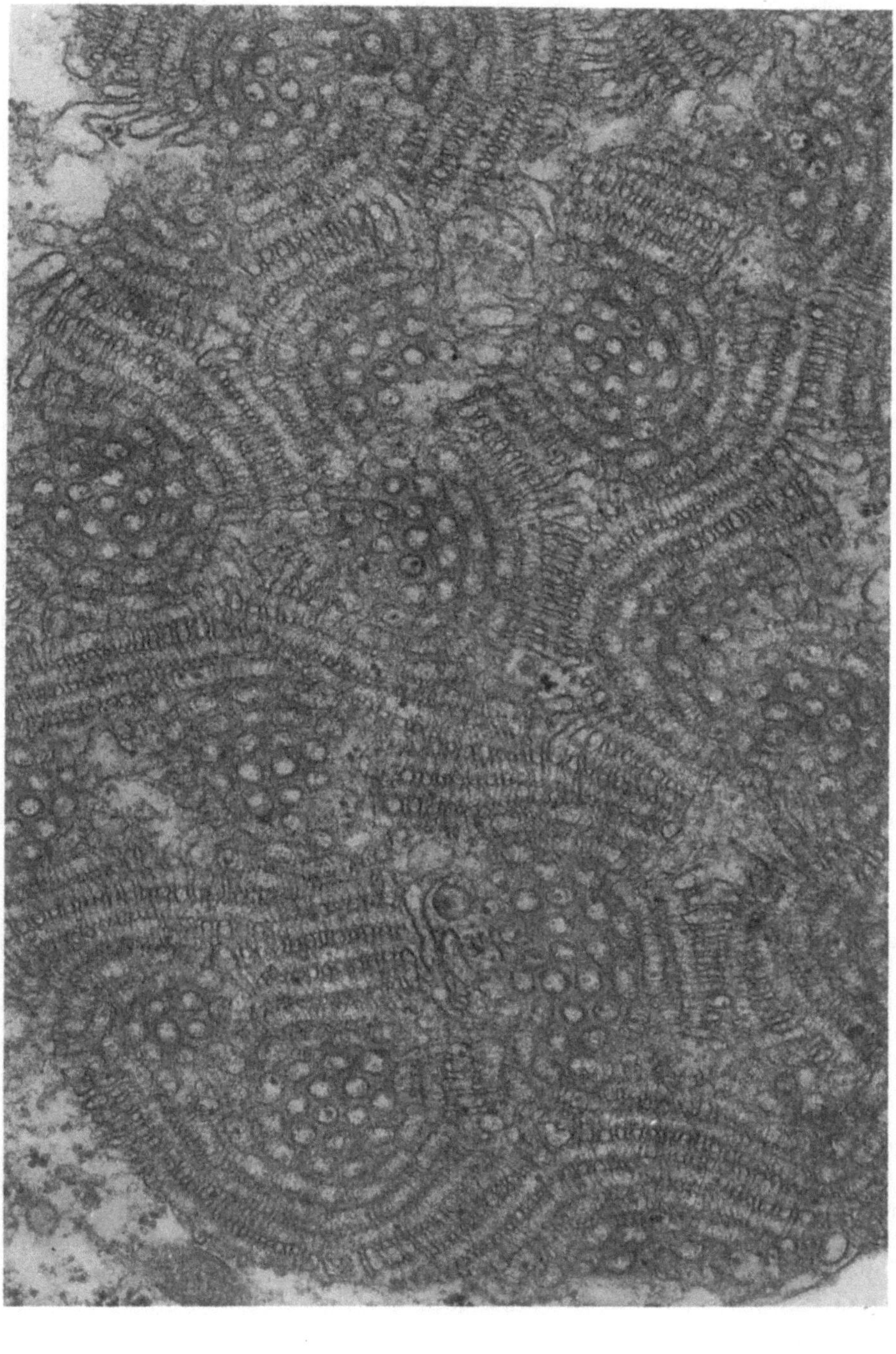

Fig. 67. Canaliculate lamellar bodies seen in rat pinealocytes. LIN, 1967

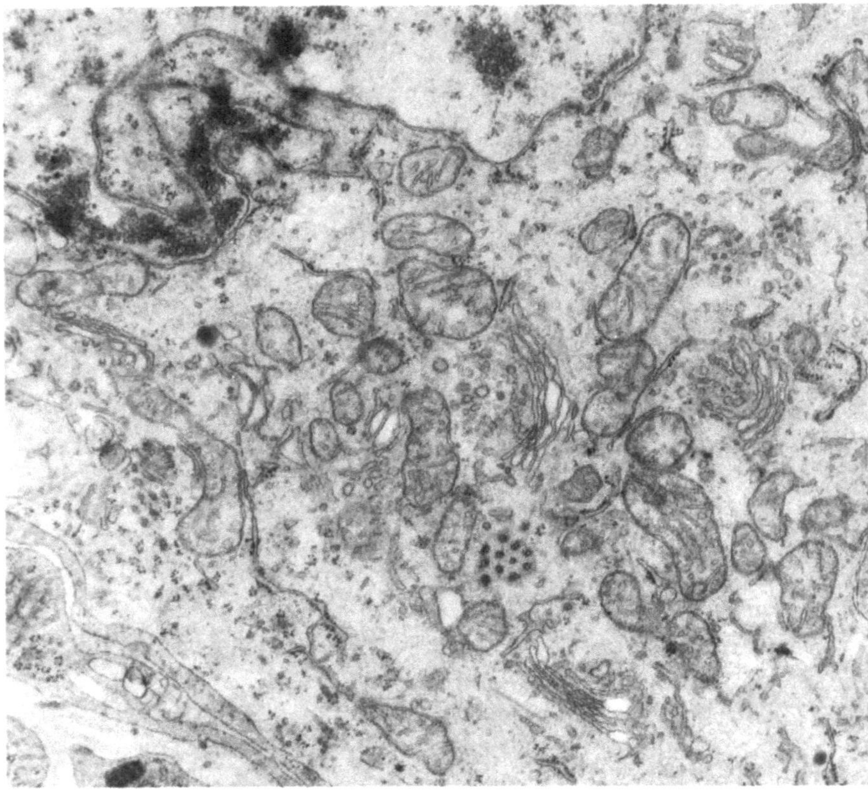

Fig. 68. Golgi zones in guinea-pig pinealocyte showing no signs of granule formation. LUES, 1971

ing puberty (BOSTELMANN, 1969a), pregnancy (LUES, 1971), after exposure to constant darkness (BOSTELMANN, 1968a; LUES, 1971), gonadectomy (GUSEK and BUSS, 1966a; KOWALZIG and GUSEK, 1971; BOSTELMANN, 1969c), Prolan administration (BUSS, 1965), oestrogen implantation (CLEMENTI et al., 1965a) and noradrenaline administration (GUSEK et al., 1965; BUSS and GUSEK, 1965; BUSS, 1965). They decreased after exposure to continuous illumination (BOSTELMANN, 1968a).

Golgi Apparatus. A Golgi apparatus, apparently consisting of several dictyosomes, is regularly present in mammalian pinealocytes. In view of the concept that the pineal is an endocrine organ, particular attention was paid to this cell organelle in the hope of obtaining information on the secretory nature of the pinealocyte. However, since in mammalian pinealocytes typical secretory granules are extremely rare, it is not surprising that very little direct evidence exists for participation of the Golgi apparatus in the formation of the secretory products (Fig. 68). Accumulation of electron-dense material in Golgi cisternae reminiscent of the production of dense-core vesicles has been noted in a few studies (ITO and MATSUSHIMA, 1968a, b; LIN et al., 1975). The actual formation

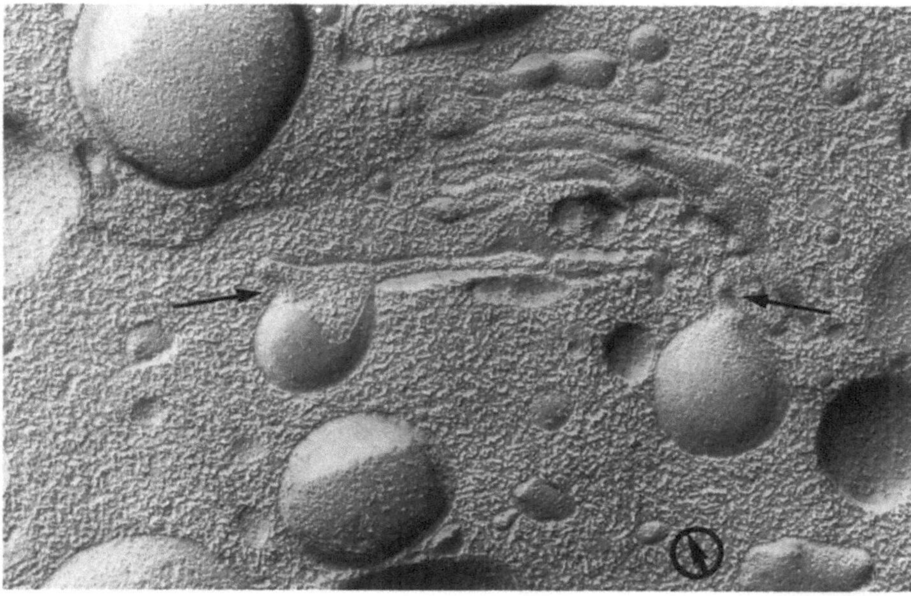

Fig. 69. Freeze-fractured replica of Golgi apparatus, showing the separation process of large granules from Golgi cisterns (*arrows*). Rat pinealocyte. KRSTIĆ, 1974; ×54,000

and separation of a large vesicle in the Golgi apparatus has been demonstrated in the rat pineal by means of freeze-etching electron microscopy (KRSTIĆ, 1974) (Fig. 69). Indirect evidence for participation of the Golgi apparatus in the formation of dense-core vesicles comes from studies showing that dense-core vesicles are present in the vicinity of dictyosomes (Fig. 70) (mouse: ITO and MATSUSHIMA, 1968a; hamster: SHERIDAN and REITER, 1970b, SHERIDAN, 1975, LIN et al., 1975; rabbit: LEONHARDT, 1967, ROMIJN, 1973, ROMIJN et al., 1976; guinea-pig: LUES, 1971; hedgehog: PEVET and SABOUREAU, 1974; cow, sheep: ANDERSON, 1965; dog: WELSER et al., 1968; dormouse: ROUX et al., 1974; noctule bat: PEVET et al., 1977a; mole-rat: PEVET et al., 1976b; golden mole: PEVET and KUYPER, 1978; gerbil: WELSH and REITER, 1978). From a functional point of view it is interesting to note that the Golgi complex-associated dense-core vesicles show daily fluctuations, with largest numbers being present at 12.00 h and lowest numbers at 24.00 h (ROMIJN et al., 1976). For further details of the granules (size etc.) and the relation to secretory products, see p. 165.

In addition to the formation of dense-core vesicles, the Golgi apparatus of the pinealocytes has been shown to be involved in the production of electron-lucent vesicles measuring between 35 and 120 nm in width, the majority lying between 40 and 100 nm (ROMIJN, 1973b), and coated vesicles (or acanthosomes) (WOLFE, 1965; LEONHARDT, 1967; ROMIJN, 1973b).

Structural appearance and intracellular localization of the Golgi apparatus do not present any unusual features (Figs. 68, 69). Individual dictyosomes consist of three to seven membranous sacs (LUES, 1971), which appear either flat (ham-

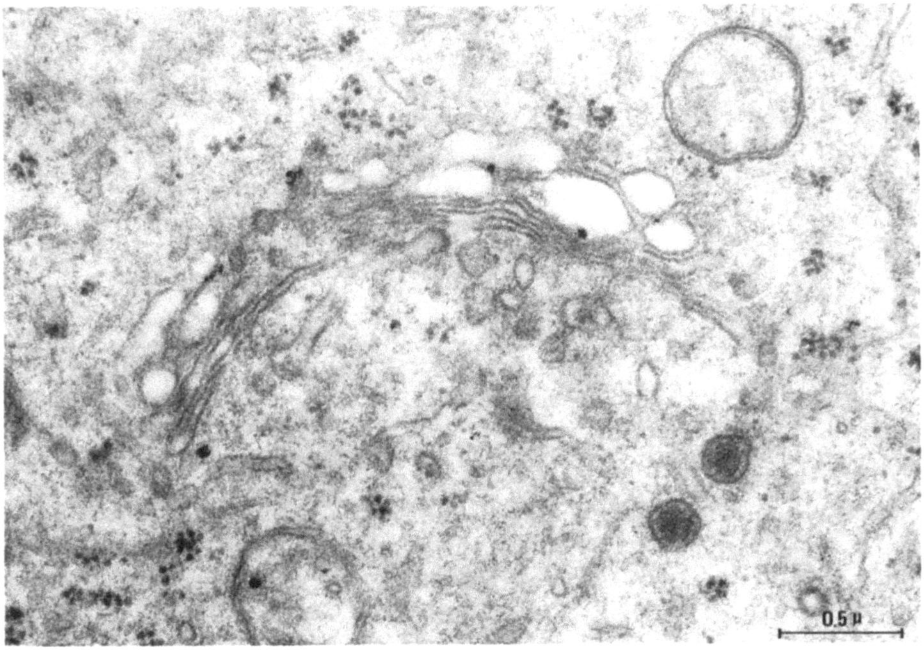

Fig. 70. Golgi apparatus with associated dense-core vesicles. Rabbit pinealocyte. ROMIJN et al., 1976

ster: SHERIDAN and REITER, 1968, 1970b; cow and sheep: ANDERSON, 1965; rat: ARSTILA, 1967) or distended, either along their entire lengths (rabbit: LEONHARDT, 1967) or mainly at their ends (guinea-pig: LUES, 1971). In cow and sheep (ANDERSON, 1965) some areas of the Golgi sacs showed evaginated fuzzy regions.

The localization of the Golgi apparatus has been reported to be juxtanuclear in dog (SANO and MASHIMO, 1966), hamster (SHERIDAN and REITER, 1968), guinea-pig (LUES, 1971), hedgehog (PEVET and SABOUREAU, 1973), mole (PEVET, 1974; PEVET and SMITH, 1975b), dormouse (ROUX et al., 1974) and mole-rat (PEVET et al., 1976b). In the noctule bat many dictyosomes are widely dispersed around the nucleus (PEVET et al., 1977a). In the rat, dictyosomes lie in the vicinity of the nucleus (ARSTILA, 1967), though WOLFE (1965) mentioned a random occurrence of Golgi membranes in the perikaryon. In the squirrel pinealocyte the Golgi apparatus is "dispersed throughout the cytoplasm" (POVLISHOCK et al., 1975). The osmiophilia of the Golgi complex in rat pinealocytes has been studied by KAPPERS et al. (1962).

The activity of the Golgi apparatus has been little studied in quantitative terms. According to visual assessments, the Golgi apparatus is strongly developed in rabbits (LEONHARDT, 1967), while in rat (ARSTILA and HOPSU, 1964) and mole (PEVET, 1974) it differs greatly in size among pinealocytes. Morphometric studies revealed that in rat pinealocytes the Golgi apparatus occupies 0.56% ±

0.02% of the cell volume (KRSTIĆ, 1977b). The only attempt at objectively assessing the size of the Golgi apparatus in relation to function is the study by PEVET and SMITH (1975a) in the mole. Here it was found that the number of dictyosomes changes in relation to the sexual activity of the animals. Maximum activity was noted in male moles during phases of high sexual activity and in females during oestrus, gestation and lactation, low activity being present during anoestrous phases.

Under a variety of physiological and experimental conditions, relatively few changes have been observed in the Golgi apparatus, as assessed visually. An increase in size was noted during puberty (BOSTELMANN, 1969a) and after exposure to darkness (LUES, 1971), gonadectomy (BOSTELMANN, 1969c; GUSEK and BUSS, 1966a; KOWALZIG and GUSEK, 1971), hypophysectomy (KARASEK, 1971a) and administration of noradrenaline, nialamide, or Prolan (BUSS, 1965; BUSS and GUSEK, 1965).

Mitochondria. Pinealocyte mitochondria show a number of features with respect to size, shape, internal structure, membrane discontinuities, frequency, location and swelling, and thus attract special attention in most species studied. Mitochondria lacking unusual features were described in dog (SANO and MASHIMO, 1966), gopher (SHERIDAN and REITER, 1973) and mole (PEVET and SABOUREAU, 1973).

Although the majority of pinealocyte mitochondria appear to be of normal *size*, extreme variations have been frequently noted. Extremely large mitochondria, epitomized as "giant" mitochondria (MILOFSKY, 1958; GUSEK and SANTORO, 1960; CLEMENTI et al., 1965a; WOLFE, 1965), measured 4–5.8 μm in length in rats (MILOFSKY, 1958; GUSEK and SANTORO, 1960, 1961; ARSTILA, 1967; BOSTELMANN, 1965; LIN, 1965), the largest mitochondria being present near the nucleus (ARSTILA and HOPSU, 1964), 3–4 μm in rabbits (WARTENBERG and GUSEK, 1965) and hamsters (BUCANA et al., 1974) and 2–3 μm in cat (WARTENBERG, 1968) and guinea-pig (LUES, 1971). In the dormouse, the mitochondria measured between 0.5 and 2.0 μm in size (ROUX et al., 1974).

In addition to the large mitochondria, relatively small ones measuring 0.2–0.25 μm have also been found (rat: GUSEK and SANTORO, 1960, 1961; hamster: SHERIDAN and REITER, 1968). The average size of pinealocyte mitochondria appears to lie between 0.5 and 1.5 μm (BOSTELMANN, 1965; SHERIDAN and REITER, 1968). The reported decrease in mitochondrial size in perivascular regions of pinealocytes (DE ROBERTIS and PELLEGRINO DE IRALDI, 1961a) could not be confirmed by WOLFE (1965).

Extreme variations in *shape* have been frequently observed. Mitochondrial profiles have been described as round, oval, elongated, rod-shaped, filamentous, tortuous, arciform, branched and club-shaped (GUSEK and SANTORO, 1960, 1961; BOSTELMANN, 1965; BUCANA et al., 1974). In the hamster, mitochondria with attenuated central portions linking two normal-appearing extremities were observed, and it has been suggested that the attenuated part is apparently responsible for encircling other structures and producing cup-shaped mitochondria (SHERIDAN and REITER, 1968). In rabbit pinealocytes the shape of the mitochondria was particularly variable near the nucleus (WARTENBERG, 1968).

Regarding the *internal structure* of mitochondria, it has been observed that the inward reflection of the inner mitochondrial membrane takes several forms. Cristiform, tubular, tubulovesicular, angular and plexiform profiles have been described. In the rat (GUSEK and SANTORO, 1960, 1961; BOSTELMANN, 1965; ARSTILA, 1967), cow, sheep (ANDERSON, 1965), mouse (ITO and MATSUSHIMA, 1968a) and guinea-pig (LUES, 1971) the majority of mitochondria exhibit cristae. In the rat the cristae show transverse, oblique or longitudinal orientations (GUSEK and SANTORO, 1960, 1961; ARSTILA, 1967), though transversely oriented cristae predominate (WOLFE, 1965). Sometimes transversely and longitudinally arranged cristae occurred in the same mitochondrion; zig-zag arrays of cristae are sparse (WOLFE, 1965). BOSTELMANN (1965) noted a particularly dense packing of the cristae.

Tubular cristae were found in pinealocyte mitochondria of rat (GUSEK and SANTORO, 1960, 1961; ARSTILA, 1967; however, WOLFE (1965) could not find any), guinea-pig (LUES, 1971), dormouse (ROUX et al., 1974) and hamster (PELLEGRINO DE IRALDI, 1966; SHERIDAN, 1967; SHERIDAN and REITER, 1968, 1970b; BUCANA et al., 1974). In the latter species mitochondria with tubular cristae appear to be fairly abundant and electron dense (Fig. 71). As shown in Fig. 72, the tubular cristae may be concentrically arranged. In longitudinal sections they are arranged like a string of beads (Fig. 72, inset). Tubular profiles were particularly prominent in large mitochondria, while cristiform profiles were typical of small mitochondria (PELLEGRINO DE IRALDI, 1966; SHERIDAN and REITER, 1968; LUES, 1971).

Tubulovesicular invaginations have been described in the rat (ARSTILA, 1967), tubulosaccular structures in the guinea-pig (LUES, 1971), angular profiles in cow and sheep (ANDERSON, 1965) and plexiform cristae in the hamster (CLABOUGH, 1971; BUCANA et al., 1974).

While in the majority of species particular types of mitochondria were not attributed to special types of pinealocytes, CLABOUGH (1971), studying the hamster pineal, noted that small mitochondria (ca. 0.2–0.4 µm in diameter) with few cristae were exclusively present in the so-called P_1 pinealocytes, while larger mitochondria measuring ca. 0.5 by 0.75 µm with a plexiform array of cristae in a dense matrix were typical of P_2 pinealocytes. To what extent these differences are functional in nature remains to be elucidated. An interesting observation was that the polymorphism of the mitochondria was more pronounced in cold-exposed golden hamsters than in control animals (BUCANA et al., 1973).

The *matrix* of pinealocyte mitochondria is often difficult to assess because of the tendency of mitochondria to swell, particularly after glutaraldehyde fixation (see p. 135). In the rat, the mitochondrial matrix varied in electron density (GUSEK and SANTORO, 1961), being high in dark pinealocytes and low in light pinealocytes (ARSTILA, 1967). In cow and sheep the matrix was fairly lucid (ANDERSON, 1965). An especially dense matrix was typical of mitochondria with tubular cristae (LUES, 1971, guinea-pig). Intramitochondrial granules do not appear to be regular features of pinealocyte mitochondria. In the rat WOLFE (1965) could not detect any and ARSTILA (1967) stated that they occur only rarely. In some rat pinealocyte mitochondria, dense intercristal bands were found, which, in tangential view, exhibited grid-like striations with a periodicity

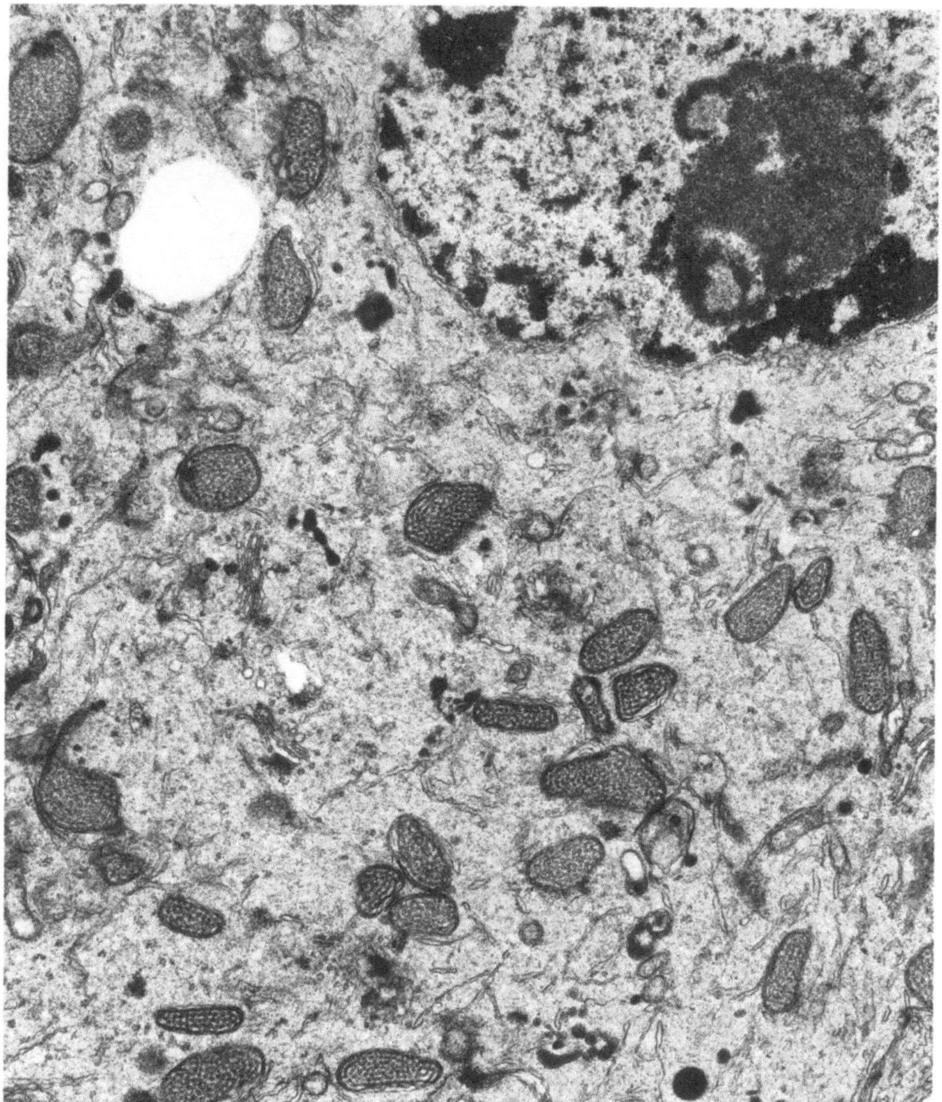

Fig. 71. Survey electron micrograph of hamster pinealocyte showing many mitochondria with tubular crests. By courtesy of Dr. C.D. BUCANA, Frederick Cancer Research Center, Frederick, Maryland

of ca. 40 nm (WOLFE, 1965). Another rare finding is the presence of microcylinders (Fig. 73) within mitochondrial cristae (LIN, 1965). These structures found in 2 out of 18 Long-Evans rats, measured 27–33 nm in diameter and consisted of probably six slender filamentous subunits surrounding a central filament. In two mitochondria with microcylinders glycogen particles were noted within enlarged intracristal spaces.

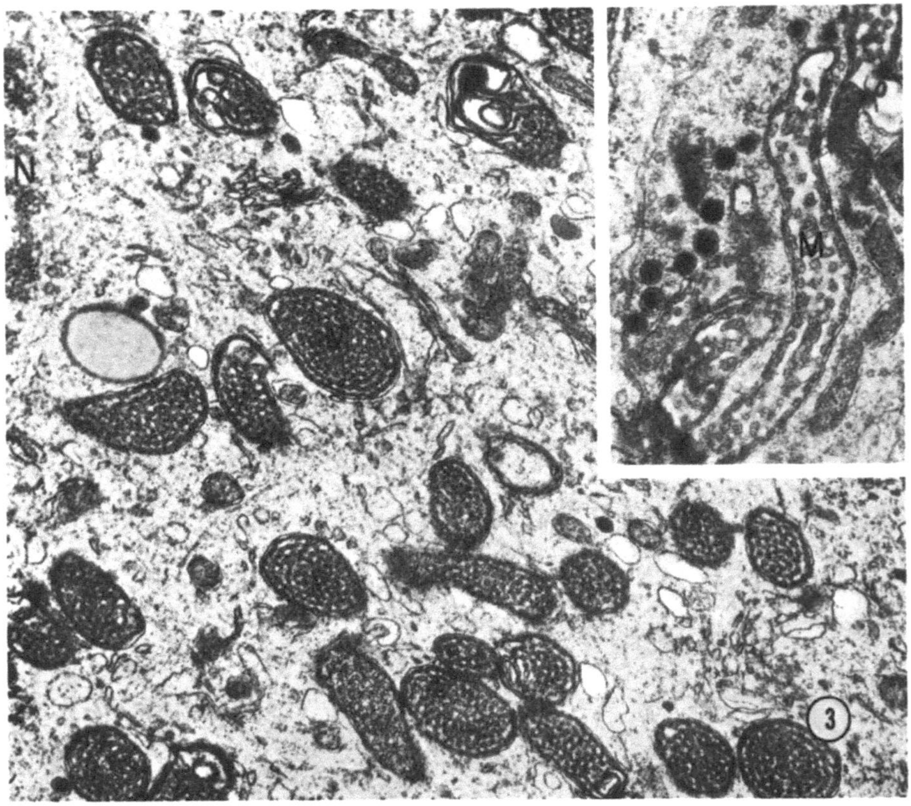

Fig. 72. High-power electron micrograph of mitochondria with tubular crests from hamster pinealocytes. The crests show concentric arrangement. *Inset* shows longitudinal section of mitochondrion (*M*) with cristae arranged like a string of beads. BUCANA et al., 1974; ×34,000; *inset*, ×44,500 (reduced to 60%)

The mitochondrial *envelope* also exhibits special features. In the rat, gaps were noted and interpreted as sites of extrusion of mitochondrial enzymes into the cytoplasm. WOLFE (1965) also found that the paired membranes of the mitochondrial envelope were difficult to trace continuously. However, he did not regard the apparent discontinuities as passageways between the mitochondrial matrix and the hyaloplasm, but felt that the membranes became invisible as they tilted toward the plane of the section. In view of the fact that mitochondria of both guinea-pig (LUES, 1971; VOLLRATH, unpublished observations) and hamster (CLABOUGH, 1971) were occasionally seen to be in continuity with the endoplasmic reticulum or the outer membrane of the nuclear envelope, the problem of apparent discontinuities of the mitochondrial envelope should be re-examined.

Mitochondrial *swelling* has frequently been noted in mammalian pinealocytes (e.g. GONZALES et al., 1969) and regarded as being peculiar to these cells in contrast to adjacent glial cells and other tissues processed in an identical manner

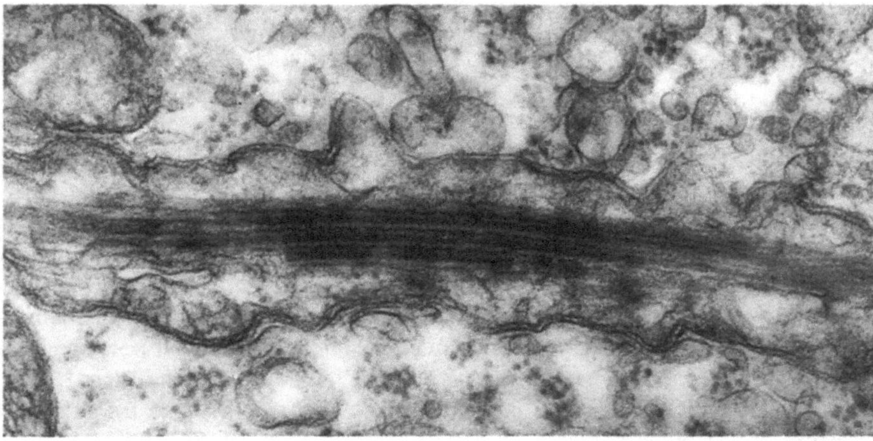

Fig. 73. Mitochondrial profile comprising microcylinders. LIN, 1965; ×72,000

(DUNCAN and MICHELETTI, 1966). Although DUNCAN and MICHELETTI found that the swelling occurred with all the fixatives used, WARTENBERG (1968) noted swelling only when the tissue had been fixed in glutaraldehyde, but it was avoided by the use of straight osmium tetroxide fixation. Nevertheless, it appears that pinealocyte mitochondria are particularly prone to swelling, as shown by the fact that swollen mitochondria were especially abundant after oestrogen implantation into the habenular region (CLEMENTI et al., 1965a; rat) or exposure to continuous illumination (rat: HALARIS, 1969, HALARIS and MATUSSEK, 1969; guinea-pig: LUES, 1971).

A *transformation* of mitochondria into strongly osmiophilic bodies has been described by GUSEK and SANTORO (1960, 1961), but this could not be confirmed (WOLFE, 1965; LUES, 1971). In the dormouse, mitochondria comprising myelin bodies in the centre have been described (ROUX et al., 1974), which may also point to a transformation of these organelles. WOLFE (1965) considered the possibility that attenuated mitochondria might develop into what he called "vernate" bodies.

Precise quantitative data on the *amounts* of mitochondria present in pinealocytes are rare. Without going into detail, it suffices to mention that in most species investigated the mitochondria have been described as being abundant. It should be noted, though, that due to an uneven distribution of mitochondria within the cytoplasm there may be areas which are particularly rich in mitochondria while others are poor in this organelle (see below). In rat pinealocytes mitochondria occupy 6.1% ± 0.06% of the cell volume (KRSTIĆ, 1977b).

Little detailed knowledge on the intracellular *location* of mitochondria is currently available. In view of the recent findings that mammalian pinealocytes may resemble photoreceptor cells more closely than has hitherto been thought (PEVET and COLLIN, 1976; PEVET et al., 1977b), the distribution of mitochondria within the pinealocytes should be carefully re-examined with respect to whether they show a localization similar to that in photoreceptor cells, where they lie accumulated in the so-called ellipsoid. A distribution of mitochondria in what could be called the supranuclear region of the mammalian pinealocyte, reminiscent of the situation in photoreceptor cells, has been clearly depicted in the mole (Fig. 56) (PEVET and COLLIN, 1976) and noctule bat (PEVET et al., 1977b) (Fig. 74). That mitochondria may be similarly distributed in rat pinealocytes is indicated in WOLFE's (1965) paper in Fig. 1, where pinealocytes surrounding a central lumen show numerous mitochondria in the apical cytoplasm. Also interesting is the observation by LUES (1971) in the guinea-pig that in some pinealocytes mitochondria and dictyosomes were particularly abundant in what he called active zones which lay close to the cell nucleus, especially where it showed invaginations. In this context one should mention the histochemical studies dealing with mitochondrial enzymes (p. 94) which support the view that ellipsoid-like regions may be fairly frequent in the guinea-pig pineal. Electron microscopically these areas may not be easily detectable due to a random orientation of the pinealocytes.

Apart from the "ellipsoid" region of the mammalian pinealocyte, mitochondria have been frequently noted to be numerous in the vicinity of the nucleus (cow, sheep: ANDERSON, 1960, 1965; rabbit: WARTENBERG and GUSEK, 1965; LEONHARDT, 1966, 1967). In randomly oriented rat pinealocytes most of the mitochondria were found in the peripheral parts of the perikarya and their processes (ARSTILA and HOPSU, 1964), some of which may correspond to relics of inner segments.

Frequently an association of mitochondria with other cell organelles or inclusions has been observed. A close association with lipid droplets has been mentioned in cow, sheep (ANDERSON, 1965), rat (GUSEK and SANTORO, 1960, 1961; WOLFE, 1965) and guinea-pig (LUES, 1971). Mitochondria near pigment inclusions were seen in rat (GUSEK and SANTORO, 1960, 1961), cow and sheep (ANDERSON, 1965). An abundance of mitochondria in the vicinity of centrioles was noted in the rabbit (WARTENBERG and GUSEK, 1965; ROMIJN, 1973b). ARSTILA and HOPSU (1964) described a close association with synaptic ribbons and WOLFE (1965) and WELSER et al. (1968) mentioned that mitochondria were frequently encompassed by branching arcades of endoplasmic reticulum.

Apart from the non-random distribution of the mitochondria, which points to a relationship of the mammalian pinealocyte to pineal photoreceptor cells of lower vertebrates, nothing substantial from a functional point of view has yet been offered. As regards the pleomorphism of the mitochondria, especially their tubular cristae, SHERIDAN and REITER (1968) have considered, in analogy to the presence of these mitochondria in steroid-producing cells, that hamster pinealocytes may be involved in steroid metabolism, but thin layer chromatography of chloroform extracts of pineal glands failed to identify a steroid which might be correlated with the presence of the tubular mitochondria. Instead they suspected that tubular mitochondria may be restricted to species demonstrating marked pineal inhibitory effects on the gonads.

Whatever the functional significance of the mitochondrial polymorphism may be, it seems to clearly indicate a functional complexity of mammalian pinealocytes similar to, for example, hepatic parenchymal cells, the mitochondria of which likewise possess lamellar and tubular cristae.

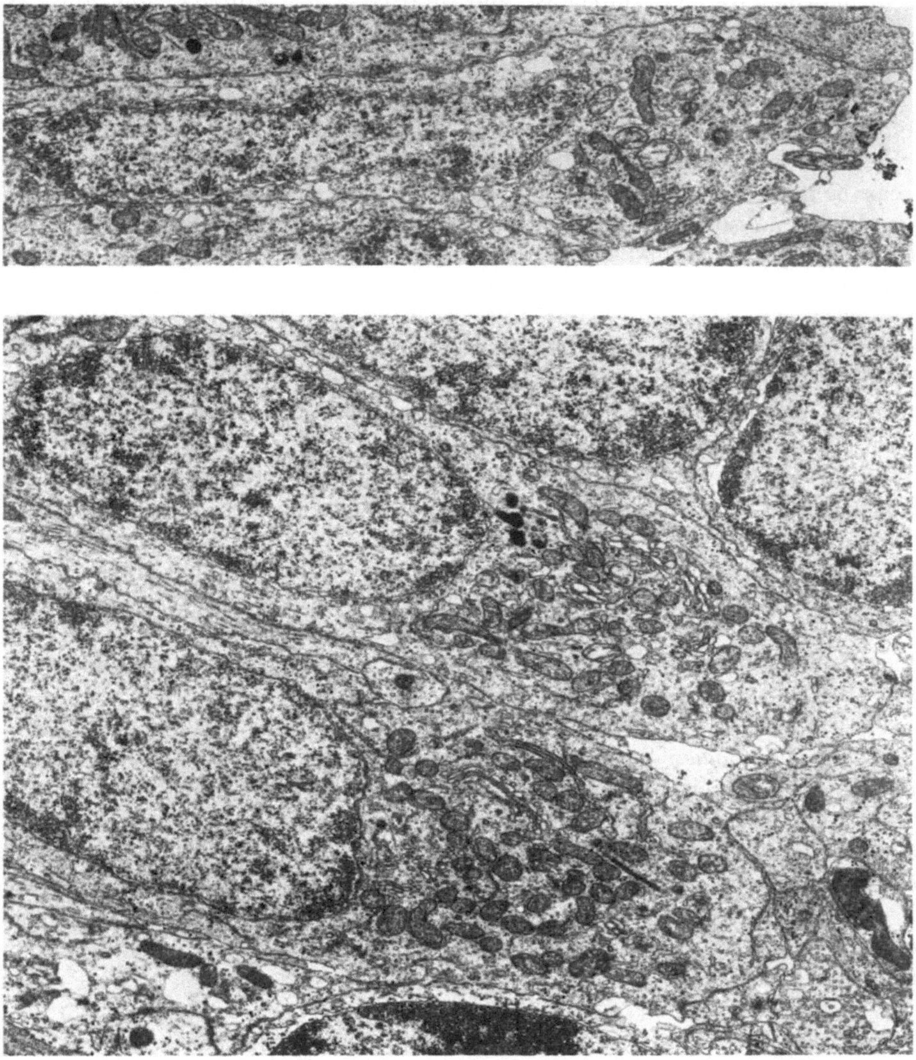

Fig. 74. Accumulation of mitochondria in certain parts of pinealocytes. Noctule bat. PEVET et al., 1977b

Lysosomes. It does not appear that lysosomes play an important role in mammalian pinealocytes when judged by their amounts. Although they are present and show changes under various physiological and experimental conditions, they are altogether too rare to be of great significance.

The most detailed study on pinealocyte lysosomes has been carried out by ARSTILA (1967) in the rat. In this species, a number of "single membrane limited inclusion bodies" were clearly identified as cytosomes and primary lysosomes by applying enzyme histochemical techniques at the electron microscopic level. Cytosomes measuring about 0.3–1.5 μm in diameter were shown to contain acid phosphatase, aryl sulphatase and thio-acetic acid esterase, and could be further subdivided according to their electron microscopic appearance. One type possessed a fairly uniform and lightly granular inner matrix and contained membranous structures, a few dense or empty vesicles or granules. The second type comprised mitochondria or large empty vesicles. Both types were rare. Type I cytosomes, more numerous than type II cytosomes, were found mainly near the nucleus, while type II cytosomes were present throughout the perikaryon. Primary lysosomes appeared as round, oval or spindle-shaped electron-dense bodies, the round ones measuring about 150 nm in diameter and the spindle-shaped ones, 1–2 μm in length and 0.4 μm in width. In contrast to the cytosomes, they lacked thio-acetic acid esterase. Primary lysosomes were altogether rare. As can be expected, they were more numerous near the Golgi apparatus, where they are formed. They were especially abundant in the terminal clubs of pinealocyte processes. In light pinealocytes they were less numerous than in dark pinealocytes.

Lysosomes have been tentatively identified in cow, sheep (ANDERSON, 1965), dog (WELSER et al., 1968), hamster (SHERIDAN and REITER, 1968; LIN et al., 1975), guinea-pig (LUES, 1971), hedgehog (PEVET and SABOUREAU, 1973), mole (PEVET, 1974), mole-rat (PEVET et al., 1976b), noctule bat (PEVET et al., 1977a), dormouse (ROUX et al., 1974, 1977) and ground squirrel (POVLISHOCK et al., 1975). In dormouse and ground squirrel they were rather common. In the dog most of the lysosomes were found near a cilium and the Golgi complex (WELSER et al., 1968).

Multivesicular Bodies. Such bodies have been reported to be numerous in rat pinealocytes, especially in their processes (ARSTILA, 1967) and near the Golgi apparatus, where they were frequently capped by one or several flattened vesicles (WOLFE, 1965). They were likewise commonly observed in the ground squirrel (POVLISHOCK et al., 1975). In the guinea-pig they were rare and lay mainly in the perikaryon (LUES, 1971). In hamsters, blinding increased their amounts (CLABOUGH, 1971; LIN et al., 1973), as did superior cervical ganglionectomy (LIN et al., 1975).

Microbodies. Oval granules measuring up to 0.15 μm in diameter and resembling microbodies have been found in some rat pinealocytes (GUSEK and SANTORO, 1961).

Centrioles. There is clear ultrastructural evidence that the centrioles of fully differentiated mammalian pinealocytes may exhibit a number of unusual features. It has frequently been noted that typical centrioles were either lacking (rat: WOLFE, 1965; LIN, 1970) or rare (hamster: SHERIDAN and REITER, 1968; BUCANA et al., 1974; guinea-pig: LUES, 1971; LIN, 1972; rat: ARSTILA, 1967; pocket gopher: SHERIDAN and REITER, 1973; hedgehog: PEVET and SABOUREAU, 1973; ground squirrel: POVLISHOCK et al., 1975). In the other species investigated centrioles have also been found, but no details regarding their frequencies have been given. Centrioles do not appear to have been mentioned in the dormouse (ROUX et al., 1974).

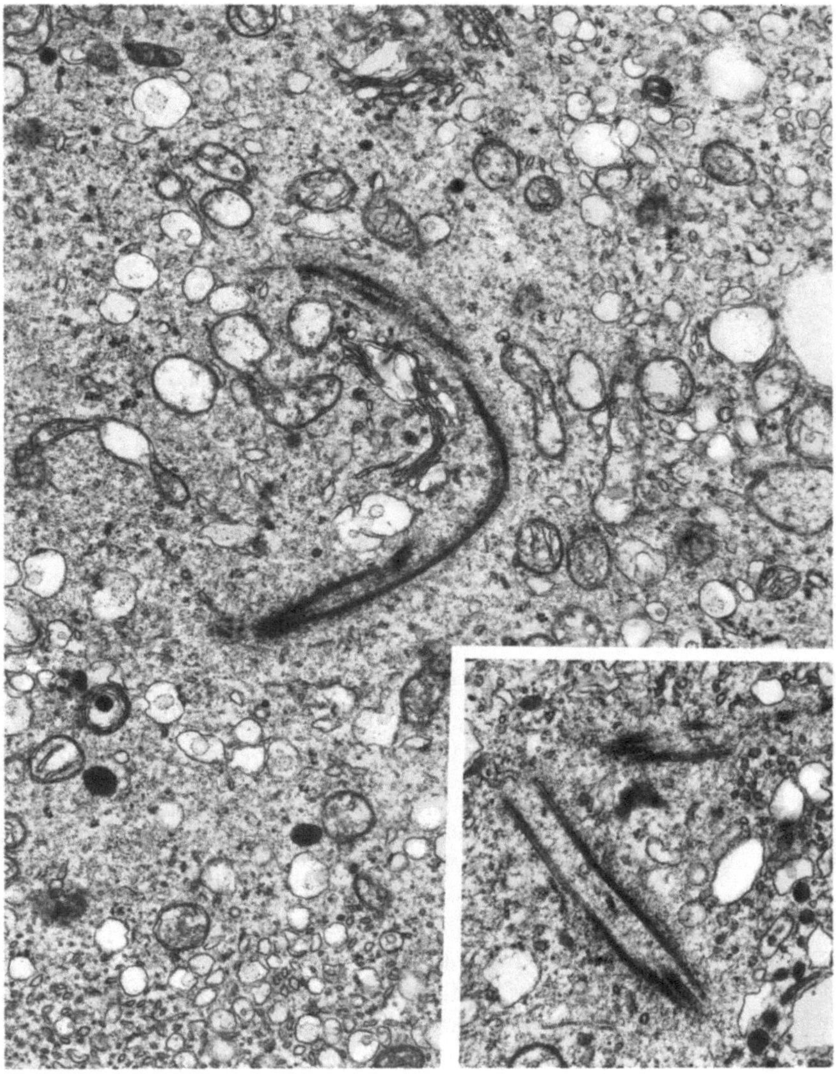

Fig. 75. Perinuclear area of hamster pinealocyte showing curved microtubular sheaf of modified centriole. ×39,000 *Inset* shows angulation of the elongated microtubular component. ×29,000. BUCANA et al., 1974

A striking feature of pinealocyte centrioles is their apparent transformation into what has been called microtubular sheaves or microtubular bouquets (Fig. 75). This conspicuous transformation, when first observed by WOLFE (1965) and subsequently studied in detail by LIN (1969, 1970, 1972), could not be satisfactorily explained in terms of function. Both authors were of the opinion that the transformation is an expression not of centriolar degeneration but rather of pinealocyte differentiation. In view of the recent findings that in

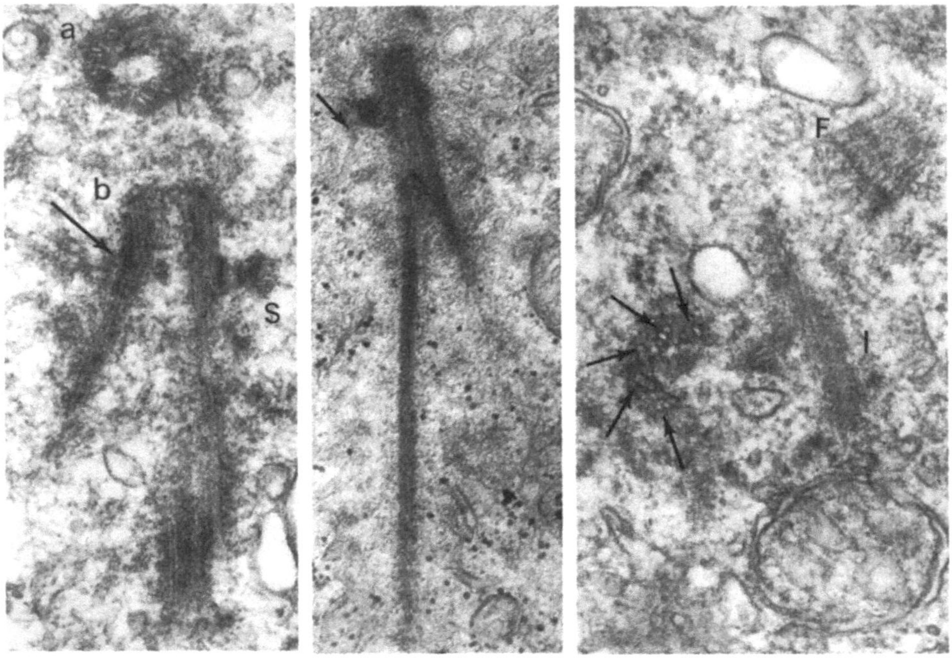

Fig. 76. Altered centrioles from 1-month-old rat. *Left:* one centriole (*a*) is cut transversely and displays a slight derangement of component structures in the wall. The other (*b*) is apparently elongated and shows a bursting of the wall at the lower end and angulation of the component microtubules (*arrow*). *S*, segmented arm; ×52,000 *Centre*: elongated centriolar tubules diverging downwards. The *arrow* indicates a cytoplasmic microtubule terminating in a lateral arm. ×38,000 *Right*: the *arrows* indicate five cross-sectional profiles of triplet tubules embedded in dense material, representing a half of a centriolar wall. A longitudinal profile of such microtubular sheaves is present nearby (*I*). *F*, cross-striated fibre. ×60,000. LIN, 1970

some species, e.g. the mole (PEVET and COLLIN, 1976) and the noctule bat (PEVET et al., 1977b), structures reminiscent of remnants of outer segments can be seen, i.e. cilia of the 9+0 pattern, it may well be that the transformation of all or some centrioles (see below) into microtubular sheaves may represent an abortive attempt to produce an outer segment or at least its cilium.

The view that the centriolar changes are progressive rather than regressive in nature has been clearly substantiated in the rat, which in the adult state lacks typical centrioles in the pinealocytes. While in fetal (CLABOUGH, 1973) and newborn (ZIMMERMAN and TSO, 1975) rats the centrioles still display the conventional fine structure, between 15 days and 6 weeks of age they become elongated, in some instances doubling their lengths (LIN, 1969, 1970). Concurrently, small amounts of dense fine-grained material accumulate at uniform intervals along the outer surface of the centriole. Moreover, portions of the centriolar wall slant laterally, due to tilting of their microtubules, giving rise to microtubular sheaves which may remain in continuity with either normal or modified centrioles (see also BUCANA et al., 1974) (Figs. 76, 77). In guinea-pig

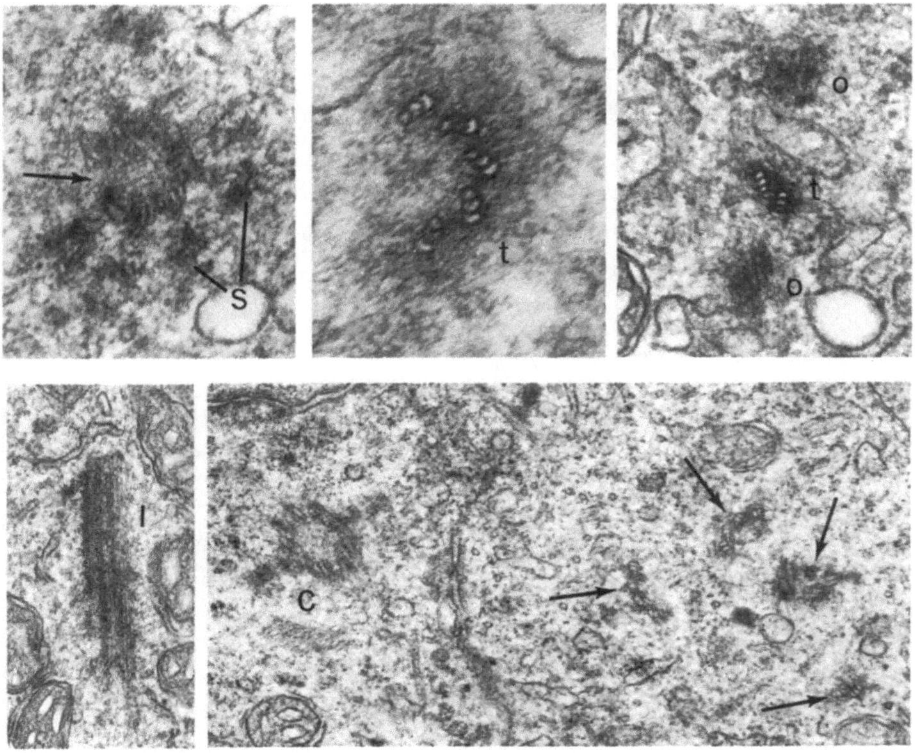

Fig. 77. *Top left:* Nearly transverse section of a centriole from a 32-day-old rat, showing a defect (*arrow*) in the wall. *S*, satellite body; ×70,000 *Top centre:* microtubular sheaves in transverse (*t*) section. ×122,000 *Top right:* microtubular sheaves in transverse (*t*) and oblique (*o*) section. ×60,000 *Bottom left:* microtubular sheaves in longitudinal (*l*) section. ×40,000 *Bottom right:* portions of two pinealocytes from a 30-day-old rat. The cell on the right contains fragments of centriolar wall or microtubular sheaves (*arrows*) in cross or oblique section. The other cell has an unaltered centriole (*C*) with satellite arms. ×44,000. Lin, 1970

pinealocytes, in which both typical and modified centrioles are present, one end of the elongated centriole may remain unbroken, while the other end may split into microtubular sheaves arranged in straight or skewed arrays (Lin, 1972). In adult guinea-pigs, the modified centrosomal area in many instances included one or more striated fibres similar to the striated rootlet fibres of the ciliary apparatus (Lin, 1972) (Fig. 78). A ciliary rootlet in continuity with a centriole has also been noted in the hedgehog (Pevet and Sabourea, 1973).

Whether or not the modification of the centrioles has something to do with the fact that the pinealocytes do not divide mitotically (see p. 75) and synthesize melatonin, which is known to arrest mitosis (Banerjee and Margulis, 1973), remains to be established. The finding that melatonin administration increased the number of procentrioles (and microtubules) in rat pinealocytes (Freire and Cardinali, 1975) points, however, to the opposite.

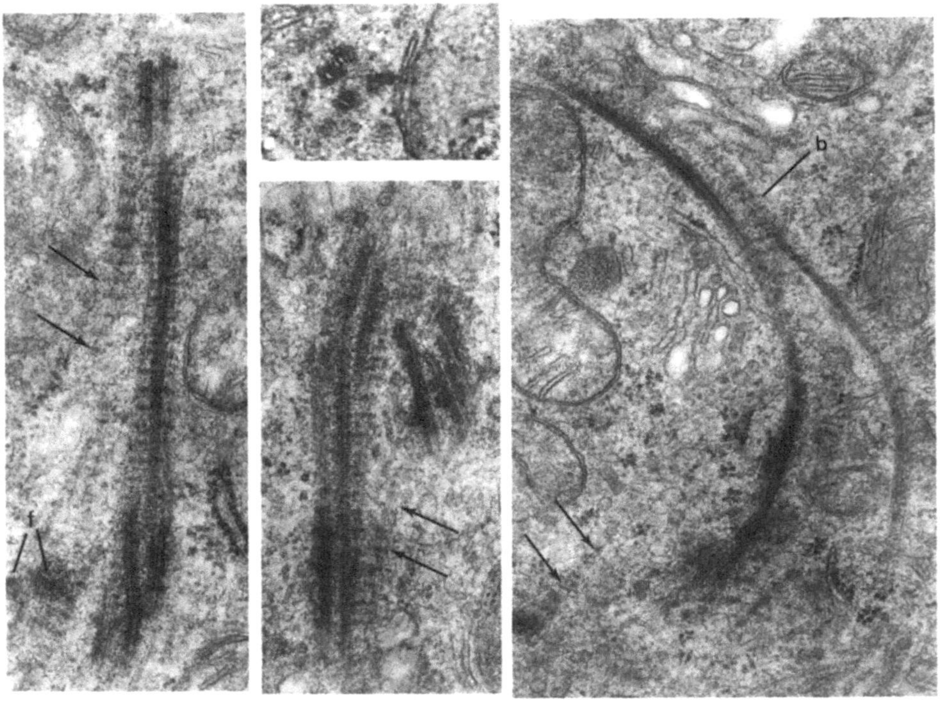

Fig. 78. *Left* and *centre bottom*: Longitudinal sections of microtubular sheaves with associated striped borders. Striated fibres are also discernible, some of them apparently continuous with the striped borders (*arrows*). Dense patches (*f*) may represent obliquely sectioned striated fibres. ×45,000 *Centre top*: cross-sectional view of a loosely arranged microtubular sheaf with dense material surrounding the tubules. ×54,000 *Right*: two curving microtubular sheaves. *b*, striped border; *arrows*, striated fibre. ×45,000. LIN, 1972 (reduced to 75%)

Cilia. The cilia of mammalian pinealocytes have recently attracted special attention as they have been thought to provide, at least in some species, morphological evidence for differentiation of mammalian pinealocytes from photoreceptor cells, whose outer segments are regarded as modified cilia showing a 9+0 pattern. Cilia have been found to be numerous in pinealocytes of noctule bats (PEVET et al., 1977b), mole-rat (PEVET et al., 1967b) and mole (PEVET, 1974), in which species it has even been claimed (PEVET and COLLIN, 1976) that *each* pinealocyte possesses a 9+0 cilium. Particularly interesting is the observation that in noctule bats the cilia occasionally expanded distally to form club-shaped structures reminiscent of rudimentary outer segments of lizards and birds (PEVET et al., 1977b). Moreover, concentric lamellar structures vaguely resembling discs of outer segments were relatively often seen in the vicinity of the cilia, though no actual connection of the two structures could be detected (PEVET and COLLIN, 1976; PEVET et al., 1977b).

Modified cilia, where the plasma membrane encompassing the rather short shaft was expanded into a bulbous tip in which a few short filaments were not in register with those of the nine peripheral pairs, have been observed also in adult but not juvenile cows and sheep (ANDERSON, 1965). In rats distally expanded cilia have been noted in fetal (CLABOUGH, 1970b, 1973) and early postnatal (ZIMMERMAN and TSO, 1975) pinealocytes. In adult rats cilia appear to be very rare as indicated by the fact that they were either not found at all (WOLFE, 1965; LIN, 1970) or were few in number (ARSTILA, 1967; KARASEK, 1970). In cow, sheep (ANDERSON, 1965), dog (SANO and MASHIMO, 1966; WELSER et al., 1968), pocket gopher (SHERIDAN and REITER, 1973), hedgehog (PEVET and SABOUREAU, 1973), golden mole (PEVET and KUYPER, 1978) and ground squirrel (POVLISHOCK et al., 1975) cilia were occasionally seen. In cats, they appear to be more numerous (DUNCAN and MICHELETTI, 1966) than in the species mentioned above.

The ultrastructural appearance of pinealocyte cilia has not always been specified, so that it is not known to which pattern most of them conform. It is clear, though, that not all the cilia observed exhibit a 9+0 pattern and that different types may be found in a given species. While in adult rats a 9+1 pattern was observed (KARASEK, 1970), neonatal rats possessed 9+0 cilia (LIN, 1970). In dogs, 9+2 cilia were seen (SANO and MASHIMO, 1966; WELSER et al., 1968).

Microtubules and Microfilaments. These are often characteristic features of mammalian pinealocytes. It has to be noted, however, that they vary greatly in amount and localization. Microtubules usually exceed microfilaments in amount and it would appear that microtubules are particularly abundant in pinealocyte processes, the ultrastructure of which will be dealt with separately (see p. 161).

The presence of fine microtubular canaliculi, resembling those in nerve fibres was already noted in what was apparently the first electron microscopic study of the mammalian pinealocyte (rat: MILOFSKY, 1957). Later it became clear that two forms of microtubules have to be distinguished, namely "ordinary" microtubules and microtubules arranged in sheaves (microtubular sheaves or bouquets, WOLFE, 1965) which, as pointed out above (see p. 140), are centriolar derivatives (WOLFE, 1965; LIN, 1970, 1972; BUCANA et al., 1974).

Microtubular sheaves (Figs. 75–78) were first noted in pinealocytes of rats (WOLFE, 1965), in which they were typically composed of seven to ten skewed microtubules. ARSTILA (1967) referred to these structures as microcylinders, apparently to distinguish them from the ordinary microtubules, which do not form similar sheaves. Microtubular sheaves were also found in guinea-pig (LIN, 1972), gerbil (WELSH and REITER, 1978) and hamster (BUCANA et al., 1974), but were apparently absent in mouse and monkey (LIN, 1970).

The gross ultrastructural appearance of the sheaves varied considerably, as shown by the fact that straight, slightly bent and strongly curved structures have been observed (WOLFE, 1965; LIN, 1970, 1972; BUCANA et al., 1974). Occasionally, the sheaves branched (LIN, 1972).

The fine structural examination of the microtubular sheaves showed that their microtubules, which measure 3–4 μm (ARSTILA, 1967) or 2.4 μm in length (LIN, 1970), exhibit roughly the same diameter as ordinary microtubules (see p. 145). They are usually arranged in doublets or triplets (ARSTILA, 1967; LIN, 1970, 1972; BUCANA et al., 1974) and close inspection reveals that the apposed microtobules of a unit share a common wall along the area of contact, similar to the triplet tubules of the centriole (LIN, 1970).

A characteristic feature of some microtubular sheaves is the presence of a denticulate border (WOLFE, 1965), consisting of tooth-, spur- or cog-like structures at regular intervals (LIN, 1970; BUCANA et al., 1974). In the guinea-pig, the denticulate border was apparently much better developed than in the rat (LIN, 1972). In many instances the denticulate border of rat microtubular sheaves displayed alternating thick and thin projections (LIN, 1970). In the guinea-pig (Fig. 78), the denticulate border possessed transverse bands at about 70 nm intervals and was made up of very fine filaments running parallel to the long axis of the microtubular sheaf (LIN, 1972). The nature of the lateral spurs of microtubular sheaves has been interpreted in different ways. Originally, they were thought to be due to the presence of short microtubules rather than centriolar satellites (WOLFE, 1965), but LIN (1970) thought that they seemed to bear some resemblance to satellite bodies in their texture.

The exact functional significance of the microtubular sheaves is obscure. As pointed out above (p. 140), they may have something to do with an abortive attempt to produce outer segments. There is no indication that the sheaves, once formed, increase or decrease in number with advancing age. In a 27-month-old rat, the pinealocytes still contained the same numbers of microtubular sheaves, whereas other organelles, especially the endoplasmic reticulum, showed a remarkable diminution (LIN, 1970). As can be expected from the fact that microtubular sheaves are derived from centrioles, they are mainly localized in the perikarya (WOLFE, 1965; ARSTILA, 1967; LIN, 1970; BUCANA et al., 1974), where they usually assemble in an ill-demarcated area of the cytoplasm around which two or more profiles of the Golgi complex are deployed (LIN, 1972). The observation that microtubular sheaves may be situated in a cytoplasmic process would suggest that these structures are not stationary but capable of migration (LIN, 1970). Other elements that may appear in proximity to sheaves include tubular or vesicular profiles of the agranular endoplasmic reticulum, free ribosomes and cytoplasmic microtubules which may converge upon a dense mass identical with centriolar satellites (LIN, 1970). Less frequently found was a feltwork of delicate filaments and cross-striated fibres reminiscent of ciliary rootlet fibres (LIN, 1970), which may have been formed by the denticulate border of the microtubular sheaves.

Ordinary *microtubules*, in contrast to microtubular sheaves, show great inter-species variation in pinealocyte perikarya. They are abundant in cow and sheep (ANDERSON, 1965) and moderately abundant in rat (ARSTILA, 1967; FREIRE and CARDINALI, 1975), deep pineal of hamster (SHERIDAN and REITER, 1970b), seal (CUELLO, 1973), mole (PEVET, 1974), ground squirrel (POVLISHOCK et al., 1975), and in dark pinealocytes of the gopher (SHERIDAN and REITER, 1973). In the rabbit, microtubules were noted near centrioles (LEONHARDT, 1967; ROMIJN, 1973b), implying that they are rare in other regions of the perikaryon. Microtubules are rare in pinealocyte perikarya of cat and monkey (WARTENBERG, 1968) and perhaps virtually non-existent in guinea-pig (LUES, 1971), hedgehog (PEVET and SABOUREAU, 1973) and light pinealocytes of the gopher (SHERIDAN and REITER, 1973).

The *location* of the microtubules within the perikarya varies. A mainly perinuclear location was found in Long-Evans rats (ARSTILA, 1967). The rare microtubules observed in cat and monkeys were situated near the plasma membrane (WARTENBERG, 1968). In some species the microtubules swerved from the perikaryon into the cell processes (ANDERSON, 1965; SHERIDAN and REITER, 1970b). In the hamster (SHERIDAN and REITER, 1968) and ground-squirrel (POVLISHOCK et al., 1975) microtubules were especially numerous at the site of origin of the processes from the perikaryon. In the other species investigated the microtubules are apparently randomly scattered in the perikaryon.

Fine structurally, the microtubules do not possess any unusual features. Their diameters were given as 20 nm in hamster (SHERIDAN and REITER, 1968) and gopher (SHERIDAN and REITER, 1973), 20–25 nm in rat (WOLFE, 1965; ARSTILA, 1967; ARSTILA and HOPSU-HAVU, 1967), about 25 nm in hedgehog (PEVET and SABOUREAU, 1973), 26 nm in the seal (CUELLO, 1973) and between 22 and 30 nm in cow and sheep (ANDERSON, 1965). Unequivocal instances of bifurcation or interanastomosis of microtubules, or of merging with other cytoplasmic organelles or with the plasmalemma were not encountered in the rat (WOLFE, 1975). However, in cow and sheep longitudinally arranged microtubules whose interior appeared to be confluent with vesicular and elongated structures were sometimes observed (ANDERSON, 1965). Moreover, they appeared to possess a banded (WELSER et al., 1968) or striated pattern, particularly following fixation in glutaraldehyde, postfixation in osmium and subsequent staining with uranyl acetate and lead citrate (ANDERSON, 1965).

Microfilaments, which are less common than microtubules, show great interspecies variation with respect to amount, location and arrangement. They are rare in the rat (WOLFE, 1965; ARSTILA, 1967; ARSTILA and HOPSU-HAVU, 1967) and distinctly more numerous in the guinea-pig (LUES, 1971), pocket gopher (SHERIDAN and REITER, 1973) and mole-rat (PEVET et al., 1976b).

Owing to the scarcity of the microfilaments, it is understandable that the results obtained regarding their intracellular location differ. In the rat, microfilaments were first noted to be restricted to the perikaryon (WOLFE, 1965). Later, ARSTILA (1967) thought that they were restricted to the pineal processes, but eventually ARSTILA and HOPSU-HAVU (1967) decided that filaments were usually located in the perikaryon and in some instances also in the cell processes. In the gopher (SHERIDAN and REITER, 1973) microfilaments were confined to the perikaryon and in the guinea-pig, usually to the processes (LUES, 1971).

Microfilaments formed either loose (WOLFE, 1965) or compact bundles, measuring 1–2 µm in diameter, in which up to 30 filaments lay parallel and tightly packed (ARSTILA and HOPSU-HAVU, 1967). In the pocket gopher (SHERIDAN and REITER, 1973) the filaments either lay singly or formed bundles. Microfilaments measured 6 nm (CUELLO, 1973, seal) or 7–8 nm in width and 0.5–1.0 µm in length (ARSTILA and HOPSU-HAVU, 1967). Some filaments resembling rootlet fibres of the ciliary apparatus showed a cross-banding with periodicities of approximately 60 nm (SHERIDAN and REITER, 1973; gopher) or 70 nm (rat: WOLFE, 1965; guinea-pig: LIN, 1972).

The functional significance of the microtubules and microfilaments of the

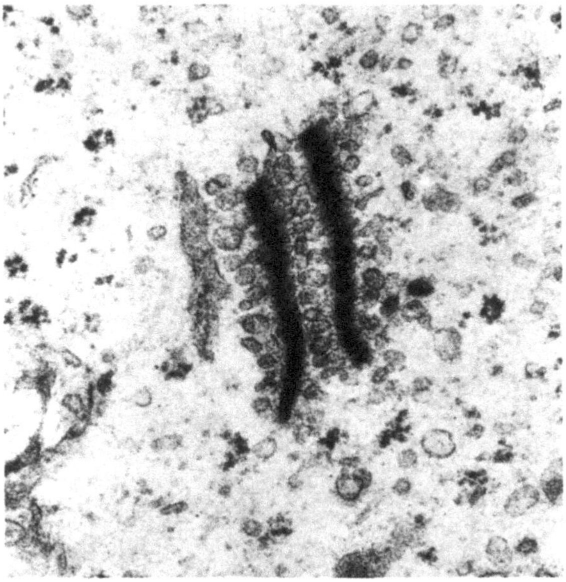

Fig. 79. Group of synaptic ribbons, each ribbon consisting of an electron-dense rod surrounded by electron-lucent vesicles. Guinea-pig pineal organ. VOLLRATH and HUSS, 1973; ×61,000

mammalian pinealocyte has not been investigated. Their presence in both perikarya and cell processes points to a similarity with nerve cells and astrocytes. In view of the secretory nature of the pinealocytes they may play a role in the transport and release of substances, similar to other mammalian secretory cells. In this context it is interesting to recall that melatonin, by interacting with microtubular protein, arrests mitosis (BANERJEE and MARGULIS, 1973). In the rat pineal, administration of melatonin led to a large increase in the amounts of microtubules and assembly areas (FREIRE and CARDINALI, 1975). That the microtubules may be somehow related to secretory processes in the pineal is suggested by the fact that they are particularly abundant during darkness, when melatonin secretion is enhanced (FREIRE and CARDINALI, 1975). Another possibility is that these structures are involved in the contraction and relaxation of individual pinealocytes, perhaps leading to widening and narrowing, respectively, of the intercellular spaces (for further details, see p. 173) in the pineal.

Synaptic Ribbons. In addition to the microtubular sheaves, mammalian pinealocytes comprise another cell organelle which is not commonly found in mammalian cells, the so-called synaptic ribbons (Fig. 79). These structures have attracted special attention in the past several years, not only because they are "possible reminders of the original neurosensory function of the cell", as suggested by KAPPERS (1971b), but also because they have been shown to react under various experimental (VOLLRATH and HUSS, 1973; VOLLRATH and HOWE, 1976; KARASEK, 1976) and physiological conditions (VOLLRATH, 1973; KURUMADO and MORI, 1977). This has led to the hypothesis that synaptic ribbons in the mammalian pineal may be involved in intercellular communication between adjacent pinealocytes (VOLLRATH and HUSS, 1973; VOLLRATH, 1973), a concept previously expressed by HOPSU and ARSTILA (1965) on purely morphological grounds. Although we are far from knowing the chemical substances involved and whether or not the ribbons are indeed synaptic in function, it has been suggested that they may be a device to synchronize the function of pinealocytes not lying close to sympathetic nerve fibres or alternatively to establish interpinealocytic circuits which could enable the pineal to act not merely as a subordinate neuroendocrine transducer governed by impulses from sympathetic nerve fibres but to play a more active and independent role in neuro-endocrine integration (VOLLRATH and HUSS, 1973).

The concept that synaptic ribbons in the mammalian pineal may be involved in intercellular communication was triggered by the fact that similar structures occur at sites where they are doubtlessly synaptic in function (retina: SJÖSTRAND, 1958; DOWLING and BOYCOTT, 1966; organ of Corti: C. SMITH and SJÖSTRAND, 1961; vestibular organ: WERSÄLL et al., 1965; electric receptor organs of teleost fishes: SZAMIER and WACHTEL, 1970; SZABO and WERSÄLL, 1970; pineal organs of lower vertebrates: see below). Some authors strictly avoid the adjective "synaptic" for the ribbons present in mammalian pinealocytes and prefer to call them "ribbons surrounded by vesicles" (COLLIN, 1969b; PEVET and COLLIN, 1976; PEVET et al., 1977b) to distinguish them from the synaptic ribbons in pineals of lower vertebrates in which they form part of the synapses between photoreceptor cells and afferent nerve fibres.

Before going into the details of synaptic ribbons, it should be noted that in the mammalian pinealocyte two types of "synaptic" structures are present, namely synaptic ribbons and synaptic spherules. Whereas the occurrence of synaptic ribbons, also termed synaptic spines (ARSTILA and HOPSU, 1964), vesicle-crowned rodlets (WOLFE, 1965), vesicle-crowned lamellae (ARSTILA, 1967) or ribbons surrounded by vesicles (COLLIN, 1969b) has been known for a long time, the presence of synaptic spherules, as distinct from the ribbons, does not appear to have been fully appreciated. One of the reasons for this may be that transversely cut synaptic ribbons bear some resemblance to synaptic spherules. That the two types should be regarded as distinct entities is suggested by comparative studies on the hair cells of the vestibular organ: At identical sites, these cells contain synaptic ribbons in mammals and synaptic bodies (=spherules) in elasmobranchs, teleosts, lizards and birds (WERSÄLL and BAGGER-SJÖBÄCK, 1974). Relatively little is currently known about the synaptic spherules of mammalian pinealocytes. In the rabbit they have been described as "rosette-like structures" (LEONHARDT, 1967; ROMIJN, 1973b, 1975a). They consist of a round or flattened mass of electron-dense material which measures 120 nm in diameter and is surrounded by a mostly single layer of 8–12 electron-lucent vesicles, 30 nm in width. Since they were sometimes seen in the immediate vicinity of the plasmalemma and the electron-dense core was apparently discharged into the intercellular space, the material was regarded as secretory substances of the pinealocytes (LEONHARDT, 1967). In the guinea-pig, synaptic spherules have been described as "vesicle-crowned balls" (LUES, 1971), the electron dense core measuring 90–130 nm in width. That the spherules are not just transverse sections through ribbons is shown by the observation that the

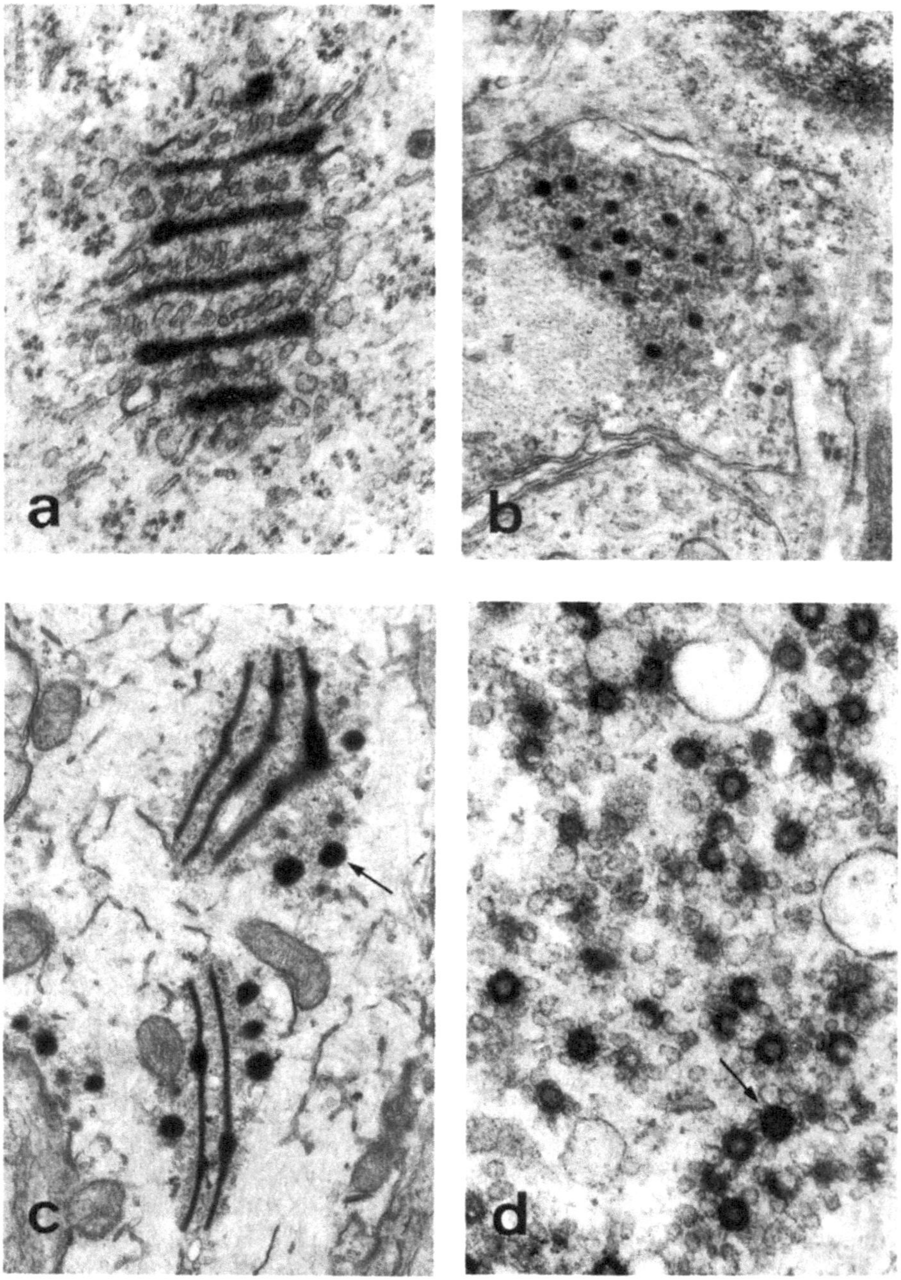

Fig. 80. a Synaptic ribbons showing terminal enlargements. ×57,000 **b** Cluster of synaptic spherules. ×24,500 **c** Pictures which have been interpreted as showing that ribbons and spherules may fuse. ×21,000. **d** Cylinder-like structures which may be precursors of spherules. ×57,000. Lues, 1971

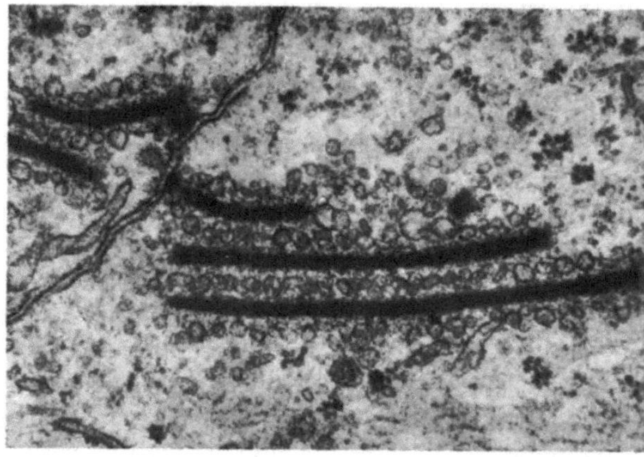

Fig. 81. Two groups of synaptic ribbons lying in the immediate vicinity of two plasmalemmata. VOLLRATH and HUSS, 1973; ×61,000

dense core of the ribbons is distinctly smaller (30–40 nm) in width (LUES, 1971). Spherules and ribbons of guinea-pig pinealocytes are depicted in Fig. 80a and b. Figure 80c shows that spherules and ribbons may fuse. Moreover, guinea-pig pinealocytes have been shown to contain cylinder-like structures (Fig. 80d) which, it has been suggested, may be precursors of the spherules (LUES, 1971).

Whether or not synaptic spherules and synaptic ribbons are functionally identical in the mammalian pineal remains to be elucidated. The observation that in pregnant guinea-pigs synaptic ribbons showed a slight increase and synaptic spherules a strong increase in number (LUES, 1971) seems to suggest a functional dissimilarity. The fact that in rabbits with sympathectomized pineals the spherules first increased in amount, followed by the synaptic ribbons, points to the same conclusion (ROMIJN, 1975a).

Synaptic ribbons, lying singly or in groups (Fig. 79) and often, but not exclusively, in the immediate vicinity of a plasmalemma (Fig. 81), consist of an electron-dense rod of varying length (see below) and of vesicles surrounding the rod. It appears that they are present in all species investigated, and although it is unclear whether they are *regularly* present in the deep pineals (p. 28), their presence in the deep pineal of the hamster has been substantiated by HEWING (1978).

The concept that synaptic ribbons may have something to do with intercellular communication was derived from an analysis of their localization in relation to cell membrane, their topographical relationships to adjacent structures and their numerical changes under experimental and physiological conditions. An analysis of the localization of the ribbons in both pinealocyte perikarya and processes in relation to the cell membrane in guinea-pigs has shown that on average between 70% and 80% of the single or clustered synaptic ribbons lie close to the cell membrane (VOLLRATH, 1973). Taking into consideration that at least some of the ribbons which do not appear to border a cell membrane are related to a plasmalemma lying in a different plane of the tissue block, it may be assumed that most of the ribbons are indeed closely linked to the cell membrane. This close association is also illustrated by the fact that occasionally the central electron-dense core of the ribbon is connected to the cell membrane by a stalk which widens before attachment (rat: ARSTILA, 1967). Direct contacts of the electron-dense rod of the ribbons with the cell membrane were usually observed in the rat (ARSTILA and HOPSU, 1964) and sometimes in the cat and monkey (WARTENBERG, 1968). Thickenings of the cell membrane were absent in most instances (WARTENBERG, 1968), but LUES (1971) noted an occasional desmosome-like appearance. At sites where subsurface cisternae (see p. 122) lie close to a synaptic ribbon, the cisternae may contain electron-dense material similar to that of the ribbon (Fig. 82).

Analysis of the topographical relationships of the ribbons of both perikarya and processes to neighbouring structures revealed that 93.1% of the ribbons faced adjacent pinealocytes, 5.3% lay near blood vessels, 1.3% faced nerve fibres and 0.3% were spatially related to collagenous fibres (VOLLRATH, 1973). In the majority of cases the ribbons lay opposite a pinealocyte process (VOLLRATH, 1973), which, when one applies the terminology used by HOPSU and ARSTILA (1965), means that "axo-axonic" and "somato-axonic" contacts are far more frequent than "axo-somatic" and "somato-somatic" contacts.

Synaptic ribbons of pinealocytes came into the limelight when it was shown that they increase strikingly in number when guinea-pigs are exposed to continuous illumination for up to 70 (LUES, 1971) or 64 days (VOLLRATH and HUSS, 1973), and that they undergo changes in shape and localization, resulting in the formation of so-called paired ribbon fields (see below). These observations, which demonstrated that synaptic ribbons are not so rare as to defy quantification (PELLEGRINO DE IRALDI and SUBURO, 1971a) or that it is difficult to ascribe a role to them (PEVET and SABOUREAU, 1973), triggered studies as to the presence or absence of circadian changes. As shown in Figs. 83 and 84, the amounts of synaptic ribbons show a clear 24-h rhythm in both guinea-pigs (VOLLRATH, 1973) and rats (KURUMADO and MORI, 1977), with decreasing values during the morning, an increase late in the afternoon prior to the onset of darkness, and high values during the dark phase, which roughly corresponds to the melatonin-forming activity of the pineal (RUDEEN et al., 1975). In the baboon, *Papio ursinus*, synaptic ribbons also increased during the night (THERON et al., 1979).

The concept that synaptic ribbons may be functionally important structures has gained support from experimental studies involving exposure to cold (KRSTIĆ, 1973), immobilization (MILINE et al., 1968), sympathectomy (ROMIJN, 1975a), orchidectomy (KARASEK, 1976), administration of drugs known to influence pineal function (ROMIJN, 1976; VOLLRATH and HOWE, 1976) and exposure to continuous illumination (rabbit: ROMIJN, 1975a; dormouse: ROUX et al., 1977), corroborating the findings obtained in guinea-pigs (VOLLRATH and HUSS, 1973).

The precise function of the synaptic ribbons in mammalian pinealocytes remains to be established; unfortunately it is also not known in other organs. For the retina it has been suggested that the electron-dense rods may guide the surrounding synaptic vesicles to the presynaptic membrane (BUNT, 1971; GRAY and PEASE, 1971). In pinealocytes they could have a similar function since a large proportion of the ribbons are arranged perpendicular to the plasmalemma (KARASEK, 1976b), favouring the transport of vesicles in a conveyor-belt fashion (BUNT, 1971).

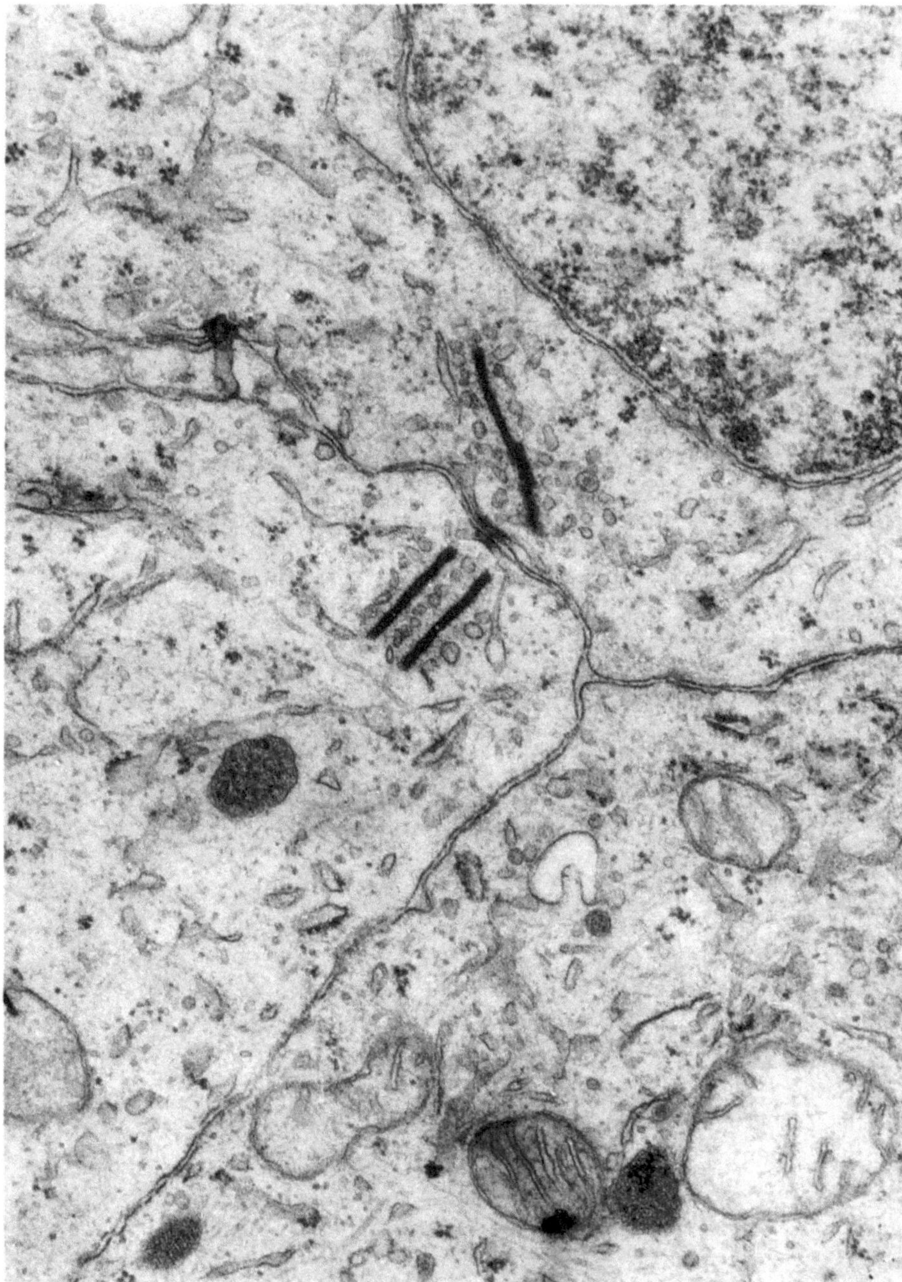

Fig. 82. Synaptic ribbons lying near subsurface cisterns which in certain areas contain electron-dense material similar to that of the ribbons. Guinea-pig pinealocytes. VOLLRATH, previously unpublished electron micrograph; × 61,500 (reduced to 87%)

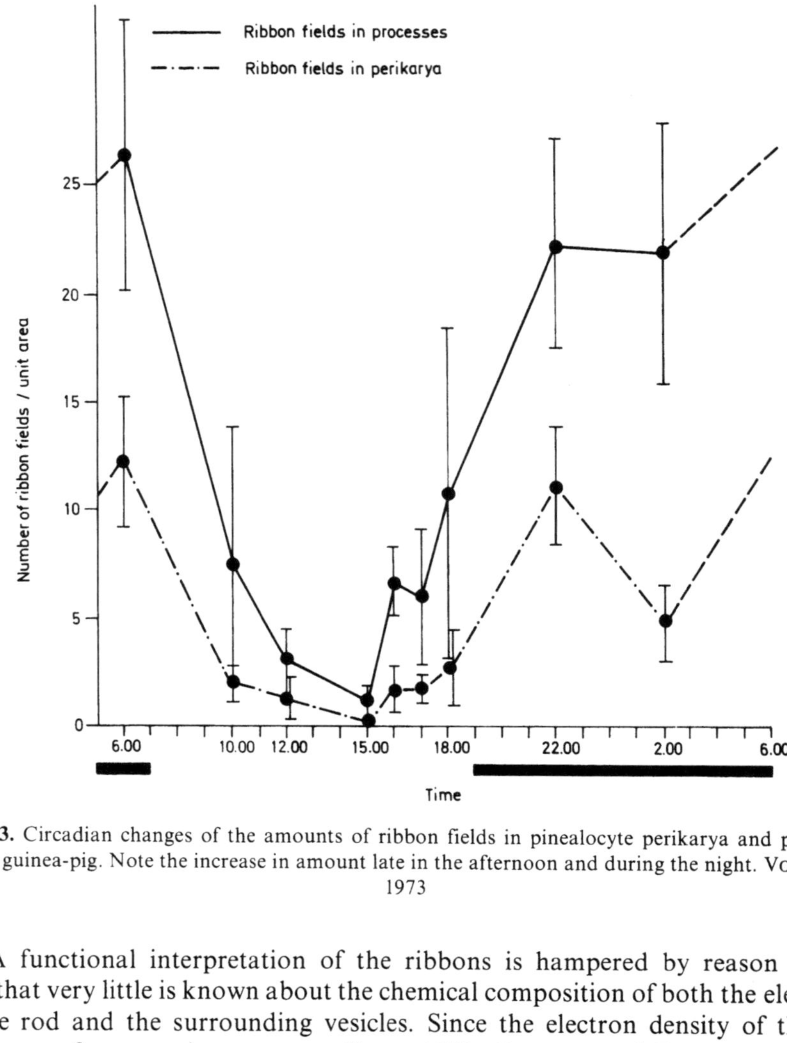

Fig. 83. Circadian changes of the amounts of ribbon fields in pinealocyte perikarya and processes in the guinea-pig. Note the increase in amount late in the afternoon and during the night. VOLLRATH, 1973

A functional interpretation of the ribbons is hampered by reason of the fact that very little is known about the chemical composition of both the electron-dense rod and the surrounding vesicles. Since the electron density of the rod decreases after reserpine treatment (LUES, 1971; OSBORNE and THORNHILL, 1972), it has been suggested that the rod may be involved in the storage of monoamines (OSBORNE and THORNHILL, 1972). It consists of proteins and lacks carbohydrates and detectable quantities of sodium and calcium ions (KRSTIĆ, 1975d, 1976a). In other organs the rod consists likewise of proteins, lacking DNA and RNA (BUNT, 1971; SZAMIER, 1974).

The vesicles, surrounding the rods in a single or occasionally a double row (LUES, 1971) and measuring between 30 and 50 nm in width (WOLFE, 1965; ARSTILA, 1967; LUES, 1971; ROMIJN, 1973b; PEVET and SABOUREAU, 1973; WELSH and REITER, 1978), resemble synaptic vesicles of cholinergic neurons. There is, however, no indication for the presence of acetylcholinesterase activity (ARSTILA, 1967; ERÄNKÖ et al., 1970a, b), which would rule out cholinergic mechanisms. Although occasionally some of the vesicles may contain a dense core (ARSTILA, 1967; ROMIJN, 1973b), cytochemical studies have clearly revealed

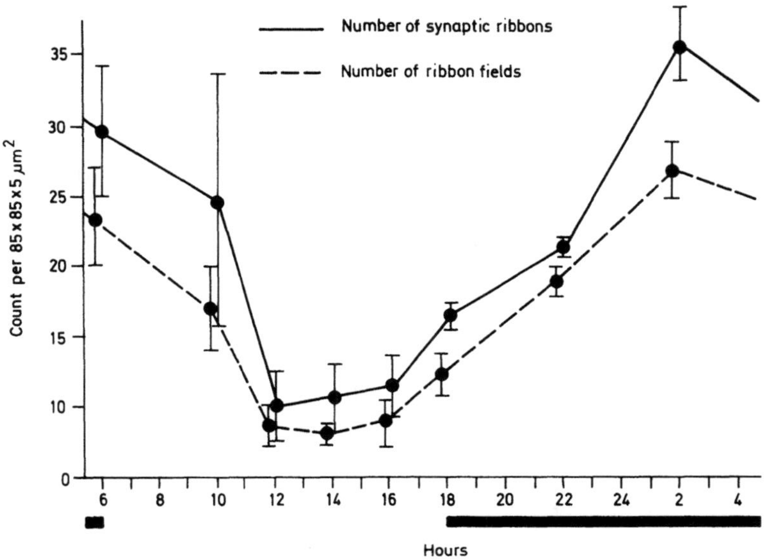

Fig. 84. Circadian changes of ribbon fields and synaptic ribbons in the rat pineal gland. Compare with Fig. 83 and note the corresponding behaviour. KURUMADO and MORI, 1977

that the vesicles differ from those present in sympathetic nerve fibres (KRSTIĆ, 1976a). Whether or not the vesicles contain GABA, as suggested by KRSTIĆ (1976a), remains to be seen.

Despite their prominence and their characteristic changes under various physiological and experimental conditions, the synaptic ribbons of mammalian pinealocytes still defy a clear-cut functional interpretation. Even if one assumes that they are somehow related to intercellular communication, the question as to whether they are excitatory or inhibitory in function cannot be answered with certainty, especially since the results obtained are not compatible: synaptic ribbons are abundant after exposure to continuous illumination and rare during the light phase of a 24-h cycle. Considering all the results, it is not unreasonable to suspect an excitatory function of the ribbons required for an enhanced secretory activity of the pineal. Synaptic ribbons increase prior to, and during, phases of high melatonin synthesis. After exposure to continuous illumination and after sympathectomy, when melatonin formation is reduced, the ribbons increase in number to perhaps counteract the inhibition of the pineal.

While it is basically not difficult to envisage that synaptic ribbons of mammalian pinealocytes may have a similar function to the ribbons in other organs, there are a number of structural features in which they differ from one another. Hence, some caution is necessary when extrapolating.

An interesting but functionally obscure finding is that synaptic ribbons of mammalian pinealocytes tend to form clusters, also termed "ribbon fields" (VOLLRATH and HUSS, 1973; KARASEK, 1976b; KURUMADO and MORI, 1977) (Figs. 79–82). In the guinea-pig, out of 1,400 ribbon sites investigated, roughly 60% comprised only one synaptic ribbon and the remainder, 2 to 12 (VOLLRATH

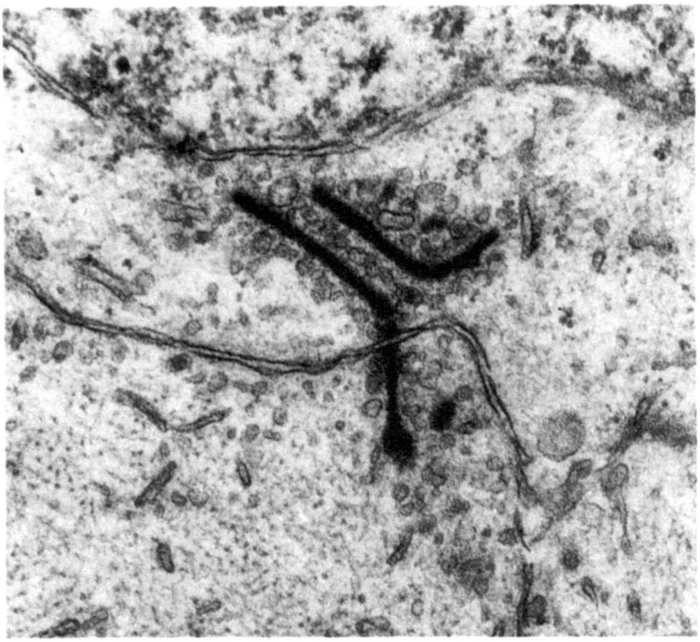

Fig. 85. Synaptic ribbons lying directly opposite one another in adjacent pinealocytes. Guinea-pig. VOLLRATH, 1974; ×61,500

and HUSS, 1973). In another study, a field with as many as 21 ribbons was seen (VOLLRATH, 1973). Although it is not known how the ribbons are formed (see below), it appears that the grouping pattern may have something to do with the age of the ribbons, as shown by the fact that singly lying ribbons are particularly abundant when the ribbons and ribbon fields increase in number. By contrast, ribbon fields comprising two and more ribbons predominate in phases when the ribbons and fields are rare and probably old (VOLLRATH, 1973). Considering the numerical changes of the ribbons over a period of 24 h (Figs. 83, 84), one may conclude that the ribbons have an average life-span of 8 h, since (i) it takes roughly this time (from 7.00 to 15.00 h) until most of the ribbons have disappeared and (ii) there is a distinct drop in the number of ribbons at 2.00 h, i.e. 8 h after they had started to increase (VOLLRATH, 1973). Whether or not the ribbons lying in groups are reserve structures which may eventually move to the cell membrane (VOLLRATH and HUSS, 1973) remains to be established.

Another striking feature of the ribbons in mammalian pinealocytes is that they may lie directly opposite one another in adjacent pinealocytes giving the impression (Fig. 85) of being a single structure intersected by the plasmalemmata of the adjacent pinealocytes. The functional significance of such an arrangement, termed "paired ribbon fields" (VOLLRATH and HUSS, 1973), is difficult to envisage. They are rare under physiological conditions (HOPSU and ARSTILA, 1965; ARSTILA, 1967; LUES, 1971; VOLLRATH, 1973) and numerous after continuous illumination (VOLLRATH and HUSS, 1973). Whether they function as a bidirectional intercellular link, in contrast to the unidirectional link of ribbons lying on one side of the plasmalemma only (VOLLRATH and HUSS, 1973), or whether they neutralize each other remains to be elucidated.

An unsolved morphological problem of the synaptic ribbons is their three-dimensional structure. In sections, the electron-dense rod shows great variation. Its length measured up to 0.2 μm in gerbil (WELSH and REITER, 1978), 0.65 μm in cat and monkey (WARTENBERG, 1968), 0.8 μm in the rat (WOLFE, 1965), 1 μm in the rabbit (ROMIJN, 1973b), 1–1.5 μm in the hedgehog (PEVET and SABOUREAU, 1973) and 2 μm in the guinea-pig (LUES, 1971). The width, by contrast, is fairly similar in all species examined, measuring between 25 and 45 nm (WOLFE, 1965; ARSTILA, 1967; WARTENBERG, 1968; LUES, 1971; ROMIJN, 1973b; PEVET and SABOUREAU, 1973; WELSH and REITER, 1978), though it should be noted that they may have circumscribed thickenings (Figs. 80c, 85). Longitudinally cut ribbons usually appear straight or slightly bent (WARTENBERG, 1968; LUES, 1971). Using serial sections, ARSTILA (1967) observed that the distal end occasionally coiled and came to lie parallel to the plasma membrane. Following long-term exposure to continuous illumination, guinea-pigs exhibited strongly bent (boomerang-shaped) ribbons (LUES, 1971; VOLLRATH and HUSS, 1973). U-shaped ribbons were found in rabbits (ROMIJN, 1973b) and in guinea-pigs under experimental conditions (LUES, 1971; VOLLRATH and HUSS, 1973).

Close inspection of the electron-dense rod at high magnifications reveals a varying appearance. In the majority of cases it is a homogeneous electron-dense structure consisting of fine-granular material. ARSTILA (1967) concluded that such an appearance is indicative of oblique or tangential sectioning, because in precisely horizontally sectioned ribbons a trilaminar structure was typical, consisting of three parallel membranes measuring ca. 7 nm in width. According to KRSTIĆ (1976d) (Figs. 86, 87), synaptic ribbons, as they appear in sections, may be profiles of different structures, namely synaptic rods and synaptic lamellae. As the terms imply, in one type of ribbon the central electron-dense core is rod-like, while it is lamella-like in the other, the lamella being either round or like a half-moon.

Origin and formation of the synaptic ribbons have not been clearly identified. Since they show a simultaneous increase in pinealocyte perikarya and processes, it has been suggested that they form in situ (VOLLRATH, 1973), although a very rapid transport from perikarya to processes cannot be excluded. KARASEK (1976) is of the opinion that the central rod forms from microtubular sheaves (p. 143), but convincing evidence is lacking. The origin of the vesicles surrounding the central rod or lamella is likewise obscure. Since the vesicles lack glucose-6-phosphatase and thiamine pyrophosphatase activities, they do not appear to originate from the endoplasmic reticulum and the Golgi apparatus (KRSTIĆ, 1975c, 1976a). However, in baboon pinealocytes it appears as though the vesicles of the synaptic ribbons originate from the smooth endoplasmic reticulum and that microfilaments may play a role in the movement of vesicles of the smooth endoplasmic reticulum to sites where new synaptic ribbons are being formed (THERON et al., 1979). The occurrence of synaptic ribbons during ontogenesis has been studied by HEWING (1979) in the hamster. For further details on

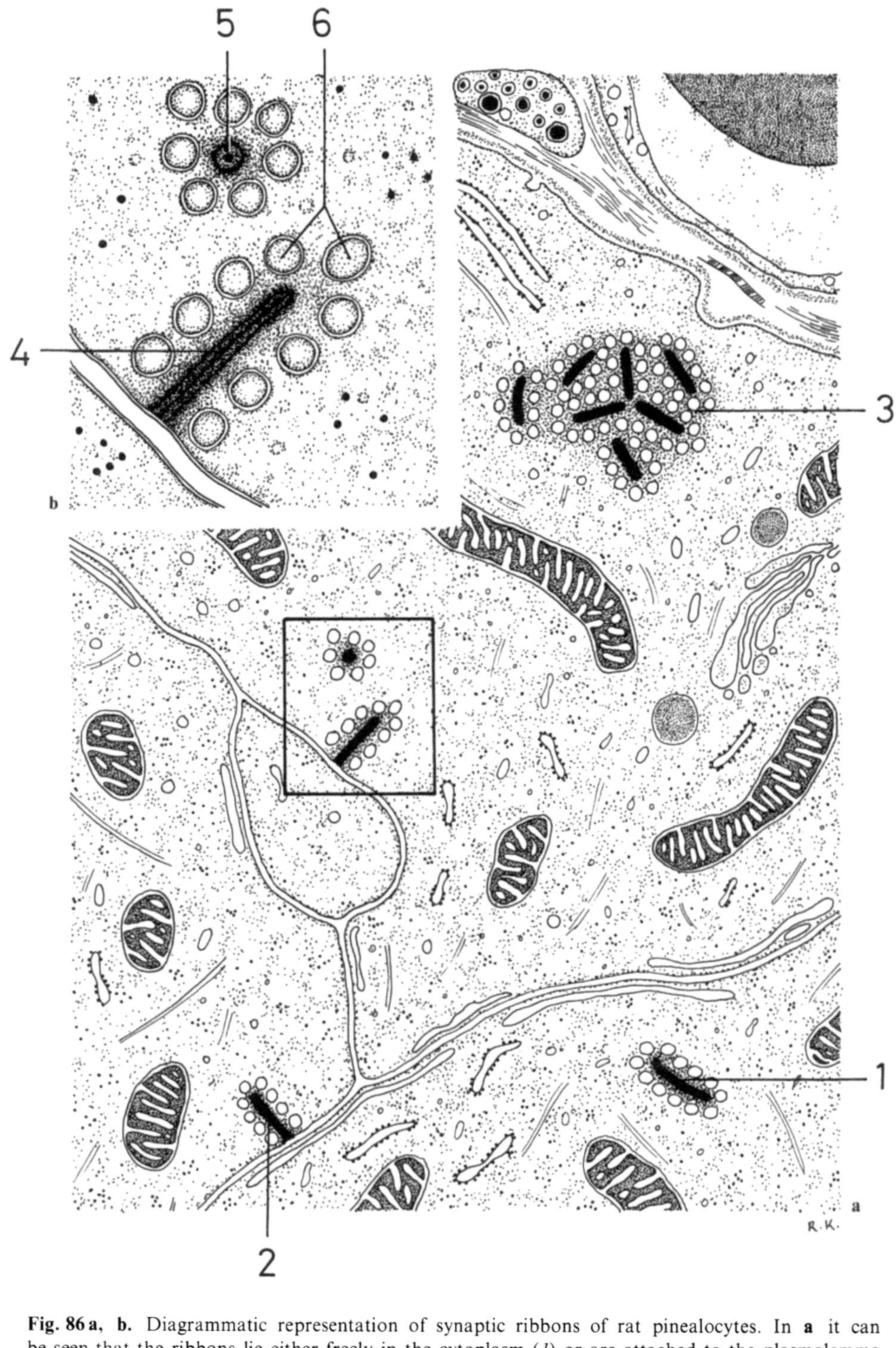

Fig. 86 a, b. Diagrammatic representation of synaptic ribbons of rat pinealocytes. In **a** it can be seen that the ribbons lie either freely in the cytoplasm (*1*) or are attached to the plasmalemma (*2*). Groups of synaptic ribbons are shown at (*3*). ×28,000. In **b** synaptic ribbons are shown as they appear in suitable sections at higher magnification (×90,000): three electron-dense lines (*4*) can be distinguished; in transverse sections they appear as an electron-dense circular structure comprising an electron-dense fibre in the centre (*5*). *6*, Synaptic vesicles. ×90,000. Krstić, 1976d

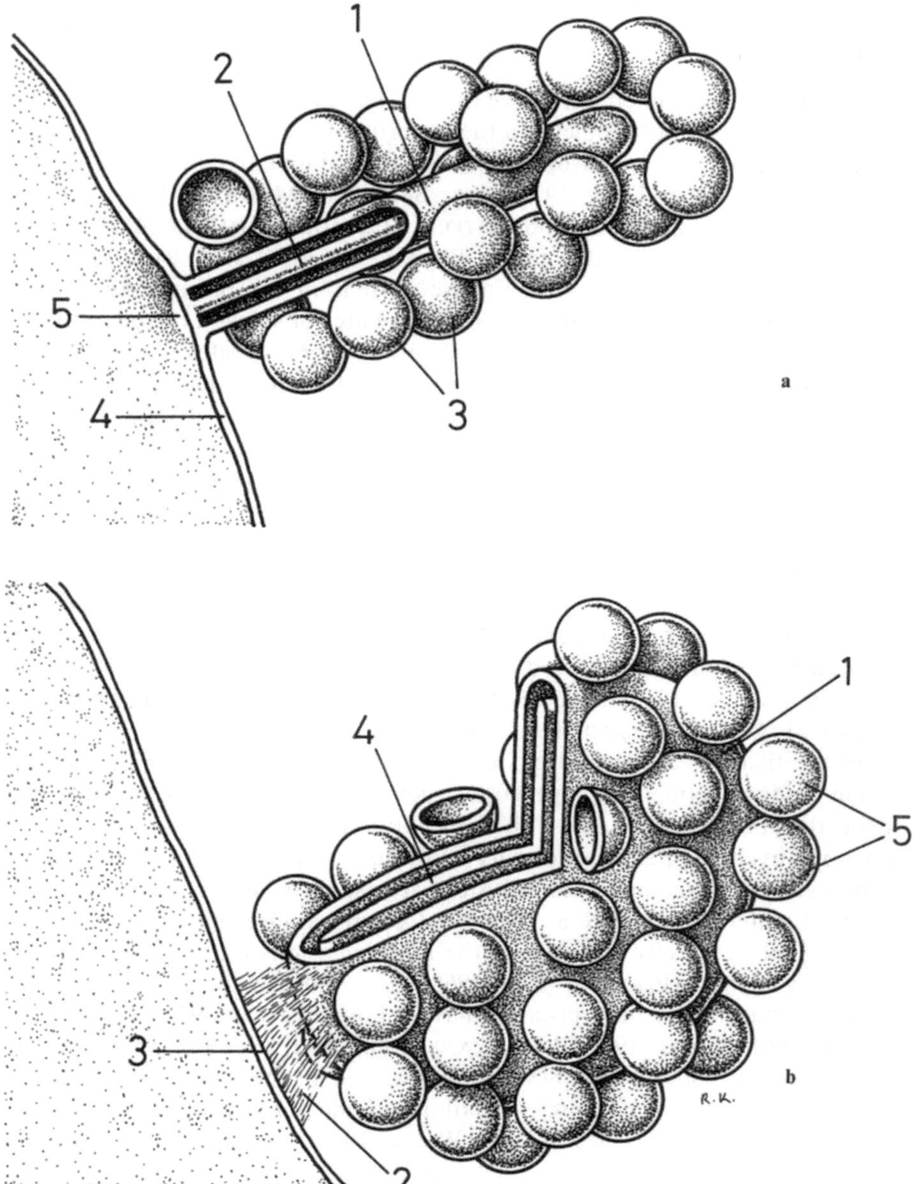

a

b

Fig. 87 a, b. Three-dimensional appearance of two types of synaptic structures, as reconstructed from sectioned profiles. **a** shows a synaptic ribbon which is basically rod-like. *1*, synaptic rod with central filament (*2*); *3*, synaptic vesicles, of which up to 50 may surround a single rod; *4*, presynaptic plasmalemma; *5*, shallow depression. ×190,000. **b** shows a synaptic lamella (*1*) comprising a central membrane ca. 7 nm thick (*4*); the lamella, which may be either round or like a half moon, is attached to the presynaptic plasmalemma (*3*) by delicate filaments (*2*); synaptic vesicles (*5*) are also present. ×190,000. Krstić, 1976d synaptic ribbons in guinea-pig pinealocytes see the paper by McNeill and Whitehead (1979).

Lipid Inclusions. As pointed out above (p. 79), mammalian pinealocytes contain varying amounts of lipid droplets. In the majority of species examined by means of electron microscopy, lipid droplets were small in amount (cow, sheep: Anderson, 1965; guinea-pig: Lues, 1971; dog: Sano and Mashimo, 1966; Welser et al., 1968; gopher: Sheridan and Reiter, 1973; mouse: Ito and Matsushima, 1968a; mole: Pevet, 1974; dormouse: Roux et al., 1974; seal: Cuello, 1973). Large to moderate amounts of lipid are present in rat pinealocytes, where they have been regarded as morphological correlates of secretory products (Gusek and Santoro, 1961; Gusek et al., 1965) or as storage sites for pineal hormones (De Martino et al., 1963; Bostelmann, 1969c; Gonzales and Blasquez, 1975), and in ground squirrel pinealocytes (Povlishock et al., 1975).

Originally regarded as lipoidal pigment granules (Gusek and Santoro, 1960, 1961), the droplets were later classified as lipid inclusions (Arstila and Hopsu, 1964; Bostelmann, 1965; Wolfe, 1965; Arstila, 1967). The lipid inclusions observed vary considerably in shape and size. Round structures were described by Gusek and Santoro (1961), Arstila and Hopsu (1964) and Arstila (1967), whereas globular to elaborately scalloped or arborescent forms were mentioned by Wolfe (1965) and bizarre-shaped inclusions by Bostelmann (1965). They measured 0.5–0.5 µm (Gusek and Santoro, 1961) and 1–4 µm (rarely 10 µm) in the rat (Wolfe, 1965) and 0.5–2 µm in the ground squirrel (Povlishock et al., 1975). As can be expected, the appearance of the inclusions varies depending on the fixative, buffer and embedding medium (Arstila, 1967). Osmium fixation yields rather pale homogeneous structures, the electron density apparently depending on the buffer used. Following glutaraldehyde fixation the inclusions show a relatively high electron density and usually a lighter zone in the centre. In Epon-embedded material, the inclusions are usually round, while in Vestopal-embedded material they have a more star-like appearance. It should be noted, however, that the electron-density of the lipid inclusions found in the pinealocytes varies anyhow (Perrelet et al., 1968; Povlishock et al., 1975).

Lipid inclusions do not appear to have a bounding membrane (Gusek and Santoro, 1961; Wolfe, 1965; Bostelmann, 1965; Arstila, 1967; Welser et al., 1968), though a limiting membrane was observed in the squirel (Povlishock et al., 1975). Occasionally a close association with membranes of the endoplasmic reticulum (Gusek and Santoro, 1961; Wolfe, 1965; Arstila, 1967) and mitochondria (Gusek and Santoro, 1960, 1961; Wolfe, 1965; Anderson, 1965; Lues, 1971) has been noted. Whereas the majority of authors do not mention a distinct distribution pattern of lipid droplets, Bostelmann (1965) observed that they were mainly localized in processes of pinealocytes and only some in the vicinity of the nucleus. According to De Martino et al. (1963), they were particularly numerous in pinealocytes located near blood vessels. In the squirrel, an even distribution with no polarity was encountered (Povlishock et al., 1975). According to S. Milcu et al. (1962) osmiophilic droplets are more numerous in males than in females.

Regarding the origin of lipid droplets in rat pinealocytes, the possibility has been discussed that lipid droplets may represent final stages of mitochondrial transformation (GUSEK and SANTORO, 1960, 1961), a view not shared by others (WOLFE, 1965; LUES, 1971).

In view of the concept that the lipid droplets may contain serotonin and melatonin (DE MARTINO et al., 1963), it is interesting to note that administration of *p*-chlorophenylalanine, a serotonin-depleting agent, led to a distinct reduction of the electron-density of the lipid inclusions (PERRELET et al., 1968; PERRELET and ORCI, 1969) (Fig. 88). GONZALES and BLASQUEZ (1975) reported that lipid droplets migrated through pinealocyte processes and that they were released into pericapillary spaces.

Changes in the amounts of lipids under various physiological and experimental conditions are mentioned in Table 16 and in Section B. XI.

Glycogen. See p. 77.

Pigments. See p. 76, 170.

Plasmalemma. The plasmalemma of the mammalian pinealocyte has received little attention, probably because it does not possess any unusual features. No attempt has been made in the present survey to scan the literature carefully with regard to the presence or absence of junctional complexes. Here it suffices to mention that relatively few precise data on this topic are currently available. According to WOLFE (1965, rat), pinealocytes form neither desmosomes (maculae adherentes) nor tight junctions (zonulae occludentes), though zonulae adherentes and fasciae adherentes were supposed to be present. Recently, a fascia adherens has been studied by means of freeze-etched electron microscopy, which showed focal contacts (KRSTIĆ, 1974). In contrast to the findings in rat, the presence of desmosomes has been reported in pinealocytes of hamster (SHERIDAN and REITER, 1968), guinea-pig (LUES, 1971), bat (STAMMER, 1972), dormouse (ROUX et al., 1974) and rabbit (ROMIJN, 1973b). Zonulae occludentes were supposedly present in guinea-pig pinealocytes (LUES, 1971).

Pinocytotic vesicles or caveolae have been mentioned in cow, sheep (ANDERSON, 1965), hamster (SHERIDAN and REITER, 1968) and guinea-pig (LUES, 1971). In cattle and sheep caveolae appeared at random over the expanse of the plasmalemma and they were supposed to be more frequent in juvenile animals than in mature ones (ANDERSON, 1965).

The close relationships of the plasmalemma to subsurface cisterns and synaptic ribbons has been mentioned on p. 122 and p. 149.

In view of the presence of intercellular cytoplasmic bridges between certain cells, it is interesting to note that discontinuities of the plasmalemma may also be present in the mammalian epiphysis cerebri. GUSEK (1968), in a review article, mentioned and illustrated unpublished results showing apparent gaps in the pinealocyte cell membrane, especially in epiphyseal areas distant from capillaries. Two types of gaps were distinguished: one was totally devoid of vesicular material and the other was characterized by the presence of a number of mainly oval vesicles approximately 60 nm in diameter. GUSEK thought it possible that these sites represent areas of exchange between adjacent pinealocytes and hypothesized that they might be involved in transferring substances

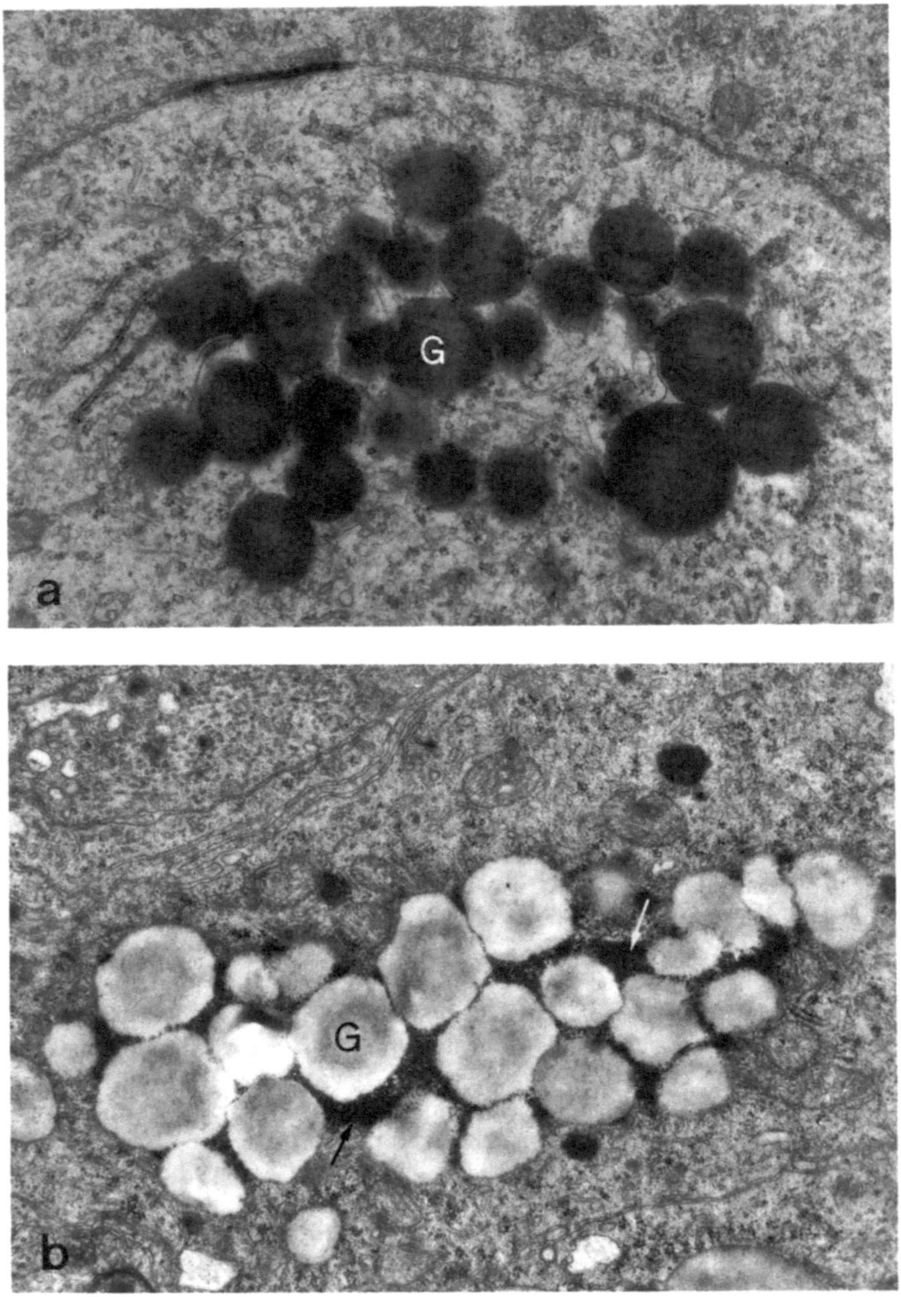

Fig. 88a, b. Lipid globules (*G*) before (**a**, ×17,000) and after (**b**, ×19,000) *p*-chlorophenylalanine treatment. Note the decrease in electron density of the globules and the occurrence of electron-dense fine granular material of unknown nature between the globules (*arrows*). PERRELET and ORCI, 1969 from pinealocytes lying distant from capillaries to those in the immediate vicinity of blood vessels.

d) Cell Processes

As pointed out above (p. 111), cytoplasmic processes are a characteristic feature of mammalian pinealocytes. Due to the great tortuosity of the processes it is virtually impossible to follow their courses and to study their ultrastructural appearance over long distances. Hence, caution is necessary when extrapolating from the findings obtained.

With respect to the number of cytoplasmic processes and their assumed relationship to processes of photoreceptor cells in some species (PEVET and COLLIN, 1976), see p. 111. Since the concept that mammalian pinealocytes may resemble pineal photoreceptor cells of lower vertebrates more closely than has been hitherto suspected is a relatively new one (PEVET and COLLIN, 1976; PEVET et al., 1977b), previous authors have not looked at the pinealocyte processes from this point of view. There is therefore a possibility that the results concerning the number of cytoplasmic processes obtained may have to be re-interpreted.

It is clear though, that pinealocyte processes, termed polar processes by WOLFE (1965), fall into two categories: (i) long processes and (ii) short processes (WOLFE, 1965; LUES, 1971; POVLISHOCK et al., 1975; PEVET and COLLIN, 1976). According to WOLFE's (1965) study in the rat, a long polar process arising from the perikaryon tapers to a neck-like or collicular segment, where it is joined to adjacent epiphyseal cells by a zonula adherens. Distal to its collicular segments, the polar process often swerves sharply and then pursues a course of variable length and tortuosity before ending as a polar terminal in the perivascular space or among clusters of pinealocytes. Short polar processes, on the other hand, were frequently observed to terminate close to their collicular segments. While in rats neck-like segments appear to be present in both long and short processes, in the guinea-pig they (Fig. 89) seem to be confined to short processes (LUES, 1971). In the ground squirrel, by contrast, a neck-like region was found in long processes (POVLISHOCK et al., 1975). The arrangement of processes was apparently less complex in this species than in the others, so that their number and lengths could be easily assessed. Here the pinealocytes possessed numerous short processes, measuring 0.5–1.0 μm in length, and a single large process measuring 5–9 μm in length and 1–2 μm in width, the long process frequently giving off ancillary processes (POVLISHOCK et al., 1975).

Ramification of pinealocyte processes has been mentioned in hamster (SHERIDAN and REITER, 1968, 1970b), rabbit (LEONHARDT, 1967), cow, sheep (ANDERSON, 1965) and guinea-pig (LUES, 1971).

A characteristic feature of pinealocyte processes is that they often possess bulbous endings which differ strikingly in appearance from the preterminal parts of the processes.

α) Preterminal Parts

Ultrastructurally it appears that the processes contain the same elements as the perikaryon, except for the Golgi apparatus, centriole and its derivatives

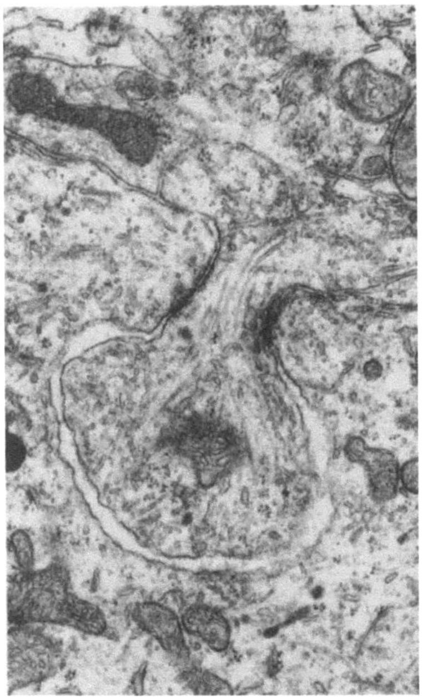

Fig. 89. Short pinealocyte process with neck-like constriction. Guinea-pig. LUES, 1971; ×21,000

(e.g. WARTENBERG, 1968; ITO and MATSUSHIMA, 1968a; LUES, 1971). Most authors mention the presence of microtubules (sheep, cow: ANDERSON, 1960, 1965; hamster, gopher: SHERIDAN and REITER, 1968, 1970b, 1973; guinea-pig: LUES, 1971; rabbit: LEONHARDT, 1967; ROMIJN, 1973b; dog: SANO and MASHIMO, 1966; cat, monkey: WARTENBERG, 1968; mole: PEVET, 1974; hedgehog: PEVET and SABOUREAU, 1973; mole rat: PEVET et al., 1976b). In some instances, microtubules were particularly numerous at sites where the processes left the perikaryon (hamster: SHERIDAN and REITER, 1968, 1970b; ground squirrel: POVLISHOCK et al., 1975). In cow and sheep, microtubules were more numerous in large processes than in small ones (ANDERSON, 1965). In the rat, microtubules were rare (ARSTILA, 1967). By contrast, in some species, the processes or at least parts of them are characterized by an abundance of microfilaments (guinea-pig: LUES, 1971; dormouse: ROUX et al., 1974), mitochondria (rat: ARSTILA and HOPSU, 1964; mole: PEVET, 1974) or smooth endoplasmic reticulum (guinea-pig: LUES, 1971; ground squirrel: POVLISHOCK et al., 1975).

Finally, it should be mentioned that in guinea-pigs synaptic ribbons are more frequent in pinealocyte processes than in perikarya (VOLLRATH, 1973).

β) Bulbous Endings

The bulbous endings of pinealocyte processes (Fig. 90) are of particular interest, at least those lying in the immediate vicinity of blood vessels, because they contain a fairly large number of both electron-lucent vesicles and dense-core vesicles, which may be a morphological correlate of secretory products. A not-

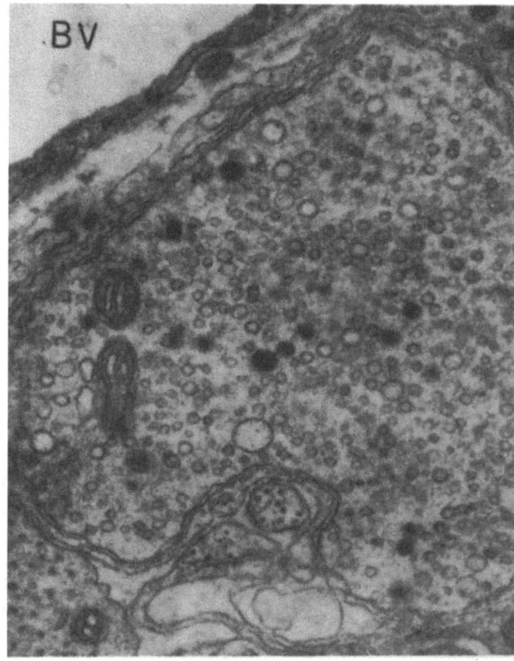

Fig. 90. Bulbous ending of pinealocyte process showing large numbers of vesicles. Hamster. *BV*, lumen of blood vessel. SHERIDAN and REITER, 1970b

able exception is the gerbil, in which dense-core vesicles could not be detected in the endings of pinealocyte processes (WELSH and REITER, 1978). Relatively little detailed knowledge on the ultrastructure of components other than vesicles is currently available. In addition to vesicles, the bulbous endings contain large amounts of smooth endoplasmic reticulum in the rabbit (LEONHARDT, 1967; ROMIJN, 1973b), many mitochondria in cow and sheep (ANDERSON, 1965), lysosomes in the rat (ARSTILA, 1967) and synaptic ribbons and spherules in the guinea-pig (LUES, 1971).

Because of an abundance of vesicular inclusions (Fig. 90) the endings have been characterized as "plurivesicular secretory processes" (DE ROBERTIS and PELLEGRINO DE IRALDI, 1961a), though apparently some of these processes were misidentified terminals of autonomic axons (WOLFE, 1965). A profusion of vesicles is typical of most species investigated. An exception is the rabbit, in which the terminals also contain large amounts of smooth endoplasmic reticulum (LEONHARDT, 1967; ROMIJN, 1973b).

The vesicles contained within the terminals are of two types, i.e. clear and dense-cored (Figs. 90, 108, 109). As can be seen from Table 18, the clear ones ranged in size between 200 and 210 nm. According to quantitative studies in the mouse, histogram peaks occur at 60 and 70 nm respectively (ITO and MATSUSHIMA, 1968a, b). Dense-core vesicles show a slightly smaller range in size (see Table 18) with a prominent peak at 110 nm (ITO and MATSUSHIMA, 1968a, b).

The relative amounts of clear and granular vesicles in the endings have apparently not been studied in detail. It is apparent, however, that clear vesicles outnumber granular ones.

Table 18. Size of vesicles in mammalian pinealocytes

Species	Clear vesicles (nm)	Dense-core vesicles (nm)	Author(s)
A. Vesicles in bulbous endings of pinealocyte processes			
Rabbit	35–65		ROMIJN (1973b)
Rat	42.5		DE ROBERTIS and PELLEGRINO DE IRALDI (1961a)
Mouse	40–210	70–190	ITO and MATSUSHIMA (1968a, b)
Hamster	40–120	70–120	SHERIDAN and REITER (1970b)
Gopher	40–120	70–120	SHERIDAN and REITER (1973)
Ground squirrel	20–150		POVLISHOCK et al. (1975)
Mole	30–35		PEVET (1974)
Hedgehog	30–40	40–150	PEVET and SABOUREAU (1973)
B. Vesicles in perikarya of pinealocytes			
Rhesus monkey		50–140	SHERIDAN and SLADEK (1975)
Sheep	80	80	ANDERSON (1965)
Cow	80	80	ANDERSON (1965)
Dog	18–120	18–120	SANO and MASHIMO (1966)
Ferret		Present	HERBERT (1971)
Seal		200–250	CUELLO (1973)
Rabbit	40–150	60–300	LEONHARDT (1967)
	35–120	55–250	ROMIJN (1973b)
Rat	40–100	100–120	ARSTILA (1967)
Mouse	60	114	ITO and MATSUSHIMA (1968a, b)
		110	PELLEGRINO DE IRALDI (1969)
Hamster	Present	70–100	SHERIDAN and REITER (1968)
	30–140	100–140	CLABOUGH (1971)
		100–120	SHERIDAN (1969), SHERIDAN and SLADEK (1975)
Guinea-pig	30–90	75–100	LUES (1971)
Gopher	40–120	70–120	SHERIDAN and REITER (1973)
Gerbil	140–280	53–79	GREGOREK (1973)
	30–50	50–120	WELSH and REITER (1978)
Ground squirrel		80–150	POVLISHOCK et al. (1975)
Mole-rat		100–200	PEVET (1976)
Mole	50–150	80–200	PEVET (1974)
Hedgehog	30–45	40–150	PEVET and SABOUREAU (1973)
Golden mole	Present	40–350	PEVET and KUYPER (1978)

Dense-core vesicles, although altogether rare, are relatively frequent in mouse (ITO and MATSUSHIMA, 1968a, b; BENSON and KRASOVICH, 1977), hamster (SHERIDAN and REITER, 1968, 1970a) (Fig. 90) and ground squirrel [in the latter species at least in some terminals (POVLISHOCK et al., 1975)]. Their occasional presence has been described in hedgehog (PEVET and SABOUREAU, 1973), guinea-pig (LUES, 1971), pocket gopher (SHERIDAN and REITER, 1973), rabbit (ROMIJN, 1973b) and dog (SANO and MASHIMO, 1966). In some species the vesicles have been shown to undergo typical numerical changes. These will be dealt with in the next section.

The origin of the vesicles has not been studied in detail. While it is most likely that they are formed in the Golgi apparatus and migrate through the pre-terminal parts of the pinealocyte processes into the bulbous endings, there are also indications that they form locally from central tubular complexes consisting of a saccular core of about 0.5 μm in diameter surrounded by a dense mass of tubules (ARSTILA, 1967, rat) or from smooth endoplasmic reticulum (ROMIJN, 1973b, rabbit).

e) Morphological Correlates of Secretory Products

As pointed out above, morphological correlates of pineal secretory products have long been searched for, which is understandable in view of the scarcity of dense-core vesicles in the pineal and their abundance in other endocrine organs. The lack of clear-cut evidence for the presence of secretory granules in rat pinealocytes led WOLFE (1965) to suspect that "the epiphyseal cell could be a secretory cell without any visible granulation of its cytoplasm". It is therefore understandable that other structures, such as lipid droplets (p. 79) and so-called grumose or vermiculate inclusions (WOLFE, 1965) were thought to be related to secretory products.

Recent studies by PEVET (1977a; PEVET and KAPPERS, 1977a, b; PEVET and KARASEK, 1977) have suggested that basically two types of morphologic correlates of secretory products may be present in the mammalian pineal, one type being the dense-core vesicles found both near the Golgi apparatus (p. 128) and in the bulbous endings (p. 162), the other type being the material comprised by the cisterns of the rough endoplasmic reticulum, e.g. in the mole-rat and noctule bat (p. 122). PEVET further concluded that in species in which two types of pinealocytes (p. 168) are present, each type is characterized by a special type of secretory mechanism, while in species with one type of pinealocytes only, such as mole, hedgehog and rat, the two mechanisms may be found in the same cell. Since the second type of secretory mechanisms has been discussed above (p. 122) and is less common than the "granular" type of secretion, it is appropriate to concentrate here on the dense-core or granular vesicles.

What evidence is there that these granules are the morphological correlate of secretory products? For the time being there is only indirect evidence for their being secretory granules.

To begin with, they are similar to organelles which in other organs contain active principles. Secondly, they appear to originate in the Golgi apparatus (see p. 128), which plays an important role in secretory cells. Thirdly, although the vesicles are altogether rare, they have been found in most species investigated (Table 18). Fourthly, the predominant localization of the vesicles within the bulbous endings of pinealocyte processes, which very often lie in the immediate vicinity of blood vessels, is compatible with the assumption that they release their active principles into the bloodstream. Finally there are the results of quantitative studies which show that the granules undergo characteristic numerical changes over a period of 24 h and under experimental conditions.

A 24-h rhythm of the number of dense-core vesicles has been clearly demonstrated in rabbit and mouse. In rabbits (ROMIJN et al., 1976) the rhythms showed slightly different patterns near the Golgi apparatus and in the bulbous endings (Fig. 91a), compatible with the assumption that the granular vesicles migrate

a
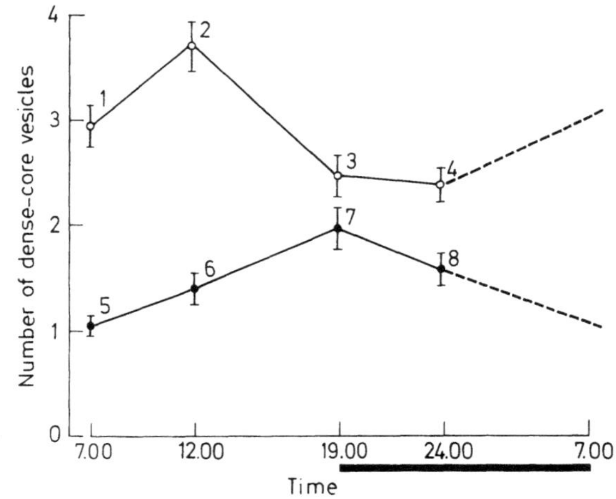

b
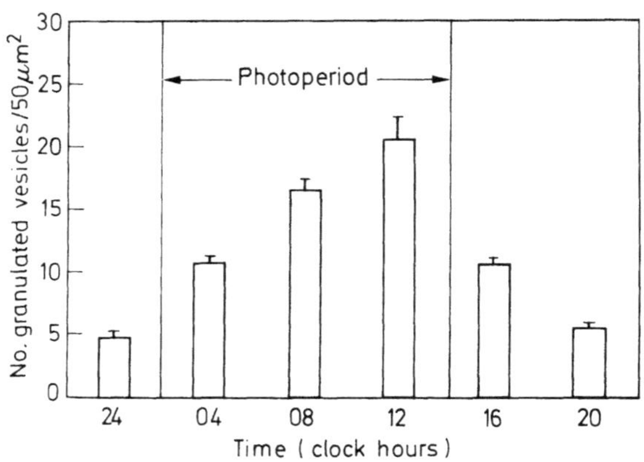

Fig. 91 a, b. Daily changes in the number of dense core vesicles in mammalian pinealocytes. **a** Rabbit. *Top*, number of dense core vesicles in perikarya near Golgi apparatus; *bottom*, number of dense core vesicles in terminals of pinealocyte processes. ROMIJN et al., 1976 **b** Mouse. BENSON and KRASOVICH, 1977

from the Golgi apparatus into the terminals. In the mouse (Fig. 91 b), the numerical changes are particularly pronounced, a four-fold difference being noted between late photoperiod maximum and late dark period minimum (BENSON and KRASOVICH, 1977). Both species have in common the fact that the granulated vesicles accumulate during the light period and decrease during the dark phase, showing that the number of granules and the melatonin-forming activity (see p. 251) of the pineal cannot be directly correlated. Since the granules are influenced by melatonin administration, BENSON and KRASOVICH (1977) suspected that they may contain the antigonadotropic (peptide) principles of the pineal (see below).

That the granules may represent secretory products is suggested by both in vitro and experimental studies. In vitro it has been shown (ROMIJN and

GELSEMA, 1976) that following addition of noradrenaline, which is known to enhance melatonin formation, the dense-core vesicles increase strikingly in number. Relevant experimental studies have been carried out in hamsters and mice only. In hamsters, blinding is followed by a strong increase in the number of electron-lucent and dense-core vesicles (CLABOUGH, 1971; LIN et al., 1975; SHERIDAN, 1975), a situation known to be accompanied by enhanced melatonin formation (p. 245). By contrast, sympathectomy, leading to a depression of the secretory activity of the pineal, is followed by a decrease in the number of dense-core vesicles in both hamster (LIN et al., 1975) and mouse (BENSON and KRASOVICH, 1977), the latter species also showing a 70% decrease of dense-core vesicles following exposure to continuous illumination (UPSON et al., 1976).

Regarding the results of sympathectomy in hamsters, it should be noted that a numerical decrease of the dense-core vesicles was not always seen. SHERIDAN (1975) found that they were more frequently seen than in control animals, but he thought that this was apparently primarily due to an increase in the number of vesicle-containing terminals rather than an increase of the vesicles within the terminals.

The chemical nature of the vesicular contents can only be surmised. Three substances, or groups of substances, are possible candidates: serotonin, melatonin and peptides. Strong advocates for the presence of serotonin or melatonin within granulated vesicles measuring 50–140 nm in diameter (hamster) have been SHERIDAN and REITER (1968, 1970b), especially after it was shown that administration of *p*-chlorophenylalanine, a serotonin-depleting agent, decreased the amounts of the electron-dense vesicles (SHERIDAN and KEPPEL, 1971). However, recently SHERIDAN reconsidered the earlier interpretation, because in combined fluorescence and electron microscopic studies an inverse relationship between the amounts of dense-core vesicles and the serotonin fluorescence was noted both in hamster and monkey pineals (SHERIDAN and SLADEK, 1975). Whether or not the granular vesicles contain peptide substances cannot yet be decided. That they may contain an antigonadotropic substance other than melatonin is suggested by the finding that melatonin administration led to an increase in the number of dense-core vesicles during the light period and to a decrease during the dark period (BENSON and KRASOVICH, 1977).

Whatever the vesicles may contain, their relative rarity is a characteristic feature of most, if not all, mammalian pinealocytes, suggesting that it is apparently not the primary function of the pinealocyte to *store* large amounts of its secretion. It would rather appear that the pinealocyte produces its active principles for more or less immediate release.

The ultrastructure of pinealocytes of transplanted and cultured pineals will be dealt with on p. 234 and p. 236. The effects of various experimental conditions on the pinealocytes are mentioned in Section B XI.

f) Different Types of Pinealocytes?

In view of the great number of different active principles present in the mammalian epiphysis cerebri, it is of particular interest whether or not different types of pinealocytes can be distinguished on ultrastructural grounds. Since this problem has been discussed in detail by PEVET (1977a) and is still far from clear, only a short account will be given here.

Very briefly, the situation is as follows. Apart from the recent study by PEVET and COLLIN (1976) in which mole pinealocytes were subdivided into three categories depending on how closely the cells were structurally related to photoreceptor cells of pineal organs of lower vertebrates (p. 111), a distinction was usually made taking into consideration the general ultrastructural aspect of the pinealocytes and their localization within the parenchyma. Frequently a distinction has been made between light and dark pinealocytes (e.g. ARSTILA and HOPSU, 1964; GUSEK et al., 1965; WARTENBERG and GUSEK, 1965; SHERIDAN and REITER, 1968, 1973; BERERHI and ABBAS-TERKI, 1970; LUES, 1971; ROMIJN, 1973b; PEVET and SABOUREAU, 1973; PEVET, 1974; UPSON et al., 1976). Although there can be no doubt that in most species investigated light- and dark-appearing cells are clearly distinguishable, the above terminology has been criticized on account of the fact that glial cells of the pineal also appear dark and that there are supposedly two distinct types of pinealocytes which do not fall into these categories (PEVET, 1977a). PEVET (1977a) therefore speaks of pinealocytes of population I and pinealocytes of population II, dark and light pinealocytes being present in both populations of pinealocytes. According to PEVET, pinealocytes of population I represent the majority of pineal parenchymal cells and contain the dense-core vesicles previously described (pp. 128, 165). Pinealocytes of population II form only a small portion of the pineal parenchymal cells and are practically always located near perivascular or extracellular spaces. In some species they lack dense-core vesicles and instead contain floccular secretory material in their endoplasmic reticulum. According to PEVET (1977a) a single population of pinealocytes (i.e. pinealocytes of population I) is present in rat, cow, dog, mole, hedgehog, dormouse, ferret and goat of Cameroun. Pinealocytes of population II, in addition to pinealocytes of population I, are supposed to be present in noctule bat, mole-rat, chinchilla, gerbil, guinea-pig, macaque, mouse, pocket gopher and rabbit.

In view of the fact that the classification of pinealocytes will probably continue to be puzzling, some of the major differences between light and dark pinealocytes in the most widely studied species will be mentioned. Apart from a pronounced electron-density of the cytoplasmic matrix (Fig. 92) and a larger number of cytoplasmic processes, the dark pinealocytes usually contain nuclei which are particularly electron-dense, comprising larger amounts of and more widely dispersed condensed chromatin than light pinealocytes. The cytoplasm comprises larger amounts of rough endoplasmic reticulum, polysomes and sometimes also of lysosomes (ARSTILA and HOPSU, 1964; GUSEK et al., 1965; ARSTILA, 1967; SHERIDAN and REITER, 1968, 1973; BERERHI and ABBAS-TERKI, 1970; LUES, 1971; UPSON et al., 1976). In hedgehog (PEVET and SABOUREAU, 1973) and mole (PEVET, 1974) dark pinealocytes show only a greater electron-density of the cytoplasm than light pinealocytes.

Often differences in the mitochondria have been reported. In both guinea-pig (LUES, 1971) and mouse (UPSON et al., 1976) the mitochondria of dark pinealocytes were smaller in size and had a more electron dense matrix than those of light pinealocytes. While in the guinea-pig the mitochondria of dark pinealo-

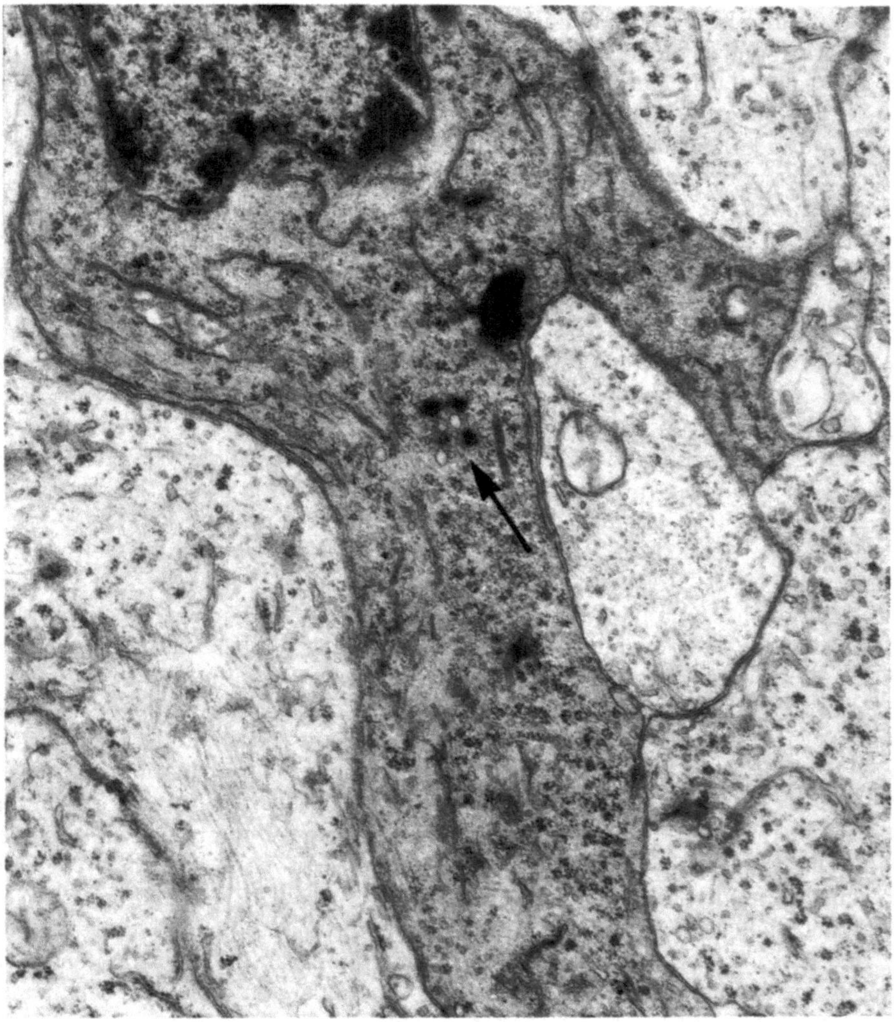

Fig. 92. Electron micrograph showing "dark" pinealocyte surrounded by profiles of "light" pinealo-cytes. Guinea-pig, exposed to continuous darkness for 30 days. Lues, 1971; ×30,000

cytes usually lacked tubular cristae which are seen in those of light pinealocytes, dark pinealocytes of mice predominantly comprised mitochondria with tubular and irregularly arranged cristae, while those of light pinealocytes were characterized by poor fixation and irregular cristae.

Differences in mitochondria have also been noted in hamsters, where the pinealocytes were classified into the more frequent P_1 pinealocytes and the less frequent P_2 pinealocytes (Clabough, 1971). Small cristiform mitochondria were noted in P_1 cells, while mitochondria with a plexiform array of cristae in a dense matrix were typical of P_2 pinealocytes. Since according to Clabough

(1971) the P_2 pinealocytes contain fewer ribosomes and larger amounts of smooth endoplasmic reticulum than P_1 pinealocytes, it appears that the P_2 cells correspond to the light pinealocytes of other species and the P_1 pinealocytes to the dark pinealocytes, with the exception that the dark pinealocytes are more frequent than light pinealocytes.

Dark and light pinealocytes also differ with respect to their lipid contents. Dark pinealocytes usually comprise distinctly less lipid than light pinealocytes (ARSTILA and HOPSU, 1964; GUSEK et al., 1965; ARSTILA 1967). In the hamster, by contrast, a greater lipid content has been attributed to dark pinealocytes (SHERIDAN and REITER, 1968) and to P_2 pinealocytes (CLABOUGH, 1971). Glycogen and vacuoles were typical of pinealocytes of population II of the noctule bat (PEVET et al., 1977a). Large intracellular vacuoles have also been described in hedgehog and mole (PEVET, 1975), but it is not clear whether they are restricted to a distinct type of pinealocyte. Glycogen is also a characteristic feature of the dark pinealocytes in rabbits (ROMIJN et al., 1977a).

Another characteristic feature of a second type of pinealocyte is the presence of pigment granules. In the rabbit singly lying pigment granules fusing to form large globules (Fig. 93) have been noted in dark pinealocytes (WARTENBERG and GUSEK, 1965; ROMIJN, 1973b). In noctule bat, pigment globules were typical of what has been termed pinealocytes of population II (PEVET et al., 1977a). Whether or not the pigmented cells reported in pocket gopher (SHERIDAN and REITER, 1973) and in chinchilla (MATSUSHIMA and REITER, 1975b) represent pinealocytes remains to be elucidated.

The functional significance of the dark pinealocytes is obscure. As pointed out above, they do not appear to be a homogeneous population of cells. They may be either dark-appearing forms of the usually light-appearing main type of pinealocyte, or a distinct cell type. With respect to the first interpretation it is interesting to note (Fig. 94), that in a given section the intrinsic pinealocytes show different shades of grey, showing that it is probably wise not to place too much emphasis on the electron density of the cells. Gradations between dark and light cells are also conspicuous in the mouse pineal (UPSON et al., 1976). In some species, it would appear that the varying electron-density is merely the expression of differences in the functional activity of the cells under question (rat: ARSTILA, 1967; gopher: SHERIDAN and REITER, 1973; hedgehog: PEVET and SABOUREAU, 1973; mole: PEVET, 1974, 1976). According to GUSEK et al. (1965) dark pinealocytes may represent quiescent or exhausted forms of light pinealocytes. This view is substantiated by the finding that dark cells are particularly abundant following conditions of enhanced pineal activity such as blinding (SATODATE et al., 1973), exposure to continuous darkness (LUES, 1971) and pregnancy (LUES, 1971). The effects of various fixation procedures have not been studied in detail. In the gerbil pineal, dark pinealocytes were demonstrable after immersion fixation but not after perfusion fixation (WELSH and REITER, 1978).

That some of the dark pinealocytes and also the pinealocytes of population II may be a distinct cell population is indicated by the fact that in some species they show an uneven distribution in the parenchyma. In some species (e.g. rabbit: WARTENBERG and GUSEK, 1965; ROMIJN, 1973b; guinea-pig: LUES, 1971)

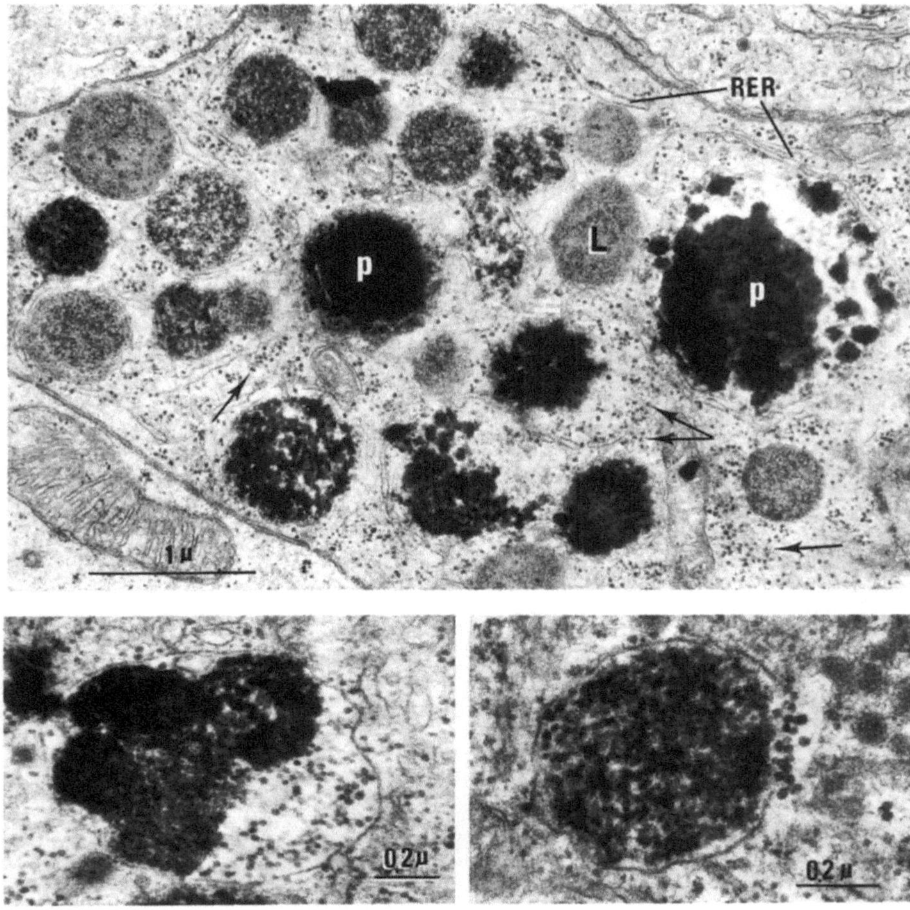

Fig. 93. Electron micrograph showing different stages of large membrane-bound pigment bodies in "dark" pinealocytes of rabbit. *Top*: survey picture; *p*, pigment body; *L*, lysosome; *RER*, rough endoplasmic reticulum; *arrows*, abundance of glycogen granules scattered throughout the cytoplasm. Note the apparent transitional forms between lysosomes and pigment bodies. *Bottom left and right*: membrane-limited pigment bodies composed of a dense granular mass and numerous solitary glycogen granules, the latter being of the same appearance as those present in the surrounding cytoplasm. ROMIJN, 1973b

they are particularly abundant near blood vessels, suggesting that they may be functionally similar to astrocytes. That they are not astrocytes is shown by the finding that in the guinea-pig they possess synaptic ribbons (LUES, 1971) and that in some species astrocytes could be identified in addition to the dark pinealocytes (PEVET, 1977a).

In addition to their being present near blood vessels, there is the interesting feature that they may be confined to certain regions of the pineal. In the rabbit (ROMIJN, 1973b) and mouse (UPSON et al., 1976) dark pinealocytes were restricted to what has been termed the medullary regions of the parenchyma,

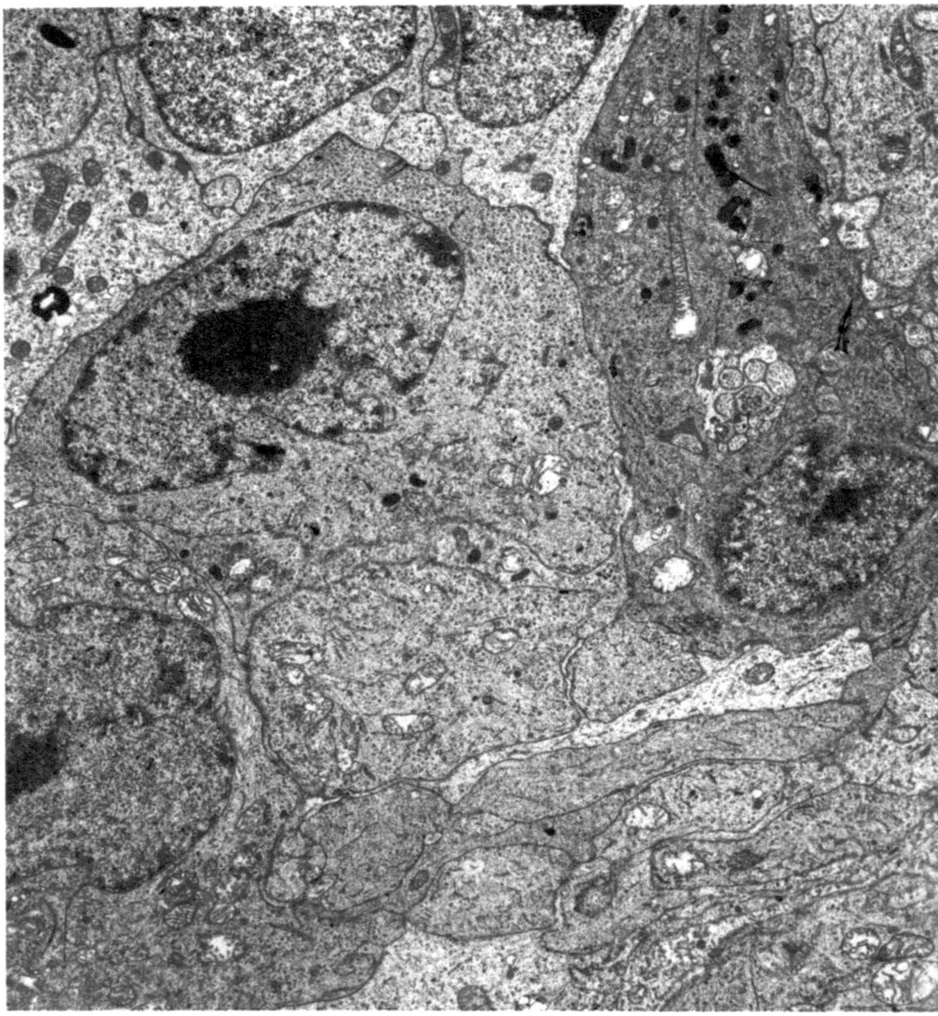

Fig. 94. Electron micrograph from a guinea-pig pineal organ showing densely packed pinealocytes, arranged in an epithelioid fashion leaving only extremely narrow intercellular spaces. VOLLRATH, previously unpublished electron micrograph

suggesting that the cells are not just different functional stages of one and the same cell type. ROMIJN (1973b) speculated that it is the function of the dark pinealocytes to take up substances previously synthesized in, and released from, light pinealocytes, to convert these precursors into pineal-specific end-products and eventually to discharge them into blood vessels. However, as yet there is little evidence to support this concept. It is also an open question whether the dark pinealocytes or the glial cells contain arginine vasotocin (BOWIE and HERBERT, 1976).

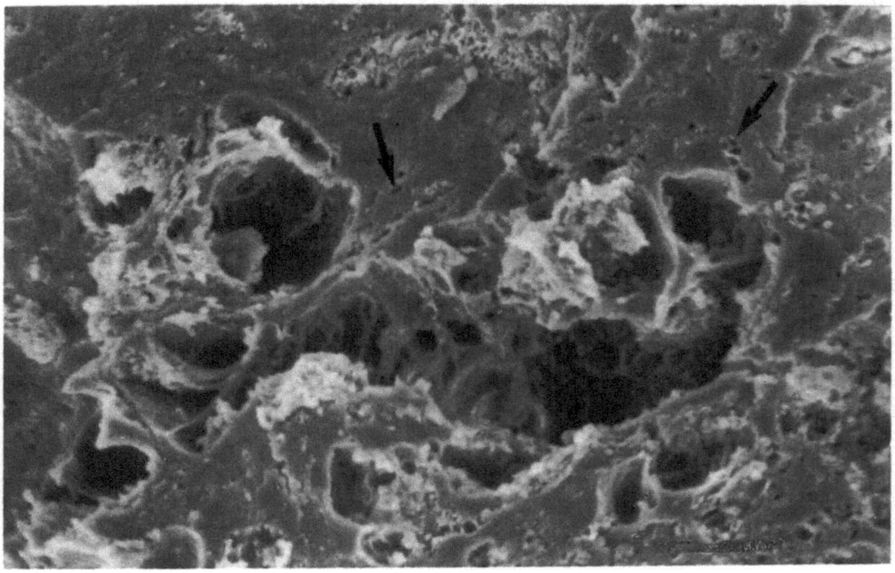

Fig. 95. Scanning electron micrograph showing intercellular canaliculi of the rat pineal gland. KRSTIĆ, 1975b

g) Intercellular Relationships

Mammalian pinealocytes are usually tightly packed in an epithelium-like fashion leaving intercellular spaces of only 15–25 nm in width (GUSEK and SANTORO, 1961; WOLFE, 1965; SHERIDAN and REITER, 1968; ROMIJN, 1973b), an example of which is shown in Fig. 94. There are areas, however, where the intercellular spaces widen considerably, and over long distances (loc. cit. and BOSTELMANN, 1965; RODIN and TURNER, 1966; ARSTILA, 1967; LUES, 1971). ARSTILA (1967) noted spaces measuring 1–2 µm in width and RODIN and TURNER (1966) found spaces extending between adjacent pinealocytes for a distance of up to 10 µm. According to KRSTIĆ (1975b), who studied the rat pineal by means of scanning electron microscopy, the spaces measure up to 100 µm in length and 20 µm in width, forming a three-dimensional network of anastomosing channels (Fig. 95). Local distentions of the intercellular spaces have also been described under the term "interfacial lakes," since they may contain electron-dense flocculent material (WOLFE, 1965; ROMIJN, 1973b). These spaces are not to be confused with flocculent material-filled cavities which are due to a radial disposition of pinealocytes (pinealocytes forming rosettes) (WOLFE, 1965).

That the channels described above may be of special functional importance has been suggested by experimental studies. QUAY (1973a, 1974b), using India ink perfusions of rat pineals, not only demonstrated that the spaces cover a large part of the parenchyma but also that the widths of the canaliculi underwent a 24-h rhythm, the canaliculi being wide during the day and narrow during the night. Moreover, it was noted that administration of serotonin reversed the nocturnal narrowing of the spaces. Since all pinealocytes lie closely related to the canaliculi (KRSTIĆ, 1975b), which in turn open into perivascular spaces (GUSEK and SANTORO, 1961; RODIN and TURNER, 1966; LUES, 1971; ROMIJN, 1973b; QUAY, 1973a, 1974b), close interactions between the pinealocytes and the blood stream and vice versa are possible. These may be important for transport of the secretory products to the blood stream and for transport of neurotransmitters from sympathetic nerve fibres, which lie preferentially in perivascular spaces (p. 220), to pinealocytes.

Occasionally the enlarged intercellular spaces were found to contain segregated cytoplasmic material with cell organelles (GUSEK and SANTORO, 1961; LUES, 1971). According to LEONHARDT (1967, rabbit) the space may contain electron-dense material stemming from extruded central cores of synaptic spherules (see p. 147).

4. Electrophysiological Properties

In view of the fact that mammalian pinealocytes have many features in common with both nerve cells and receptor cells, it is pertinent to inquire into their electrophysiological properties.

Following initial observations by TAYLOR and WILSON (1970), SCHAPIRO and SALAS (1971) studied the electrical activity of the rat pineal in detail. They noted that the pineal gland shows spontaneous electrical activity consisting of low amplitude, high frequency waves, varying between 13 and 22 cycles/s, which are interrupted periodically at intervals of 5–15 s by burst activity of high amplitude waves of 6–12 cycles/s, the duration of the bursts varying from 1–3 s. In view of the circadian activity changes of the pineal, it is worthy of note that burst activity was more predominant during periods of daytime recordings than during night-time. Moreover it was found that after photic stimulation a following response occurs.

In the rat pineal, evidence was obtained that differences in the electrical properties exist between cortical and medullary regions, spontaneous spike activity being restricted to the cortical region (DAFNY and MCCLUNG, 1975; MCCLUNG and DAFNY, 1975). Moreover it was shown that the pineal responds to a number of stimuli which reach the organ via central fibres (DAFNY et al., 1975; DAFNY, 1977). This point will be discussed in Section B VII 2, dealing with the central input of the pineal.

Recently single cell recordings have been carried out in the guinea-pig pineal (SEMM, 1978; SEMM and VOLLRATH, 1978, 1979). In this species spontaneous spike activity is also demonstrable (Fig. 96), corroborating previous findings obtained in rats. Moreover it was shown that the elements responsible for the generation of action potentials (probably pinealocytes) can be subdivided into groups depending on the stimuli by which they can be influenced, namely cells with photic, acoustic and olfactory inputs. In view of the concept that the mammalian pineal gland is phylogenetically derived from a photoreceptor

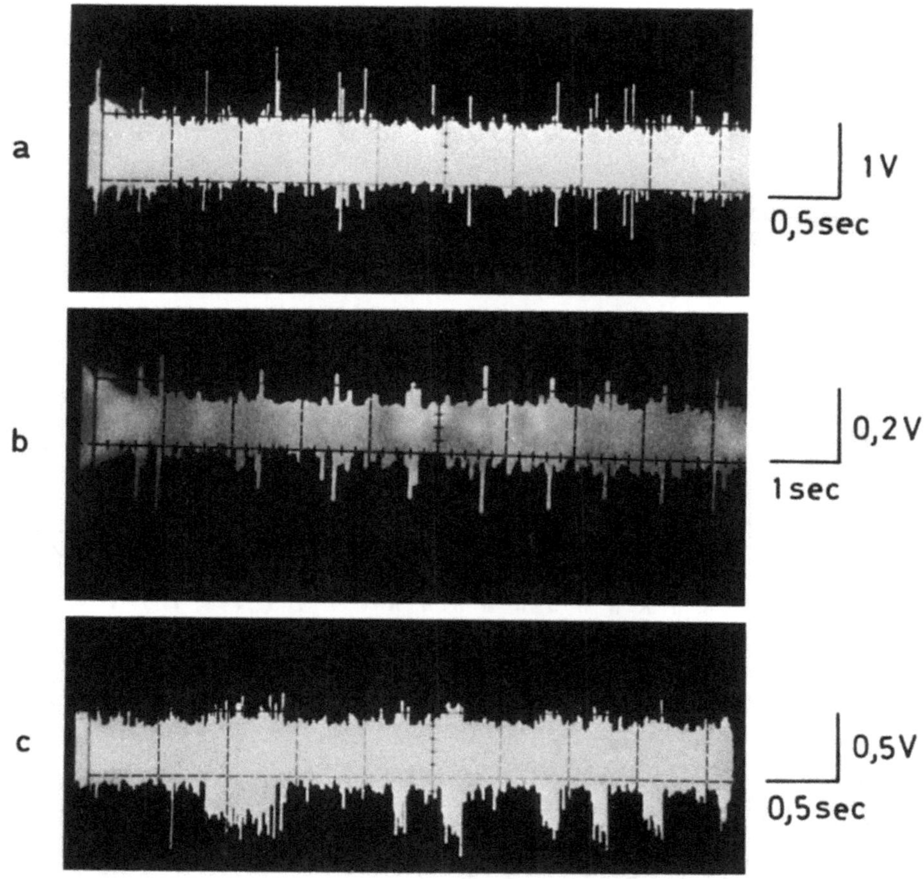

Fig. 96a–c. Direct oscillographic recordings of the electrical activity of pineal cells (guinea-pig) showing various patterns of spontaneous activity: **a** irregular activity; **b** regular activity; **c** burst activity. SEMM and VOLLRATH, previously unpublished recordings

organ, it was most interesting to find that the cells responding to optic stimuli can be further subdivided into ON-cells, OFF-cells, ON-OFF-cells, and cells showing a graded response similar to that shown by retinal ganglion cells. Long-term recording from single cells showed that the spontaneous firing rate increases steadily during the night and decreases early in the morning (Fig. 97), the firing rate running parallel to the secretory activity of the pineal gland.

Finally, pinealocytes have also been studied in vitro (SAKAI and MARKS, 1972). These studies have shown that the parenchymal cells have a resting membrane potential resembling that of other secretory systems studied in vitro, such as liver, and that administration of catecholamines such as noradrenaline or isoproteronol leads to a hyperpolarization which in the pineal appears to involve β-adrenergic receptors, in contrast to the liver, where α-adrenergic receptors appear to play a role. In view of the fact that the pineal appears to

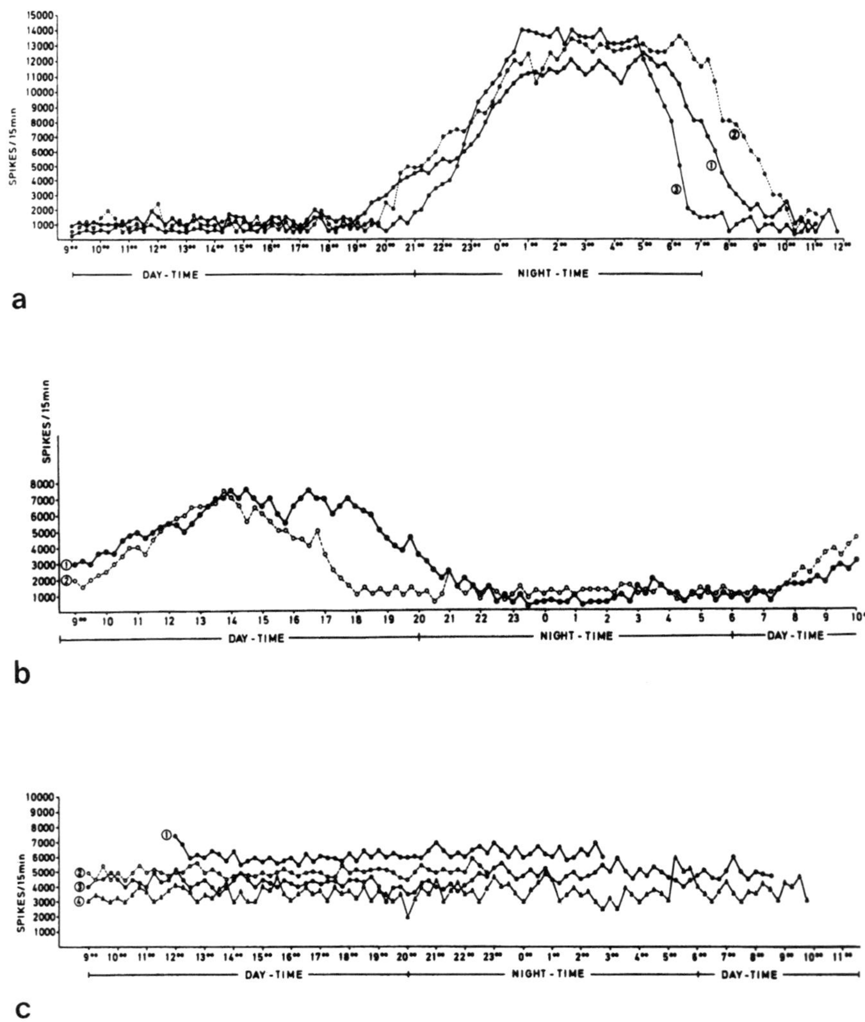

Fig. 97a–c. Long-term recordings from guinea-pig pineal cells, showing three different behavioural patterns. **a** Cells showing low electrical activity during the day and increasing activity during the night, a behaviour similar to that of melatonin formation; note the close correspondence between the length of the night at different seasons (*1*, autumn; *2*, winter; *3*, summer) and the nocturnal activation. **b** Two cells showing increased electrical activity during the day and diminished activity during the night; this activity pattern can as yet not be correlated with any known pineal functions. **c** Pineal cells whose activity remains fairly constant over long periods; note the absence of a clear circadian rhythm. Semm and Vollrath, previously unpublished recordings

influence the gonads and vice versa (Section B XI 11), it is interesting to note that administration of oestradiol inhibits the noradrenaline effect. In vitro, electric field stimulation of pinealocytes has been shown to release cyclic GMP (rat, Zatz and Romero, 1978b).

VI. Glial Cells

Besides pinealocytes, the epiphyseal parenchyma comprises a varying number of glial cells, mainly astrocytes. In addition to intraparenchymal glial cells, ependymal cells may be present, provided that the pineal lies closely related to the third cerebral ventricle. Schwann cells will not be considered here.

1. Oligodendrocytes

Oligodendrocytes are not a regular feature of the mammalian epiphysis cerebri. They have been found in the guinea-pig (LUES, 1971), gopher (SHERIDAN and REITER, 1973) and ground squirrel (POVLISHOCK et al., 1975), the latter species being an exception in so far as the oligodendrocytes outnumber the astrocytes.

2. Astrocytes and Interstitial Cells

Although in the pineals of some species astrocytes have been clearly identified, often no attempt was made to classify the glia-like cells, either because the cells were too rare to be studied in detail by electron microscopy or because they were structurally insufficiently similar to any of the known glial cells. Consequently, either non-committal terms such as interstitial cells were used, or terms which point to certain characteristic features of the cells, such as fat-laden cells.

Unfortunately, the issue has become very complicated by reason of the fact that, in the most widely studied species, the rat, the cells in question differ in many respects from the classic astrocytes and that they have many features in common with the so-called dark pinealocytes, the relationships of which to the main type of pinealocytes and to glial cells is largely obscure (p. 168). A great step forward in the classification of the interstitial cells in the rat pineal has recently been made by means of immunohistochemical studies (MØLLER et al., 1978a). Using antibodies against the glial marker proteins, S-100 protein and glial fibrillary protein (GFA), at the light microscpoic level, results were obtained which suggest that the interstitial cells are indeed astrocytes. It therefore seems appropriate to deal with astrocytes and interstitial cells under the same heading.

Astrocytes and their processes show great interspecies variations. This is partly due to the fact that the pineal bodies have different localizations and that superficially situated pineals contain few glial cells and processes, while deeply placed pineals may have an abundance of these elements (HÜLSEMANN, 1967; SHERIDAN and REITER, 1973). Precise data on the number of glial cells in relation to pinealocytes are available for few species. In the superficially placed rat pineal, glial cells amounted to 12% of the cells scanned (WALLACE et al., 1969), while in cat and monkey the ratio of glial cell perikarya to pinealocyte perikarya was less than 1:10 (WARTENBERG, 1968).

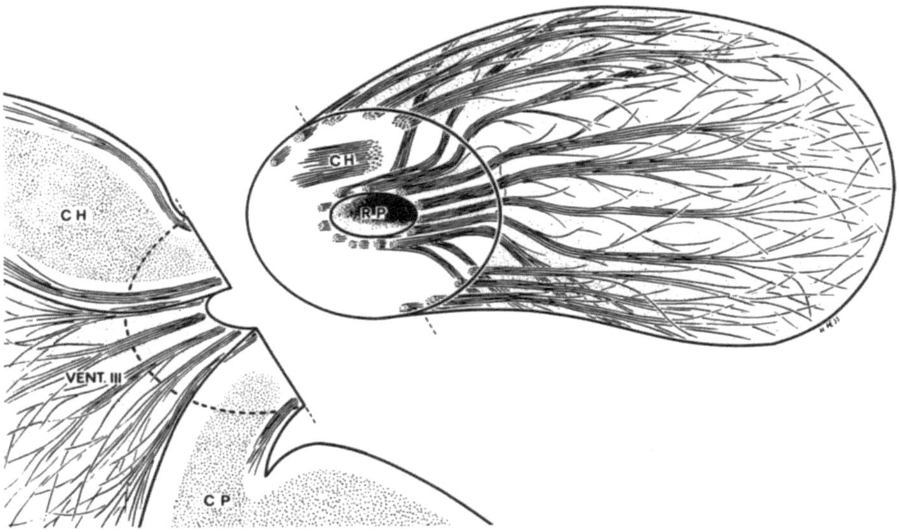

Fig. 98. Schematic representation of glial fibres in the pineal region of *Macaca mulatta*. *CH*, habenular commissure; *CP*, posterior commissure; *RP*, pineal recess; *VENT. III*, third ventricle. ECKHARDT, 1972

Like in the brain, astrocytes are present in two forms, namely fibrous astrocytes and protoplasmic astrocytes (FASSBENDER, 1962; HOLLMANN, 1963; ANDERSON, 1965; WARTENBERG and GUSEK, 1964, 1965; WARTENBERG, 1966, 1968; DUNCAN and MICHELETTI, 1966). Fibrous astrocytes appear to outnumber protoplasmic astrocytes.

Fibrous astrocytes and their processes, the glial fibres, are abundant in the human epiphysis (HÜLSEMANN, 1967). The fibres show a net-like arrangement and surround groups of pinealocytes, especially in the basal third of the organ. A particularly dense network was also present in the vicinity of trabeculae and in the periphery of the organ, occasionally forming glial patches. In other primates, glial fibres show regional differences as well. In the rhesus monkey (HÜLSEMANN, 1967; ECKHARDT, 1972), glial fibres were rare in the distal areas of the organ. Proximally they formed long, distinct bundles running just beneath the surface of the organ (Fig. 98) or in the vicinity of blood vessels. Moreover prominent bundles were found subependymally. By contrast in the pineals of *Pan* and *Cebus* the glial fibres formed short bundles (MERKER, 1968).

An important point is that not all the glial fibres found in the pineal parenchyma originate from within the pineal. There is clear evidence that glial fibres may originate in structures lying outside the pineal (rabbit: WARTENBERG and GUSEK, 1964, 1965; cat: WARTENBERG, 1966, 1968; KUSCHE, 1966; rhesus monkey: HÜLSEMANN, 1967; squirrel monkey: WARTENBERG, 1968; *Pan, Ateles*: MERKER, 1968). In the rhesus monkey, these fibres reach the epiphysis via the stalk (Fig. 99) and originate from the subependymal layer of the 3rd ventricle (ECKHARDT, 1972). In non-primate mammals the amounts of glial fibres vary characteristically from species to species. No glial fibres could be demonstrated

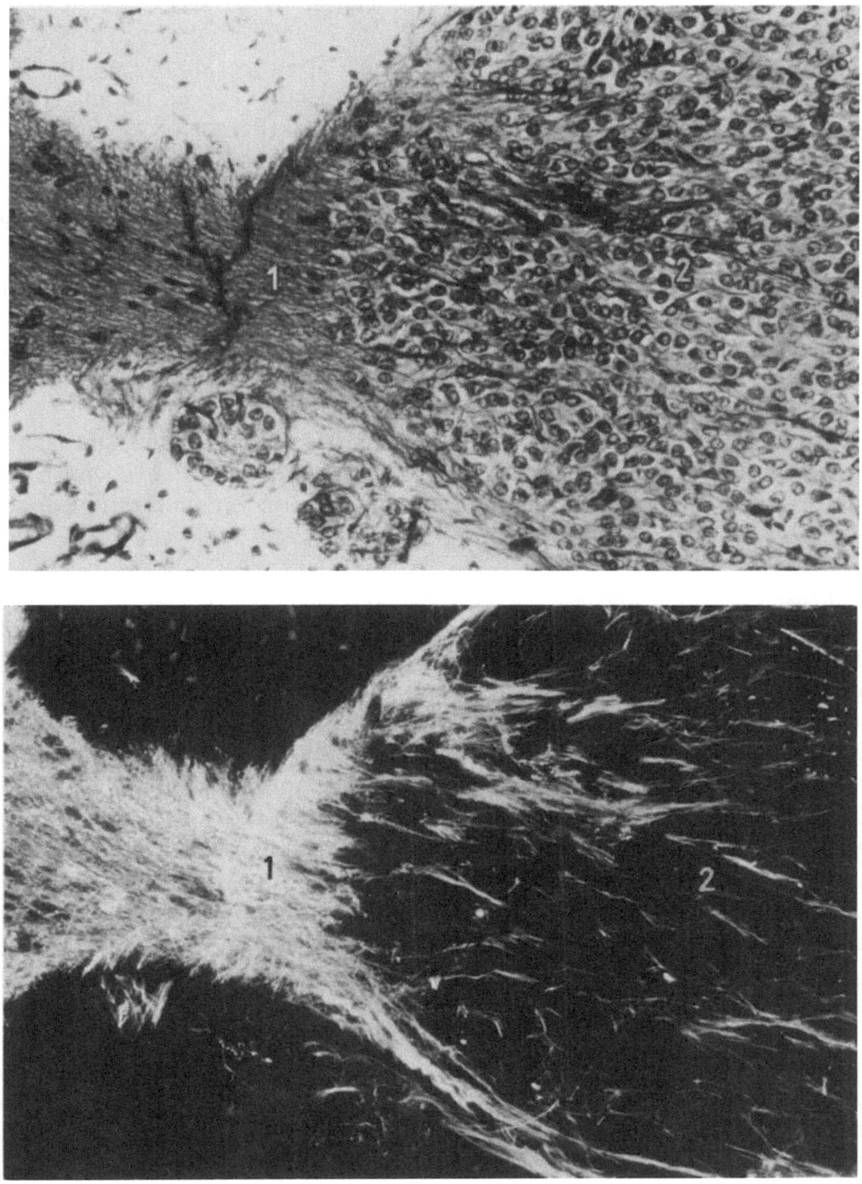

Fig. 99. Light micrographs showing that in the pineal bodies of Rhesus monkeys glial fibres reach the organ via the stalk region. *Top*: conventional light microscopy; *bottom*: fluorescence microscopy. HÜLSEMANN, 1967

in guinea-pig, mouse, *Jaculus* (HÜLSEMANN, 1967) and rat (GUSEK and SANTORO, 1960, 1961; WOLFE, 1965). They were rare in rabbit (WARTENBERG and GUSEK, 1965; HÜLSEMANN, 1967) and gopher (SHERIDAN and REITER, 1973), but regularly present in hedgehog, *Tupaia*, dog, horse, sheep, dormouse (HÜLSEMANN, 1967) and cat (WARTENBERG, 1965; KUSCHE, 1966). In the hamster, they were rare in the superficial epiphysis (SHERIDAN and REITER, 1968) and frequent in the deep pineal (SHERIDAN and REITER, 1970b).

The localization of the glial fibres also varied. Species exhibiting a prominent subcapsular or marginal distribution include hedgehog, *Tupaia*, rabbit (HÜLSEMANN, 1967), cat (ZACH, 1960; KUSCHE, 1966), horse (FASSBENDER, 1962), cow, sheep (LANG, 1959) and pig (HEINIGER, 1965a). Moreover, accumulations of glial fibres have been noted near blood vessels and connective tissue septa in dormouse (HÜLSEMANN, 1967) and horse (FASSBENDER, 1962). In some species, the central areas of the epiphysis contained an abundance of glial fibres (cat: ZACH, 1960; KUSCHE, 1966; horse: FASSBENDER, 1962). Glial fibres surrounding individual or groups of pinealocytes have been mentioned in pig (HEINIGER, 1965a), sheep and cattle (LANG, 1959; HOLLMANN, 1963; ANDERSON, 1965; HÜLSEMANN, 1967). Glial patches have been observed by HOLLMANN (1963), but could not be detected by LANG (1959).

Relatively little detailed information is currently available on the light microscopy of protoplasmic astrocytes. Electron microscopic studies have shown that in the cat and monkey protoplasmic astrocytes form an uninterrupted glial barrier separating pinealocytes from perivascular and subarachnoidal spaces (WARTENBERG, 1968).

Ultrastructurally epiphyseal astrocytes differ from pinealocytes in having larger amounts of condensed chromatin, a smaller number of nuclear indentations and larger amounts of both endoplasmic reticulum and lysosomes (WARTENBERG, 1968). The mitochondria are smaller and more electron-dense than those of pinealocytes (ANDERSON, 1965; WARTENBERG, 1968). In the deep pineal of the hamster they possess triangular-shaped hollow tubules ("prismatic" mitochondria) (Fig. 100) similar to astrocytes of the cerebral cortex (SHERIDAN and REITER, 1970b). Sometimes large amounts of glycogen are present (ANDERSON, 1965; ROMIJN, 1973b). The processes of astrocytes, attached to one another by desmosomes, intermediate junctions, tight junctions (ANDERSON, 1965; WARTENBERG, 1968) or half-desmosomes (CUELLO, 1973), contain abundant filaments 4–8 nm thick (ANDERSON, 1965; WARTENBERG, 1968; CUELLO, 1973), as illustrated in Fig. 101. They terminate among pinealocytes, on other glial cells or near perivascular spaces (ANDERSON, 1965; ROMIJN, 1973b).

The interstitial or fat-laden cells of the rat pineal, which, as pointed out above, appear to be astrocytes, differ from pinealocytes in that they are smaller in size, have nuclei which lack indentations and comprise large amounts of rough endoplasmic reticulum, many fat droplets and a particularly large Golgi complex (ARSTILA and HOPSU, 1964; ARSTILA, 1967).

Virtually nothing is known concerning the function of the astrocytes or astrocyte-like interstitial cells in the epiphysis cerebri, other than what is known about astrocytes in general. It has been suggested that, in the rat, they may contain arginine vasotocin (BOWIE and HERBERT, 1976), which would explain why these cells contain large amounts of rough endoplasmic reticulum. BOSTELMANN (1968b) is of opinion that following pineal lesions they differentiate into macrophages and hence do not represent typical astrocytes. SCHON et al. (1975) have shown that rat pineal gliocytes, but not pinealocytes and sympathetic nerve fibres, take up GABA. For further details on GABA, see MATA et al. (1976).

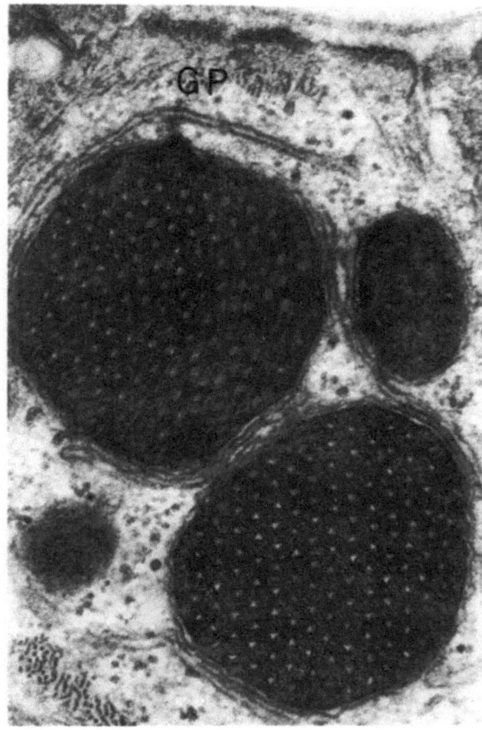

Fig. 100. Mitochondria with prismatic crests from astrocytes of a hamster pineal. SHERIDAN and REITER, 1970b

3. Ependymal Cells

Depending on the position of the pineal body in relation to the third ventricle, it shows a more or less intimate relationship to ependymal cells. Information on the light and electron microscopic appearance of these ependymal cells is relatively rare, but highly desirable in view of the concept that functional interrelationships may exist between the pineal body and the cerebrospinal fluid of the third ventricle (p. 266).

In *man,* the ependymal lining covering the pineal parenchyma in certain regions consists of flat to cuboidal cells in the region of the suprapineal recess and similar cells together with some columnar cells in the pineal recess (BARGMANN, 1943). Electron microscopic studies in human fetuses have shown that the apical parts of the ependymal cells in the pineal recess are joined by zonulae adherentes, showing a dense filamentous network on the cytoplasmic side, of the junction. As indicated by studies involving ruthenium red, these junctions are apparently impermeable (MØLLER, 1976).

Underlying the ependymal layer of the suprapineal recess is a rather compact layer of connective tissue (BARGMANN, 1943), so that pinealocytes do not appear to come into direct contact with the ependymal lining in this region. Occasionally rosettes probably formed by ependymal elements of the subcommissural organ are incorporated into the pineal parenchyma (PALKOVITS et al., 1962a, b, c).

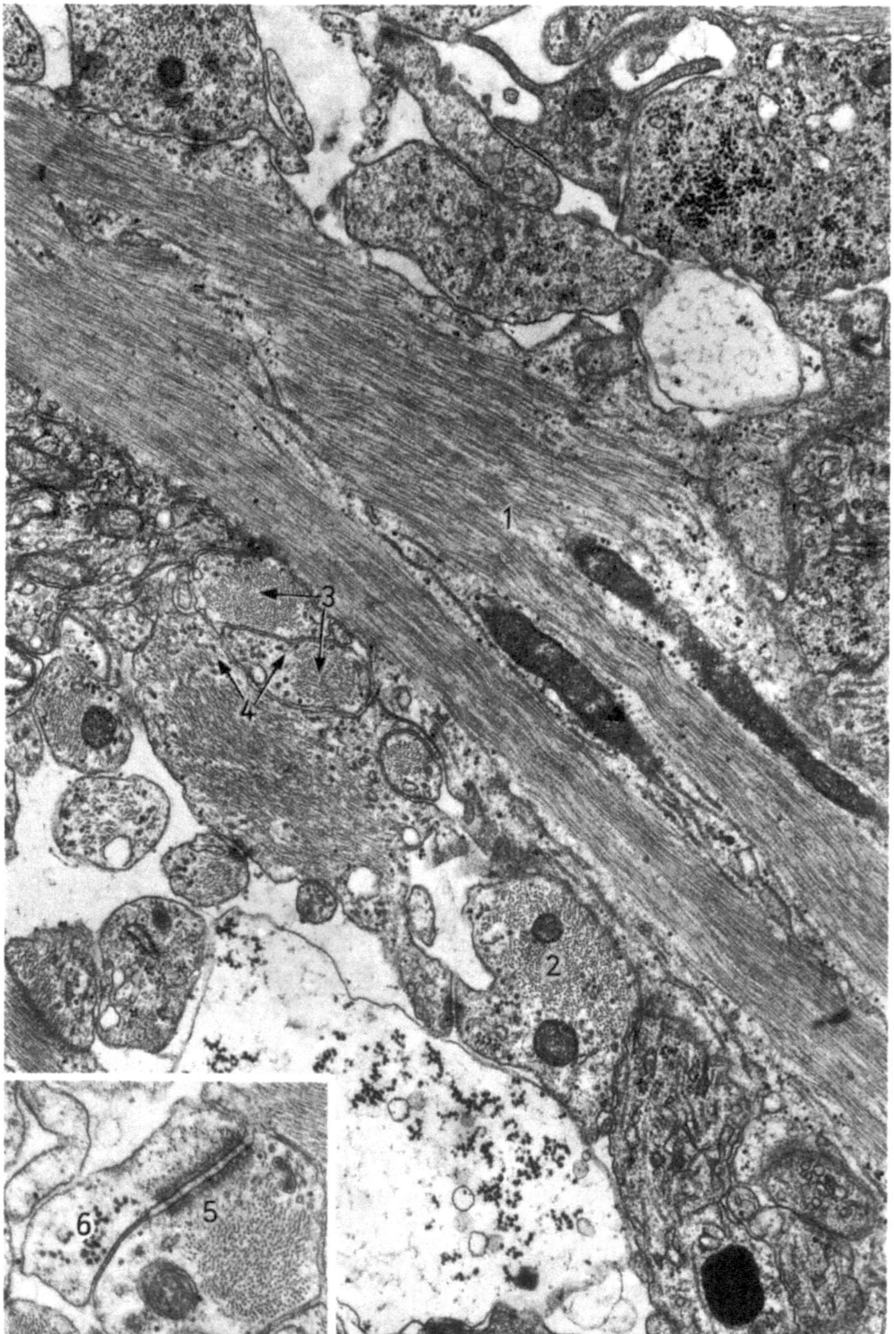

Fig. 101. Glial processes showing an abundance of fine filaments. Squirrel monkey. *1*, process cut longitudinally; *2*, cross section, containing many filaments (*3*), and some tubules (*4*); *5*, desmosome (macula adhaerens); *6*, tight junction (zonula occludens). WARTENBERG, 1968

In *primates* other than man, the tall ependymal cells decreased in height in the pineal recess (HOSAKA et al., 1957; MERKER, 1968; QUAY, 1970b), the ependymal lining being discontinuous in some areas in *Macaca mulatta* (Eckhardt, 1972). In the orangutan, *Pongo pygmaeus*, the ependyma bordering the recess is irregular and extensively perforated by complexly branching channels which run for 100–200 µm within the pineal body and become progressively attenuated (QUAY, 1970b). As in man, a subependymal fibrous layer is present in non-human primates. It is very dense in *Macaca* and *Cebus* and loose and net-like in *Ateles* (MERKER, 1968). Ependymal cells arranged in rosettes are seen in the pineal recess, but not within the epiphysis cerebri. Ependymal cells resembling those of the subcommissural organ have been found incorporated into the lining of the pineal recess of *Macaca* and *Pan* (MERKER, 1968). WARTENBERG (1968), using the electron microscope, noted that in rhesus and squirrel monkey pineals the tall ependymal cells and their nuclei do not have the regularity seen in other regions of the wall of the third ventricle. The nuclei, irregular in shape, possess deep invaginations. The cytoplasm is studded with ribosomes and a Golgi apparatus is well developed. The apical cell surface is packed with microvilli and cilia. Basally processes emerge extending into hypendymal layers.

In *cow, sheep* (LANG, 1959), and *horse* (FASSBENDER, 1962) the ependymal lining tends to be flat and simple in the suprapineal recess. In the pineal recess it is pseudostratified, columnar and discontinuous in some areas. Occasionally, pinealocytes seem to be interspersed among the ependymal cells. Electron microscopically both microvilli and cilia are seen in the pineal recess (ANDERSON, 1965). In the pineal recess of the horse, stereocilia are poorly developed, in contrast to the suprapineal recess, which also contains kinocilia (FASSBENDER, 1962). As in man, the cattle pineal contains rosettes formed by ependymal cells. According to OLSSON (1958) two types of rosettes may be distinguished. One type contains granulated ependymal cells similar to ependymal cells of the subcommissural organ and is restricted to the ventral and caudal areas of the organ. The second type consists of non-granular cells and can be found anywhere in the organ.

Information on ependymal cells in carnivores is restricted to *cats* (FLEISCHHAUER, 1964; KUSCHE, 1966; WARTENBERG, 1968), in which the cells show great variation in shape, ranging from flat to cylindrical (Fig. 102). In addition to long cilia, the cells are characterized by the presence of pleomorphic nuclei possessing phloxinophilic, PAS-negative inclusions. Subependymal fibres are not present, but occasionally processes of tanycytic ependymal cells extend into the pineal parenchyma. As regards the transport of chemical substances across the ependyma, see p. 267. In the *dog* pineal gland a network of yellow fluorescent nerves (formaldehyde-induced fluorescence) extends to the deepest part of the pineal recess, forming an intimate connection with the ependymal cells in this region (BERTLER et al., 1963).

Recent studies in rodents by means of scanning electron microscopy have shown that the ependymal lining of the pineal is not uniformly built. According to WEINDL and SCHINKO (1977), in the hamster it consists of non-ciliated cells in both the suprapineal and pineal recesses, although according to SHERIDAN and REITER (1970b) transmission electron micrographs clearly showed a few

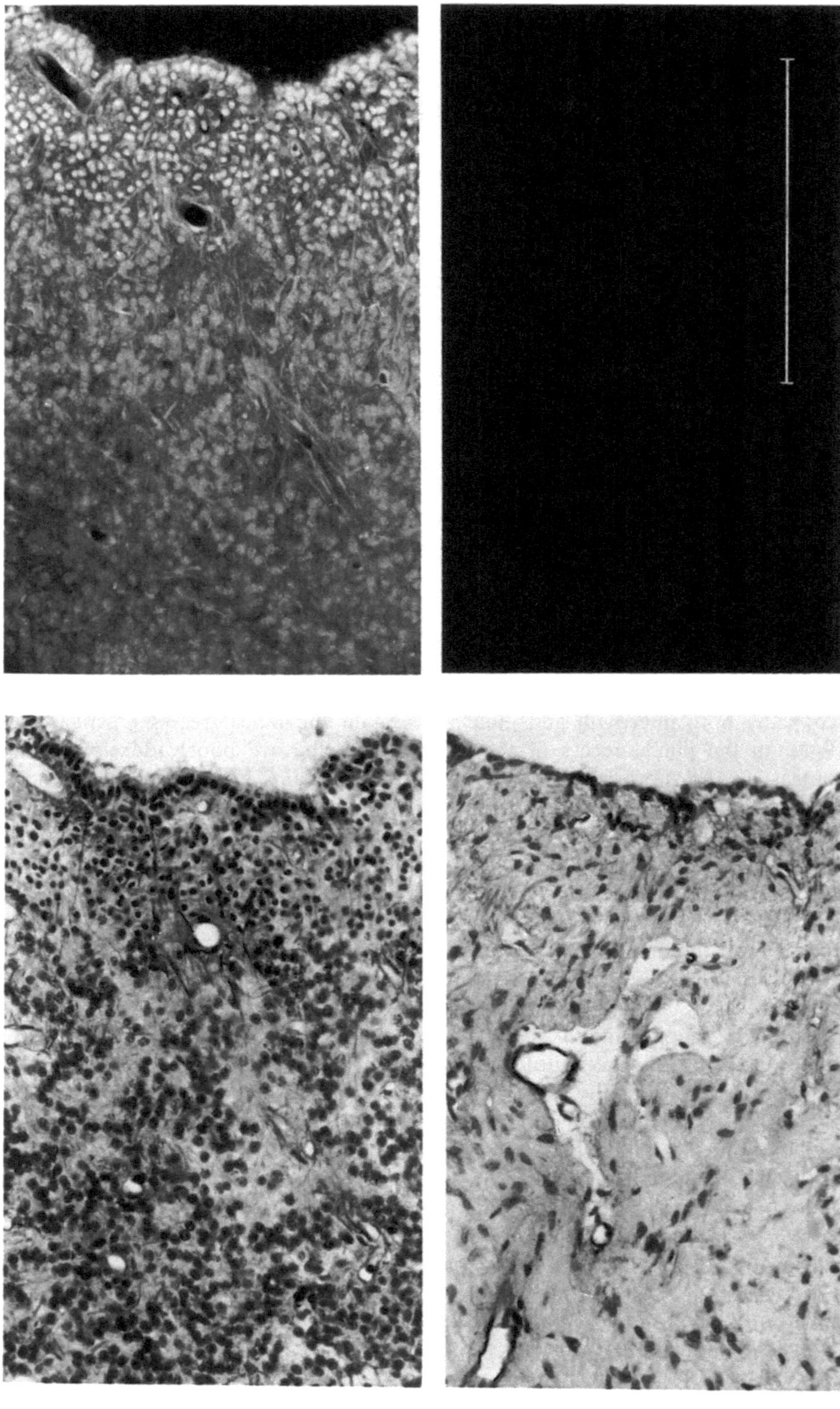

Fig. 102

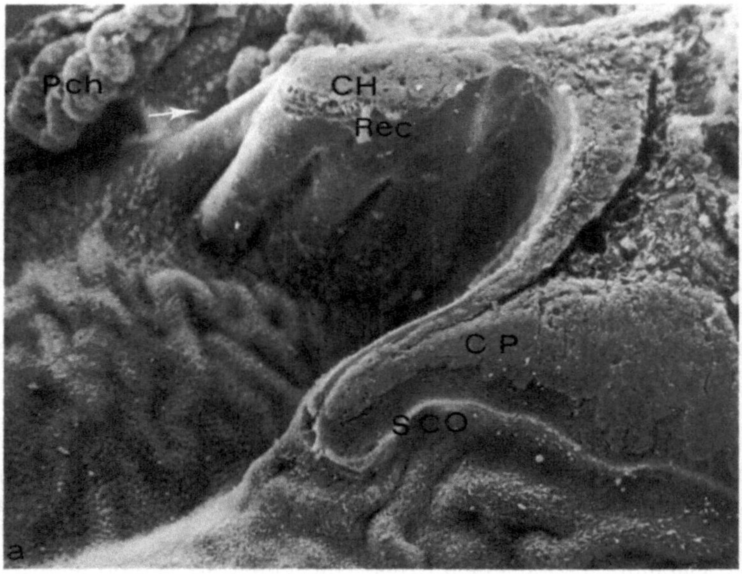

Fig. 103 a, b. Scanning electron micrographs of a mid-sagittal section through a hamster brain. **a** Survey to show the pineal recess (*Rec*), habenular (*CH*) and posterior (*CP*) commissures, choroid plexus (*Pch*) and subcommissural organ (*SCO*). × 100 **b** Suprapineal recess showing supra-ependymal cells, which are macrophage-like. The cells below are covered by microvilli. WEINDL and SCHINKO, 1977

Fig. 102. Light micrographs showing the ependymal covering of the cat epiphysis cerebri (*left*). Penetration of 3′, 6′-diaminoacridine trihydrochloride into the epiphysis (*top left*) but not into the neurohypophysis (*top right*), when injected into the lateral ventricles. *Bottom right:* section of the neurohypophysis adjacent to that shown above. FLEISCHHAUER, 1964 cilia in the pineal recess, adjacent ependymal cells being joined by maculae adherentes. Most interesting is the finding that the ependymal lining of the pineal recess may be lacking in a fairly large area, so that pinealocytes can come into direct contact with the cerebrospinal fluid (HEWING, 1978). Scanning electron microscopic studies have shown the presence of macrophage-like supra-ependymal cells in the suprapineal recess of the hamster (Fig. 103).

In the guinea-pig (KRAPP, 1978), the pineal parenchyma was found to be separated from the cerebrospinal fluid of the third ventricle by a continuous ependyma consisting of cells bearing either microvilli alone or microvilli plus kinocilia, the latter measuring 2–25 μm in length. Moreover, supra-ependymal neurons have been observed.

In a marsupial, *Trichosurus vulpecula* (TULSI, 1977, 1978a, b), the pineal recess is characterized by an abundance of morphologically heterogeneous supra-ependymal cells. Several of these cells look remarkably like neurons with as many as eight processes and not uncommonly form a network extending over several hundred microns.

In the elephant, *Loxodonta africana*, the epiphysis cerebri is covered by ependyma consisting of ciliated columnar cells, some of which bear pear-shaped apical protrusions. Ependymal cells arranged in rosettes or cell buds are rare in the floor of the pineal recess and relatively frequent in its roof (HAUG, 1972).

VII. Innervation

The mammalian epiphysis cerebri contains principally three groups of nerve fibres, namely sympathetic, commissural and parasympathetic fibres. Whereas sympathetic nerve fibres have been clearly shown to be of great importance for the proper functioning of the epiphysis cerebri, relatively little is known about the functional significance of both the commissural and the parasympathetic nerve fibres. Occasionally, neurosecretory fibres have been found. In some species, pineal ganglia are present.

1. Sympathetic Innervation

The presence of sympathetic nerve fibres within the epiphysis cerebri, which had long been suspected (see BARGMANN, 1943; HARTMANN, 1957; KAPPERS, 1960a), was clearly established by KAPPERS (1960a, b, 1961, 1964) in the albino rat. Since the pineal was subsequently regarded as an exclusively sympathetically innervated organ, it became a highly favoured object of pharmacological research. The importance of the sympathetic nerve fibres resides in the fact that they release noradrenaline, which in turn stimulates melatonin formation via β-adrenergic receptors and the adenyl cyclase–cAMP system (p. 10). In addition to noradrenaline, the fibres contain serotonin released from pinealocytes (p. 190) and octopamine (MOLINOFF and AXELROD, 1972; SAAVEDRA, 1974; JAIM-ETCHEVERRY and ZIEHER, 1975; AXELROD and SAAVEDRA, 1977).

a) Establishment of the Sympathetic Nature of the Sympathetic Nerve Fibres in the Epiphysis Cerebri

The sympathetic nature of the nerve fibres in question has been established by a number of means, such as interference with the superior cervical ganglia, autoradiography, administration of false transmitters, chemical sympathectomy, fluorescence histochemistry, electron microscopy and pharmacological studies.

That the mammalian pineal organ is innervated by sympathetic nerves was shown as early as 1947, when THIÉBLOT et al. (1947a) found that in cats electrical stimulation of the superior cervical ganglion was followed by histological changes in the pineal, namely a greater polymorphism of pinealocyte nuclei when compared with controls. In a similar approach VOLKMAN and HELLER (1971) found that the NAT activity of the pineal increased more than threefold, whereas HIOMT activity decreased (HELLER, 1972). Preganglionic electrical stimulation in *blinded* animals, by contrast, led to an 18% reduction of pineal HIOMT activity (BROWNSTEIN and HELLER, 1968).

The sympathetic innervation of the mammalian epiphysis cerebri came into the limelight when it was shown that after superior cervical ganglionectomy, intrapineal nerve fibres began to fragment after 8 days and were almost completely lacking after 2 months or more (KAPPERS, 1960a). These findings were later amply confirmed by electron microscopy (rat: RODIN and TURNER, 1965; ARSTILA, 1967; JAIM-ETCHEVERRY and ZIEHER, 1968; cat: DUNCAN and MICHELETTI, 1966; hamster, mouse: PELLEGRINO DE IRALDI, 1966, 1969; rabbit: ROMIJN, 1975a, b; ferret: JOHNSON et al., 1972) and histochemistry involving formaldehyde-induced fluorescence (BERTLER et al., 1963; OWMAN, 1964a, b; HÅKANSON and OWMAN, 1965; ERÄNKÖ et al., 1970; BJÖRKLUND et al., 1972) and the acetylcholinesterase (AChE) reaction (A. MACHADO and LEMOS, 1971). A disappearance of the sympathetic nerve fibres from the epiphysis cerebri after superior cervical ganglionectomy is also suggested by the fact that the noradrenaline content of the organ decreases by 84.9% (PELLEGRINO DE IRALDI and ZIEHER, 1966).

The occasional persistence of nerve fibres has been attributed to incomplete removal of the superior cervical ganglia, rare leptomeningeal fibres entering the pineal (KAPPERS, 1960a) and central fibres coming from the habenular nuclei (DAVID et al., 1975).

In some studies the sympathetic nature of the nerve fibres under consideration was confirmed by severing the preganglionic nerve fibres supplying the superior cervical ganglia, which apparently takes longer than ganglionectomy to produce demonstrable changes in the pineal (ROMIJN, 1975a). This may explain OWMAN'S (1964a, b) finding that no effects were noticeable 4 days after decentralization.

Moreover, sympathectomy by means of the false transmitter 6-hydroxydopamine has provided evidence for the sympathetic nature of the fibres under consideration (rat: ERÄNKÖ and ERÄNKÖ, 1971; rabbit: ROMIJN, 1975b, 1976).

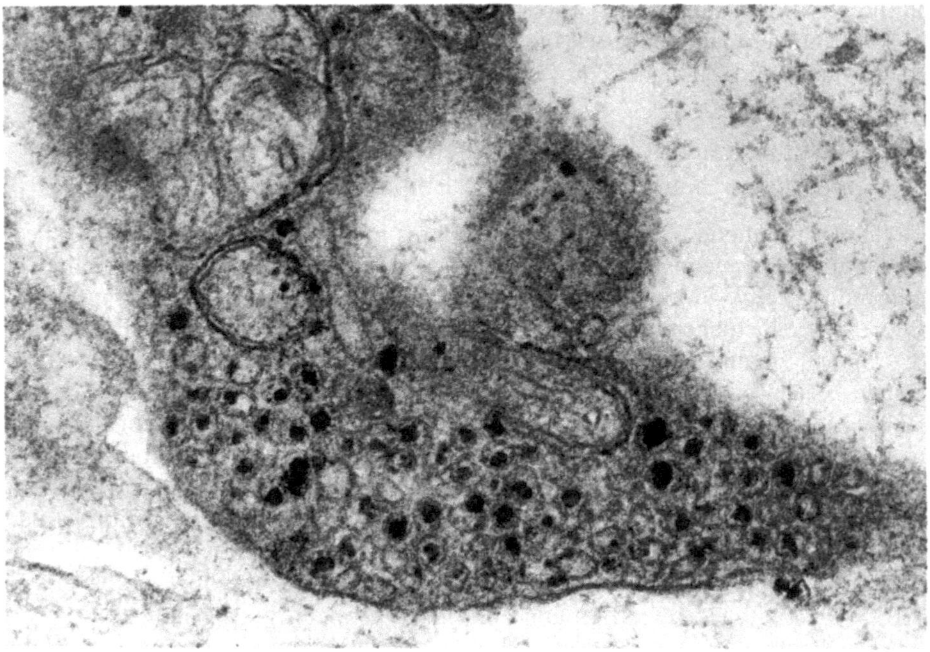

Fig. 104. Sympathetic nerve fibres of the rat pineal gland, showing some dense-core vesicles and some clear vesicles. Compare with Fig. 105. RICHARDS and TRANZER, 1969; ×75,000

Sympathectomy by means of nerve growth factor antiserum has been described by C. MACHADO et al. (1969a). The effects of sympathectomy on pinealocytes will be mentioned on p. 204.

The aminergic nature of the fibres under question can also be demonstrated by the application of false transmitters. BONDAREFF (1966a), using the noradrenaline analogue α-methylnoradrenaline following the depletion of the dense-core vesicles by means of metaraminol, noted that the analogue occurred in vesicles which are normally associated with the storage of noradrenaline. 5-Hydroxydopamine uptake (Figs. 104, 105) was clearly demonstrated by RICHARDS and TRANZER (1969) and by PELLEGRINO DE IRALDI et al. (1971). In the ferret (JOHNSON et al., 1972) nerve fibre vesicles were not labelled by 5-hydroxydopamine, but by 5-hydroxydopa.

The uptake of amines into terminals of the nerve fibres under consideration has been corroborated by autoradiography at the electron microscopic level. WOLFE et al. (1962) were the first to show a selective uptake of tritium-labelled noradrenaline by the nerve fibre terminals. That the uptake is restricted to the nerve terminals and does not occur in preterminal areas was subsequently confirmed in the mouse (BUDD and SALPETER, 1968, 1969). A labelling was also obtained following injection of ^{3}H-dopamine, ^{3}H-adrenaline, ^{3}H-Dopa and ^{3}H-5-hydroxytryptophan (TAXI and DROZ, 1966; TAXI, 1969). The uptake of ^{3}H-5-hydroxytryptophan (Fig. 106), confirmed by BAK et al. (1970), is of

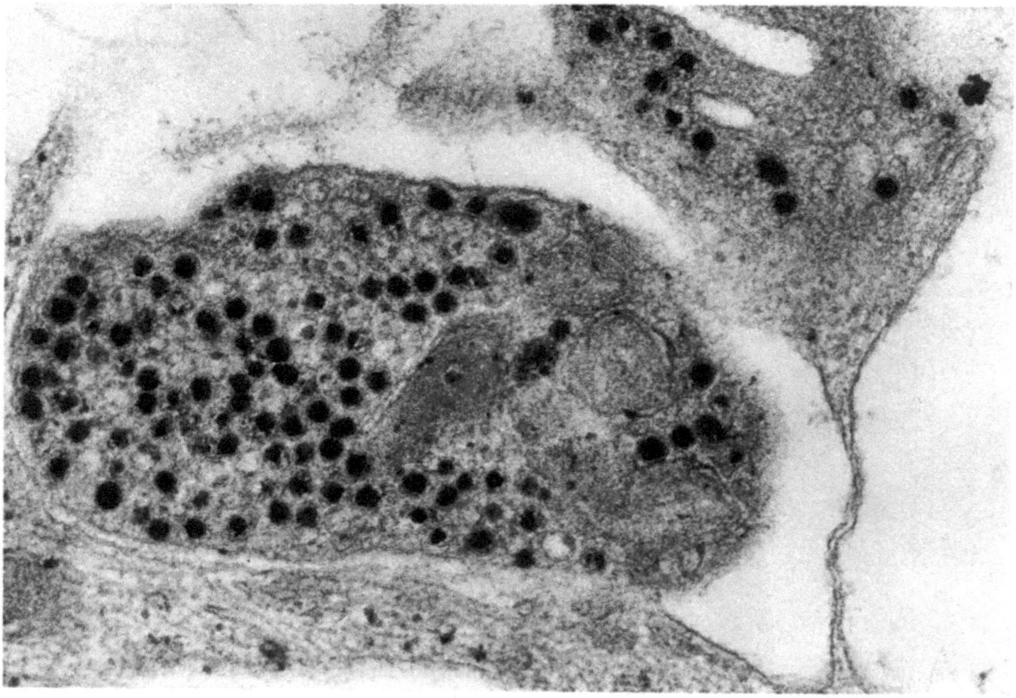

Fig. 105. Sympathetic nerve fibres from the pineal of a rat treated with 5-hydroxydopamine to demonstrate aminergic nerve fibres. Note the intense labelling of vesicles. Compare with Fig. 104. RICHARDS and TRANZER, 1969; × 75,000 (reduced to 85%)

particular interest since it points towards an uptake of the 5-hydroxytryptophan derivative, serotonin (p. 190).

A widely used technique for demonstrating the aminergic nature of the nerve fibres in the pineal has been the formaldehyde-induced fluorescence method of FALCK and HILLARP (FALCK et al., 1962). Except in man (HÅKANSON and OWMAN, 1966), the method clearly reveals the amines present in these fibres (Fig. 107) and the distributional patterns of the aminergic fibre elements (p. 192).

Green fluorescence due to the presence of noradrenaline has been demonstrated in pineal nerve fibres of cat, rabbit (BERTLER et al., 1963; HÅKANSON and OWMAN, 1966), sheep, pig, cattle (OWMAN, 1964a), ferret (TRUEMAN and HERBERT, 1970a, b), mole (PEVET et al., 1976a), Mongolian gerbil (NIELSEN and MØLLER, 1978), rhesus monkey (SHERIDAN and SLADEK, 1975) and hamster (OWMAN, 1964a), though in the latter species a yellow serotonin fluorescence has been detected by SHERIDAN and SLADEK (1975). Excitation and emission maxima of the green fluorescent fibres are at 415 and 475 μm respectively (NIELSEN and MØLLER, 1978), while fibres containing both noradrenaline and serotonin have two emission maxima at 480 and 520 μm, respectively (SHERIDAN and SLADEK, 1975).

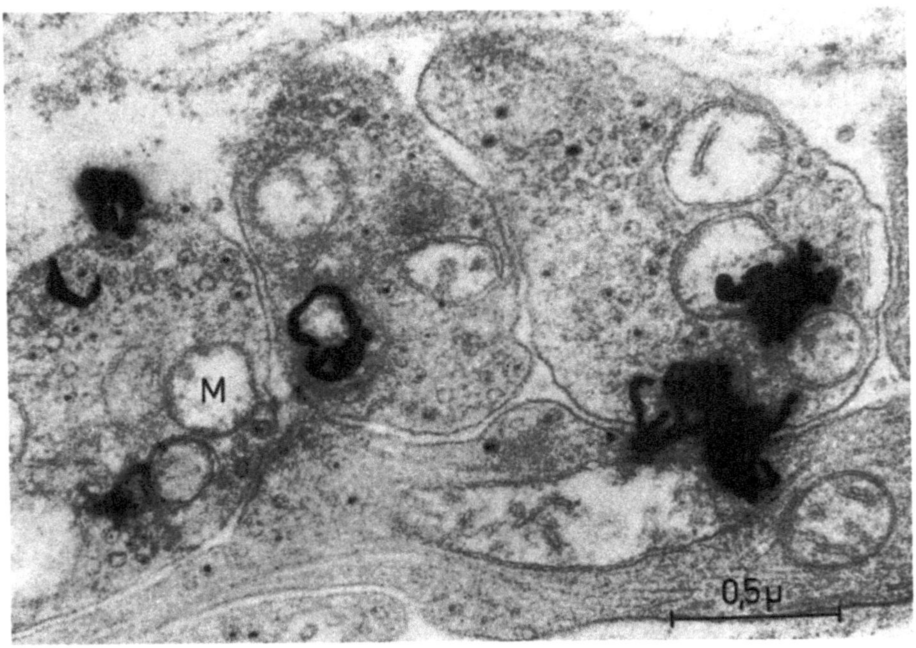

Fig. 106. Electron microscopic autoradiogram to show the uptake of 5-hydroxytryptophan into sympathetic nerve fibres of the rat pineal gland. *M*, mitochondrion. BAK et al., 1970

In many species, such as rat, mouse, guinea-pig and dog (BERTLER et al., 1963; OWMAN, 1963c, 1964a), the sympathetic fibres take up and retain serotonin (ZWEIG and AXELROD, 1969) amounting to about 30% of the total pineal serotonin content (NEFF et al., 1968, 1969) and resulting in a yellow fluorescence (OWMAN, 1963c, 1964a; ERÄNKÖ et al., 1970; BJÖRKLUND et al., 1972), with an emission peak at 520 nm (BJÖRKLUND et al., 1972). That the fibres are basically noradrenergic has been demonstrated by various means. To begin with, it has been shown that the sympathetic fibres penetrating the capsule of the epiphysis cerebri exhibit an intense green fluorescence and turn abruptly yellow when they enter the epiphyseal parenchyma (BERTLER et al., 1963, 1964; BJÖRKLUND et al., 1972; SHERIDAN and SLADEK, 1975). The same effect has been noted in pineal grafts transplanted into the anterior chamber of the eye (OWMAN, 1964b). In accordance with the assumption of an uptake of serotonin released from the pinealocytes is the observation that in newborn rats the nerve fibres showed a green fluorescence as long as the pinealocytes did not exhibit a yellow serotonin fluorescence (OWMAN, 1964a; C. MACHADO et al., 1968). Finally, there is the observation that by exposing the pineal sections for several minutes to UV light the yellow fluorescence of the nerve fibres is quenched and a greenish one persists (OWMAN, 1964b).

The fluorescence studies have also shown that distinct varicosities are present in intra-epiphyseal sympathetic nerve fibres, in contrast to those lying outside

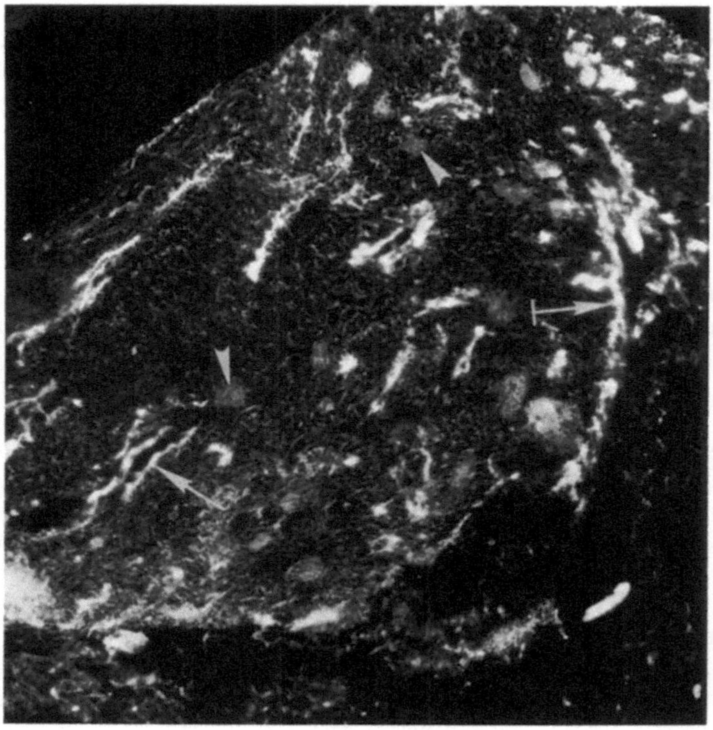

Fig. 107. Hamster pineal. Note the intense histofluorescence of the sympathetic nerve fibres inside the parenchyma (*arrow*) and the capsule (*barred arrow*). In addition, a small number of pinealocytes (*arrowheads*) show a distinct serotonin fluorescence. SHERIDAN and SLADEK, 1975; ×200

the parenchyma (BERTLER et al., 1963, 1964; OWMAN, 1964b; HÅKANSON and OWMAN, 1965; C. MACHADO et al., 1968).

Frequently, the sympathetic nerve fibres of the pineal have been studied fluorescence microscopically after the administration of various substances. Following intraperitoneal administration of *dl*-noradrenaline (5 mg/kg), the yellow serotonin fluorescence turned green, typical of catecholamines (BERTLER et al., 1963). A disappearance or decrease of the fluorescence was observed following administration of reserpine (BERTLER et al., 1963, 1964; BARRETT et al., 1969; BJÖRKLUND et al., 1972), α-methyl-*m*-tyrosine (BERTLER et al., 1964) or aramine (BERTLER et al., 1963). A marked enhancement of the fluorescence was obtained after treatment with nialamide, a MAO inhibitor (BERTLER et al., 1963; BJÖRKLUND et al., 1972). No changes in the intense yellow fluorescence of the intra-epiphyseal sympathetic fibres following nialamide were noted by FALCK et al. (1966).

Finally, the aminergic nature of the nerve fibres under consideration has been corroborated by electron microscopy, either alone or in combination with a number of cytochemical and pharmacological procedures. These results will be dealt with separately (p. 193).

b) Location of Sympathetic Nerve Fibres

The sympathetic nerve fibres innervating the epiphysis cerebri reach the organ by two routes. According to KAPPERS (1960a), in the rat the majority of sympathetic fibres reach the organ as nervi conarii, whereas a small number accompany pial vessels (KAPPERS, 1965a). An interesting exception appears to be the macaque monkey, in which the nervus conarii has been claimed to be parasympathetic (KENNY, 1961). That the nervi conarii do indeed comprise the sympathetic fibres innervating the pineal has been shown by the fact that after bilateral transection of the nervus conarii most of the fluorescent nerve fibres of the epiphysis cerebri disappear (OWMAN, 1964b).

The nervus conarii, which according to studies in the rabbit consists of some 150 axons (ROMIJN, 1973a), is basically a paired nerve, as can be expected from its origin from the paired superior cervical ganglia of the sympathetic trunk. In some species the nervi conarii reach the epiphysis cerebri as a single bundle (e.g. cat, bat: KENNY, 1965a). Two nervi conarii have been described in rat (KAPPERS, 1960a; KENNY, 1965a), rabbit (ROMIJN, 1973a) and man [in both adult (KENNY, 1965a) and fetal specimens (MØLLGARD and MØLLER, 1973)]. Since in earlier studies (see KAPPERS, 1965a) the presence of a single nervus conarii was described in the human, there is the possibility of interindividual variation.

In man (KENNY, 1965a), the nervi conarii were observed to:

run anteriorly beneath the floor of the great cerebral vein from the dura mater, at the junction of the vein and the floor of the straight sinus, through the suprapineal arachnoid body towards the posterior pole of the pineal body, passing laterally to a prominent median superior cerebellar vein that drains into the great cerebral vein. In its course each nerve formed part of a neurovascular bundle, but the fibres of the nerve which penetrated the posterior pole at a number of sites in many instances enter the organ independently of vessels.

In animal species the sites of entrance of the nervi conarii vary. In the rat, the nervi conarii penetrate the epiphysis symmetrically at its dorsolateral surface (KAPPERS, 1960a). Fluorescence microscopically the fibres were shown to enter the organ in the form of two or more thick bundles of non-varicose fibres, which split up into smaller fascicles (BERTLER et al., 1963, 1964; OWMAN, 1964b; HÅKANSON and OWMAN, 1965; C. MACHADO et al., 1968). In the bat, the unpaired nerve extends to the posterior pole of epiphysis and the majority of fibres enter the organ at the superior surface of the posterior one-third, whereas a small number run along the inferior surface (KENNY, 1965a). In the cat, most of the fibres enter at the posterior pole, accompanied by a small arteriole (HARTMANN, 1957), the remaining fibres continuing anteriorly along the inferior surface of the organ as far as the anterior one-third (KENNY, 1965a). Unilateral removal of the superior cervical ganglion revealed that most of the fibres do not cross to the other side of the organ (RODRIGUEZ-PEREZ, 1962).

The sympathetic fibres which do not form part of the nervi conarii have been reported to enter the epiphysis cerebri either at its inferior surface (HARTMANN, 1957: hedgehog, cat, dog, calf, *Macaca mulatta*) or along its entire surface (rabbit: ROMIJN, 1973a; deer: HARTMANN, 1957).

The intra-epiphyseal distribution of sympathetic fibres varies from species to species. Moreover, slightly divergent results have been obtained depending on the methods used. By fluorescence microscopy or ordinary light microscopy, sympathetic nerve fibres were seen to have an almost exclusively perivascular localization in the guinea-pig (OWMAN, 1964a). In the Mongolian gerbil, most of the fluorescent fibres lie perivascularly (NIELSEN and MØLLER, 1978). In rabbit and sheep they form a rich network mainly related to pinealocytes. In rat, hamster (see also SHERIDAN and SLADEK, 1975) and cat they occupy an intermediary position in that they are distributed both around blood vessels and in the parenchyma. In porcine and bovine pineals very few isolated nerve fibres were present (OWMAN, 1964a). In the hooded rat (GARDNER, 1953) the nerve fibres were apparently uniformly distributed throughout the gland. In the laboratory rat, perivascular nerves were densest in the very periphery of the organ (BERTLER et al., 1964).

Electron microscopically a mainly perivascular localization of the sympathetic nerve fibres has been described in rabbit (ROMIJN, 1975b), mouse (MATSUSHIMA and ITO, 1972), cat (DUNCAN and MICHELETTI, 1966), hamster, gopher (SHERIDAN and REITER, 1968, 1973), rat, ground squirrel (MATSUSHIMA and REITER, 1977) and others. In the chinchilla, by contrast, they lie mainly among parenchymal cells (MATSUSHIMA and REITER, 1977).

In species in which the pineal complex consists of different parts (p. 28), the sympathetic nerve fibres are not restricted to the pineal proper or superficial pineal. In the rat sympathetic fibres leave the superficial pineal, forming a green fluorescent nerve plexus which can be followed through the pineal stalk without interruption into the lamina intercalaris, or deep pineal, and thence into the medial habenular nucleus on both sides (BJÖRKLUND et al., 1972). In the deep pineal the green fluorescent fibres enclosed the intrinsic cells (BJÖRKLUND et al., 1972; WIKLUND, 1974). Following bilateral cervical ganglionectomy the fibres disappeared either partly (HÅKANSON and OWMAN, 1965) or completely (BJÖRKLUND et al., 1972). In the hamster (SHERIDAN and SLADEK, 1975; REITER and HEDLUND, 1976) and Mongolian gerbil (NIELSEN and MØLLER, 1978), as in the rat, sympathetic nerve fibres extend to the deep pineal.

c) Enzyme Histochemistry

In addition to MAO activity (p. 102), the sympathetic nerve fibres of the pineal are characterized by histochemically demonstrable AChE activity (ERÄNKÖ et al., 1970a, b; TRUEMAN and HERBERT, 1970a, b; A. MACHADO and LEMOS, 1971; DAVID and ANAND KUMAR, 1978). Following superior cervical ganglionectomy (A. MACHADO and LEMOS, 1971) or chemical sympathectomy with 6-hydroxydopamine (ERÄNKÖ and ERÄNKÖ, 1971) the enzyme activity disappeared. AChE-positive nerve fibres of unknown origin have been observed near blood vessels of cattle, sheep and pig pineals (ARVY, 1961). In bovine pineals, acetylcholine (0.6 μm/g) has been demonstrated (GIARMAN et al., 1959).

d) Ultrastructure

α) *Vesicle Content*

The numerous electron microscopic studies dealing with the innervation of the epiphysis cerebri have concentrated on the functionally most important

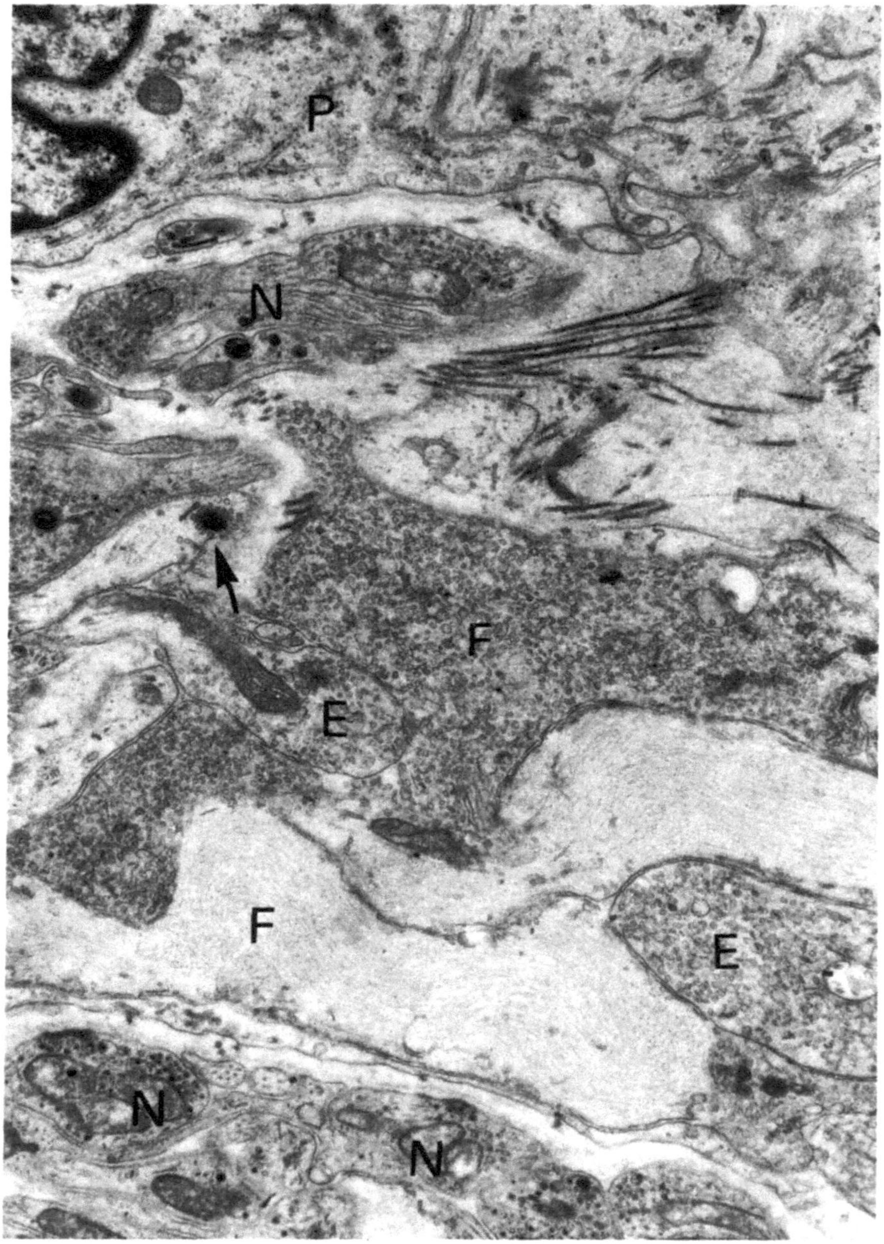

Fig. 108. Electron micrograph showing sympathetic nerve fibres (*N*), pinealocytes (*P*) and pinealocyte processes (*F*), some of which contain large amounts of clear vesicles not unlike those seen in nerve fibres. *E*, end-bulbs of pinealocyte processes. *Arrow*, synaptic spherule. Adult male guinea-pig. LUES, 1971; ×36,000

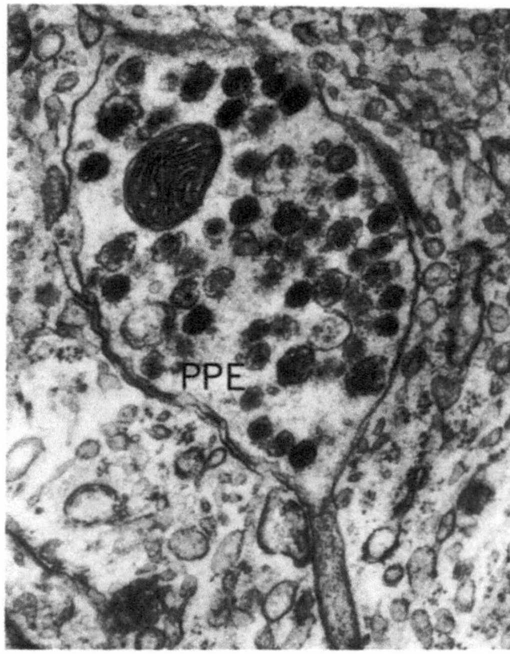

Fig. 109. Terminal of pinealocyte process (*PPE*) showing a large amount of dense-core vesicles. On first sight this may be mistaken as aminergic nerve fibre. SHERIDAN and REITER, 1968

parts of the sympathetic nerve fibres, the nerve fibre terminals. Since they are structurally similar to terminals of pinealocytes and often have an identical perivascular localization, the two can be easily confused (DE ROBERTIS and PELLEGRINO DE IRALDI, 1961a, b; PELLEGRINO DE IRALDI and DE ROBERTIS, 1961; rectification: PELLEGRINO DE IRALDI and DE ROBERTIS, 1963). Due to great interspecies variation of the vesicular contents of pinealocyte terminals (p.164), generally applicable criteria to distinguish them from nerve fibre terminals cannot be given. The complexity of the issue is illustrated in Figs. 108 and 109. In species where the terminals of pinealocyte processes contain a relatively large amount of dense-core vesicles (Fig. 109), these are generally larger than those of the nerve terminals.

Frequently it has been shown that the endings of the sympathetic nerve fibres of the pineal contain the plurivesicular content which is typical of adrenergic nerves (PELLEGRINO DE IRALDI et al., 1971), namely small and large dense-core vesicles as well as electron-lucent vesicles (Table 19).

Regarding the proportions of the dense-core and electron-lucent vesicles present, it has to be taken into consideration that they vary diurnally. The dense-core vesicles decrease in number between 08.00 and 24.00 h, after which there is a striking increase, reaching a maximum at 08.00 h; the electron-lucent vesicles show a roughly inverse behaviour (Figs. 110, 111) (MATSUSHIMA and ITO, 1972). Moreover the type of fixation matters: following fixation in glutaraldehyde-OsO_4 a greater amount (81%) of vesicles exhibited dense cores than after OsO_4 fixation alone (40%) (A. MACHADO, 1967a). In intact rats and mice 30%–60% of the vesicles examined contained dense cores (PELLEGRINO

Table 19. Size of vesicles in sympathetic nerve fibres

Species	Dense-core vesicles (DCV)				Clear vesicles (nm)	Author(s)
	Small DCV		Large DCV			
	Overall size (nm)	Size of dense core (nm)	Overall size (nm)	Size of dense core (nm)		
Monkey	40	20	100–120	70		WARTENBERG (1968)
Cat	40	20	100–120	70		WARTENBERG (1968)
Rabbit	40	25–30			40	WARTENBERG and GUSEK (1965)
	45–50	25	100		45–50	ROMIJN (1975b)
	25[a]		60[a]			ROMIJN (1975b)
Rat	40–50		90–120			DUFFY and MARKESBERY (1970)
	48.5; 51.6[b]		84.5; 85.8[b]	55.4; 54.6[b]		A. MACHADO (1967a)
	40	20				RODIN and TURNER (1965a, 1966)
	45		100			ARSTILA and RINNE (1967)
	50	20	100			BONDAREFF (1966a, b), BONDAREFF and GORDON (1966)
	45		100			HASSLER and BAK (1966)
	50		80–100		50	JAIM-ETCHEVERRY and ZIEHER (1969)
	50		100		50	BAK et al. (1970)
	45	25	100–120	65	45	ARSTILA (1967)
	42.5	21				PELLEGRINO DE IRALDI and DE ROBERTIS (1963)
	23.4[c]					KRSTIĆ (1972a)
	20–30[a]		50–70[a]			PELLEGRINO DE IRALDI and GUEUDET (1969)
	40–60[a]		80–100[a]			JAIM-ETCHEVERRY and ZIEHER (1968)
	25.5[a]		60[a]			PELLEGRINO DE IRALDI and GUEUDET (1968)
Mouse	65	25–30	100	60	65	MATSUSHIMA and ITO (1972)
	60	25				ITO and MATSUSHIMA (1968a)
	40–60				40–60	PELLEGRINO DE IRALDI (1969)
Hamster		20	Present			SHERIDAN and REITER (1968)
	40–60	20	60–120			SHERIDAN and REITER (1970b)
Guinea-pig	60	20–25	80	50		LUES (1971)
Gopher	50	20				SHERIDAN and REITER (1973)
Ground squirrel	60–80				20–60	POVLISHOCK et al. (1975)

[a] After $K_2Cr_2O_7$ staining.
[b] Depending on fixative used.
[c] After ZIO staining.

DE IRALDI and DE ROBERTIS, 1963; RODIN and TURNER, 1965; BONDAREFF, 1965a; PELLEGRINO DE IRALDI and GUEUDET, 1969; BAK et al., 1970; DUFFY and MARKESBERY, 1970; MATSUSHIMA and ITO, 1972). In the rabbit this figure was 75%–90% (ROMIJN, 1975b).

As mentioned above, dense-core vesicles fall into two categories, small and large, the small ones exceeding the large ones by a factor of 10–15 (Fig. 110) (MATSUSHIMA and ITO, 1972). In rat and ground squirrel about 2%–5% and

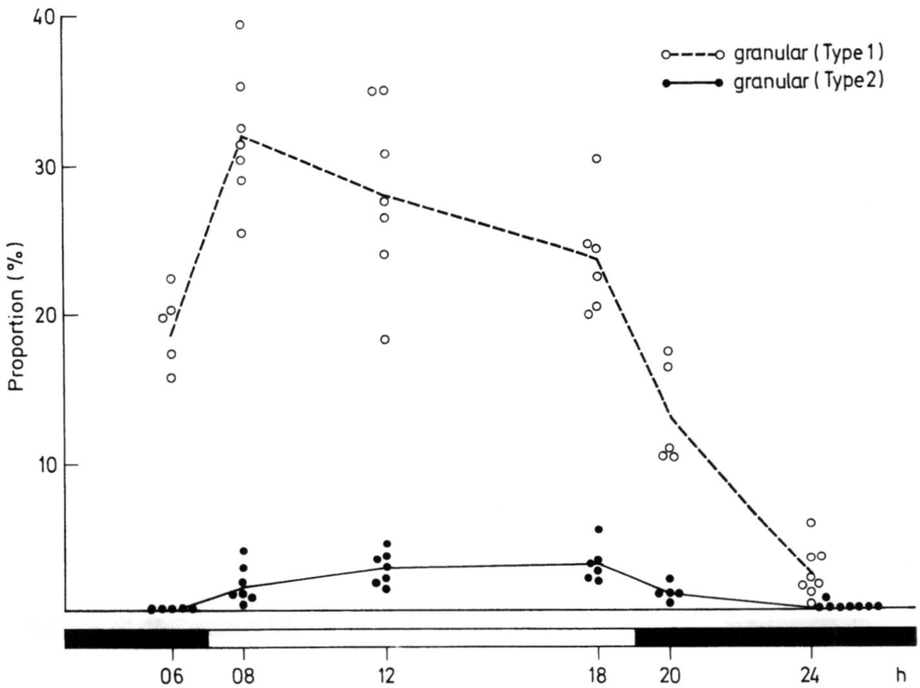

Fig. 110. Proportions and circadian behaviour of dense-core vesicles in terminals of sympathetic nerve fibres. Mouse pineal gland. Type 1 granules are characterized by a halo between the dense core and the bounding membrane. In Type 2 granules the dense core fills almost the entire vesicle. MATSUSHIMA and ITO, 1972

in the chinchilla about 10% of the total number of vesicles are large dense-core vesicles (MATSUSHIMA and REITER, 1977). As can be seen from Table 19, the small dense-core vesicles measure between 40 and 60 nm in diameter and possess an often eccentrically positioned dense core of some 25 nm. According to DUFFY and MARKESBERY (1970) a fairly large proportion (85%) of the vesicles examined contain delicate grains which are distinctly smaller than the dense cores. In contrast to the large dense-core vesicles, the small ones possess a fairly wide clear space between the central core and the bounding membrane. Since with appropriate methods the space can be shown to contain reactive substances it has been thought that it comprises a matrix (see p. 199).

The relatively rare large dense-core vesicles measure 60–120 nm (Table 19). They do not possess a clear halo between the dense core and the bounding membrane (DUFFY and MARKESBERY, 1970) and are more electron dense than the small ones (MATSUSHIMA and ITO, 1972). Histograms of vesicle sizes have been published by PELLEGRINO DE IRALDI and DE ROBERTIS (1961), PELLEGRINO DE IRALDI et al. (1965), ITO and MATSUSHIMA (1968a, b) and MATSUSHIMA and REITER (1977). It should be noted that vesicle size depends on the fixation procedure used. According to A. MACHADO (1967a) the mean size of the small dense-core vesicles is 48.5 nm in OsO_4-fixed material and 51.6 nm after glutaralde-

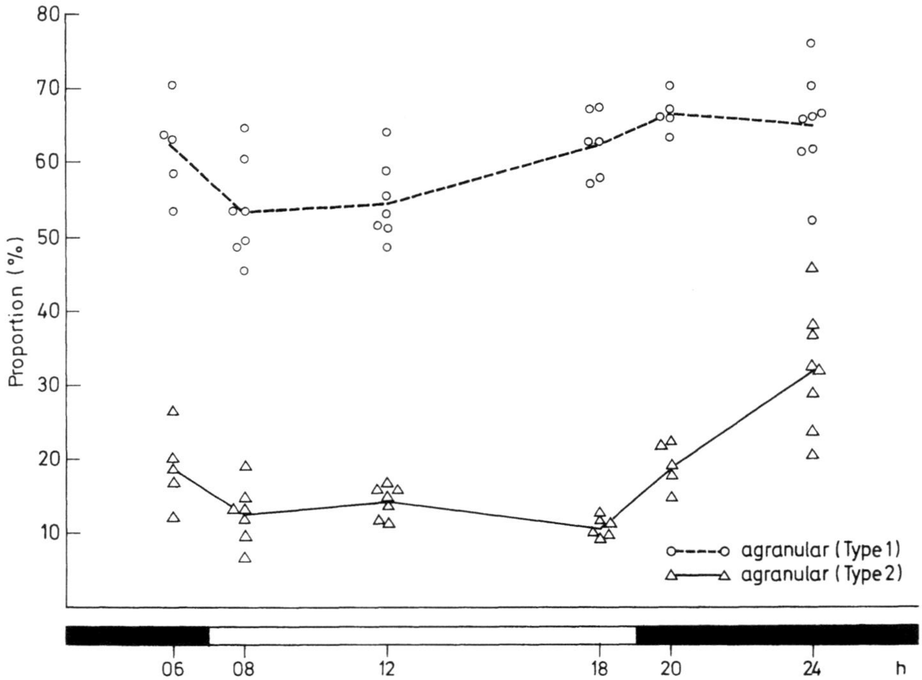

Fig. 111. Proportions and circadian behaviour of agranular vesicles in terminals of sympathetic nerve fibres. Mouse pineal gland. Type 1 agranular vesicles are spherical; Type 2 vesicles are elliptical or flat. MATSUSHIMA and ITO, 1972

hyde-OsO_4, the large dense-core vesicles measuring 85.8 nm and 84.5 nm respectively.

The clear vesicles can be subdivided into round and flat ones, but only some authors have done so (e.g. MATSUSHIMA and ITO, 1972). The clear, round vesicles have roughly the same size as the dense-core vesicles (for details see Table 19), and the flat ones measure 80 by 30 nm (MATSUSHIMA and ITO, 1972). Frequently pharmacological studies have been carried out to study the effects of certain drugs on the ultrastructural appearance of the dense-core vesicles. Repeatedly it has been shown that the administration of the amine-depleting agent, reserpine (1.0–5.0 mg/kg), was followed by an almost complete disappearance of the dense cores of the granulated vesicles (PELLEGRINO DE IRALDI and DE ROBERTIS, 1961, 1963; PELLEGRINO DE IRALDI et al., 1965; PELLEGRINO DE IRALDI and SUBURO, 1971a; PELLEGRINO DE IRALDI and GUEUDET, 1968; PELLEGRINO DE IRALDI, 1969; JAIM-ETCHEVERRY and ZIEHER, 1969; DUNCAN and MICHELETTI, 1966; ARSTILA, 1967; DUFFY and MARKESBERY, 1970; LUES, 1971; WOOD, 1973; ROMIJN, 1976). Morphologically demonstrable effects are apparent as early as 10 min after injection and maximal effects can be noted after 24 h (PELLEGRINO DE IRALDI and DE ROBERTIS, 1961). A decrease in the amounts of dense-core vesicles was also noted after oxypertine (ARSTILA, 1967; PELLEGRINO DE IRALDI and SUBURO, 1971a; DUFFY and MARKESBERY, 1970), aramine,

5-hydroxytryptophan, α-methyl-3-hydroxyphenylalanine (PELLEGRINO DE IRALDI et al., 1965; PELLEGRINO DE IRALDI and DE ROBERTIS, 1965), metaraminol (BONDAREFF, 1966b; BONDAREFF and GORDON, 1966), *p*-chlorophenylalanine (BLOOM and GIARMAN, 1967, 1970; PELLEGRINO DE IRALDI and GUEUDET, 1969; PELLEGRINO DE IRALDI and SUBURO, 1971a, b; PELLEGRINO DE IRALDI et al., 1971), α-methylmetatyrosine (PELLEGRINO DE IRALDI and DE ROBERTIS, 1965; RICHARDS and TRANZER, 1969), α-methyl-*p*-tyrosine (JOHNSON et al., 1972; see, however, ROMIJN, 1976) or desmethylimipramine (DMI) (DUFFY and MARKESBERY, 1970), which is in accordance with the view that the vesicles contain both noradrenaline and serotonin (see p. 190). Dense-core vesicles increased in amount following iproniazid (PELLEGRINO DE IRALDI and DE ROBERTIS, 1965; PELLEGRINO DE IRALDI et al., 1965; HASSLER and BAK, 1966) and angiotensin II (PANAGIOTIS and HUNGERFORD, 1966).

Cytochemical methods have frequently been applied at the ultrastructural level to analyse the contents of both the dense-core and the electron-lucent vesicles.

The presence of both noradrenaline and serotonin has been demonstrated in both dense-core and electron-lucent vesicles by means of the WOOD technique using potassium dichromate following different fixation procedures (PELLEGRINO DE IRALDI, 1969; PELLEGRINO DE IRALDI and GUEUDET, 1969; JAIM-ETCHEVERRY and ZIEHER, 1968, 1969, 1971; PELLEGRINO DE IRALDI et al., 1971; ROMIJN, 1975b). That the potassium-dichromate technique differentiates very accurately between aminergic and non-aminergic neurons has been clearly revealed in the rabbit, in which the triple fixation after TRANZER and SNIPES (1968) stained the vesicles in sympathetic nerve endings but not parasympathetic terminals (ROMIJN, 1975b).

A frequently applied staining method has been the zinc iodide – osmium tetroxide technique of CHAMPY-MAILLET. Although a positive reaction was first thought to be typical of cholinergic nerve fibres only, it was subsequently shown that in the epiphysis cerebri the sympathetic nerve fibres are stained by this procedure (PELLEGRINO DE IRALDI and GUEUDET, 1968; PELLEGRINO DE IRALDI and SUBURO, 1970, 1971a). With this technique it was even possible to stain differentially the core and the matrix of the dense-core vesicles, short-term impregnation (1/2 h) staining the central core and impregnation for 2 h and more staining the matrix, resulting in a pale dense core (PELLEGRINO DE IRALDI and SUBURO, 1970, 1971a). A differential staining of matrix and core can also be obtained if potassium iodide is used instead of zinc iodide, the matrix being more reactive to zinc iodide and the dense-core having a greater affinity to potassium or cadmium iodide (PELLEGRINO DE IRALDI and SUBURO, 1971a, b). Following administration of *p*-chlorophenylalanine, a serotonin-depleting agent, the zinc iodide-reactive material of both core and matrix disappears (PELLEGRINO DE IRALDI and SUBURO, 1971b). Potassium iodide-reactive material, on the other hand, remains unaffected and it has been suggested that it may represent a substance of unknown nature. Further studies (PELLEGRINO DE IRALDI and LETZEN, 1973) have shown that zinc iodide may reveal different substances in the matrix, but not the core: zinc iodide used at 4° C for 2 h demonstrates serotonin in the matrix of the vesicles in the pineal but not in the vas deferens;

the dense cores are stained in both organs when zinc iodide is applied at 20° C for 15 h. PELLEGRINO DE IRALDI and SUBURO (1971 a, 1972) have suggested that the tyramine-releasable pool of noradrenaline is present in the matrix and the tyramine-resistant pool in the dense-core.

KRSTIĆ (1972a) demonstrated that exposure of rats to low ambient temperatures leads to a distinct increase in both the amount and size of zinc iodide-positive vesicles.

From both the pharmacological and cytochemical studies it is clear that the *small* dense-core vesicles as well as the electron-lucent vesicles are intimately related to storage and release of noradrenaline and serotonin or their precursors. Whether or not all the electron-lucent vesicles are functionally identical is difficult to decide. In view of the current concept that the flat electron-lucent vesicles contain GABA, this possibility should also be considered for the pineal. Caution is necessary, however, since flat vesicles have been noted to occur after reserpine administration (PELLEGRINO DE IRALDI and SUBURO, 1971 a; DUFFY and MARKESBERY, 1970; WOOD, 1973). Furthermore, as their amount shows an inverse relationship to that of the dense-core vesicles (MATSUSHIMA and ITO, 1972), it is perhaps more likely that in the epiphysis cerebri they represent depleted dense-core vesicles. Another line of thought, triggered by the observation that round electron-lucent vesicles, in contrast to flat ones, do not change in amount over a period of 24 h, was that some, if not most, of the round clear vesicles are cholinergic in nature, representing a cholinergic link in postganglionic adrenergic transmission (MATSUSHIMA and ITO, 1972). In this context it is interesting to recall that epiphyseal sympathetic nerve fibres exhibit a positive reaction for acetylcholinesterase (ERÄNKÖ et al., 1970; A. MACHADO and LEMOS, 1971).

The functional significance of the *large* dense-core vesicles measuring 60–120 nm in diameter has not been unequivocally solved. Although they seem to contain noradrenaline in sympathetic fibres of organs other than the epiphysis cerebri, they have been reported to be unaffected by drugs including reserpine, metaraminol, *p*-chlorophenylalanine, oxypertine, DMI and pargyline (BONDAREFF, 1965a; BONDAREFF and GORDON, 1966; PELLEGRINO DE IRALDI and SUBURO, 1971 a; C. MACHADO et al., 1968; JAIM-ETCHEVERRY and ZIEHER, 1969; DUFFY and MARKESBERY, 1970), suggesting that they do not contain amines. However, JAIM-ETCHEVERRY and ZIEHER (1969), using the WOOD technique and phosphotungstic acid in ethanol, have suggested that the large vesicles contain, in addition to noradrenaline and serotonin, a substance stainable with phosphotungstic acid in ethanol and resistant to reserpine, which would explain the negative results observed by other workers. According to this concept the large dense-core vesicles are functionally basically similar to the small ones.

Information regarding morphological changes in sympathetic nerve fibres *in relation to function* is scanty. MATSUSHIMA and ITO (1972), studying the relative proportions of different types of vesicles in nerve endings of the mouse epiphysis cerebri over a period of 24 h, found that after the onset of darkness the dense-core vesicles showed a particularly striking decrease (see also MATSUSHIMA et al., 1979) while the clear flat vesicles increased in amount. This finding is in accordance with the concept that the release of noradrenaline, the content of which increases during darkness (WURTMAN et al., 1967 a), is responsible for the chemical changes leading to an increased formation of melatonin during darkness (see WURTMAN et al., 1968a).

An atrophy of the sympathetic fibres was observed after blinding (SATODATE et al., 1973), which deprives the epiphysis cerebri of its photic input.

β) *Other Constituents*

As pointed out above, the non-vesicular components of the sympathetic nerve fibres have received little attention. In addition to dense-core vesicles and electron-lucent vesicles, the endings of nerve fibres may contain mitochondria (e.g. WARTENBERG and GUSEK, 1965; BONDAREFF, 1965a; ARSTILA, 1967) as well as vacuoles (ARSTILA, 1967) and various saccular configurations of (smooth) endoplasmic reticulum (BONDAREFF, 1965a). Moreover, neurotubules and neurofilaments are present, which show a positive reaction for biogenic amines following treatment with sodium molybdate; it has been suggested that this reflects transport of biogenic amines from the perikarya to nerve fibre terminals (WOOD, 1973).

Nerve fibre terminals measured 0.56 μm (mean diameter) in perivascular spaces and 0.78 μm in the parenchyma, while non-terminal parts measured 0.47 and 0.45 μm respectively (ROMIJN, 1975b).

Large *preterminal* swellings of adrenergic fibres have been found in the pineals of rat, ground squirrel and chinchilla. In the latter species they lay predominantly within the parenchyma and only occasionally in perivascular spaces (MATSUSHIMA and REITER, 1977); they measured up to 15 μm or more in diameter and contained mitochondria, membrane-bound dense and lamellated inclusion bodies, multivesicular bodies, electron-lucent vesicles and dense-core vesicles, the whole structure resembling Herring bodies of neurosecretory fibres. Regarding their function, it has been suggested that the inclusion bodies may represent degenerative changes which occur in the adrenergic nerve fibres of the pineal in relation to the functional activity of the pineal gland. In this context it is perhaps relevant to mention that multilamellar bodies have been described in the rat; they were thought to represent transformed nerve fibres (PAWLIKOWSKI and KARASEK, 1967). Varicosities as seen by light microscopy are shown in Fig. 112. The question as to what extent the sympathetic nerve fibres are ensheathed by Schwann cells has received relatively little attention. From the few published remarks it would appear that sometimes the fibres are only incompletely surrounded by Schwann cell cytoplasm (BONDAREFF, 1965a; ARSTILA, 1967; KAPPERS, 1969). The basal lamina which usually surrounds the nerve fibre bundles also appears to be incomplete (KAPPERS, 1969).

γ) *Developing Nerve Fibres*

As pointed out above, sympathetic nerve fibres penetrate the epiphysis cerebri in relatively early stages of development (see p. 52). In man (HÜLSEMANN, 1971), the nerve fibres first possessed few varicosities. Occasionally granules with "laminated structures" were present. In 150-day-old fetuses the axons were clearly shown to contain both clear and dense-core vesicles.

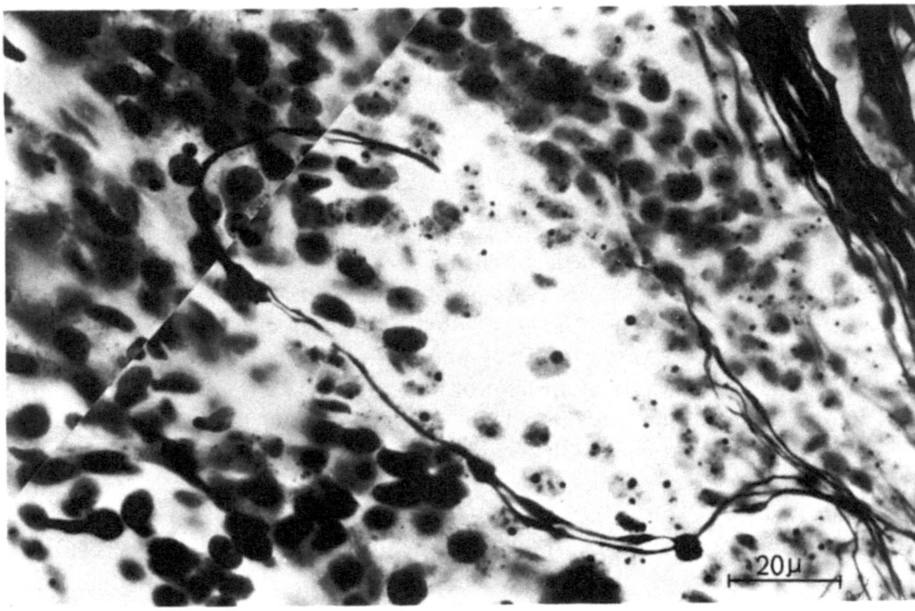

Fig. 113

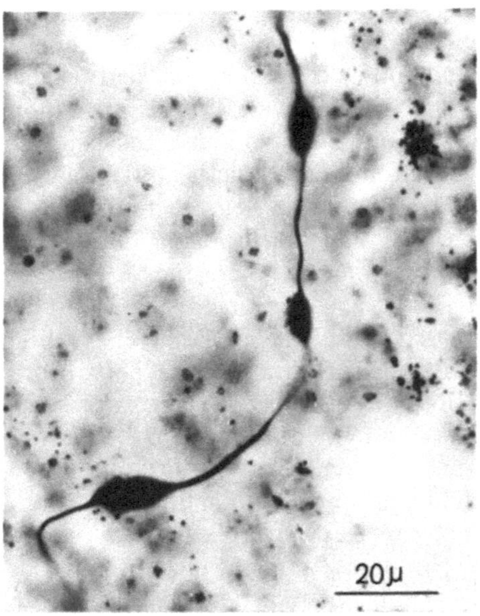

Fig. 112

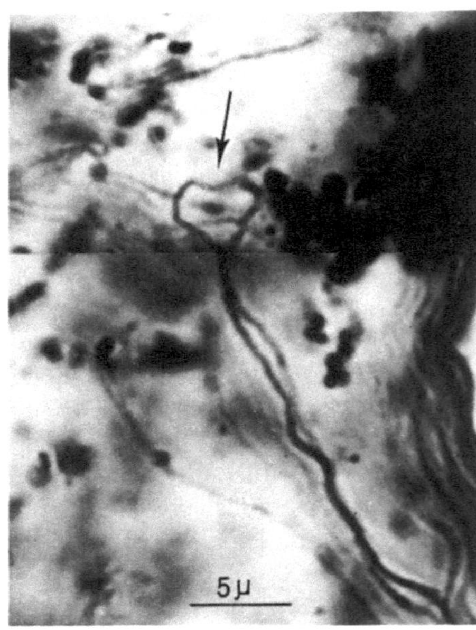

Fig. 11

Fig. 112. Intrapineal nerve fibre showing three prominent varicosities. Richardson stain. ROMIJN, 1973a

Fig. 113. Aberrant habenular commissural fibres making a loop of 180° in the dorsorostral part of a rabbit pineal organ. Richardson stain. ROMIJN, 1973a

Fig. 114. Aberrant posterior commissural fibre penetrating the rostral part of the rabbit pineal organ and making a hairpin loop of 180° (*arrow*). ROMIJN, 1973a

In the rat (A. MACHADO, 1971) a typical feature of early sympathetic nerve fibres is the presence of dilated microtubules forming saccular or tubular profiles. Glycogen is demonstrable until the 1st month of life. Coated vesicles, dense-core vesicles and electron-lucent vesicles are noticeable directly after birth. As in adults, the dense-core vesicles can be subdivided into small and large ones. Small dense-core vesicles were virtually absent at birth, but with $KMnO_4$ fixation a few were demonstrable. With increasing age they increased in amount. Moreover, the size of the dense cores, but not the vesicles as a whole, increased from 16 nm on day 1 to 27.2 nm on day 15 after birth. Large dense-core vesicles with a mean diameter of 82 nm were more numerous than in adult rats. Electron-lucent vesicles amounted to 62% of the total vesicle population in rats aged 1–3 days. The amount was halved by 7 days and decreased to 22% at 15 days.

As regards the origin of the different types of vesicles, it was concluded that electron-lucent vesicles may perhaps be formed from tubular or saccular profiles of axonal endoplasmic reticulum. Small dense-core vesicles were thought to develop from electron-lucent vesicles and large ones from accumulations of electron-dense material within sacs of smooth endoplasmic reticulum (A. MACHADO, 1971). Individual or bundled axons were invested by Schwann cell processes by day 21 (A. MACHADO, 1967b).

e) Ultrastructural Relationships Between Sympathetic Nerve Fibres and Pinealocytes

In view of the fact that sympathetic nerve fibres and their transmitter substance, noradrenaline, appear to be of utmost importance for the regulation of pinealocyte function (see p. 9), the ultrastructural relationships between these fibres and the target cells have been carefully examined. Since this aspect has been extensively discussed by KAPPERS (1969), only a brief account will be given.

As mentioned above (p. 192), in most species investigated the majority of sympathetic nerve fibres have a perivascular location and hence do not come into direct contact with pinealocytes. Exceptions are apparently the chinchilla, in which nerve fibres usually come in close contact with pigment-containing cells (MATSUSHIMA and REITER, 1975b), and the rat, in which a fairly large proportion of fibres terminate among pinealocytes. In the rat WOLFE (1965) noted that axons, directly opposed to pinealocytes, frequently contained aggregations of dense-core vesicles. Thickenings of the axolemma were observed by WOLFE (1965), RODIN and TURNER (1965a, b, 1966) and KAPPERS (1969), but ARSTILA (1967) could not detect any. KAPPERS (1969) noticed that even the plasmalemma of the pinealocytes showed thickenings.

Concerning the functional interpretation, it has to be taken into consideration that these contacts are altogether rare. Nevertheless, it is clear that at sites where these contacts are present the target cells may be affected more rapidly than elsewhere (KAPPERS, 1969). KAPPERS (1969) even hypothesized that although a simple close apposition of the axolemma to the pinealocyte cell membrane may suffice for the propagation of neural impulses, the propagation may be more efficient when specific synaptic contacts are present.

On the whole it would appear that in the majority of species, and partly also in the rat, the neurotransmitter reaches the pinealocytes by way of diffusion over more or less wide distances. According to KAPPERS (1969), the neurotransmitter released from nerve fibre terminals and varicosities diffuses through the perivascular space and reaches the target cells more or less directly, depending on the architecture of the perivascular space and the adjacent tissue. Principally, four different possibilities were distinguished: (1) The neurotransmitter may reach the pinealocyte directly if pinealocyte processes lie freely in the perivascular space. (2) If the pinealocyte processes are covered by a basal lamina, they are supposedly less readily influenced by the transmitter. (3) A fairly slow influence on the pinealocytes is thought to be exerted by the transmitter if a glial barrier exists between nerve fibres and parenchyma. (4) In addition to the possibilities mentioned by KAPPERS, it has to be taken into consideration that in some species (e.g. guinea-pig: LUES, 1971) the perivascular space has fairly large outpocketings penetrating into the parenchyma. These outpocketings may well represent channels by which the transmitter reaches the effector cells, the accessibility depending on the presence or absence of a basal lamina. As shown by light microscopy (QUAY, 1974b) and electron microscopy (KRSTIĆ, 1975b), the epiphysis cerebri possesses an intercellular system of continuous microscopic channels enmeshing individual pinealocytes (p. 173), which may well transport the neurotransmitter from the perivascular space to pinealocytes.

f) The Pinealocyte Deprived of Its Sympathetic Input

In view of the fact that the sympathetic innervation is of great functional importance for the pinealocyte, it is pertinent to examine in what way this cell is affected by sympathectomy.

Morphological effects have been little studied. QUAY (1971a), studying the composition of the rat pineal after cutting the nervi conarii and the tentorium cerebelli, noted that pineal volume was not significantly changed when compared with sham-operated controls (Table 20). Both parenchymal cytology and vascularity were normal and equivalent in both groups. Electron microscopy (LIN et al., 1973, 1975; ROMIJN, 1975a) revealed signs of atrophy, including a decrease in the amount of rough endoplasmic reticulum, a decrease in the number of dense-core vesicles and a vesiculation of the smooth endoplasmic reticulum. A striking increase of lysosomes was noted by LIN et al. (1973, 1975), and an increase in the number of synaptic ribbons and synaptic spherules by ROMIJN (1975a). SHERIDAN (1975) found an increase of dense-core vesicles, which he attributed primarily to an increase in the number of pinealocyte process terminals that contain these vesicles rather than to an increase in vesicles with dense core per terminal. Sympathectomy by means of 6-hydroxydopamine led to vesiculation of the smooth endoplasmic reticulum in the terminals of pinealocyte processes and a striking increase in the numbers of synaptic ribbons and synaptic spherules (ROMIJN, 1976). SHERIDAN and KEPPEL (1971) noted a decrease in the number of pinealocyte dense-core vesicles, but more recently (SHERIDAN,

Table 20. Estimate of volumes of pineal organs (mm^3), derived from planimetry of complete sections at intervals through each pineal organ after sham-operation or severance of nervi conarii (QUAY, 1971 a)

	No.[a]	Mean ± SEM	Range
Experiment 1			
Sham-operated	8	0.514 ± 0.050	0.300–0.692
Nerve-cut	9	0.520 ± 0.043	0.348–0.690
Experiment 2			
Sham-operated	5	0.607 ± 0.79	0.412–0.840
Nerve-cut	5	0.472 ± 0.045	0.358–0.598

[a] Number of animals (pineal organs).

1975) more pinealocyte processes and terminals containing dense-core vesicles have been seen.

Histochemically, the changes appear to be rather subtle. A slight decrease in serotonin fluorescence was found in some studies (BERTLER et al., 1963; OWMAN, 1964a, b, 1965; HÅKANSON and OWMAN, 1966), which is corroborated by a chemically determined decrease of pineal serotonin (PELLEGRINO DE IRALDI et al., 1963; FISKE, 1964). No changes in serotonin histofluorescence were seen by CSILLIK (1963) and TRUEMAN and HERBERT (1970a, b). MORGAN and HANSEN (1978), besides noting that it takes 12–24 h for pineal noradrenaline to disappear after superior cervical ganglionectomy, found that the serotonin fluorescence diminished and that some residual fluorescence remained deep within the pineal.

The enzymes involved in serotonin and melatonin formation react differently to sympathectomy. Tryptophan hydroxylase does not change (DEGUCHI, 1977a). 5-Hydroxytryptophan decarboxylase activity increases (PELLEGRINO DE IRALDI and RODRIGUEZ DE LORES ARNAIZ, 1964; SNYDER and AXELROD, 1964; SNYDER et al., 1965a). HIOMT activity decreases (WURTMAN et al., 1964a; EICHLER and MOORE, 1971; MOORE and RAPPORT, 1971; NAGLE et al., 1973a; DE GALLARDO and TRAMEZZANI, 1975), as does NAT activity, especially during the night (KLEIN et al., 1971a; DEGUCHI and AXELROD, 1972b). Moreover, denervation results in supersensitivity of β-adrenergic receptors of pinealocytes (DEGUCHI and AXELROD, 1972b, 1973a, b), and oestradiol uptake by the pineal decreases (NAGLE et al., 1973b). Moreover, testosterone administration stimulates pineal protein synthesis only if the sympathetic innervation is intact (NAGLE et al., 1975). Likewise, an intact innervation is required for follicle-stimulating hormone (FSH), luteinizing hormone (LH) or prolactin to increase pineal HIOMT activity (CARDINALI et al., 1976a). The importance of an intact innervation is also reflected in the finding that a transplanted pineal does not function properly (REITER, 1967a).

The influence of the sympathetic nervous system on the pinealocytes can be demonstrated also by administering drugs which deplete the sympathetic fibres of their noradrenaline content. Reserpine administration is followed by an increase in the number of both dense-core vesicles in the vicinity of the

Golgi complex and synaptic spherules, the occurrence of vacuoles, grumose and vermiculate bodies, and a vesiculation of smooth endoplasmic reticulum (ROMIJN, 1976). Synaptic ribbons are unaffected by short-term administration of reserpine, but decrease in number after long-term administration (VOLLRATH and HOWE, 1976). LUES (1971) noted a decrease in the electron density of the ribbon rod. Rat pinealocytes showed no apparent changes after a single injection of reserpine (5 mg/kg, IP) (ARSTILA, 1967).

Administration of *dl*-α-methyl-*p*-tyrosine methylester-HCl, a specific inhibitor of tyrosine hydroxylase, which is rate-limiting with regard to noradrenaline synthesis in postganglionic sympathetic neurones, led to distension of Golgi cisternae, distension and vesiculation of smooth endoplasmic reticulum and a striking increase in the number of synaptic spherules and synaptic ribbons, the latter often lying in clusters and showing a conspicuous increase in length (ROMIJN, 1976).

As regards the fine structure of pinealocytes after administration of 5-hydroxy-L-tryptophan, 5-hydroxy-DL-tryptophan, (*S*)-(+)-*p*-chloramphetamine (2R:3R) hydrogenetartrate, and L-propranolol, see ROMIJN (1976). The effects of noradrenaline, L-dopa, propranolol and *p*-chlorophenylalanine on synaptic ribbons have been studied by VOLLRATH and HOWE (1976), and the effects of *p*-chlorophenylalanine on hamster pinealocytes by SHERIDAN (1975).

Fluorescence microscopically detectable changes of pinealocyte serotonin levels following administration of a number of different drugs have been mentioned above.

The effects of drugs on pinealocyte enzymes have frequently been investigated. Following administration of *p*-chlorophenylalanine, pineal tryptophan hydroxylase activity decreases (DEGUCHI and BARCHAS, 1972b), leading to a serotonin depletion (DEGUCHI and BARCHAS, 1972a). Administration of the MAO inhibitor, pargyline, led to conflicting results. While ILLNEROVÁ (1973b) noted a decrease of pineal NAT, BADE et al. (1977) found an increase during darkness and no detectable effect during light. The antidepressant, desmethylimipramine, causes an increase of both NAT and melatonin (PARFITT and KLEIN, 1977) when given to isolated pineals. Further information may be obtained from the papers by MILIN and STERN (1958, 1960), AIRAKSINEN et al. (1970) and others listed in the references.

Taking all these findings together, there is clear evidence that the sympathetic fibres play an important role in pinealocyte functioning, by both releasing noradrenaline at certain times of the 24-h cycle followed by a reuptake and by synthesizing and storing the transmitter at other times. Another important aspect is that the sympathetic nerve endings apparently protect the pineal against stress-induced effects, such as an increase of NAT, by taking up catecholamines stemming from extra-epiphyseal sources (PARFITT and KLEIN, 1976).

g) Central Pathways Linking the Sympathetic Nerve Fibres with the Optic System

As the sympathetic fibres convey the information of the environmental lighting conditions to the pineal, the question has to be considered how the sympathetic fibres are centrally linked to the optic system. This problem was approached by lesioning distinct areas of the brain and by studying the effects of these lesions on pineal HIOMT and NAT activities under various experimental conditions. The first such experiments led to the concept that the two enzymes studied are controlled by different neuronal pathways, namely (a) HIOMT by the accessory optic tract running through the medial forebrain bundle and terminating in the medial terminal nucleus of Bochenek in the rostral tegmentum of the mid-brain (AXELROD et al., 1966; MOORE et al., 1967, 1968; MOORE, 1969; WURTMAN et al., 1967b), and (b) NAT by the retinohypothalamic projection innervating the suprachiasmatic nucleus (MOORE and KLEIN, 1974; MOORE, 1974; see also NISHINO, 1976; MOORE and TRAYNOR, 1976). More recent results, however, led MOORE (1978a) to conclude that "the original analysis of the effect of medial forebrain bundle lesions as being equivalent to blinding was incorrect and that their effect should be viewed as equivalent to sympathectomy." Here it is relevant to note that central sympathetic fibres appear to run through the medial forebrain bundle (MOORE, 1978a, b). It now appears that the retinohypothalamic projection is essential both for maintaining the HIOMT response to light and for the entrainment of the NAT rhythm. The proposed neural pathways are as follows (MOORE, 1978a, b): Input from the ganglion cell layer of the retina projects bilaterally upon the neurons of the suprachiasmatic nuclei. The suprachiasmatic nuclei project caudally into the periventricular hypothalamus and the ventral tuberal area and neurons of the ventral tuberal area into the lateral hypothalamic nucleus. Neurons from this nucleus course directly to the intermediolateral cell column of the upper thoracic cord, the source of preganglionic fibres to the superior cervical ganglion.

A direct link of the cervical sympathetic nerves to the optic nerve on the one hand and the epiphysis cerebri on the other has been confirmed by electrophysiology (BROOKS et al., 1975): Electrical stimulation of the optic nerve leads to a greatly reduced tonic activity within the cervical sympathetic fibres, and stimulation of the cervical sympathetic nerve evokes a discharge in the pineal in the form of action potentials. Electrical stimulation of the medial forebrain bundle was likewise effective in altering the electrical activity of the pineal (DAFNY and MCCLUNG, 1975; MCCLUNG and DAFNY, 1975). That the effects of light on the pineal are not mediated by endocrine organs has been shown by FISKE et al. (1962), who found that the decrease in pineal weight following exposure to continuous illumination also occurred after hypophysectomy, gonadectomy, adrenalectomy or thiouracil-feeding. A direct influence of light on mammalian pineal function has been ruled out by studies using organ culture (CARDINALI et al., 1973a).

In new-born rats evidence has been obtained for the presence of an as yet unidentified non-retinal pathway of light to the pineal gland (ZWEIG et al., 1966). For an effect of chiasmotomy on pineal serotonin levels in *Macaca irus*, see KENNY et al. (1973).

2. Commissural Fibres

In addition to sympathetic fibres, the mammalian epiphysis cerebri is regularly supplied by fibres coming from the posterior and habenular commissures.

A frequently expressed view is that these fibres are of no significance for pineal function and that those which appear to lie closely related to parenchymal cells are merely aberrant commissural fibres (e.g. KAPPERS, 1960a, 1965a, 1969; KENNY, 1965a; WURTMAN et al., 1968a). Whether or not this view is tenable is discussed below.

With the exception of the bat (KENNY, 1965a), commissural fibres invading the pineal appear to be present in all species investigated (GARDNER, 1953; HARTMANN, 1957; HOSAKA et al., 1957; KAPPERS, 1960a; KENNY, 1965a). While there can be no doubt as to the presence of these fibres, it is unclear which commissure supplies most of them. In the albino rat (KAPPERS, 1960a) and hooded rat (GARDNER, 1953) the majority of fibres stem from the habenular commissure. PFISTER et al. (1975) claim that, in addition to fibres from the habenulae, fibres also originate in the superior colliculus. In the dog, one author found that most of the fibres come from the posterior commissure (HARTMANN, 1957), while YAMADA et al. (1957) are of the opinion that the prominent commissural innervation stems from the habenular ganglia. A similar situation is found in cats: according to KENNY (1965a) the majority of fibres stem from the habenular commissure; NIELSEN and MØLLER (1975), on the other hand, described a prominent bundle of nerve fibres measuring 20–25 μm in thickness that comes from the posterior commissure and also numerous nerve fibres from the habenular region. In *Cercopithecus aethiops* a nerve tract consisting of 10–15 fibres extends from the posterior commissure into the pineal, while single fibres reach the organ from the habenular commissure (NIELSEN and MØLLER, 1975; HAYES et al., 1970). The degree of penetration of the commissural fibres into the epiphysis varies considerably. In the human epiphysis (KENNY, 1965) only a few commissural fibres intermingle with the pineal parenchyma, but none of these fibres can be traced deeply into it. In other primates the fibres reach the distal areas of the organ (HARTMANN, 1957; HOSAKA et al., 1957; NIELSEN and MØLLER, 1975). A similar situation is apparently present in the hamster (PFISTER et al., 1975). The centre of the epiphysis comprises commissural fibres in cat (HARTMANN, 1957; KENNY, 1965a; NIELSEN and MØLLER, 1975) and calf (HARTMANN, 1957). Fibres are restricted to the proximal part of the epiphysis in rat (KAPPERS, 1960a), rabbit (ROMIJN, 1973a), horse and roebuck (HARTMANN, 1957).

For a functional interpretation of the commissural fibres it has to be taken into consideration that, as pointed out by LE GROS CLARK (1939/40) in the monkey and corroborated by KAPPERS (1960a) and KENNY (1965a) in the rat and by ROMIJN (1973a) in the rabbit, commissural fibres form loops (Figs. 113, 114) and leave the organ contralaterally. The other fibres, which do not return to the commissures, have been regarded as aberrant for a number of reasons. To begin with, the majority of fibres in the rat do not penetrate deeply into the superficial epiphysis cerebri. Some merely cross the parenchyma to reach adjacent leptomeningeal tissue, while others which seem to be in contact with parenchymal cells do not exhibit terminal thickenings (KAPPERS, 1960a). The fact that in the dog the myelinated nerve fibres neither loose their myelin sheaths nor form end-bulbs also seems to support the concept that the fibres are aberrant ones (HARTMANN, 1957). There are, however, some observations, both morphological and functional, which strongly urge reconsideration of this interpretation.

Thus it was shown in the hooded rat (GARDNER, 1953) that following lesioning the habenular and posterior commissures the nerve fibre terminals on parenchymal cells underwent complete degeneration while the intrapineal arterial sympathetic nerve fibres remained unchanged. A word of caution is necessary, however, since the same author found no changes in the nervous elements of the pineal following bilateral removal of the superior cervical ganglia, an operation which has been clearly shown to destroy the largest part of the intrapineal nerve fibres (see p. 187). Moreover, there is the observation that in cat and dog pineals commissural fibres represent the main nervous component, whereas in the rat sympathetic fibres predominate (TUBAHARA, 1955). In the cat and monkey, where commissural fibres penetrate deeply into the parenchyma, some fibres possess end-bulbs close to parenchymal cells (NIELSEN and MØLLER, 1975). Also of interest is the finding that interruption of the pineal stalk in rats is followed by a reduction in pineal size and a transitory increase in initial rate of phase shift of locomotor rhythms resembling that after complete pinealectomy (QUAY, 1971b, c).

The concept that the habenulo-epiphyseal complex is a functional entity (MILIN and STERN, 1960; SOULAIRAC and SOULAIRAC, 1963; FAURE et al., 1966; and others) has gained support from electrophysiological studies. Following observations by SCHAPIRO and SALAS (1971) that photic stimuli reach the pineal even after complete sympathectomy, DAFNY and MCCLUNG (1975), MCCLUNG and DAFNY (1975) and RØNNEKLEIV et al. (1978a) electrically stimulated the habenular nuclei and noted clear responses in the pineal. Moreover, connections have been shown to exist between the amygdala, ventromedial hypothalamus, olfactory and acoustic centres and the pineal (DAFNY et al., 1975; DAFNY, 1977). In view of the fact that the habenular nuclei, which receive input from stria medullaris fibres, appear to play an important role in olfactory reflex mechanisms (RAUSCH and LONG, 1971), and since some of the strial fibres reach the epiphysis cerebri directly by means of the habenular commissure (KAPPERS, 1960a), it is interesting to note that in rats stimulation of the olfactory system leads to a pronounced atrophy of both nerve cells in the internal habenular nucleus and pinealocytes (MILINE et al., 1963a). It is also relevant that removal of both eyes and olfactory bulbs results in a severe retardation in the growth of the reproductive organs and in the onset of puberty in both male and female rats and that removal of the pineal reverses these changes nearly completely (REITER et al., 1969; REITER and ELLISON, 1970; REITER and SORRENTINO, 1971a).

A direct link between the brain and the pineal has been clearly demonstrated in the ferret, in which lesions in the habenulae or the posterior commissure result in degenerative changes in the intrapineal ganglion (see below) (DAVID and HERBERT, 1973). Another interesting observation is that in the ferret, bilateral lesions of the stria medullaris lead to degenerative changes in the habenular nuclei (DAVID and HERBERT, 1973).

In the cat, commissural fibres terminate among ependymal cells of the pineal recess (HARTMANN, 1957).

3. Parasympathetic Innervation

A parasympathetic innervation of the epiphysis cerebri, in contrast to the sympathetic supply, is not a regular feature in mammals (KAPPERS, 1960a, 1965a, 1969). Species with a clearly identified parasympathetic innervation are the macaque monkey and the rabbit. In the rat, parasympathetic fibres may also be present.

In the macaque monkey the presence of parasympathetic fibres in the pineal has been established by lesioning the greater superficial petrosal nerve, leading to degeneration of the parasympathetic (sic!) nervus conarii, intra-epiphyseal nerve fibres and intra-epiphyseal nerve cells (KENNY, 1961). The parasympathetic innervation of the rabbit pineal has been studied in great detail (ROMIJN, 1972, 1973a, c, 1975b). It is relatively complicated because it consists of both intra- and extra-epiphyseal nerve cells; consequently pre- and postganglionic parasympathetic fibres are present in the pineal. The intra-epiphyseal nerve cells amount to approximately 40 prominent perikarya (Fig. 115) showing both axosomatic (Fig. 116) and axodendritic (Fig. 117) contacts. As depicted diagrammatically in Fig. 118, the intra-epiphyseal ganglion cells are innervated by both preganglionic parasympathetic fibres and postganglionic sympathetic fibres, the epiphyseal parenchyma being innervated by postganglionic parasympathetic fibres, the ganglion cells of which lie either intra- or extrapineally.

In the rat pineal, cholinergic nerve terminals have been found by electron microscopy (WOOD, 1973). Interestingly, these fibres do not occur in perivascular spaces, where the sympathetic fibres lie, but rather within the parenchyma, where they apparently make synaptic contacts with pinealocyte perikarya and processes, showing distinct thickenings of the pre- and postsynaptic membranes.

On biochemical grounds a parasympathetic innervation of the mammalian epiphysis cerebri has been denied because choline acetyltransferase is barely (LA BELLA and SHIN, 1968) or not at all detectable (SCHRIER and KLEIN, 1974). Moreover, SCHRIER and KLEIN (1974) have pointed out that the choline acetyltransferase activity measured by other authors is in fact due to the presence of carnitine acetyltransferase.

Altogether there is little experimental evidence to support the notion that the parasympathetic fibres have a profound effect on pinealocyte function.

In the rabbit (ROMIJN, 1975a) parasympathectomy of the epiphysis cerebri by means of bilateral stereotaxic coagulation of the facial nerves was followed by minor changes of the pinealocytes; the rough endoplasmic reticulum lost some of its ribosomes, became irregularly dilated and tended to form spiral structures, while the smooth endoplasmic reticulum showed signs of vesiculation and of transformation into lamellar whorls (p. 122). Comparable changes occurred after administration of the parasympatholytic drugs, atropine methylbromide and oxyphenecycline (ROMIJN, 1976).

WARTMAN et al. (1969a, b), studying HIOMT activity in rat pineals, noted that atropine methylbromide inhibited the high levels of HIOMT in constant darkness, while the cholinomimetic, oxotremorine oxalate, restored HIOMT activity in darkness.

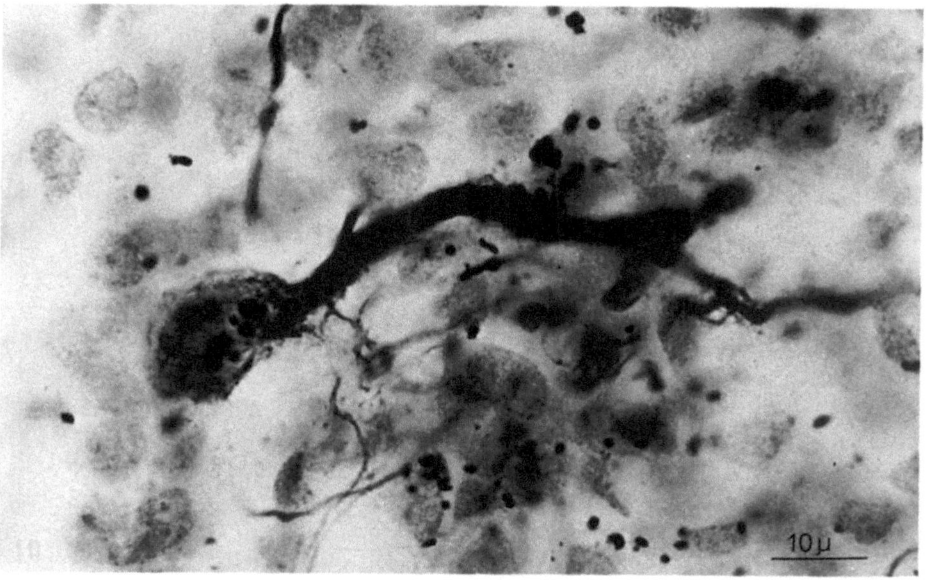

Fig. 115

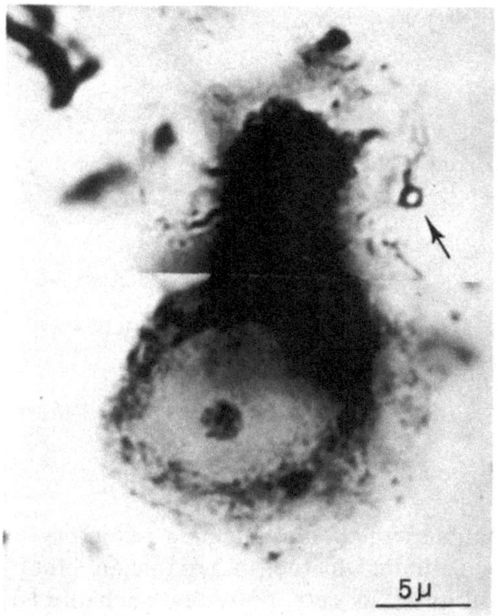

Fig. 116

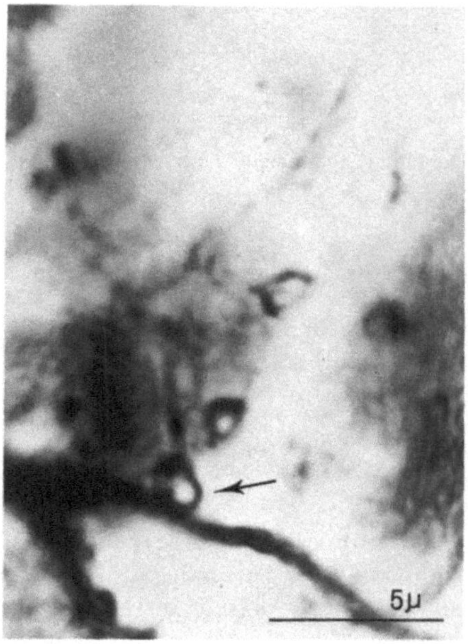

Fig. 117

Fig. 115. Intrapineal neuron with a thick process branching between pinealocytes. Richardson stain. ROMIJN, 1973a

Fig. 116. Annular nerve ending (*arrow*) close to the soma of an intrapineal neuron. Richardson stain. ROMIJN, 1973a

Fig. 117. Annular nerve ending (*arrow*) close to a dendrite of an intrapineal neuron in the rabbit. Richardson stain. ROMIJN, 1973a

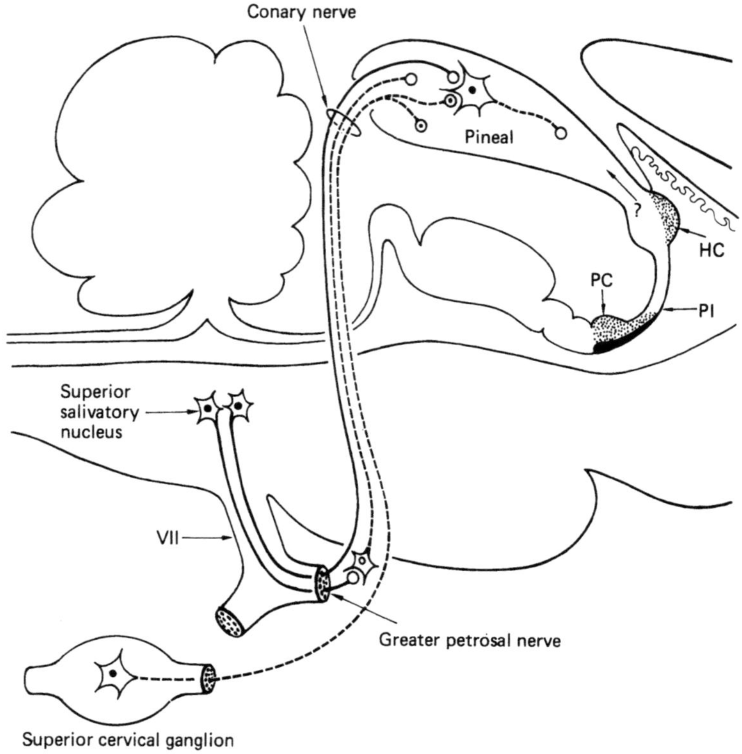

Fig. 118. Diagrammatic representation of the parasympathetic innervation of the rabbit pineal gland. ROMIJN, 1975b

Finally, it should be recalled that the sympathetic fibres of the pineal have parasympathetic-cholinergic characteristics (p. 106).

4. Neurosecretory Fibres

Neurosecretory fibres (i.e. fibres stainable by the common "neurosecretory" stains such as chrome alum haematoxylin, aldehyde fuchsin, or alcian blue) are irregularly present in the mammalian epiphysis cerebri. A few such fibres have been described in the hedgehog (BARGMANN, 1954; HARTMANN, 1957; OKSCHE, 1965a), the rhesus monkey (LUKASZYK and REITER, 1974, 1975) and the Mongolian gerbil (JAPHA et al., 1974). The origin of these fibres is unknown. In the monkey the fibres are present in the vicinity of the pineal recess, between groups of pinealocytes and occasionally in perivascular spaces. Whether their tinctorial similarity with the secretory product of the subcommissural organ has any relevance is obscure. The absence of neurosecretory material in intra-epiphyseal ganglion cells suggests an extra-epiphyseal origin of these fibres.

5. The So-called Fetal Pineal Nerve

Recently attention has been drawn to the fact that in fetal specimens a so-called pineal nerve is present, which may be homologous to the anuran pineal nerve (MØLLER et al., 1975; see also p. 48). In human fetuses, the nerve was found in the median plane, extending from the posterior part of the pineal body to a small mesodiencephalic cluster of nerve cells situated dorsal to the rostral-most area of the subcommissural organ (MØLLGAARD and MØLLER, 1973). An identical nerve has been found in sheep and rabbit fetuses (MØLLER et al., 1975). In view of the present concept that the pineal appears to contain efferent nerve fibres only (KAPPERS, 1965a), it is interesting to note that in sheep and rabbit there are indications for the presence of both pinealofugal (afferent) and pinealopetal (efferent) fibres in this nerve (MØLLER et al., 1975). The functional significance of this nerve is obscure. Since it is restricted to fetal stages, the possiblity has been discussed that it may (a) simply represent a vestigial homologue of the pineal nerve of anurans, (b) play a role during development only or (c) have been incorporated into the pineal stalk and the parenchyma (MØLLER et al., 1975). Closely related to this nerve are clusters of nerve cells (see p. 214).

6. Intra-epiphyseal Ganglion Cells

The presence of intra-epiphyseal ganglion cells is not a regular feature in mammals (see BARGMANN, 1943; KAPPERS, 1965a). According to QUAY (1965a) and KAPPERS (1965a) they appear to be restricted to primates and in these they are very variable in amount. Recently ganglion cells have also been found in non-primates. KENNY (1965a), studying two bats, found one nerve cell only. In rabbits (ROMIJN, 1973a, c) some 40 nerve cells are present, lying scattered near the caudal tip of the epiphysis in the transitional area of the medulla. A distinct intra-epiphyseal ganglion has been found in the ferret (TRUEMAN and HERBERT, 1970a, b; DAVID and HERBERT, 1973; DAVID et al., 1973). Recently neuron-like cells have also been found in the distal part of the ground squirrel pineal (MATSUSHIMA and REITER, 1978).

Relatively little is known about the functional significance of these ganglion cells. According to KAPPERS (1965a) they do not represent sensory cells and afferent nerve fibres are lacking in the mammalian epiphysis cerebri. In the rabbit (ROMIJN, 1973a, c, 1975b) they have been regarded as intramural parasympathetic neurons which, according to electron microscopy (ROMIJN, 1975b), show axosomatic and axodendritic synapses formed by both postganglionic noradrenergic and preganglionic cholinergic nerve fibres. In the ferret, as mentioned above (p. 209), the ganglion appears to receive input from the habenular nuclei (DAVID and HERBERT, 1973). Electron microscopically (DAVID et al., 1973) axodendritic synapses are commonly seen, in contrast to axosomatic and axo-axonic synapses. Since the axons involved contain both electron-lucent and dense-core vesicles it would appear that they are aminergic in nature. The ganglion cells contain electron-lucent vesicles measuring 50–80 nm in diameter, as well as granulated vesicles measuring 60–120 and 200–300 nm in diameter, but they do not exhibit a specific amine fluorescence. A characteristic feature is the presence of AChE, which has been demonstrated by means of both light (TRUEMAN and HERBERT, 1970a, b) and electron microscopy (DAVID et al., 1973; DAVID and ANAND KUMAR, 1978).

7. Extra-epiphyseal Ganglion Cells

Occasionally the presence of pineal-associated ganglia has been described (see BARGMANN, 1943). In the recent literature they have been mentioned (a) as parietal ganglion or WARBURG's ganglion and (b) as PASTORI's ganglion. A small WARBURG's ganglion has been identified in a young adult human, where it lay between the suprapineal recess and the superior surface of the pineal; in older individuals it was not seen (KENNY, 1965a). In human fetuses of 30–168 nm CRL (Fig. 119) it was found on the dorsal surface of the great cerebral vein, which separates the ganglion from the epiphysis (MØLLGAARD and MØLLER, 1973). In the same material, PASTORI's ganglion (Fig. 120) was found at the posterior pole of the epiphysis cerebri, beneath the great cerebral vein (MØLLGAARD and MØLLER, 1973).

The functional significance of these ganglia is unknown. According to KENNY (1965a), WARBURG's ganglion contains multipolar nerve cells and seems to form part of the autonomic nervous system. MØLLGAARD and MØLLER (1973) are of the opinion that both ganglia belong to the parasympathetic nervous system.

Moreover, ganglion cells have been found associated with the recently discovered fetal pineal nerve linking the pineal and an area of the CNS closely related to the subcommissural organ (see p. 213). In human fetuses the nerve possesses two ganglia. One lies just above the external limiting membrane and comprises bipolar nerve cells. The other is located close to the hypendyma of the rostral-most part of the subcommissural organ (MØLLGAARD and MØLLER, 1973). In sheep and rabbit fetuses (MØLLER et al., 1975) the ganglion cells do not form a cluster but are located at various intervals along the length of the nerve.

Fig. 119. Sagittal section of a human pineal gland (fetus, 100 mm CRL) showing the Marburg ganglion lying above the great cerebral vein. Toluidine blue staining. MØLLGAARD and MØLLER, 1973; ×265

Fig. 120. Sagittal section of a human pineal gland (fetus, 168 mm CRL) showing the Pastori ganglion in a connective tissue septum at the posterior pole. Toluidine blue staining. MØLLGAARD and MØLLER, 1973; ×600

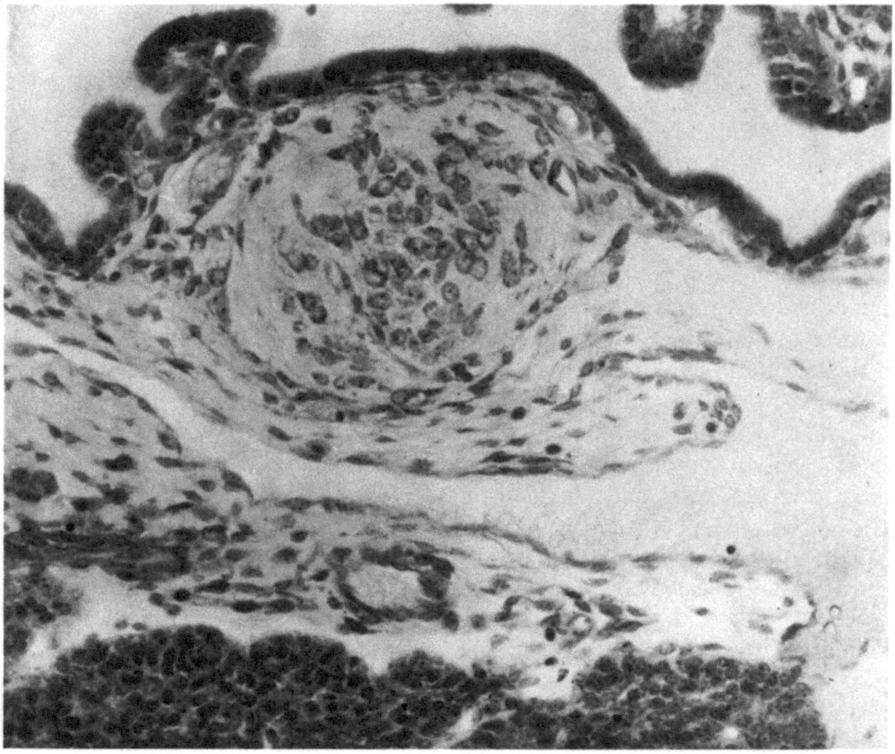

Fig. 119

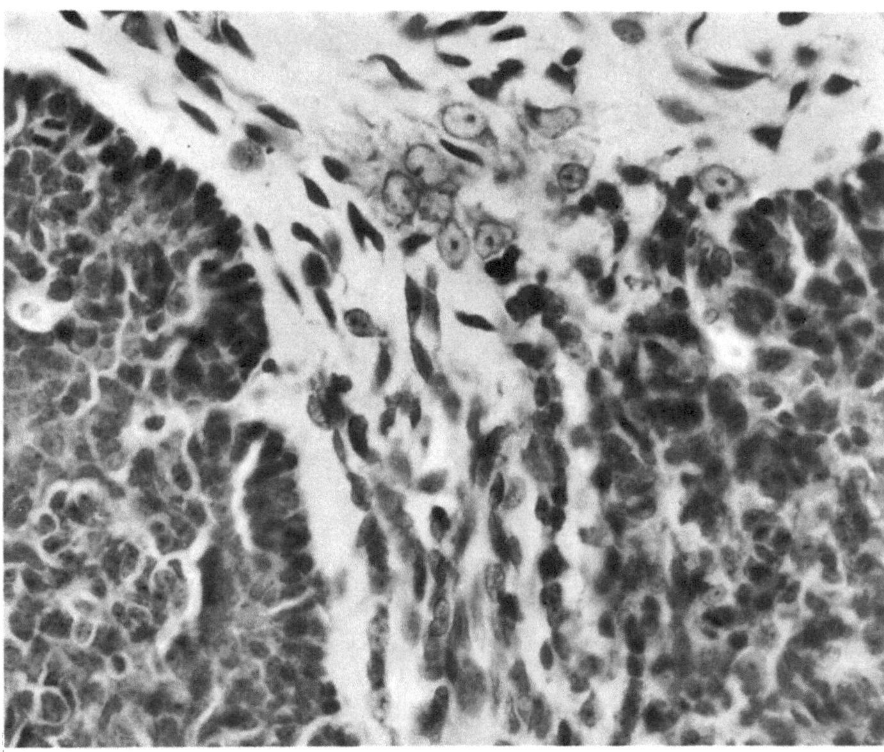

Fig. 120

VIII. Blood Supply

1. General Aspects

The mammalian epiphysis cerebri appears to have a copious blood supply. GOLDMAN and WURTMAN (1964) concluded that the "... minimum rate of rat pineal blood flow per gram exceeds that of most endocrine organs, equals that of neurohypophysis and is surpassed only by that of the kidney." According to these authors the true blood flow amounted to 4 ml/min per gram. In view of the fact that the blood content of the rat is only one-fifth that of the anterior lobe of the pituitary and one-third that of the neurohypophysis (QUAY, 1958a), it would appear that the blood flow is exceedingly high (QUAY, 1974a). However, the oxygen consumption of pineal tissue is relatively low (QUAY, 1962c, 1963d) and decreases with age. In pineals of young goats (1–3 months of age) it is almost twice as high as in animals over 6 years of age (HELLMAN and LARSSON, 1961). In rats, calculation of number of erythrocytes per mm^3 of pineal tissue revealed that the mean blood content was lowest between 3 and 5 weeks of age (QUAY, 1958a). While in a first study (QUAY, 1958a) the blood content of the pineal cortex exceeded by 1.2–2.0 times that of the medulla, more recent calculations have yielded virtually identical results in cortex and medulla (QUAY, 1972b). From a functional point of view it is interesting to note that at the daily onset of light a pronounced vasoconstriction occurs, as assessed by counting the number of red blood cells. In the rat pineal cortex (medulla), red blood cells amounted to 563 ± 34 (520 ± 47) cells/mm^3 tissue/10^3 at 3.00 h and to 296 ± 36 (225 ± 27) cells/mm^3 tissue/10^3 at 6.00 h (lights on at 5.00 h) (QUAY, 1972b).

Attempts have been made to alter the blood supply. Castration or administration of oestradiol fail to produce significant effects in either sex. Significant vasodilatation is obtained by injecting adenosine or picrotoxin and by ether anaesthesia. Vasoconstriction has never been observed, but noradrenaline administration, even in small doses, inhibits the vasodilatation occurring during ether anaesthesia (QUAY, 1958a). Removal of the major nervous input to the epiphysis cerebri by decentralization or superior cervical ganglionectomy reduces the blood flow by one third and the ability of the epiphysis to make a characteristic compensatory vasomotor response to pressor drugs is lost (GOLDMAN, 1976). ROLLAG et al. (1978b), using radioactively labelled microspheres, found that 5 min after lights on pineal blood flow in ewes is significantly diminished when compared with 1 min before lights on.

2. Arterial Supply and Venous Drainage

The main blood supply to the epiphysis cerebri is provided by branches of the posterior choroidal arteries arising from the posterior cerebral arteries (GLADSTONE and WAKELEY, 1940; LE GROS CLARK, 1940; BARGMANN, 1943; VON BARTHELD and MOLL, 1954; WARWICK and WILLIAMS, 1973; QUAY, 1974a).

These branches are apparently so thin that it is not justified to call them epiphyseal arteries (VON BARTHELD and MOLL, 1954). Before entering the epiphysis, the arteries run inside the pia mater investing the epiphysis. They give off small arterioles which may spread in the capsule before penetrating the epiphysis. In the human pineal the arterioles follow the trabeculae where they give off capillary vessels (GLADSTONE and WAKELEY, 1940; BARGMANN, 1943). In the rodent pineal, arterioles are apparently not regularly present. In the mouse pineal the arteriolar character of the blood vessels vanishes in the capsule so that the entire organ appears to be supplied by capillaries; occasional arterioles entering the pineal parenchyma leave the organ after a longer or shorter course through the organ (VON BARTHELD and MOLL, 1954). In rat, chinchilla and ground squirrel there are likewise no conspicuous arterioles (MATSUSHIMA and REITER, 1975a, b). Whether or not the presence or absence of arterioles in the pineal body is just a question of organ size remains to be established.

Relatively little is known about the intra-epiphyseal distribution of capillaries. In cow, sheep and goat a capillary network is evenly distributed throughout the organ (BARGMANN, 1943). In rabbits, blood vessels are particularly abundant in the transitional area of cortex and medulla (ROMIJN, 1973a).

The capillary net of the epiphysis is drained by venules which, passing through the trabeculae and capsule, unite to form a vessel which joins the great cerebral vein of GALEN (vena cerebri magna), the latter terminating in the anterior part of the straight sinus (GLADSTONE and WAKELEY, 1940). In man, blood is also drained into the bifurcation of the internal cerebral veins (BARGMANN, 1943). COOPER (1960) stresses that dorsally the human pineal is related to the two internal cerebral veins and not to the great cerebral vein

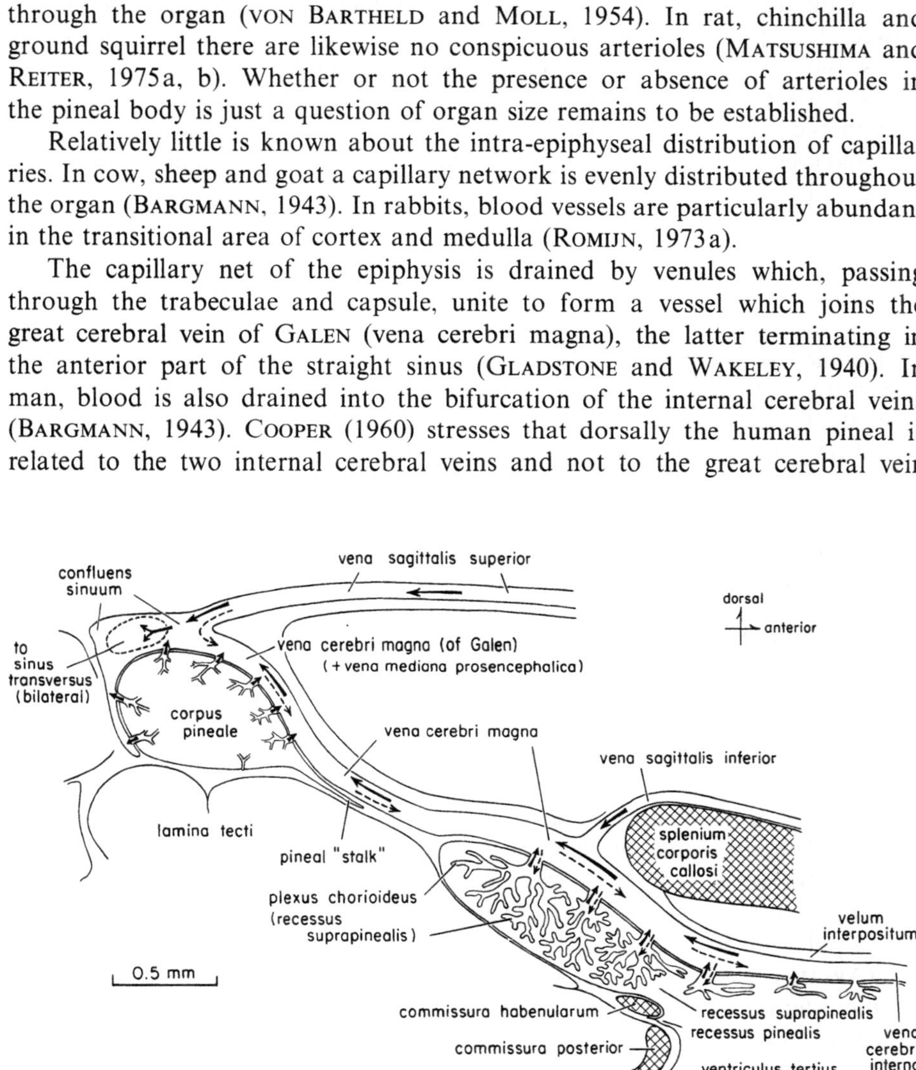

Fig. 121. Diagram of the blood vessels in the pineal region of the rat. *Heavy arrows* indicate the usually presumed direction of the blood flow; *broken arrows* indicate sites of anatomically possible reversed blood flow on the basis of perfusion studies. QUAY, 1973b

and that by elevation or depression the pineal might influence the flow of blood from the two internal cerebral veins into the great cerebral vein and hence into the straight sinus.

In rodents (VON BARTHELD and MOLL, 1954; KAPPERS, 1960a; GREGOREK et al., 1977) venous drainage is via the great cerebral vein or corresponding veins. In the mouse, the epiphyseal stalk lacks veins, venules and arterioles (VON BARTHELD and MOLL, 1954).

In view of the fact that the brain appears to be a major target of pineal substances, the question has been studied as to whether there is a direct vascular link between the epiphysis cerebri and the brain. By means of India ink injections carried out in rats, QUAY (1973b) demonstrated that the only such link is between the epiphysis cerebri and the choroid plexus of both the suprapineal recess and the third ventricle, via the great cerebral vein, and that such a pathway would require a reversed blood flow, which is apparently not uncommon in the region of the great cerebral vein (see also REITER et al., 1975f; GREGOREK et al., 1977). The vascular relationships of the rat pineal are shown in Fig. 121.

3. Development

Detailed studies on the development of the epiphyseal blood supply have apparently not been carried out. QUAY (1958a) noted that in rats the blood content of the epiphysis was largest between 3 and 5 weeks of age. In the cow (MEYBURG, 1965) blood vessels penetrated the tip of the epiphysis at 8 cm CRL. At 9 cm CRL the capsule revealed a prominent vascular network, from which more and more vessels penetrated the organ between 10 and 13 cm CRL. A maximal vascularization was noted at 14 cm CRL. At this stage the vessels ran parallel to the organ surface, before penetrating further into the parenchyma. A diagrammatic representation of the pineal-related blood vessels in the human has been given by VON BARTHELD and MOLL (1954).

4. Histochemistry of Blood Vessels

Histochemical data are apparently restricted to histochemically demonstrable enzymes. The *endothelial* cells have been shown to possess high alkaline phosphatase activity (ARVY, 1961; BOSTELMANN, 1963; GUSEK and BUSS, 1966; BAYEROVÁ and BAYER, 1967; ARSTILA, 1967; TAPP et al., 1973), which increases strikingly during puberty (BOSTELMANN, 1969a). Moreover, acid phosphatase, arylsulphatase (ARSTILA, 1967), ATPase (ARVY, 1961; BAYEROVÁ and BAYER, 1967), nonspecific esterase (BOSTELMANN, 1963), NADH diaphorase (BOSTELMANN, 1963; BAYEROVÁ and BAYER, 1967) and succinate dehydrogenase activities (BAYEROVÁ and BAYER, 1967) have been demonstrated.

In the *pericytes*, both acid phosphatase and arylsulphatase have been noted (ARSTILA, 1967).

5. Ultrastructure of Capillaries

Systematic studies on the ultrastructure of epiphyseal capillaries are rare. The only such study appears to be that by MATSUSHIMA and REITER (1975a, b), which deals with four rodent species. From these reports and other scattered data it would appear that pineal capillaries do not exhibit any unusual features, though species differences are apparent.

Although no actual measurements are given, it is apparent that the thickness of the endothelial cells varies considerably. According to MATSUSHIMA and REITER (1975a, b) the endothelial lining is often attenuated in rat (see also GUSEK and SANTORO, 1961), mouse and ground squirrel and relatively thick in chinchilla. A particularly thin endothelium was also noted in the gopher (SHERIDAN and REITER, 1973).

In some species the endothelial cells possess characteristic surface specializations. Cytoplasmic processes on the luminal side of the endothelium are found in rat and squirrel monkey, but they are rarely present in the mouse. In the chinchilla true microvilli are seen in addition to cytoplasmic folds. Marginal folds, lying near a junction of two adjacent cells, are present in chinchilla (MATSUSHIMA and REITER, 1975a), cow and sheep (ANDERSON, 1965). While the basal surface of endothelial cells is generally smooth in rat, mouse and chinchilla, it shows small cytoplasmic processes in the ground squirrel (MATSUSHIMA and REITER, 1975a).

In rat, mouse and ground squirrel the junctions occurring between adjacent endothelial cells appear quite variable. Usually, the plasma membranes abut each other bluntly or overlap for short distances. Occasionally, interdigitations are encountered. Plasma membranes opposing at the junction usually run parallel to each other with a gap of about 20 nm, which narrows to about 4 nm near the luminal end of the junction. Occasionally, the outer leaflets of adjacent plasma membranes show signs of fusion. Typical tight junctions are absent in rat, mouse and ground squirrel, but zonulae occludentes are seen in the chinchilla (MATSUSHIMA and REITER, 1975a). Maculae adhaerentes are present in cow, sheep (ANDERSON, 1965) and dormouse (ROUX et al., 1974).

Fenestrations or pores of the endothelium are not present in sheep, cow (ANDERSON, 1965), cat (MASHIMO, 1965; DUNCAN and MICHELETTI, 1966; WARTENBERG, 1968), monkey (WARTENBERG, 1968), guinea-pig (LUES, 1971), gopher (SHERIDAN and REITER, 1973), dog (MASHIMO, 1965; SANO and MASHIMO, 1968; WELSER et al., 1968), chinchilla (MATSUSHIMA and REITER, 1975a), dormouse (ROUX et al., 1974) or gerbil (WELSH and REITER, 1978). In the hamster no pores have been noted (SHERIDAN and REITER, 1968), but, according to unpublished results by SHERIDAN (SHERIDAN and REITER, 1973), fenestrated vessels may be present on or near the surface of the epiphysis. In the rabbit pineal, fenestrations are scarce (ROMIJN, 1973b), which would explain why WARTENBERG and GUSEK (1965) could not find any. They are likewise rare in the rat (MATSUSHIMA and REITER, 1975a): while their presence has been clearly established by various authors (MILOFSKY, 1957; GUSEK and SANTORO, 1961; GUSEK et al., 1965; BUSS, 1965; WOLFE, 1965; MASHIMO, 1965), negative results have also been reported (RODIN and Turner, 1965a, 1966).

Fenestrations are frequent in mouse (ITO and MATSUSHIMA, 1968a, b; MATSUSHIMA and REITER, 1975a; MØLLER et al., 1978b) and ground squirrel (MATSUSHIMA and REITER, 1975a; POVLISHOCK et al., 1975). The fenestrations, most of which appear to be bridged by a delicate membranous diaphragm roughly 5 nm thick (ITO and MATSUSHIMA, 1968a, b; MATSUSHIMA and REITER, 1975a), measure between 50 and 70 nm in diameter (GUSEK and SANTORO, 1961; BUSS, 1965; ITO and MATSUSHIMA, 1968a, b; MATSUSHIMA and REITER, 1975a).

The cytoplasm of the endothelial cells is endowed with the regularly occurring organelles such as mitochondria, Golgi complex, cisternae of the rough endoplasmic reticulum and ribosomes (GUSEK and SANTORO, 1961; ANDERSON, 1965; MATSUSHIMA and REITER, 1975a). Occasionally, fine filaments are present (ANDERSON, 1965; MATSUSHIMA and REITER, 1975a), in addition to microtubules and multivesicular bodies (MATSUSHIMA and REITER, 1975a). Lysosome-like bodies are particularly abundant in the chinchilla (MATSUSHIMA and REITER, 1975a).

A regular feature of capillary endothelial cells is the presence of a large number of pinocytotic vesicles, some of which are coated vesicles (MATSUSHIMA and REITER, 1975a). These vesicles have also been described by GUSEK and SANTORO (1961, rat), ANDERSON (1965, sheep, cow), LUES (1971, guinea-pig) and SANO and MASHIMO (1968, dog). In the dog their size has been given as 21 nm (SANO and MASHIMO, 1968), but subsequent studies have revealed distinctly larger sizes. According to MATSUSHIMA and REITER (1975a), in the rat, mouse and ground squirrel 68%–75% of the total number of vesicles measured between 70 and 120 nm in diameter, whereas in the chinchilla 65% of the vesicles ranged between 40 and 70 nm in diameter. BUSS (1965, rat) found them to measure between 60 and 200 nm.

The basal lamina, on which the capillary endothelium rests, does not appear to present any unusual features. As pointed out below (p. 221) the capillary basal lamina represents the inner boundary of the prominent perivascular spaces.

Little is known about pineal pericytes, apart from the fact that they show both acid phosphatase and arylsulphatase activities (ARSTILA, 1967).

6. Perivascular Spaces

The perivascular spaces of the mammalian epiphysis cerebri are of particular interest in different respects. To begin with, they are predominant sites of sympathetic nerve fibres. Secondly, species differences are prominent. Thirdly, pinealocyte processes may enter the perivascular spaces, making possible a close interaction between sympathetic nerve fibres and pinealocytes. Finally, perivascular spaces together with their extensions into the parenchyma may provide important pathways for the transport of neurotransmitters to, and hormones from, pinealocytes which do not lie in the immediate vicinity of blood vessels.

The key papers on perivascular spaces in the mammalian epiphysis are those by RODIN and TURNER (1966) and WARTENBERG (1968). Whereas the paper by RODIN and TURNER gives a detailed account of perivascular spaces in rats only, WARTENBERG'S paper compares the spaces in different species and stresses their importance for the general architecture of the epiphysis.

According to RODIN and TURNER (1966) each vessel within the rat pineal is provided with a perivascular space, varying in width from 5–30 μm, and only one vessel is seen in each space. ARSTILA (1967) noted that the perivascular spaces were wider after fixation by perfusion than after immersion. The inner boundary of the space is consistently formed by the capillary basal lamina. The outer boundary consists of the epiphyseal parenchyma with a more or less reduced basal lamina. This latter point is of particular importance because it may explain the partly conflicting results obtained by various authors. Whereas RODIN and TURNER (1966) were of the opinion that the outer cytoplasmic membrane of pinealocytes formed the immediate boundary of the space without an additional delimiting membrane, GUSEK and SANTORO (1961) felt that the pineal parenchyma was separated from the perivascular space by an incomplete basal lamina. A similar conclusion was arrived at by ARSTILA (1967) and WARTENBERG (1968). The differences of opinion are understandable when one realizes that pinealocyte processes devoid of a basal lamina penetrate the perivascular spaces, giving rise to smooth to sharp undulations of the parenchymal surface (RODIN and TURNER, 1966). Discontinuous outer basal laminae have also been found in hamster (SHERIDAN and REITER, 1968), guinea-pig (LUES, 1971), gopher (SHERIDAN and REITER, 1973) and ground squirrel (POVLISHOCK et al., 1975).

An interesting feature of the perivascular spaces is that they are highly complex and not uniform (Fig. 122). According to WARTENBERG (1968) the simplest type is present in squirrel monkey (WARTENBERG, 1968), cow, sheep (ANDERSON, 1965) and rabbit (WARTENBERG and GUSEK, 1965), in all of which the spaces are narrow and contain but a few nerve fibres. Sometimes a discontinuous glial sheath formed by astrocytes prevents the pinealocyte processes from entering the perivascular spaces. In the cat (WARTENBERG, 1968) the perivascular spaces are fairly wide and contain large amounts of nerve fibres. Occasionally pinealocyte processes, covered by a thin glial sheath and a continuous basal lamina, bulge into the space. In rhesus monkey (WARTENBERG, 1968) the space is even wider than in the cat, and pinealocyte processes devoid of a glial covering but surrounded by a basal lamina penetrate deeply into the space. An even closer topographical relationship between pinealocyte processes and perivascular spaces is present in the rat (WARTENBERG, 1968), in which the perivascular space is extremely wide and contains large amounts of pinealocyte processes devoid of both a glial sheath and a basal lamina, in addition to endings of non-myelinated nerve fibres, a few collagenous fibres and processes of interstitial cells (ARSTILA and RINNE, 1967). A characteristic feature of rat perivascular spaces is that they emit extensions measuring up to 10 μm (RODIN and TURNER, 1966) between adjacent pineal cells and that they are in continuity with enlarged intercellular spaces or cyst-like formations of pineal parenchyma (GUSEK and SANTORO, 1961). Similar extensions have been noted in the guinea-pig (LUES, 1971) and rabbit (ROMIJN, 1973b), and their functional significance has been discussed (p. 173). They may represent an important route for transmitter substances to reach pinealocytes (KAPPERS, 1969), though it should not be overlooked that nerve fibres are not restricted to perivascular spaces, but may terminate within pineal parenchyma.

The precise situation in other species is difficult to assess from the papers

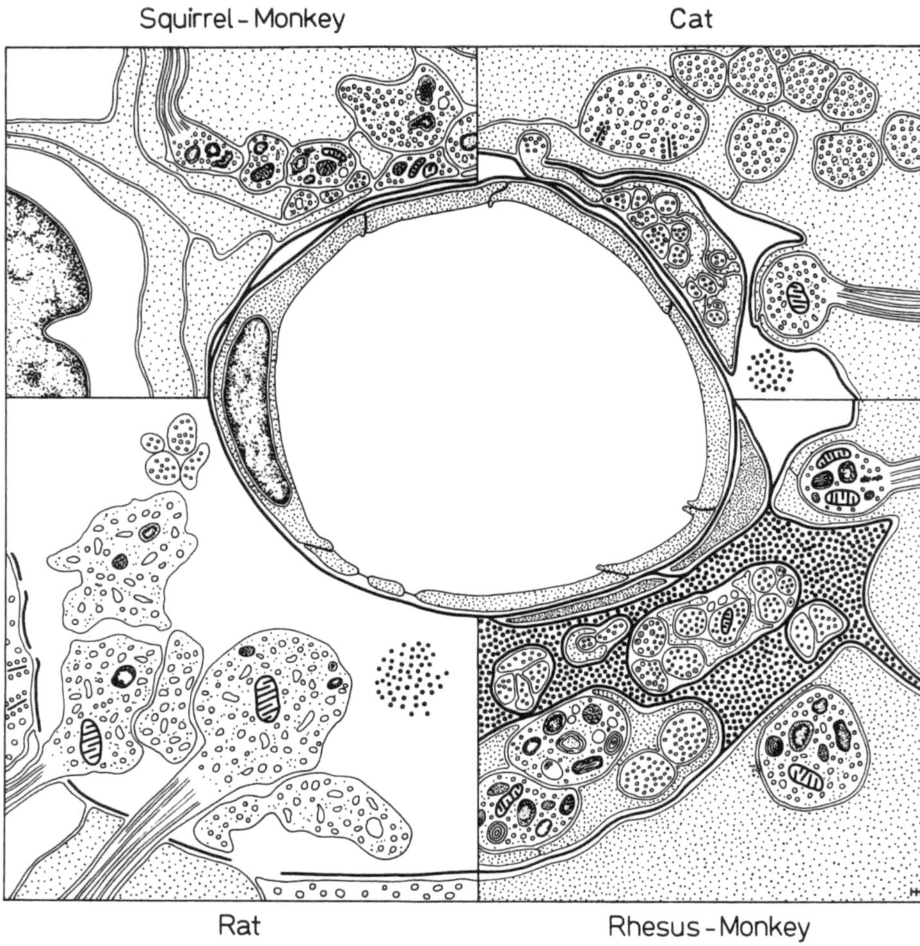

Fig. 122. Topography of perivascular spaces in the pineals of four mammalian species. WARTENBERG, 1968

published. That pinealocyte processes terminate in the perivascular space has also been found in dog (SANO and MASHIMO, 1966), hamster (SHERIDAN and REITER, 1968), guinea-pig (LUES, 1971), mouse (ITO and MATSUSHIMA, 1968a, b; PELLEGRINO DE IRALDI, 1969) and gopher (SHERIDAN and REITER, 1973). Invariably, in these species the perivascular spaces also contained nerve fibres.

Experimental studies have shown that the perivascular spaces with the terminals of both pinealocyte processes and nerve fibres undergo changes of potential importance. In hamsters blinding is followed by an increase in size of pinealocyte terminals, and after superior cervical ganglionectomy there is an increase in the number of pinealocyte terminals (SHERIDAN, 1975). In rats the interesting observation has been made that the glial cells surrounding the perivascular space retract after bilateral adrenalectomy, so that only 34% of the space, instead of 60% in intact control animals, is covered by glial elements (DEUSSEN-SCHMITTER et al., 1976).

7. Pineal Blood-Brain Barrier

Ever since the demonstration by WISLOCKI and LEDUC (1952) that after administration of silver nitrate via the drinking water for 9–15 months the epiphysis cerebri showed silver deposits in the immediate vicinity of blood vessels, it has been commonly accepted that the epiphysis cerebri lacks a blood-brain barrier. However, as pointed out by DRETZKI (1971), this conclusion is not entirely justified, since according to the above authors neither silver nor trypan blue are taken up by parenchymal cells: silver is deposited on strands of reticular fibres, whereas trypan blue is taken up by perivascular macrophages and is occasionally found in the cytoplasm of endothelial cells. Experimental studies by DRETZKI (1971) using Myofer (an iron-polysaccharide complex) as a marker have shown that this substance has to be administered over a period of at least 6 days before it occurs in some of the pinealocytes lying adjacent to perivascular spaces, most of the Myofer being present in endothelial cells and perivascular spaces. After longer periods of treatment, Myofer is uniformly taken up by pinealocytes and glial cells, macrophages showing the strongest labelling. It thus appears that, as far as short-term experiments are concerned, a definite blood-pineal barrier exists for trypan blue and Myofer which may be overrun after longer time intervals. Trypan blue given over periods of between 9 and 28 days was not taken up by rat pinealocytes (COLLIER, 1943). In rabbits trypan blue was found only in the vicinity of blood vessels where adventitial cells were present (WEINDL, 1967). In guinea-pigs it takes longer for Myofer to reach the pinealocytes than in rats (WARTENBERG et al., 1972). As in rats, most of the substance resides in the narrow perivascular spaces and macrophages, a small amount also being present in glial cells. It thus appears that distinct interspecies differences exist with regard to the leakiness of the blood-pineal barrier.

Recently, this problem has been studied using horseradish peroxidase (HRP) and microperoxidase (MP) given intravenously to mice (MØLLER et al., 1978b). In all experiments a pronounced extravasation of HRP and MP occurred, either by means of transendothelial vesicular transport (vesicle diameter, 50–150 nm) or via intercellular junctions, endothelial pores and channels formed by fusion of vesicles with the luminal and abluminal cell membranes. However, to what extent these substances are taken up by pinealocytes was not specified.

IX. Other Constituents

1. Connective Tissue

For a detailed review of epiphyseal connective tissue, the reader is referred to BARGMANN'S (1943) monograph. Here it suffices to mention that the amounts of connective tissue present show distinct interspecies variation. It is regularly present in the capsule of the epiphysis cerebri. The thickness of the capsule is apparently not directly related to the size of the epiphysis cerebri, though actual measurements are lacking. In contrast to previous studies (BARGMANN, 1943), the pineal capsule of cattle and sheep has been described as being relatively thick (ANDERSON, 1965). A relatively thick capsule was also noted in the horse (HEINIGER, 1965a), corroborating previous observations (see BARGMANN, 1943). In the pig pineal the capsule is of medium thickness, while in carnivores it is relatively thin (HEINIGER, 1965a).

As mentioned above (p. 67), more or less well developed connective tissue septa extend from the capsule into the parenchyma and surround groups of pinealocytes. Again, most of the relevant information may be found in BARGMANN's (1943) monograph. A new finding appears to be that in the rat pineal the interstitial tissue contains small numbers of elastic fibres in addition to collagenous and reticular fibres (CHANG, 1969). Basophilic granules in the interstitial tissue of the pineal body of *Macaca mulatta,* interpreted as secretory material, have been demonstrated after aldehyde fuchsin staining following performic acid oxidation (HOLMGREN, 1958a).

2. Mast Cells

Pineal mast cells have attracted special attention as the pineal contains substantial amounts of histamine (GIARMAN and DAY, 1958/59; K. TAYLOR et al., 1972; GARBARG et al., 1974), though it is not entirely clear in what kind, or kinds, of cell histamine is contained (QUAY, 1974a).

According to A. MACHADO et al. (1965) and HÅKANSON and OWMAN (1966) using toluidine blue or Astrablau staining, the number of mast cells in the epiphysis cerebri varies considerably from species to species (Table 21). In man the amount of mast cells present does not correlate with the age of the individuals, and in cattle there are no statistically significant differences between males and females. Mast cells appear to occur relatively early during ontogenesis as shown by the fact that in the cynomolgus monkey they were already present in pineals of 1-day-old animals and in cats at 18 days of age (CAMMERMEYER, 1973). Other species containing mast cells in the pineal are chimpanzee, *Macaca mulatta,* bushbaby, capybara and agouti (CAMMERMEYER, 1973).

The localization of mast cells varies considerably. In man, they are concentrated in, or close to, the capsule and the connective tissue septa, especially along blood vessels. An even distribution throughout the parenchyma is characteristic of cattle, goat, sheep, monkey and guinea-pig, in which the cells lie either singly or in small groups. In marmoset and mouse, mast cells are exclusively present in the capsule (A. MACHADO et al., 1965). In the rat, mast cells are rarely present inside the pineal, but are regularly observed in the meninges near the pineal (BERTLER et al., 1964). In capybara and agouti they lie in the dorsal parts of the pineal (CAMMERMEYER, 1973).

Whether or not pineal mast cells differ from those of other parts of the body has apparently not been investigated. Due to a close correspondence of mast cell number and histamine content of the epiphysis cerebri it has been concluded that pineal mast cells contain histamine (A. MACHADO et al., 1965).

Table 21. Incidence of mast cells in the pineals of different mammalian species

Species	Precise estimate[a]	Rough estimate[b]
Cattle	1,375.0	Large number
Goat	805.0	
Monkey	783.0	
Guinea-pig	680.0	
Man	300.0	
Sheep	194.0	Moderate number
Marmoset	55.0	
Mouse	43.0	
Rabbit	18.0	Small number
Cat	9.9	Small number
Fox	7.6	
Syrian hamster	7.3	
Dog	4.7	
Pig	3.9	Very few
Rat	2.2	Not specified
Horse	1.0	

[a] Number of mast cells/100 microscopical fields; A. MACHADO et al., 1965.
[b] HÅKANSON and OWMAN, 1966.

HÅKANSON and OWMAN (1966), using the fluorescence microscope following treatment with formaldehyde vapour, found that dopamine is present in cattle mast cells and serotonin in humans, the latter observation suggesting that pineal mast cells may differ from those in other parts of the body. In pig, cat and rabbit mast cells were not visible by fluorescence microscopy and in the rat their fluorescence is masked by that of the pinealocytes (HÅKANSON and OWMAN, 1966).

As hinted at above (p. 90), epiphyseal histamine may not be restricted to mast cells. That pinealocytes may contain some has been concluded on the grounds that in the rat pineal the histamine content exhibits a characteristic 24-h-rhythm with low levels during the day and a striking increase after the onset of darkness (GARBARG et al., 1974). BERNAD and CSABA (1970), using tissue culture and the supposedly specific *O*-phthaldialdehyde reaction, claimed that a special type of pinealocyte contains histamine, though it should be noted that according to HÅKANSON et al. (1974) the reaction employed is not specific.

3. Concretions (Corpora Arenacea)

One of the best known yet functionally poorly understood features of the mammalian epiphysis cerebri is that the organ is subject to calcification, the basic aspects of which have been dealt with by BARGMANN (1943), KITAY and ALTSCHULE (1954a) and QUAY (1974a). Contrary to common belief, pineal calcification does not appear to be a sign of functional inactivation. Recently, the concept has been put forward that the formation of concretions, at least

Table 22. Incidence of pineal calcification in humans (KITAY and ALTSCHULE, 1954a)

No. of cases	Decade									Average
	1st %	2nd %	3rd %	4th %	5th %	6th %	7th %	8th %	9th %	%
666	2	25	38	59	59	58	69	–	–	45
1271	3	26	47	58	58	67	65	73	–	49
114	0	60	78	71	75	86	95	70	100	76
70	0	25	55	61	69	83	–	–	–	56
2724	5	28	49	61	62	66	70	69	100	51
375	0	11	42	36	72	45	83	–	–	33

in some species, is directly related to the secretory activity of the epiphysis cerebri (for details see below).

A characteristic feature of the calcareous concretions in the pineal is that they are not universally present. They are typical of human and ungulate pineals; QUAY (1974a) suspected that their development is related to the stromal architecture and transport activities of the epiphysis cerebri, since species with concretions are characterized by prominent stroma and interlobular tissue. This view seemed reasonable at the time when it was expressed, since concretions had not been found in rodents, carnivores, insectivores, bats and primates other than man. However, recently concretions have also been detected in the rhesus monkey (LUKASZYK and REITER, 1975), Mongolian gerbil (JAPHA et al., 1974, 1976; REITER et al., 1976a; NIELSEN and MØLLER, 1978) and rat (ERDINÇ, 1977). In contrast to ERDINÇ, who found concretions in only one rat epiphysis cerebri, DIEHL (1978a) found concretions in 77 out of 86 Sprague-Dawley rats of 450–550 g body weight, which suggests that corpora arenacea may regularly be present in aged rats.

The problem of a possible age-dependency of the occurrence of pineal concretions has been thoroughly studied in humans. There is clear evidence that their incidence increases with age (Table 22), the incidence being negligible in the first decade of life, approximating 25% in the second decade, followed by a gradual increase from the third to the eighth decade (KITAY and ALTSCHULE, 1954a). The incidences reported vary depending on the technique used for detection. As can be expected, histological techniques are superior to roentgenographic studies. While WILDI and FRAUCHIGER (1965) detected concretions in 84% of 298 histologically investigated pineals, KITAY and ALTSCHULE (1954a) reported an incidence of between 33% and 76% among subjects of all ages. BAYEROVÁ et al. (1962) found fluorescence and polarizing microscopical techniques useful for the study of delicate concretions in the human pineal. There is also great interindividual variation, both in the old (ARIETI, 1954) and young. In one study dealing with subjects of up to 20 years of age, 1 out of 16 pineals contained concretions, which led to a concept that calcification of the pineal in children strongly suggests the presence of a tumour, in particular a pinealoma or teratoma (LÖFGREN, 1958). However, when one considers that

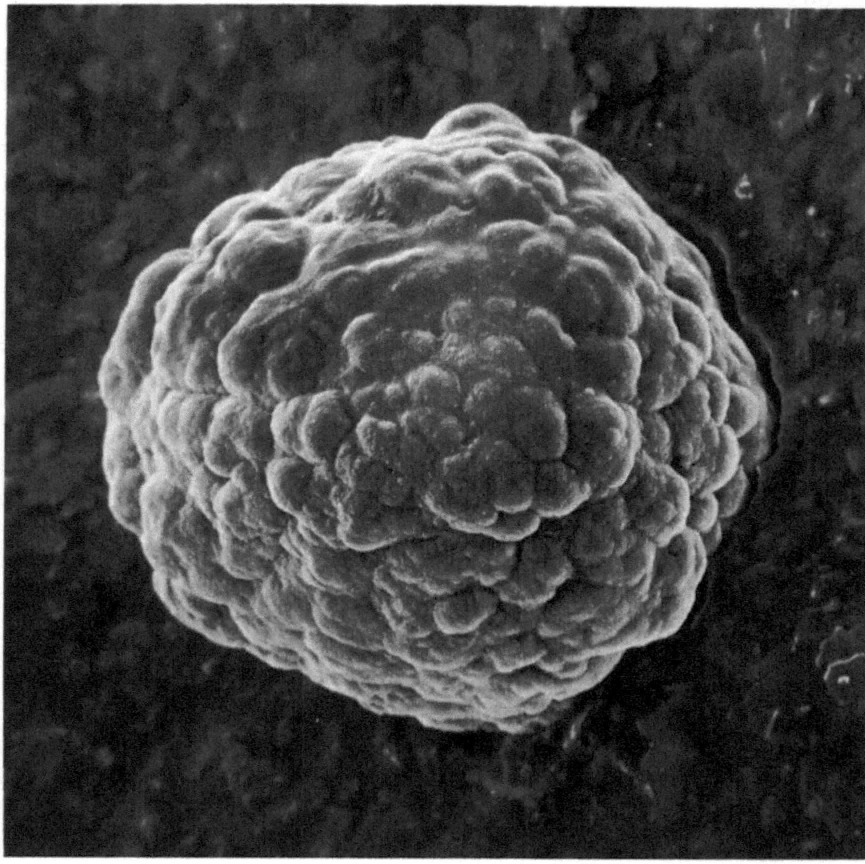

Fig. 123. An isolated acervulus from a human pineal gland. Note the lobulated surface. Scanning electron micrograph. KRSTIĆ, 1976b; ×275

in 55 subjects of 0–12 years of age 27 pineals contained concretions (HEIDEL, 1965), the above concept is not very likely to be correct.

The concretions are very variable in shape and appearance (BARGMANN, 1943; ARIETI, 1954; QUAY, 1974a). They range from small microscopically hardly visible structures to large bodies showing concentric lamination or coralliform appearance, both in humans and animal species. Under the scanning electron microscope human acervuli have a mulberry-like appearance (Fig. 123) and are composed of a varying number of spherical particles measuring 135–800 μm, between which groups of smaller globules (4–14 μm in diameter) are seen (KRSTIĆ, 1976b; see also MICHOTTE et al., 1977). By transmission electron microscopy the acervuli have been shown to consist of needle-shaped, randomly oriented crystals, 30–50 nm long and 2.5–3.5 nm wide (KRSTIĆ, 1976b; KRSTIĆ and GOLAZ, 1977).

Pineal concretions are found primarily in the glial and stromal compartments of the organ, very often lying in the immediate vicinity of blood vessels (BARGMANN, 1943; QUAY, 1974a). PILLING and HAWKINS (1977) found that the centre of calcification in 72 out of 73 glands studied was within 2 mm of the midline of the gland; in one organ it lay 2.6 mm from the midline.

In the Mongolian gerbil they are often seen in chromophobic but not in chromophilic pinealocytes (NIELSEN and MØLLER, 1978). They are present in both the superficial pineal and the pineal stalk, but are absent in the deep pineal (JAPHA et al., 1974, 1976). In the rat, intrapineal concretions are restricted to subcapsular regions of the gland (DIEHL, 1978a). In the pineal region, concretions are not restricted to the pineal organ proper. In humans, a juxta-epiphyseal position near the habenulae has been described (VOSS, 1957), and in rats the leptomeningeal tissue adjacent to the superficial pineal is often the site of concretions (DIEHL, 1978a). Technical details for the radiological localization of the pineal gland may be obtained from the paper by MOREAU (1966).

The *chemistry* of the concretions has been examined by various means. Histochemical studies have revealed that concretions and colloid droplets have identical properties and that the concretions consist of both inorganic and organic substances (VOSS, 1957; KHELIMSKY, 1969). According to VOSS (1957), the concretions give positive PAS and HALE reactions and show metachromasia following toluidine blue staining, pointing to the presence of both proteoglycans and glycoproteins (see also PALLADINI et al., 1965). Similar results have been obtained in the gerbil (JAPHA et al., 1976). The presence of calcium salts has been demonstrated by various staining reactions, including the alizarin red and VON KOSSA techniques (BARGMANN, 1943; QUAY, 1974a; SALIICHUK, 1964; EARLE, 1965; JAPHA et al., 1976; REITER et al., 1976a). X-ray diffraction and X-ray fluorescent spectroscopical studies reveal the pattern of a poorly resolved apatite (EARLE, 1965). Whereas EARLE is of the opinion that the pattern obtained matches that of hydroxyapatite or carbonate apatite, ANGERVALL et al. (1958) concluded that no other calcium compounds besides hydroxyapatite can be detected in X-ray diffractograms. In addition to calcium, iron (EARLE, 1968) and magnesium (SALIICHUK, 1964) appear to be present in calcospherites. By means of electron probe microanalysis, Ca and P as well as small amounts of Mg, S and Sr have been detected (KRSTIĆ, 1976b (Fig. 124). Gerbil pineal acervuli have a chemical composition similar to those in the human (KRSTIĆ and GOLAZ, 1977). A very high content of trace elements has been reported by MICHOTTE et al. (1977).

The genesis of the concretions has been thoroughly discussed by BARGMANN (1943), who concluded that colloid droplets serve as precursors (see also KHELIMSKY, 1958). Electron microscopic studies have revealed that concretions may originate from degenerating myelin or connective tissue ground substance (BAYER and BAYEROVÁ, 1967). LUKASZYK and REITER (1975) suspect that microvesicles stemming from exocytotic processes eventually coalesce to form colloid deposits which then might become secondarily calcified. They also suspect that a carrier protein may be important (see below). According to JAPHA et al. (1976), the concretions develop from within pinealocytes, but using the scanning electron microscope KRSTIĆ (1976b) always found the concretions in the extracellular

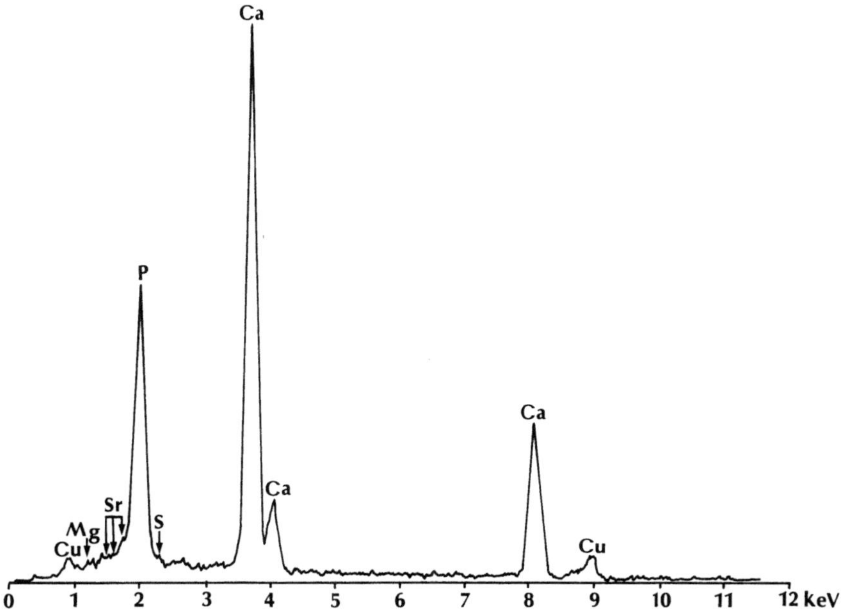

Fig. 124. X-ray energy dispersive analysis of pineal concretions (human). Note the large peaks of *Ca* and *P*. In addition, small amounts of *Mg*, *S* and *Sr* can be detected. The *Cu* peaks stem from the specimen supporting grids made of copper. KRSTIĆ, 1976b

space and suspected that globular bodies fuse to form the lobe-like structures of the concretions.

The functional meaning of the concretions in the human pineal is unclear. KITAY and ALTSCHULE (1954a) concluded that no relationship has ever been established between the occurrence of calcification and any functional status of the human pineal gland. In view of the fact that the metabolic activity of the pineal remains unaltered in old age (WURTMAN et al., 1964c) and that there is no correlation between the degree of calcification and either the histological structure (CASSANO et al., 1961) or enzyme levels (5-hydroxytryptophan decarboxylase, MAO, HIOMT) (OTANI et al., 1968a) of the epiphysis cerebri, it is apparently unjustified to correlate the presence of concretions with atrophic processes. In view of the high incidences of pineal calcification it is not very likely that a relationship exists between calcification and endocrinopathy, including disturbances of libido, homosexuality etc., as suggested by PENDE (1953).

A currently discussed view (Fig. 125) is that the concretions have something to do with exocytotic processes in the pineal. LUKASZYK and REITER (1975) propose that they are the result of a binding process of calcium and (hypothetical) pineal carrier protein ("neuroepiphysin"), forming a complex which is then deposited on adjacent multilayered concretions. That the concretions may indeed be related to secretory processes is suggested by the finding that in the gerbil the pineal is almost devoid of corpora arenacea after a depression of pineal function by sympathetic denervation (REITER et al., 1976a). DIEHL (1978a) attributes the formation of concretions to a reduced drainage of tissue fluid in the pineal.

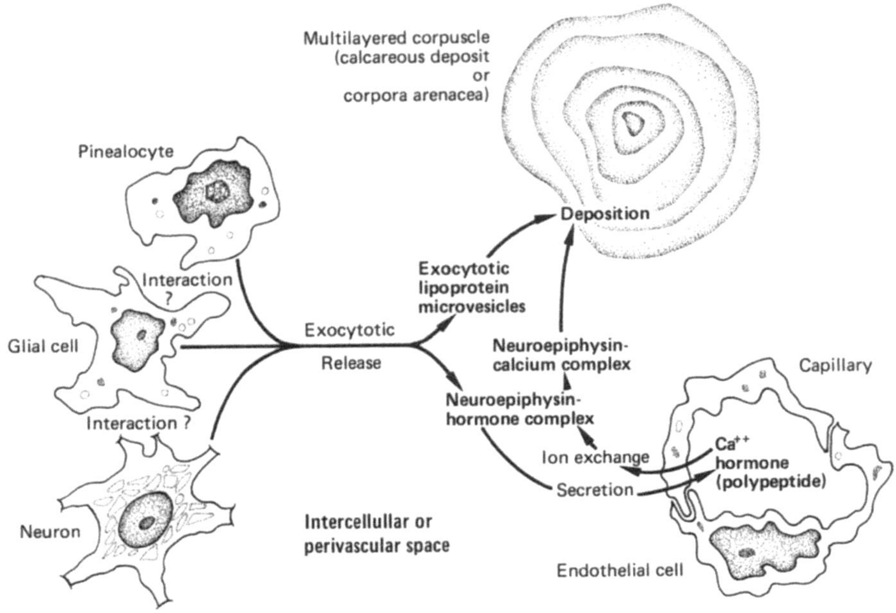

Fig. 125. Schematic representation of the formation of pineal concretions (hypothetical), which is self-explanatory. LUKASZYK and REITER, 1975

4. Cysts

According to BARGMANN (1943, 1958), cysts are a regular feature of the mammalian epiphysis cerebri. Recent studies have contributed little to this aspect of pineal morphology. In human pineals, WILDI and FRAUCHIGER (1965) described the presence of cysts in the centre of glial patches in 33% of the cases investigated. Some were already found at birth. In a total of 104 pineals (MEGYERI, 1960), 37% had cysts whose walls consisted of glial tissue and 4% had cysts with walls consisting of parenchymal cells. In pigs, cysts appeared to be particularly numerous in young sows, boars and hermaphrodites (HEINIGER, 1965a). In rats, cysts lying beneath the pineal have been described (GIROUD and MARTINET, 1965).

5. Melanophores

In addition to pigment-containing pinealocytes and glial cells, there are special pigment-containing stromal cells, the melanophores. BARGMANN (1943) has thoroughly reviewed the subject and stated that melanophores are found in the pineal stroma of man, horse (see also FASSBENDER, 1962), ass, cow, monkey, goat, sheep, dog, rabbit and cat. The dark pigment present has been identified as melanin by means of chemical, histochemical and spectrophotometric techniques (SANTAMARINA and MEYER-ARENDT, 1956; SANTAMARINA, 1958).

In electron microscopic studies epiphyseal melanophores have not been unequivocally identified. SHERIDAN and REITER (1973) noted a regular occurrence of pigment-containing cells in the gopher pineal, but the authors were not sure whether they represented melanophores or mast cells. It is also unclear what sort of cells the pigment-containing interstitial cells of horse and goat represent (MIKAMI, 1951). Pigment-containing cells are apparently not restricted to intraepiphyseal connective tissue. SANTAMARINA (1958) noted pigment granules in the pia mater of the pineal capsule, especially in the vicinity of blood vessels. As regards a possible correlation of melanin pigmentation and gonadal function in cattle, see SANTAMARINA (1958).

6. Striated Muscle

A peculiar feature of the mammalian epiphysis cerebri is the occasional presence of striated muscle fibres (see BARGMANN, 1943). Originally only found in man and cattle, striated muscle has recently been detected in rats. In this species the incidence of striated muscle is very variable (Table 23), being present in up to more than 30% of the animals examined.

Striated muscle fibres lie primarily near the surface of the epiphysis (BARGMANN, 1943; DERENBACH, 1951; DILL, 1963; TAPP and BLUMFIELD, 1970; KRSTIĆ, 1972b; DIEHL, 1978b); sometimes they occur in the region of the pineal stalk (QUAY, 1959b; KRSTIĆ, 1972; DIEHL, 1978b). While according to BARGMANN (1943) and QUAY (1959b) the fibres are not related to blood vessels, such a relationship has been observed by DILL (1963), KRSTIĆ, (1972b) and DIEHL (1978b). Figure 126 shows that the amounts of striated muscle fibres in the pineal may be substantial.

Electron microscopic studies (KRSTIĆ, 1972b) have confirmed the impression gained by light microscopy (DERENBACH, 1951; DIEHL, 1978b) that the striated muscle fibres are embryonic in appearance, insofar as they possess abundant sarcoplasm, scarce myofibrils, helically arranged polysomes, differently located

Table 23. Incidences of striated muscle fibres in the rat pineal (DIEHL, 1978b)

Strain	No. of pineals investigated	No. of pineals containing striated muscle fibres	Sex	Total incidence (%)	Author(s)
Mixed and Long-Evans	1,200	2	Male	0.25	QUAY (1959b)
		1	Female	–	
Unspecified	Unspecified	1	–	–	KAPPERS (1960a)
Holzman	Unspecified	2	Female	–	DILL (1963)
Unspecified	4	2	–	50.00	KENNY (1965b)
Sprague-Dawley	144	2	Mixed	1.39	TAPP and BLUMFIELD (1970a)
Wistar	35	8	Mixed	22.86	KRSTIĆ (1972b)
Sprague-Dawley	96	31	Male	32.39	DIEHL (1978b)

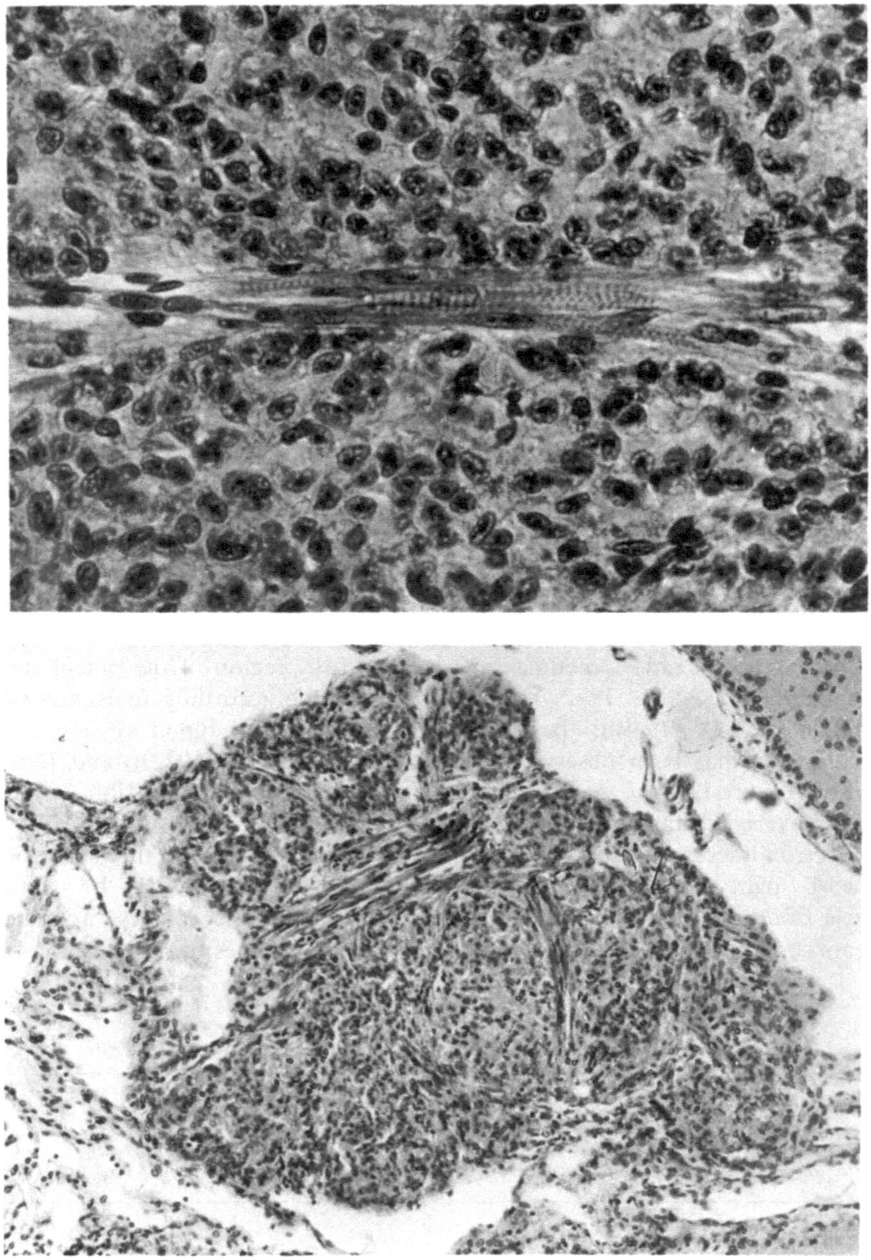

Fig. 126. Light micrographs showing the presence of striated muscle fibres in the rat pineal gland. *Top*, fibres exhibiting a clear cross striation. ×1270. *Bottom*, sagittal section through the marginal zone of a pineal gland, showing an extensive network of muscle fibres. ×310. By courtesy of B.J.M. Diehl, Department of Anatomy, University of Mainz. Previously unpublished light micrographs

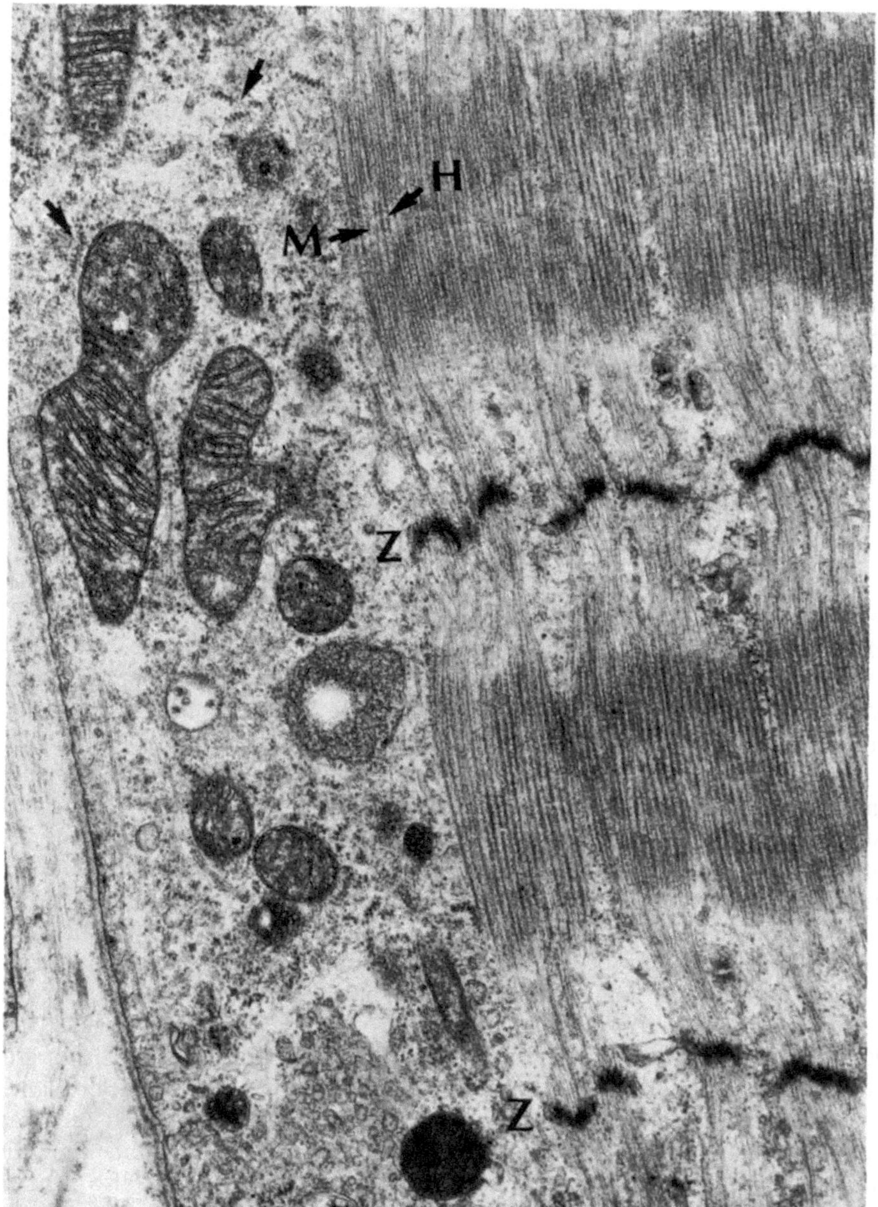

Fig. 127. Electron micrograph of striated musculature found in rat pineal gland. The *unlabelled arrows* point to helically arranged polyribosomes. *H*, H-band; *M*, M-band; *Z*, Z-line. KRSTIĆ, 1972b; × 30,000

triads, transition of rough into smooth endoplasmic reticulum and primitive A-bands (Fig. 127); H- and M-bands are mostly lacking.

Origin and functional significance of the epiphyseal striated muscle are obscure. DILL (1963) discussed the possibility that the myogenic tissue may have come from blood vessels or the leptomeninx. In this context it is interesting that in the bat KENNY (1965b) found striated muscle in the leptomeninges adjacent to the pineal body. Comparing pineal striated musculature with musculature of mesodermal and ectodermal origins, DIEHL (1978b) concluded that the pineal musculature is of mesodermal origin. QUAY (1959b) and TAPP and BLUMFIELD (1970a) discuss that the epiphyseal striated muscle fibres are a rare anomaly, possibly an example of atavism. KRSTIĆ (1972b) remarked that during pinealectomy the pineal occasionally retracts when it is touched, but he was hesitant to ascribe this to the presence of the muscle fibres.

X. The Transplanted and Cultured Epiphysis Cerebri

1. Transplantation

In view of the concept that the pineal gland is a neuro-endocrine transducer (WURTMAN and ANTON-TAY, 1969) for which the neural input is essential, one cannot expect that structure and function of a transplanted pineal are identical to those of a pineal left in situ. Nevertheless it would seem that the transplanted pineal has a fairly normal histological appearance. HOLMES (1957), using both autografts and homografts, noted that autografting was more successful than homografting. Autografts of adult rat epiphyseal tissue, transplanted into the anterior chamber of the eye and examined up to 10 weeks after operation, showed good vascularization, and the pineal cells were apparently healthy and normal, showing no mitotic figures. Homografts were either resorbed or became fibrosed. Normal histological appearance of pineal tissue was also noted when the organ had been transplanted to muscle or under the renal capsule (IFFT, 1962; GITTES and CHU, 1965; REITER, 1967a). In a comparative study SIMIONESCU and GOLDSTEIN (1962) found that pineal autografts took best when transplanted into muscular tissue, ensuring viability of the tissue for up to 60 days; subcutaneous adipose tissue and retroperitoneal tissue were less favourable sites and intermuscular fascial tissue was least favourable. Pineal tissue also remains viable in growing cartilage, where it stimulates the growth of cartilage (SIMIONESCU et al., 1962).

Transplanted pineals lack sympathetic nerve fibres (e.g. OWMAN, 1964b; WRAGG et al., 1967) until surrounding nerve fibres re-innervate the gland. According to MOORE (1975a) pineal transplants to the anterior chamber of the eye show functional re-innervation after 4 weeks.

Pinealocytes from pineal transplants look ultrastructurally fairly normal up to 1 month (KRSTIĆ, 1976c; AGUADO et al., 1977) and exhibit yellow fluorescence after formaldehyde vapour treatment, indicative of serotonin (OWMAN, 1964b), the intensity being less than in control pineals left in situ (WRAGG et al., 1967). HIOMT is present, but shows a relatively low activity (GITTES and CHU, 1965; WRAGG et al., 1967). As long as the pineal transplants are not re-innervated they do not respond to changes in environmental lighting (WRAGG et al., 1967). MOORE (1975a), studying NAT activity in pineal transplants after reinnervation had occurred, noted that although the overall enzyme activity is reduced, there are clear day-night differences, as in intact animals. In view of the common origin of the ocular and pineal sympathetic fibres from the superior cervical ganglia it is interesting to note that in pineals transplanted to the anterior chamber of the eye of ganglionectomized animals the day–night differences of NAT activity are no longer demonstrable.

Since the transplanted pineal retains only a little of its secretory activity, it is not surprising that no effect can be demonstrated if one pineal or a small number of pineal transplants are used in any one animal (e.g. REITER, 1967a), while if sufficient numbers of transplants are grafted, an effect is clearly present (IFFT, 1962; MOSZKOWSKA, 1947, 1958a).

2. Tissue Culture

As pointed out above, pineal tissue cultured in vitro has been widely used to study pineal function. Technical details have been given by SHEIN (1975). TROWELL (1959), culturing the whole pineal of rats weighing 60 g, found that the tissue is regularly healthy and unaltered for 6 days and is usually so for 9 days. SHEIN et al. (1967), studying cells of explanted pineal *fragments*, described that they have a healthy appearance throughout a 10-day period of incubation. After the first 2–3 days the cell population consists almost entirely of pinealocytes. There is little cellular outgrowth from the explants until the 4th day. An extensive growth appears at 8–10 days, when occasional scattered, more elongate cell forms can be observed in addition to pinealocytes.

The most detailed morphological studies of cultured rat pineal tissue are those by HUNGERFORD and POMERAT (1962, 1965), who cultured the tissue for up to 29 days and studied the outgrowing cells by means of phase-contrast, time-lapse cinematography and histological techniques. With these techniques at least three distinct parenchymal cells types can be distinguished. Chief cells with three to four processes are characterized by marked pinocytosis at the end of the cell processes. Fat droplets, stained by oil red 0, are small in number in the chief cells and particularly abundant in what has been termed fat-laden cells, which exhibit rapid nuclear rotation. The third type of cell, "basophilic cells", is characterized by prominent cytoplasmic granules which stain deep blue with MAY-GRÜNWALD-GIEMSA, brown to black with HORTEGA's silver carbonate but only faintly with HEIDENHAIN's haematoxylin. Moreover, the granules are weakly stained with PAS and strongly stained with aldehyde fuchsin. They do not exhibit metachromasia with methylene blue, Azure II or toluidine blue. In addition, a large epithelioid element is observed together with macrophages and astrocytes. Mitoses are frequent in both chief cells and astrocytes. However,

Quay and Kahn (1963) noted that mitoses of parenchymal cells are extremely rare.

Histochemically, pinealocytes in vitro are similar to those in vivo. The enzyme patterns are identical (Taşcă et al., 1968; S. Milcu et al., 1968), although basophilic pinealocytes show greater enzyme activities than the main type of pinealocyte. Pinealocytes in vitro exhibit the typical yellow formaldehyde-induced fluorescence due to serotonin (Bernad and Csaba, 1970; Arstila et al., 1971).

The *ultrastructure* of cultured pineal tissue has been extensively studied and it would appear that the organelles of the pinealocytes basically do not differ from those of non-cultured cells (Arstila et al., 1971; Karasek, 1974, 1976; Romijn and Gelsema, 1976; Rowe et al., 1977; Buda and Klein, 1978; Piekut and Knigge, 1978). That the cultured pinealocytes continue to form melatonin has been mentioned above. Further details may be obtained from the papers by Nathanson et al. (1977), Rowe et al. (1977) and Piekut and Knigge (1978).

The question of granule formation and storage of indoleamines has also been studied in vitro. Combined pharmacological and autoradiographic studies at both the light and electron microscopic levels suggest that in the rabbit pineal indoleamine synthesis and storage do not take place within dense-core vesicles but more diffusely in the cytosol of light pinealocytes, the initial action of tryptophan 5-hydroxylase occurring in mitochondria (Romijn et al., 1977b).

In a number of studies drugs which are known to play a role in the regulation of pineal function have been tested for their effects on pinealocytes in vitro. In view of the important role of noradrenaline as neurotransmitter responsible for the initiation of the increased activity of the pineal gland after the onset of darkness, it is not surprising that administration of noradrenaline to the culture medium is followed by an increase in the amounts of rough endoplasmic reticulum, mitochondria, granulated vesicles (see also Romijn and Gelsema, 1976; Romijn et al., 1977b), often lying in the immediate vicinity of synaptic ribbons (Romijn and Gelsema, 1976), and a strong development of the Golgi apparatus (Karasek, 1974). Administration of dibutyryl cAMP leads to the occurrence of mitochondria with tubular cristae (Karasek, 1974) and light microscopically the pinealocytes are seen to develop processes within less than 4 h which retract when the effect of the drug has passed (Wilkinson, 1975). *p*-Chlorophenylalanine has virtually no effect on light and dark pinealocytes apart from the fact that in light pinealocytes the mitochondria show an increase in both the size of the organelle and the number of cristae (Romijn et al., 1977b).

Tissue culture has also been used to study the relationships between the pineal and assumed target organs. Rat pineal tissue cultured in the presence of adenohypophyseal tissue shows signs of enhanced activity (Karasek, 1974). The addition of oestradiol to the incubation medium is followed by a reduction of the chromophilia of the pinealocytes (Quay and Kahn, 1963), and electrophysiologically a decrease of the membrane potential is noted (Sakai and Marks, 1972). In histochemical studies thyroxine administration increases the activities of a number of enzymes, while cortisol (except arylsulphatase), testosterone propionate and stilboestrol have no demonstrable effects (S. Milcu et al., 1968).

For the effects of noradrenaline, cAMP etc., on biochemical parameters of cultured pinealocytes, see p. 10. Here it suffices to mention some additional points which may be of importance. Administration of morphine apparently has no effect on the synthesis of pineal indoleamines (SHEIN et al., 1970). The MAO inhibitor, harmine, inhibits ^{14}C-serotonin oxidation, but stimulates ^{14}C-melatonin production (KLEIN and ROWE, 1970). Ouabain or elevated potassium ion concentration block the noradrenaline- and cAMP-induced increase of pineal NAT activity (PARFITT et al., 1975).

XI. Functional Aspects

The foregoing sections have made clear that we have a fairly clear picture of the morphology, biochemistry and pharmacology of the mammalian epiphysis cerebri. As regards its target organs, we are in an only slightly better situation than in 1969, when WURTMAN and ANTON-TAY (1969) stated that one knows what the pineal does, but that it is much less clear what the organ is for. Hence it is appropriate to begin with what the pineal does. It is clearly established that the pineal synthesizes melatonin and that this process is influenced by light and darkness. In addition to melatonin, the epiphysis cerebri contains a number of other highly active substances of perhaps hormonal nature.

1. Melatonin Formation

As pointed out above (p. 5), the biosynthesis of melatonin in the pineal has been clearly established. While it was previously thought that melatonin formation is restricted to the epiphysis cerebri, it has recently become evident that it also occurs in the retina (CARDINALI and ROSNER, 1971; NAGLE et al., 1972a, 1973a, 1974a, c; HYYPPÄ et al., 1973; BUBENIK et al., 1978; PEVET et al., 1978), Harderian gland (CARDINALI et al., 1972a; CARDINALI and WURTMAN, 1972; BUBENIK et al., 1976b, c 1978), enterochromaffin cells (RAIKHLIN et al., 1975, 1978; RAIKHLIN and KVETNOY, 1976, 1977; BUBENIK et al., 1977) and human erythrocytes (ROSENGARTEN et al., 1972). Immunoreactive material in the cerebellum which was thought to be melatonin (BUBENIK et al., 1974) has subsequently been interpreted as *N*-acetylserotonin (BUBENIK et al., 1976a, b). Very small amounts of melatonin have been detected in peripheral nerves of man (BARCHAS and LERNER, 1964), monkey and cow (LERNER et al., 1959b). After pinealectomy, melatonin is still demonstrable in blood (OZAKI and LYNCH, 1976; KENNAWAY et al., 1977), urine (OZAKI and LYNCH, 1976) and hypothalamus (KOSLOW, 1974). The daily melatonin rhythm persists, yet with reduced amplitude (LYNCH et al., 1975b).

Melatonin formation in the epiphysis cerebri appears to have been demonstrated in all mammalian species in which it has been searched for. Until fairly recently, melatonin in the epiphysis cerebri had been demonstrated in relatively

Table 24. Pineal HIOMT activities in different mammalian species (QUAY and SMART, 1967)

Species	No. of animals	Mean ± SEM[a]
Monkey (*Macaca mulatta*)	3	42,524.0 ± 8,481.0
Sheep (*Ovis aries*):		
Castrate males	6	36,473.0 ± 12,003.0
Intact females	30	23,579.0 ± 1,351.0
Beef (*Bos taurus*)	4	20,602.0 ± 2,012.0
Raccoon (*Procyon lotor*)	3	13,518.0 ± 3,851.0
Rat (*Rattus norvegicus*)	13	4,139.0 ± 334.0
Cat (*Felis catus*)	2	871.0 ± 29.0

[a] Results expressed as net CPM of ^{14}C-melatonin formed per mg tissue per hour.

few species by direct assay (cow: LERNER et al., 1958, 1959a, b, 1960; rat: PROP and KAPPERS, 1961; QUAY, 1964a; LYNCH, 1971; DEGEN et al., 1972; dog: MILLER and MAICKEL, 1970; kangaroo: QUAY and BAKER, 1965; rhesus monkey, langur: QUAY, 1974a; man: LERNER and CASE, 1959, quoting unpublished results by LERNER, CASE and MORI). A large body of indirect evidence for the formation of melatonin in the epiphysis, namely the demonstration of the universal presence of the enzymes HIOMT and NAT, has been beautifully corroborated by recently developed radio-immunoassay techniques for the demonstration of melatonin, which appear to be fairly reliable (WETTERBERG, 1977). One great advantage of these techniques is that they are sensitive enough to measure melatonin in serum, allowing melatonin determinations also in humans. Together with other techniques, e.g. gas chromatography-mass spectrometry (KENNAWAY et al., 1977), it has been clearly established that in the human melatonin is present in both epiphysis cerebri and serum and is excreted in the urine, all showing corresponding day–night changes of the melatonin concentrations (PELHAM et al., 1972, 1973; LYNCH et al., 1975a; ARENDT et al., 1975, 1977; MATTHEWS et al., 1977; J.A. SMITH et al., 1977; G.M. VAUGHAN et al., 1976; GREINER and CHAN, 1978).

Although the formation of melatonin in the mammalian epiphysis cerebri appears to be a common phenomenon, there is great interspecies variation, as assessed by HIOMT measurements (Table 24) and radio-immunoassay of serum melatonin levels (Table 25). Species-related differences in the circadian rhythm of melatonin formation will be mentioned on p. 255. Interspecies differences of other biochemical parameters have been summarized by WURTMAN et al. (1968a) and QUAY (1974a) (see also SELLEI and FRAUCHIGER, 1968). There is also great intraspecies variation of melatonin formation in both humans (ARENDT et al., 1977; GREINER and CHAN, 1978) and animal species (HEARSE et al., 1969). Sex differences have apparently not been studied in detail and no attempt has been made in the present review to scan the large amount

Table 25. Serum melatonin levels (pg/ml) as determined by radio-immunoassay

Species	Day	Night	Author(s)
Man	47–481		WETTERBERG and ARENDT (1975)
	< 20	80–100	MATTHEWS et al. (1977)
	50	100–130	ARENDT et al. (1977)
	30–50	179 ± 26	G. VAUGHAN et al. (1978 b)
	21.88		G. VAUGHAN et al. (1978 a)
	ca. 100–700[a]	ca. 100–720[a]	WEINBERG et al. (1979)
Rat	6.3 ± 3	75 ± 27	MATTHEWS et al. (1977)
		120	WILKINSON et al. (1977)
Camel	29	221	MATTHEWS et al. (1977)
Calf	19 ± 4	121 ± 24	HEDLUND et al. (1977)
Sheep	10–30	100–300	ROLLAG and NISWENDER (1976)
Ewe	140 ± 16.8	297 ± 46.5	ROLLAG et al. (1978 c)
Pig	22	76	MATTHEWS et al. (1977)
Skink	35 ± 5.9	240 ± 43	MATTHEWS et al. (1977)
Donkey	24.5	108	MATTHEWS et al. (1977)

[a] Although the day/night values appear to be very similar, the individuals studied always showed higher night and lower day values.

of literature with respect to possible sex differences in pineal function. Hence only some incidental findings are mentioned. According to ARENDT et al. (1977) the mean melatonin *serum* concentration is higher in human females than males, while cerebrospinal fluid concentrations are similar. In rabbits, pineal NAT activity shows no sex differences (HEARSE et al., 1969). HYYPPÄ et al. (1973) administered low doses of 6-hydroxydopamine to new-born rats and noted a 106% increase in HIOMT activity at 30 days of age in males, but not females. In cattle, the pineal contained an anti-oestrogenic substance in females, but not in males (JUSZKIEWICZ and RAKALSKA, 1963). For morphological differences of the pineal complex in *Peromyscus*, see QUAY (1956 a).

The figures given in parentheses denote the pages where relevant information may be found regarding melatonin formation during ontogenesis (257), puberty (260), oestrous cycle (296), pregnancy (301), old age (261) and different seasons (299). Distribution and fate of melatonin in the body will be mentioned on p. 262.

2. Formation of Indoleamines Other than Melatonin and Related Substances

Although a great deal of pineal research has concentrated on melatonin, there are other indoleamines or related substances present in the mammalian epiphysis cerebri which may be equally important. Whether or not serotonin is released from the pineal and exerts an effect on other organs has apparently not been studied. Recently 5-methoxytryptophol has come into the limelight (WURTMAN and OZAKI, 1978; HOOPER et al., 1979), details of its synthesis having been given by QUAY (1974 a). Other substances with biological activity are 5-

hydroxytryptophol and *N*-acetylserotonin (see REITER and VAUGHAN, 1977). Finally there are a number of β-carbolines (QUAY, 1974a), one of which was originally thought to influence adrenocortical function (p. 306).

3. Presence of Peptides, Polypeptides and Proteins with Possible Hormonal Properties

In addition to indoleamines and related substances the mammalian epiphysis cerebri appears to contain a great number of peptides, polypeptides and proteins which may have hormonal properties. Whether or not they are actually synthesized in the epiphysis cerebri is an open question.

a) Vasotocin

It took approximately a decade until the significance of vasotocin, extracted as early as 1963 (S. MILCU et al., 1963; NEACŞU, 1972a, b), was generally appreciated, although the active principles having pressor and oxytocic properties were soon identified as lysine vasotocin in the pig (PAVEL, 1965) and arginine vasotocin in cow (EBELS et al., 1965b; PAVEL and PETRESCU, 1966; PAVEL, 1969, 1971; CHEESMAN, 1970; CHEESMAN and FARISS, 1970; ROSENBLOOM and FISHER, 1975), man (PAVEL et al., 1973a, 1973/74; ROSENBLOOM and FISHER, 1975; LEGROS et al., 1976; REINHARZ and VALLOTTON, 1977), mouse (PAVEL et al., 1973b), rat (BOWIE and HERBERT, 1976) and sheep (ROSENBLOOM and FISHER, 1975). From a functional point of view it is interesting that these substances not only have pressor, antidiuretic and oxytocic activities but also exert antigonadotropic (PAVEL and PETRESCU, 1966; PAVEL, 1969; PAVEL et al., 1973b, 1973/74; M. VAUGHAN et al., 1974; REITER and VAUGHAN, 1977, lit.) and anticorticotropic effects (PAVEL et al., 1973c; PAVEL, 1975). Although the possibility has been discussed that arginine vasotocin influences melatonin synthesis via NAT (BINKLEY et al., 1973a, 1976) and that melatonin influences arginine vasotocin (p. 265), it is still not clear whether arginine vasotocin is indeed a pineal substance. BENSON et al. (1976a) have demonstrated that in bovines and rats vasotocic activity is not present in pineal parenchyma, but rather in tissues of what has been termed the stalk region. These authors suspect that the vasotocin detected resides in (a) ependymal cells attached to the pineal (p. 181), (b) the subcommissural organ, in which arginine vasotocin has been demonstrated by means of immunohistochemistry, or (c) cells of the subcommissural organ incorporated into the pineal organ proper (p. 66). One of the strongest proponents of the arginine vasotocin concept, PAVEL, also made quite clear that in the bovine pineal region arginine vasotocin is only contained in those parts which contain secretory ependymal cells (PAVEL, 1971). An absorption of arginine vasotocin from the CSF has been considered unlikely (PAVEL, 1971). Demonstration that cell cultures from pineal glands of human fetuses produce vasotocin-like peptides (PAVEL et al., 1973a) suggests a local synthesis. In rat fetuses, evidence for vasotocin biosynthesis by cultured pineal ependymal cells has been obtained by chromatography (PAVEL et al., 1977d). Whether or not the "ependymal" concept of vasotocin can be maintained is an important point.

An exclusive presence of vasotocin in cells of the ependyma and subcommissural organ seems unlikely because, in the rat pineal, elements dissimilar from those mentioned above contain immunoreactive arginine vasotocin (p. 180) (BOWIE and HERBERT, 1976) and because the posterior half of the rat pineal, which is devoid of ependymal cells, synthesizes vasotocin (PAVEL et al., 1978). PEVET et al. (1979) assume that a vasotocin-like peptide, rather than vasotocin itself, is present in the pineal. The release of arginine vasotocin into the CSF will be mentioned on p. 267. According to REINHARZ and VALLOTTON (1974) pineal glands contain proteolytic enzyme activity which is able to attack arginine vasotocin and arginine vasopressin.

Since in the hypothalamo-neurohypophyseal system the peptide hormones are closely associated with carrier proteins, the neurophysins, it was only logical to search for similar substances in the epiphysis cerebri. Biochemical studies in the cow have suggested that the pineal indeed contains neurophysin-like proteins (REINHARZ et al., 1974, 1975), but, as shown by the discussion following the paper of REINHARZ et al. (1975), the issue is far from clear. Neurophysins in the pineal have also been demonstrated by means of immunochemistry (LEGROS et al., 1975, 1976). The presence of two neurophysins in the adult human epiphysis cerebri has been described by REINHARZ and VALLOTTON (1977). Recently, cultured pineal ependymal cells have been reported to release neurophysin (REINHARZ et al., 1978).

b) Hypothalamic Releasing Hormones

The presence of hypothalamic releasing hormones in the pineal may be also of importance, but unfortunately little detailed knowledge is currently available. There are indications for the presence of LHRH, TRH and somatostatin. Indirect evidence points to the occurrence of PIF (p. 278). According to WHITE et al. (1974) gonadotrophin-releasing hormones are present in ovine, bovine and porcine pineals. They amount to 185 ng/gland in pig, 500 ng/gland in cow and 490 ng/gland in sheep, these values being 4–10 times higher than those reported in the hypothalami of the corresponding species. Immunohistochemical studies have shown that LHRH in the pineal is localized arround blood vessels, the reaction being variable from one animal to the other. Ultrastructurally a diffuse positive reaction was clearly demonstrable in as yet unidentified perivascular cells and, most interestingly, also over some nerve endings (PELLETIER, 1976). The fact that radioactively labelled LHRH is taken up by the pineal (DUPONT et al., 1974) has been interpreted as indicating that an exchange of the substances may take place between blood and pineal (PELLETIER, 1976). It should not be overlooked, though, that negative results have also been obtained. No evidence for the presence of LHRH was obtained in ovine pineals (CARSON et al., 1977). Gonadotrophin-releasing hormone activity was absent in the mouse pineal (GROSS, 1976) and WEINDL and SOFRONIEW (1978) failed to detect immunostaining for LHRH, oxytocin, arginine vasopressin and neurophysin.

TRH has been found in pineals of pig, cow and sheep, in the same amounts as in the hypothalamus (WHITE et al., 1974). In female rats TRH was not detected in samples from pineal gland and habenula; if present, being below 0.3 μg/mg protein (BROWNSTEIN et al., 1974). However, TRH has been detected in rat pineals by BRAMMER et al. (1979), who made the interesting observation that the TRH content was highest at 12.00 h and lowest at 24.00 h.

Somatostatin has been demonstrated in both rat (PELLETIER et al., 1975) and guinea-pig pineals (DUBÉ et al., 1975), where it had a similar localization to LHRH (see also PELLETIER, 1976). By contrast, WEINDL and SOFRONIEW (1978) were unable to detect this substance in a variety of species.

Hypothalamic releasing hormones may be also some of the more or less well defined substances mentioned in the following section.

c) Structurally Unidentified Presumptive Polypeptides and Proteins

Without going into great detail it suffices to mention that a large number of more or less well defined fractions of mostly low molecular weight have been extracted from the pineal, some of which have a strong influence on reproductive functions (for a review see REITER and VAUGHAN, 1977; BENSON, 1977; for further details see p. 284). For chemical details of the various fractions see JOUAN et al. (1965), THIÉBLOT and BLAISE (1966, 1967), THIÉBLOT et al. (1966a, b), MOSZKOWSKA (1967a, b), MOSZKOWSKA and EBELS (1971), MOSZKOWSKA et al. (1965, 1971, 1973, 1974, 1976), EBELS (1967, 1975, 1976), EBELS et al. (1965a, 1970, 1973, 1975, 1978a), DAMIAN and IANAS (1971), DAMIAN (1973), BENSON and MATTHEWS (1971), BENSON et al. (1971, 1972a, b, 1976b), BENSON and ORTS (1972), CITHAREL et al. (1972), ORTS and BENSON (1973), ORTS et al. (1974a, b) and ROSENBLUM et al. (1979).

The observation that a gonadotrophin-inhibiting substance other than melatonin (perhaps a polypeptide) disappeared from the urine in pinealectomized rats (OTA et al., 1971) strongly suggests that the peptides gain access to the systemic circulation and fulfil some of the criteria of true hormones.

4. Exogenous Factors Influencing Pineal Function

As pointed out above, the epiphysis cerebri is a neuro-endocrine transducer that is greatly influenced by exogenous factors, of which only a few have been thoroughly studied. Clearly established are the effects of light and darkness on melatonin formation. To what extent pineal substances other than melatonin are photically influenced has not been investigated in detail. Since the effects of light and darkness were first studied in long-term experiments it is appropriate to first deal with this facet.

a) Photic Influences

α) Pineal Morphology and Function Under Constant Illumination

Exposure of laboratory animals to continuous illumination has been clearly shown to depress pineal function. The first study to demonstrate this was probably that by MILINE (1957a, b), who showed that after several weeks' exposure the pinealocytes of rats and rabbits exhibited distinct signs of atrophy as indicated by a clear decrease in pineal weight (FISKE et al., 1960, 1962; WURTMAN et al., 1961, 1963a; QUAY, 1961a, b, 1962b; ZBORAY, 1965; SINGH and TURNER, 1969; ROWE et al., 1970; REITER, 1973e; MAS and ALONSO SOLIS, 1977; and others). At the light microscopic level pinealocytes of light-exposure rats are small in size, contain few nucleoli and the cytoplasm shows little basophilia (ROTH et al., 1962; ROTH, 1965). Quantitative studies by QUAY (1963c) show that after 11 weeks, in addition to a decrease in the total volume of the pineal, the largest nucleolar eosinophilic mass was significantly reduced, especially in the medulla of the organ. The depressive effect of light has been light microscopically confirmed in hamster (QUAY, 1961a, b) and mouse (KACHI et al., 1971a).

Histochemically, a decrease of the pineal lipid content is seen (QUAY, 1961a; ZBORAY, 1965; PROP and EBELS, 1968). Enzyme studies reveal a reduced activity of cytochrome oxidase, succinate dehydrogenase (see also QUAY, 1963b), NADH diaphorase, hydroxybutyrate dehydrogenase, α-glycerophosphate dehydrogenase, alkaline phosphatase and lipase, while acid phosphatase and non-specific esterase activities increase and glutamate dehydrogenase and MAO activities are unaffected (BOSTELMANN, 1968a). Glycogen decreases in rats (QUAY, 1963b) but increases in mice (KACHI et al., 1971a). Lactic acid increases (NIR et al., 1970b), which, taken together with the activity of some of the oxidative enzymes investigated, indicates suppression of oxidative cell processes (NIR et al., 1970b). Pineal RNA and protein synthesis decrease, but DNA levels are apparently not affected (NIR et al., 1969a). However, an *increased* RNA content has been found by NOVÁKOVÁ et al. (1972) (see p. 94).

Electron microscopic studies have yielded results which differ to a certain extent, but altogether they seem to confirm the concept of reduced pineal activity following exposure to continuous illumination. Quantitative studies carried out in mice (UPSON et al., 1976) have revealed that the concentration of pinealocyte nuclei increases 40% ($P<0.005$), indicating a decrease in pinealocyte volume. Similar observations have been made in rats, in which nuclear size remains unchanged (BOSTELMANN, 1968a). Altogether it would appear that a decrease in nuclear size is not the rule. In guinea-pigs even a small increase (15%) occurs (VOLLRATH, 1975). Nevertheless, the nuclei do appear to be affected. In the rat the pinealocyte nuclei show less folding (BOSTELMANN, 1968a), but in the mouse they appear more folded and more irregular than in the controls (UPSON et al., 1976). A decrease in both the number of nucleoli and the amount of cell body cytoplasm are typical of the rat (BOSTELMANN, 1968a). Mitochondria decrease in amount (BOSTELMANN, 1968a; LUES, 1971), swell and become larger, showing an almost complete loss of cristae together with a rarified matrix (HALARIS and MATUSSEK, 1969; HALARIS, 1969; LUES, 1971; UPSON et al., 1976). The Golgi complex shows a 35% decrease in the mean number of Golgi stacks per pinealocyte (UPSON et al., 1976). In guinea-pigs a roughly 20-fold increase in the amount of synaptic ribbons occurs accompanied by distinct morphological changes of the ribbons (LUES, 1971; VOLLRATH and HUSS, 1973; VOLLRATH, 1974); similar effects are present in *Eliomys quercinus* (ROUX et al., 1977).

Continuous illumination also influences the cytoplasmic processes of pinealocytes. In mice the space occupied by pinealocyte terminals decreases to half that of the control (UPSON et al., 1976). In guinea-pigs, the processes increase strikingly in number, but the total area occupied by these processes increases only slightly, resulting in a large number of relatively small processes. Since the number of processes radiating from the perikarya remain virtually unchanged, this may mean that the processes must have branched and/or become longer (VOLLRATH, 1975). The finding that in the mouse the number of dense-core vesicles counted per terminal decreases by 70% points to a decreased secretory activity of the pinealocyte (UPSON et al., 1976). As can be expected from the fact that the effect of continuous illumination is mediated by sympathetic nerves, these fibres are affected as well. KUROSUMI and KAWABATA (1966) noted that under these conditions most vesicles lack a dense core.

In addition to morphological parameters, a great number of biochemical and pharmacological parameters have been determined following exposure of animals to continuous illumination. That melatonin formation is depressed has been shown by direct melatonin assay (TOMATIS and ORIAS, 1967) as well as by measuring enzymes involved in melatonin production, namely HIOMT (WURTMAN et al., 1963a, 1964a; AXELROD et al., 1965a, 1966; QUAY, 1967a, 1968c; WARTMAN et al., 1969; ROWE et al., 1970; REITER and KLEIN, 1971; REITER, 1973e; YOCHIM and WALLEN, 1974a; and others) and NAT (REITER and KLEIN, 1971; KLEIN and WELLER, 1970c; DEGUCHI and AXELROD, 1972a), though REITER and KLEIN (1971) noted decreased NAT activities during night time only and elevated levels in the afternoon.

Since melatonin is formed from serotonin, it is not surprising that serotonin and serotonin-forming enzymes are also affected by continuous illumination. Serotonin concentrations show a decrease, whereas the activity of 5-hydroxytryptophan decarboxylase increase (QUAY and HALEVY, 1962; SNYDER and AXELROD, 1964; SNYDER et al., 1965a). Succinate dehydrogenase (QUAY, 1963b) and tyrosine hydroxylase activities (MCGEER and MCGEER, 1966) decrease.

Monoamine oxidase activity, P^{32}-phosphate uptake and ATP content (QUAY, 1963b) are unaffected, while pineal free amino acid levels decrease (NIR et al., 1973a), accompanied by changes of the water-soluble proteins (NIR et al., 1973b). Pineal antidiuretic activity decreases (KÖNIG and MEYER, 1971). The electrical activity of the pineal is apparently not altered by exposure of animals to continuous illumination (SALAS and SCHAPIRO, 1970).

Finally it should be mentioned that exposure to continuous illumination not only reduces the overall activity of the pineal gland, but that it also abolishes the *circadian rhythm* of melatonin formation (p. 251), as assessed by NAT measurements (KLEIN and WELLER, 1970c; REITER and KLEIN, 1971) and melatonin radio-immunoassay (ROLLAG and NISWENDER, 1976). In new-born rats the rhythm is not abolished by continuous light (DEGUCHI, 1975c; for further details see p. 259).

β) *Pineal Morphology and Function Under Constant Darkness*

The effects of light deprivation on the epiphysis cerebri have been studied by various means, e.g. exposure to continuous darkness, blinding or sympathectomy, the effects of the latter procedure having been dealt with on p. 204.

The results obtained after constant darkness and blinding are often less pronounced and less consistent than those after continuous illumination, one of the reasons being that under light deprivation pineal rhythms become free-running (see below).

Exposure to constant darkness may result in an increase in pineal weight (WURTMAN and AXELROD, 1964a; see, however, MAS and ALONSO SOLIS, 1977), but when rats are kept in constant darkness from birth onwards, a decrease occurs (RELKIN, 1967). Pinealocyte size, nucleoli and cytoplasmic basophilia increase (ROTH et al., 1962; MILINE, 1965). The nuclei, light microscopically either unaffected (ROTH et al., 1962) or smaller in size (MILINE, 1965), are slightly irregular and strongly indented, as confirmed ultrastructurally (BOSTELMANN, 1968a). Electron microscopy shows that the size of pinealocyte nuclei increases following light deprivation in rats (BOSTELMANN, 1968a) and is little affected in guinea-pigs (VOLLRATH, 1975).

Cytophotometric studies (NOVÁKOVÁ et al., 1971) reveal a decrease in RNA, but chemical determinations of RNA, DNA and protein content do not exhibit any differences in comparison to controls (NIR et al., 1969a). Histochemical studies reveal enhanced activities of cytochrome oxidase, succinate dehydrogenase, NADH diaphorase, hydroxybutyrate dehydrogenase, α-glycerophosphate dehydrogenase and lipase (BOSTELMANN, 1968a). The lipid content is apparently not affected when studied by light microscopy (ZBORAY, 1965).

Ultrastructurally, the cytoplasm is particularly rich in ribosomes and shows larger amounts of both mitochondria and lipid droplets (BOSTELMANN, 1968a). Golgi apparatus (LUES, 1971; BENSON and SATTERFIELD, 1975; ROUX et al., 1977; UPSON and BENSON, 1977), Golgi vesicles and saccules (CLABOUGH, 1971) as well as smooth endoplasmic reticulum (LIN et al., 1973, 1975) show signs of slight or pronounced hypertrophy. Terminals of pinealocyte processes are larger in size than in controls (SHERIDAN, 1975), and the dense-core vesicles they contain are more numerous in hamsters (CLABOUGH, 1969, 1971; SHERIDAN, 1975), but show a 55% decrease in number in blinded mice when compared with intact controls (UPSON and BENSON, 1977). Moreover, an increase in multivesicular bodies, membranous lamellar structures of probably mitochondrial origin – apparently more numerous in blinded males than females – and macrophages has been noted (hamster: CLABOUGH, 1971). Synaptic ribbons show a strong increase in some guinea-pigs, but not in others (VOLLRATH and HUSS, 1973).

Chemical investigations reveal that the lactic acid content of the pineal gland is unaffected (NIR et al., 1970b). The daily fluctuations of basic amino acids and taurosine are not abolished, but those of aspartic acid and methionine are (NIR et al., 1973a). Day–night differences in soluble pineal proteins disappear in immature but not in mature rats (NIR et al., 1973b). Pineal proteins decrease in amount (GARWEG et al., 1976).

Very variable results have been obtained regarding the formation of *melatonin* after light deprivation. This may be partly due to the fact that under this experimental condition enzyme studies apparently do not correctly reflect the amounts of melatonin present, as shown by the observation that in rats the melatonin rhythm persists while that of HIOMT is abolished (LYNCH and RALPH,

1970). Pineal melatonin concentrations have been found to be higher than in intact controls (TOMATIS and ORIAS, 1967). As the melatonin-forming rhythm becomes free-running (see below), it is understandable that HIOMT measurements have yielded variable results: No effect was noted by QUAY (1967a) and GARWEG et al. (1976) (see also FREHN et al., 1973); increases of HIOMT activity have been reported by WURTMAN and AXELROD (1964a), AXELROD et al. (1965a), ANTON-TAY and WURTMAN (1968), WARTMAN et al. (1969a), REITER and ELLISON (1970), REITER et al. (1971), NAGLE et al. (1973a) and FARISS and BOTZ (1975), while ROWE et al. (1970) described a decrease. In congenitally anophthalmic guinea-pigs, in contrast to surgically blinded guinea-pigs, HIOMT activity is lower than in intact controls (FARISS and BOTZ, 1975).

There is clear evidence that under constant darkness the rhythm of melatonin formation persists (YOCHIM and WALLEN, 1974a; FARISS and BOTZ, 1975; ROLLAG and NISWENDER, 1976) and becomes free-running, as shown by the fact that an inversion of the NAT rhythm occurs (KLEIN and WELLER, 1970c; KLEIN et al., 1971c). The circadian rhythm of NAT is regularly delayed by 1 or 2 h per week and is still present after 65 days in constant darkness (DEGUCHI, 1975a). According to YOCHIM and WALLEN's (1974a) HIOMT studies, exposure to constant darkness significantly depresses the amplitude of the daily fluctuations of melatonin formation. In congenitally anophthalmic guinea-pigs there are no statistically significant differences between day and night levels of HIOMT (FARISS and BOTZ, 1975). The electrical activity of the pineal is essentially unchanged under constant darkness (SALAS and SCHAPIRO, 1970).

γ) Effects of Short-term Photic Changes on Pineal Morphology and Function

In view of the fact that pineal function is strongly influenced by photic changes, it was only logical to examine whether the organ also responds to short-term photic changes and, if so, how long it would take for such a response to occur.

Morphological parameters have been little investigated with regard to this question. VOLLRATH and HOWE (1976), studying the amount of synaptic ribbons in the guinea-pig pineal, found that extension of the dark period by 7 h and extension of the light phase by 2–3 h completely prevented the changes of synaptic ribbons which normally occur. The concept that the pineal also responds to short-term photic changes has gained strong support from *biochemical* studies. Interestingly, the pineal response is faster to light than to darkness. In rats kept under diurnal lighting conditions, there is relatively little increase in NAT activity in the first hour after the onset of darkness, while after 4 h in darkness a 50-fold increase is seen (DEGUCHI and AXELROD, 1972a). When rats are brought out of darkness to light at 22.00 h, NAT activity disappears very rapidly: while in the first 5 min there is almost no change, in the second 5 min the enzyme activity decreases rapidly to less than 10% of the initial activity and remains at this level as long as the rats are kept in light. When rats are returned to darkness after 10 min in the light, NAT activity gradually returns to the initial level after 3 h (DEGUCHI and AXELROD, 1972a). Recently it has been shown (ILLNEROVÁ et al., 1979) that at 23.00 h 1 min exposure to light suffices to cause the rapid decrease of pineal NAT and melatonin.

According to KLEIN and WELLER (1972a) the halving time of pineal NAT after exposure to light is about 3 min. When lights are switched on during the night, pineal serotonin concentrations increase about twofold within 14 min (ILLNEROVÁ, 1971a; see also BROWNSTEIN et al., 1973a, b, c), while serum melatonin concentration drop precipitously (ROLLAG and NISWENDER, 1976). A parallel decline of melatonin in pineal and serum has been demonstrated by ILLNEROVÁ et al. (1978). The signal for "turnoff" of melatonin formation appears to be an abrupt decrease in adenosine 3′,5′-monophosphate (KLEIN et al., 1978). Short-term studies in ewes have shown that the nocturnal increase of serum melatonin concentrations can be delayed by continued illumination. After lights off there is a lag period of 5–10 min before the melatonin concentrations rise within 2–10 min. When the lights are turned on after 1 h, melatonin concentrations decline rapidly to baseline levels (ROLLAG et al., 1978c).

While extension of the daily light period for a few hours abolishes the changes of pineal metabolism normally occurring after the onset of darkness (QUAY, 1963c, 1966/67), exposure of animals to darkness leads to an increase of NAT activity only during the expected dark-time, but not during the normal light-time, which apparently represents a refractory period for any effects of darkness (BINKLEY et al., 1973b). According to RUDEEN and REITER (1977), a definite NAT rhythm persists when animals are subjected to as little as 15 min of light in a 24-h period and there is a refractory period of ca. 8 h before darkness is capable of inducing an increase in NAT activity.

Electrophysiological studies have demonstrated a fourfold electrical depression of the pineal in light when compared to darkness (NEWMAN TAYLOR and WILSON, 1969).

δ) *Light Spectra and Intensities Most Effective in Influencing Pineal Function*

Recent studies have clarified the light spectra and intensities that are most or least effective in influencing pineal function. Pineal activity as assessed by HIOMT measurements is most strongly suppressed by green light. Blue or yellow lights are less effective. Exposure to red light has virtually no effect, while UV light produces a transient stimulation. For experimental studies it is important that common fluorescent light (cool-white) is more effective than a light source that simulates natural sunlight (Vita Lite) in decreasing pineal HIOMT activity (CARDINALI et al., 1972a, c). As far as the light intensities are concerned, it has been found that rats exposed to as little as 0.5 $\mu W/cm^2$ exhibit a greater than 50% decrease in NAT activity, maximum inhibition being obtained with 15 $\mu W/cm^2$ (MINNEMAN et al., 1975).

b) Ambient Temperature

In addition to photic stimuli the epiphysis cerebri is dependent on the ambient temperature. Exposure to cold (3–10° C) for 1 day to several weeks leads to hypertrophy and hyperplasia of pinealocytes, accompanied by a decrease of sudanophilic substances, occurrence of pleomorphic and enlarged mitochondria, enlargement of Golgi apparatus, increase in number of ribosomes, degranulation of vesicles surrounding synaptic ribbons, an increase in electron-density in the basal part of the rod and prominent mebraneous whorls, especially in the hamster (MILINE, 1969; MILINE et al., 1970; BUCANA et al., 1973). Recent studies (MATSUSHIMA and MORISAWA, 1978) have shown that short-term exposure of mice to cold for 30–60 min results in a strong increase in the number of dense-core vesicles in pinealocytes, accompanied by an increase of lysosome-like bodies. Nerve terminals increase in size, resembling HERRING bodies (MILINE et al., 1970), and show a distinct increase in both the amount and size of zinc iodide-positive vesicles (KRSTIĆ, 1972a).

Histochemical studies reveal increases of lactate, glucose-6-phosphate and succinate dehydrogenase activities (DEVEČERSKI, 1972b), and biochemical evidence also exists for the increase of the latter enzyme (QUAY, 1958b). Melatonin formation increases in adult rats (URRY et al., 1976), but is depressed in suckling rats (ULRICH et al., 1973/74). Cold-induced testicular regression in hamsters is apparently not mediated by the pineal (LIU and FREHN, 1973). For the role of the pineal in the cold acclimatization of the lemming, see FOLK et al. (1977).

In contrast to cold exposure, the effects of exposure to heat on the pineal have been little investigated. The available evidence suggests that heat depresses pineal function, as indicated by decreases of pineal protein, RNA (NIR et al., 1972b), HIOMT (NIR et al., 1975) and NAT activities (NIR and HIRSCHMANN, 1978a). In suckling rats, measurements of HIOMT and NAT have yielded different results: while HIOMT is not temperature-sensitive, NAT activity decreases (ULRICH et al., 1973/74).

c) Noise and Vibration

Exposure of rabbits to noise and vibration leads to an atrophy of pinealocytes, followed by pinealocyte hyperplasia, especially in the marginal zone of the epiphysis cerebri (MILINE, 1957a). Auditory stimuli are followed by electrical short-latency responses of the pineal in rats (DAFNY, 1977), pinealocyte atrophy in hibernating bats (MILINE et al., 1969c) and pyknosis of pinealocyte nuclei in rats (FONTANA and ALVES DE AZEVEDO, 1978). Monoamine oxidase activity increases (J. MILIN, 1976).

d) Olfactory Stimuli

Long-term olfactory stimulation has been shown to result in pinealocyte atrophy (MILINE et al., 1963a). Additional evidence comes from studies showing that the effects of light-deprivation on pineal and reproductive organs are accentuated by removal of the olfactory bulb (REITER et al., 1969; REITER and ELLISON, 1970; REITER and SORRENTINO, 1971a). Anosmia alone is not sufficient to bring about statistically significant differences in enzyme activities (HIOMT: REITER and ELLISON, 1970, REITER et al., 1971; NAT: KLEIN et al., 1971c). The fact that electrical stimulation of the olfactory system results in short-latency responses of pineal elements indicates connections between the olfactory system and the epiphysis cerebri (DAFNY, 1977).

e) Nutritional Factors

Nutritional factors may also affect pineal function (REITER and SORRENTINO, 1971a). KŘEČEK and PALATÝ (1967) reported that in young rats pinealectomy prevents the decrease in weight and citric acid content of seminal vesicles caused by premature weaning. Underfeeding renders rats highly sensitive to the antigonadotropic action of the pineal (SORRENTINO et al., 1971a). Feed restriction to 53% (but not to 33%) for 21 days (but not for 28 days) depresses pineal weight; HIOMT and NAT activities increase when feed intake is restricted to 33% that of controls, but not to 53% (URRY et al., 1976). The effects of hypoglycaemia are discussed on p. 312. In pinealectomized rats a melatonin rhythm persists, though with reduced amplitude, the rhythm being absolished by fasting (LYNCH et al., 1975b).

In addition to the quantity, the quality of the feed also appears to play a role. In the vole, *Microtus montanus,* dietary supplement of fresh green plant food (green lettuce) decreases pineal weight and increases the size of the reproductive organs, suggesting that the pineal may be involved in an evaluation of the quality of the nutrition (NEGUS and BERGER, 1971). In young rats, a high tryptophan diet increases pineal serotonin levels by 250%, while a high phenylalanine diet leads to a reduction by 30% (QUAY, 1963e). Some of the nutritional effects on the pineal may be mediated by the sympathetic nervous system, as sucrose feeding has been shown to stimulate the sympathetic nervous system (YOUNG and LANDSBERG, 1977). Water deprivation for 23 h daily has apparently no effect on the daily NAT cycle in the pineal (MOORE and TRAYNOR, 1976).

f) Isolation, Crowding, Captivity, Immobilization and Other Stressful Stimuli

Juvenile rats, isolated in small cages for ca. 2 months starting at 3–4 weeks of age, have pineals which are 7%–10% heavier than in controls; while the ratio of pineal weight to brain weight is also greater, pineal mid-day serotonin concentrations and HIOMT activities are unaffected (QUAY et al., 1969a).

Crowding has been found to be without effect on pineal melatonin formation in the Uinta ground squirrel (FREHN et al., 1973). Captivity of hares for 3 days is followed by pineal oedema, especially in the centre, haemorrhages and atrophic pinealocytes, followed by a reversal to normal after 1 week and hyperplasia of pinealocytes after 2 weeks (MILINE, 1957a). In the 2nd week of captivity the pinealocytes show an increase in the activity of a number of histochemically demonstrable enzymes, and ultrastructurally virtually all cell organelles increase in size and number (MILINE et al., 1972).

Physical immobilization, which in rats is one of the most stressful stimuli, structurally results in regressive changes of the pineal followed by signs of enhanced activity (MILINE et al., 1966, 1968). After only 2 h of immobilization, pineal NAT activity shows a 100-fold increase and pineal melatonin content a 6-fold increase (LYNCH et al., 1973a; LYNCH and WURTMAN, 1973). This increase is apparently due to an action of the adrenal medulla, as shown by the fact that it can be prevented by bilateral adrenalectomy (LYNCH et al., 1977).

CCl_4 poisoning of mice is followed by a decrease of pineal volume from 4.65 to 3.88 mm^3 and a reduction of size of pinealocyte nuclei from 7.65 to 6.3 µm (LEGAIT et al., 1978). Ether stress for 20 min leads to a fourfold rise of pineal NAT activity at 7.00 h, but not at 1.00 and 19.00 h (G. VAUGHAN et al., 1978d). Urethane has no effect on pineal NAT activity (SHIVERS, 1978).

That stressful stimuli of apparently low magnitude can influence pineal function is illustrated by the observation that subcutaneous saline injections increase pineal melatonin levels by 30%–50% (LYNCH et al., 1973b).

Swimming stress increases pineal NAT activity (PARFITT and KLEIN, 1976).

In formalin-stressed rats, pinealectomy increases the mortality (HOLBAN et al., 1965). The effects of emotional stress on pineal morphology have been studied by R. MILIN (1979).

g) Irradiation

It appears to be mainly of academic interest that the epiphysis cerebri is influenced also by irradiation. In animal species whole-body irradiation (6–8 Gy) results in pinealocyte hypertrophy and hyperplasia after 4–9 days, followed by signs of pinealocyte involution after 8–10 weeks (MILINE et al., 1960, 1963b, 1969a). Irradiation (30 Gy) to the chest is not followed by histological changes in the pineal (RODIN and TURNER, 1965a). Pineal HIOMT activity is depressed during the first 24 h after irradiation, beginning as early as 1 h after irradiation (BARFUSS et al., 1969; ELLIS, 1970), increases from days 4 to 19 and returns to control levels by day 22 (ELLIS, 1970; ELLIS et al., 1970).

The effect of X-rays on the pineal is apparently not direct, as shown by the observation that (a) X-irradiation (4.5 Gy) to the head increases HIOMT activity while that to the body diminishes it and (b) in vitro extremely high doses (38.70 Gy) are required before an effect is noticeable (ELLIS et al., 1973).

A radioprotective role of melatonin has been suggested by MILINE (1960b) and others (see ELLIS et al., 1973).

h) Drug Action

The effects of drugs on pineal structure and function are mentioned where they appear to be most appropriate. Here a few results are reported which do not appear to fit into other sections.

In vitro, mescaline has been found to stimulate pineal serotonin synthesis (SHEIN et al., 1971). *p*-Chloroamphetamine given intraperitoneally depletes pineal serotonin levels (FULLER and PERRY, 1977a), as does *p*-chlorophenylalanine (RUBIO et al., 1977), the decrease in serotonin content being followed by an elevation of pineal noradrenaline levels for 24 h (RUBIO et al., 1977).

Pineal noradrenaline content increases after administration of desipramine

(FULLER and PERRY, 1977b). Electrophysiological studies have shown that morphine increases the amplitudes of the evoked potentials in the pineal body (DAFNY and BURKS, 1977). Administration of chlorpromazine enhances the accumulation of melatonin in various organs (WURTMAN and AXELROD, 1966b). Central depressants have been tested by ZATZ and BROWNSTEIN (1979a): nocturnal NAT activity in rat pineals is depressed by pentobarbital, thiopental, morphine sulphate, lidocaine and diazepam and is not affected by phenobarbital, chloral hydrate and ethyl alcohol. Carbachol given intraventricularly mimics the effects of light on the circadian rhythm of NAT in the rat pineal, perhaps by acting via the suprachiasmatic nucleus (ZATZ and BROWNSTEIN, 1979b). Low concentrations of lithium inhibit the synthesis of cAMP and cGMP in the rat pineal gland (ZATZ, 1979).

5. Rhythms of Pineal Function

There is clear evidence that pineal function undergoes typical rhythms, at least melatonin formation. To what extent substances not directly related to melatonin synthesis show similar rhythms largely remains to be elucidated. Daily rhythms of pineal function are very pronounced in some species and have been studied in great detail.

a) Daily Rhythms

Morphology and function of the mammalian pineal show day–night changes which appear to be due to circadian rhythms synchronized by changes in environmental lighting conditions.

The morphological and histochemical parameters showing circadian changes have been dealt with above. For ease of comparison with biochemical and other data, a brief summary will be given. During the day the following parameters show a maximum: pineal weight (p. 39), nuclear size (p. 73), nucleolar size (p. 73), mitotic activity (p. 75), intercellular canaliculi (p. 174) and dense-core vesicles (p. 165). During the night, maxima are found with respect to pinealocyte size (p. 75), succinate dehydrogenase activity (p. 102), glycogen (p. 78) and synaptic ribbons (p. 150).

It is now clearly established that in man and other mammalian species melatonin formation shows circadian changes with generally a low activity during the day and high activity during the night (Table 25). In man, melatonin measurements by means of bio-assay (PELHAM et al., 1973; G. VAUGHAN et al., 1976, 1978b) and radio-immunoassay (ARENDT et al., 1977; J.A. SMITH et al., 1977; KENNAWAY et al., 1977; MATTHEWS et al., 1977; FEVRE et al., 1978) in plasma and serum samples taken from individuals over single (Fig. 128) or four consecutive 24-h cycles (Fig. 129) clearly demonstrate the circadian rhythmicity, showing a mean peak at mid-dark phase. For future work it is interesting to note that in post-mortem pineals melatonin measurements by gas chromatography with electron capture detection (GREINER and CHAN, 1978) and measurements of NAT and HIOMT activities (J.A. SMITH et al., 1977) have shown

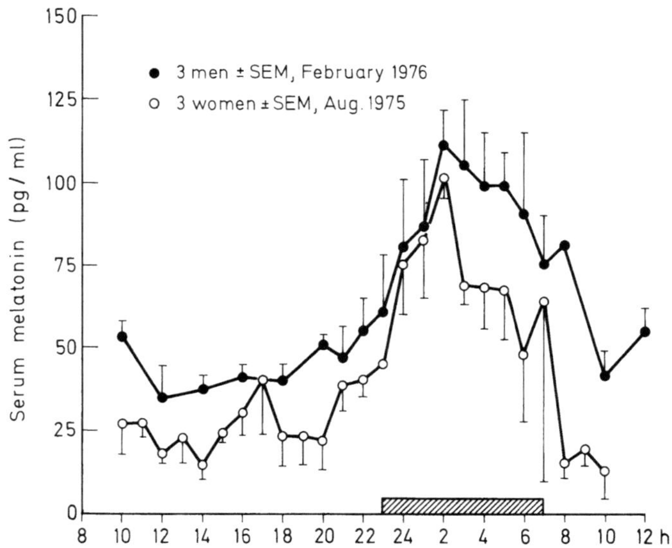

Fig. 128. Twenty-four hour variations in human serum melatonin as assessed by radioimmunoassay; after ARENDT et al. (1977). This diagram differs from the one originally published in that the blank values have been subtracted. By courtesy of Dr. J. ARENDT, Department of Biochemistry, University of Surrey, Guildford, United Kingdom

that melatonin levels and enzyme activities are apparently directly related to the time of death of the subject. Daily rhythms in human urinary melatonin (LYNCH et al., 1975a, b; JIMERSON et al., 1977) and 5-hydroxyindoleacetic acid (5-HIAA) excretion (G. VAUGHAN et al., 1976) conform to those present in blood.

That neither changes in environmental lighting conditions nor the sleep–wake cycles of individuals are solely responsible for driving the rhythm of melatonin formation is shown by the fact that in man, unlike in rats, exposure to continuous illumination and sleep deprivation or changes in the sleep–wake cycle under a diurnal lighting schedule do not markedly affect melatonin rhythms and concentrations (G. VAUGHAN et al., 1976; JIMERSON et al., 1977), in contrast to 5-HIAA (G. VAUGHAN et al., 1976). As regards the effects of autonomic drugs on melatonin and 5-HIAA, see G. VAUGHAN et al. (1976).

The various metabolic processes accompanying and driving the circadian rhythm of melatonin formation in the pineal have been studied in great detail in laboratory animals, especially rats.

As mentioned above (p. 9), an increase in pineal melatonin formation is triggered by an increased release of noradrenaline from intrapineal sympathetic nerve fibres, which in rats occurs immediately after the onset of darkness. Pineal noradrenaline levels in rats, apparently in contrast to hamsters (MORGAN et al., 1976), and gerbils (MORGAN and REITER, 1977), show marked diurnal variations, being lowest at the end of the daylight period and highest 12 h later, the nocturnal rise being in phase with the rise in melatonin (WURTMAN and AXELROD, 1966a; WURTMAN et al., 1967a). While ^{3}H-noradrenaline *uptake*

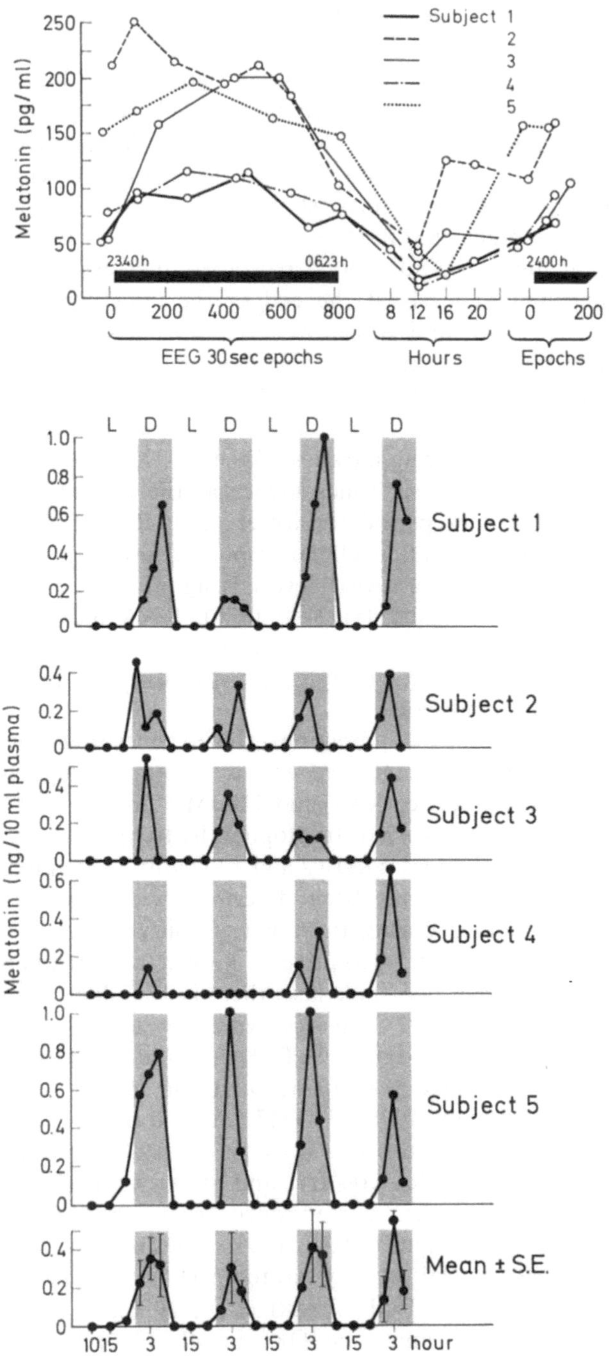

Fig. 129. Interindividual variation of plasma melatonin levels in humans. *Top*, note the relatively low melatonin levels during the day and the increase late in the afternoon in all the individuals tested; although the daily secretion patterns are similar, the concentration of the hormone in the blood differs strikingly between individuals. G. VAUGHAN et al., 1978b *Bottom*, melatonin levels in five individuals (not the same as those in the *top* figure) measured on four consecutive days (bio-assay). Note that the melatonin levels may differ strikingly between days. G. VAUGHAN et al., 1976

by pineal nerve endings during the day is not significantly different from that at night, the ^{3}H-noradrenaline *turnover* shows a characteristic 24-h rhythm, the half-life being more than twice as long during the day as at night (BROWNSTEIN and AXELROD, 1974). Catechol-*O*-methyltransferase, an enzyme involved in the metabolism of catecholamines, shows high levels during the day and low ones during the night (BÄCKSTRÖM and WETTERBERG, 1972). The binding sites of noradrenaline, the β-adrenergic receptors, increase in number during the day, reaching highest levels at the end of the light period, and decrease during the night (ROMERO and AXELROD, 1974, 1975; ROMERO et al., 1975a). Adenyl cyclase, somehow coupled to the β-adrenergic binding sites (WEISS, 1968a, 1969, 1970, 1971; WEISS and COSTA, 1967a, b, 1968a, b; WEISS and CRAYTON, 1970c; WEISS and STRADA, 1972), increases in activity during the light period and decreases during darkness (WEISS, 1970), the product of adenyl cyclase, cAMP, being about six times higher at the end of the light period than at the end of the dark period (EBADI et al., 1970b, c). Cyclic nucleotide phosphodiesterase, which modifies cellular responsiveness to β-adrenergic stimulation by modifying intracellular cAMP levels, is significantly more active during night than day (MINNEMAN and IVERSEN, 1976). RNA induced by cAMP is synthesized during darkness and declines during light (ZATZ et al., 1976a). For further studies on pineal RNA synthesis, see MORISSEY and LOVENBERG (1977, 1978a, b) and RATLIFF et al. (1979).

As a result of these complex interactions, the enzymes involved in melatonin synthesis are stimulated. As mentioned above, tryptophan hydroxylase, 5-hydroxytryptophan decarboxylase, NAT and HIOMT are engaged in melatonin synthesis. The first enzyme involved, tryptophan hydroxylase, was long thought not to undergo circadian changes (SNYDER and AXELROD, 1965; DEGUCHI, 1977a), but recently a clear rhythm has been described with low activities during the day and high activities during the night (SHIBUYA et al., 1978; SITARAM and LEES, 1978). While nothing definitive is known on circadian rhythms of pineal 5-hydroxytryptophan decarboxylase, there is abundant evidence for such a rhythm in NAT and HIOMT activities, with low activities during the day and high activities during the night (NAT: KLEIN and WELLER, 1970c; KLEIN et al., 1971a; MOORE and TRAYNOR, 1976; DEGUCHI, 1977b; WILKINSON et al., 1977; STEPHENS and BINKLEY, 1978; HIOMT: AXELROD et al., 1965a; WALLEN and YOCHIM, 1974a; DE GALLARDO and TRAMEZZANI, 1975; for differing HIOMT results, see QUAY, 1967a, 1968c; MERRITT and SULKOWSKI, 1969).

As a result of the enzyme actions mentioned, serotonin levels are high during the day and low during the night (QUAY, 1963c, 1964b; FISKE, 1964; MERRITT and SULKOWSKI, 1969). Pineal *N*-acetylserotonin (KLEIN and WELLER, 1972a), melatonin in pineal (LYNCH, 1971; OZAKI et al., 1976), blood (OZAKI et al., 1976; WILKINSON et al., 1977) and urine (OZAKI et al., 1976), and 5-methoxytryptophol (WILSON et al., 1978) show inverse rhythms. Plasma tryptophan, in contrast to plasma melatonin, does not exhibit the typical day–night differences (sheep: KENNAWAY et al., 1978).

Considerable effort has been devoted to clarifying whether the daily rhythms are exogenous or endogenous. While it was previously held that some parameters follow an endogenous and others an exogenous rhythm, it is now becoming apparent that the rhythms are endogenous in nature. The issue has been greatly clarified by studying the pineal rhythms of serotonin, NAT and noradrenaline. That the daily rhythm of serotonin (QUAY, 1963c, 1964b; HEDLUND and RALPH, 1967) is endogenous and persists in the absence of diurnal lighting cues (SNYDER et al., 1964a, 1965b) but depends on an intact sympathetic innervation (FISKE, 1964; SNYDER et al., 1965b) has been known for a long time. This rhythm depends on NAT, the rhythm of which persists in complete darkness and in blinded animals (KLEIN and WELLER, 1970c). In agreement with the concept of an endogenous nature of pineal rhythms is the finding that in rats exposed to constant darkness the NAT rhythm is regularly delayed by 1 or 2 h every week (DEGUCHI, 1975a). Further support comes from studies showing that the circadian NAT rhythm of rats kept under an ultradian lighting schedule of LD 6:6, persists, increasing once a day during the dark period closest to the previous time of darkness (DEGUCHI, 1977b). Also of interest is the finding that a circadian rhythm of NAT *inducibility* exists and that this rhythm persists under continuous illumination for at least 36 h (ALPHS and HELLER, 1978).

Incompatible with the concept of an endogenous nature of the pineal rhythm was the finding that the pineal noradrenaline rhythm was abolished by light deprivation or continuous illumination for as little as 1 day (WURTMAN et al., 1967a). This apparent discrepancy has recently been clarified by the demonstration that although the pineal noradrenaline levels are exogenously triggered there is an apparently endogenous 24-h rhythm in noradrenaline *turnover*, as indicated by the fact that during the day the disappearance of noradrenaline is slow in both intact and blinded rats and that during the night it is fast in intact animals as well as blinded animals exposed to light (BROWNSTEIN and AXELROD, 1974).

As mentioned above (p. 207) pineal circadian rhythms appear to be controlled by the suprachiasmatic nucleus (MOORE and EICHLER, 1976). According to RUDEEN and REITER (1977) 15 min of light per 24 h suffice for the maintainance of a distinct circadian rhythm of NAT. It is also interesting that if intact and blinded animals are kept in the same cage under diurnal lighting conditions, the NAT rhythm persists in a normal manner in the intact animals and is shifted in the blinded rats (DEGUCHI, 1975a). The circadian rhythm of NAT is apparently not influenced by the feeding schedule (DEGUCHI, 1977b).

Complete inversion of pineal rhythms following reversal of the lighting schedule has been observed after 6 days (serotonin rhythm, SNYDER et al., 1967a) and 3 days (NAT, NIR et al., 1974b). For an inversion of the number of pineal synaptic ribbons, see VOLLRATH and HOWE (1976). In humans it took 5–7 days before 50% or more of the total daily melatonin excretion occurred during the new 8-h period of darkness and sleep (LYNCH et al., 1978).

An important point is that the rhythm of melatonin formation shows distinct differences between species and apparently also in different regions of the pineal complex. Interspecies differences apply to both day–night ratios and the onset of an enhanced melatonin formation. According to RUDEEN et al. (1975) the day–night ratios of melatonin formation are 1:60 in rat, 1:3 in hamster and Mongolian gerbil and 1:1.5 in guinea-pig. A strongly enhanced melatonin synthesis begins to occur with the onset of darkness in rat (AXELROD et al., 1965a) and sheep (KENNAWAY et al., 1977) and between 24.00 h and 4.00 h in the hamster (RUDEEN et al., 1975; PANKE et al., 1979), while in *Macaca mulatta* (QUAY, 1966a) and guinea-pigs it already increases late in the afternoon (RUDEEN et al., 1975). In view of the concept that in rodents the pineal is a complex organ, sometimes consisting of completely separated superficial and deep pineals (p. 28), it is interesting to note that in rats the region of the deep pineal does not show rhythmic changes of NAT activities (MOORE, 1975b).

Other pineal parameters showing rhythmic changes that may be of a circadian nature are 5-hydroxyindole-3-acetic acid (QUAY, 1964a), antidiuretic activity being highest 3 h after the onset of darkness and lowest in the middle of the light period (KÖNIG and MEYER, 1967, 1971), testosterone uptake (NAGLE et al., 1974b), 8 S receptors for binding sex steroids (CARDINALI et al., 1975a) and ^{32}P uptake (KARASEK et al., 1972). No diurnal changes have been noted for MAO activity (SNYDER and AXELROD, 1965) and 5-methoxyindole acetic acid (QUAY, 1969a).

b) Weekly Rhythm

Recently it has been suggested that in the melatonin formation, as assessed by HIOMT measurements, undergoes a weekly rhythm, in the pineal rat with highest levels on Saturdays and lowest levels on Thursdays (VOLLRATH et al., 1975). Further studies have to be carried out to determine whether or not the concept of a weekly rhythm can be maintained (see REITER, 1976a).

c) Annual Rhythms

The presence of annual pineal rhythms has been reviewed by REITER (1975a, 1976a). Since most of the relevant studies are related to gonadal function, this point will be dealt with below (p. 299). Here it suffices to give a few additional data. Morphological studies have shown that in *Glis glis* and *Eliomys quercinus* both pineal volume and nuclear size of pinealocytes show a distinct decrease in July and August (LEGAIT et al., 1975b). In hibernating red-cheeked sousliks (*Citellus erythrogenys major* Brandt) the boundary between the peripheral and central parts of the pineal gland disappears. The pinealocyte nuclei are pyknotic and homogeneously stained, and condensed and degenerating forms are often seen. The blood supply is reduced and the serotonin level is halved (6.7 ± 0.6 µg) (POPOVA et al., 1975). Seasonal ultrastructural variations in the pinealocytes of the woodchuck, *Marmota monax*, a rodent hibernator, point to an increased synthetic and secretory activities in autumn (FRINK et al., 1978). Seasonal fluctuations of hydroxyindole levels have been mentioned in the Northern fur seal (KEYES et al., 1971). Moreover, it has been shown that after *l*-noradrenaline or K^+ stimulation in rats, the pineal cGMP efflux, but not that of cAMP, exhibits a minimum in January and February (O'DEA et al., 1978). In *humans*, serum melatonin levels show two peaks, one in January and the other in July (ARENDT et al., 1978), but in ewes no statistically significant differences in plasma melatonin levels have been detected between February, November, June and August (ROLLAG et al., 1978c). Pinealectomy in white-tailed deer did not alter the annual *pattern* of serum androgen levels, though the mean levels were depressed (R. BROWN et al., 1978). In hibernating *Eliomys quercinus* kept under constant darkness, elevation of the ambient temperature from 6° to 22° C leads to a strong increase of the serotonin fluorescence of the pineal (CORDONNIER and ROUX, 1978).

d) Other Rhythms

Rhythmic changes associated with the oestrous cycle will be dealt with on p. 296.

6. Onset of Pineal Function During Ontogenesis

a) Rat

Onset of pineal function during ontogenesis has been most thoroughly studied in the rat. In fetuses relatively little is known about pineal function. Whereas melatonin formation starts only after birth, there are indications that the fetal pineal is involved in the secretion of peptidergic substances. Between 16 and 20 days of intra-uterine life pineal cells contain material which stains brilliantly with paraldehyde fuchsin, especially in the cell processes (OWMAN, 1960/61, 1964a, c). In view of the close spatial relationships between the fetal pineal and the subcommissural organ (SCO) and the observation that cells from the SCO may be incorporated into the pineal organ (OLSSON, 1958) the secretory material may be identical with that of the SCO. Whether or not the stainable material is identical with vasotocin, which has also been identified in human fetal pineals (PAVEL et al., 1973a, 1973/74), remains to be established. The occurrence of marked histological changes in the lower ileum after pinealectomy in utero may also point to a secretory activity of the fetal epiphysis cerebri (OWMAN, 1962, 1963a, b, 1964a).

Melatonin formation appears to start only after birth, though a prenatal onset cannot be excluded (p. 258). Two aspects will be considered: (a) the time of onset of melatonin formation and (b) the time of the occurrence of a circadian rhythm.

The enzymes involved in the formation of serotonin and melatonin appear to occur, or are first demonstrable, at different stages of development. The first enzyme involved, tryptophan hydroxylase, already shows a high activity at birth (HÅKANSON et al., 1967). 5-Hydroxytryptophan decarboxylase activity is not detectable at birth (ILLNEROVÁ, 1969), but is clearly demonstrable on the 3rd postnatal day, exhibiting $^1/_{12}$ that of the adult levels (ILLNEROVÁ and TOLAR, 1973). Over the next 2 weeks there is a sharp rise (ZWEIG and SNYDER, 1968; HERNANDEZ and ILLNEROVÁ, 1970). Adult enzyme levels are reached on the 18th day (ZWEIG and SNYDER, 1968) or 20th day if values are expressed per mg pineal and by about 40 days if values are expressed per pineal (ILLNEROVÁ, 1969).

NAT activity, showing identical developmental patterns in males and females, is already detectable 4 days prior to birth. One day after birth, daytime and nocturnal values are about equal. A statistically significant night–day difference is first noted at 4 days. At 7 days, the daytime values decrease sharply by one-half to one-third and in the 3rd postnatal week nocturnal values reach adult levels (KLEIN et al., 1971b; ELLISON et al., 1972). Most recently VANEČEK and ILLNEROVÁ (1979) have shown that in new-born rats NAT activity begins to increase within 1 h after birth, followed by a decrease during the first night after birth. However, in contrast to the results reported above, nocturnal values were already higher than diurnal values on the 2nd day of life.

Very variable results have been obtained for the time of occurrence of HIOMT activity. According to TAIT et al. (1969), high activity is already present 1 day after birth, which was thought to be due to the lack of photic stimulation in utero. On day 16 a marked decrease, as related to body weight, occurs, coinciding with the opening of the eyes. ILLNEROVÁ (1972) and ILLNEROVÁ and TOLAR (1973) first detected HIOMT activity on the 3rd postnatal day. In other studies it was rarely measurable at 6 days, exhibiting 10% of the adult levels at 12 days (ZWEIG and SNYDER, 1968). According to KLEIN and LINES (1969) enzyme activity occurs only on day 10, then rising fairly sharply, attaining 40% of the 8- to 10-week values on the 14th day. Full activity is reached at 34 days (ZWEIG and SNYDER, 1968), but according to KLEIN and LINES (1969) the enzyme activity increases up to 6–7 weeks of age. However, if HIOMT activity is calculated in relation to body weight, a peak is present at about 3 weeks of age, a decrease occurring thereafter (KLEIN and LINES, 1969). With this method of calculating HIOMT activity, the enzyme activity decreases to one-third of the 3-week values at 8–10 weeks of age, and it has been suggested that the relative decrease in melatonin production after 21 days might permit previously inhibited sexual maturation to develop. An elevation of HIOMT levels during darkness was first noted in 39-day-old rats (KLEIN and LINES, 1969). However, BALEMANS et al. (1978a), studying the methylation of a number of pineal 5-hydroxyindoles, found clear day–night differences in 10-day-old Wistar rats. Pineal acetyl-CoA hydrolase is measurable 4 days before birth. It increases rapidly during the 1st week after birth, thereafter gradually decreasing and stabilizing at adult levels 3–4 weeks after birth (NAMBOODIRI et al., 1979).

Serotonin concentrations are 10% of the adult levels on day 1 and 50% at day 6, exhibiting adult levels at 12 days (QUAY and HALEVY, 1962; ZWEIG and SNYDER, 1968; ZWEIG et al., 1966). In accordance with these findings ILLNEROVÁ (1971b) found that serotonin was barely detectable in new-born rats and was clearly present at 3 days postnatally. Histochemically serotonin is demonstrable in some pineal cells around day 3 (C. MACHADO, 1967; C. MACHADO et al., 1968). Strong fluorescence is present at 6–7 days and the adult state is reached at 10–11 days. HÅKANSON et al. (1967) found serotonin at 1 week after birth and adult levels at 3 weeks of age.

The development of *melatonin* concentrations has apparently not been studied, but it is clear from the foregoing sections that melatonin formation appears to increase slowly during the 1st postnatal week, reaching adult levels at about the end of the 3rd week of life. The importance of the 3rd week for pineal development is also shown electrophysiologically: the pineal begins to respond to light at 18 days, the adult pattern being reached at 20 days (SCHAPIRO and SALAS, 1971). According to RELKIN (1968) days 19 and 20 after birth represent a critical time for the development of pineal function, as rats exposed to 20 and 21 days of diurnal lighting before being placed in total darkness mature later and have greater pineal weights and lower ovarian weights than rats kept for 18 and 19 days in diurnal lighting before being transferred into darkness.

Considerable effort has been devoted to elucidating the age at which serotonin and melatonin formation begin to show their circadian rhythms and the factors responsible for setting the biological clock. Studies dealing with the metabolic mechanisms underlying pineal circadian rhythmicity (p. 254) have shown that although at birth a completely developed postsynaptic adrenergic control system for pineal NAT is present (YUWILER et al., 1977; VANEČEK and ILLNEROVÁ, 1979), a circadian rhythm of NAT is not established before the end of the 1st week of life (ELLISON et al., 1972). This is in agreement with studies showing that a circadian rhythm of serotonin levels is present at 6 days after birth (ZWEIG et al., 1966). Adenyl cyclase activity does not change significantly during the first 30 days after birth (WEISS and CRAYTON, 1970c), but phosphodiesterase converting cyclic 3′,5′-AMP to 5′AMP increases abruptly between 8 and 16 days of age (WEISS and STRADA, 1972). Between 4 and 17 days of age the sensitivity to noradrenaline increases, as shown by the fact that in vitro NAT stimulation increases from between 2- and 3-fold to 15-fold (ELLISON et al., 1972).

A functioning adrenergic innervation coupled to the retina appears to be important for the development of the circadian rhythm of pineal melatonin formation. Pineal circadian rhythmicity does not develop in immunosympathectomized rats (ILLNEROVÁ, 1973a). Conflicting results reported by WRAGG et al. (1968) and C. MACHADO et al. (1969b) have been thought to be due to a decrease of serotonin levels during storage of pineal tissue (ILLNEROVÁ and ŠKOPKOVÁ, 1976). In agreement with the concept of a neuronal control of pineal circadian rhythmicity is the fact that rhythmicity occurs when the sympathetic nerve fibres have reached the pineal, which is accompanied by a striking increase in pineal noradrenaline content (MOORE and SMITH, 1971).

Other parameters of the catecholamine metabolism investigated show that the pineal dopamine content is not detectable at 1 and 4 days, amounting to 4 ng at 10 days, 10 ng at 30 days and 12 ng at 60 days (HYYPPÄ, 1971). Dopa decarboxylase has been found to occur on day 3 (ILLNEROVÁ and TOLAR, 1973) and at 1 week after birth (HÅKANSON et al., 1967). Monoamine oxidase activity was found as early as 1 day after birth (ZWEIG and SNYDER, 1968), but HÅKANSON et al. (1967) could not detect it until the end of the 1st week of postnatal life. According to ZWEIG and SNYDER (1968) the activity on day 1 is 50% of the adult level; between 6 and 12 days a rapid increase occurs, adult levels being reached by 12 days.

Most interesting studies have been carried out to determine how the lighting pattern influences the development of circadian pineal NAT rhythms. While it was first thought that a rhythm does not develop in animals born and raised under continuous illumination (ILLNEROVÁ, 1973a), it was later demonstrated that it does develop, the rhythm being delayed by 3 h every week (DEGUCHI,

1975c). The rhythm also develops under constant darkness (DEGUCHI, 1975c). The *serotonin* rhythm does not develop in constant light, but appears when the animals are subsequently reared in constant darkness (ILLNEROVÁ, 1975a). In a first such study (ILLNEROVÁ, 1972), the rhythm did develop under constant light.

More detailed studies were undertaken to investigate whether the biological clock that controls the NAT rhythm is set inborn, independently of environmental lighting, or is set after birth (DEGUCHI, 1975c). When female rats were mated in darkness and maintained in darkness throughout pregnancy and after delivery, the pups developed a typical circadian rhythm. When females were mated under reversed lighting conditions and transferred into darkness or light, the rhythmic patterns in their pups was inverted 180° from that of pups born of mothers kept under normal lighting conditions. The fact that the NAT rhythms of pups raised in darkness were similar to those of their mothers was taken as evidence that in the absence of light–dark cycles the mother rat sets the rhythm of the pups in such a way that it is synchronous with her own rhythm.

Rearing pups in the presence of a foster mother exposed to a lighting regimen different from the original mother revealed that the rhythm in the pups was closer to the original mother than to the foster mother, suggesting that the original mother plays the predominant role (DEGUCHI, 1975c). Further studies using original and foster mothers under reversed ultradian lighting schedules of LD 6:6 showed that the pups synchronized with the NAT rhythm of the foster mother (DEGUCHI, 1977b; see also DEGUCHI, 1978).

Finally it should be noted that in the developing rat pineal not all parameters show an increase. Arginine vasotocin decreases from fetal to adult stages (PAVEL et al., 1975b).

b) Sheep

A few data are available for pineal functional development in sheep. KENNAWAY and SEAMARK (1975), studying the endocrine changes preceding parturition in this species, noted that HIOMT activity is present as early as 130 days of gestation (duration of gestation, 147–150 days). A sudden rise in HIOMT activity occurs in the last 4–5 days of pregnancy, together with changes in plasma cortisol, and it has been suggested that changes in fetal pineal–hypothalamic–pituitary–adrenal interactions may induce parturition in sheep. From these results it is clear that prenatal pineal function may differ according to species investigated. To extrapolate from the findings obtained in rats, in which the pineal matures relatively late, may lead to erroneous conclusions.

7. Pineal Morphology and Function During Puberty and in Old Age

a) Puberty

In view of the concept that the pineal may be involved in endocrine changes occurring during puberty and that it may help to prevent precocious sexual maturity, studies have been carried out to examine pineal morphology and function during this phase of development. While experimental studies involving pinealectomy or injection of pineal substances resulting in slight acceleration or delay of sexual maturity (p. 288) seem to suggest enhanced pineal activity before puberty and diminished activity after puberty, morphological studies have not provided unequivocal data.

CLEMENTI et al. (1965a) described a greater abundance of mitochondria before puberty than in adult animals, accompanied by an abundance of synaptic ribbons and a well developed Golgi apparatus actively engaged in producing electron-lucent vesicles and dense-core vesicles, and concluded that pinealocytes of prepuberal rats are more active than those of adult animals. Completely opposing results have been obtained by other authors. CASSANO et al. (1961) and DE MARTINO et al. (1962) noted a linear increase in pinealocyte nuclear volume from prepuberal via adult to aged rats, pineal ^{32}P uptake showing a similar tendency. Osmiophilic granules which may contain melatonin and serotonin are rare in impuberal rats, but numerous in adult rats (DE MARTINO et al., 1963). Prepuberally pinealocytes are small in size, possess few processes and contain little plurivesicular material (DE MARTINO et al., 1964; BOSTELMANN, 1969a).

After puberty, rat pinealocytes enlarge, showing an increase in mitochondria, smooth endoplasmic reticulum, ribosomes, Golgi apparatus, lipid droplets and a number of histochemically demonstrable enzymes; this led BOSTELMANN (1969a, b) to suggest that the rat pineal gland acquires its full functional capacity only after puberty.

b) Old Age

Considerable effort has been devoted to assessing human pineal morphology and function in old age. Biochemical studies of HIOMT, MAO, histamine-*N*-methyltransferase (WURTMAN et al., 1964c) and hydroxytryptophan decarboxylase activities (VON STUDNITZ, 1967; OTANI et al., 1968a) show no decrease with age, suggesting that the pineal may be active throughout the entire life-span.

Gravimetric studies reveal no clear signs of a decrease of pineal weight beyond 50 years of age (RODIN and OVERALL, 1967; TAPP and HUXLEY, 1971; VON EYL and GUSEK, 1975; LEGAIT and LEGAIT, 1977) (Table 5). On the contrary, there is sometimes a tendency for an increase, even after decalcification of the pineal (TAPP and HUXLEY, 1971). It is interesting that up to 60 years of age pineal weight is increased in females compared with males, the glands of females reaching a maximum in the 49–59 year age group (TAPP and HUXLEY, 1971).

Histological studies (ARIETI, 1954; S. MILCU and VREJOIU, 1959) do not provide evidence for functional inactivity of the human pineal in old age. Histological changes from 0–90 years of age are minimal (WILDI and FRAUCHIGER, 1965). Lobulation of pineal tissue appears to slightly increase with age (Table 11) (see also TAPP and HUXLEY, 1972), as does fibrosis (TAPP and HUXLEY, 1972). Cyst formation does not appear to be age dependent and large amounts of calcium present in the glands of young adults (TAPP and HUXLEY, 1972) challenge an assumption that calcification is merely an age-dependent process. Mean percentage cellularity in females is constant from 5–90 years of age, while there is a slight decrease of this parameter in males after 44 years of age. The amount of intracellular pigment in females is highest in the 30–44 year age group, falling slightly in older women. In males, maximum pigment is found in the 60–74 year age group. However, the nuclei of pinealocytes have mean diameters of 6 μm in all age groups in both sexes (TAPP and HUXLEY, 1972). There is an increase in pineal lipid with age (QUAY, 1957a, b), but this change does not lend itself to a meaningful functional interpretation (QUAY, 1974a).

In animal species the problem of pineal ageing has also been studied. In 1,000-day-old rats, the pineal does not show any signs of degeneration (DE MARTINO et al., 1961), the nuclei of pinealocytes being distinctly larger than in 190-day-old animals (CASSANO et al., 1961). When pineals of 25-month-old Sprague-Dawley rats are compared under the electron microscope with those of young adult specimens they show a slight increment of connective tissue and lysosomes, but neither signs for an age-associated change in biogenic amine content nor a loss of amine synthesizing or storage capacity (BONDAREFF, 1965b). Pineal glycogen content decreases with age (KACHI et al., 1975, mice). Succinate dehydrogenase activity decreases in pineals of 1-year-old rats (QUAY, 1958b, 1959a), but ^{32}P incorporation increases up to an age of 1,000 days (CASSANO et al., 1961; DE MARTINO et al., 1962). Oxygen consumption and conversion of glucose into CO_2 in goat pineals decreases with age (HELLMAN and LARSSON, 1961). Pineal HIOMT activity in 3-year-old Uinta ground squirrels is higher than in juvenile, 1- and 2-year-old specimens (ELLIS and BALPH, 1976). Recently it has been demonstrated that 24-month-old rats have fewer β-adrenergic receptors in the pineal, than 3-month-old rats and hence have a reduced capacity to develop supersensitivity (GREENBERG and WEISS, 1978).

All these findings taken together do not indicate that in old age there are any drastic changes in pineal morphology and function; there are no consistent signs for a significantly depressed pineal function; even an enhanced pineal function cannot be ruled out.

8. Target Organs of Pineal Substances: General Aspects

Before considering the target organs of epiphyseal substances in detail, a few general points of fundamental importance have to be discussed. The first is the mode of release, distribution and fate of melatonin within the body. The second point deals with the effects of pineal substances on pineal morphology and function. In this section important data will be presented as to whether melatonin is indeed a pineal *hormone* whose primary site of action lies outside the pineal or whether it is perhaps just a pineal transmitter substance whose primary site of action is within the pineal. The third point concerns effects of pineal substances not related to any particular organ.

a) Release, Distribution and Fate of Pineal Substances

As pointed out above (pp. 237, 240), the epiphysis cerebri contains inter alia two well defined substances of hormonal nature, melatonin and arginine vasotocin, whose release, distribution and fate within the body have to be considered. While we have a fairly complete picture concerning melatonin, virtually nothing is known about arginine vasotocin apart from the fact that it is released into the CSF (PAVEL, 1973). Since it is still unclear whether vasotocin is actually a pineal substance (p. 240), this substance will not be discussed here.

As will be recalled, there are corresponding changes in melatonin concentrations in pineal, blood and urine over a period of 24 h (p. 251) which strongly suggest that melatonin synthesized in the pineal gains access to the systemic circulation (the question of a possible release of melatonin into the third cerebral ventricle will be discussed on p. 266).

A word of caution is necessary, however, since melatonin is also synthesized in extrapineal sites (p. 237) and may show rhythmic changes even in pinealectomized animals (rats: LYNCH et al., 1975b; ewes: KENNAWAY et al., 1977). Experimental studies using radioactively labelled melatonin show that ^{3}H-melatonin binds to plasma proteins in both man and rat (CARDINALI et al., 1972b) and is rapidly taken up by all tissues, including the brain (KOPIN et al., 1961). In cats, 1 h after intravenous administration the various organs examined show clear differences in ^{3}H-melatonin uptake (Table 26) (WURTMAN et al., 1964e, 1968a). It is particularly interesting that the epiphysis cerebri shows the highest uptake. Apart from the iris, ovary, pituitary and other endocrine organs, peripheral nerves show fairly high ^{3}H-melatonin concentrations. Other studies pointing to the presence of melatonin in peripheral nerves are those by LERNER et al. (1959b) and BARCHAS and LERNER (1964). It should be noted that nerves apparently do not synthesize melatonin (AXELROD and WEISSBACH, 1961). Studies on the subcellular distribution of ^{3}H-melatonin in rat ovary and adrenals by means of sucrose density gradients show that ^{3}H-melatonin is present mainly in the soluble supernatant fraction, small amounts also being present in the nuclear and mitochondrial fractions (WURTMAN et al., 1964e). Specific cytoplasmic melatonin receptors have been demonstrated in ovary (also human), testis, uterus, skin, liver and eye, but not in brain by COHEN et al. (178).

For experimental studies using melatonin, it is important to note that there is a rapid decrease of melatonin within the first 10 min after injection, approxi-

Table 26. Tissue distribution of ^{3}H-melatonin (WURTMAN et al., 1968a)[a]

Tissues	μCi/100 g	μg/100 g	Tissues	μCi/100 g	μg/100 g
Plasma	0.5	0.6	Kidney	1.1	1.3
Pineal	20.2	25.3	Uterus	0.9	1.1
Iris choroid	5.0	6.3	Pancreas	0.6	0.8
Ovary	4.6	5.6	Salivary glands	0.6	0.8
Pituitary	2.3	2.9	Spleen	0.6	0.8
Sympathetic chain	2.0	2.6	Heart	0.5	0.6
Peripheral nerve	1.8	2.2	Skin	0.4	0.5
Testis	1.5	1.9	Brain	0.4	0.5
Thyroid	1.4	1.8	Diaphragm	0.2	0.3
Adrenal	1.2	1.5	Adipose tissue	0.2	0.3

[a] Cats were given 1 mCi ^{3}H-melatonin (200 μCi/μmol) intravenously; the animals were killed 1 h later, and tissues were assayed for ^{3}H-melatonin.

mating a half-life of 2 min. After 40 min melatonin disappears more slowly, corresponding to a half-life of 35 min (KOPIN et al., 1961). According to MAICKEL et al. (1974) melatonin has a half-life of 12 min in plasma and up to 18 min in various rat organs. The importance of the time of day of melatonin injections is discussed on p. 283; for different modes of melatonin administration, see pp. 283, 284.

The major metabolic pathway for melatonin is 6-hydroxylation in the liver, followed by conjugation mainly with sulphate (70%–80%) or to a lesser degree (5%) with glucuronic acid and excretion in urine (80%) and faeces (20%) (KOPIN et al., 1960/1961; KVEDER and MCISAAC, 1961; TABORSKY et al., 1965).

Melatonin reaching the CSF may not affect the brain in its original form. According to HIRATA et al. (1974) the mammalian brain converts intracisternally injected ^{14}C-melatonin to two new metabolities, namely N^{γ}-acetyl-N^2-formyl-5-methoxykynurenamine and N^{γ}-acetyl-5-methoxykynurenamine.

b) Effects of Pineal Substances on Pineal Morphology and Function

According to QUAY (1974a), one of the neglected pathways by which melatonin may possibly have its effects is inside the pineal gland itself, where methoxyindoles could possibly function as chemical messengers (QUAY, 1965f, 1969a; BENSON, 1977).

There is clear evidence that administration of pineal extracts or well defined pineal substances is not without effect on pineal morphology and function, though it should be noted that when these experiments are carried out in vivo these effects cannot necessarily be taken as evidence that the substances given act directly on the pinealocytes. From the results obtained it appears that administration of pineal extracts and melatonin have opposing effects. Pineal extracts appear to depress pineal activity, as shown by the fact that their administration is followed by a statistically significant decrease in both the number of pinealocyte nucleoli and the amount of granular material in the nuclei together with a slight decrease in size of the pinealocytes and a lightening of the cytoplasm (HOLMGREN et al., 1960). Depressive effects of pineal extracts were noted also by MILINE and ŠĆEPOVIĆ (1959). Electron microscopy reveals smaller pinealocytes containing smaller and less numerous mitochondria, accompanied by a decrease of lipid droplets and an increase in the number of dark pinealocytes, when compared to intact controls (BUSS, 1965; GUSEK et al., 1965).

Melatonin treatment in rats for 14 days, on the other hand, is followed by an increase in the number of ribosomes, procentrioles and microtubules accompanied by prominent nucleoli, Golgi apparatus and annulate lamellae (FREIRE and CARDINALI, 1975). In mice, melatonin given for 5 days increases the number of dense-core vesicles in the pericapillary pinealocytes and it has been suggested that these vesicles represent the stored or secretory form of a pineal antigonadotropic factor the synthesis and/or secretion of which is stimulated by melatonin (BENSON and KRASOVICH, 1977). Moreover, melatonin treatment partially counteracts the light-induced depletion of pineal lipids in rats exposed to constant illumination, but not to natural photoperiods (EBELS and PROP, 1965). Pineal protein synthesis is apparently not altered by melatonin

(ORSI et al., 1973). When administered at the 8th hour of the daily photoperiod, melatonin suppresses the diurnal rhythm of pineal serotonin, but not when given 6 h later (FISKE and HUPPERT, 1968). Studies dealing with the effects of melatonin on pineal HIOMT have yielded variable results. In vitro melatonin inhibits HIOMT (WEISS, 1968b), while in vivo it stimulates both HIOMT and NAT (FREIRE and CARDINALI, 1975). In vitro melatonin, but not serotonin, enhances pineal thyroxine uptake (CADY and DILLMAN, 1971). Most interesting in vivo findings are that melatonin induces the release of arginine vasotocin into cerebrospinal fluid (PAVEL, 1973) and that melatonin inhibits reproduction in the male hamster provided that the animals are neither pinealectomized nor have sympathetically denervated pineal glands (REITER et al., 1977a). Arginine vasotocin, in turn, decreases pineal melatonin synthesis (SARTIN et al., 1978).

It is therefore not unlikely that pineal substances, by way of as yet unknown mechanisms, influence pineal function either before they are released from the pineal or after they have passed through the systemic circulation. In this context it is relevant to recall that after ^{3}H-melatonin injections the pineal shows by far the highest radioactivity of all organs examined (WURTMAN et al., 1964e) (Table 26).

c) Effects of the Pineal Not Related to Any Particular Organ

Before considering the possible target organs of the epiphysis cerebri, it is relevant to note that pineal substances may affect basic cellular processes.

Frequently, it has been noted that the pineal is not without effect on mitosis. In rats, pinealectomy increases the mitotic rate in liver, intestinal mucosa and spleen (BINDONI and CAMBRIA, 1966, 1968). By contrast, in healing wounds pinealectomy depresses mitotic activity (WEICHSELBAUM et al., 1975, mice). Which one of the pineal substances is responsible for this effect is not yet clear. In the wound-healing studies mentioned melatonin administration does not lead to statistically significant differences when compared with control and sham-pinealectomized mice (WEICHSELBAUM et al., 1975). In HeLa and ICB cells melatonin antagonizes colchicine-induced mitotic arrest, as shown by the observation that melatonin is without effect when given alone or with colchicine, but effective when given 1 h prior to colchicine (FITZGERALD and VEAL, 1976). Melatonin appears to act on microtubules (QUAY, 1968d), as exemplified by the findings that in plant cells melatonin arrests mitosis by interacting with microtubular proteins (BANERJEE and MARGULIS, 1973) and that light concentrations of melatonin can displace colchicine from tubulin, the protein subunit of microtubules (WINSTON et al., 1974). Since melatonin decreases brain microtubule protein, leading to a strong inhibition of the fast axoplasmic flow (CARDINALI and FREIRE, 1975), and increases pinealocyte microtubules (FREIRE and CARDINALI, 1975), it is not unlikely that the pineal also influences microtubules in other organs where they are known to be involved in the release of secretory products.

The fact that in adult rats pinealectomy increases ^{32}P incorporation into DNA in liver, intestinal mucosa and spleen, whereas it decreases it in the prepuberal rat liver, indicates metabolic effects of the epiphysis cerebri (BINDONI and CAMBRIA, 1966). Following pinealectomy RNA synthesis in intestinal mucosa and spleen is strongly stimulated in prepuberal, but not in adult rats (CAMBRIA and BINDONI, 1966). A significant decrease in muscle RNA has been noted at 30 days after pinealectomy provided that the animals had fasted for 12 h, but not for 48 h, before they were killed (I. MILCOU et al., 1967). In nephrectomized rats pinealectomy increases blood urea levels, while administration of pineal extracts decreases it (I. MILCOU et al., 1967; I. MILCU et al., 1968, 1969; NANU-IONESCU et al., 1969). An increase in intercellular permeability in connective tissue following pinealectomy has been described by MARCEAN et al. (1968).

Smooth musculature is apparently not directly affected by melatonin, though it inhibits serotonin-induced contractions (ARUTYUNYAN et al., 1964; RAHAMIMOFF et al., 1965).

The pineal may influence phosphate metabolism (THIÉBLOT et al., 1949a; SIMONNET and STERNBERG, 1951) and has been shown to affect nerve growth factor, the levels of which are depressed by pinealectomy and by melatonin treatment (PEREZ-POLO et al., 1978).

9. Pineal–Brain Interactions

a) Pathways

As pointed out below, there are clear indications that the pineal influences brain structure and function. It is therefore of interest to discuss the pathways mediating these effects. Since there is no convincing evidence as yet that afferent nervous pathways link the pineal with the brain, one must assume that it is by way of humoral mechanisms that the pineal influences the brain. Two pathways may be involved, namely the systemic circulation and the CSF. Since in the epiphysis cerebri the alleged morphological correlates of the secretory products are predominantly located in club-shaped endings of pinealocyte processes often lying in the immediate vicinity of blood vessels, it would appear that melatonin and other pineal products are released into the systemic circulation. That melatonin readily reaches the brain has been demonstrated by KOPIN et al. (1961) and WURTMAN et al. (1964e).

As the hypothalamus appears to be an important target organ of the epiphysis cerebri, it has frequently been discussed whether or not the pineal releases its secretory products into CSF. Unfortunately this important problem has not been studied in detail. Morphological studies alone are not sufficient to provide clear-cut answers. As described in Section B II 2, the topographical relationships of the pineal to the third ventricle are fairly variable. While in some species, e.g. the hedgehog, the pineal organ projects into the third ventricle, in the majority of species examined only a part of the pineal is in contact with the third ventricle. A contact is most regularly established in the area lying between the habenular and posterior commissures, though it should be noted that in this area the epiphyseal parenchyma is separated from the CSF by a fairly variable, but sometimes partly reduced ependyma (see p. 186). Another site of contact may be between the pineal and the suprahabenular, or suprapineal, recess, not to forget that in man the entire epiphysis cerebri appears to be bathed in CSF (p. 20). In rodents in which the bulk of pineal tissue may lie distant from the third ventricle the situation is very variable. In the guinea-pig (KRAPP, 1978) only the proximal part of the elongate pineal is in contact with the third ventricle. In the hamster, in which the pineal consists of a superficial and a deep pineal, only the deep pineal is in contact with the third ventricle (SHERIDAN and REITER, 1970a). Here there is a constant and fairly large contact area, where the ependyma may be greatly reduced enabling direct contact of pineal parenchyma with the CSF (HEWING, 1978). In the most widely studied species, the rat, the contact site of the deep pineal with the third ventricle is very variable, ranging from virtually no contact to a situation comparable to that in hamsters (VOLLRATH and BOECKMANN, 1978). However, when the deep pineal extends in a posterior direction, beyond the intercommissural region, there is the possibility that functional interrelations exist between the deep pineal and the large suprahabenular recess of the third ventricle (VOLLRATH and BOECKMANN, 1978). The superficial pineal, by contrast, does not come into direct contact with the CSF of the third ventricle.

At present there is little information available as to a possible release of pineal substances into the third ventricle. There can be no doubt that melatonin is present in the CSF, as shown in humans (I. SMITH et al., 1976) and calves (HEDLUND et al., 1977). Release of melatonin directly into the third ventricle is suggested by the fact that during the night the melatonin levels in calves increase 17-fold in the CSF and only 6-fold in the plasma (HEDLUND et al., 1977). However, in humans the melatonin levels are usually lower in the CSF than in the serum (ARENDT et al., 1977; G. VAUGHAN et al., 1978a). Arginine vasotocin has been shown to be released into the third ventricle following melatonin injections (PAVEL, 1973), but since the exact site of origin of vasotocin is not clear (p. 240), these findings cannot be taken as evidence that pineal substances are actually released into the third ventricle. According to MESS the choroid plexus concentrates melatonin from the peripheral circulation and secretes it into the CSF (MESS et al., 1978, references!).

A possible interaction between the CSF and the pineal is suggested by intense staining of the whole pineal body, in contrast to the hypophysis, following injection of bromophenol blue into the cerebral ventricles (FELDBERG and FLEISCHHAUER, 1960; FLEISCHHAUER, 1961). A complete penetration of the epiphysis is also demonstrable after intraventricular injections of 3,6-diamino-acridinetrihydrochloride (FLEISCHHAUER, 1964). Fluorescein-labelled bovine serum proteins also reach the pineal gland (KLATZO et al., 1964). The interactions between the CSF and the pineal are fairly rapid as it takes as little as 1.5 min for *l*-^{14}C-phenylalanine injected into the lateral cerebral ventricle to reach the pineal gland and to be metabolized (BAGCHI and ZARYCKI, 1971a, b).

In this context it is also relevant to note that, when judged by the amount of arginine vasotocin released into the third ventricle, intraventricular injections of melatonin are 50 times more potent than intravenous injections (PAVEL, 1973).

b) Hypothalamus

Although pineal substances have been shown to affect various parts of the brain, it would appear that they act primarily on the hypothalamus (CARDINALI, 1974). Melatonin, which appears to be demonstrable in the hypothalamus of intact animals under physiological conditions (A. GREEN et al., 1973), becomes concentrated within the hypothalamus after systemic or intraventricular administration (WURTMAN et al., 1964e; ANTON-TAY and WURTMAN, 1969; CARDINALI et al., 1973b, 1978) and induces a variety of metabolic changes, including elevation of both serotonin (ANTON-TAY et al., 1968) and GABA levels (ANTON-TAY, 1971), and a decrease in protein synthesis (ORSI et al., 1973; CARDINALI et al., 1974a) and microtubule protein (CARDINALI and FREIRE, 1975). The uptake of noradrenaline, serotonin, dopamine and glutamate by hypothalamic synaptosomes was first found to be suppressed by melatonin in vitro (CARDINALI et al., 1975b), but in a later study smaller doses of melatonin inhibited the serotonin accumulation only (CARDINALI, 1975). Melatonin stimulates hypothalamic Δ^4-reductase activity in the rat and depresses it in the hamster (FREHN et al., 1974). It reduces hypothalamic MAO activity (URRY and ELLIS, 1975). In ferrets melatonin produces differential serotonin rhythms in hypothalamus and pineal (YATES and HERBERT, 1976).

Pinealectomy increases protein synthesis (CARDINALI et al., 1974a), hypothalamic autofluorescence (A. SMITH, 1974) and MAO activity (URRY and ELLIS, 1975), blocks the isoproterenol-induced decrease in hypothalamic tubulin levels (CARDINALI et al., 1975e), but is without effect on hypothalamic serotonin and melatonin levels (KOSLOW, 1974). Hypothalamic synaptosomes from pinealectomized rats, or rats which had undergone pinealectomy combined with superior cervical ganglionectomy, show a depressed serotonin uptake, while uptakes of noradrenaline, dopamine and glutamate are not affected (CARDINALI, 1975). No effect on hypothalamic serotonin levels was observed after pinealectomy by VERMES et al. (1974). However, pinealectomy was found to increase the uptake of 17β-oestradiol in the hypothalamus of mature castrated female rats (GOGAN et al., 1972).

Moreover, there is indirect evidence that melatonin and serotonin influence hypothalamic releasing hormones. In vitro studies have shown that in the presence of serotonin, hypothalamic tissue is no longer capable of stimulating FSH and LH secretion (MOSZKOWSKA, 1964, 1965b). Administration of melatonin or serotonin changes FSH, LH and prolactin levels when injected into the third ventricle, but is without effect when injected into pituitary portal vessels (KAMBERI et al., 1970, 1971; KAMBERI, 1973).

In view of these results it is not surprising that pinealectomy alone or combined with administration of pineal extracts affects parvocellular hypothalamic neurons, which are likely to be involved in the production of LHRH. In both the arcuate and ventromedial hypothalamic nuclei serotonin, as assessed by formaldehyde-induced fluorescence, shows distinct changes (SMITH and KAPPERS, 1975). Mean nuclear diameters are not affected by pinealectomy (CARNICELLI et al., 1963). According to LEONARDELLI et al. (1978), melatonin treatment increases the number of LHRH immunoreactive cells in the hypothalamus. In hamsters bearing melatonin antibodies LHRH stores are reduced (KNIGGE and SHERIDAN, 1976). There is also biochemical evidence that the pineal influences the hypothalamic releasing hormones LHRH, PIF and/or PRF (BLASK and REITER, 1975a, 1978; EBELS et al., 1975). Hypothalamic LHRH content depressed by blinding has been shown to be unaffected by pinealectomy, in contrast to FSHRH (BLASK et al., 1979). Although many findings concerning the releasing hormones suggest an action of the pineal at the hypothalamic level, there is also evidence for an effect at the pituitary level: in organ culture, melatonin (1 n*M*) inhibits the action of LHRH on LH release (neonatal rats, MARTIN and KLEIN, 1976). Moreover, it has been shown that intracarotid injection of melatonin is without effect on 17-oxosteroid secretion by the testes; however, when LHRH was given after melatonin, then the testicular response to LHRH was diminished, but not that to HCG (dog: YAMASHITA et al., 1978; MIENO et al., 1978). In the suprachiasmatic nucleus pinealectomy does not drastically influence serotonin uptake (MEYER and QUAY, 1976) and mean nuclear diameter (CARNICELLI et al., 1963).

Interactions also appear to exist between the pineal and the supraoptic and paraventricular nuclei of the hypothalamus. MILINE (1969) discusses a functional antagonism between the pineal and these nuclei since in rats exposed to cold the supraoptic nucleus shows signs of involution when the pineal is intact and signs of hypertrophy together with a decrease of neurosecretory material after pinealectomy (MILINE et al., 1969b). Pinealectomy results in an increase in mean nuclear diameters in both supraoptic and paraventricular nuclei (CARNICELLI et al., 1963), the neurosecretory material showing variable results: a decrease in both nuclei (AZZALI et al., 1963, 1964) or the paraventricular nuclei (MILINE, 1963) has been described by one group of authors, while BUGNON and MOREAU (1961b) report on a striking increase of neurosecretory material in the paraventricular nuclei, but not in the supraoptic nuclei. TPPase activity, an enzyme of the Golgi apparatus used as a marker of the neurosecretroy activity of the nerve cells, decreases after pinealectomy (DE VRIES and KAPPERS, 1971; DE VRIES, 1972a, b), this effect being negated by hypothalamic pineal implants (DE VRIES, 1972c).

The reduced diuresis after pinealectomy can be normalized by administration of pineal extracts (BARBAROSSA et al., 1961), which are known to have diuretic effects (MILIN and NESIĆ, 1958). According to ARUTYUNYAN et al. (1964) both melatonin and serotonin have antidiuretic effects. No effect of pinealectomy on daily urine volumes could be demonstrated by KARPPANEN et al. (1970a).

The administration of pineal extracts to non-pinealectomized animals has led to variable results: no effects on neurosecretory material of the paraventricular and supraoptic nuclei could be demonstrated by BUGNON and MOREAU (1961b), while the nerve cells of the paraventricular nuclei hypertrophy (ARON et al., 1961; MILINE, 1963). A twofold increase in the oxytocin content of the paraventricular nucleus after intravenous injection of bovine pineal extracts has been described by S. MILCOU and PAVEL (1960). MILINE (1960a, 1968) described diminished neurosecretory activity in the supraoptic nucleus after administration of pineal extract or melatonin.

Other hypothalamic nuclei showing an increase of the mean nuclear diameters after pinealectomy are preoptic and ventral and dorsal premamillary nuclei, while the lateral mamillary nucleus is unaffected (CARNICELLI et al., 1963).

The effects of hypothalamic releasing hormones on pineal function have also been investigated. In vitro, TRH, LRH and SRIF do not alter cAMP levels in the rat pineal, but TRH and, to a lesser extent, LHRH are potent inhibitors of noradrenaline-stimulated cAMP accumulation (TSANG and MARTIN, 1976). The ultrastructure of cultured pinealocytes is apparently not altered by the addition of LHRH (KARASEK et al., 1978b). In vivo, LHRH, TRH and somatostatin induce vasotocin release into the CSF (GOLDSTEIN and PAVEL, 1977). Intracarotid injection of MIF reduces pineal vasotocin stores (PAVEL et al., 1977a).

Another line of evidence for an interaction between the hypothalamus and the pineal body is that in immature rats electrolytic lesions of the median eminence lead to higher NAT and melatonin contents in the pineal when compared to controls (WILKINSON et al., 1976). As regards the role of serotonin and dopamine in hypothalamo-pituitary function, see the review article by SMYTHE (1977).

c) Extrahypothalamic Actions

α) *Effects of Melatonin Administration in Humans*

The administration of melatonin to human subjects has been reviewed and commented upon by LERNER and NORDLUND (1975). From the data published it would appear that relatively few patients have been treated with melatonin and over relatively short periods of time. For short-term studies LERNER and NORDLUND (1975) think it feasible to administer 100 mg/day intravenously or 1.2 g/day orally, which are fairly large doses in comparison with physiological levels of melatonin. In humans, melatonin treatment does not appear to have very striking effects. According to LERNER and NORDLUND melatonin administration for a few days up to 6 weeks is followed in some patients by mild sedation, cutaneous flush, abdominal cramps, diarrhoea, scotoma lucidum and migraine headaches. There is also an increase in EEG alpha activity, EEG synchronization, induction of sleep, more frequent REM cycles, the occurrence of vivid dreams and moderate feelings of elation (ANTON-TAY et al., 1971). An effect in relation to sleep has also been documented by CRAMER et al. (1974), who showed that after an intravenous injection of 50 mg in daytime, sleep occurs within 15–40 min and lasts for 26–60 min.

In schizophrenic patients injection of a pineal extract was found to cause no local reaction and no change in pulse, blood pressure, body temperature and leucocyte count (ALTSCHULE, 1957), but in a later study an increase in the number of eosinophils was noted (ALTSCHULE and GIANCOLA, 1960). Melatonin does not affect the signs of parkinsonism, but in some patients it facilitates sleep. Some patients become tranquil, whereas they had frequently been anxious, nervous or depressed prior to the administration of melatonin (COTZIAS et al., 1971a; PAPAVASILIOU et al., 1972).

β) Sedation and Sleep

A sedative effect of melatonin has been corroborated in animal species. In cats melatonin implantation into the preoptic region induces sleep (MARCZYNSKI et al., 1964). It also results in an increase of hexobarbital-induced sleeping time (ARUTYUNYAN et al., 1964; BARCHAS et al., 1967). In addition to melatonin, vasotocin has been shown to induce sleep (PAVEL et al., 1977b).

In rats, following pinealectomy a dissociation develops between the nyctihemeral variations of paradoxical sleep and slow-wave sleep. While the amounts of paradoxical sleep and slow-wave sleep do not change, as assessed on a 24-h basis, paradoxical sleep decreases during the light phase and increases during the dark phase. The diurnal changes of slow-wave sleep remain normal, but the circadian periodicity of paradoxical sleep is almost completely abolished (MOURET et al., 1974). In humans, melatonin does not lead to significant changes in the percentage and frequency of REM sleep or in dream activity (CRAMER et al., 1974).

A sedative effect of the pineal can also be demonstrated in experimental studies involving *seizures* and pinealectomy. Electrically induced seizures in cats are remarkably suppressed by pineal extracts (ROLDÁN and ANTÓN-TAY, 1968). Melatonin reduces or suppresses epileptiform activity when the focus produced by convulsant agents is localized in primary sensory areas, but not in other cortical areas or limbic structures (FARIELLO and BUBENIK, 1976). Following pinealectomy both the voltage and the duration of interhippocampal evoked potentials increase (BINDONI and RIZZO, 1964, 1965), and seizure-like discharges occur in the EEG (NIR et al., 1969b). Actual seizures are observed when parathyroidectomized rats are pinealectomized (REITER and MORGAN, 1972; REITER et al., 1973a, b, 1975a). Adventitious movements induced by L-dopa are blocked by melatonin (COTZIAS et al., 1971b). Pinealectomy-induced convulsions occur in Mongolian gerbils (PHILO and REITER, 1978a); and in pentazole-induced convulsing mice, propagation of convulsions to other mice is inhibited by pinealectomy (HATA and KITA, 1978). Moreover, convulsions induced by shaking mice up and down are inhibited by pinealectomy (HATA and KITA, 1978).

Frequently the effects of the pineal on locomotor activity have been studied, some of the results supporting the notion of a sedative effect by the pineal. Thus it was found that pinealectomy increases locomotor activity (REISS et al., 1963a; AIRAKSINEN and KARPPANEN, 1970; for a depressive effect see KINCL et al., 1970), the increase apparently being restricted to hypoactive rats (KRAPP, 1977). An additional peak of locomotor activity only has been observed by KARPPANEN et al. (1973). Administration of pineal substances reduces locomotor activity (REISS et al., 1963a; R. WONG and WHITESIDE, 1968). In some studies pinealectomy was without apparent effect on locomotor (REMLEY et al., 1969) and treadwheel activity (RELKIN, 1970).

That melatonin may play an important role in regulating locomotor activity is also shown by the interesting observation that in rats kept in continuous darkness for 16–18 days locomotor activity rhythms can be used as predictors of the pineal melatonin content: animals showing high locomotor activity exhibit ten fold higher pineal melatonin contents than those exhibiting low locomotor activity (RALPH et al., 1970, 1971). Recently similar data have been obtained with respect to pineal NAT (POHL and GIBBS, 1978). Also of interest is the observation that in the central region of the pineal nuclei are densely packed in slow-running rats and widely spaced in fast-running rats (REISS et al., 1967).

γ) Locomotor Rhythms

As the pineal affects locomotor activity (see above), QUAY has thoroughly investigated to what extent the *rhythms* of locomotor activity are influenced by the epiphysis cerebri. In a first study of a series QUAY (1965b) noted that the free-running circadian rhythm in the male rat's running activity in both standard daily photocycle and continuous light is similar in intact and in pinealectomized animals. More detailed investigations revealed that pinealectomy does not affect the timing, resetting and number of active 10-min periods (QUAY, 1968f). However, an effect of the pineal becomes apparent after reversal of the lighting schedule: pinealectomized animals initially shift more quickly in starting time and activity mid-point, have less increase post $\Delta\Phi$ in activity phase length and amount of activity, and have a different distribution of activity per transient cycle (QUAY, 1970d, e). These differences observed led QUAY (1970f) to encourage clinicians to examine the possibility of using forced phase-shifting of circadian or 24-h rhythms as a diagnostic test for destruction of the pineal, as pineal-deprived patients will phase-shift their daily time of primary behavioural and physiological activity more quickly than normal. In young animals initial mean shift in activity starting time following reversal of photoperiod is even greater than in adult animals, namely 6–7 h ahead in pinealectomized rats as compared to the controls (QUAY, 1970g).

Intact *nervi conarii* and *tentoria cerebelli* are apparently unnecessary for pineal-based delay in activity phase-shifting (QUAY, 1971a). Since when the pineal stalk is severed the initial rate of phase-shift shows a transitory increase resembling that after complete pinealectomy, it has been suggested that transmission via the dura mater to the pineal body of periodic changes in intracranial pressure may contribute to the physiological mechanism for timing of end-point in the circadian rhythm in running activity during resetting (QUAY, 1971b, c). Studies in hamsters have shown that after photoperiod reversal pinealectomized animals re-entrain their activity faster than intact controls (FINKELSTEIN et al., 1978).

δ) Somatic Brain Effects

The effects of the epiphysis cerebri on brain structure and chemical composition have been little investigated. In *developing* rats pinealectomy results in lighter brain weight (QUAY, 1974c) and a decrease in myelin formation accompanied in both sexes by significantly lower values for total lipid, cerebroside, sulphatide, phosphatidyl inositol, phosphatidyl serine and sphingomyelin, an increase in phosphatidyl choline in males and a decrease in proteolipid protein in females (RELKIN et al., 1973; RELKIN and SCHNECK, 1975). In *adult* rats, pinealectomy decreases ^{32}P uptake in the acid-soluble fraction of the brain (REISS et al., 1963b) and is without effect on brain serotonin and noradrenaline levels (KARPPANEN et al., 1970a, 1973; KARPPANEN, 1974). However, pinealectomized rats on a sodium-deficient diet have a lower cerebral potassium content than sham-operated or unoperated controls, which led QUAY (1965c, 1969c) to suggest that the pineal participates in homeostasis of brain composition. This notion is supported by results showing that melatonin enhances protein synthesis in cerebral cortex slices in vitro (CARDINALI et al., 1973c) and increases dopamine, noradrenaline (WENDEL et al., 1974), serotonin (COTZIAS et al., 1971b), 3′, 5′-GMP (RUDMAN, 1976; CSF) and pyridoxal phosphokinase (ANTON-TAY et al., 1970).

The LD_{50} (mg/kg) for melatonin, as assessed in cats and mice, is 125 when given intravenously and 930 when given orally. Slight ataxia, unrest and assumption of a lateral position for a brief period occur after intravenous doses of 100–150 mg/kg, while after 200–250 mg/kg clonic or clonicotonic spasms leading to death occur. Melatonin given subcutaneously lowers body temperature by 2°–3° C (ARUTYUNYAN et al., 1964).

A depressive effect of pineal extracts on the habenular nuclei has been described by MILINE and ŠĆEPOVIĆ (1959).

The *subcommissural organ* (SCO) is apparently not affected by pinealectomy (LENYS, 1966) or administration of pineal homogenate, while administration of SCO homogenate is supposedly followed by a 12.1% decrease of pinealocyte nuclear volume (PALKOVITS, 1968). NOMOKONOVA and SHAPIRO (1965), studying the SCO and the pineal organ after modifications of the water-electrolyte metabolism and following catalepsy produced by intermittent light flashes, noted that the SCO and the pineal differed in their behaviour, the SCO resembling more closely the neurosecretory nuclei of the anterior hypothalamus than the pineal.

ε) *Behavioural Studies*

Pineal functioning and behaviour has been reviewed by SAMPSON (1975). Here it suffices to mention that in female rats exploratory behaviour as studied in the open field test increases after pinealectomy, is depressed by administration of pineal extracts in both intact and pinealectomized animals and is not affected by melatonin (SAMPSON and BIGELOW, 1971). In male rats pinealectomy increases exploratory behaviour in emotional but not in non-emotional rats (KRAPP, 1977). An increase in exploratory movements after pinealectomy has also been observed in mice (HATA and KITA, 1978).

Inter-male aggression in adult house mice is reduced when one or both pair members are pinealectomized, but it is increased in chronically maintained groups when all males are pinealectomized (MCKINNEY et al., 1975). PATERSON et al. (1978a) found that pinealectomy increases attacking latency and that administration of melatonin shortens the latency. As administration of MSH reduces aggression, but not when given jointly with melatonin, it has been concluded that the reduction of aggression observed in MSH-treated groups may, to a large extent, depend on an intact pineal gland, and that MSH may act by inhibiting melatonin secretion (PATERSON et al., 1978b).

Locomotor activity and rhythms have been mentioned above (p. 271, p. 272). Feeding rhythms will be discussed on p. 304, sexual and maternal behaviours on p. 304.

d) Neurological and Mental Diseases

No attempt has been made to scan the literature with respect to changes of pineal function in relation to neurological and mental diseases. Hence, only incidental findings will be reported.

Parkinsonism, despite some encouraging reports, is apparently not greatly improved by melatonin (SHAW, 1977, review). In epileptic patients melatonin apparently raises the convulsive threshold (ANTON-TAY, 1971) and may therefore provide potential therapy for epilepsy (SHAW, 1977).

In schizophrenic patients administration of pineal extracts sometimes has a beneficial effect (ALTSCHULE et al., 1954; ALTSCHULE, 1957, 1975b; ALTSCHULE and GIANCOLA, 1960; ELDRED et al., 1960; BIGELOW, 1975), reversing a non-specific biochemical abnormality, i.e. a rise in blood pyruvate, α-ketoglutarate and citrate concentrations after ingestion of glucose, to normal (ALTSCHULE, 1975b).

According to JONES et al. (1969) the metabolism of β-^{14}C-melatonin may differ in schizophrenic and non-schizophrenic patients in so far as schizophrenic patients may excrete more of an unidentified acidic metabolite.

In depressed patients melatonin excretion apparently does not differ from that in normal patients (JIMERSON et al., 1977). Desmethylimipramine, an antidepressant drug, causes an increase of pineal NAT and melatonin (PARFITT and KLEIN, 1977).

In mice made experimentally phenylketonuric, subnormal maze-learning ability was largely prevented by administration of melatonin or 5-hydroxytryptophan (WOOLLEY and VAN DER HOEVEN, 1964).

Convulsions occur in Mongolian gerbils after pinealectomy (PHILO and REITER, 1978a).

The effects of central depressants on pineal function have been mentioned above (p. 251).

10. Pineal-Pituitary Interactions

Considerable effort has been devoted to studying the influence of the pineal on the pituitary and vice versa. Despite many conflicting and negative findings a functional antagonism between the two organs, as suggested by earlier workers (REISS et al., 1949a; MOSZKOWSKA, 1951b, 1958a, b, 1963, 1966; TRENTINI et al., 1969), appears to be present. Recent relevant review articles are those by REITER (1974a, b, 1976b, 1977) and REITER and VAUGHAN (1977).

a) Effects of Hypophysectomy on the Pineal

That hormonal changes may affect the epiphysis cerebri has been clearly demonstrated by studies involving hypophysectomy. Hypophysectomy results in an increased uptake of ^{32}P and ^{131}I in the pineal (REISS et al., 1949a, b, 1963b: for a negative finding see SKAUG et al., 1958) and a decrease in pineal weight (ALEXANDER et al., 1970a; URRY et al., 1972), though in one study no change was noted (REISS et al., 1963b). Quantitative morphological studies reveal a statistically significant decrease in pineal volume correlated with marked atrophy of the pinealocytes throughout the entire organ (ITO and MATSUSHIMA, 1968c), which explains the observation that the cell content and the nuclear density (REISS et al., 1949a, 1963b) of the pineal increase. The lipid content also decreases (PANAGIOTIS and HUNGERFORD, 1962; HUNGERFORD and PANAGIOTIS, 1962; ZWEENS, 1964, 1965; LUPULESCU, 1968; SATODATE et al., 1970a, b; KARASAK, 1971a, b). Electron microscopic studies (LUPULESCU, 1968; SATODATE et al., 1970a, b; KARASEK, 1971a, b) have yielded somewhat contradictory results. On the whole it would appear that pinealocyte function is depressed following hypophysectomy, as indicated by an atrophy of endoplasmic reticulum, small and collapsed ergastoplasmic vesicles, small mitochondria showing tigroid appearance, increase of autophagic vacuoles and smaller nuclei and nucleoli than in control animals. The Golgi apparatus was found to be strongly developed by KARASEK (1971a, b), but rarely seen by LUPULESCU (1968). SATODATE et al. (1970a, b) noted a decrease of dense-core granules in sympathetic nerve endings.

Pineal adenyl cyclase shows either a proportional decrease in the basal and noradrenaline-stimulated activity (WEISS and CRAYTON, 1970a) or no change (WEISS, 1970). HIOMT activity decreases in immature rats (ALEXANDER et al., 1970a; URRY et al., 1972, 1976) and also in mature rats at 6 and 7 weeks after hypophysectomy (URRY et al., 1976), but not at 1 and 3 weeks (URRY et al., 1972), pointing to long-term rather than short-term effects of hypophysectomy on the pineal.

b) Effects of Pinealectomy and Pineal Substances on Non-hormonal Parameters of the Pituitary

α) *Pinealectomy*

The effects of pinealectomy on the pituitary are very variable, one of the reasons for the discrepancies being the stage of development of the animals at which they are operated upon. Pituitary *weight* is usually not affected by pinealectomy in *adult* animals (RENNELS and DILL, 1961; REISS et al., 1963b; ROWE et al., 1970; KINSON and LIU, 1973a; SEIBEL and SCHWEISTHAL, 1973; REITER et al., 1974a; REITER, 1975b), though a decrease has been reported (RØNNEKLEIV and MCCANN, 1975b). In *young* animals pinealectomy has been found to increase pituitary weight (WURTMAN et al., 1959; QUAY, 1974c) or not to affect it (WRAGG, 1965, 1967; LOMBARD DES GOUTTES, 1967b). Relevant studies in experimentally affected animals have shown that pinealectomy depresses posterior pituitary weight elevated by anosmia (RØNNEKLEIV and MCCANN, 1975b), increases pituitary weight in NIH black rats kept under LD 1:23 for 18 weeks (REITER and HESTER, 1966) and in rats subjected to blinding combined with anosmia (SHIINO et al., 1974) and is without effect on pituitary weight elevated by castration (TALBOT and REITER, 1973/74). That sham-pinealectomy may decrease posterior pituitary weight has been demonstrated by QUAY (1974c).

Despite the fact that in adult animals pituitary weight is not greatly affected by pinealectomy, there are some indications that this operation is not without effect on pituitary *histology*. While HOLMES (1956) did not observe any changes, other authors have found an increase in the number of acidophilic cells (SIMONNET et al., 1954; THIÉBLOT and BLAISE, 1963) showing signs of hypertrophy (CLEMENTI et al., 1966) and a decrease of chromophobes (THIÉBLOT and BLAISE, 1963). Authors using a terminology based on functional rather than tinctorial criteria describe hyperplasia, hypertrophy and degranulation of corticotropic and prolactin cells, hypertrophy of FSH and LH cells (hyperplasia of the latter cells has been described by GIROD et al. [1964] and CZYBA et al. [1965]) and signs of reduced activity in FSH cells (BUGNON and MOREAU, 1961a, 1964). Electron microscopically gonadotropic cells look normal after pinealectomy, occasionally showing dilatation of the rough endoplasmic reticulum (CLEMENTI et al., 1969). Pituitary mitotic rate is significantly increased by pinealectomy (BINDONI and RAFFAELE, 1968).

Uptake of ^{32}P by the pituitary is not affected by pinealectomy (MALM et al., 1959; REISS et al., 1963b), but is increased by pinealectomy in adrenalectomized animals (LINGJAERDE et al., 1959). Monoamine oxidase activity in the pituitary (URRY and ELLIS, 1975) and uptake of oestradiol in the neurohypophysis (PEDROZA GARCIA et al., 1974) are enhanced by pinealectomy.

β) Pineal Substances

Administration of pineal substances is probably more often without effect on pituitary weight (REISS et al., 1963b, c; BARCHAS et al., 1969; COLLU et al., 1971a; TIGCHELAAR and NALBANDOV, 1971; DE PROSPO and HURLEY, 1971a; SEIBEL and SCHWEISTHAL, 1973; REITER et al., 1974a; SACKMAN et al., 1977) than that it decreases it (MOSZKOWSKA, 1964, 1965b; NABER et al., 1969; DEBELJUK et al., 1970a; KINSON and LIU, 1973b). Pituitary weight in *experimentally affected* animals is also not uniformly influenced by pineal substances. In adult male rats melatonin administration inhibits pituitary hypertrophy which follows castration (DEBELJUK et al., 1970b; TALBOT and REITER, 1974/74), but is without detectable effect on anterior pituitary weight in females following unilateral ovariectomy (RUBIN and TRAUM, 1971). In prepuberal male rats kept under constant illumination, anterior pituitary weight is reduced following melatonin administration (DEBELJUK, 1969).

Histological studies have shown that administration of pineal extracts results in an increase in the number of eosinophilic cells (NOTARIO, 1956), hyperactivity of TSH cells (BUGNON and MOREAU, 1963, 1964) and FSH cells showing a stronger PAS-positive reaction than after pinealectomy (BUGNON and MOREAU, 1961a). Administration of pineal extracts normalizes the amounts of eosinophilic and chromophobic pituitary cells altered by pinealectomy (THIÉBLOT and BLAISE, 1963).

Metabolic changes in the pituitary after administration of pineal substances are suggested by (a) a decrease of both protein synthesis (ORSI et al., 1973) and MAO activity (URRY and ELLIS, 1975) after melatonin and (b) increases of leucine, thymidine (PERESS et al., 1976) and ^{32}P incorporation (REISS et al., 1963c) after pineal extracts. Pineal extracts are not uniform with respect to ^{32}P incorporation. TCA-soluble extracts decrease the specific activity by 50%, whereas both TCA-precipitate and deproteinized pineal extracts increase the specific activity by 44% (REISS et al., 1963b). In the pars intermedia of the rat pituitary, administration of melatonin leads to an increase of serotonin levels; pars distalis and pars nervosa, by contrast, are not affected (PIEZZI and WURTMAN, 1970).

c) Growth Hormone and Body Weight

Although studies dealing with the relationship between the pineal and growth are relatively new and scarce, it would appear that the data obtained are compatible with the idea that some pineal constituents diminish growth hormone synthesis or secretion and consequently restrict growth (REITER, 1974a).

Pinealectomy has been little used to evaluate the pineal-growth hormone relationship. In rats, pinealectomy abolishes the dark-induced decrease of *growth hormone* in both pituitary and plasma (RELKIN, 1972a) and prevents the decrease in growth hormone levels which occurs after light deprivation alone (SORRENTINO et al., 1971a, b) or in combination with anosmia (SORRENTINO et al., 1971c, e; SHIINO et al., 1974). By contrast, RØNNEKLEIV and MCCANN (1978) noted that pinealectomy significantly reduces growth hormone release during the day, but does not significantly further reduce the release at night.

Body weight is usually not significantly affected by pinealectomy (LOMBARD DES GOUTTES, 1967b; BICK et al., 1968, 1969; ROWE et al., 1970; M. SMITH et al., 1973; RELKIN, 1973; REITER et al., 1974a; W. BROWN et al., 1977), though an increase has been noted by MALM et al. (1959) and J. HOFFMANN and POMERANTZ (1971). Body growth retarded by blinding can be accelerated by pinealectomy (SORRENTINO et al., 1972). In young animals pinealectomy stimulates weight increase (KINCL and BENAGIANO, 1967).

More data are available regarding the effects of pineal *substances* on growth hormone levels. In humans, oral administration of 1 g melatonin causes a rapid and significant decrease of serum growth hormone levels elevated by insulin-induced hypoglycaemia (SMYTHE and LAZARUS, 1974a). However, melatonin treatment alone leads to an elevation of serum GH levels (SMYTHE and LAZARUS, 1974b). In rats, growth hormone release is blocked by melatonin and stimulated by serotonin (SMYTHE and LAZARUS, 1973a; COLLU et al., 1972). A reduced growth-promoting activity is present in pituitaries of rats injected with bovine pineal extracts (GRACHEV et al., 1972). The intermediary of growth hormone action, somatomedin, is likewise elevated by serotonin and inhibited by roughly 35% by melatonin, as assessed by the uptake of $^{35}SO_4$ into costal cartilage (SMYTHE et al., 1974). A blockage of serotonin-induced growth hormone secretion has been described by SMYTHE and LAZARUS (1973b).

Studies dealing with the effects of pineal substances on *body weight* have produced very variable results, ranging from no effect (NABER et al., 1969; DEBELJUK et al., 1970a; KINSON and PEAT, 1971; COLLU et al., 1971a; KINSON and LIU, 1973b; REITER et al., 1974a) to depression (TILSTRA and PROP, 1962; TALBOT and REITER, 1973/74; K. HOFFMANN, 1973; M. VAUGHAN and VAUGHAN, 1974), statistically insignificant increases (TINACCI and FAZZINI, 1964; KUNKEL, 1969) and significant increases (SACKMAN et al., 1977).

Effects on body weight due to differences in food intake will be mentioned on p. 313.

d) Prolactin and Milk Yield

Prolactin secretion which is under the control of a hypothalamic inhibitory factor (PIF) is clearly influenced by the pineal. It appears as though the active pineal principles inhibit the release or synthesis of PIF, leading to an increased release of prolactin from the pituitary resulting in depressed pituitary stores and increased blood prolactin levels (RELKIN and ADACHI, 1972; REITER, 1974a).

Consistent with the above concept is that *pinealectomy* is followed by an increase in pituitary prolactin levels (RELKIN, 1972b, d, 1973; PAVEL et al., 1975c) and a decrease in blood prolactin concentrations (RELKIN, 1972b, d, 1973; RØNNEKLEIV and MCCANN, 1975a; W. BROWN et al., 1977; NILES et al., 1977a). Despite a drop in blood prolactin levels, the diurnal *rhythm* of this hypophyseal hormone is not greatly affected by pinealectomy (NILES et al., 1977a), though in nocturnal samples depressed concentrations (RØNNEKLEIV et al., 1973; BARRELL and LAPWOOD, 1978) and increased prolactin levels (KIZER et al., 1975) have been reported. Pinealectomy has no demonstrable effect on prolactin levels in the afternoon of pro-oestrus (BLAKE, 1976), but in conscious freely moving male rats pinealectomy reduced the plasma prolactin titres at 17.30 h, and there were often very low values during the day and higher values during the night (RØNNEKLEIV et al., 1978b). In animals exposed to darkness alone (RELKIN et al., 1972; REITER and JOHNSON, 1974a, b, c; REITER, 1975b, 1976b) or darkness combined with anosmia, pinealectomy (DONOFRIO and REITER, 1972; SHIINO et al., 1974; BLASK and REITER, 1975a, b) or sympathetic denervation of the pineal (BLASK et al., 1977) elevate the depressed pituitary prolactin stores, though sometimes the restoration is incomplete. An initial depression of pituitary prolactin levels by pinealectomy has been described by REITER (1975b).

Compatible with the concept that the pineal inhibits PIF via melatonin is that *administration of melatonin* results in depressed pituitary prolactin stores (REITER et al., 1976b, 1977a; SACKMAN et al., 1977) and elevated blood prolactin levels (KAMBERI et al., 1971; LU and MEITES, 1973; M. VAUGHAN et al., 1978) and that immunization against melatonin depresses blood prolactin levels (NILES et al., 1977a). In light-deprived animals and in hamsters during winter months, administration of melatonin *elevates pituitary* prolactin levels (REITER et al., 1975c, 1976c, 1978).

That among the various pineal compounds the indoleamines are not the only substances which influence prolactin secretion is shown by the fact that arginine vasotocin injected into the mouse third ventricle on the day of pinealectomy (a) prevents the increase in pituitary prolactin which occurs 7 days after pinealectomy (PAVEL et al., 1975c) and (b) increases plasma prolactin levels, even in the absence of the pineal (M. VAUGHAN et al., 1978, 1976d). However, in pro-oestrous rats arginine vasotocin drastically suppresses the prolactin surge occurring in the afternoon (CHEESMAN et al., 1977a). Involvement of substances other than melatonin is supported by the fact that in *pinealectomized* rats melatonin does not alter pituitary prolactin levels (SACKMAN et al., 1977) and that in vitro rat hemipituitaries incubated with bovine pineal extracts release up to 285% more prolactin than the corresponding control halves, suggesting that a PRF may be present in the pineal (BLASK and VAUGHAN, 1976; BLASK et al., 1976). Melatonin is without effect on prolactin release in vitro (BLASK and VAUGHAN, 1976), but pineal *extracts* have been shown to reduce both the basal and TRH stimulated prolactin release (DEMOULIN et al., 1977a).

Milk ejection is not significantly affected by pinealectomy or administration of melatonin in the rat (MIZUNO and SENSUI, 1970), but in rabbits melatonin slightly decreases the milk yield, leaving the composition of the milk unchanged (SHANI et al., 1971).

e) Adrenocorticotrophic Hormone (ACTH)

Although there is considerable evidence that the pineal influences the adrenal cortex (see p. 305), relatively little is known about the role of the pineal in regulating ACTH secretion. According to NUZZI et al. (1962), KITAY (1963) and JOUAN and SAMPEREZ (1965a) the pituitaries of pinealectomized rats show hypersecretion of ACTH. Melatonin has no effect on pituitary ACTH (JOUAN and SAMPEREZ, 1965b; BARCHAS et al., 1969), but pineal extracts increase pituitary ACTH (NUZZI et al., 1962). Administration of ACTH, in contrast to FSH, LH and prolactin, does not affect pineal HIOMT activity (CARDINALI et al., 1976a).

f) Thyroid-Stimulating Hormone (TSH)

Relatively few studies have dealt with the influence of the pineal on TSH secretion. ROWE et al. (1970) found no effect on TSH levels in pituitary or plasma at 27–28 days after pinealectomy, but RELKIN (1972c, 1978) noted that in rats pinealectomized at 21 days of age there was a short-lived decrease in pituitary TSH accompanied by elevated plasma TSH and protein bound iodine levels. In young rats melatonin administration was followed by a decrease of pituitary TSH, while plasma TSH increased (PANDA and TURNER, 1968). However, RELKIN (1978) found that melatonin decreases plasma TSH.

g) Follicle-Stimulating Hormone (FSH)

Regulatory influences of the epiphysis cerebri on FSH secretion and gonads have been thoroughly reviewed by REITER (1973a, b; 1974b; 1976b, 1977), REITER and VAUGHAN (1977) and REITER et al. (1975e). Hence it suffices to give a brief survey.

Pinealectomy has been relatively little used to study the influence of the pineal on FSH secretion, and the results obtained are very variable. No effects have been noted by TALBOT and REITER (1973/74), SLAMA-SCEMAMA (1976) and BLAKE (1976). A threefold *increase* in pituitary FSH levels after pinealectomy has been described by FRASCHINI (1969) and FRASCHINI and MARTINI (1970). That pituitaries of pinealectomized rats contain more uterine weight-stimulating activity than those of intact controls points in the same direction (BARBAROSSA et al., 1959). In adult rats, pinealectomy does not affect pituitary FSH levels, but prevents the late nocturnal decrease in *serum* FSH levels (RØNNEKLEIV and MCCANN, 1975b).

Melatonin given to normal male subjects and postmenopausal women has no demonstrable effect on FSH secretion (FIDELEFF et al., 1976). In animal species pituitary FSH levels are reduced by administration of melatonin (DEBELJUK et al., 1970a; KAMBERI et al., 1971), serotonin (KAMBERI et al., 1970, 1971) or 5-methoxytryptophol (TALBOT and REITER, 1973/74). *Blood* FSH levels are also reduced by melatonin (SORRENTINO, 1968), provided that melatonin treatment is carried out in the afternoon and over several weeks (TAMARKIN et al., 1976). Decreased serum FSH levels have also been noted after administration of melatonin-free pineal extract (DAMIAN et al., 1978). Under experimental conditions such as castration (THIÉBLOT et al., 1966f; TALBOT and REITER, 1973/74) or continuous illumination (DEBELJUK, 1969) melatonin is not capable of lowering FSH levels. Arginine vasotocin is apparently without effect on FSH, both in vivo (CHEESMAN et al., 1977a) and in vitro (DEMOULIN et al., 1977b). However, MOSZKOWSKA and EBELS (1968) demonstrated a depressive effect.

In vitro, FSH release is blocked when pituitary tissue is incubated in the presence of either pineal fragments or extracts (EBELS et al., 1965a; MOSZKOWSKA, 1965a, 1966). An anti-FSH effect of pineal extracts in humans has been described by BRET and LEGROS (1959).

h) Luteinizing Hormone (LH)

Studies dealing with the effects of the pineal on LH secretion have yielded very variable results, especially those involving pinealectomy.

In intact adult *male* rats *pinealectomy* increases *pituitary* LH levels (CLEMENTI et al., 1966, 1969; FRASCHINI and MARTINI, 1970) and prevents the nocturnal decrease of *serum* LH (RØNNEKLEIV and MCCANN, 1975b), but in immature male rats this operation has virtually no effect on both pituitary and serum LH levels (HUS-CITHAREL et al., 1977). In rams, blood samples collected at 20 min intervals for 26 h showed that pinealectomy increased LH levels (BARRELL and LAPWOOD, 1978). In adult female rats (J. HOFFMANN and CULLIN, 1975) and ewes (ROCHE et al., 1970b) pinealectomy is without effect on LH. Studies in relation to the oestrous cycle show no effect of pinealectomy on LH levels on the afternoon of pro-oestrus (BLAKE, 1976), but point to an elevation of serum LH levels at dioestrus (M.L. SMITH et al., 1973). However, in rats kept under LD 8:16 serum LH levels at pro-oestrus are reduced by pinealectomy (JACOBSON and MANN, 1978). Pinealectomy in 9-day-old rats abolishes the plasma LH peak occurring at 14.00 h in springtime, but not in other seasons (SLAMA-

SCEMAMA and L'HÉRITIER, 1978). In prepuberal male rats pinealectomy alone or in combination with melatonin administration increases pituitary and serum LH levels (ALONSO et al., 1978). At *puberty* pinealectomized female rats show decreased pituitary LH levels when compared to intact controls (SLAMA-SCEMAMA, 1976). Pinealectomy cannot further elevate the high LH levels following castration (CLEMENTI et al., 1966, 1969; TALBOT and REITER, 1973/74; RØNNEKLEIV and McCANN, 1975b), but can elevate LH levels previously lowered by frontal hypothalamic de-afferentation (TRENTINI et al., 1976), light deprivation alone (REITER and JOHNSON, 1974a, b, c; 1975b) or light deprivation combined with anosmia (BLASK and REITER, 1975b). In young male rats, pinealectomy increases plasma LH levels at 16 and 18 days of age (HUS-CITHAREL et al., 1977). In blinded hamsters, pinealectomy depresses the elevated pituitary LH levels and increases the unchanged serum LH levels (REITER et al., 1975c).

Administration of *pineal substances* shows fairly clearly that the pineal may suppress LH secretion in animal species. In normal male subjects and postmenopausal women melatonin has no effect on LH levels (FIDELEFF et al., 1976). Following melatonin administration in rats, serum LH levels decrease (KAMBERI et al., 1970; MOGUILEVSKY et al., 1976), while pituitary LH levels increase (ADAMS et al., 1965). It should be noted that interspecies differences appear to exist. In intact hamsters melatonin administration *depresses* pituitary LH concentration (REITER et al., 1977a), while it increases pituitary LH levels lowered by darkness (REITER et al., 1975b, 1976c). No effect of melatonin could be demonstrated by DEBELJUK et al. (1970b). In addition to melatonin (DOMANSKI et al., 1975), arginine vasotocin is a potent suppressor of the pre-ovulatory surge of LH (CHEESMAN et al., 1977b; OSLAND et al., 1977). In vitro arginine vasotocin is without effect on LH levels (DEMOULIN et al., 1977b). While melatonin appears to act at the hypothalamic level (melatonin does not abolish the effect of LHRH but suppresses the effect of clomiphene) (MOGUILEVSKY et al., 1976), it has been suggested that arginine vasotocin acts at a higher neuronal level than where the prostaglandin E_2-induced release of LHRH takes place (OSLAND et al., 1977). A reduction of LH in the culture medium when bovine pineal substances are added to a culture system consisting of hypothalamic and adenohypophyseal tissue indicates an effect at the hypothalamic level (HAYES et al., 1973). Melatonin implanted into the pituitary is without effect on LH levels (FRASCHINI et al., 1968). Different fractions of a partially purified bovine pineal extract also reduce serum LH in rats (ORTS, 1977), as do melatonin-free pineal extracts (DAMIAN et al., 1978).

An inhibitory effect of pineal substances on LH release has also been demonstrated under certain *experimental conditions*. In castrated animals pineal indoleamines (FRASCHINI et al., 1968, 1971; ROCHE et al., 1970a; DEBELJUK et al., 1970b) as well as melatonin-free pineal extracts (ORTS and BENSON, 1973; ORTS et al., 1974a, b) reduce the post-castration rise of serum and pituitary LH. Melatonin also prevents the rise in plasma LH which occurs after administration of pregnant mare serum (REITER and SORRENTINO, 1971b). In prepuberal rats exposed to continuous illumination no change in LH levels is observed following melatonin administration (DEBELJUK, 1969). Concerning the role of the pineal gland in the regulation of LH release in rats with different types of the anovulatory syndrome, see RUZSAS et al. (1977).

i) Melanocyte-Stimulating Hormone (MSH)

Since in lower vertebrates melatonin is an antagonist of pituitary MSH it was only logical to examine whether or not in mammals functional interrelationships exist between the pineal and the pituitary pars intermedia. The results obtained suggest that such relationships may exist. In pinealectomized rats the MSH content of the pituitary is considerably higher than in controls (KASTIN et al., 1967a, b, 1969; PAVEL et al., 1975a), but plasma MSH levels are not affected (KASTIN et al., 1969). Exposure to continuous light increases while exposure to continuous darkness decreases pituitary MSH, the effect of darkness being negated by pinealectomy (KASTIN et al., 1967c).

The fact that radioactively labelled MSH-inhibiting factor (REDDING et al., 1972) and MSH accumulate in the pineal gland has been considered as evidence that each component of the pineal – hypothalamus-pituitary pars intermedia axis may affect every other component by a complex system of interactions (KASTIN et al., 1972, 1975). However, this concept has been challenged by TILDERS and SMELIK (1975), who found that the diurnal rhythm of pituitary MSH levels persists after either superior cervical ganglionectomy or pinealectomy and that melatonin, in contrast to previous studies (KASTIN and SCHALLY, 1967), is without effect on pituitary MSH. PEASLEE et al. (1969) noted that melatonin increases pituitary MSH levels in 70-day-old female rats, but not in weanling rats. PAVEL et al. (1975a) attribute the effect of the pineal on MSH to arginine vasotocin, which is thought to be at least a hundred million times more active than melatonin in decreasing pituitary MSH levels, by inhibiting the synthesis and/or release of MSH-release inhibiting factor.

Finally it should be mentioned that the pineal contains a peptidase which cleaves MIF (MARKS et al., 1974; GRIFFITHS, 1976) and that melatonin enhances the serotonin levels of the pars intermedia of the rat pituitary (PIEZZI and WURTMAN, 1970).

11. Pineal and Reproductive Organs

a) General Considerations

Although it had been suspected for a long time that functional interrelationships exist between the pineal and the reproductive organs, the actual proof for this assumption was not easily obtained. Under certain experimental conditions an antigonadotropic effect of the pineal could be clearly established in some studies, but most experiments carried out have failed to provide clear-cut answers. At the time of writing this review we are perhaps witnessing why most of the previous studies have not yielded the expected results.

Recent research shows clearly that pineal–gonadal relationships are very complex and that it is absolutely essential to use suitable animals for experimentation and to keep them under a lighting schedule which is similar to that occurring under natural conditions. REITER (1973a) emphasizes that exposure of nocturnal rodents to light–dark schedules routinely used in laboratories where the duration of artificial light is 12–16 h per day may severely bias or completely negate any pineal effect. According to REITER, pinealectomizing rodents that experience "long" days may be analogous to ovariectomizing animals that have been previously hypophysectomized.

A suitable laboratory animal for demonstrating the antigonadal effect of the pineal under physiological conditions is the hamster. In this species, in which the gonads regress in both males and females with decreasing day-length in the autumn, the regression takes place only if the pineal is functioning properly. Gonadal atrophy does not occur if the hamsters have previously been pinealectomized (CZYBA et al., 1964; HOFFMAN and REITER, 1965b; REITER, 1975c) or when the pineal has been deprived of its neuronal input by either superior cervical ganglionectomy or decentralization (REITER, 1967a, 1968a; REITER and HESTER, 1965), enabling the hamsters to reproduce even during the winter (REITER, 1973/74).

The complexity of the pineal–gonadal relationships can be illustrated in several respects. A most interesting recent finding is that melatonin, which for a long time was thought to be an important *anti*gonadotropic substance, is actually *pro*gonadotropic under certain conditions. This was detected after discovering a novel mode of administering melatonin over a wide range of concentrations at relatively constant rates for protracted periods, namely the subcutaneous implantation of polydimethylsiloxane capsules of varying lengths, filled with crystalline melatonin. With this method it was shown that in hamsters kept under 14 h light and 10 h dark, implantation of melatonin-filled capsules measuring 50–200 mm in length and releasing between 26 ± 1.1 and 101 ± 1.6 µg melatonin per day leads to the expected striking decrease of weight of seminal vesicles and testes, accompanied by a marked reduction in the number of spermatocytes and spermatids. However, in hamsters kept under 6 h of light and 18 h of darkness, implantation of capsules measuring between 25 and 100 mm in length and releasing between 14 and 50 µg melatonin/day results in an increase in size of these organs (TUREK et al., 1975). Progonadotropic effects of melatonin have also been demonstrated by K. HOFFMANN (1972, 1974), REITER et al. (1975c), REITER and BANKS (1976) and others (see below).

When designing experiments involving melatonin administration it should be borne in mind that melatonin (a) may exert an inhibitory effects on pineal antigonadotropic peptidergic substances (ORTS et al., 1975), (b) influences pineal melatonin formation (p. 265) and (c) is effective only when given during a suitable phase of the 24-h cycle. TAMARKIN et al. (1975, 1976) found that in male hamsters melatonin results in testicular regression only when given in the afternoon; in females it takes 7–8 weeks to induce acyclicity when melatonin is given at 13.30 h and 2.5 weeks when given at 19.45 h, intermediate effects being present when melatonin is administered between 16.00 and 18.00 h. Similar results have been obtained by REITER's group (REITER et al., 1976b, 1977b; SACKMAN et al., 1977), who additionally showed that melatonin is effective only when the pineal is intact. The importance of correct timing of injections when studying pineal – gonadal relationships is also illustrated by the findings that (a) the pineal 8 S protein receptors for sex steroids are likely to show a circadian rhythm (CARDINALI et al., 1975a), (b) uptake of testosterone by pineal tissue shows a maximum at 6.00 h and a minimum at 14.00 h (NAGLE et al., 1974b) and (c) testosterone propionate given at 6.00 h yields a 150% increase in ^{3}H-leucine incorporation into pineal proteins, as measured 24 h later, whereas given at 14.00 h it produces only a 54% increase (NAGLE et al., 1975).

In females, the stage of the oestrous cycle has to be taken into account, as shown by the observation that inhibition of compensatory ovarian hypertrophy is most effective when melatonin is given on the day of dioestrus and least effective on the day of pro-oestrus (M. VAUGHAN et al., 1971b).

In addition to the time of day and the stage of the oestrous cycle, the mode of administering pineal substances matters. Melatonin is somewhat more effective in inhibiting ovulation in immature pregnant mare serum (PMS)-treated rats when administered intra-ocularly than subcutaneously, and more effective subcutaneously than intravenously (POMERANTZ and SORRENTINO, 1973a). A most perplexing finding is that melatonin restricts its own action when administered simultaneously by two different modes: in the male hamster subcutaneous melatonin implants inhibit reproductive atrophy induced by daily melatonin injections (REITER et al., 1977b). Also relevant are the observations (a) that bovine pineal extract is capable of blocking ovulation induced in mice by HCG if the extract and HCG are administered simultaneously by intraperitoneal injection and (b) that it is ineffective when given at different sites, since the extract deactivates HCG due to the presence of an enzyme of the neuraminidase type (CHEESMAN and FORSHAM, 1974a, b).

Finally, variations in the effects of melatonin injections can be expected when different commercial preparations are used. According to TIGCHELAAR and NALBANDOV (1971) melatonin obtained from Regis and Sigma decreases ovarian and uterine weights, while melatonin obtained from Calbiochem and Nutritional Biochemicals does not alter these parameters in immature rats.

Furthermore, it should be taken into consideration that gonadal function may be influenced by pineal substances other than melatonin, namely arginine vasotocin (p. 240), gonadotrophin-releasing hormones (p. 241) and a number of structurally unidentified presumptive polypeptides and proteins (p. 242) with diverse functions. According to EBELS et al. (1975, see also EBELS, 1975, 1976) one fraction inhibits the release of gonadotrophin-releasing hormone (GRH) in the hypothalamus. A second fraction appears to decrease GRH activity without inhibiting its release. A third fraction stimulates both the activity and the release of GRH, while a fourth fraction acts on the pituitary itself and represents a GRH. It is therefore not surprising that if crude pineal extracts are used very variable results may occur, depending on the ratios of the various fractions present.

Finally, the pineal may affect reproductive function by altering the rate at which sex steroids are metabolized. Δ^4-Reductase, which is important for the elimination of testosterone and is present in liver and hypothalamus, is stimulated by melatonin in the rat and depressed in the hamster (FREHN et al., 1974).

b) The Pineal as a Target Organ of Sex Hormones

As the epiphysis cerebri possesses receptors for sex steroids (CARDINALI et al., 1975a), takes up sex steroids (CARDINALI et al., 1974b; NAGLE et al.,

1974b) and converts androgens into α-reduced metabolites (CARDINALI et al., 1974b) and oestrogens (CARDINALI et al., 1974d), it is appropriate to regard the epiphysis cerebri inter alia as a target organ of sex hormones.

α) Effects of Gonadectomy on the Pineal

Despite some conflicting results it would appear that gonadectomy has a stimulatory effect on the epiphysis cerebri, resulting in an initial or permanent increase of pineal weight (DES GOUTTES, 1964; WURTMAN et al., 1965a; DAS GUPTA, 1968a; no effect: ALEXANDER et al., 1970a; HOUSSAY and BARCELO, 1972b; decrease: DES GOUTTES, 1964, between 6 and 25 days of age; DAVID et al., 1975) and an increase in pineal volume (ITO and MATSUSHIMA, 1968c). Histologically, the pineals of gonadectomized mice show a slight reduction in nuclear density in the periphery of the organ, but not in the centre (ITO and MATSUSHIMA, 1968c), and a loose arrangement together with a foamy appearance of pinealocytes in the periphery of the rat pineal, while in the centre the cells remain densely packed and granular (GUSEK and BUSS, 1966a, b). Histochemically, castration is followed by a slight to moderate increase of the activity of oxidative enzymes, acid phosphatase and non-specific esterase (DAS GUPTA, 1968a; BOSTELMANN, 1969b) and an increase of pineal cells containing a yellow autofluorescent material; serotonin levels are not affected (PEVET et al., 1975a; KARASEK et al., 1976b).

Electron microscopically it has been demonstrated that the pinealocyte nuclei increase in size and exhibit deeper invaginations and more and larger nucleoli than in control rats (GUSEK and BUSS, 1966a; BOSTELMANN, 1969c; KOWALZIG and GUSEK, 1971). The cytoplasm is characterized by a strong increase in the amount of both types of endoplasmic reticulum, an increase in both the number of ribosomes and mitochondria, the latter showing polymorphism and increase in size, an enlarged Golgi apparatus and an increase in the amount of lysosomes (CLEMENTI et al., 1965a, b; GUSEK and BUSS, 1966a; BOSTELMANN, 1969c; KOWALZIG and GUSEK, 1971; KARASEK et al., 1976a, b). Synaptic ribbons, microtubules, microtubular sheaves (KARASEK, 1976; KARASEK et al., 1976a), dense-core vesicles and clear vesicles (CLEMENTI et al., 1965a) increase as well. Lipid droplets show varying behaviours depending on whether gonadectomy is carried out in immature or in mature rats. In immature rats an increase of lipid droplets occurs (ZWEENS, 1963, 1964, 1965; DAS GUPTA, 1968a; BOSTELMANN, 1969c). In mature rats they have been noted to increase (KARASEK et al., 1976a) or to decrease (S. MILCU et al., 1962; BOSTELMANN, 1969c). Moreover, dark pinealocytes increase in number, at the expense of light pinealocytes (SATODATE et al., 1970a, b). The changes described above are more pronounced when gonadectomy is combined with administration of chorion gonadotrophin (GUSEK and BUSS, 1966a) or LHRH (KARASEK et al., 1976a).

In contrast to the morphological findings, which point to an enhanced activity, melatonin formation, as assessed by HIOMT measurements, decreases in mature rats (NAGLE et al., 1974a; CARDINALI et al., 1974c; WALLEN and YOCHIM, 1974b), while it increases in immature rats (ALEXANDER et al., 1970a). Pineal ^{32}P uptake (BREWER and QUAY, 1958) and basic levels of adenyl cyclase activity (WEISS and CRAYTON, 1970a, b) are not affected by gonadectomy. However, the stimulatory effect of noradrenaline on adenyl cyclase activity increases (WEISS and CRAYTON, 1970a, b), while pineal cAMP concentrations decrease (KARASEK et al., 1978a). Recently, melatonin levels, as assessed by radio-immunoassay, have been shown to increase in serum and to remain unaltered in the pineal after oophorectomy (OZAKI et al., 1978).

β) Effects of Gonadotrophins on the Pineal

Administration of chorion gonadotrophin increases pineal weight (TRENTINI et al., 1973), the pinealocytes showing signs of hypertrophy (TRENTINI et al., 1973; BUSS, 1965; GUSEK and BUSS, 1966a; KARASEK and MAREK, 1978) and enhanced colloid production (THIÉBLOT et al., 1949b). In the rat pineal, the lipid content (ZWEENS, 1964) and leucine aminopeptidase as well as MAO activities increase (TRENTINI et al., 1969), while in the guinea-pig pineal histochemically demonstrable enzymes show no changes (VOLLRATH and SCHMIDT, 1969). Since in orchidectomized rats, FSH, LH and prolactin all increase pineal HIOMT activity, it has been suggested that gonadotrophins may have a direct effect on the pineal (CARDINALI et al., 1976a). In *hypophysectomized* rats neither pineal weight nor HIOMT activity are affected by HCG (URRY et al., 1972).

In vitro, addition of HCG or pregnant mare serum gonadotrophin (PMSG) leads to a marked increase in rough endoplasmic reticulum, Golgi cisternae, mitochondria, multivesicular bodies, lipid droplets, microtubules, microtubular sheaves, and synaptic ribbons of pinealocytes (KARASEK et al., 1978b).

γ) Effects of Oestrogens and Progesterone on the Pineal

Oestrogens accumulate readily in the pineal (NAGLE et al., 1972b, 1973b; MARKS et al., 1972; ANAND KUMAR, 1973; STUMPF and SAR, 1975a; DAVID et al., 1975), apparently forming a nuclear receptor oestrogen complex, which is depressed by denervation of the pineal (CARDINALI, 1977). The uptake shows no circadian changes (NAGLE et al., 1974b), but it is depressed by superior cervical ganglionectomy (MARKS et al., 1972; NAGLE et al., 1973b). On the whole it would appear that oestrogens exert an inhibitory influence on pineal activity, though it should be noted that the effects may differ, depending on the route of administration and whether intact or castrated animals are used. Oestradiol-17-β may be methylated in the pineal by means of HIOMT (WEISZ et al., 1978).

In *intact* animals, systemic administration of oestrogens depresses pineal weight (WURTMAN et al., 1965a) and pinealocyte cell organelles (CLEMENTI et al., 1965a, b), whereas it enhances pineal protein, RNA and DNA levels (NIR et al., 1970a), ^{3}H-leucine incorporation (CARDINALI et al., 1976b), MAO activity (URRY et al., 1976) and noradrenaline turnover (CARDINALI et al., 1975c). Pineal HIOMT activity is depressed by oestrogens (WURTMAN et al., 1965a), but oestradiol given to immature female rats has no effect on noradrenaline-induced NAT activity (WILKINSON and ARENDT, 1978). 17-β-Oestradiol does not affect pineal melatonin levels, as assessed by RIA (OZAKI et al., 1978).

In vitro studies show that base line levels of adenyl cyclase activity are apparently not affected by oestrogens, while they inhibit the noradrenaline-induced activation of the enzyme by over 80% (WEISS and CRAYTON, 1970a, b). Electrophysiologically it can be seen that oestradiol inhibits the noradrenaline-induced hyperpolarization of the pinealocyte cell membrane (SAKAI and MARKS, 1972).

Administration of oestrogens by means of stereotactic implantation into the habenular region, like systemic administration, is followed by morphological signs of depressed pinealocyte activity, while implantation into the median eminence has a stimulatory effect (CLEMENTI et al., 1965a).

In *ovariectomized* rats, pineal weight is depressed by oestrogens (HOUSSAY and BARCELO, 1972a). Pineal protein content (CARDINALI et al., 1974c) and catechol *O*-methyltransferase activity (SCARDAPANE and CARDINALI, 1977) are unaffected, while ^{3}H-leucine incorporation into pineal protein is enhanced (CARDINALI et al., 1974c). Pineal HIOMT activity shows opposing results: an increase has been noted by HOUSSAY and BARCELO (1972a), CARDINALI et al., (1974c) and WALLEN and YOCHIM (1974b), while NAGLE et al. (1972b) report on a decrease. NAT activity shows no change (ILLNEROVÁ, 1975b). Serum melatonin levels elevated by oophorectomy are partially suppressed by 17-β-oestradiol (OZAKI et al., 1978). In vitro, oestradiol has been shown to induce, in the pineal, a specific protein of unknown functional significance (MIZOBE and KUROKAWA, 1978).

In contrast to oestrogens, *progesterone* appears to have little effect on pineal function, despite the fact that the hormone readily accumulates in the pineal (ANAND KUMAR, 1973; LUTTGE and WALLIS, 1973; CARDINALI et al., 1975d; DAVID et al., 1975). Histochemically demonstrable enzymes (VOLLRATH and SCHMIDT, 1969), adenyl cyclase activity (WEISS and CRAYTON, 1970a, b), HIOMT activity in pregnant rats (KANTARJIAN, 1976) and ^{32}P uptake of the pineal (BREWER and QUAY, 1958) are unaffected by progesterone.

In *ovariectomized* rats, progesterone does not affect pineal weight, but depresses pineal HIOMT activity (HOUSSAY and BARCELO, 1972a) and serum melatonin levels (OZAKI et al., 1978). In vitro progesterone inhibits the release of melatonin from pineals of adult ovariectomized rats (WILKINSON and ARENDT, 1978).

δ) *Effect of Androgens and Anti-androgens on the Pineal*

Like oestrogens and progesterone, testosterone is avidly taken up by the pineal (ANAND KUMAR, 1973; ANAND KUMAR et al., 1973; NAGLE et al., 1974a; CARDINALI et al., 1974b, 1975d), the uptake being maximal at 6.00 h and minimal at 14.00 h (NAGLE et al., 1974b). Testosterone is then bound to cytoplasmic proteins and converted into 5α-reduced metabolites (CARDINALI et al., 1974b).

Relevant *morphological* studies are rare. THIÉBLOT et al. (1947b) report that pinealocyte nuclei become hyperchromatic after testosterone administration. Lipid droplets increase in number (S. MILCU et al., 1962; S. MILCOU and PETREA, 1964). In female rats in testosterone-induced constant oestrus the pinealocytes exhibit a higher incidence of large mitochondria, lipid inclusions and a hypertrophy of the Golgi apparatus when compared to controls (CLEMENTI et al., 1965b).

Testosterone enhances pineal ^{3}H-leucine incorporation in both intact (CARDINALI et al., 1976b) and castrated rats (NAGLE et al., 1975), the increase 24 h later being more pronounced (150%) when the hormone is injected at 6.00 h than when injected at 14.00 h (54%) (NAGLE et al., 1975). Testosterone accelerates pineal noradrenaline turnover (CARDINALI et al., 1975c) and decreases MAO activity significantly in dosages of 0.3 and 1.0 mg, but not at 3.5, 15 and 30 mg (URRY et al., 1976). That the effects of testosterone are dose-dependent applies also to melatonin formation. In doses ranging from 0.1 to 1 mg/day, testosterone increases pineal HIOMT activity, whereas a 5-mg dose decreases it significantly (NAGLE et al., 1974a). HOUSSAY and BARCELO (1972b) describe a decrease of pineal HIOMT activity in testosterone-treated castrated rats when compared to intact male rats. Adenyl cyclase activity (WEISS and CRAYTON, 1970b), the serotonin rhythm (FANTINI et al., 1966) and the pineal catechol *O*-methyltransferase activity of castrated rats (SCARDAPANE and CARDINALI, 1977) are not affected by testosterone. A possible feedback of dehydroepiandrosterone upon the pineal in mice has been suggested by FARMER (1974).

Administration of anti-androgens (cyproterone acetate) results in morphological changes indicative of increased pineal function (GUSEK, 1971, 1976; KOWALZIG and GUSEK, 1971).

c) Effects of Pinealectomy and Pineal Sympathectomy on Reproductive Organs

As already concluded by KITAY and ALTSCHULE (1954a) and THIÉBLOT and LEBARS (1955), pinealectomy has a progonadotropic effect which can be most clearly demonstrated in maturing animals and in animals subjected to suitable experimentation. In intact adult animals the effects of pinealectomy are relatively slight.

α) *Prepuberal Animals*

Pinealectomized animals mature slightly earlier than intact ones. In *female* rats vaginal opening has been a frequently studied parameter. Premature vaginal opening following pinealectomy has been reported a number of times (SIMONNET and THIÉBLOT, 1951; SIMONNET et al., 1952b, 1954; KINCL and BENAGIANO, 1967; HENZL et al., 1970), though sometimes no effect could be demonstrated (WRAGG, 1965, 1967). Effects on the ovaries are variable. As regards ovarian *weight* after pinealectomy, some authors have found an increase (SIMONNET et al., 1951a; KITAY, 1954a; WURTMAN et al., 1959, 1961; THIÉBLOT and BLAISE, 1963; M. VAUGHAN et al., 1973), whereas others were unable to detect any difference (WRAGG, 1965, 1967; AIRAKSINEN and KARPPANEN, 1970; HENZL et al., 1970; PITIS et al., 1970; DEFRONZO and ROTH, 1971). Increase in the number of corpora lutea has been reported by SIMONNET et al. (1952b). Ovarian lactic acid dehydrogenase activity (HENZL et al., 1970), RNA content, RNA/DNA ratio, histology (DEFRONZO and ROTH, 1971) and phosphate uptake (THIÉBLOT et al., 1963) are apparently not affected by pinealectomy, while uterine weights increase (AIRAKSINEN and KARPPANEN, 1970). The development of the mammary gland is not accelerated by pinealectomy (MISHKINSKY et al., 1966).

Frequently pinealectomy has been combined with other experimental procedures, and most of these studies show that the pineal exerts an inhibitory effect on the gonads. Thus pinealectomy negates the delay of vaginal opening caused by light deprivation (REITER and ELLISON, 1970; OSMAN et al., 1972). In unilaterally ovariectomized immature rats pinealectomy was found to be without effect on the weight of the remaining ovary (PITIS and SCHERZER, 1969), but it inhibits ovarian and uterine atrophy due to blinding in androgen-sterilized rats (REITER et al., 1968a; REITER, 1967c, 1970d). Moreover, pinealectomy leads to the occurrence of compensatory ovarian hypertrophy in unilaterally ovariectomized, blinded, anosmic rats, in which the remaining ovary normally shows no increase in weight (REITER, 1972b). Rats stimulated by PMS ovulate to a greater extent and earlier when they have been pinealectomized (DUNAWAY, 1968, 1969; DUNAWAY and O'STEEN, 1967).

In prepubertal *male* animals a progonadotropic effect of pinealectomy is illustrated by earlier testicular descent (RELKIN, 1970), stimulated spermatogenesis (PITIS and SCHERZER, 1969; PITIS et al., 1970) as well as more frequent mounting and earlier pelvic thrusting (BAUM, 1968). Testicular weight is usually unaffected (rat: RENNELS and DILL, 1961; KINCL and BENAGIANO, 1967; BAUM, 1968; ROTH, 1964, 1965; KINSON and LIU, 1973a; mouse: M. VAUGHAN and REITER, 1972), though occasionally permanent (SIMONNET and THIÉBLOT, 1950, 1951; ASAGOE and HAMAMOTO, 1959) or transient increases (LOMBARD DES GOUTTES, 1967b) have been noted. Development of reproductive organs delayed by blinding is accelerated by pinealectomy (REITER, 1968b).

The weights of the accessory sex glands may be unaffected (KINCL and BENAGIANO, 1967; KŘEČEK and PALATÝ, 1967) or elevated (ROTH, 1964, 1965). One reason for this apparent discrepancy could be the time of weaning, as seminal vesicle weight and citric acid levels are unaffected in normally weaned rats and elevated in prematurely weaned rats (KŘEČEK and PALATÝ, 1967). Moreover, the androgen levels may be of importance since in young rats pinealectomy does not alter the response of the sex organs to darkness unless the animals have been pretreated by androgens (MOSZKOWSKA and SCEMAMA, 1968a, b). Pinealectomy is without effect on compensatory testicular hypertrophy following hemicastration (HUS-CITHAREL et al., 1977).

In the Djungarian hamster, pinealectomy partially abolishes the inhibitory effect of short photoperiods on the gonads, while in long photoperiods pinealectomy retards gonadal development slightly but significantly (BRACKMANN and HOFFMANN, 1977).

The problem of precocious puberty due to pineal tumours in humans has been reviewed by KITAY (1954b). Pineal tumours are frequently accompanied by premature sexual maturation, especially in males.

β) Adult Animals

Pinealectomy in adult animals supports the concept that the epiphysis cerebri has an inhibitory influence on the gonads.

Pinealectomy in *females* results in an increased incidence of oestrus (CHU et al., 1964; GITTES and CHU, 1965; M. SMITH et al., 1973), prolonged oestrus

(Albertazzi et al., 1966), a shorter interval between two oestrous phases (M. Smith et al., 1973) or permanent oestrus (Nir et al., 1969b). No effect on vaginal oestrous smears could be detected by Mizuno and Sensui (1970). Apparently cycle length is not affected by pinealectomy (Alleva et al., 1970; Roche et al., 1970b; Blake, 1976), though some authors describe significantly more 5-day and fewer 4-day cycles than in intact controls (J. Hoffmann and Pomerantz, 1971; J. Hoffmann and Cullin, 1975). Irregular cycles after superior cervical ganglionectomy have been described by Guerillot et al. (1973). Ovulation is little affected by pinealectomy (Alleva et al., 1970; Roche et al., 1970b), and pinealectomized rats easily become pregnant (Mitchell and Yochim, 1970; Tigchelaar and Nalbandov, 1975; Kantarjian, 1976).

The effects of pinealectomy on ovarian and uterine weights are not very pronounced. Ovarian weight was either not affected (Hoffman and Reiter, 1966; Bick et al., 1968, 1969; Quay, 1969b; Reiter et al., 1966; M. Smith et al., 1973; J. Hoffmann and Cullin, 1975) or temporarily increased, followed by a significant decrease to below control levels (Pazo et al., 1968). Increased luteinization of the ovaries has been observed by Girod et al. (1967) in the hamster. In the rat the corpora lutea present are apparently not active enough to induce the formation of deciduomas following traumatization of the uterus (Kehl et al., 1962). Uterine weight is either increased (Quay, 1969b; Airaksinen and Karppanen, 1970) or unchanged (Bick et al., 1969). Serum oestrogen levels increase after pinealectomy, but progesterone levels are unaffected (G. Vaughan et al., 1978c).

Frequently the effects of pinealectomy have been investigated in animals previously subjected to experimentation expected to be better suited to demonstrate pineal-gonadal relationships. In rats bearing an anovulatory syndrome due to frontal de-afferentation, pinealectomy or superior cervical ganglionectomy are followed by a resumption of ovulation (Tima et al., 1973; Mess et al., 1973). Pinealectomy or superior cervical ganglionectomy lower ovarian weight enhanced by exposure of animals to continuous illumination (Bick et al., 1969), stimulate compensatory ovarian hypertrophy (Dickson et al., 1971), prevent the acceleration of the oestrus by artificial light in ferrets (Herbert, 1967, 1968, 1969) and abolish the decrease in size of the reproductive organs which occurs after light deprivation alone or in combination with anosmia (Reiter, 1967b, 1968c, 1969a, b, c, 1970b; Reiter et al., 1971). It should be noted that in hamsters, in which the ovaries become larger when they show functional regression (Reiter, 1969a), pinealectomy lowers ovarian weight which had increased due to blinding (Reiter and Johnson, 1974a, b). While no effect of sympathectomy on the incidence of cornified vaginal smears after exposure to continuous illumination could be demonstrated by Moore and Rapport (1971) and Brown-Grant and Östberg (1974), Guerillot and DaLage (1968) noted a decrease in the incidence. If eyeless testosterone-treated rats have either their pineal gland or the superior cervical ganglia removed, the incidence of oestrous smears is equivalent to that in testosterone-treated rats with eyes (Reiter, 1969d).

In adult *males*, except hamsters (see below), testes and accessory sex organs usually increase in size after pinealectomy, though differential effects are sometimes present between the two groups of organs (CARNICELLI et al., 1963; ROTH, 1964; THIÉBLOT and BLAISE, 1965; HOUSSAY et al., 1966a, b; HOUSSAY and PAZO, 1966; MOTTA et al., 1967; PAZO et al., 1968; AIRAKSINEN and KARPPANEN, 1970; KINSON and LIU, 1973a; NAGLE et al., 1974c; CHARLTON et al., 1976). Hyperplasia of interstitial cells has sometimes been noted (LOMBARD DES GOUTTES, 1967b; TIMMERMANS and DEVEČERSKI, 1972a), the hydroxylase activity of the interstitial cells being unchanged (TIMMERMANS and DEVEČERSKI, 1972c). According to LOSADA (1977) and LOSADA and SARRAT (1977) the initial activation of the gonads is followed by degenerative processes due to anterior pituitary hyperfunction. Testosterone levels are enhanced (KINSON and PEAT, 1971; NAGLE et al., 1974c), but depending on the time of day and the interval between pinealectomy and the determination of the testosterone levels the results may vary, ranging from a decrease (6.00 h; after 4 weeks), through no change (18.00 h; 4 and 12 weeks) to an increase (6.00 h; 8 and 12 weeks; 18.00 h, 8 weeks) when compared to controls (KINSON and LIU, 1973a).

In hamsters pinealectomy yields demonstrable effects only if the animals are not maintained in long days of, for example, 12–16 h of light per 24 h, which effectively "pinealectomizes" animals since photic stimulation inhibits pineal function (REITER, 1973a). That pinealectomy alone has no effect on testis and seminal vesicle weight in hamsters has been demonstrated by BUSH and SEIBEL (1977). From a physiological point of view it is most interesting that pinealectomy (CZYBA et al., 1964, 1965; HOFFMAN and REITER, 1965b; REITER, 1975c; SMIT-VIS, 1972), superior cervical ganglionectomy or decentralization (REITER, 1967a) prevent gonadal atrophy in winter and hibernation, enabling hamsters to reproduce even during the winter (REITER, 1973/74). The progonadotropic effect of pinealectomy or sympathectomy has been amply confirmed in light-deprived hamsters (HOFFMAN and REITER, 1965b; REITER, 1967a, 1968a, 1969a, 1972c, d; EICHLER and MOORE, 1970, 1971; SORRENTINO and REITER, 1971; SEIBEL and SCHWEISTHAL, 1973; REITER and JOHNSON, 1974c; REITER et al., 1974a; BUSH and SEIBEL, 1977). In light-deprived, unilaterally castrated hamsters pinealectomy results in compensatory testicular enlargement, showing that the activated pineal gland of the blind hamster is capable of suppressing compensatory testicular hypertrophy (HOFFMAN and REITER, 1965c; REITER, 1970c). In the Djungarian hamster, *Phodopus sungorus*, pinealectomy inhibits the acceleration of testicular development during long photoperiods (K. HOFFMANN and KÜDERLING, 1975).

Electron microscopic studies in golden hamsters have shown that pinealectomy elicits an enhanced activity of Leydig cells (LIN and WING, 1978).

In experimental animals other than hamsters pinealectomy or sympathectomy are likewise capable of activating depressed gonads and accessory organs (rat: REITER, 1968a, b, c; ALLIOT et al., 1975; vole, *Microtus agrestis*: CLARKE and FARRAR, 1975). In pinealectomized white-tailed deer mean serum androgen levels are lower than in intact controls (R. BROWN et al., 1978), whereas in pinealectomized rams there is a trend towards higher plasma testosterone levels (BARRELL and LAPWOOD, 1978).

d) Effects of Pineal Substances on Reproductive Organs

As pointed out above (p. 283), considerable progress has recently been made concerning the mode and timing of administering pineal substances. Moreover, it should be borne in mind that the substances used do not appear to act directly on the reproductive organs, but rather via hypothalamus (p. 268) and anterior pituitary (p. 274).

α) Females

In *maturing* females, vaginal opening, which is accelerated by pinealectomy (p. 288), is usually delayed by administration of serotonin (CORBIN and SCHOTTELIUS, 1961; ROBSON and BOTROS, 1961; O'STEEN, 1965; MOSZKOWSKA et al., 1971), melatonin (WURTMAN et al., 1963b; MOTTA et al., 1967; COLLU et al., 1971a), methoxytryptophol (COLLU et al., 1971a) or pineal extracts (MOSZKOWSKA, 1947, 1951b, 1965b; ORTS et al., 1974a) as well as by grafting extra pineals (MOSZKOWSKA, 1947, 1958a). While some authors noted no effect of pineal substances on the time of vaginal opening (MCISAAC et al., 1964; MOSZKOWSKA, 1965b), JÖCHLE (1956), using the pineal extract "Glanepin", found an acceleration by 7 days.

Very variable effects of pineal substances have been noted with respect to growth of the reproductive organs. Retrospectively this is probably not surprising when one considers that, as shown for melatonin (M. VAUGHAN, 1970) and arginine vasotocin (M. VAUGHAN et al., 1976a) in males, the substances accelerate gonadal growth when given immediately after birth and have a depressant effect when injected at later stages. However, differential effects may occur also when melatonin is given at later stages: melatonin enhances uterine weight and uterine and ovarian DNA when given between 25 and 35 days of age and depresses these parameters when administered between 45 and 55 days of age (SINGH and TURNER, 1971). Progonadal effects, not necessarily demonstrable simultaneously in ovaries and uteri, have also been demonstrated by LENTI et al. (1957), THIÉBLOT and BLAISE (1966), THIÉBLOT et al. (1966d) and ORTS et al. (1974a). However, in the majority of studies either a decrease in weight of the reproductive organs has been noted (KITAY and ALTSCHULE, 1954b; CORBIN and SCHOTTELIUS, 1961; ROBSON and BOTROS, 1961; WURTMAN et al., 1963b, 1964f; FRANCHIMONT, 1964; MCISAAC et al., 1964; MOSZKOWSKA, 1965b; ADAMS et al., 1965; MOTTA et al., 1967; SORRENTINO, 1968; and others) or there was no effect (TILSTRA and PROP, 1962; CHU et al., 1964; EBELS and PROP, 1965; COLLU et al., 1971a).

The effects of pineal substances have also been studied in *experimentally affected* immature rats. In PMS-treated animals ovulation is blocked or depressed by melatonin (LONGENECKER and GALLO, 1971; REITER and SORRENTINO, 1971b; POMERANTZ and SORRENTINO, 1973; POMERANTZ and REITER, 1974; SORRENTINO, 1975) or pineal extracts (OTA et al., 1975, 1977), though sometimes no effects are demonstrable (OTA and HSIEH, 1968; M. SMITH et al., 1972). Ovarian and uterine hypertrophy due to HCG or PMS can be inhibited by 5-methoxytryptophol, 6-hydroxymelatonin, melatonin, *N*-acetylserotonin, 5-

methoxyindole-3-acetic acid, arginine vasotocin (M. VAUGHAN et al., 1975, 1976b; CHEESMAN and FARISS, 1970) or melatonin-free bovine pineal powder (YAMASHITA et al., 1965; M. SMITH et al., 1972). Compensatory ovarian hypertrophy after unilateral castration at 6 weeks of age can be prevented by a single dose of melatonin given at birth, but not when it is given at 17 days of age (M. VAUGHAN and VAUGHAN, 1974).

In *adult* females *ovulation* is usually completely or partially inhibited by administration of either serotonin (ENDERSBY et al., 1970), melatonin (COLLU et al., 1971b; YING and FISKE, 1972; YING and GREEP, 1973; FARRELL et al., 1968; FISKE and MACDONALD, 1973; ANDRE and PARRISH, 1979) or bovine pineal extracts (BENSON et al., 1976b; ORTS et al., 1977), melatonin being without effect in ewes (ROCHE et al., 1970a). To demonstrate the inhibitory effect of pineal indoleamines on ovulation, the injection schedules are important. In rats, melatonin blocks ovulation when the substance is given between 13.30 h and 16.30 h on the day of pro-oestrus (YING and GREEP, 1973). On the day of pro-oestrus, serotonin is effective when given at 16.00 h, 17.00 h or 18.00 h, but not at 14.00 h or 19.00 h (ENDERSBY et al., 1970). In the rabbit the optimum time of injection of melatonin is about 2 h before ovulation (FARRELL et al., 1968). In the autumn, when spontaneous ovulation is low, 2 mg melatonin *increases* the ovulation rate, whereas 30 mg does not (FARRELL et al., 1968). In hamsters, melatonin injections given at 13.30 h require 7–8 weeks to induce acyclicity; however, when given at 19.45 h all the females become anovulatory after only 2.5 weeks, injections given at 16.00 h and 18.00 h having intermediate effects (TAMARKIN et al., 1976). Further studies dealing with melatonin injections at different times during the night reveal that 10 μg melatonin given at 20.00 h or 4.00 h stops oestrous cyclicity, while the same dosage given at 24.00 h and 1.00 h has no effect (TAMARKIN et al., 1977a). Administration of a number of synthetic melatonin analogues has revealed that ovulation-blocking activity greatly depends on the 5-methoxy group and that position 1 is not substituted (FLAUGH et al., 1979).

The oestrous cycles of animals treated with melatonin, 5-methoxytryptophol (WURTMAN et al., 1963b; CHU et al., 1964; MCISAAC et al., 1964), a pineal substance termed anestrina (BIANCHINI and OSIMA, 1960) or the bovine pineal extract Glanepin (JÖCHLE, 1956) are characterized by decreased incidences of oestrous phases, apparently accompanied by a prolongation of dioestrus, as shown by administration of melatonin, arginine vasotocin (M. VAUGHAN et al., 1976c) or serotonin (ALBERTAZZI et al., 1966). No effect on the oestrous cycle could be demonstrated by PROP and EBELS (1968) and MEYER et al. (1961), though the latter authors reported that in rats of post-reproductive age a protein-free bovine pineal extract arrests the persistent state of oestrus.

The effects of pineal substances on ovarian and uterine weights in intact animals are variable, ranging from a decrease (ROBSON and BOTROS, 1961; MEYER et al., 1961; M. VAUGHAN et al., 1976e), through no effect (EBELS and PROP, 1965; DEPROSPO and HURLEY, 1971a) to a slight increase (REISS et al., 1963c). Oestrogen and progesterone secretion patterns are not affected (*Macaca fascicularis:* FISKE et al., 1975). ^{32}P uptake of the ovaries decreases (REISS et al., 1963c), but NORRIS (1971) noted a significant increase of ovarian ^{32}P uptake after melatonin administration. Ovarian ascorbic acid content is not affected by administration of beef pineal extracts (SCHNEIDER et al., 1968).

Depressant effects of pineal substances on the ovaries can be readily demonstrated under *experimental conditions* in which the ovaries are stimulated. One such condition is after unilateral ovariectomy when the remaining ovary hypertrophies (compensatory ovarian hypertrophy, COH). This hypertrophy has frequently been inhibited by substances stemming from human postmortem pineals (MATTHEWS et al., 1971), pineal extracts from animal species or pineal indoleamines (MOSZKOWSKA, 1951a, 1963; BIANCHINI and OSIMA, 1962; REISS et al., 1963c; SORRENTINO, 1968; NORRIS, 1970; M. SMITH, 1971; BIGELOW et al., 1972; RUBIN and TRAUM, 1971; M. VAUGHAN et al., 1970, 1971a, b, 1972a; BENSON et al., 1971, 1972a, b, 1976b; MATTHEWS and BENSON, 1973; EBELS et al., 1973; BENSINGER et al., 1973; ORTS et al., 1974a; EBELS, 1975, 1976). Whereas in adult rats melatonin has to be given over relatively long periods (9 days) to be effective (RUBIN and TRAUM, 1971), in mice a single injection on the day of unilateral ovariectomy suffices to inhibit COH observed 10 days later (M. VAUGHAN et al., 1971a, b). Compensatory ovarian hypertrophy is also prevented by *N*-acetylserotonin, 5-hydroxytryptophol, 5-methoxytryptophol and vasopressin, but not by 5-hydroxytryptophan, serotonin, 6-hydroxymelatonin, 5-hydroxyindole acetic acid, 5-methoxyindole acetic acid and oxytocin (M. VAUGHAN et al., 1972a, b). Substantially more effective than melatonin in inhibiting COH are bovine pineal extracts (BENSON et al., 1972a) and arginine vasotocin, which has been claimed to be a million times more potent than the extracts (PAVEL et al., 1973/74). Pinealectomy shows that the epiphysis cerebri is not required for arginine vasotocin to inhibit COH (M. VAUGHAN, 1976). In view of the fact that arginine vasotocin is released into the third ventricle (PAVEL and COCULESCU, 1972), it is interesting that intraventricular injections of arginine vasotocin inhibit COH (PAVEL et al., 1973b), whereas intravenous or intraperitoneal injections do not.

In addition to COH, pineal substances inhibit stimulatory effects of gonadotrophins (SOFFER et al., 1965; HIPKIN, 1970b) and continuous illumination (JÖCHLE, 1956; WURTMAN et al., 1961; IFFT, 1962; TAKAHASHI et al., 1973; for negative results see EBELS and PROP, 1965). In pinealectomized animals, pineal substances inhibit ovarian hypertrophy (WURTMAN et al., 1959) and normalize the sexual cycle (ALBERTAZZI et al., 1966). Moreover, they counteract the ovulation-inducing effect of pinealectomy in rats which are anovulatory due to frontal de-afferentation (TIMA et al., 1973). However, there are also indications for progonadotropic effects of pineal substances, as shown in hypophysectomized rats (PETRONIO and TAVAZZA, 1958a; NORRIS, 1971), light-deprived animals (REITER et al., 1975c, d, 1977b) and rats treated with HCG, under which condition melatonin potentiates the effects of HCG on uterine weight (HIPKIN, 1970a).

Finally the question has to be discussed whether the pineal substances act directly on the reproductive organs. While the majority of results point to an indirect action via the hypothalamus (p. 268) and the anterior pituitary (p. 274), there are some indications from in vitro studies that pineal substances may have a direct effect. Melatonin, which is selectively taken up by the ovary (WURTMAN et al., 1964e), increased progesterone synthesis in vitro in a dose-related manner in a human corpus luteum of the menstrual cycle (MacPhee et al., 1975). In the cow, no direct effect on ovarian progesterone synthesis is demonstrable (Ewig and Wickersham, 1968). In rat ovarian homogenates, melatonin decreases oestrogen production (Balestreri et al., 1969).

β) Males

Recent studies have made clear that the effects of pineal substances, especially melatonin, on male gonads and accessory organs vary depending on the dosage used (Turek et al., 1975), the time of day (Tamarkin et al., 1976, 1977a) and the season at which the substances are given (Turek et al., 1976a) and whether photoperiodic or non-photoperiodic rodents are studied (Turek et al., 1976b). As in females, melatonin may have antigonadotropic or progonadotropic properties.

In *young* animals pineal substances have been found to be mainly antigonadotropic, though often differential effects existed between testes, seminal vesicles, ventral prostate and coagulating glands, which will not be mentioned in detail. Not infrequently one or two of these organs show retardation in growth following administration of pineal substances (Moszkowska, 1951b, 1965b; Moszkowska et al., 1971; Kappers, 1962a; Motta et al., 1967; Debeljuk, 1969; Debeljuk et al., 1970a; M. Vaughan et al., 1974, 1976e; Brackmann, 1977). Moreover, decreases have been observed with respect to the number of interstitial cells, cell height in the epididymis (Milin et al., 1959), testicular DNA synthesis, fructose content of seminal vesicles and prostate (Kinson and Robinson, 1970) as well as testosterone secretion (Kinson and Peat, 1971; Kinson and Peat, 1971; Kinson and Liu, 1973b; S. Milcu et al., 1975). In a substantial number of papers pineal substances have been described as being without effect (Tilstra and Prop, 1962; Ebels and Prop, 1965; Thiéblot and Blaise, 1967; Kunkel, 1969), while in some a stimulatory effect has been documented (Lenti et al., 1957; Nasr et al., 1961; Thiéblot and Blaise, 1966; Thiéblot et al., 1966d).

Despite some rare findings showing that in *intact adult* animals testicular weight is reduced by melatonin (short-tailed weasel, *Mustela erminea*: Rust and Meyer, 1969; Djungarian hamster, *Phodopus sungorus*: K. Hoffmann, 1973; K. Hoffmann and Küderling, 1977) there was a substantial amount of evidence against an antigonadotropic effect of melatonin (golden hamster: Reiter, 1974b; Reiter et al., 1975b, c) until it was found that interspecies differences exist. While in non-photoperiodic animals such as rats and mice melatonin has no effect on testicular size, melatonin has a clear inhibitory effect in photoperiodic rodents such as golden hamsters and grasshopper mice (Turek et al., 1976b), provided that melatonin is given at the right time of the 24 h cycle, namely late in the afternoon (Tamarkin et al., 1976; Reiter et al., 1976b, 1977b; Sackman et al., 1977) and during the night at 20.00 h or 4.00 h, but not between 24.00 h and 1.00 h (Tamarkin et al., 1977a). That the effect is due to melatonin itself, rather than to another antigonadotropic substance of the pineal released by the administered melatonin has been demonstrated by the observation that melatonin induces gonadal quiescence in pinealectomized hamsters (Tamarkin et al., 1977b).

A most interesting phenomenon is that melatonin may exert progonadal effects provided that gonadal function is depressed, as shown in Djungarian hamsters (K. HOFFMANN, 1972, 1974), golden hamsters (TUREK et al., 1975; REITER et al., 1975b, c, 1976c, 1977b; BITTMAN, 1978a; TUREK and LOSEE, 1978) and rats (REITER et al., 1975d; REITER and BANKS, 1976). A progonadotropic effect of melatonin has also been suggested by the observation that in hamsters bearing melatonin antibodies blinding leads to an even greater testicular atrophy than in sighted animals (SHERIDAN, 1976). The fact that hamsters exposed to short days undergo gonadal collapse followed by recrudescence and are insensitive to the regressive effects of such photoperiods has been shown experimentally to be due to insensitivity of pineal target tissues (BITTMAN, 1978b). According to TUREK and LOSEE (1978) melatonin is inhibitory when the photoperiod stimulates neuroendocrine-gonadal activity and is stimulatory when the photoperiod inhibits reproductive activity. However, studies involving immunization against melatonin cast some doubt on the role of melatonin in blinded rats (NILES et al., 1977b) and hamsters kept under LD 1:23 (G. BROWN et al., 1976).

It is still an open question whether melatonin acts directly or indirectly on the testes and accessory organs or whether both ways apply. A direct peripheral action of pineal substances is suggested by the facts that melatonin decreases testis weight in hypophysectomized animals (DEBELJUK et al., 1971) and that administration of suspensions of dried cattle pineal glands increases testicular and accessory organ weight in hypophysectomized but not in intact rats (REISS et al., 1963a). In vitro, pineal substances depress gonadal steroid production (BALESTRERI et al., 1969; ELLIS, 1969, 1972) and oxygen consumption (I. MILCU et al., 1960; DAMIAN et al., 1962). However, transplantation of the pineal into rat testes has no visible effect on the testes (COSNIER, 1960). Melatonin given subcutaneously diminishes the frequency of contractions but not the tonus of seminiferous tubules (rat: ELLIS and BUHRLEY, 1978).

Precisely what role the pineal plays when melatonin is administered in high doses is not known. REITER et al. (1976b) point out that the pineal and its sympathetic innervation have to be intact for melatonin to exert its inhibitory effect and that subcutaneous melatonin implants inhibit reproductive atrophy induced by daily melatonin injections (REITER et al., 1977a). However, K. HOFFMANN and KÜDERLING (1977) and TUREK (1977) describe testicular inhibition by melatonin in both intact and pinealectomized animals.

Arginine vasotocin has no effect on ventral prostate weight in adult mice (BENSON et al., 1976a). In patients suffering from prostate cancer administration of *pineal extract* led to sclerotic degeneration of seminiferous tubules and interstitial tissue (ALTIERI and SORRENTINO, 1956). Pineal extracts given to adult rabbits diminish testicular weight (NASR et al., 1961). Testicular function and histology after serotonin administration have been studied by BOCCABELLA et al. (1962).

e) Pineal Function and the Oestrous Cycle

The results mentioned in Sections B XI 11c β and B XI 11d α show that the oestrous cycle may be severely affected by pinealectomy or administration

Table 27. Pineal weight and activities of some histochemically demonstrable enzymes of the rat pineal gland during the oestrous cycle (adapted from FASANO et al., 1971a, b)[a]

	Oestrus		Metoestrus	Dioestrus		Pro-oestrus	
	Early	Late		Early	Late	Early	Late
Pineal weight	2	1	7	3	6	5	4
Succinate DH[b]	6	7	1	4	5	3	2
Lactate DH[b]	2	6	5	4	1	3	7
Alkaline phosphatase	7	1	2	5	4	6	3
Acid phosphatase	7	2	2	5	6	4	3

[a] Using an arbitrary scale where 1 denotes lowest values and 7 highest values.
[b] DH, dehydrogenase activity.

of pineal substances. However, it has not yet been discussed what role, if any, the pineal plays in regulating the oestrous cycle under physiological conditions. Moreover, it has not been mentioned which morphological, histochemical and functional changes occur in the epiphysis cerebri during the oestrous cycle.

Relevant studies in humans are rare. The results of one such study (WETTERBERG and ARENDT, 1975) showing that serum melatonin levels are highest at the time of menstrual bleeding and lowest at ovulation suggest that melatonin may be somehow involved in the regulation of the menstrual cycle.

In small laboratory mammals a variety of parameters change in relation to the oestrous cycle. In the albino rat, pineal weight shows a maximum at metoestrus, followed by a more or less steady decrease, reaching a minimum at late oestrus (FASANO et al., 1971b) (Table 27). Pinealocyte size, as assessed by measuring nuclear density in the mouse, is slightly larger in metoestrus/dioestrus than in pro-oestrus/oestrus during the day, whereas during the night the reverse is found (Table 13) (KACHI et al., 1973). Histochemically, the amounts of glycogen are usually slightly, but statistically insignificantly, higher in metoestrus/dioestrus than pro-oestrus/oestrus (KACHI et al., 1973). In the rat pineal the phospholipid content is lowest just before ovulation; it increases after ovulation and reaches a maximum at dioestrus (ZWEENS, 1963, 1964, 1965). Pineal protein and RNA levels are high at dioestrus, intermediate at pro-oestrus and metoestrus and low at oestrus (NIR et al., 1972a). CARDINALI and VACAS (1978a), by contrast, noted a striking decrease of pineal protein synthesis on the day of pro-oestrus. Histochemical studies reveal (Table 27) that some enzyme activities appear to change very rapidly (FASANO et al., 1971a, 1977; FASANO, 1972).

Biochemical studies dealing with pineal serotonin and melatonin formation during the oestrous cycle have revealed a very complex situation, which is still far from clear. Pineal serotonin levels are given in Table 28, showing that between 17.00 h and 18.00 h the levels are identical at all stages; between 4.00 h and 5.00 h at oestrus serotonin levels are distinctly lower than at metoestrus, while between 21.00 h and 21.30 h at pro-oestrus the levels are significantly higher than at dioestrus (QUAY, 1963c, 1964b). The day–night rhythm of pineal NAT activity is not altered during the oestrous cycle (CARDINALI and VACAS,

Table 28. Pineal serotonin at different times during the oestrous cycle in Long-Evans rats 13–20 weeks old when killed (QUAY, 1963c)[a]

Time	Pro-oestrus	Oestrus	Metoestrus	Dioestrus
4.00–5.00	56.9 ± 5.2 (10)	51.1 ± 2.7 (16)[b]	65.4 ± 3.9 (14)[b]	63.4 ± 2.7 (10)
11.30–13.00	79.7 ± 5.2 (16)	77.4 ± 5.3 (19)	70.2 ± 3.8 (21)	69.8 ± 3.4 (15)
17.00–18.00	62.0 ± 3.9 (13)	62.8 ± 3.3 (14)	65.6 ± 3.1 (14)	61.7 ± 4.0 (11)
21.00–21.30	31.5 ± 3.3 (6)[b]	23.3 ± 4.5 (6)	25.6 ± 1.2 (7)	15.4 ± 1.3 (5)[b]

[a] Nanograms of serotonin base per pineal; means ± SEM (number of samples) are listed.
[b] Probability that the means of these populations are actually equal or different in the opposite direction is <0.01, based on "Student"-Fisher t.

1978a). Melatonin excretion, as assessed by radio-immunoassay, is depressed at proestrus (OZAKI et al., 1978). Plasma melatonin levels in ewes do not show statistically significant differences between the various stages of the oestrous cycle (ROLLAG et al., 1978c). HIOMT measurements have yielded variable results. According to WURTMAN et al. (1965a, 1968a) HIOMT activity is twice as high in metoestrus and dioestrus than during pro-oestrus and oestrus. CARDINALI et al. (1974c), by contrast, found maximum HIOMT levels at oestrus and minimum levels during dioestrus and pro-oestrus. More recent studies have not confirmed these observations (WALLEN, 1972; WALLEN and YOCHIM, 1971, 1972, 1974a; VOLLRATH et al., 1975). WALLEN and YOCHIM (1974a) demonstrated that there is no change in the mean rate of HIOMT activity throughout the oestrous cycle, but that there are striking differences in the *amplitudes* of the diurnal oscillations, the oscillations being greatest during dieoestrus and pro-oestrus and very small in the oestrous/metoestrous interval so that no statistically significant differences in enzyme activity are present between 9.00 h and 23.00 h.

These differences are thought to be due to the fact that during the oestrous cycle at least two rhythms interact to create a complex waveform with a beat frequency. Mathematical calculations, based on the assumption that one of the rhythms is the 24-h photoperiodic rhythm, reveal that the second rhythm has a duration of 19.2 h in 4-day-cyclic rats and 20.0 h in 5-day-cyclic rats. Further studies designed to test these results showed that when in 5-day-cyclic rats the 24-h solar day was changed to an artificial 20-h day many animals developed a persistent oestrous syndrome 4–5 days after the onset of the experiment and this, together with the fact that pinealectomized control rats no longer exhibited the relationship between vaginal cornification and phase of the cycle, was taken as evidence that the second rhythm did indeed have a duration of 20 h (WALLEN and YOCHIM, 1974c). Since the characteristic rhythmicity of HIOMT activity persisted after ovariectomy, it was concluded that the ovary and its hormones are not the controlling factors and that either the pineal gland generates its own mixed-wave rhythm which in turn modifies reproductive cyclicity, or that the rhythm is generated elsewhere in the central nervous system, the information being simultaneously transmitted to the pineal gland and the hypothalamus–pituitary axis (WALLEN and YOCHIM, 1974b). The main role attributed to the pineal by YOCHIM and WALLEN (1974c) is that the organ uses its rhythm to adjust its daily output of antigonadotropic principles, which in turn modulates the rhythmic activity in the hypothalamus, thus altering the basal rates of secretion of hypophyseotropic factors.

The results obtained by WURTMAN et al. (1965a, 1968a) that HIOMT activity in metoestrus and dioestrus is distinctly higher than in pro-oestrus and oestrus may be due to the fact that the day of the week at which the sampling is done was not taken into consideration. KANTARJIAN (1976), working on the assumption that the HIOMT activity of the pineal undergoes a 7-day cycle (see also VOLLRATH et al., 1975), found that when the sampling is done on Wednesdays only, there are virtually no differences of HIOMT activity between the different stages of the oestrous cycle. However, when HIOMT activity is studied irrespective of the day of the week, then results similar to those described by WURTMAN et al. (1965a, 1968a) are obtained.

Finally, there are other findings pointing to interrelationships between the pineal and the oestrous cycle. In vitro it has been shown that the ^{3}H-oestradiol uptake of the pineal varies as a function of the stage of the oestrous cycle, maxima being present at dioestrus and minima at pro-oestrus (NAGLE et al., 1973b). Nuclear receptor oestrogen complex, by contrast, shows a maximum at pro-oestrus (CARDINALI, 1977). Pineal adenyl cyclase activity is stimulated by noradrenaline at all stages of the oestrous cycle, except pro-oestrus (WEISS and CRAYTON, 1970a, b; DAVIS, 1978).

f) Pineal Function in Seasonal Breeders

In seasonal breeders the pineal may play a vital role by ensuring that the young are born during the appropriate season (REITER, 1973a, 1974c, d, 1975a; REITER et al., 1975e). In view of this concept it is relevant to note that in certain species the structure of the pineal undergoes characteristic seasonal changes. In the golden hamster, pinealocyte nuclei are largest from December to February and smallest during the summer months, the nuclei of the testicular interstitial cells showing an inverse relationship (PFLUGFELDER, 1956; MOGLER, 1958) (Fig. 130). An identical behaviour is seen in ewes (NEŠIĆ, 1962). Since there are indications that in certain species the pineal parenchyma can be subdivided into medullary and cortical zones (see p. 69), it is interesting that when the ewe pineal is particularly active this subdivision is not conspicuous: In March, April and May cortical and medullary regions are clearly distinguishable. In June, July and August, glial patches and small cysts occur and from September to November cortical and medullary pinealocytes are identical in size, some of them containing PAS-positive substances (NEŠIĆ, 1962).

Electron microscopically, 60%–80% of *hedgehog* pinealocytes show lamellar bodies from the end of September to mid-December, when the animals are sexually quiescent (PEVET and SABOUREAU, 1973). In the preparatory phase of the sexual cycle (mid-December to mid-March), the pinealocytes are characterized by increasing numbers of electron-dense bodies and lysosomes, while in the phase of sexual activity (mid-March to the end of September) almost all the pinealocytes exhibit areas of "cytoplasmic condensations" characterized by somewhat more electron-dense cytoplasm comprising the usual organelles,

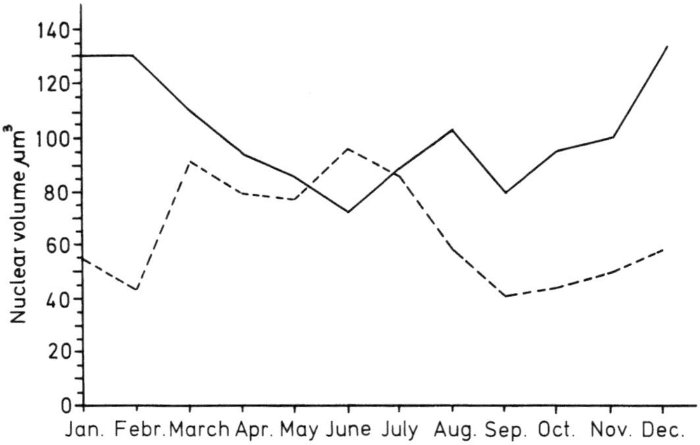

Fig. 130. Nuclear volumes of pinealocytes (*solid line*) and interstitial cells of testes (*broken line*) in hamsters throughout a 1-year period. Modified from MOGLER, 1958

some of which undergo lysis (PEVET and SABOUREAU, 1972, 1973). Quantitative studies carried out in *moles* killed between the beginning of January and May reveal that in the pinealocytes of males paracrystalline structures (Fig. 61) increase in number and the Golgi apparatus enlarges from January to March, followed by decreases (PEVET and SMITH, 1975a; PEVET et al., 1975c). Secretory granules, which are altogether rare, apparently do not change in amount (PEVET and SMITH, 1975a, b). The pinealocytes of *Eliomys quercinus* show an abundance of lysosomes, lipid droplets and myelin bodies during hibernation, while in the spring synaptic ribbons increase in amount (RICHOUX et al., 1975).

Pineal indoleamine synthesis has been little investigated. Seasonal fluctuations of the pineal hydroxyindole content have been reported for the Northern fur seal (KEYES et al., 1971) and *Eliomys* (BARRY, 1971). HIOMT activity in Uinta ground squirrels, *Spermophilus armatus*, is lowest in the breeding season and highest just before the animals enter hibernation (ELLIS and BALPH, 1976).

Experimental evidence for interrelationships between annual reproductive rhythms and the pineal comes from studies in two species in particular, namely the hamster and ferret. In hamsters, the gonads atrophy during the winter if the animals are kept under natural photoperiodic and temperature conditions. That this effect is mediated by the pineal has been demonstrated by showing that pinealectomy prevents gonadal atrophy (REITER, 1973c, d) and that if pinealectomized females are caged with pinealectomized males they ovulate, become pregnant and successfully deliver young during the winter months (REITER, 1973/74).

The pineal and light are not solely responsible for the annual reproductive rhythms. REITER (1972a) noted that if adult hamsters are kept long enough under short daily photoperiods, the initial atrophy of the testes is followed after 30 weeks by spontaneous restitution which persists up to 80 weeks. A second involution of the testes due to light deprivation can be achieved only after 30 weeks provided that the hamsters had been exposed to a lighting schedule of LD 14:10 for 22 weeks. Since 1 week or 10 weeks are not sufficient, it was thought that either a transient refractoriness of the brain to the antigonadotropic principle of the pineal existed or that there was a temporary failure of the pineal to be activated. Reactivation of the reproductive system in the spring is not a function of increasing photoperiodic length, since it also occurs in hamsters blinded at the beginning of February (REITER, 1975c). According to STETSON et al. (1975) the annual testicular cycle depends primarily on the position of light relative to the circadian system and not on the absolute amount of light or darkness on each day. A so-called resonance experiment in which light cycles consisting of a 6-h period coupled with dark periods of various lengths to give cycles of 24, 36, 48 and 60 h duration were used has shown that a circadian rhythm of *photosensitivity* determines whether or not light will stimulate the hypothalamus–pituitary–gonadal axis.

In the ferret, in which the females normally come into oestrus once a year in the spring, the onset of oestrus can be advanced to December or January by exposure to artificial light, the effect of light being abolished by pinealectomy or superior cervical ganglionectomy (HERBERT, 1967, 1968, 1969, 1971). Pinealectomy combined with diurnal lighting conditions does not affect the oestrus occurring in the first postoperative spring (HERBERT, 1971), but delays oestrus by 20–30 weeks in the second postoperative year (HERBERT, 1972; HERBERT and VINCENT, 1973). In the summer, oestrus can be curtailed by exposing ferrets to artificial short days, provided that the pineal is present (THORPE and HERBERT, 1974). Melatonin administration shortens the duration of oestrus and induces photosensitivity in ferrets made photorefractory by prolonged exposure to long photoperiods (THORPE and HERBERT, 1976).

g) Pineal Function During Pregnancy

In view of the fact that functional interrelations exist between the pineal and the gonads it was only logical to examine pineal structure and function during pregnancy. Yet systematic and detailed studies are rare. As regards the older literature on this topic, the reader is referred to the publications by BARGMANN (1943) and KITAY and ALTSCHULE (1954a). Recent studies have confirmed results reviewed by BARGMANN (1943) that in pregnant animals the pineal undergoes structural changes. Pineal weight in the rat decreases significantly in the second half of gestation provided that the mother carries more than ten fetuses (HUANG and EVERITT, 1965). A decrease in pineal weight in the last 4–5 days of pregnancy has been noted by FASANO (1972), but the relationship between the number of fetuses carried and pineal weight could not be confirmed. While in rats there is a tendency for pineal weight to decrease during pregnancy, pregnant ewes (SANTAMARINA and VENZKE, 1953) show an increase in pineal weight, accompanied by increased vascularity.

Light microscopy reveals that towards the end of pregnancy pinealocyte nuclear size increases by 24.5% in the rat, by 22.5% in the hamster and by 10.8% in the mouse. In rats the increase in size is most pronounced at 5 days of pregnancy, a second increase occurring between 20 days of pregnancy and 24 h after delivery. Five days after delivery nuclear size is still high (KANTARJIAN, 1976).

Histochemical studies at the light microscopic level show that NADPH- as well as NADH-diaphorase activities are greatly enhanced in the second half of gestation in guinea-pigs (VOLLRATH and SCHMIDT, 1969). The results obtained in rats are variable. RZESZOWSKA-TOKARSKA (1969) noted that in the second half of pregnancy ATPase, thiamine pyrophosphatase, glucose-6-phosphatase, alkaline phosphatase and lipase activities are enhanced. FASANO (1972) found that lactate and succinate dehydrogenases increase at 10 days of gestation, alkaline and acid phosphatase activities being highest between the 15th and 18th days. DEVEČERSKI (1972a), on the other hand, noted a moderate decrease of lactate and isocitrate dehydrogenase activities at 7 days, and a strong decrease of isocitrate dehydrogenase activity at 15 days of gestation.

Electron microscopy reveals that when compared to non-pregnant controls, the pinealocytes in pregnant guinea-pigs (LUES, 1971) and rats (KANTARJIAN, 1976) possess more strongly lobulated nuclei, larger amounts and more polymorphic mitochondria as well as increased numbers of lysosomes. Moreover, the smooth endoplasmic reticulum increases in amount. In view of the problem of the functional significance of the so-called dark pinealocytes it is interesting that the electron-density of the pinealocyte cytoplasm increases in the rat and that more dark pinealocytes are present in the guinea-pig. Quantitative data obtained in rats (KANTARJIAN, 1976) show that rough endoplasmic reticulum increases by some 100% ($P < 0.001$) and the Golgi apparatus by 25.6% ($P < 0.1$). An increase in size of the Golgi apparatus has also been demonstrated in pregnant moles (PEVET and SMITH, 1975a).

Data on melatonin formation during pregnancy are rare. YOCHIM and WALLEN (1974b), studying HIOMT activity in *pseudo*pregnant rats, noted that the diurnal oscillations of HIOMT activity become smaller, resembling the oestrous-metoestrous interval of the cycle. Since the nocturnal activity is greater than in controls, the mean rate of enzyme activity is identical to that measured during the oestrous cycle. In pregnant rats HIOMT activity decreases after the onset of pregnancy and remains low until the 10th day; at the 15th day it shows a distinct peak, followed by a decrease from the 20th to the 22nd day (KANTARJIAN, 1976). HIOMT activity was found to be high in pregnant moles (PEVET et al., 1978). Melatonin measurements in humans in the third trimester have revealed a normal circadian pattern and no differences to non-pregnant controls (C. MATTHEWS et al., 1977), but during labour the melatonin levels were higher than prior to labour (MITCHELL et al., 1978). That the pineal is not unaffected during pregnancy is also shown by the fact that sham-pinealectomy in pregnant rats, in contrast to non-pregnant ones, leads to a distinct increase in pineal HIOMT activity (KANTARJIAN, 1976).

Experimental studies have not provided clear-cut answers as to the function of the pineal during pregnancy. Following pinealectomy, or melatonin administration, rats become pregnant easily (MITCHELL and YOCHIM, 1970; MESS et al., 1973; TIGCHELAAR and NALBANDOV, 1975; KANTARJIAN, 1976), the time of ovulation, number of ova released, number of implantation sites and litter size being unaffected (KINCL and BENAGIANO, 1967; TIGCHELAAR and NALBANDOV, 1975; M. VAUGHAN et al., 1976c), though occasionally smaller litter size has been noted (GUERRA et al., 1973; KANTARJIAN, 1976). In the spotted skunk,

Spilogale putorius latifrons, which is characterized by a long pre-implantation period of some 220 days, neither the pre-implantation period nor postimplantation physiology are adversely affected by pinealectomy (MEAD, 1972). Different results can be obtained when other species and pineal substances other than melatonin are studied. In contrast to rats (see above), melatonin administration reduces the incidence of fertile matings in mice (M. VAUGHAN et al., 1976c), and pineal peptides appear to be more potent than indoleamines, as shown by the fact that arginine vasotocin reduces the incidence of fertile matings and that those animals which have become pregnant do not deliver viable pups (M. VAUGHAN et al., 1976c). A reduced incidence of pregnancy and a reduced mean number of fetuses is observed after administration of fractions F3 and F4 of bovine pineal extracts (ORTS et al., 1977). Indirect evidence for the importance of indoleamines in relation to pregnancy is provided by the finding that resorption of fetuses and cannibalism is induced by *p*-chlorophenylanine (*p*-CPA), which inhibits serotonin synthesis (W. MOORE and HAMPTON, 1974).

In view of the fact that melatonin crosses the placenta and is demonstrable in fetal brain, heart, skin, muscle, skeleton, liver and other viscera (KLEIN, 1972), there is a possibility that maternal melatonin influences fetal organs. Following daily melatonin administration from the 17th to 20th days of gestation, the female offspring show significantly reduced pituitary and adrenal weights, while in the male offspring kidney weight is significantly lower (M. VAUGHAN et al., 1971c). Sex distribution of the offspring is apparently normal after pinealectomy (KINCL and BENAGIANO, 1967).

Finally, the effects of the pineal on the duration of gestation have been investigated. According to GRISHCHENKO and ZUBKOV (1974) the pineal may be involved in the timing of parturition insofar as prior to parturition a reduction in melatonin synthesis leads to a discharge of hypothalamic neurosecretory substances resulting in the onset of labour. In vitro melatonin does not induce uterine contractions (KAPPERS, 1962a; TILSTRA and PROP, 1962), but induces a slight inhibition of uterine contractility (HERTZ-ESHEL and RAHAMINOFF, 1965). Although according to some authors (MIZUNO and SENSUI, 1970; MITCHELL and YOCHIM, 1970) pinealectomy does not affect the duration of pregnancy, MESS et al. (1973) and KANTARJIAN (1976) did find a prolongation of ca. 1 day, not only during that gestation in which pinealectomy had been carried out, but also in successive pregnancies (KANTARJIAN, 1976). Melatonin injections do not affect the duration of gestation (KANTARJIAN, 1976). In view of the fact that the pineal responds to changes in environmental lighting conditions, it is relevant to note that the duration of pregnancy can be extended by a prolongation of the daily illumination (MITCHELL and YOCHIM, 1970) and by exposure to a lighting schedule of LD 4:4 (CHEN and BESCH, 1975). As regards the question of multiple pregnancies and photoperiodicity, see TIMONES and CARPEN (1968).

Altogether it would appear that the pineal plays a subtle role during pregnancy, probably by influencing hormonal parameters. One train of thought is that the pineal may be responsible for cessation of pituitary gonadotrophin secretion in the second half of pregnancy (VOLLRATH and SCHMIDT, 1969), while

KANTARJIAN (1976) is of the opinion that the pineal controls the critical levels of gonadotrophins, especially at the beginning and near the end of pregnancy. In this context it is interesting to note that melatonin administration increases pituitary LH levels on the 1st and the 23rd days of pregnancy, abolishes the increase of serum LH levels occurring between the 22nd and 23rd days of pregnancy and lowers pituitary FSH levels on the 1st day of pregnancy only (TIGCHELAAR and NALBANDOV, 1975).

The milk ejection response is apparently not significantly affected by pinealectomy (MIZUNO and SENSUI, 1970). For further details regarding milk yield see p. 279.

h) Maternal and Sexual Behaviours

No attempt has been made in the present review to scan the literature concerning the influence of the pineal on maternal and sexual behaviours. Hence, only some incidental findings will be reported. With respect to maternal behaviour, it is relevant to recall (p. 303) that depletion of serotonin by *p*-CPA leads to cannibalism in rats (W. MOORE and HAMPTON, 1974), but it is doubtful whether this effect can be attributed to the pineal organ. According to TAGLIAMONTE et al. (1969), *p*-CPA elicits sexual excitation in both normal and pinealectomized rats and pinealectomy alone does not lead to episodes of sexual excitation in rats.

12. Pineal–Adrenal Interrelationships

Attempts to establish the role of the epiphysis cerebri in the regulation of the hypothalamus–pituitary–adrenal axis have been fraught with inconsistent results (M. VAUGHAN et al., 1972b). Nevertheless there are some hard facts available which support the concept that the pineal may not be without effect on adrenal function.

Size and weight of the adrenals do not appear to be ideal parameters for demonstrating an influence of the pineal. *Pinealectomy* often does not lead to statistically significant differences (YAMADA, 1961; RENNELS and DILL, 1961; PANAGIOTIS and HUNGERFORD, 1961a; BUGNON and MOREAU, 1962; ROTH, 1964; TRENTINI et al., 1965; HOUSSAY et al., 1966a; WRAGG, 1965, 1967; LOMBARD DES GOUTTES, 1967b; KINSON et al., 1968a, b; HENZL et al., 1970; DE FRONZO and ROTH, 1972; SEIBEL and SCHWEISTHAL, 1973; OGLE and KITAY, 1976). An increase in adrenal weight was noted by WURTMAN et al. (1959), ASAGOE and HAMAMOTO (1959), REITER et al. (1966), FRASCHINI et al. (1968), M. VAUGHAN et al. (1972b) and G. VAUGHAN and VAUGHAN (1972). A *decrease* of adrenal weight following pinealectomy has rarely been observed (HOFFMAN and REITER, 1966; CHERDYNTSEV, 1967). *Pineal substances* are not much better in altering adrenal weight. In the majority of studies pineal substances were without detectable effects (KITAY, 1963; JOUAN and SAMPEREZ, 1965b; DEPROSPO and HURLEY, 1971a; COLLU et al., 1971a; RUBIN and TRAUM, 1971; FRASCHINI et al., 1971). In some studies an increase in adrenal weight was observed (BUGNON and MOREAU, 1962; GROMOVA et al., 1967; NARANG et al., 1967; KINSON and LIU, 1973b); in others a decrease occurred (MASSA, 1950; FRANCHIMONT, 1964; SINGH and TURNER, 1971).

In addition to adrenal weight, other parameters have been investigated which cannot be differentially attributed to either the adrenal cortex or the adrenal medulla. Pinealectomy decreases the adrenal ascorbic acid concentration (SIMONNET et al., 1951b; ASAGOE and HAMAMOTO, 1959), and melatonin administration is without effect on adrenal DNA levels (SINGH and TURNER, 1971).

There is some morphological evidence that adrenalectomy affects the pineal. Ultrastructurally an increase in the number of pinealocyte processes abutting the perivascular spaces can be seen; this occurs at the expense of glial cells and is accompanied by an increase of HIOMT activity (DEUSSEN-SCHMITTER et al., 1976). Lipase activity increases (MARKINA, 1970), while the pineal lipid content decreases, the decrease being more pronounced when the animals are kept on a salt-free diet (PANAGIOTIS and HUNGERFORD, 1961b). A sodium-high diet, by contrast, increases pineal lipids (MARKINA, 1970). The pineal serotonin content is not significantly altered by adrenalectomy (MCKENNY et al., 1966).

a) Adrenal Cortex

Histology and histochemistry of the adrenal cortex are inconsistently affected by pinealectomy and administration of pineal substances. Following *pinealectomy*, WURTMAN et al. (1960) noted a slight and statistically insignificant increase in the ratio of the thickness of the zona glomerulosa to total cortical thickness, while ŠĆEPOVIĆ (1963a) found that the glomerular zone was smaller and the zonae fasciculata and reticularis larger than in controls. However, other authors were unable to detect any clear histological differences (PANAGIOTIS and HUNGERFORD, 1961a; BUGNON and MOREAU, 1962; BUGNON et al., 1963). At the cellular level, a decrease in nuclear size in the zone glomerulosa has been noted (GIACOMELLI, 1962). Histochemically, there are no detectable changes in the lipid content (RENNELS and DILL, 1961), but glucose-6-phosphate dehydrogenase, TPN diaphorase, DPN diaphorase, 3β-ol steroid dehydrogenase (TRENTINI et al., 1965) and esterase activities (GIACOMELLI, 1962) decrease in the zona glomerulosa after pinealectomy.

Administration of pineal extracts results in insignificant changes in zona glomerulosa thickness (WURTMAN et al., 1960), a decrease in size of the zona fasciculata (NOTARIO, 1956) accompanied by a decrease of DNA and RNA levels (SUARDI et al., 1959), an increase in the cholesterol content (FAZIO and GIORDANO, 1953) and an increase in the activity of a number of histochemically demonstrable enzymes in both intact (TRENTINI et al., 1964) and pinealectomized animals (VASSANELLI et al., 1965; serotonin). Nuclear volumes of the zona glomerulosa do not show consistent changes after administration of pineal body extract, whereas a 12.4% increase is demonstrable in the zona fasciculata (PALKOVITS and FÖLDVARI, 1963).

Pineal–adrenal interrelationships can be readily demonstrated under certain *experimental* conditions. Compensatory adrenal hypertrophy (CAH) is blocked by melatonin, which when given to neonatal rats has a long-term effect on

CAH (M. VAUGHAN and VAUGHAN, 1974), serotonin, *N*-acetyl-5-hydroxytryptamine, 5-hydroxytryptophol, 5-methoxytryptophol, 6-hydroxymelatonin (female mice: G. VAUGHAN and VAUGHAN, 1972; M. VAUGHAN et al., 1972b), bovine pineal extract (rat: ANTON-TAY et al., 1962; DICKSON and HASTY, 1972) and arginine vasotocin (when injected into the third ventricle; it is without effect when given IV or into the pituitary; mouse: PAVEL et al., 1973c; PAVEL, 1975). 5-Hydroxytryptophan and 5-methoxyindoleacetic acid have no effect on CAH (G. VAUGHAN and VAUGHAN, 1972; M. VAUGHAN et al., 1972b). Moreover, the castration- and cold-induced increases of adrenal weight are counteracted by melatonin (M. VAUGHAN et al., 1972b). In blind anosmic rats CAH occurs only when the animals have been pinealectomized (REITER, 1972b). In hypophysectomized rats, treatment with pineal extracts has been claimed to accelerate the involution of the adrenals (PETRONIO and TAVAZZA, 1958b).

Regeneration of the adrenal cortex is not influenced by pinealectomy (C. MACHADO, 1965, 1966). Pineal tissue autografted to enucleated adrenals stimulates adrenal cortex regeneration, causing a hyperplastic response of the zona glomerulosa, while the two other zones regenerate in a normal manner (SIMIONESCU and SCHERZER, 1964a, b). MILINE (1965) suspects an influence of the epithalamo-epiphyseal complex on the adrenals since in darkness both the pineal and the glomerular zone hypertrophy.

Finally, *hormone* synthesis in the adrenal cortex has been studied, and it would appear that the pineal affects both mineralo- and glucocorticoid secretion. The concept that the pineal regulates *aldosterone* secretion has been greatly promoted by FARRELL and collaborators. Following initial observations pointing to the presence of a diencephalic centre regulating aldosterone secretion (FARRELL, 1958), it was later found that both the diencephalon and the epiphysis cerebri contained a regulatory substance (NEWMAN TAYLOR, 1960; NEWMAN TAYLOR and FARRELL, 1962, 1963; NEWMAN TAYLOR et al., 1962), first termed glomerulotropin (FARRELL, 1959a, b). This term was subsequently replaced by adrenoglomerulotropin since renal physiologists felt that the original term implied an effect on the glomerulus of the kidney (FARRELL, 1960a, b). Further analyses revealed that the active principle was 1-methyl-6-methoxy-1,2,3,4-tetrahydro-2-carboline (FARRELL and MCISAAC, 1961). The concept that a specific carboline accounts for the aldosterone-stimulating property of the epiphysis cerebri was soon abandoned, since FARRELL's results could not be replicated by a number of authors (see REITER et al., 1975e, for a review) and since FARRELL himself felt that adrenoglomerulotropin was not present in the pineal in sufficient amounts to have a physiological significance (cited after REITER and FRASCHINI, 1969).

Subsequent studies provided clear evidence that the pineal affects hormone synthesis by substances other than adrenoglomerulotropin, though the effects have often been variable. Pineal extracts inhibit adrenal beta-hydroxylase activity, thus blocking the synthesis of aldosterone, cortisol and cortisone, but not desoxycorticosterone (LOMMER, 1966), and melatonin reduces the production of Δ^4-3-ketonic corticosteroids from endogenous precursors in vitro (GIORDANO et al., 1970) and suppresses aldosterone production (GROMOVA et al., 1967). The fact that pinealectomy decreases aldosterone levels is in keeping with an inhibitory influence of the pineal (KINSON and SINGER, 1967; KINSON et al., 1967, 1968a, b). However, contradictory results have been obtained as well. In some studies pinealectomy was without effect on aldosterone secretion (DAVIS, 1961; BLAIR-WEST et al., 1964; VAN DER WAL et al., 1965; KARPPANEN et al., 1975), while in others it was followed by a decrease (JOUAN and SAMPEREZ, 1965a), the latter observation being in agreement with findings showing that pineal extracts elevate aldosterone production (GIORDANO and BALESTRERI, 1963; JOUAN, 1963; JOUAN and MENDELL, 1964; JOUAN and SAMPEREZ, 1965a). Which one of the pineal substances is responsible for these diverse effects is not known. According to FABRE et al. (1965) it could be ubiquinone, but JOUAN, who among others demonstrated that adrenoglomerulotropin is without effect on aldosterone secretion (JOUAN, 1962), suggested that it was serotonin (JOUAN, 1963; JOUAN and SAMPEREZ, 1964, 1965a).

The results concerning the secretion of other adrenocortical hormones have been slightly more consistent than those concerning aldosterone production. Pinealectomy may be without demonstrable effect on *corticosterone* production (KINSON and SINGER, 1967; VERMES et al., 1974; JACOBS, 1974; TAKAHASHI et al., 1976), but if the animals are studied at suitable times after the operation, an effect of pinealectomy may well be present. In adult rats a significant increase in plasma corticosterone levels is seen 30 days after pinealectomy, but not 90 and 180 days after surgery (KINSON et al., 1967, 1968a, b). By contrast, in young female rats pinealectomy increases plasma corticosterone levels after 10 but not after 30 days (NIR et al., 1971b). Long-term and sex-dependent differences are indicated by the fact that pinealectomy carried out at 1 day of age results in an elevated corticosterone concentration at 50 days of age in female but not in male rats (HENZL et al., 1970). According to OGLE and KITAY (1976) pinealectomy decreases dihydrocorticosterone and tetrahydrocorticosterone secretion, but not that of corticosterone. Melatonin, but not serotonin, stimulates 5α-reductase activity, resulting in an increase of dihydrocorticosterone and tetrahydrocorticosterone secretions with a concomitant decline of corticosterone secretion (OGLE and KITAY, 1978). In ovariectomized rats pinealectomy diminishes the post-castration rise in adrenal 5α-reductase activity (OGLE and KITAY, 1976). In blinded rats the disturbed corticosterone rhythm was restored to normal by pinealectomy in one study (JACOBS, 1974) but not in another (TAKAHASHI et al., 1976). In rams, pinealectomy abolished the early morning rise of *cortisol* in two out of four animals (BARRELL and LAPWOOD, 1978).

Administration of pineal substances has shown that in *humans* urinary 17-ketosteroid excretion is depressed by aqueous bovine pineal extract (MORGANO et al., 1956). In *animal* species plasma corticosterone is reduced after administration of bovine pineal extract (DICKSON and HASTY, 1972; PAVEL et al., 1977c) and glomerulotropin (FRANCHIMONT, 1964); melatonin increased corticosterone production in one study (GROMOVA et al., 1967), was without effect in another (BARCHAS et al., 1969) and decreased it in a third (MOTTA et al., 1971). Arginine vasotocin depressed plasma cortisol levels (PAVEL et al., 1977c). ACTH-mediated corticosterone release by isolated adrenal showed variable responses: it was depressed by aqueous bovine pineal extract, enhanced by 5-methoxytryptophol and not affected by melatonin (PORTER and HEIMANN, 1977a, b). A complicating feature when administering pineal substances with the aim of studying adrenocortical function is that these substances affect gonadal function and that interrelationships exist between the gonads and adrenals (OGLE and KITAY, 1976; RAMALEY and BARTOSIK, 1975; MANN et al., 1975).

The question as to which effects adrenocortical hormones have on the pineal has been little studied. There is clear autoradiographic evidence that ^{3}H-dexamethasone is taken up by pinealocyte cytoplasm and nucleoplasm (REES et al., 1975; WAREMBOURG, 1975). Aldosterone administration reduces the number of dense-core vesicles in pinealocytes (CLEMENTI et al., 1962), while desoxycorticosterone increases the pineal lipid content (HUNGERFORD and PANAGIOTIS, 1962). Cortisol is without effect on pineal serotonin levels (MCKENNY et al., 1966) and methylprednisolone is ineffective in changing the circadian serotonin rhythm during development (CAGNONI et al., 1966).

b) Adrenal Medulla

In view of the key role of noradrenaline in regulating pinealocyte function (p. 9) it is surprising that interrelationships between the pineal and the adrenal medulla have been little investigated. The effects of pinealectomy on adrenal weight have been mentioned above (p. 304), but the results obtained do not lend themselves to a meaningful interpretation because the changes in volume were not studied separately for adrenal medulla and cortex. More specific studies have shown that pinealectomy leads to a decreased serotonin uptake of the medulla (BARÁTH and CSABA, 1973), an increase in ^{3}H-tyrosine incorporation (CSABA and NAGY, 1973) and depression of the nocturnal rise of dopamine-β-hydroxylase activity (BANERJI and QUAY, 1976). Plasma levels of dopamine-β-hydroxylase are not affected by either pinealectomy or adrenalectomy, but when the two procedures are combined there is a decrease of the enzyme levels, but only during the middle of the dark phase (BANERJI and QUAY, 1978). Recent electron microscopic studies have shown that pinealectomy increases the amplitude of rhythmic changes of the numbers of small clear vesicles in nerve endings on adrenal medullary adrenaline cells and simultaneously leads to a phase shift of the diurnal peak by about 3 h (KACHI et al., 1979).

The fact that the stress-induced increase in pineal NAT activity is blocked by bilateral adrenalectomy, in contrast to the dark-induced rise of pineal NAT activity (LYNCH et al., 1977), suggests an influence of the adrenal medulla on the pineal. Electron microscopically, administration of the noradrenaline derivate, aludrin, results in an increase in pinealocyte endoplasmic reticulum, free ribosomes, mitochondria, Golgi apparatus and nucleoli (BUSS, 1965).

13. Pineal–Thyroid Interrelationships

Pineal – thyroid interrelationships have been frequently studied, and it would appear that the pineal usually exerts an inhibitory influence on thyroid function, though some studies have pointed to stimulatory effects.

Thyroid *weight* is either increased by pinealectomy (HOUSSAY et al., 1966a; HOUSSAY and PAZO, 1966, 1968; LOMBARD DES GOUTTES, 1967b; CSABA et al., 1968a; SINGH and TURNER, 1971; DEFRONZO and ROTH, 1972) or not affected (BICK et al., 1968, 1969; MESS, 1969; ROWE et al., 1970), but it does not show a decrease. A decrease of thyroid weight results after administration of pineal substances in adult animals (DE LUCA et al., 1961; BASCHIERI et al., 1963; NABER et al., 1969), while in immature rats melatonin administration increases thyroid weight (PANDA and TURNER, 1968; SINGH and TURNER, 1971).

Morphologically there are clear signs that pinealectomy influences the thyroid. Hypertrophy and hyperplasia of thyroid follicular cells have been noted by ŠĆEPOVIĆ (1960, 1961, 1963b) and BUGNON and MOREAU (1963). Nuclear size of follicular cells is decreased at 12 and 21 days after pinealectomy, but increased at 45 days after the operation (PESCHKE et al., 1974; PESCHKE, 1976). Histochemically, alkaline and acid phosphatase, 5-nucleotidase as well as carboxylic esterase activities are enhanced (ŠĆEPOVIĆ and SCHIPPERS, 1965). Electron microscopically, crystalloid structures are seen in the endoplasmic reticulum of follicular cells, accompanied by an increase of connective tissue (CSABA and BARATH, 1975).

Administration of crude pineal extracts results in an increase in size of the thyroid follicles (SUARDI et al., 1959; NOTARIO, 1956; BARBAROSSA et al., 1962a), accompanied by a diminished cytoplasmic basophilia (SUARDI et al., 1959). Morphological signs of depressed thyroid function were observed by BUGNON and MOREAU (1963). In immature rats the DNA content of the follicular cells is increased by melatonin (PANDA and TURNER, 1968; SINGH and TURNER, 1971).

The investigation of other and more specific parameters of thyroid function reveals fairly clearly an inhibitory effect of the pineal on thyroid follicular cells. After pinealectomy the following parameters show an enhancement: ^{3}H-tyrosine incorporation rate (CSABA and NAGY, 1973), ^{131}I-uptake (CSABA et al., 1968a, 1970; SEIBEL and SCHWEISTHAL, 1973) and thyroid hormone secretion rate (ISHIBASHI et al., 1966; NEGOESCU et al., 1968; RELKIN, 1978). Pineal substances depress thyroid hormone secretion rate (ISHIBASHI et al., 1966; NARANG et al., 1967; SINGH and TURNER, 1972; VRIEND and REITER, 1977) and usually inhibit the uptake of radioactive iodine (DE LUCA et al., 1961; ANTON-TAY et al., 1963; BARBAROSSA et al., 1962b, 1963; BASCHIERI et al., 1963; REITER et al., 1965; DE PROSPO et al., 1968, 1969; DE PROSPO and HURLEY, 1971b), though occasionally no such effect could be demonstrated (NABER et al., 1969; GOLDMAN et al., 1969). According to CSABA and NAGY (1976) melatonin *increases* blood thyroxine levels provided that it is given together with TSH, melatonin apparently potentiating the effects of TSH. NIR et al. (1978b) found that 5-methoxytryptophol, 5-hydroxytryptophol and serotonin, but not melatonin, produced an increase in serum thyroxine levels, whereas *thyroid* thyroxine content decreases after melatonin or serotonin and was not affected by 5-methoxytryptophol and 5-hydroxytryptophol; moreover, it was shown that thyroid cAMP content was elevated by melatonin and serotonin. Administration of CPA, a drug which increases pineal serotonin levels, increases ^{131}I uptake and thyroid weight (DE PROSPO and MELGAR, 1975).

Moreover, the effects of pinealectomy and pineal substances have been tested in *experimentally affected* animals. In propylthiouracil-treated animals, pinealectomy produces no change in thyroid enlargement and radio-iodine accumulation (YAMADA, 1961), but in blinded male golden hamsters pinealectomy elevates the plasma thyroxine levels lowered by blinding (VRIEND et al., 1977). According to WERNER et al. (1971) the thyroid of pinealectomized rats is more sensitive to cold stress than that of intact controls. Pineal substances have been shown to accelerate thyroid involution in hypophysectomized rats (PETRONIA and TAVAZZA, 1958b) and to depress thyroid ^{131}I uptake elevated by blinding (SEIBEL and SCHWEISTHAL, 1973).

Since pineal tissue avidly takes up thyroid hormones both in vivo (FORD and GROSS, 1958; FORD, 1965; STUMPF and SAR, 1975b) and in vitro (CADY et al., 1971) and since melatonin increases the uptake of thyroxine by bovine pineal slices in vitro (CADY and DILLMANN, 1971), it was only logical to examine whether or not thyroid hormones influence pineal function. Following earlier observations that administration of potassium iodide leads to pinealocyte hyperplasia and hypertrophy (MILINE and ŠĆEPOVIĆ, 1959; MILIN and ŠĆEPOVIĆ, 1960) and that thyroxine in vitro stimulates histochemically demonstrable enzymes of pinealocytes (S. MILCOU et al., 1968; S. MILCU et al., 1968; and p.104) and leads to a disappearance of pinealocyte fluorescence (CSABA and BERNAD, 1972), recent systematic in vitro studies have shown (NIR and HIRSCHMANN, 1978b) that thyroid hormones, but not TSH, increase the *N*-acetylserotonin and melatonin contents of pineal tissue. This has been interpreted as indicating that a direct positive feedback exists between the thyroid and epiphysis cerebri. According to DILLMAN and CADY (1971) pineal tissue de-iodinates thyroxine.

There is also evidence that the epiphysis cerebri influences thyroid *parafollicular* or *C cells*. In the rat, pinealectomy leads to a numerical increase of C cells (KRSTIĆ, 1969; CSABA and BARATH, 1974a). The nucleocytoplasmic ratio changes in favour of the cytoplasm, and electron microscopically an increase in the number of electron-dense secretory granules is observed (CSABA and BARATH, 1974a), together with an increase in the amount of endoplasmic reticulum (CSABA and BARATH, 1975). Administration of calcitonin depresses the activities of a number of histochemically demonstrable enzymes in the pineal, pointing to a reduced metabolic activity (KRSTIĆ and TARSOLY, 1972).

Finally, two miscellaneous findings should be mentioned. According to CABASSA and SCIASCIA (1965) melatonin diminishes uptake of tri-iodothyronine by the red blood cells. CSABA and BARATH (1974b) noted that pinealectomy retards uptake and depletion of ^{3}H-serotonin by thyroid mast cells.

14. Pineal–Parathyroid Interrelationships

Relatively few studies have dealt with possible interrelationships between the epiphysis cerebri and the parathyroid. The available evidence suggests that the pineal may inhibit parathyroid function, as shown by the fact that pinealectomy leads to hypertrophy and hyperplasia of parathyroid cells, accompanied by the usual changes of the cell organelles (KRSTIĆ, 1965a, 1966, 1967; MILINE and KRSTIĆ, 1966) and slightly elevated blood calcium levels (KRSTIĆ, 1967), while administration of pineal substances is followed by atrophy of parathyroid cells (KRSTIĆ, 1965b; MILINE and KRSTIĆ, 1966). KISS et al. (1969) observed a decrease of ^{3}H-methionine incorporation into the parathyroid after pinealectomy.

Very dramatic effects of the epiphysis cerebri can be demonstrated in parathyroidectomized rats. While parathyroidectomy alone is not followed by any severe changes, pinealectomized animals experience convulsive attacks and most of them die within 14 h (REITER and MORGAN, 1972; REITER et al., 1972a, b, 1973a, b, 1975a; POMERANTZ and REITER, 1973). The underlying mechanisms are apparently not completely understood, but it has been noted that differences exist between different rat strains (PHILO and REITER, 1978b).

15. Pineal, Islets of Langerhans and Carbohydrate Metabolism

That the epiphysis cerebri may affect the carbohydrate metabolism was insinuated above (p. 306) when interrelationships between the pineal and adrenal glucocorticoid production were discussed. The concept that the pineal influences the carbohydrate metabolism and the islets of Langerhans has been greatly promoted by Romanian workers who have shown that bovine pineal extracts exert an insulin-like action leading to hypoglycaemia and increased glucose tolerance and an increase in hepatic and muscular glycogenesis after glucose loading, while pinealectomy produced a biochemical syndrome characterized by diminished glucose tolerance, a decrease in hepatic and muscular glycogenesis and an increase in blood pyruvate concentration (S. MILCU et al., 1971; S. MILCU and MILCU, 1958; S. MILCOU et al., 1963; I. MILCOU et al., 1957, 1966a; I. MILCU et al., 1961, 1962, 1963, 1964a, b, 1965, 1967; PARHON et al., 1952; NANU-IONESCU and IONESCU, 1969). Corroborative evidence for a participation of the pineal in the regulation of the carbohydrate metabolism comes from other groups of researchers. ALCOZER et al. (1956) noted a drop in blood sugar levels in humans after administration of pineal extract and ATKINS (1973) reported that melatonin significantly reduced the glucose-induced insulin release in mice. Constant infusion of melatonin in vivo to rats undergoing intravenous glucose tolerance tests did not lead to any changes in blood sugar levels (BAILEY et al., 1974). CSABA and BARATH (1971) noted that pinealectomy is followed by a decrease in blood sugar levels and that the rise in blood sugar levels after alloxan treatment is far higher in pinealectomized rats than in non-pinealectomized control rats. GORRAY and QUAY (1978), by contrast, noted a *slight* decrease in blood glucose levels in alloxan-treated rats and a *strong* decrease after sham-pinealectomy, when compared to intact controls.

Studies dealing directly with the islets of Langerhans are scarce. After pinealectomy and ^{3}H-tyrosine injection the islets of Langerhans are more heavily labelled than in normal controls (CSABA and NAGY, 1973). In alloxan-treated animals pinealectomy leads to necrosis of B cells 2 h later, but when "pinealine" is given prior to alloxan the B cells are not altered (S. MILCOU et al., 1957). Using microdissected islets in vitro, melatonin has been shown to reduce basal insulin secretion in rats, but not in mice (BAILEY et al., 1974). In vivo, melatonin did not decrease glucose-induced insulin secretion at all (FELDMAN and LEBOVITZ, 1972), or did so only marginally (BAILEY et al., 1974).

In the adipose tissue of the rabbit, melatonin decreases glucose oxidation (MURTHY and MODESTO, 1975). In schizophrenic patients who show excessive rises in blood pyruvate, α-ketoglutarate and citrate concentrations after ingestion of glucose, administration of a pineal extract reverses these changes to normal (ALTSCHULE, 1975b).

Finally there are findings showing that changes in the carbohydrate metabolism affect the pineal. According to LYNCH et al. (1973a) insulin-induced hypoglycaemia is followed by a more than 300-fold increase in pineal NAT activity.

16. Miscellaneous Effects of the Epiphysis Cerebri

a) Adipose Tissue, Lipid Metabolism and Liver

In most of the relevant studies *adipose tissue* has been clearly shown to be affected by pineal substances. The lack of effect reported by TRAUTMANN and SILBER (1947) is not surprising in view of the small number of pigs studied. Administration of an antigonadotropic pineal extract to young rats leads to an increase in the amount of adipose tissue in females and a decrease in males (DE SMET and VAN DE VELDE, 1954) and to a shift of the adipose tissue from the abdomen to the subcutis in adult females, but not males (VAN DE VELDE and DE SMET, 1960). Melatonin administration decreases ^{14}C-glucose oxidation in rabbit adipose tissue (MURTHY and MODESTO, 1975), enhances the release of glycerol and free fatty acids in the presence of noradrenaline, but not in its absence (M. VAUGHAN and BARCHAS, 1966; rat adipose tissue), and stimulates growth of brown adipose tissue in the hamster (HELDMAIER and HOFFMANN, 1974). Pinealectomy is without effect on intrascapular brown adipose tissue in hamsters kept under LD 14:10, but prevents its hypertrophy due to exposure of the animals to LD 1:23 (REITER, 1975b).

Precisely which pineal substance is responsible for the above effects is not entirely known. According to M. VAUGHAN and BARCHAS (1966) 6-hydroxymelatonin has similar effects as melatonin in enhancing the release of glycerol and free fatty acids from adipose tissue in the presence of noradrenaline. That substances other than melatonin are important has been shown by the observation that while melatonin does not affect serum cholesterol, liver cholesterol and 17-ketosteroid excretion in the urine (DAMIAN, 1972; DAMIAN et al., 1973a), administration of pineal polypeptides for 30 days decreases the serum levels of free and esterified cholesterol, phospholipids, free fatty acids, total lipids and β-lipoproteins while α-lipoproteins increase by 854% (DAMIAN et al., 1974). According to RUDMAN's work (RUDMAN et al., 1970, 1971, 1972, 1973; RUDMAN and SCOTT, 1975) the pineal activates lipase activity in fat cells by a lipolytic substance residing in a fraction which has a molecular weight of 1,000–3,000, lacks tyrosine and methionine, and is inactivated by trypsin.

According to DAMIAN et al. (1973b, 1974) pineal polypeptides reduce arteriosclerosis, especially atheroma lesions of the aorta.

Effects of the epiphysis on the *liver* have been little studied, but it would appear there may well be interrelationships between the two organs. Pinealectomy increases liver protein levels, RNA content (PETRELLI, 1964), lactate, citrate, malate, glutamate concentrations (VEECH et al., 1975) and biliary cholesterol (I. MILCOU et al., 1966b), these effects usually being reversed by administration of pineal extracts (loc. cit). Moreover, it has been shown that pineal extracts lead to a 50% increase of the free methionine content of the liver (CURZIO et al., 1967) and to an increase of liver and plasma alkaline phosphatase activity, accompanied by a decrease in the intestine (PETRONIO, 1958). Melatonin stimulates liver Δ^4-reductase in the rat and depresses it in the hamster (FREHN et al., 1974). In vitro melatonin inhibits rat liver *N*-acetyltransferase activity (HOWD et al., 1976).

b) Food Intake, Water Intake, Alcohol Consumption and Electrolyte Metabolism

Food intake is relatively little affected by the pineal. Following *pinealectomy* a 5.77% increase of food intake has been observed (ISHIBASHI et al., 1966), while after aministration of melatonin food intake shows a reduction of between 8% and 16% (ISHIBASHI et al., 1966; NARANG et al., 1967; SINGH and TURNER, 1972). However, according to TAKAHASHI et al. (1976) mean values of food intake are not affected by pinealectomy. Pinealectomy is also without effect on changes in feeding *rhythms* induced by changes in environmental illumination (BAUM, 1970).

Water intake is influenced by pinealectomy neither under normal lighting schedules (YAMADA, 1961; BLISS and BATES, 1972; DONOVICK et al., 1973) nor under continuous illumination (STEPHAN and ZUCKER, 1972).

Interesting behavioural studies have been carried out with respect to *alcohol consumption*. If adult male hamsters are given a choice between water or 5% ethanol, they show a preference for the alcohol solution, this preference being exaggerated by blinding and reduced by pinealectomy (BLUM et al., 1973a; REITER et al., 1974b). Melatonin increases ethanol consumption in the rat (GELLER, 1971), but not in the hamster (REITER et al., 1974b). Also of interest is the finding that ethanol consumption increases during darkness, when the pineal is known to actively synthesize melatonin (BLUM et al., 1973a). In rats, the ratio of ethanol intake to water intake, on a weekly basis, is greater in intact that in pinealectomized animals (BLUM et al., 1973b).

Studies dealing with *sodium* intake have yielded variable results. While according to BATES and BLISS (1972) pinealectomized rats ingest more saline than sham-operated controls, the reverse was found by JELÍNEK and KŘEČEK (1968). In *neonatal* rats pinealectomy increases the saline intake (KŘEČEK et al., 1975). Serum sodium and potassium concentrations show no consistent changes (YAMADA, 1961; RELKIN, 1973; KARPPANEN, 1974) although lowered potassium levels with unchanged sodium levels (KARPPANEN et al., 1970b) and sodium, but not potassium, retention (TANNER and HUNGERFORD, 1962; PALKOVITS, 1975) have been reported after pinealectomy. According to QUAY (1965c) pinealectomized rats kept on a sodium-deficient diet have a lower cerebral potassium content than sham-operated or unoperated controls, while no consistent and significant differences in the electrolyte contents of other organs, serum and urine are detectable. A diminished calcium excretion in urine and faeces after pinealectomy has been described by MENIGOT et al. (1970), and after administration of pineal polypeptides Mg^{2+} levels increased in blood and brain and decreased in urine (NEACŞU and GHEORGHIU, 1966).

In intact rats *melatonin* administration decreases serum Ca^{2+} and Mg^{2+} levels and increases phosphate values (CSABA and BÓKAY, 1977). Melatonin administration in adrenalectomized–thymectomized male rats on 1% saline is without effect on inorganic phosphate excretion, fluid intake, food intake and urine volume, the latter parameter being stabilized and synchronized with respect to its diurnal rhythm (HALPERN and MARTIN, 1972). After administration of pineal body extract a strong sodium retention lasting for at least 9 h has been observed (A. MACHADO and DA SILVA, 1962, 1963). In view of the presence of arginine vasotocin in the pineal gland it may also be relevant that [4-leucine, 8-arginine]-vasotocin has a distinct natriuretic action in rats under mild water load or isotonic saline load (MACHOVÁ and JOŠT, 1975).

Electrolyte imbalance is reflected in pineal structure and function. Rats maintained on a low salt diet show a decrease in pineal weight and fat content (PANAGIOTIS and HUNGERFORD, 1961a, b, 1962). Following excess sodium administration, compact fat droplets in rat pinealocytes change into ring-like structures (PROP, 1965). The pineal serotonin content decreases (BEAUVALLET et al., 1966), while the melatonin content increases (LYNCH et al., 1973b).

c) Urinary System

There is as yet little evidence for interrelationships between the epiphysis cerebri and the urinary system. Following pinealectomy, kidney weight has been reported as being lower than in controls (KINSON et al., 1968b) or as being unaffected (BRUINVELS et al., 1964). The renin content of the kidney was not affected (BRUINVELS et al., 1964).

Studies dealing with diuresis have yielded variable results. In view of the fact that the epiphysis contains antidiuretic activity (p. 256), one would expect that pinealectomy increases diuresis. However, diuresis was found to be unaffected (KARPPANEN et al., 1970a) or to be reduced, in which case diuresis reversed to normal after administration of pineal extracts (BARBAROSSA et al., 1960, 1961), which are known to have diuretic effects (MILIN and NEŠIĆ, 1958). Melatonin and serotonin appear to be antidiuretic (ARUTYUNYAN et al., 1964).

The urea content of the blood increases after pinealectomy, and administration of pineal powder reduces the amount of urea present in nephrectomized rats (I. MILCU et al., 1969).

In the bladder pinealectomy results in a thickening of the epithelial lining accompanied by an increase in lactate dehydrogenase activity (TIMMERMANS and DEVEČERSKI, 1972b).

d) Blood Pressure and Haematocrit

A regularly occurring effect of *pinealectomy* is a distinct and statistically significant increase in arterial *blood pressure* (ZANOBONI and ZANOBONI-MUCIACCIA, 1967; KARPPANEN et al., 1970a, b, 1973; KARPPANEN, 1974; NEUVONEN and KARPPANEN, 1970) which can be depressed by melatonin (S. HOLMES and SUGDEN, 1976). Both an increased activity of the vasomotor centres (KARPPANEN et al., 1973) and the renin–angiotensin–aldosterone system (KARPPANEN et al., 1970a; KARPPANEN and VAPAATALO, 1971) may be involved in this process. Which role a renin-like activity of the pineal (HAULICA et al., 1975) plays remains to be elucidated. In view of a possible involvement of angiotensin, it is interesting that administration of angiotensin II in the rat is followed by a marked loss of stainable fat (PANAGIOTIS and HUNGERFORD, 1962) and an increase in the amount of dense-core granules of the sympathetic nerve fibres in the pineal (PANAGIOTIS and HUNGERFORD, 1966). In spontaneously hypertensive rats the number of pineal β-adrenergic receptors may be higher than in normotensive rats, as shown by the fact that in the former isoproterenol, a β-adrenergic agonist, has a more pronounced effect on pineal NAT than in the latter (ILLNEROVÁ and ALBRECHT, 1975).

Under certain experimental conditions pinealectomy lowers the blood pressure, e.g. after adrenalectomy combined with unilateral nephrectomy, provided that the adrenals have been removed before puberty (JELÍNEK and KŘEČEK, 1968).

The haematocrit was not influenced by pinealectomy in one study (RELKIN, 1973) and was slightly increased from 42.5 ± 0.4 ml/100 ml to 44.5 ± 0.7 ml/100 ml in another (KARPPANEN, 1974).

e) Retina and Harderian Gland

In the *retina* melatonin formation, as assessed by HIOMT measurements, increases after *pinealectomy* (NAGLE et al., 1973a, 1974c). Increased melatonin levels after pinealectomy in the retina, as studied immunohistochemically, have been reported by BUBENIK et al. (1978). In aged rats, melatonin formation in the retina, as assessed by HIOMT measurements, is increased when compared to young adult ones (CREMER-BARTELS, 1978). Rod outer segment shedding is not affected by pinealectomy (TAMAI et al., 1978). Light-induced photoreceptor degeneration is likewise not affected by pinealectomy at 40 and 50 days of age, but pinealectomy has a beneficial effect when carried out at 60 days of age (RUDEEN and O'STEEN, 1979). As retina and pineal share a common antigen (KALSOW and WACKER, 1973, 1977), the role of the pineal in retina-induced experimental allergic uveitis has been investigated. These studies have shown that the pineal shows moderate to heavy lymphoid infiltrates (KALSOW and WACKER, 1978). In retinal pigment epithelium of *Cavia porcellus* melatonin has been described as leading to pigment aggregation (PANG and YEW, 1979).

That the *Harderian* glands and the epiphysis are functionally interrelated has been suggested by a number of studies. According to WETTERBERG et al. (1970a, b) in blinded suckling rats the Harderian glands act as photoreceptor organs mediating the depressive effect of light on pineal HIOMT activity. In *adult* female rats removal of the Harderian glands does not affect pineal NAT and HIOMT activities (REITER and KLEIN, 1971). Melatonin injections briefly elevate the porphyrin content of the Harderian gland (JOÓ and KAHÁN, 1975), and in blinded animals the pineal is responsible for the decrease in the formation and storage of porphyrin granules which occurs after blinding (CLABOUGH and NORVELL, 1974). Pinealectomy reverses the development of a female type of Harderian gland in castrated, blinded male hamsters (CLABOUGH and NORVELL, 1973).

f) Body Temperature

The daily rhythm of body temperature in rats is not affected by removal of the epiphysis cerebri (FIORETTI et al., 1974). Melatonin given subcutaneously lowers body temperature by 2°–3° C (ARUTYUNYAN et al., 1964).

g) Hair Growth, Hair Colour, Skin and Wound Healing

According to HOUSSAY et al. (1966a, b) and HOUSSAY and PAZO (1966), *hair growth* in mice is stimulated by pinealectomy and retarded by melatonin.

As regards *hair colour*, in guinea-pigs it was found that melatonin administration for 1 month had no significant effect on the relative proportion of dark to light coloured hair. However, 1 month after cessation of treatment this ratio was reduced below the normal control level, but only on the anterior abdominal wall of the black animals (CLIVE and SNELL, 1969).

The pineal organ clearly influences hair colour and moult in some photoperiodic mammals. Following the initial experiments in the weasel, *Mustela erminea*, showing that melatonin administration in the summer leads to the development of a new white winter coat (RUST and MEYER, 1969), HOFFMANN (1972, 1973, 1974, 1977) has thoroughly studied this aspect in the Djungarian hamster (*Phodopus sungorus*). In this species pelage change in the winter is accelerated by exposure to long photoperiods and this acceleration is delayed by administration of melatonin. In the summer, melatonin was effective only when the animals were kept under short photoperiods. Under these experimental conditions melatonin inhibited the moult into the winter pelage.

In canine *melanosis*, melatonin administration has been found to lead to a visible lightening of the melanotic areas within 48 h and to a desquamation of melanin deposits (RICKARDS, 1965). As psoralen is used in psoriatic skin disorders and leads to pigmentation, the effect of this substance on melatonin formation has been studied: in organ culture it was found to stimulate melatonin synthesis (rat: BÄCKSTRÖM and WETTERBERG, 1977).

Wound healing is retarded by pinealectomy due to a decrease in mitotic rate (WEICHSELBAUM et al., 1975).

h) Intestine

In the intestinal mucosa pinealectomy results in an increase of ^{32}P incorporation into DNA and RNA as well as an increase in mitotic rate (BINDONI and CAMBRIA, 1966, 1968; CAMBRIA and BINDONI, 1966).

In the isolated rat duodenum, melatonin (1.25×10^{-9} moles/ml) inhibits spontaneous contractions and suppresses the motile response of the duodenal segment to serotonin (QUASTEL and RAHAMIMOFF, 1965).

In pinealectomized rat fetuses remaining in utero, the epithelial cells of the distal part of the ileum are filled with large acidophilic, PAS-positive globules, which has been interpreted as meaning that protein absorption is shifted from postnatal to prenatal life (OWMAN, 1963a, b, 1964d).

i) Immune System

There is some evidence that the epiphysis cerebri affects the immune system. According to CSABA et al. (1966) pinealectomy in rats diminishes the primary response of antibody production and increases the secondary response.

The *thymus* responds to pinealectomy with an increase in weight (LOMBARD DES GOUTTES, 1967b; CHERDYNTSEV, 1967). In pinealectomized–hypophysectomized rats a 50% *reduction* of thymus weight occurred (M. VAUGHAN et al., 1973). Thymus weight is depressed by administration of a purified bovine pineal fraction (M. VAUGHAN et al., 1974). Histologically the thymus of pinealectomized rats is characterized by a disproportional increase of epithelial and connective tissue elements, accompanied by a decrease of lymphoid elements (CSABA and BARATH, 1975). Administration of pineal extract, by contrast, leads to hyperplasia and hypertrophy of thymocytes, hypertrophy of epithelial cells, and an increase in the number of Hassall's bodies (NEŠIĆ, 1961). A stimulatory effect on the reticular cells has been confirmed by DEVEČERSKI (1963). Pinealectomy increases $^{35}SO_4$ uptake of the thymus (CSABA et al., 1967), the ^{45}Ca (CSABA et al., 1967) and the ^{131}I uptakes (CSABA et al., 1970) being reduced.

Lymph nodes show a depressed proliferation after administration of an aqueous pineal extract (PETRONIO and DORI, 1957).

Spleens of pinealectomized rats exhibit a higher iron content and a greater accumulation of trypan blue in the histiocytes than do those of controls; in addition there are slight increases of acid phosphatase, TPN diaphorase, glucose-6-phosphate dehydrogenase and glutamate dehydrogenase activities (BARBOLINI et al., 1966). After removal of the spleen (TRENTINI et al., 1968) or in states of methyl-cellulose hypersplenism (TRENTINI et al., 1971) the pinealocytes show an increase in nuclear and nucleolar size accompanied by an increase in both nucleolar and cytoplasmic RNA.

The administration of pineal extracts in cases of immunological disorder has been suggested by KARLER (1962/63).

j) Mast Cells

Pinealectomy decreases mast cell numbers in the thymus (CSABA et al., 1968b) and retards ^3H-serotonin uptake of mast cells in the thyroid gland (CSABA and BARATH, 1974b). Pinealectomy combined with administration of cortisone or DOCA stimulates mast cell production (CSABA et al., 1965).

k) Cancer and Other Diseases

The possibility that the epiphysis cerebri might have some influence on the growth of tumours has been recognized for some time (DOBROVOLSKAIA-

ZAVADSKAIA and ZÉPHIROFF, 1940; AUBERT et al., 1970; LAPIN, 1976, review; 1978). The available evidence suggests that the epiphysis exerts an inhibitory effect on tumour growth, as shown by the fact that pinealectomy accelerates tumour growth in rats bearing Walker 256 carcinoma (RODIN, 1963; BARONE and DAS GUPTA, 1970) and in hamsters bearing melanomas (DAS GUPTA and TERZ, 1967a, b), while implantation of pineal tissue or administration of pineal extracts lead to some alleviation of symptoms and temporary regression of metastases (HOFSTÄTTER, 1950, 1959; BERGMANN and ENGEL, 1950; ENGEL and BERGMANN, 1951/52; SANDER and SCHMID, 1952).

In several studies it has been claimed that the pineal shows structural changes in subjects and animals bearing malignomas. A correlation was thought to exist between calcification of the pineal body and the presence of cancers of the head, neck, breast and metastatic carcinoma of the skull and brain (DREXLER et al., 1957), but later DREXLER (1963) arrived at different conclusions. RODIN and OVERALL (1967) noted that pineal glands from malignant cases were significantly larger than from non-malignant cases. TAPP and BLUMFIELD (1970b), on the other hand, found that the pineal glands of patients dying with malignant disease had significantly *lower* weights than those of patients dying from other causes. Since the differences in weight between the two groups of pineals diminished after decalcification of the pineal it was thought that the higher weight of the glands from patients without malignomas was due, at least in part, to the presence of a greater amount of mineral in these glands.

In rats bearing fibrosarcomas, the pineal shows a higher lipid content than in normal controls (HUXLEY and TAPP, 1972). Whether or not the pineal has anything to do with spontaneous adenomata in the rat anterior pituitary is not known: QUAY (1967b) found them in both pinealectomized and non-pinealectomized animals.

In two cases of Tay-Sachs disease the pineals showed a diffuse infiltration with cytosomes, accompanied by precocious puberty (RELKIN et al., 1971). In five cases of xeroderma pigmentosum an atrophy of the pineal gland has been described (MORI, 1973).

Parasite growth appears to be affected by the pineal as shown by the observation that pinealectomy abolishes the growth and division synchrony of *Plasmodium berghei* (ARNOLD et al., 1969a, b, c).

XII. Tumours of Pineal Tissue

It is beyond the scope of the present review to discuss in detail the occurrence and histological appearance of pineal tumours, which are rare in humans and extremely rare in animals (FRAUCHIGER, 1961; THOMPSON et al., 1961). According to ZÜLCH (1956, cited after JELLINGER, 1973) 0.4% of brain tumours are pineal neoplasms. In Japan they amount to 4.5% (ARAKI and MATSUMOTO, 1969; cited after JELLINGER, 1973). In America, among 4,000 neoplasms of the central nervous system 35 were pineal tumours (WURTMAN et al., 1968a). KITAY and ALTSCHULE (1954a), reviewing the literature on 437 pineal tumours, found that over 50% of the tumours have been found in patients 20 years of age and younger and that the incidence is three times as frequent in men as in women. In children 1–16 years of age parenchymal and non-parenchymal pineal tumours are equally frequent, the tumours being about five times more frequent in males than in females (KITAY, 1954b).

A characteristic feature of pineal tumours is their great variability: KITAY and ALTSCHULE (1954a) distinguish as many as 30 different types. Two types are of particular interest, namely pinealomas and pineocytomas: pinealomas because they are the commonest type of pineal tumours and pineocytomas because they supposedly represent a true pinealocyte tumour.

Pinealomas do not represent an homogeneous group of tumours. Ectopic pinealomas, frequently lying near the optic chiasm, structurally resemble pinealomas and may have an extrapineal origin, in particular from the pituitary. Electron microscopically, pinealomas have been shown to consist of lymphocyte-like cells and another type of cell resembling a pinealocyte in some respects (RAMSEY, 1966; CRAVIOTO and DART, 1973; TABUCHI et al., 1973). HIOMT activity has been detected in both a metastasis from a parenchymal pinealoma (WURTMAN and AXELROD, 1964b; WURTMAN et al., 1964d, 1965b) and an ectopic pinealoma (WURTMAN and KAMMER, 1966). A lack of HIOMT activity has been noted in a spinal cord metastasis of an ectopic pinealoma, but since the tumour tissue contained methanol-forming enzyme activity, which is characteristic of pituitary tissue, the "pinealoma" was probably of pituitary origin (SNYDER et al., 1967b). The ectopic pinealoma described by MISUGI et al. (1967) appears to belong to the same category of tumours. Since the neoplastic cells possessed an abundance of membrane-bound electron-dense granules measuring 300–500 nm in diameter, which have never been observed in normal pineal tissue, a pituitary origin is more likely than a pineal origin. The question whether or not melatonin can be used as a tumour marker in patients with a pineal tumour has not yet been settled (BARBER et al., 1978; ARENDT, 1978; TAPP, 1978; BARBER, 1978).

Enzyme histochemical studies of a pinealoma have revealed high activities of alkaline phosphatase, succinate dehydrogenase and β-glycerophosphate dehydrogenase and moderate activities of lactic dehydrogenase, glucose-6-phosphate dehydrogenase, isocitrate dehydrogenase, β-hydroxybutyric dehydrogenase, acid phosphatase, non-specific esterase, β-glucuronidase, glutamate dehydrogenase and malate dehydrogenase; leucine aminopeptidase activity could not be demonstrated (KAWASHIMA et al., 1965).

Pineocytomas (NIELSEN and WILSON, 1975) comprise cells resembling pinealocytes. These cells contain abundant microfilaments and microtubules, large mitochondria and empty alveolate vesicles. Dense-core vesicles measuring 250–300 nm in diameter are extremely rare and lie in cell processes near capillaries, similar to "normal" pinealocytes. Synaptic ribbons have been found neither in pinealomas nor pineocytomas.

According to MARSHALL and DAYAN (1964) the histological nature of the stromal reaction of pinealomas suggests an immune reaction by the body against this type of tumour. A totally calcified pinealoma has been investigated by MØLLER et al. (1979). The main constituent was calcium phosphate.

In patients with disseminated neoplasms, 4% show a metastasis to the pineal body (ORTEGA et al., 1951). Metastases to the pineal have been described in patients bearing tumours in stomach (JOYNER, 1962), bronchi and kidney (OUYANG and ROZDILSKY, 1966).

Further details of pineal tumours may be obtained from the following papers: FRIEDMAN (1947), MCGOVERN (1948), WALTON (1949), KALM and MAGUN (1950), RINGERTZ et al. (1954), HANSEN (1955), KAGEYAMA and BELSKY (1961), IWATA et al. (1962), WEBER (1963), M. DAVID et al. (1963), NISHII (1963), RUBIN and KRAMER (1965), LISS and SCHARENBERG (1965), SUZUKI and IWABUCHI (1965), DAYAN et al. (1966), POPPEN (1966), BESTLE (1968), MISUGI et al. (1967), JELLINGER et al. (1970), RUBINSTEIN and OKAZAKI (1970), KOMATSU et al. (1971), SCHINDLER et al. (1977), JENKIN et al. (1978), SOLARSKI et al. (1978), SUNG et al. (1978), DIAZ-FLORES et al. (1978), PECKER et al. (1978), CAMINS and SCHLESINGER (1978), DONAT et al. (1978), MCCORMACK et al. (1978), ONOYAMA et al. (1979) and others.

Neoplastic transformation of hamster pineal cells by oncogenic DNA viruses in vitro has been described by ORME et al. (1968).

C. The Non-mammalian Epiphysis Cerebri: Introductory Remarks

The non-mammalian epiphysis cerebri, especially that of lampreys, fishes, amphibians and reptiles, is structurally not directly comparable to that of mammals. Hence it is appropriate to make some general remarks before going into structural and functional details.

Several striking differences exist between the epiphysis cerebri of mammals and that of lower vertebrates. One is that the epiphyseal complex of the latter consists of two components, namely the pineal organ proper and the parapineal organ in fishes, the frontal organ in amphibians and the parietal, or third, eye in lacertilian reptiles. The second distinguishing feature is the presence of true photoreceptor cells in the lower vertebrates (see Fig. 190). Thirdly, there are major differences in the innervation pattern: in mammals afferent or pinealofugal fibres are lacking, while efferent or pinealopetal sympathetic fibres seem to predominate, at least functionally. By contrast, in lower vertebrates, especially fishes and amphibians, the pinealofugal or afferent fibres are prominent. The avian pineal organ resembles structurally that of mammals more closely than that of anamniotic vertebrates. Despite these differences, the pineal organs of all groups of vertebrates have in common that they are light sensitive and that they produce melatonin.

D. The Pineal Complex in Lampreys and Fishes

I. General Aspects

In lampreys and fishes the pineal complex consists basically of two distinct organs, the pineal organ and the parapineal organ, the latter being variable. There is as yet no evidence to relate the paraphysis cerebri to pineal structure and function. In contrast to the pineal complex, it is a separate telencephalic circumventricular organ (HOLMGREN, 1959a; WURTMAN et al., 1968a; LEONHARDT, 1980).

Structure and function of the pineal complex in lampreys and fishes have been reviewed by GLADSTONE and WAKELEY (1940), BARGMANN (1943), HOLMGREN (1959a), VAN DE KAMER (1965), KAPPERS (1965a), OKSCHE (1965a), FENWICK (1970a), HOFFMAN (1970), UECK (1974) and RALPH (1975).

II. Gross Anatomy of the Pineal Organ

A notable difference between the epiphysis cerebri of mammals and lampreys and fishes is that in mammals the pineal organ extends in a posterior direction whereas in lampreys and fishes the epiphyseal complex points anteriad. In both groups of vertebrates the distal part of the epiphysis cerebri may lie close to the skull. In lampreys and fishes either part of, or the entire, pineal organ may be hollow.

The pineal organ is basically an elongate structure consisting of a thin proximal stalk and a distal thickening. In some species the proximal part of the stalk also shows a thickening which, when compared with the distal thickening, is rather inconspicuous. As in mammals, the pineal organ originates from the area lying between the habenular and posterior commissures.

In *cyclostomes* the situation is not uniform with regard to the presence of a pineal organ. Among the myxinoids, *Myxine glutinosa* lacks a pineal in the adult state, but a pineal anlage is present in an early stage of development (ADAM, 1965). In the frequently investigated lamprey, *Lampetra planeri*, the pineal organ has the form of a dorsoventrally flattened vesicle (COLLIN, 1969a). The pineal stalk connecting the pineal vesicle with the epithalamus consists mainly of nerve fibres forming the pineal tract (MEINIEL and COLLIN, 1971) (Fig. 131). Projected onto the skull, the pineal vesicle is situated close behind

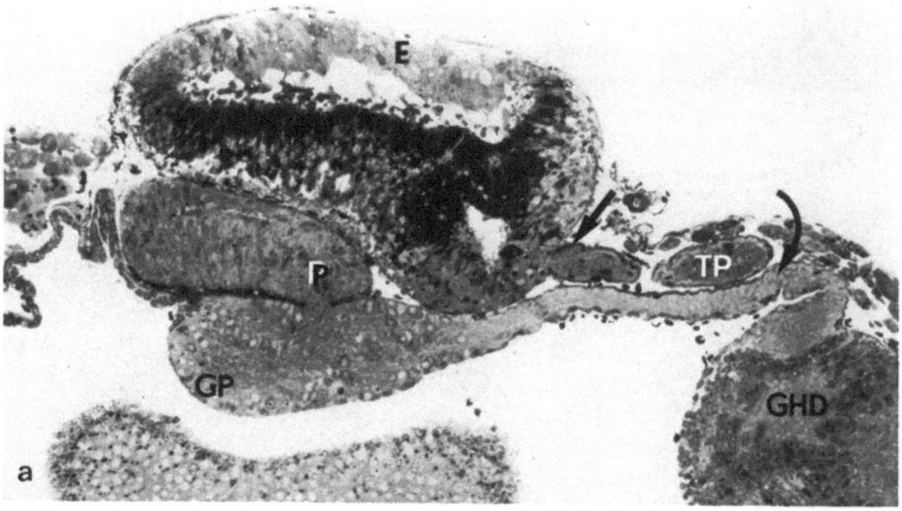

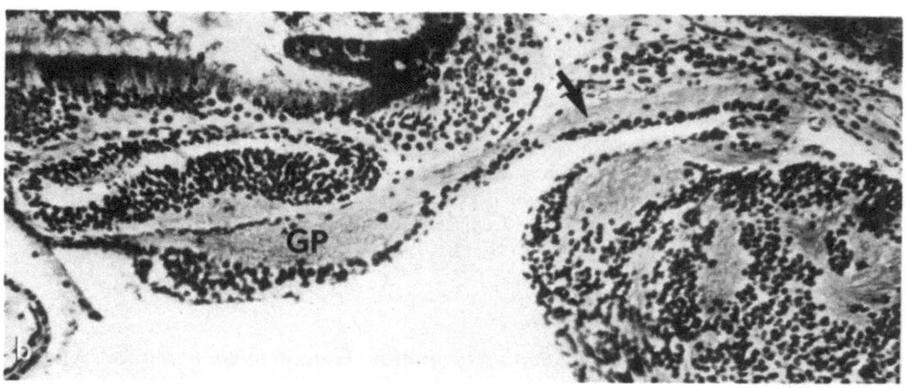

Fig. 131 a, b. Pineal region of *Lampetra planeri*. *E*, pineal organ proper; *GHD*, left habenular ganglion; *GP*, accumulation of ganglion cells; *P*, parapineal organ; *TP*, pineal tract, emerging from the base of the pineal organ (*straight arrow*). **a** Toluidine blue staining; ×110. **b** Staining according to Bodian; ×125. MEINIEL and COLLIN, 1971

the nasal orifice where it lies below a light spot characterized by typical epidermis and dermis and lacks cartilage (COLLIN, 1969a). The pineal organ appears to be highly differentiated also in *L. fluviatilis*, *L. zanandreai* (BERTOLINI and MANGIA, 1966) and *Geotria australis* Gray (EDDY and STRAHAN, 1968, 1970). It is poorly developed in *Mordacia mordax* Richardson (EDDY and STRAHAN, 1968, 1970).

In *elasmobranch* fishes, as in cyclostomes, the pineal organ is variably developed. It is apparently lacking in *Torpedo ocellata* and *T. marmorata* (BARGMANN, 1943; however, see UECK and KOBAYASHI, 1978, 1980). A frequently observed pattern is that of a long tubular structure which may be subdivided into an end-vesicle, a delicate stalk with a narrow lumen and a dilated part in the

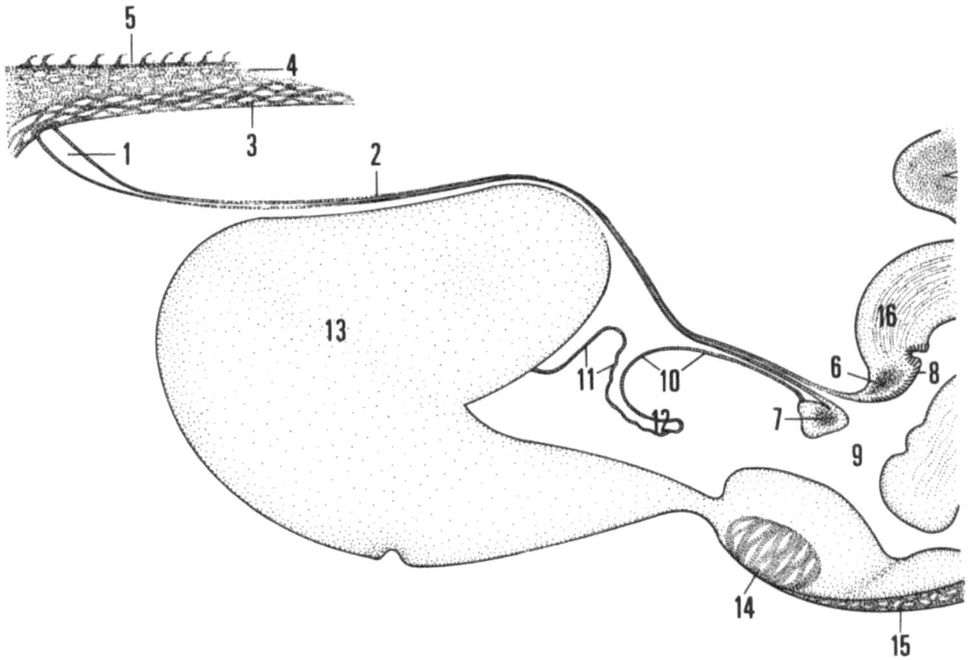

Fig. 132. Pineal region of the dogfish, *Scyliorhinus canicula*. Midsagittal section. *1*, pineal end-vesicle; *2*, pineal stalk; *3*, connective tissue of braincase; *4*, dermis; *5*, epidermis; *6*, posterior commissure; *7*, habenular commissure; *8*, subcommissural organ; *9*, third ventricle; *10*, dorsal sac; *11*, paraphysis; *12*, velum transversum; *13*, telencephalon; *14*, chiasma opticum; *15*, hypophysis; *16*, tectum opticum. RÜDEBERG, 1969a

proximal region of the stalk (*Squalus acanthias, Galeus melanostomus*: ALTNER, 1965; *Scyliorhinus canicula*: RÜDEBERG, 1969a). As in some teleost fishes, the pineal organ may be more or less solid, the intra-epiphyseal lumen being confined to the proximal part of the stalk (*Etmopterus spinax*: ALTNER, 1965).

Figure 132 shows the pineal organ of a dogfish, *Scyliorhinus canicula*, measuring 40–60 cm in length. In this specimen the total length of the organ is 13–16 mm, 2 mm of which are occupied by the end-vesicle. The dorsoventral diameter of the end-vesicle measures 0.3–0.4 mm and the horizontal diameter, 0.5–0.6 mm. The thin middle part of the stalk is only 0.05–0.07 mm in diameter. Proximally the stalk measures 0.3 mm in width over a distance of 0.2 mm. The distal end of the pineal end-vesicle is located in a shallow depression of the connective tissue layer that forms the roof of the braincase in the prefrontal foramen. The end-vesicle and the distal part of the pineal stalk are not covered by cartilage, and a pigment-free pineal spot is lacking in the epidermis (for further details see RÜDEBERG, 1969a).

In *teleost fishes*, as in elasmobranchs, the pineal organ consists basically of a slender stalk extending into a prominent distal thickening. Deviations from this basic pattern have occasionally been described. In *Labeo umbratus* (STEYN and WEBB, 1960) a pineal organ proper appears to be lacking though parapineal components are present. The absence of a pineal organ in *Thynnus thynnus*, as reported by RIVAS (1953), was later not confirmed (HOLMGREN,

1958b; MURPHY, 1971). Rod-shaped pineal organs with no distal thickening, or an inconspicuous one, have been found in *Symphosodon aequifasciatus* (HAFEEZ, 1971) and *Pterophyllum scalare* (BERGMANN, 1971). A slight thickening of the proximal part of the stalk is present in *Clarias lazera* (RIZKALLA, 1970). In the blind, cave-dwelling fish, *Typhlichthyes subterraneous*, the pineal organ is a small (140 μm × 40 μm) structure which lies in the immediate vicinity of, and dorsal to, the habenula and does not extend anterodorsally to contact the roof of the cranium (McNULTY, 1978c).

Typically shaped pineal organs have been found in *Rasbora daniconius*, *Chela bucaila*, *Puntius ticto*, *Garra mullya*, *Barilius bendelisis* (BELSARE, 1974a), *Phoxinus phoxinus* (BHARGAVA, 1973), *Oncorhynchus nerka* (HAFEEZ and FORD, 1967), *Gambusia affinis holbrooki* (CHÈZE and LAHAYE, 1969), *Carassius auratus* (TAKAHASHI, 1969), *Esox lucius* (OWMAN and RÜDEBERG, 1970), *Anguilla anguilla*, *Lebistes reticulatus* (RÜDEBERG, 1971), *Symphodus melops* (CHÈZE, 1969), 13 other species of teleosts (OMURA and OGURA, 1969), 12 species of deep sea fish (HOLMGREN, 1959a), *Astyanax mexicanus* (HERWIG, 1976), *Triphoturus mexicanus* (McNULTY and NAFPAKTITIS, 1976) and others. According to HOLMGREN (1969a) distal thickenings of the pineal organ in deep sea fishes are better developed and lie more closely related to the skull than in other teleosts. A particularly large pineal organ has been described in *Pleuronectes platessa*, in which it measures more than half the length of the entire brain (CHÈZE, 1966). The localization of the distal thickening of the pineal organ may depend on the size of the dorsal sac, which lies in the immediate vicinity of the pineal organ: HAFEEZ (1971), studying 15 marine and freshwater teleosts, has pointed out that in the majority of species in which the vault of the dorsal sac is high the thickening of the pineal organ lies close to the cranial roof, whereas in species lacking a dorsal sac the pineal lies close to the brain. *Gambusia affinis* (HAFEEZ, 1971) and *Lebistes reticulatus* (RÜDEBERG, 1971) apparently do not conform to this concept.

While the shapes of the pineal organs in teleosts are fairly uniform, there is great interspecies variation with regard to the internal structure of the pineal organ, ranging from hollow and extremely thin-walled organs to compact structures. This aspect will be dealt with on p. 327. The presence or absence of the so-called pineal window will be referred to on p. 356.

In the *lung fishes*, *Protopterus annectens* (DORN, 1957; HOLMGREN, 1959b) and *P. dolloi* (KAPPERS, 1965a; UECK, 1969), the pineal organs are elongated hollow structures with an end-vesicle, the lumen of the organ communicating with the third ventricle (Fig. 133).

III. Development of the Pineal Complex

There is little recent information available on the development of the pineal complex in lampreys and fishes. The older literature has been carefully reviewed by BARGMANN (1943) and HOLMGREN (1965). As can be expected from the

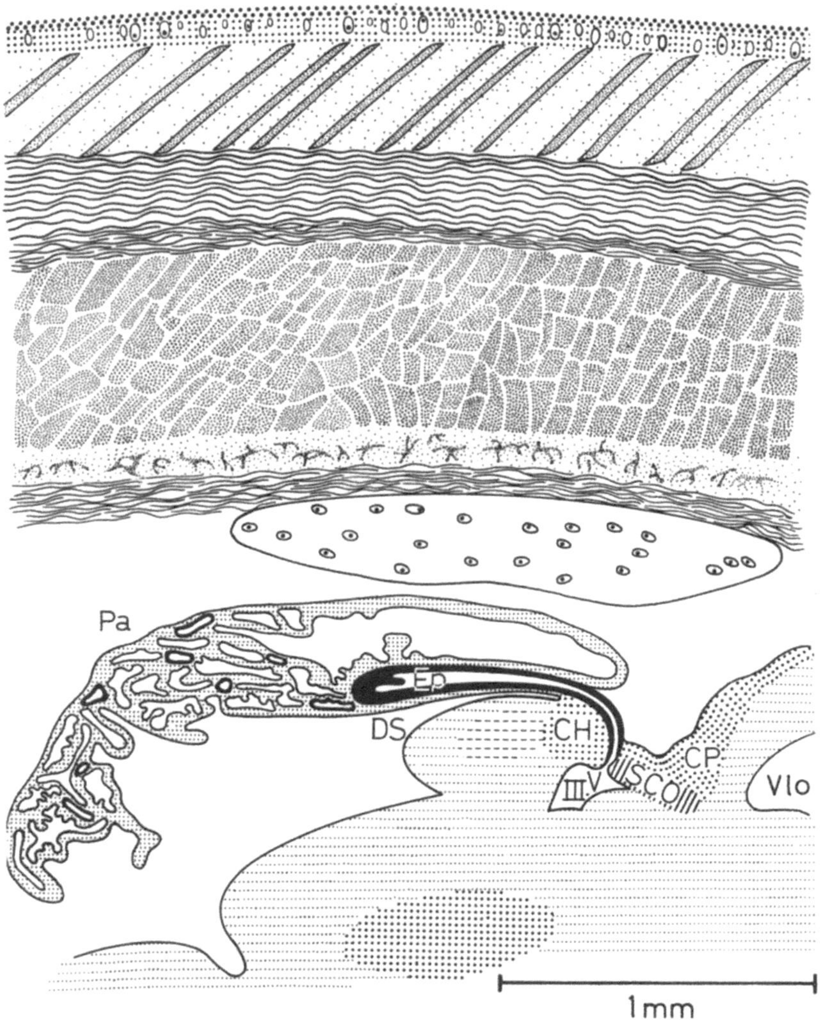

Fig. 133. Schematic representation of a midsagittal section through the pineal region of *Protopterus dolloi*. *CH*, habenular commissure; *CP*, posterior commissure; *DS*, dorsal sac; *Ep*, epiphysis cerebri; *Pa*, paraphysis; *SCO*, subcommissural organ; *Vlo*, ventriculus lobi optici; *III. V*, third ventricle. UECK, 1969

gross anatomy, the pineal complex develops from the diencephalic roof rostral to the posterior commissure. In both *Clarias batrachus* (BELSARE, 1974b) and *Triphoturus mexicanus* (MCNULTY and NAFPAKTITIS, 1976) pineal development starts with the formation of a solid cellular mass. Following elongation of the anlage accompanied by a narrowing of the stalk and a thickening of the distal part, a central lumen develops. The lumen is retained in *Triphoturus* and disappears in *Clarias*. In *Salvinellus fontinalis* embryos of 10 mm length (HOLMGREN, 1965) the pineal anlage is represented by two vesicles, posterior and anterior, giving rise to the pineal organ and the parapineal anlage respectively, the latter of which degenerates very soon. Whether the formation of the pineal anlage begins with the occurrence of a solid tissue mass, as in *Triphoturus* and *Clarias*, or with tubular evaginations is not known.

Virtually nothing is known about the development of the parapineal organ in other species. In *Clarias* a bud of cells lying rostral to the pineal organ on the left-hand side has been regarded as the parapineal anlage at 25 h after hatching, but at 40 h this mass was no longer visible (BELSARE, 1974b). In *Triphoturus*, which lacks a parapineal organ in the adult state, no parapineal anlage could be detected in larvae measuring 5 mm in length (MCNULTY and NAFPAKTITIS, 1976).

IV. Light Microscopy of the Pineal Organ

The light microscopic appearance of the pineal organ of lampreys and fishes has been described in detail by BARGMANN (1943). Since most of the findings reported are still valid, only a brief survey is given, supplemented by some important new observations.

The pineal parenchyma consists basically of three cell types: photoreceptor cells, supporting cells and ganglion cells, the histochemistry and ultrastructure of which will be dealt with on pp. 330, 333, 345, 347. While the composition of the pineal parenchyma shows relatively little variation, there are pronounced differences in the arrangement of the parenchyma.

In *lampreys* the dorsoventrally flattened epiphyseal vesicle is not uniformly built. The thickened basal part, but according to COLLIN (1968c, 1969a, b: *Lampetra planeri*) also the lateral regions, are differentiated into a retina. The dorsal part termed the pellucida differs from the retina in that ganglion cells are lacking and that it is thinner. In addition to the relatively large lumen of the epiphyseal vesicle, which can be regarded as a primary lumen, there is a secondary lumen in the caudal region of the retina, which appears to represent a vestige of the lumen of the pineal stalk. According to COLLIN (1969a) the primary lumen communicates with the secondary lumen.

In *elasmobranch fishes* the pineal organ is usually an elongated hollow structure with a prominent end-vesicle. The wall is relatively thin, giving rise to folds which project into the lumen. In *Scyliorhinus canicula*, RÜDEBERG (1969a) distinguished outer fibrous and inner nuclear zones in the pineal wall. In the pineal vesicle and in the proximal part of the pineal stalk the nuclear zone comprises four to five layers of nuclei arranged in the fashion of a pseudostratified epithelium, whereas in the remainder of the pineal stalk a single nuclear layer is seen. The fibrous appearance of the outer zone, where nuclei are only occasionally present, is due to the presence of cytoplasmic extensions of the parenchymal cells accompanied by nerve fibres running parallel to the longitudinal axis of the pineal stalk.

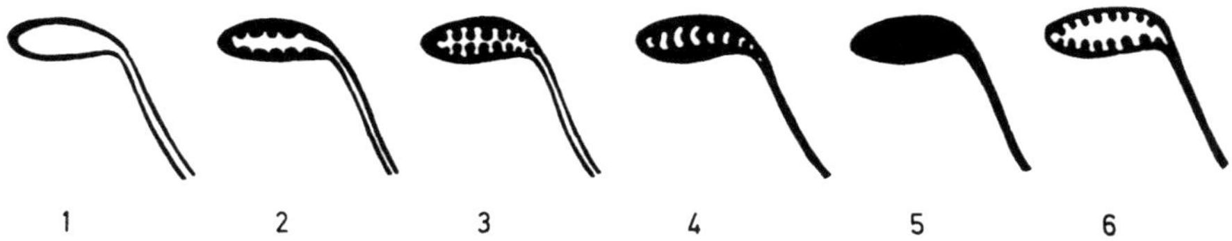

Fig. 134. Diagrammatic representation of six types of parenchymal arrangement in the teleost pineal. *1*, flat type; *2*, fold type; *3*, convoluted type; *4*, small space type; *5*, compact type; *6*, intermediate type. After OMURA and OGURI, 1969

The light microscopic appearance of the *teleost* pineal organ is of special interest in view of the concept that phylogenetically the epiphysis cerebri develops from a mainly sensory organ into a secretory one (for further details see p. 476). According to HAFEEZ (1971) tubular or saccular pineals with an extremely simple wall are photoreceptor-rich and hence mainly photoreceptive, whereas pronouncedly parenchymal, compact, highly vascularized pineal organs are poor in photoreceptors and involved in non-sensory functions, possibly secretion. Whether or not this concept is generally applicable in teleosts remains to be established, expecially since HAFEEZ's results are based on light microscopy alone. For further research along these lines a classification of pineal parenchyma into six types (OMURA and OGURI, 1969) may be useful (Fig. 134). In the *flat* type (1) there is a primitive flat epithelium surrounding a wide central lumen (*Carassius auratus, Cyprinus carpio*). The *fold* type (2) is characterized by a slightly thickened epithelium showing infoldings in places (*Plecoglossus altivelis*). In the *convoluted* type (3), thickness and infolding of the epithelium have increased to such an extent that a pineal lumen is reduced to a narrow space (*Salmo gairdneri irideus, Salvelinus fontinalis, Diaphus coeruleus, Zacco temmincki, Acanthogobius flavimanus*). Pineals belonging to the *small space* type (4) lack a continuous central lumen; consequently, there is no communication with the third ventricle; however, in some sites of the epithelium small luminal remnants can be seen (*Oryzias latipes, Lebistes reticulatus*). *Compact* type (5) pineal organs lack a central lumen and small spaces (*Pelteobagrus undiceps, Chrysophrys major*). Finally, there is an *intermediate* type (6) characterized by a wide lumen, epithelial infolding in some places and convoluted epithelium in the pineal stalk with some narrow spaces (*Mugil cephalus, Lateolabrax japonicus, Choerodon azurio*).

Since it is not always possible, on the basis of data in the literature, to classify the species investigated according to the above scheme, it appears more appropriate to distinguish just three main types of parenchymal arrangement, namely wide-spaced types, intermediate types and compact types. In the wide-spaced type the parenchymal wall is relatively thin; it may be flat or thrown into short folds. The purest of this type of pineal has been observed in *Symphysodon aequifasciatus* (HAFEEZ, 1971) and *Garra mullya* (BELSARE, 1974a). In these species the pineal walls consist of a simple flat to cuboidal epithelium devoid of infoldings. Parenchyma characterized by a few folds is found in *Gyrinocheilus aymoneri* and *Barbus tetrazona* (HAFEEZ, 1971). Species whose pineal walls are highly folded still leaving a fairly wide lumen include *Pterophyllum scalare*

(BERGMANN, 1971), *Nezumia liolepis* (McNULTY, 1976), *Triphoturus mexicanus* (McNULTY and NAFPAKTITIS, 1976), *Esox lucius* (OWMAN and RÜDEBERG, 1970), *Thynnus thynnus* (HOLMGREN, 1958) and *Salmo gairdneri* (KORF, 1974). *Engraulis mordax* (HAFEEZ, 1971) shows some degree of regional differentiation insofar as the distal part of the end-vesicle is distinctly saccular with a wide lumen, whereas proximally, where the vesicle merges with the stalk, the epithelium becomes more folded and compact, appreciably reducing the size of the central lumen. Wide-spaced type pineals are also present in *Sphyraena barracuda, Astyanax mexicanus, Cyprinodon baconi, Amierus nebulosus, Gymnosarda pelamis, Thunnus albacares, Nomeus gronovii, Bathygobius soporator, Haemulon sciurus, Spheroides spangler, Hippocampus reidi* and *Monacanthus ciliatus* (RASQUIN, 1958).

Intermediate type of pineal organs characterized by parenchyma with high folds resulting in a relatively narrow central lumen have been found in many species, including *Puntius ticto, Rasbora daniconius, Barilius bendelisis, Chela buccaila* (BELSARE, 1974a), *Phoxinus phoxinus* (OKSCHE, 1965a; BHARGAVA, 1973), *Anguilla anguilla* (OKSCHE, 1965a; RÜDEBERG, 1971), *Gambusia affinis* (CHÈZE and LAHAYE, 1969), *Bathylagus wesethi* (McNULTY, 1976), *Xiphophorus helleri, Hesperoleucus symmetricus, Zalembius rosaceus, Xiphister mucosus, Roccus saxatilis* (HAFEEZ, 1971), *Jenkinsia lamprotaenia, Sardinella macrophthalmia, Strongylura notata, Mollinesia sphenops, Gerres cinereus, Pseudupeneus maculatus, Irideo bivittata, Histrio histrio* (RASQUIN, 1958) and others.

Pineals of the compact type, which lack a central lumen or the lumen of which is hardly visible, are characteristic of a great number of deep sea fishes (HOLMGREN, 1959a), the blind cave fish, *Anoptichthys jordani*, and its ancestor, the eyed river fish, *Astyanax mexicanus* (HERWIG, 1976; see, however, RASQUIN, 1958), *Atherinopsis californiensis, Atherinops affinis, Corydoras aneus, Gibbonsia metzi* (HAFEEZ, 1971), *Lebistes reticulatus* (RÜDEBERG, 1971), *Atherina (Hepsetia) stipes, Carapus bermudensis* and *Belonesox belizanus* (RASQUIN, 1958).

Concerning the arrangement of the various cell types in the parenchyma, the paper by HAFEEZ (1971) should be consulted. Here it suffices to mention that the commonest type of pineal parenchymal organization and composition is one in which both the primary sensory cells and the supporting cells are abundant. The *sensory cells* lie predominantly close to the luminal aspect of the epithelium and possess a large basally located nucleus of oval or round shape. Two processes may be seen: an apical process projecting into the lumen and a basal process apparently representing sensory cell axons. With some exceptions, the apical process shows the usual outer and inner segments characteristic of vertebrate photoreceptor cells in general.

In *supporting cells* the cytoplasm is not readily discernible under the light microscope. They comprise nuclei of varying shapes ranging from irregular via elongate to round profiles. The supporting cells occur interspersed throughout the epithelium, but are predominant in its basal part. Supporting cells do not appear to be a homogeneous cell population. Most of these cells are ependyma-like, possessing a basal process extending to the underlying basement membrane. Another class of cells shows generally smaller, irregular nuclei, several cytoplasmic processes and a more basal localization.

Nerve cells are often difficult to identify in the pineal parenchyma by conventional light microscopy, one of the reasons being their scarcity. In some species (e.g. *Engraulis, Gambusia, Atherinopsis, Roccus*: HAFEEZ, 1971), in which the nerve cells are particularly large and comprise a large nucleus with a prominent nucleolus, they are generally located in the basal region of the epithelium. OMURA and OGURI (1969), by contrast, often noted them in the central region of the pineal epithelium. A notable difference from other nerve cells is a lack of Nissl substance (OMURA and OGURI, 1969). For further details see p. 347.

An important point of teleost pineal histology is that the sensory cells vary in amount and appearance. As regards the incidence, no precise quantitative data are available. According to OMURA and OGURI (1969) photoreceptor cells are abundant in *Carassius auratus, Cyprinus carpio, Mugil cephalus* and *Choerodon azurio*, all of which are characterized by a relatively large pineal lumen. In species with compact pineals (see above) photoreceptors are poorly developed (OMURA and OGURI, 1969; HAFEEZ, 1971). In *Symphosodon* regional differences in the amounts of photoreceptor cells have been observed: in the distal parts of the pineal, sensory cells are particularly numerous, whereas supporting cells predominate in the proximal part (HAFEEZ, 1971).

Frequently differences in the shape of the photoreceptors have been noted. These differences apply to the apical processes projecting into the pineal lumen. OMURA and OGURI (1969) described long, thick apical processes in *Carassius auratus* and short, round processes in *Choerodon azurio*. In *Osmerus mordax*, in which primary sensory cells of a secretory type are thought to be present, the apical process possesses a club-shaped ending of variable size, containing argyrophilic granules (HOLMGREN, 1959a). In some species outer and inner segments can be easily seen (HAFEEZ, 1971; TAKAHASHI, 1969; HAFEEZ and FORD, 1967). For further details on the structure of sensory cells, see p. 33.

Additional points of interest are that in *Corydoras* subcommissural-type cells may be present in the pineal, especially in its proximal part (HAFEEZ, 1971). In *Engraulis* (HAFEEZ, 1971) groups of cells lying in the proximal part of the end-vesicle, where the vesicle merges with the stalk, contain chrome alum haematoxylin-positive granules, resembling those found in ependymoglial cells. Material with identical staining properties is found in several species (*Grynocheilus, Barbus, Xiphophorus, Zalembius, Roccus*), especially in the proximal part of the pineal, where it lies in both intact or fragmented outer segments of the sensory cells.

The pineal organ of *lung* fishes, as represented by *Protopterus annectens*, does not differ from that in teleosts. According to HOLMGREN (1959b) supporting cells are rare, whereas secretory sensory cells are abundant.

V. Histochemistry of the Pineal Organ

The pineal organ of lampreys and fishes has been little studied by histochemical techniques. *Glycogen* has been demonstrated in *Fundulus heteroclitus* (HOLM-

GREN, 1959a), *Clarias lazera* (RIZKALLA, 1970), *Oncorhynchus nerka* (HAFEEZ and FORD, 1967), *Atherina stipes* (RASQUIN, 1958) and *Esox lucius* (OWMAN and RÜDEBERG, 1970). In *Esox* it is conspicuously concentrated in a small area in the ventral wall of the apical part of the end-vesicle. It is seen in lesser concentration along the basal surface of the entire ventral wall, and traces can be detected in various parts of the parenchyma. In *Clarias* glycogen is localized in cells lying near the outer and inner borders of the organ. The innermost pineal cells usually possess accumulations of glycogen granules at their free ends which are, most probably, sloughed off into the pineal lumen. In *Thunnus thynnus*, glycogen is found in supporting cells (MURPHY, 1971).

In some studies the PAS reaction was not very revealing (TAKAHASHI, 1969; BHARGAVA, 1973). In *Oncorhynchus* PAS-positive granules occur near the inner boundaries of pineal cells and in the pineal lumen; since acetylation prevents the PAS reaction, it was thought that the granules contain 1:2 glycol groups (HAFEEZ and FORD, 1967). In *Clarias batrachus* (BELSARE, 1974b) peripheral pineal cells show PAS-positive granules, whereas aldehyde fuchsin-positive cells lie in the central region of the organ. The performic acid alcian blue-PAS technique gives negative results in both juvenile and adult *Oncorhynchus* (HAFEEZ and FORD, 1967).

Lipid droplets are not detectable in *Clarias*; the occurrence of minute granules and rodlets after Sudan black B staining has been attributed to the presence of mitochondria (RIZKALLA, 1970).

Metachromatically reacting material (toluidine blue staining) has been described in the pineal lumen of *Anguilla anguilla* (HARTWIG and KORF, 1978).

Of great functional significance is the finding that the fish pineal, like that of mammals, contains serotonin. There is therefore a possibility that, like in mammals, serotonin is converted into melatonin (see below). In *Salmo gairdneri* and *Atherinopsis californiensis* a diffuse yellow (serotonin) fluorescence can be detected by means of the Falck-Hillarp technique, without suggestion of any regional differences or of localization to any particular cell type, provided that the animals had been pretreated with either 5-hydroxytryptophan or the MAO inhibitor, iproniazid, or both (HAFEEZ and QUAY, 1969). In *Esox lucius* an intense yellow fluorescence is obtained without pharmacological pretreatment, the fluorophore being evenly distributed in the cytoplasm of both sensory and supporting cells (Fig. 135; see also FALCON, 1978); microspectrofluorometric analysis reveals an excitation spectrum with a maximum at 380 nm and an emission curve with a maximum at 515 nm, pointing to the presence of serotonin (OWMAN and RÜDEBERG, 1970) (Fig. 136). In *Lampetra planeri* yellow serotonin fluorescence has been noted in both the atrium and along the pineal tract, showing a decrease in intensity in animals killed in January when compared with those killed in September and October (MEINIEL, 1978). This fluorescence is restricted to (a) a particular type of rudimentary photoreceptor cells and (b) to true pinealocytes. These rudimentary photoreceptor cells probably establish a second line of elements of the receptor type. They display a specific lamellar organelle serving the storage of 5-hydroxytryptophan (MEINIEL, 1980).

Corroborative evidence for a participation of the fish pineal in serotonin metabolism comes from studies showing that radioactive 5-hydroxytryptophan

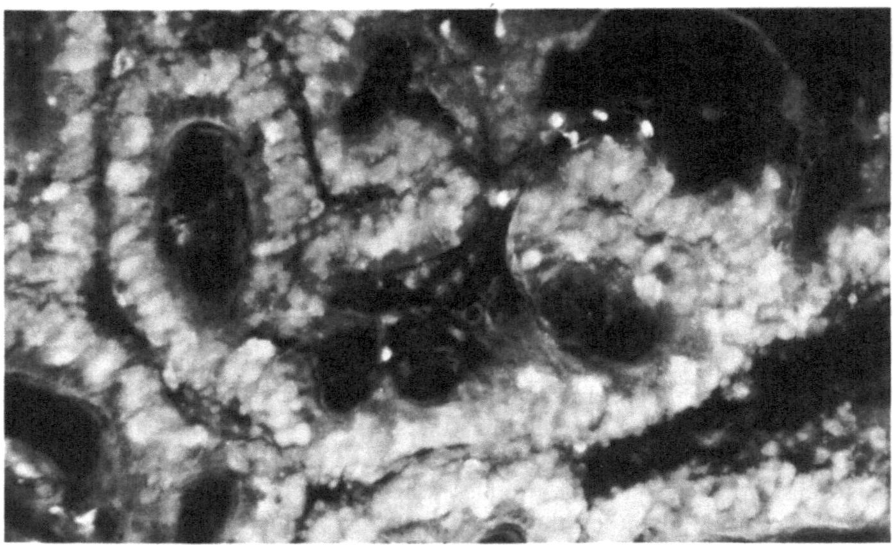

Fig. 135. Fluorescence photomicrograph (Falck-Hillarp technique) of pineal organ of *Esox lucius*. Yellow indole (serotonin) fluorescence is seen in both sensory and supporting cells throughout the gland. OWMAN and RÜDEBERG, 1970; ×220

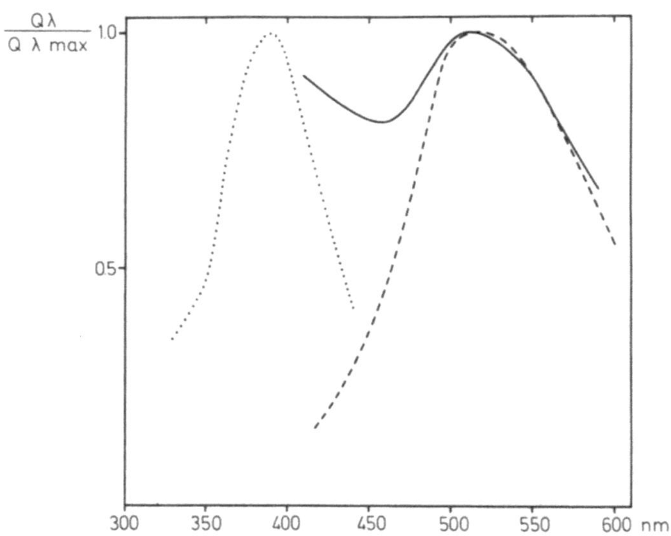

Fig. 136. Microspectrofluorometric recordings from the pineal organ of *Esox lucius*. The readings were made from the most intensely yellow fluorescent structures (cf. Fig. 135). The formaldehyde-induced fluorophore has an excitation maximum at 380 nm (*stippled line*); the emission curve has a maximum at 515 nm with a tendency of a second peak at 420 nm (*solid line*); the second peak disappears after ozone treatment of the sections (*broken line*). OWMAN and RÜDEBERG, 1970

and 5-hydroxytryptamine are rapidly and very strongly incorporated into the pineal organ of *Salmo gairdneri* (OGURI et al., 1968; HAFEEZ and ZERIHUN, 1976). Autoradiographic studies at the light microscopic level confirm the finding that serotonin is localized in both sensory and supporting cells; nerve cells are not labelled (HAFEEZ and ZERIHUN, 1976). Electron microscopically it has

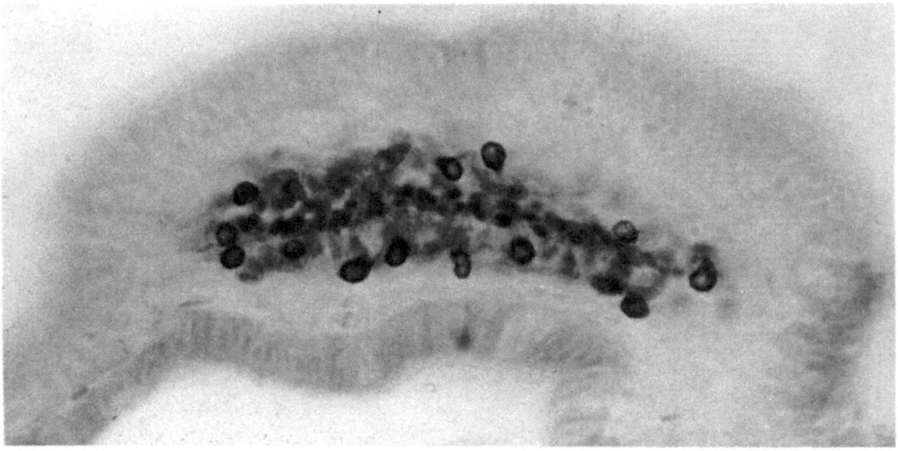

Fig. 137. Longitudinal section of the pineal stalk of a goldfish. Acetylcholinesterase reaction. Note the intensely labelled nerve cells, and, in the centre of the specimen, the reaction product in the inner segments of sensory cells. WAKE, 1973; ×600 (reduced to 80%)

been shown that the labelling of the sensory cells is more intense than that of the supporting cells (FALCON, 1978; *Esox*).

Histochemical *enzyme* studies have turned out to be an excellent means of demonstrating the presence of nerve cells in the fish pineal (WAKE, 1973; KORF, 1974), details of which are to be described on p. 347. Here it suffices to mention that, apart from nerve cells, the acetylcholinesterase reaction results in a moderately strong staining reaction of the inner segments of sensory cells of the pineal stalk in *Carassius* (WAKE, 1973) (Fig. 137), whereas in *Salmo gairdneri* (KORF, 1974) the inner segments stain only occasionally, particularly in the rostral part of the pineal organ. For the demonstration of MAO activity, see FALCON (1978).

VI. Ultrastructure of the Pineal Organ

Electron microscopic studies have clearly confirmed the light microscopic impression that the pineal parenchyma of fishes consists of photoreceptor cells, supporting cells and nerve cells. In this section only the photoreceptor and supporting cells are considered. As regards the nerve cells, the reader is referred to p. 349.

1. Photoreceptor Cells

For a detailed description of the structure and function of vertebrate photoreceptors the publications by RODIECK (1973), DARTNALL (1972), FUORTES (1972)

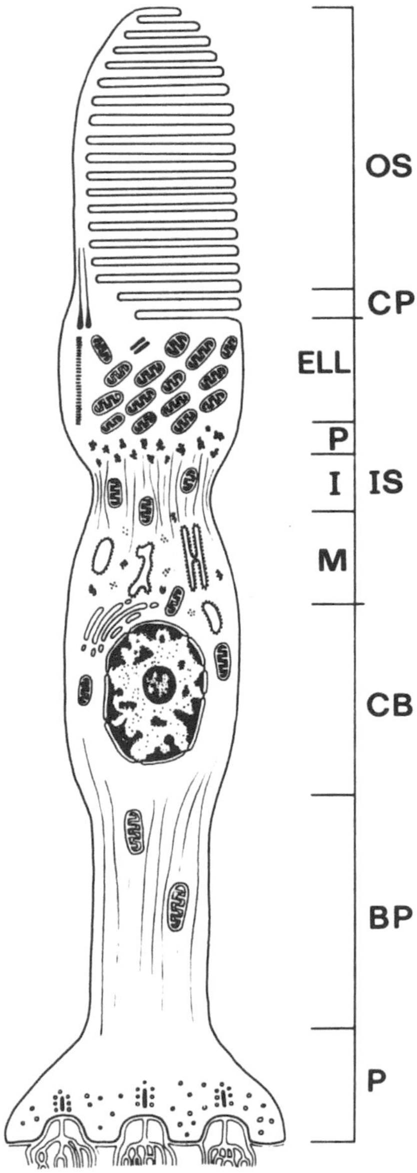

Fig. 138. Schematic representation of a photoreceptor cell. *BP*, basal process; *CB*, cell body; *CP*, connecting piece; *ELL*, ellipsoid; *I*, isthmus; *IS* inner segment; *M*, myoid; *OS*, outer segment; *P*, paraboloid. Note that presence and localization of isthmus and paraboloid may vary considerably

and ROHEN (1964) should be consulted. Here a brief survey of the basic structure of photoreceptors suffices in order to see the data reviewed below in the right perspective.

Vertebrate photoreceptors consist basically of an outer segment, a connecting stalk, an inner segment, a cell body with the nucleus and a basal process terminating with a pedicle in cones and with a spherule in rods (see Fig. 138).

A diagrammatic representation of the pineal photoreceptor cells and their derivatives is given in Fig. 190.

Under the electron microscope the outer segments are seen to be filled by lamellar structures arranged perpendicular to the photoreceptor axis. These lamellae or discs are derived from infoldings of the plasma membrane covering the outer segment and undergo renewal. The outer segment is connected to the inner segment by an eccentrically situated narrow cytoplasmic stalk, the centre of which is occupied by a 9+0 cilium. The cilium arises from one of the two centrioles of the cell which lie in the apical part of the inner segment. The cilium extends part way into the outer segment, giving rise to a narrow region devoid of discs. Ciliary rootlets originating from the basal body of the cilium pass through the inner segment and split above the nucleus. An oil droplet is occasionally present in the region below the centriole and the basal body. A regular feature is the presence of what is called an ellipsoid. It lies in the distal half of the inner segment and consists of an accumulation of mitochondria. Just below the ellipsoid a paraboloid which contains glycogen in high concentration may be present, intermingled with membranes of the endoplasmic reticulum stemming from the myoid. The myoid region of the inner segment comprises free ribosomes, rough endoplasmic reticulum and the Golgi apparatus. The cell body with the nucleus does not present any unusual features. It gives rise to a basal process which terminates with a flat conical foot, the pedicle, in cones, whereas a rounded swelling termed a "spherule" is seen in rods. A typical feature of the terminals is the presence of synaptic contacts characterized by synaptic vesicles and synaptic ribbons.

In the pineal of *lampreys* and *fishes* typical and atypical photoreceptor cells are present, the latter group comprising degenerating typical photoreceptor cells and rudimentary photoreceptor cells. This aspect has been most thoroughly discussed by FALCON (1978, 1979a, b). Precise quantitative data on the proportions of typical photoreceptor cells and rudimentary photoreceptors are apparently not available. In *Esox lucius* the interesting observation has been made that in the proximal and distal regions of the pineal typical photoreceptor cells predominate, whereas in an intermediate zone the rudimentary photoreceptors are particularly numerous (FALCON, 1978, 1979a, b). Very little is known about the exact numerical relationships between photoreceptor cells and supporting cells. For some species the ratio of photoreceptor cells to nerve cells ("convergence ratio") has been calculated, but this aspect will be dealt with in another section (see p. 351).

Apart from the general ultrastructure of the photoreceptors, which is basically similar to that described above (compare Figs. 138 and 190), three main aspects are of particular interest, namely the ultrastructural appearance of the outer segments, the presence or absence of synaptic contacts between photoreceptors and second order neurones and the question whether morphological correlates of secretory products are present in photoreceptors.

a) Outer Segments

Relatively little is known about the shape of the outer segments. Scanning electron microscopic studies carried out in a few species only have shown that

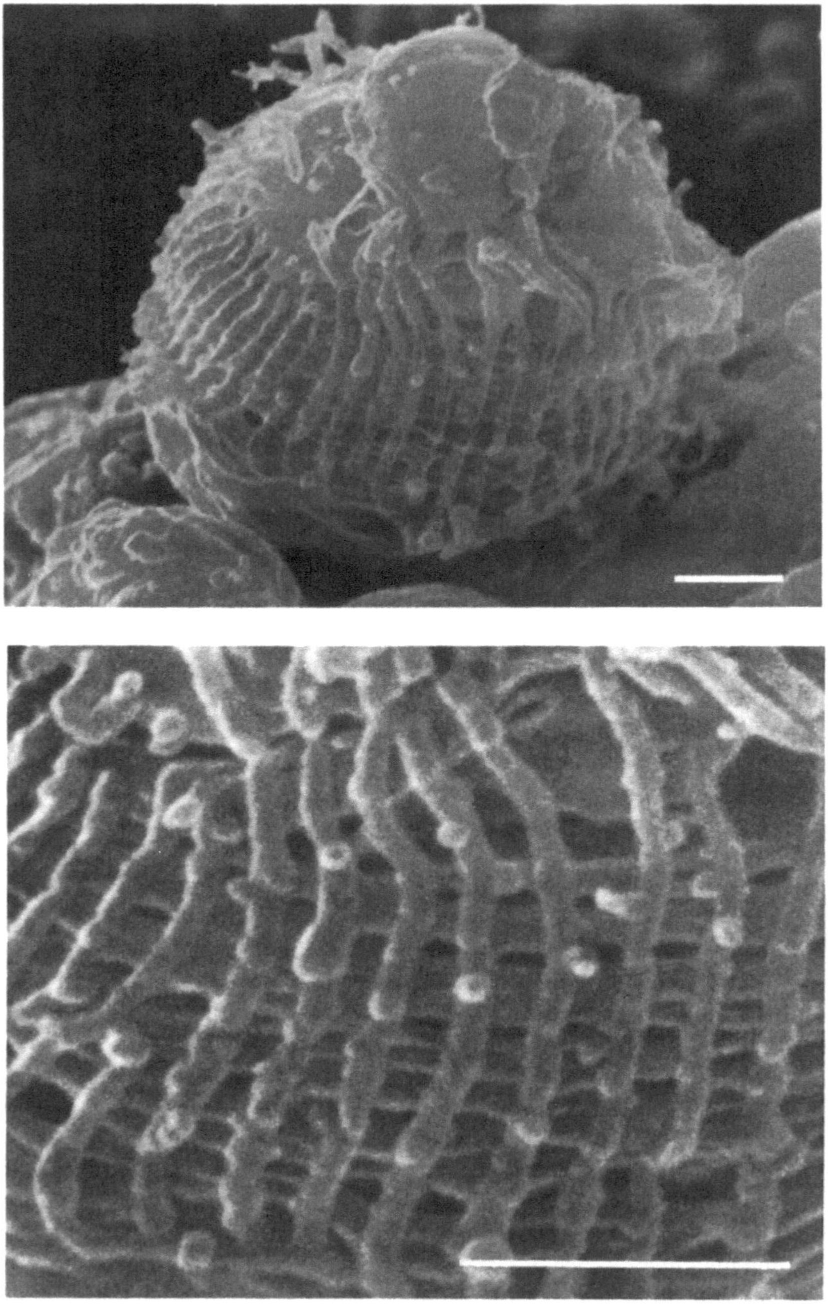

Fig. 139. Scanning electron micrographs of outer segment of a pineal photoreceptor cell of *Carassius gibelio langsdorfi*. *Top*, outer segment partly surrounded by filamentous processes which end freely at the top of the outer segment or are connected by a transverse cytoplasmic bridge. At the top of the outer segment degenerated and partly segregated lamellar material is seen. ×14,500. *Bottom*, high magnification of a part of an outer segment showing the peripheral border of the lamellar discs and the filamentous processes in a rectangular arrangement. Note that the filamentous processes show small protrusions. ×44,000. UECK et al., 1978

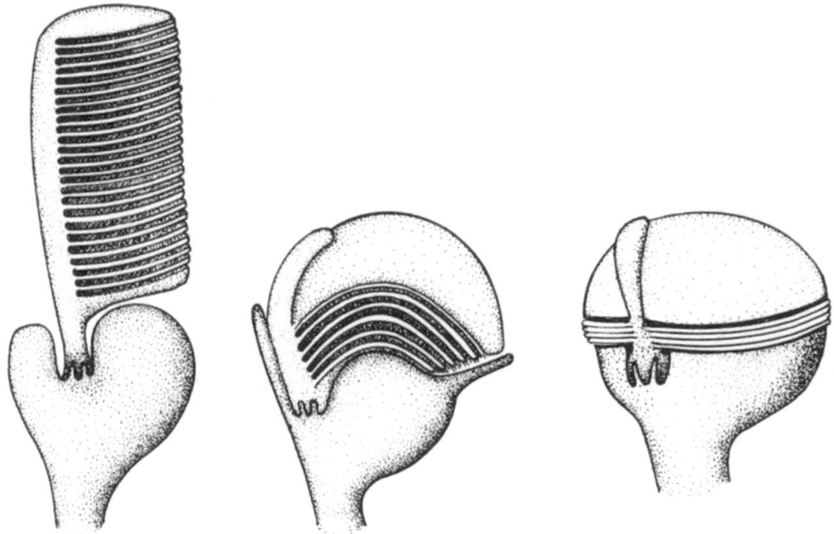

Fig. 140. Three different types of outer segments of pineal photoreceptor cells. *Pterophyllum scalare.* BERGMANN, 1971

a number of different types can be distinguished. In *Salmo gairdneri*, HARTWIG and PFAUTSCH (1973) described slender outer segments deflected in a semicircular fashion (type A), cap-like segments (type B) and cap-like segments possessing a club-shaped structure (probably corresponding to the region of the cilium) lying in a furrow extending almost to the vertex of the outer segment (type C). In *Carassius gibelio langsdorfi* (UECK et al., 1978) a slender type, a dome-like type, a cap-like type and a helical type were demonstrated. Detailed analysis revealed that some of the outer segments are covered by filamentous material comparable to calycal processes of retinal photoreceptors (Fig. 139). The tips of the outer segments may be concave or convex. Based on *transmission* electron microscopic findings, BERGMANN (1971) classified the outer segments into the forms shown in Fig. 140. Whether and how these different forms relate to pineal photoreceptor function is not known.

A striking difference to retinal photoreceptors is the presence of small numbers of lamellar discs in fish pineal photoreceptors. While the number of discs in fully differentiated retinal rods lies in the region of 1,000–2,000 per outer segment, teleost pineal outer segments usually contain not more than 50 discs, the highest reported number being 275; in the cyclostomes, elasmobranchs and lung fishes investigated, outer segments comprise between 30 and 150 discs (for further details, see Table 29). Typical outer segments are shown in Figs. 141–144. In *Typhlichthyes subterraneous*, a blind, cave-dwelling fish, the outer segments are apparently completely degenerated (MCNULTY, 1978c).

The functional significance of the variable number of discs is not exactly known. MCNULTY (1976) and MCNULTY and NAFPAKTITIS (1977), studying deep sea fishes, observed a trend for increased numbers of discs in deeper-

Table 29. Number of saccules ("discs") in outer segments of pineal photoreceptor cells in fishes

Species	Incidence	Author(s)
Lampetra planeri	30–130	Collin (1968a, c, 1969a, b)
Petromyzon marinus	High, as judged from electron micrographs	Eakin (1963)
Scyliorhinus canicula	40–150	Rüdeberg (1968a, 1969a)
Astyanax mexicanus	30–50	Herwig (1976), Omura (1975)
Phoxinus phoxinus (= *laevis*)	Up to 70	Oksche and Kirschstein (1971)
	30–40, up to 70	Oksche and Kirschstein (1967)
Pterophyllum scalare	50–70, rarely more	Bergmann (1971)
Esox lucius	20–30	Owman and Rüdeberg (1970), Falcon (1978, 1979b)
Anguilla anguilla	20–30, occasionally 10–50	Rüdeberg (1971)
Lebistes reticulatus	10–50	Rüdeberg (1971)
Triphoturus mexicanus	12 ± 0.48[a]	McNulty (1976)
Bathylagus wesethi	30 ± 1.62	McNulty (1976)
Nezumia liolepis	38 ± 2.34	McNulty (1976)
Sardina pilchardus sardina	20–50	Rüdeberg (1968b)
Latimeria chalumnae	275	Hafeez and Merhige (1977)
Carassius auratus	40–50	Takahashi (1969)
Plecoglossus altivelis	Not counted	Omura et al. (1969)
Tarletonbeania crenularis	14.03 ± 0.38[b]	McNulty and Nafpaktitis (1977)
Diaphus theta	14.90 ± 0.59	McNulty and Nafpaktitis (1977)
Lampanyctus ritteri	13.35 ± 0.52	McNulty and Nafpaktitis (1977)
Parvilux ingens	21.23 ± 0.43	McNulty and Nafpaktitis (1977)
Chologaster agassizi	17	McNulty (1978a)
Protopterus dolloi	100–130	Ueck (1969)

[a] Mean $\pm$ 2 SEM
[b] Mean $\pm$ 1 SEM.

dwelling species. In the blind cave fish, *Astyanax mexicanus*, typical outer segments have been reported to disappear after exposure to either continuous illumination or constant darkness (Omura, 1975). Herwig (1976), in the same species, described an age-dependent gradual regression of the outer segments. However, Oksche and Kirschstein (1971) were unable to detect any striking morphological changes between young and adult *Phoxinus phoxinus*.

Atypical outer segments appear to fall into three main categories. Outer segments characterized by a cytoplasmic protrusion only or small number of short discs have been described in *Lampetra planeri* and compared to *rudimentary* photoreceptors found in Sauropsida (Collin, 1968a, c, 1969a, b; Meiniel, 1980). Cytoplasmic projections which are devoid of discs and are clearly attached to cilia fall into the same category, though it is not known whether they are regenerating or regressed outer segments (Oksche and Kirschstein, 1967, *Phoxinus phoxinus*). Vesiculation of discs (see Breucker and Horstmann, 1965) has been attributed to poor fixation (Oksche and Kirschstein, 1971), whereas Rüdeberg (1969a) considers vitamin A deficiency to be another possible cause of these alterations. Finally, the shape of the outer segments and the plane of sectioning have a bearing. Oksche and Kirschstein (1971) pointed out that

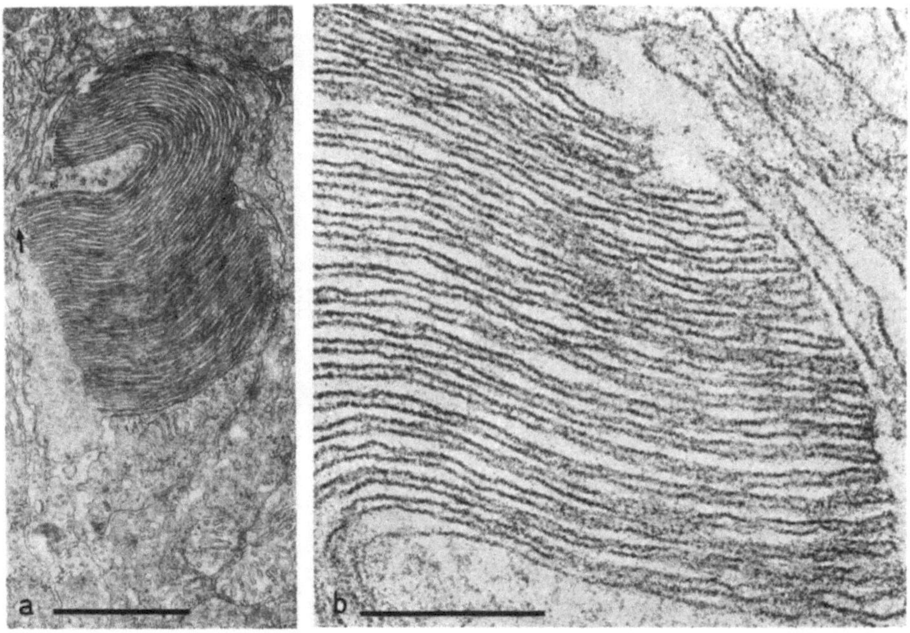

Fig. 141 a, b. Outer segments with lamellar discs of pineal photoreceptor cells of *Phoxinus phoxinus* (=*laevis*). **a** *Arrow*, invagination of plasmalemma. Bar=1 μm. ×22,000. **b** Bar=0.25 μm. ×99,000. After OKSCHE and KIRSCHSTEIN, 1967

a circular arrangement of discs may be due to sectioning deflected outer segments and may not necessarily represent degeneration.

Another complicating feature is that the outer segments of pineal photoreceptors may undergo cyclic renewal, as has been demonstrated for the photoreceptors of lateral eyes. HERWIG (1976) also considers the possibility that discs may be continuously shed, rather than cyclically.

b) Presence of Secretory Granules

As fish pineal cells show light microscopically, what can be interpreted as signs of apocrine secretion and in view of the fact that the pineal of higher vertebrates synthesizes melatonin, morphological correlates of secretory products have been carefully searched for. What looks like apocrine secretion is apparently nothing more than disintegrating outer segments which are shed into the pineal lumen (OKSCHE and VAUPEL-VON HARNACK, 1965b; TAKAHASHI, 1969) where the remnants are taken up by macrophages (RÜDEBERG, 1971; HERWIG, 1976).

Morphological signs of an endocrine activity of fish photoreceptor cells are virtually non-existent. Frequently secretory granules have been searched for, but these were either not present (OKSCHE and KIRSCHSTEIN, 1966a, 1967; RÜDEBERG, 1966, 1971; BERGMANN, 1971; TAKAHASHI, 1969) or so small in number (*Sardina*: RÜDEBERG, 1968b; *Lampetra*: COLLIN, 1969a; *Scyliorhinus*: RÜDE-

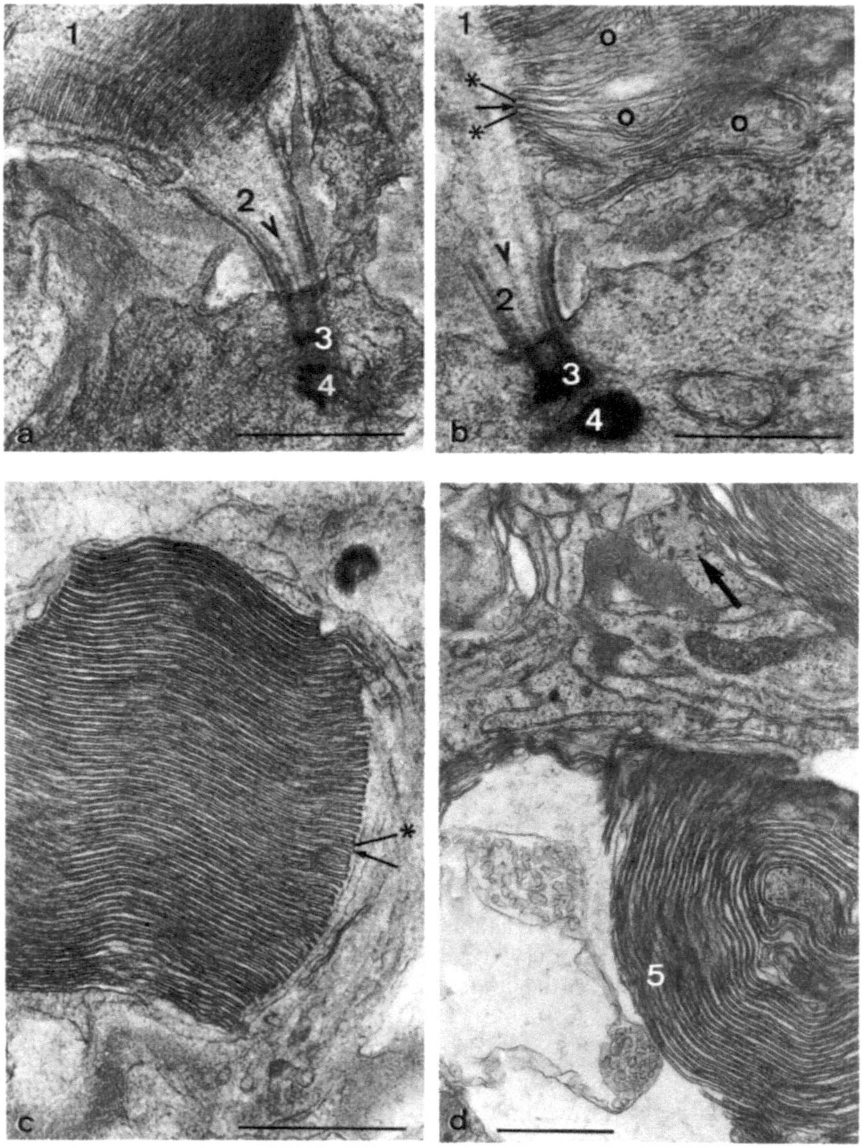

Fig. 142a–d. Regularly lamellated outer segments of pineal photoreceptor cells of *Phoxinus laevis*. **a, b** Axially sectioned outer segment (*1*) with connecting piece (*2*), centrioles (*3, 4*) and fibrils of cilia (*V*). In **a** the invaginated lamellae are regularly arranged; in **b** a few tubular and vesicle-like (*o*) structures are interspersed. *Asterisks* denote invaginations which are in continuity with the free surface of the outer segment. *Arrow*, lamellae of the outer segments. **a** ×22,400; **b** ×22,500. **c** Typical picture of lamellated outer segment. ×22,500. **d** Lamellated outer segment, sectioned vertically to its longitudinal axis and partly tangentially (*5*). *Arrow*, transverse section through cilium of the receptor type (9+0 pattern). ×15,600. OKSCHE and KIRSCHSTEIN, 1971

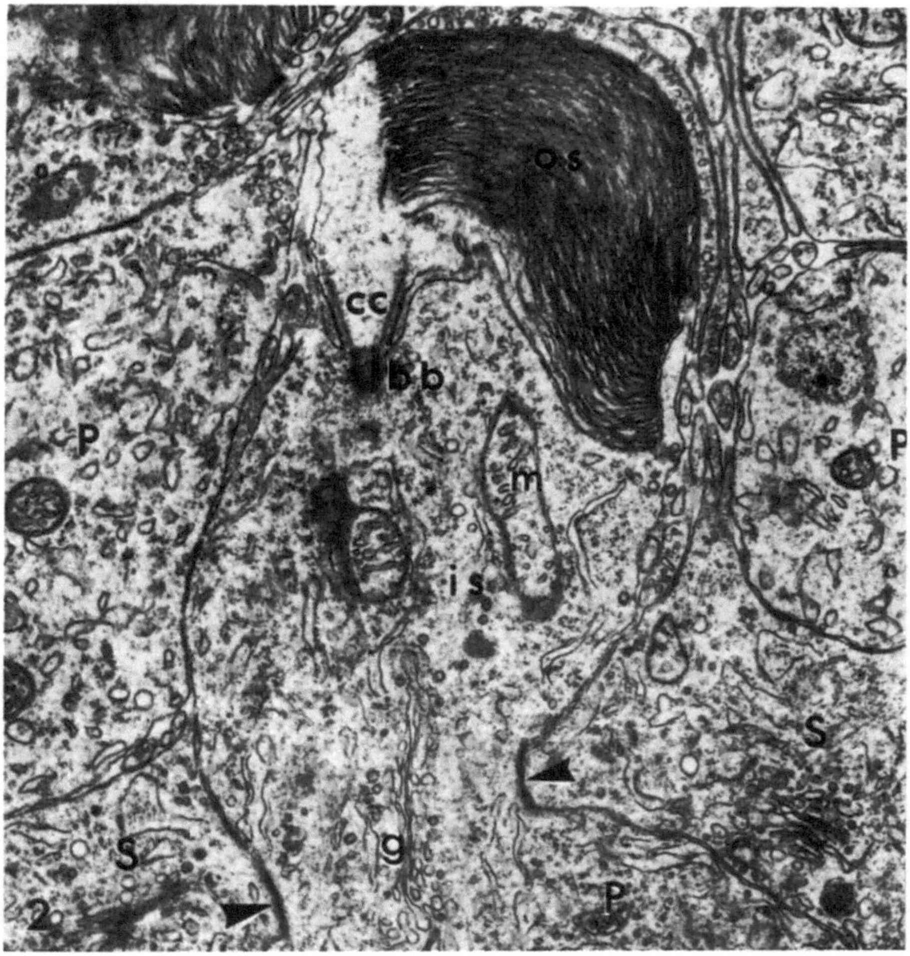

Fig. 143. Typical pineal photoreceptor cell of the blind cave fish, *Astyanax mexicanus*. *P*, photoreceptor cell; *bb*, basal body; *cc*, connecting cilium; *g*, Golgi apparatus; *is*, inner segment; *os*, outer segment; *m*, mitochondrion; *S*, supporting cell; *arrow*, terminal bars. OMURA, 1975; ×13,000

BERG, 1969a; *Astyanax*: HERWIG, 1971; *Latimeria chalumnae*: HAFEEZ and MERHIGE, 1977; *Typhlichthyes subterraneous*: MCNULTY, 1978c) that they appear to be of little functional significance. Their size, lying mostly between 50 and 100 nm, suggests that they are amine storage granules (COLLIN, 1969a). Pictures like that published by BREUCKER and HORSTMANN (1965) showing large amounts of dense-core vesicles located in a structure which resembles a transformed outer segment have never again been seen. The localization of the small number of dense-core vesicles in basal processes of photoreceptors (COLLIN, 1969a; HERWIG, 1976) suggests that the vesicles are involved in synaptic transmission rather than secretion.

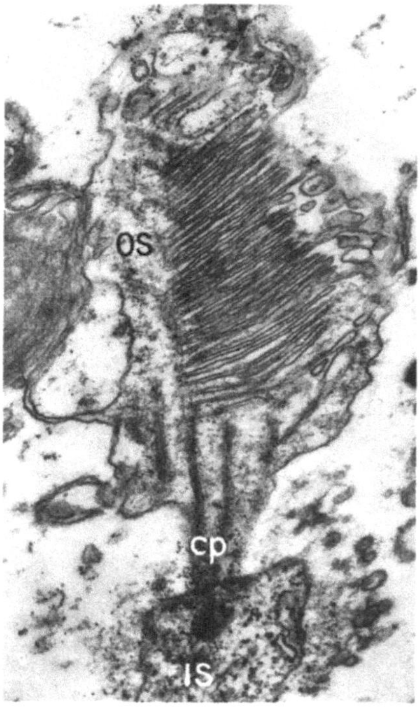

Fig. 144. Apical part of a pineal photoreceptor cell of a dogfish, *Scyliorhinus canicula*. *IS*, inner segment; *OS*, outer segment; *cp*, connecting piece. RÜDEBERG, 1968a; ×36,000

c) Synaptic Connections

As the fish pineal is characterized by the presence of fairly well developed photoreceptor cells showing practically no morphological signs of endocrine secretion, the connections of photoreceptor cells with afferent nerve fibres are of special interest. As will be recalled, the sites of synaptic transmission in retinal photoreceptors are characterized by the presence of conspicuous synaptic ribbons.

In fish pineal photoreceptors synaptic ribbons (Fig. 145) are fairly variable in amount. Whether or not this variability is due to the presence of circadian changes of synaptic ribbons as has been shown in the fish retina (WAGNER, 1973, 1975a, b) and in mammalian pineal organs (VOLLRATH, 1973; KURUMADO and MORI, 1977) remains to be seen. Synaptic ribbons have not been detected in *Uranoscopus*, *Mugil* (RÜDEBERG, 1966), *Typhlichthyes subterraneous* (MCNULTY, 1978c) and *Triphoturus mexicanus* (MCNULTY and NAFPAKTITIS, 1976); they are rare in *Scyliorhinus* (RÜDEBERG, 1969a), *Lebistes* (RÜDEBERG, 1971), *Thunnus* (MURPHY, 1971), *Latimeria* (HAFEEZ and MERHIGE, 1977), *Esox* (OWMAN and RÜDEBERG, 1970) and *Pterophyllum* (BERGMANN, 1971); moderate amounts have been seen in *Sardina* (RÜDEBERG, 1968b) and large amounts in *Lampetra* (COLLIN and MEINIEL, 1968a, b; COLLIN, 1968c), *Phoxinus* (OKSCHE and KIRSCHSTEIN, 1971) and especially *Anguilla* (RÜDEBERG, 1971). Table 30 gives a comparison of the dimensions of synaptic ribbons. In *Latimeria* the

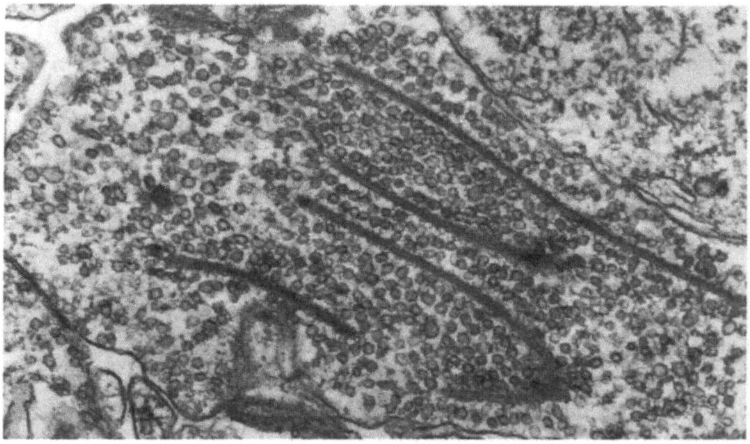

Fig. 145. Synaptic ribbons in pineal photoreceptor cells of *Sardina*. Note the great length of the ribbons and the large number of synaptic vesicles. RÜDEBERG, 1968b; ×39,000

Table 30. Synaptic ribbons in teleost fishes

Species	Ribbons		Size of vesicles (nm)	Author(s)
	Length (µm)	Width (nm)		
Phoxinus phoxinus (= *laevis*)	0.5–1.0	40	35	OKSCHE and KIRSCHSTEIN (1971)
Sardina	1.5	30–80	40–60	RÜDEBERG (1968b)
Astyanax mexicanus	0.75	55	30–60	HERWIG (1976)
Anguilla anguilla	0.5–1.8	40–60 local thickenings up to 200	40–60	RÜDEBERG (1971)
Carassius auratus	Up to 1.0	30–35	50	TAKAHASHI (1969)

interesting observation has been made that the synaptic ribbons studied lacked the vesicles which normally surround the electron-dense ribbon (HAFEEZ and MERHIGE, 1977).

Although the synaptic ribbons are conspicuous landmarks of prospective sites of synaptic contacts, the synaptic connections are usually difficult to trace, a notable exception being *Lampetra*. In *Lampetra planeri* and in ammocoetes of *Lampetra* (COLLIN, 1968d, 1969b; COLLIN and MEINIEL, 1968a, b), in which photoreceptors outnumber nerve cell perikarya, the postsynaptic elements have been clearly identified as dendrites which are usually not invaginated into the basal processes of photoreceptors. The connectivity between the two elements appears to be elaborate: a single basal process was frequently seen to be in contact with more than one dendrite, and a single dendrite has often been found to contact three basal processes of photoreceptors, but it could not be distinguished whether the processes belonged to *one* photoreceptor cell or to a number of them. In 75% of the synaptic contacts examined, the postsynaptic (dendritic) membrane exhibited thickenings characteristic of chemical synapses. Occasionally synaptic vesicles were found postsynaptically.

In the *elasmobranch, Scyliorhinus*, ribbon-bearing synapses are altogether rare; since receptor cell processes and postsynaptic elements frequently comprise synaptic vesicles without ribbons, RÜDEBERG (1969a) concluded that ribbon synapses were not frequent enough to represent the only site of transmission and postulated that "ordinary" synapses are responsible for the afferent transmission.

In *teleost* fishes the synaptic contacts between receptor cells and afferent fibres are also poorly understood. The only point that appears to have been clarified is that the postsynaptic elements are usually not invaginated into the receptor cell (OKSCHE and KIRSCHSTEIN, 1971; RÜDEBERG, 1971). The difficulty of clearly defining the synaptic contacts in teleosts is due to a number of reasons. Synaptic contacts often occur in neuropil areas, in which a great number of different and highly tortuous processes are intermingled so that it may be difficult to distinguish dendrites, axons and processes of supporting cells. Also the neuropil areas themselves are variable. In *Phoxinus laevis* the neuropil area of older specimens is more prominent than that of younger ones (OKSCHE and KIRSCHSTEIN, 1971). In some species typical neuropil areas are lacking, e.g. in *Nezumia liolepis* (MCNULTY, 1976). As mentioned above, in some species synaptic ribbons are either not present or are so inconspicuous that they may have been overlooked. An additional complicating feature is that synaptic ribbons may lie closely related to supporting cells (*Chologaster agassizi*; MCNULTY, 1978a), which are not very likely to play a role in afferent transmission. That not every synaptic contact observed in the fish pineal may be one between a photoreceptor and a dendrite is highlighted by the observation that in *Astyanax mexicanus* synapse-like structures are present between adjacent photoreceptor cell processes (HERWIG, 1976). Finally, identification of the direction of transmission is sometimes rendered difficult because both the basal processes of photoreceptor cells and the supposedly postsynaptic elements contain synaptic vesicles (e.g. *Esox lucius*, OWMAN and RÜDEBERG, 1970). In view of this complexity it is not surprising that the postsynaptic elements often could not be identified with certainty (e.g. *Phoxinus laevis*: OKSCHE and KIRSCHSTEIN, 1971; *Esox lucius*: OWMAN and RÜDEBERG, 1970) or that no attempt was made to identify them (*Sardina*: RÜDEBERG, 1968b; *Bathylagus, Nezumia*: MCNULTY, 1976; *Anguilla, Lebistes*: RÜDEBERG, 1971). A fairly convincing picture of a ribbon synapse between a photoreceptor cell basal process and a nerve cell has been published by HERWIG (1976).

Although typical synaptic contacts between photoreceptor cells and second order neurons are difficult to demonstrate, it is probably safe to conclude, from a comparative point of view and from electrophysiological experiments (see below), that sites where synaptic ribbons lie closely related to the plasma membrane of photoreceptors are strong candidates for sites of synaptic transmission. Moreover, transmission to dendrites may occur by means of tight junctions (OWMAN and RÜDEBERG, 1970).

According to BERGMANN (1971), in *Pterophyllum scalare* basal processes of photoreceptors may terminate, with or without synaptic contacts, on *axons*.

In *lung fishes* synaptic contacts have apparently not been studied.

d) Miscellaneous Findings in Photoreceptors

As pointed out above, fish pineal photoreceptors show a structure similar to that of retinal photoreceptors. In this section some findings which appear to be peculiar to *pineal* photoreceptors will be mentioned. In *outer* segments of *Scyliorhinus* abundant mitochondria have been observed (RÜDEBERG, 1969a), while in those of *Protopterus* a fairly large amount of endoplasmic reticulum and microtubules is present (UECK, 1969). A 9+2 cilium has been seen in *Parvilux ingens* (MCNULTY and NAFPAKTITIS, 1977). The connecting piece of *Clevelandia ios* gives rise to cytoplasmic processes measuring 50–80 nm in thickness, which contain what could be ribosomes (MCNULTY, 1978b). Indications that the inner segment may be involved in phagocytosis have been obtained in *Carassius* (TAKAHASHI, 1969). Cytoplasmic sheets projecting into neighbouring cells were seen in *Triphoturus* (MCNULTY and NAFPAKTITIS, 1976).

Neck-like constrictions are found at different sites of the photoreceptors: in *Carassius* (TAKAHASHI, 1969) and *Scyliorhinus* (RÜDEBERG, 1969a) they lie between perikaryon and inner segment; in *Lampetra* (COLLIN, 1969a) between ellipsoid and myoid. In photoreceptor cells of *Typhlichthyes* (MCNULTY, 1978c) mitochondria with a crystalline inclusion have been seen.

2. Supporting Cells

The regularly occurring supporting cells which lie interspersed between photoreceptor cells generally do not show any unusual features. They are slender cells extending throughout the pineal wall. Frequently cytoplasmic processes project into neighbouring cells. Apically the cells are often studded with microvilli and occasionally cilia are seen. The cells comprise the usual cell organelles in moderate amounts. For further details the papers by COLLIN (1969a), RÜDEBERG (1969a), BREUCKER and HORSTMANN (1965), OKSCHE and KIRSCHSTEIN (1967), RÜDEBERG (1968b), OWMAN and RÜDEBERG (1970), TAKAHASHI (1969), BERGMANN (1971), MURPHY (1971), MCNULTY and NAFPAKTITIS (1976, 1977), MCNULTY (1978a, c) and UECK (1969) should be consulted.

Very little is currently known about the function of these cells. apart from the fact that they may be supportive. In *Lampetra* the interesting observation has been made that in the supranuclear region the supporting cells comprise crystalline structures probably containing guanine, which may serve to reflect light. In some species supporting cells have been seen to be involved in phagocytosis of degenerated discs of outer segments (Fig. 146) (UECK, 1969; COLLIN, 1969a). Enough lysosomes are definitely present to digest the material taken up by phagocytosis. A secretory function of the supporting cells has also been considered, most recently by MCNULTY (1978a, b), who found a well developed apparatus for protein synthesis in *Chologaster* and *Typhlichthyes*. However,

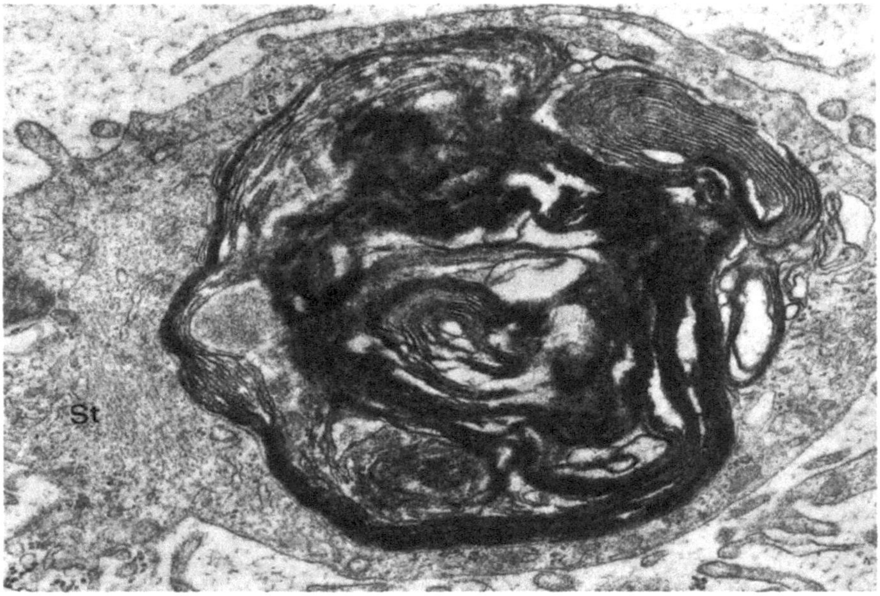

Fig. 146. Altered outer segment taken up by supporting cells (*St*). UECK, 1969; ×40,000

since dense-core vesicles are present in small numbers only, there is little direct morphological evidence for an endocrine activity of these cells. It should be recalled (p. 331) that both supporting and photoreceptor cells appear to be involved in serotonin metabolism. However, according to FALCON (1978) supporting cells apparently do not possess the capacity to *synthesize* indoleamines. The occurrence of a fluorophore or autoradiographic labelling may rather depend on an uptake of material produced in elements of the receptor line.

The ultrastructure of *nerve cells* is dealt with in the next section.

VII. Innervation of the Pineal Organ

The older literature on pineal innervation in fishes has been thoroughly reviewed by KAPPERS (1965a, b, 1971a). Hence, we can concentrate on the more recent findings. The situation in fishes is characterized by the fact that, in contrast to mammals, afferent pinealofugal fibres are prominent, whereas the presence of efferent (=pinealopetal) fibres is doubtful. The afferent fibres form the pineal tract. Although the synaptic connections of the photoreceptor cells are poorly understood (p. 343), there can be little doubt that the fibres of the pineal tract are functionally related to the photoreceptor cells, as has been shown by electrophysiological studies (DODT, 1963a; MORITA, 1966a; HANYU and NIWA, 1970).

1. Afferent Innervation

In most species of fishes relatively little is known about the *nerve cell perikarya*, the axons of which form the afferent pineal tract. There are two main reasons for this. First, the number of nerve cell perikarya is distinctly smaller than that of photoreceptors (see below: convergence ratio). Secondly, light and electron microscopically the nerve cells do not differ strikingly from neighbouring cells and often lack the staining properties typical of nerve cells, which is due to the fact that they are poor (TAKAHASHI, 1969) or lacking (OKSCHE and KIRSCHSTEIN, 1967; RÜDEBERG, 1966, 1968b) in Nissl substance. Whether or not the perikarya of the pineal tract lie outside the wall of the pineal organ is an unexplored field.

Staining reactions which have failed to demonstrate ganglion cells in the fish pineal organ include: gallocyanin, toluidine blue, methylene blue (*Scyliorhinus*: RÜDEBERG, 1969a) and iron haematoxylin (*Phoxinus laevis*: OKSCHE and KIRSCHSTEIN, 1971). Gallocyanin gave poor results in *Sardina*, whereas with Bodian's protargol method nerve cells could be clearly demonstrated (RÜDEBERG, 1968b). There are also clear interspecies differences, as shown by the fact that with the Bodian-Ziesmer technique clearly identified nerve cells could be demonstrated in *Atherinopsis* but not in *Salmo* (HAFEEZ, 1971).

The method of choice for detecting intrapineal nerve cells and for the study of their localization is the application of the histochemical method for acetylcholinesterase (AChE) according to KARNOVSKY and ROOTS (1964) (WAKE, 1973; KORF, 1974). With this technique nerve cells can be demonstrated so elegantly (Figs. 137, 147) that all the results obtained by conventional staining methods should be viewed with great care. For example, while with conventional staining methods a small number of nerve cells were found in *Carassius* (TAKAHASHI, 1969), the application of the AChE technique revealed a fairly large number of these cells (Fig. 137) (WAKE, 1973). Table 31 shows in which species nerve cells have not been identified or where they are apparently small in number. It can be seen that we are far from fully understanding the neuronal network and its various components. In view of the fact that in many species nerve cells have not been clearly identified in the pineal organ, special emphasis should be laid on the problem as to whether (a) processes of sensory cells do not form synapses in the pineal but rather extend directly to the epithalamus (see below), (b) the perikarya of the pineal tract are perhaps mainly located in the pineal stalk or in the epithalamus, and (c) the afferent connections are present at all (see below), which may well be the case in species in which fully functioning photoreceptors are not present in the pineal.

Studies by RÜDEBERG (1968b), WAKE (1973) and KORF (1974) have clearly demonstrated that the nerve cells present in the fish pineal are not evenly distributed in the organ. *Carassius* (WAKE, 1973) and *Salmo* (KORF, 1974), which have been examined with the AChE technique, have in common the fact that in the stalk region nerve cells are more densely packed than in the end-vesicle. In the end-vesicle nerve cells are prominent in the rostrolateral regions (*Carassius*) and the rostromedial part (*Salmo*). In *Sardina* (RÜDEBERG, 1968b) the protargol method revealed a group of ten nerve cells which lie

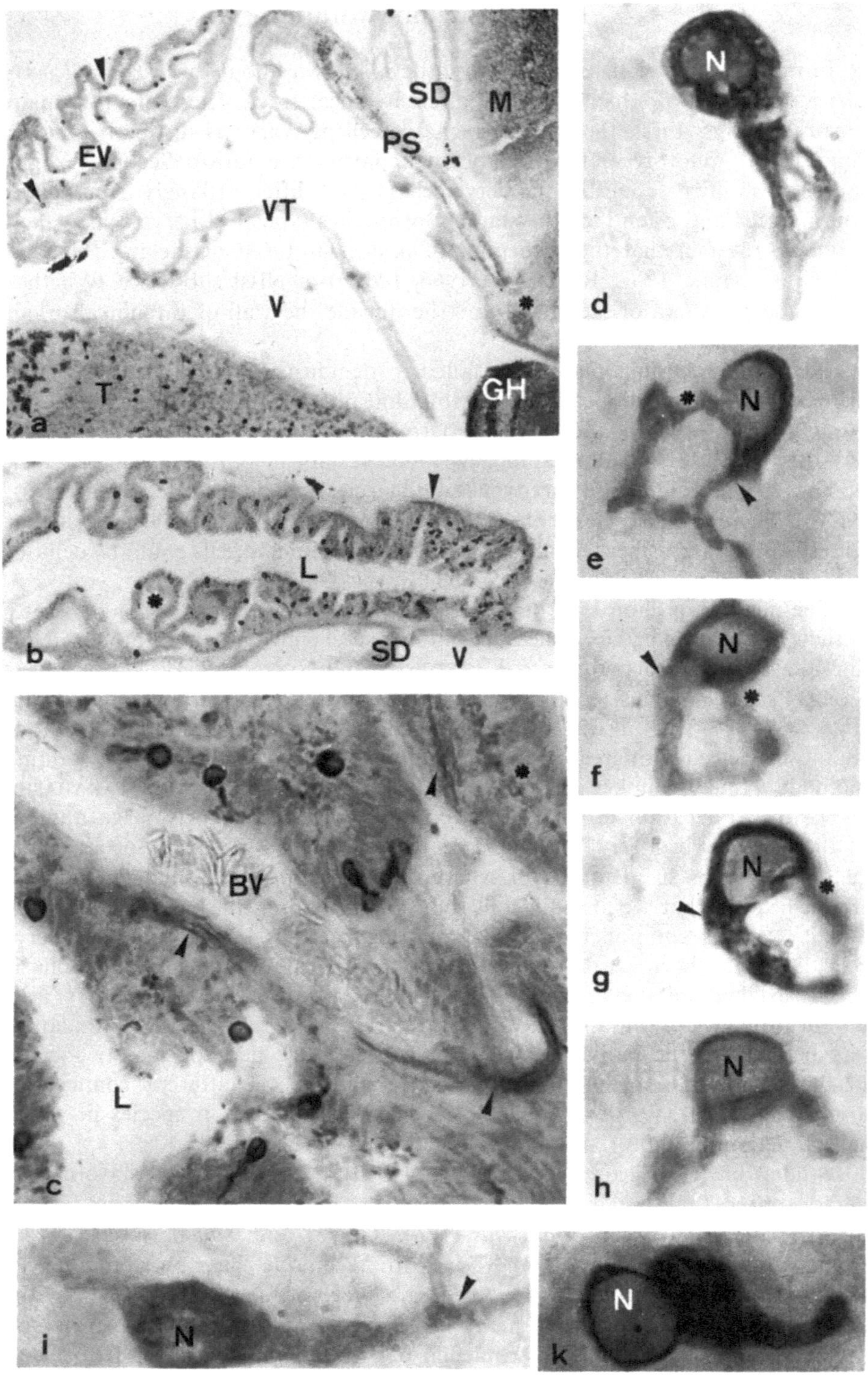

Fig. 147a–k

medially in the dorsal wall at about one-third of the vesicle's length from the rostral tip of the organ, and two symmetrical clusters of three to five nerve cells laterally.

By means of the AChE technique two (occasionally three) types of nerve cells can be distinguished: multipolar nerve cells measuring 15–20 µm in diameter in *Carassius* (WAKE, 1973) and 10–15 µm in *Salmo* (KORF, 1974) and pseudo-unipolar nerve cells measuring 5–11 µm and 5–8 µm in diameter respectively. The latter type of cell is more widely distributed than the former. Bipolar nerve cells are present in the basal region of the pineal parenchyma of *Salmo*. The AChE activity shows interspecies differences: it is high in the multipolar nerve cells of *Salmo* and the pseudo-unipolar nerve cells of *Carassius*, whereas it is low in the pseudo-unipolar nerve cells of *Salmo* and the multipolar cells of *Carassius*.

Relatively little is known about the ultrastructure of pineal nerve cells. They contain the usual cell organelles (TAKAHASHI, 1969; MCNULTY and NAFPAKTITIS, 1976). In addition, subsurface cisternae (COLLIN, 1969c; RÜDEBERG, 1969a), glycogen (COLLIN, 1969c), crystalline inclusions (HERWIG, 1976) and a small number of dense-core vesicles measuring ca. 50–85 nm in diameter (COLLIN, 1969c; HERWIG, 1976) have been found.

Little is known about intrapineal connections of nerve cells. Contacts between nerve cells and photoreceptor cells have been mentioned above (p. 342). At present it is not known whether the neuronal circuitry resembles that in the retina. In *Salmo gairdneri* a rostral pineal ganglion which appears to be interconnected with smaller adjacent neurons is present (KORF, 1974).

Only in some instances has it been shown that processes of intrapineal nerve cells contribute to the formation of the pineal tract (e.g. KORF, 1974). Nevertheless, it can be implied that in species in which the pineal comprises a fairly large number of nerve cell perikarya, these are a major, if not the sole, source for the *afferent* fibres of the pineal tract. The question as to whether the pineal tract comprises also *efferent* fibres will be discussed below. Relevant is the observation that in *Salmo* the numbers of intrapineal AChE-positive nerve cell perikarya and nerve fibres in the pineal tract are in the same order

---

Fig. 147a–k. Acetylcholinesterase-positive nerve cells in the pineal organ of *Salmo gairdneri*. **a** Parasagittal section through the epithalamic region. The pineal end-vesicle (*EV*) comprises few nerve cells (*arrowheads*) in contrast to the pineal stalk (*PS*); *GH*, habenular ganglion; *M*, mesencephalon; *SD*, dorsal sac; *T*, telencephalon; *V*, third ventricle; *VT*, velum transversum; *asterisk*, parapineal organ. ×45. **b** Parasagittal section through the pineal end-vesicle. Note that the number of nerve cells increases towards the pineal stalk and that the diameter of the nerve cell perikarya decreases in a rostrocaudal direction. *Asterisk*, parenchymal fold; *arrowhead*, initial portion of the pineal tract; *L*, lumen of pineal organ; *SD*, dorsal sac; *V*, third ventricle. ×80. **c** Horizontal section showing nerve cells in the pineal end-vesicle with processes directed towards the basal lamina; *asterisk*, selectively stained inner segment of pineal photoreceptor cells; *arrowhead*, AChE-positive nerve fibre bundle; *BV*, blood vessel; *L*, pineal lumen. ×300. **d–k** Nerve cells found in the pineal end-vesicle. **d–e** Pseudo-unipolar nerve cells; *N*, nuclei; *arrowhead*, initial segment of the process; *asterisk*, recurrent branch of the process pointing to the pineal lumen. **f–g** Nerve cells near the pineal lumen. **h–i** Bipolar nerve cells exhibiting ramification (*arrowhead*). **k** Two nerve cells; *N*, nucleus. ×1,100. KORF, 1974

Table 31. Presence of nerve cells in the pineal organ proper of fishes

A. Nerve cells not found, not mentioned or not clearly identified in	
Scyliorhinus	RÜDEBERG (1969a)
Etmopterus	ALTNER (1965)
Squalus	
Galeus	
Triphoturus	McNULTY and NAFPAKTITIS (1977)
Tarletonbeania	
Symbolophorus	
Diaphus	
Ceratoscopelus	
Lampanyctus	
Stenobrachius	
Parvilux	
Garra mulya	BELSARE (1974a)
Barilius bendelensis	
Claria batrachus	RIZKALLA (1970)
Gyrinocheilus	HAFEEZ (1971)
Barbus	
Hesperoleucus	
Corydoras	
Xiphophorus	
Symphysodon	
Zalembius	
Gibbonsia	
Xiphister	
Atherinops	
Oncorhynchus	HAFEEZ and FORD (1976)
Phoxinus	BHARGAVA (1973)
Pterophyllum	BERGMANN (1971)
Chologaster agassizi	McNULTY (1978a)
Typhlichthyes subterraneus	McNULTY (1978c)
Protopterus dolloi	KAPPERS (1965a), UECK (1969)
B. Nerve cells scarce, or no numbers given	
Puntius ticto	BELSARE (1974a)
Rasbora daniconius	
Chela bucaila	
Osmerus mordax	HOLMGREN (1959a)
Fundulus heteroclitus	
Engraulis	HAFEEZ (1971)
Gambusia	
Roccus	
Carassius auratus	TAKAHASHI (1969)
Triphoturus	McNULTY and NAFPAKTITIS (1976)
Phoxinus	OKSCHE and KIRSCHSTEIN (1967)
Astyanax	OMURA (1975)
C. Nerve cells frequent	
Squalus	KAPPERS (1965a)
Lampetra	COLLIN (1969a)
Atherinopsis	HAFEEZ (1971)
Salmo	KORF (1974)
Carassius	WAKE (1973)
Sardina	RÜDEBERG (1968b)

Table 32. Nervous apparatus of the pineal organ of *Salmo gairdneri* (KORF, 1974)

No. of animals	No. of AChE-positive pineal nerve cells	No. of nerve fibres in the pineal tract	No. of type 1 fibres	No. of type 2 fibres
1	1325	1817	1436	381
2	1629	1945	1634	311
3	1515	1816	1495	321
4	1725	2051	1740	311
5	1395	1728	1399	329

Table 33. Number of fibres forming pineal tract in fishes

Species	No. of fibres	Author(s)
Lampetra planeri	600	COLLIN (1969c)
Salmo gairdneri	1728–2051[a]	KORF (1974)
Astyanax mexicanus	350	HERWIG (1976)
Triphoturus mexicanus	1400	McNULTY and NAFPAKTITIS (1976)
Tarletonbeania crenularis	3000	McNULTY and NAFPAKTITIS (1977)
Lampanyctus ritteri	1300	
Parvilux ingens	250	
Pterophyllum scalare	360	BERGMANN (1971)
Bathylagus wesethi	530	McNULTY (1976)
Nezumia liolepis	120	

[a] For further details see Table 32.

Table 34. Convergence ratios of photoreceptor cells and nerve fibres of the pineal tract in fishes

Species	Convergence ratio	Author(s)
Tarletonbeania crenularis	3.0	McNULTY and NAFPAKTITIS (1977)
Triphoturus mexicanus	3.3	McNULTY (1976)
Bathylagus wesethi	8.9	
Nezumia liolepis	17.0	
Parvilux ingens	15.0	McNULTY and NAFPAKTITIS (1977)

of magnitude (KORF, 1974) (Table 32), which supports the notion that it is the intrapineal nerve cells which give rise to the pineal tract.

As can be seen from Table 33, the pineal tract comprises a fairly variable number of mostly non-myelinated nerve fibres, ranging from 120 to more than 2,000. These numerical differences may be partly due to the fact that the counts might have been made at different levels of the pineal stalk and that the number of fibres increases from distal to proximal. The interesting observation that *Thynnus thynnus* (HOLMGREN, 1958) and *Clarias lazera* (RIZKALLA, 1970) lack a pineal tract needs confirmation by electron microscopy.

The convergence ratios of photoreceptor cells and nerve fibres of the pineal tract, which are obtained by dividing the number of photoreceptor cells by the number of pineal tract fibres, vary considerably (Table 34). According to

Table 35. Diameters of nerve fibres present in pineal tract of fishes

Species	Diameter (μm)	Author(s)
Lampetra planeri	0.25–1.5	COLLIN (1969a)
Scyliorhinus canicula	0.2–0.7	RÜDEBERG (1969a)
Triphoturus mexicanus	0.35	MCNULTY and NAFPAKTITIS (1976)
Parvilux ingens	0.2	MCNULTY and NAFPAKTITIS (1977)
Lampanyctus ritteri	0.2–0.4	
Tarletonbeania crenularis	0.2–0.4	
Salmo gairdneri	Type 1 fibres: 0.5	KORF (1974)
	Type 2 fibres: 2.0	KORF (1974)

MCNULTY (1976) a higher convergence ratio is associated with a higher photosensitivity.

The nerve fibres are not uniform in size (Table 35). In *Salmo gairdneri*, two distinct types have been distinguished (Table 32), one of which may represent pinealopetal (=efferent) fibres (KORF, 1974). Some fibres (1%–2% in *Astyanax*, HERWIG, 1976) contain dense-core vesicles measuring 50–140 nm in diameter (OKSCHE and KIRSCHSTEIN, 1971; RÜDEBERG, 1969a, 1971; BERGMANN, 1971; MCNULTY, 1978a). Structures indicating the existence of axo-axonal contacts have been found in *Pterophyllum* (BERGMANN, 1971) and *Astyanax* (HERWIG, 1976). A few Gomori-positive neurosecretory fibres have been seen in the pineal tract of *Etmopterus spinax* (ALTNER, 1965).

The *central projections* of the pineal tract have been studied in detail by HAFEEZ and ZERIHUN (1974) using the cobalt chloride iontophoresis technique. In *Salmo gairdneri* fibres could be clearly traced to the parapineal organ, lateral habenular nucleus, pretectal area, di- and mesencephalic periventricular grey matter, dorsomedial and dorsolateral thalamic nuclei, nucleus of Darkschewitsch, cells of origin of the medial longitudinal fasciculus and the dorsal tegmentum. No fibres innervating the subcommissural organ, optic tectum and the interpeduncular nucleus were seen. With conventional histological techniques the pineal tract could not be followed further than to the posterior commissure in *Protopterus* (KAPPERS, 1965a), *Sardina* (RÜDEBERG, 1968), *Lampetra* (COLLIN, 1969c; COLLIN and MEINIEL, 1968a, 1971), *Esox* (OWMAN and RÜDEBERG, 1970), *Anguilla* (RÜDEBERG, 1971), *Carassius* (WAKE, 1974), *Bathylagus* (MCNULTY, 1976) and *Triphoturus* (MCNULTY and NAFPAKTITIS, 1976). In *Phoxinus* (BHARGAVA, 1973) the pineal tract is connected to the right habenular ganglion.

2. Efferent Fibres

There is as yet no unequivocal evidence for the presence of pinealopetal (=efferent) fibres in fishes. In analogy to the situation in mammals the efferent fibres may be aminergic in nature. In *Esox* green fluorescent fibres (FALCK-HILLARP technique) were seen in meningeal connective tissue surrounding the pineal organ, but not actually inside the pineal parenchyma (OWMAN and RÜDEBERG, 1970). By means of electron microscopy fibres containing dense-core vesicles have been clearly identified (p. 352), but it is not known whether they are afferent or efferent fibres. The observation that in some species intrapineal nerve cells contain dense-core vesicles (p. 349) shows that not all the fibres containing dense-core vesicles may necessarily be efferent. A prominent efferent innervation has been described in the sockeye salmon (HAFEEZ and FORD, 1967), but later observations by means of the cobalt chloride iontophoresis technique showing that apparently all fibres in the pineal tract of *Salmo gairdneri* are afferent (HAFEEZ and ZERIHUN, 1974) casts considerable doubt on the above interpretation.

VIII. Blood Supply of the Pineal Organ

Very little appears to have been published about the blood supply of the fish pineal organ. In *Phoxinus phoxinus* (BHARGAVA, 1973) the pineal organ receives its arterial blood supply via the epiphyseal artery which in the region of the torus longitudinalis leaves the larger of the two branches of the posterior cerebral artery. It runs below the pineal stalk and emits smaller branches which are connected with blood vessels at the posterior margin of the pineal end-vesicle. The vessels then form plexuses surrounding the upper and lower surfaces of the end-vesicle. They project into intrapineal connective tissue folds, but do not penetrate the parenchyma. The venous drainage is shown diagrammatically in Fig. 148; what is not shown in this figure is that there is another vein lying beneath the pineal stalk which drains the pineal plexus into blood vessels running on the dorsal surface of the dorsal sac which in turn reach the superior orbital sinuses.

Occasionally a larger blood vessel (MCNULTY and NAFPAKTITIS, 1976) or a number of vessels (MCNULTY and NAFPAKTITIS, 1977) have been observed in the pineal stalk. In myctophids, the blood vessels of the pineal end-vesicle are usually confined to the periphery, exceptions being *Tarletonbeania* and *Symbolophorus*, in which blood vessels frequently penetrate the central portions of the pineal epithelium (MCNULTY and NAFPAKTITIS, 1977).

Endothelial cells show fenestrations in *Bathylagus* (MCNULTY, 1976), *Sardina* (RÜDEBERG, 1968b), *Esox, Mugil, Salmo gairdneri* (OWMAN and RÜDEBERG, 1970) and *Triphoturus* (MCNULTY and NAFPAKTITIS, 1976). They lack fenestrations in *Nezumia* (MCNULTY, 1976), *Chologaster* and *Typhlogobius* (MCNULTY, 1978a).

IX. The Function of the Pineal Organ

There is clear evidence that the fish pineal is both photoreceptive and secretory in nature.

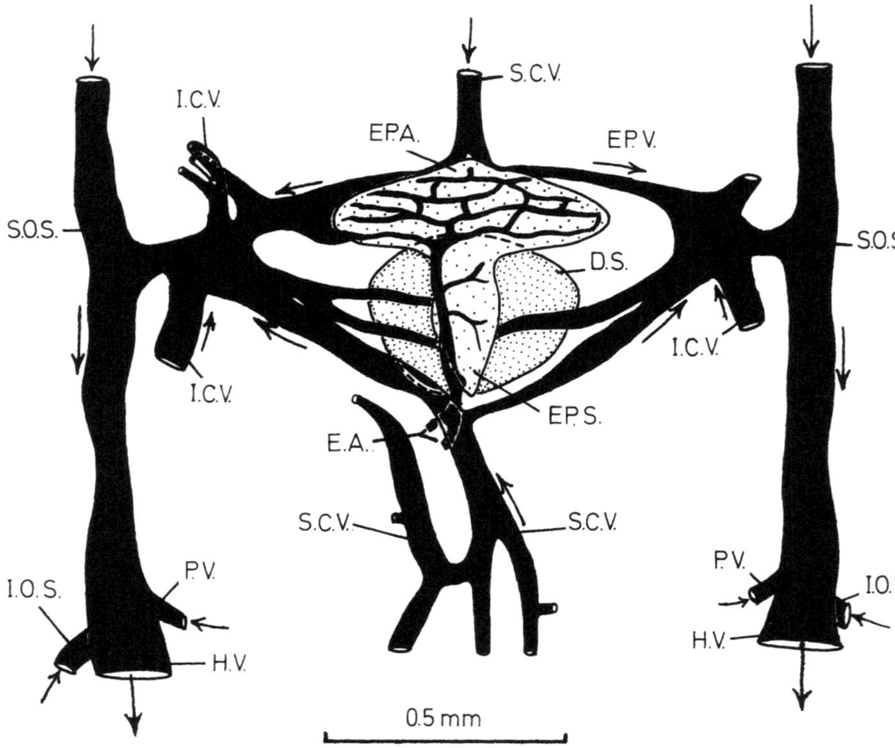

Fig. 148. Pineal organ and its vascularization, as reconstructed from serial transverse sections of a *Phoxinus phoxinus* 4.5 cm long. *D.S.*, dorsal sac; *E.A.*, epiphyseal artery; *EP.A.*, anterior expanded part of the pineal body; *EP.S.*, pineal stalk; *EP.V.*, epiphyseal vein; *H.V.*, head vein; *I.C.V.*, intracranial vein; *I.O.S.*, inferior orbital sinus; *P.V.*, pituitary vein; *S.C.V.*, supracerebral vein; *S.O.S.*, superior orbital sinus. BHARGAVA, 1973

1. Photoreception

The photosensitivity of the fish pineal organ has been directly demonstrated by electrophysiological techniques. In addition, there are morphological findings showing that the organ is affected by light and darkness.

a) Electrophysiology

Electrophysiological recordings from the pineal end-vesicle and the stalk have provided clear-cut evidence that the organ is involved in light perception (for a review see DODT, 1973). In view of the situation in mammals, in which melatonin formation of the pineal is suppressed by light, it is most interesting to note that in cyclostomes and fishes light inhibits the spontaneous electrical activity of the pineal organ. This type of response has been demonstrated in lampreys (DODT, 1973; MORITA and DODT, 1973), elasmobranchs (HAMASAKI and STRECK, 1971) and teleosts (DODT, 1963a; MORITA, 1965a, 1966a, 1975; MORITA and BERGMANN, 1971; HANYU and NIWA, 1970; TABATA et al., 1971).

However, it should be noted that two types of responses have to be distinguished. Usually an achromatic response is obtained, in which the effect of stimulation depends solely on luminance and not on the wavelength of light. A few cells show a chromatic response, i.e. they are inhibited by light of short wavelengths and show excitation after exposure to longer wavelengths.

For further details, see HANYU (1967), HANYU et al. (1969, 1978), DE LA MOTTE (1963, 1964) and TABATA et al. (1975).

b) Morphological Findings

In some studies the effects of light and darkness on pineal morphology have been examined. As early as 1939 it was noted (KNOWLES, 1939) that in the lamprey pineal organ the cell nuclei were in line in sunlight and lay scattered in darkness. The interpretation that in *Gambusia affinis holbrooki* exposure to continuous illumination for 20 days leads to an enhanced apocrine secretion in the pineal organ (CHÈZE and LAHAYE, 1969) should be reconsidered in view of the electron microscopic findings reported above that what looks like apocrine secretion is nothing more than protruding outer segments of photoreceptor cells. In the blind cave fish, *Astyanax mexicanus*, there are corresponding changes in the pineal after exposure to continuous illumination and darkness: under both conditions there is an increase in the width of the pineal lumen, a decrease in the depth of the epithelial folds, followed by a gradual disorganization, and a decrease in the amounts of glycogen (GRUNEWALD-LOWENSTEIN, 1956). Accumulation of glycogen during hours of darkness and a decrease during hours of light has been observed in *Atherina stipes* (RASQUIN, 1958), and in *Symphodus melops* disappearance of aldehydefuchsin positive perinuclear material in some epiphyseal cells after exposure to darkness has been described (CHÈZE, 1969). Electron microscopically there are signs of degeneration of outer segments: the discs are irregular in arrangement, show a decrease in number or are lacking; moreover, there is an increase in the number of phagocytic cells in the immediate vicinity of the outer segments; both photoreceptor and supporting cells show a decrease of Golgi complexes, mitochondria and rough endoplasmic reticulum, whereas no alterations are visible in ganglion cells and the neuropil (OMURA, 1975).

Most interesting results have been obtained by morphometric studies of nuclei and nucleoli of photoreceptor and supporting cells of the pineal of *Salmo gairdneri* under various experimental conditions (HAFEEZ et al., 1978): nuclei and nucleoli of photoreceptor and supporting cells are largest in fish raised in continuous darkness. Continuous light, natural photoperiods and light of either narrow or broad spectral range are not notably different in their chronic effects on receptor cell nuclei and nucleoli. However, nuclei and nucleoli of supporting cells are smallest under continuous illumination. By blinding and pineal masking it can be demonstrated that photoreceptor cells are affected by chronic incident light in the pineal region, but not by photic and visual influences via the lateral eyes; the supporting cells are affected by chronic photic input only via the lateral eyes, a fact which seems to suggest the presence of a cellular comparator system in which photic inputs from lateral eyes and pineal affect separately and respectively the adjacent supporting and receptor cells of the pineal organ.

Biochemical studies have shown that exposure to continuous light for 7 days decreases pineal taurine content, increases ornithine, lysine, tryptophan, tyrosine and 5-hydroxyindole and renders serotonin undetectable (MEISSL et al., 1978).

c) The So-called Pineal Window

The amount of light that reaches the pineal organ proper depends on the composition of the tissue covering the pineal region. In certain species of fish there is a specialization termed "pineal window" or "pineal spot" which allows the light to reach the pineal organ more easily than in species lacking such a specialization. No attempt has been made in the present review to scan the literature thoroughly as to which species have a pineal window and which have none. Incidentally obtained data are given in Table 36. Also, no attempt

Table 36. Pineal window

	Author(s)
Present	
Bathylagus wesethi	McNULTY (1976)
Triphoturus mexicanus	McNULTY and NAFPAKTITIS (1976)
Thynnus thynnus	RIVAS (1953); HOLMGREN (1958b)
Scomber	RIVAS (1953)
Auxis	
Katsuwonus	
Euthynnus	
Sarda	
Scomberomus	
Acanthocybium	
Jenkinsia lamprotaenia	BREDER and RASQUIN (1950)
Sardinella macrophthalmus	
Leptocephalus larvae Sp.?	
Brachydanio rerio	
Atherina stipes	
Carapus bermudensis	
Mugil trichodon	
Chauliodus sloanei	HOLMGREN (1959a)
Gonostoma elongatum	
Gonostoma bathyphylum	
Cyclothone microdon	
Argyropelecus olfersi	
Sternoptyx diaphana	
Triplophos elongatum	
Chlorophthalmus agassizi	
Myctophum glaciale	
Tarletonbeania crenularis	McNULTY and NAFPAKTITIS (1977)
Symbolophorus californiensis	
Diaphus theta	
Ceratoscopelus townsendi	
Lampanyctus ritteri	

Table 36 (continued)

	Author(s)
Stenobrachius leucopsaurus	
Parvilux ingens	
Typhlogarra widdowsoni	MARSHALL and THINES (1958)
Controllable pineal window present	
Sphyraena barracuda	BREDER and RASQUIN (1950)
Strongylura notata	
Hemirhamphus brasiliensis	
Synodus synodus	
Species showing only an indication of a pineal window	
Phoxinus laevis	OKSCHE and KIRSCHSTEIN (1967)
Species lacking a pineal window	
Nezumia liolepis	MCNULTY (1976)
Pterophyllum scalare	BERGMANN (1971)
Anguilla anguilla	RÜDEBERG (1971)
Oncorhynchus nerka	HAFEEZ and FORD (1967)
Protopterus annectens	HOLMGREN (1959 b)
Astyanax mexicanus	BREDER and RASQUIN (1950)
Ameiurus nebulosus	
Haemulon melanurum	
Thalassoma bifasciatum	
Eucinostomos gula	
Pseudupeneus maculatus	
Apogonichthys stellatus	
Cyprinidon baconi	
Bathygobius soporator	

has been made to review which structures overlying the pineal region are responsible for preventing light from reaching the pineal organ or for permitting it to do so. In some species there is a controllable pineal window, the degree of light penetration being regulated by pigment dispersion and aggregation in chromatophores of *Strongylura notata* (BREDER and RASQUIN, 1950).

A pineal foramen or a pineal fontanelle are not regularly present in fishes. This aspect has been dealt with by GLADSTONE and WAKELEY (1940), BARGMANN (1943) and HOLMGREN (1959a). The presence of a pineal foramen has been described in *Clarias lazera* (RIZKALLA, 1970). In *Astyanax mexicanus* a pineal fontanelle is present (HERWIG, 1976).

2. Melatonin Secretion

In addition to photoreception the fish pineal organ synthesizes and secretes melatonin. The various steps in the formation of melatonin have been described

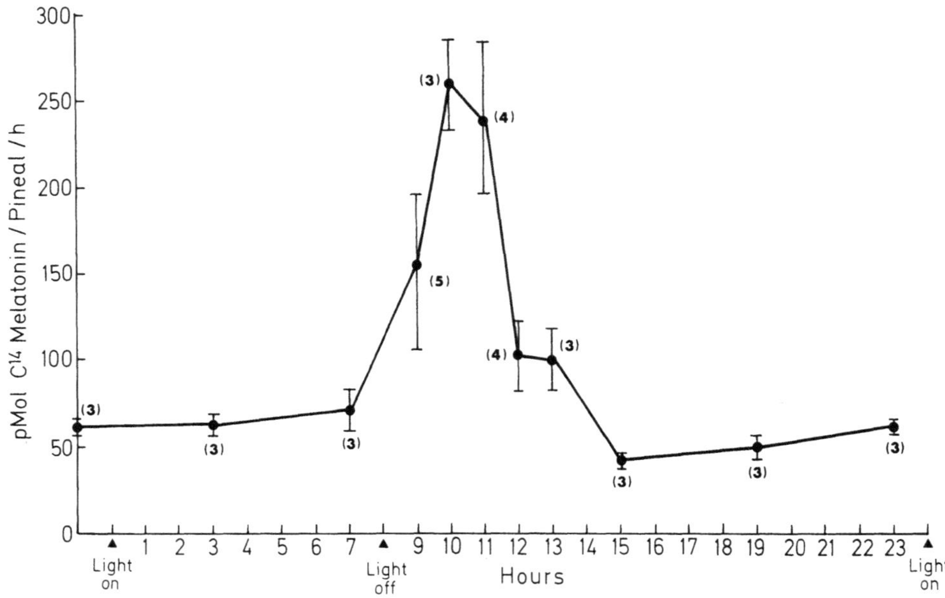

Fig. 149. HIOMT activity in the pineal bodies of ammocoetes measured over a period of 24 h. The figures in brackets are the number of animals studied. The vertical lines through the means represent 1 S.E.M. Joss, 1977

above (p. 5). There is no indication that melatonin formation in fishes differs from that in mammals. The fish pineal has been shown to contain one of the melatonin precursors, serotonin (*Salmo gairdneri, Atherinopsis californiensis*: HAFEEZ and QUAY, 1969; *Esox lucius:* OWMAN and RÜDEBERG, 1970) and to strongly incorporate radioactive 5-hydroxytryptophan and 5-hydroxytryptamin (OGURI et al., 1968; HAFEEZ and ZERIHUN, 1976). It also contains the melatonin-forming enzyme HIOMT (*Salmo gairdneri*: QUAY, 1965d; HAFEEZ and QUAY, 1970a; J.R. SMITH and WEBER, 1974; *Hesperoleucus symmetricus:* HAFEEZ and QUAY, 1970a; *Geotria australis:* JOSS, 1977). Melatonin itself has been identified in the pineal of *Oncorhynchus tshawytscha* by means of paper chromatography and fluorimetric measurements, the melatonin content being considerably (six times) higher in immature than in mature salmon (FENWICK, 1970b). Recently, melatonin has been demonstrated in trout plasma by means of radio-immunoassay (GERN et al., 1978a), which shows that melatonin gains access to the systemic circulation. Whether or not all of the melatonin stems from the pineal organ is an open question in view of the fact that the fish retina contains HIOMT and synthesizes melatonin (QUAY et al., 1969b; GERN et al., 1978b; GERN and RALPH, 1979) and that a day-night rhythm of plasma melatonin persists after pinealectomy (GERN et al., 1978c).

As in mammals, melatonin formation in fishes shows prominent day–night changes. In the lamprey, *Geotria australis*, pineal HIOMT activity reaches a peak 2–3 h after the onset of darkness (Fig. 149); altogether the burst in activity is short and confined to the first 4 h of darkness (JOSS, 1977). In the teleost fish, *Salmo gairdneri*, HIOMT activity shows a day–night rhythm with a maximum at 4.00 h and a minimum at 20.00 h (J.R. SMITH and WEBER, 1974). Plasma melatonin levels at 24.00 h are almost twice as high as those measured at 12.00 h 4 days previously (GERN et al., 1978a, c).

Little is known as to whether the rhythm of melatonin formation can be altered by changes in environmental lighting conditions. According to HAFEEZ and QUAY (1970a) pineal HIOMT activity is not affected by exposing *Hesperoleucus* and *Salmo gairdneri* to continuous illumination or darkness for 1 week. GERN et al. (1978c) noted that the melatonin rhythm persists after sectioning of the optic tract.

3. Functional Significance of the Pineal Organ

As in mammals, we are far from knowing which organs are the primary targets of the pineal organ in fishes. The centripetal course of the pineal tract suggests that the brain is a major site of action, but which of the many brain functions is influenced is not known. The functional significance of melatonin also remains to be elucidated. As in mammals (p. 263), radioactive melatonin strongly accumulates in the pineal organ. High uptake is also found in the pituitary, whereas other organs show relatively little activity (JOSS, 1977) (Table 37). Since all the major organs take up and dispose of it in several hours, JOSS (1977) speculated that the dark-induced surge of melatonin is the timer for the circadian rhythm of activities within these organs. One of the earliest effects of the pineal to be noted in fishes was a participation in the regulation of the colour change.

a) Regulation of Colour Change

Although there is little doubt that the pineal organ in cyclostomes, elasmobranchs and teleosts is somehow involved in the regulation of colour change, the precise role of the pineal is not known. This is mainly due to the fact that there are interspecies differences, age-related differences and differences

Table 37. Uptake of ^{3}H-melatonin (μmol/g tissue) into selected tissues of ammocoetes (JOSS, 1977)

Tissue	0.5 h	1 h	2 h	4 h
Pineal complex	88.40	183.60	570.0	289.0
Pituitary	4.42	14.28	94.52	19.49
Gonads	0.72	0.63	0.37	0.12
Endostyle	0.79	0.48	0.80	0.21
Brain	0.42	0.27	0.28	0.72
Liver	2.14	0.50	0.33	0.38
Gut	1.52	0.24	0.31	0.16
Tail muscle	0.064	0.61	0.27	0.10

between various types of chromatophores. An important point to be taken into consideration is that in fish melatonin is not a universal paling agent. Hence, if pinealectomy does not result in a change in body colouration, this does not necessarily mean that the pineal organ in this species does not produce melatonin.

Pinealectomy is followed by a dispersion of melatonin granules in ammocoetes of *Lampetra planeri* (YOUNG, 1935; JOSS, 1973b), *L. fluviatilis* (JOSS, 1973b) and *Geotria australis* (EDDY and STRAHAN, 1968). Pinealectomy is without visible effects on pigmentation in adult *Lampetra planeri* (JOSS, 1973b), in metamorphosing *Mordacia mordax* (EDDY and STRAHAN, 1968), in *Scyliorhinus canicula* as far as adaptation to black or white backgrounds is concerned (WILSON and DODD, 1973), in *Gasterosteus* (SCHÖNHERR, 1955) and in *Salmo gairdneri* (HAFEEZ and QUAY, 1970b).

In some species an effect of pinealectomy could be demonstrated under certain experimental conditions only. In *Scyliorhinus* pinealectomy abolished the normal paling response in darkness (WILSON and DODD, 1973) and in *Salmo gairdneri* pinealectomy was effective only in blinded animals (HAFEEZ and QUAY, 1970b).

Melatonin administration may induce pigment *aggregation*, may have no effect or may lead to pigment dispersal. Pigment aggregation has been noted in *Geotria austalis* (EDDY and STRAHAN, 1968), *Lampetra fluviatilis* when melatonin was given in very large doses (100 µg/animal) (JOSS, 1973b), *Chasmichthys gulosus* (FUJII, 1961), *Scardinius erythrophthalmus* (MIRA, 1963), *Phoxinus phoxinus* (HEALEY and ROSS, 1966), *Salmo gairdneri* (HAFEEZ, 1970) and the species listed in Table 38. This table also lists species which show no observable response to melatonin. Melatonin is also without observable effect in *Mordacia mordax* (EDDY and STRAHAN, 1968) and in *Lampetra fluviatilis* when given in "physiological" doses (JOSS, 1973b). Species showing pigment dispersal after melatonin are listed Table 38. It can be seen that pigment dispersal after melatonin appears to be a relatively rare phenomenon. The complexity of the issue is illustrated by the observation that in two of the four species showing pigment dispersal melatonin administration via the swimming water leads to pigment *aggregation* (REED et al., 1969).

In the pencil fish, *Nannostomus beckfordi anomalus*, the melanophores respond differently to melatonin (REED, 1968). Melanophores responsible for day colouration show pigment aggregation, whereas those responsible for night colouration exhibit pigment dispersal.

A paling effect of *pineal extracts* (JOSS, 1973b) and *transplanted pineals* (EDDY and STRAHAN, 1968) has also been demonstrated. When melatonin is administered in conjunction with mammalian MSH to hypophysectomized ammocoetes, the normal reaction to MSH is inhibited (JOSS, 1973b).

Finally attempts have been made to correlate the melatonin-forming activity of the pineal with the state of the melanophores. In ammocoetes of *Geotria australis* there is a fairly clear inverse relationship between pineal HIOMT activity and the melanophore index (JOSS, 1973b). However, in *Salmo gairdneri* no relationship appears to exist between plasma melatonin and background adaptation (OWENS et al., 1978).

Table 38. Effects of melatonin on fish melanophores (REED et al., 1969)[a]

Species	Dosage added to water			Injected
	0.05 μg/ml	0.5	5	2 μg
Acanth. Kuhlii sum.	–	–	–	–
Botia macracanthus	–	–	–	–
Gyrinocheilus aymonieri	0	–	–	–
Aequidens pulcher	0	–	–	–
Hyphessobrycon serpae	0	–	–	–
Rasbora heteromorpha	0	0	–	–
Brachydanio rerio	0	0	–	–
Hyphessobrycon scholzei	0	–	–	0
Moenkhausia oligolepis	0	0	+	+
Hemigrammus ocellifer	0	0	+	+
Nannostomus ocellatus	–	–	–	+
Hyphess. herbertaxelrodi	–	–	–(+)	+
Nannostomus marginatus	–	0	0	0
Thayeria boehlkei	0	0	0	0
Barbus tetrazona tetraz.	0	0	0	0
Leporinus fasciatus	0	0	0	0
Rasbora einthoveni	0	0	0	0(–)
Xiphophorus hybrid black	0	0	0	0(–)
Pterophyllum eimeki	0	0	0	0
P. eimeki (black)	0	0	0	0
Mollienes. latipinna (black)	0	0	0	0

[a] Symbols used: 0, no observable response; +, pigment dispersal (darkening); –, pigment aggregation (paling). Responses shown by less than half of the population studied are in parentheses.

b) Phototaxis

In some species phototaxis appears to be influenced by the pineal organ. Two types of phototaxis have to be distinguished; *negatively* phototactic animals try to avoid light, whereas *positively* phototactic animals do not try to escape light or are attracted by it. A clear-cut effect of the pineal on phototactic behaviour has been obtained in *Fundulus heteroclitus*, in which species the positive phototaxis of blinded control animals is changed to negative phototaxis or to an indifferent behaviour by pinealectomy (PANG, 1965, 1967). When blinded *Oncorhynchus nerka* are subjected to a destruction of the pineal region, they become almost completely insensitive to light (HOAR, 1955). In *Carassius auratus* the negative phototaxis changes to an indifferent behaviour after pinealectomy or blinding, showing a tendency towards positive phototaxis in blinded plus pinealectomized animals (FENWICK, 1970c). No effect of the pineal on phototaxis could be demonstrated in *Salmo gairdneri* (HAFEEZ and QUAY, 1970b). In *Anguilla anguilla*, pinealectomy did not change the negative phototaxis, but delayed the response time: it is 3–6 s in intact controls, 20–30 s after bilateral enucleation and 30–80 s after bilateral enucleation combined with pinealectomy (VAN VEEN et al., 1976).

c) Locomotor Activity

That the locomotor activity of fishes may be affected by the pineal organ is shown by the fact that pinealectomy is followed by an increase of the spontaneous swimming activity, accompanied by a decrease of brain serotonin levels (FENWICK, 1970e, *Carassius auratus*), and that administration of melatonin inhibits swimming activity (*Oncorhynchus nerka*: BYRNE, 1970; *Salmo gairdneri*: HAFEEZ, 1970). In *Salmo* pinealectomy and sham-pinealectomy did not significantly affect swimming activity (HAFEEZ, 1970). The circadian swimming activity *patterns* do not differ strikingly in ophthalmectomized and ophthalmectomized-pinealectomized *Anguilla anguilla* (VAN VEEN et al., 1976).

d) Gonadal Function

In view of the situation in mammals (p. 282) it was only logical to search for pineal–gonadal interrelationships in lower vertebrates. In fishes the situation appears to be similar to that in mammals: the results obtained have been very variable, but it would appear that if the right experimental approach is used an involvement of the pineal in reproduction is demonstrable. This aspect of pineal physiology has recently been reviewed (RALPH, 1978).

One group of earlier studies had shown that pinealectomy or sectioning of the pineal stalk has little, if any, effect on reproduction and the gonads (SCHÖNHERR, 1955; RASQUIN, 1958; PETER, 1968, 1970). Other studies yielded controversial results after pinealectomy: an acceleration of sexual maturation was noted by PFLUGFELDER (1954a, *Lebistes reticulatus*), whereas a delay was observed in *Fundulus heteroclitus* (PANG, 1967) and *Lampetra fluviatilis* (JOSS, 1973a).

More recent studies have shown that the pineal organ may either stimulate or inhibit gonadal maturation depending on the time of year in which the experiments are being carried out and on the photoperiod-temperature regime to which the fish are exposed.

Very briefly, the situation is as follows: females of *Oryzias latipes* pinealectomized in winter fail to lay eggs during the following breeding season. In pinealectomized fish maintained under continuous light or natural daylight ovarian weight is significantly smaller than in intact fish (URASAKI, 1972a). The mean gonosomatic index (GSI), i.e. the gonadal weight expressed as a percentage of the body weight, is larger in pinealectomized male and female *Oryzias* kept under continuous darkness or natural daylight conditions for 76 days from October to December than in intact fish. Under continuous illumination the mean GSI is smaller after pinealectomy than in intact animals, suggesting that the pineal organ stimulates gonadal growth under continuous illumination and that it inhibits it under continuous darkness or short photoperiods (URASAKI, 1972b). Finally it was shown in *Oryzias* that pinealectomy decreases ovarian weight when the animals are exposed to LD 16:8, whereas a slight increase occurs under LD 8:16 and under natural daylight conditions in November

(URASAKI, 1972c). Moreover, pinealectomy shortens the period of oviposition (URASAKI, 1973).

In *Notemigonus crysoleucas* (DE VLAMING, 1975) pinealectomy causes testicular regression during the gonadal preparatory and prespawning season and retards final testicular maturation during the spawning season when fish are exposed to a long photoperiod-warm temperature (LD 15.5/8.5; 25° C) regime. Under these conditions pinealectomy retards ovarian development or causes ovarian involution during all the phases of the gonadal maturation cycle. Under a long photoperiod-low temperature (12° C) regime pinealectomy has no apparent effects on gonadal development during the preparatory and spawning seasons. During the prespawning season pinealectomy represses ovarian development and causes testicular regression in fish exposed to a long photoperiod-low temperature regime. During the prespawning and spawning seasons pinealectomy stimulates gonadal activity in fish maintained under a short photoperiod (LD 9/15)-warm temperature regime. Under the same conditions pinealectomy fails to alter gonadal activity during the gonadal preparatory season. No obvious changes can be detected in gonadal activity in pinealectomized fish exposed to a short photoperiod-low temperature regime during any phase of the gonadal maturation cycle. Details on the effect of pinealectomy on pituitary and hypothalamus in *Notemigonus* have been given by DE VLAMING and VODICNIK (1977).

The concept that the pineal may be either antigonadotropic or progonadotropic has gained support from studies in *Carassius auratus*: From late winter to early summer pinealectomy retards gonadal activity in fish maintained on long photoperiods, but stimulates reproductive function in animals exposed to short photoperiods; no effects can be observed when pinealectomy is carried out from late summer to early winter (DE VLAMING and VODICNIK, 1978). The important role of different lighting conditions are also highlighted at the hypothalamic level: MAO activity in the medial basal lobe is not affected by pinealectomy when animals are kept under LD 16:8, but this operation elevates the enzyme activity under LD 8:16 (OLCESE and DE VLAMING, 1977).

Frequently it has been shown that melatonin administration depresses gonadal activity (*Carassius auratus:* FENWICK, 1970b; *Oryzias latipes*, when kept under LD 16/8 in January: URASAKI, 1972c; *Fundulus similis*, in January or May on a long photoperiod, but not in January on a short photoperiod: DE VLAMING et al., 1974a; *Heteropneustes fossilis*, not only melatonin, but also prolactin effective: SUNDARARAJ and KESHAVANATH, 1976; *Mystus tengara*: SAXENA and ANAND, 1977).

All these results taken together suggest that the pineal organ may play a role in regulating gonadal function. As in mammals, the pineal organ appears to exert its effects via the hypothalamus–pituitary axis (see below). The effects of sex steroids on the pineal organ have been little studied, in contrast to mammals (p. 286). According to PFLUGFELDER (1957, quoting unpublished results by GESKE) the administration of male or female sex hormones leads to a distinct decrease of both the weight of the pineal organ and the nuclear volume of pineal cells in *Lebistes reticulatus*.

e) Pituitary Function

Although there is a fairly large body of indirect evidence suggesting that the pineal organ influences anterior pituitary function, this aspect has not been studied in detail. An influence of the pineal on the gonadotropic cells is suggested by the fact that they have been found to increase in size after administration of melatonin (*Carassius auratus*: FENWICK, 1970b). In *Heteropneustes* melatonin administration was followed by a reduction of the number of gonadotrophs (SUNDARARAJ and KESHAVANATH, 1976). An increase in the number of gonadotrophs in *Notemigonus* after pinealectomy when compared with sham-operated controls (DE VLAMING and VODICNIK, 1977) points to an inhibitory influence of the pineal.

Pituitary *prolactin* levels are reduced following melatonin administration (*Fundulus similis*: DE VLAMING et al., 1974b). Pinealectomy affects pituitary prolactin levels differently depending on the photoperiod under which the animals are kept and the time of sampling: when goldfish are exposed to short photoperiods and killed 0–2 h after the onset of the light phase, pinealectomized animals have significantly lower pituitary prolactin levels than sham-operated controls; if sampling occurs 4–5 h after the onset of a short photoperiod, prolactin levels are significantly higher; when experimental animals are exposed to long photoperiods and are killed 0–2 h after the onset of light, sham-operated goldfish have significantly less pituitary prolactin than their pinealectomized counterparts; if fish are killed 2–3 h later no difference is detectable (VODICNIK and DE VLAMING, 1978).

f) Thyroid Function

Studies dealing with the influence of the pineal organ on thyroid function have yielded equivocal results. Pinealectomy was found to be without effect on the thyroid in *Carassius auratus* (PETER, 1968, 1970; FENWICK, 1970d) und *Astyanax mexicanus* (RASQUIN, 1958). In *Fundulus heteroclitus* pinealectomy was followed by thyroid hyperplasia provided that the experiment was carried out from December to March, and not from February to June (PANG, 1967). Thyroid hyperplasia was also noted in *Carassius gibelio auratus* (PFLUGFELDER, 1964) and *Lebistes reticulatus* (PFLUGFEDER, 1953, 1954, 1956). In the latter species administration of a pineal extract led to a decrease in the size of follicles which had become hyperplastic after pinealectomy, the nuclear volume decreasing from 50–60 μm^3 to 10–15 μm^3 (PFLUGFELDER, 1956, 1957).

g) Inter-Renal Function

Judging from pinealectomy experiments, the inter-renal tissue appears to be little affected by the pineal organ (FENWICK, 1970d; PETER, 1968). In *Fundulus heteroclitus* (PANG, 1967) the nuclear diameters of inter-renal cells were not changed after pinealectomy, but in one experiment (from February to June, in contrast to from December to March) the height of the inter-renal zone was greater than in controls.

h) Miscellaneous Effects

Plasma levels of Ca^{2+}, Na^{+} and Cl^{-} (FENWICK, 1970d) and serum osmolality (PANG, 1967) are apparently not affected by pinealectomy.

Metamorphosis has been found to be delayed after pinealectomy in *Lampetra*, *Geotria* and *Mordacia* (EDDY, 1969), but in *Fundulus* pinealectomy was followed by a faster growth rate (PANG, 1967). Impairment of skeletal development has been described by PFLUGFELDER (1953, 1954, 1957, 1967).

Lipid depletion following melatonin administration has been found in *Fundulus similis* (DE VLAMING et al., 1974b).

A recent study on the effects of pinealectomy and melatonin treatment on liver and plasma metabolites in *Carassius auratus* has shown how important is the time of year when the experiments are carried out (DELAHUNTY et al., 1978): pinealectomy increases plasma glucose levels in summer and depresses them in autumn; plasma lipid levels increase only in winter, spring and summer. Moreover it has been found that pinealectomy abolishes the daily changes of liver glycogen and that it decreases the liver glycogen concentration during spring and summer. Melatonin administration increases liver glycogen when the animals had been kept on long photoperiods during spring, and also leads to an increase of plasma glucose levels.

Finally it should be mentioned that the trout *retina* synthesizes melatonin, exhibiting high levels during the light phase and low levels during the night (in contrast to the situation in the pineal!), and that pinealectomy does not perturb either the rhythm or the titre of retinal melatonin (GERN et al., 1978b). Scotophobin A, a peptide isolated from brains of rats trained with electric shock to avoid the dark, is also effective in goldfish and has been suggested to act via the pineal (SATAKE and MORTON, 1979).

X. Structure and Function of the Parapineal Organ

As mentioned above (p. 322), the parapineal organ of fish is fairly variable. In older studies the absence of a parapineal organ was usually not specifically mentioned (e.g. HOLMGREN, 1959a), because the authors were under the impression that it is always lacking in the adult state. Hence, no complete list can be given as to which of the species studied possess, or lack, a parapineal organ. Table 39 lists the papers (and also the species, if they are not too numerous) in which the absence of a parapineal organ has either been definitely stated or can be deduced from the context. Species possessing a parapineal organ are mentioned in the text. *Labeo umbratus* (STEYN and WEBB, 1960) is a notable exception among fishes in that it possesses a parapineal organ, but not a pineal organ. As might be expected from its rare occurrence, the parapineal organ has been carefully studied in a few species only. The most thoroughly investigated parapineal organ is that of *Lampetra planeri*.

Table 39. Species specifically mentioned as lacking a parapineal organ in the adult state

Species	Author(s)
Great number of deep sea fishes	HOLMGREN (1959a)
Protopterus annectens	HOLMGREN (1959b)
Thunnus thynnus	MURPHY (1971)
Lebistes reticulatus	RÜDEBERG (1971)
Phoxinus phoxinus	BHARGAVA (1973)
Nezumia liolepis	MCNULTY (1976)
Bathylagus wesethi	
Tarletonbeania crenularis	MCNULTY and NAFPAKTITIS (1977)
Symbolophorus californiensis	
Diaphus theta	
Ceratoscepelus townsendi	
Lampanyctus ritteri	
Stenobrachius leucopsaurus	
Parvilux ingens	

In the cyclostome *Lampetra planeri* (BERTOLINI and MANGIA, 1966; MEINIEL, 1969a, b, 1970, 1971) the parapineal organ consists of an end-vesicle with a stalk and underlies the pineal organ. In the end-vesicle a dorsal inverted retina and a ventral direct retina can be distinguished. Both retinas are similar to the sensory cupule of the pineal organ proper and consist of photoreceptor cells, ependyma-like supporting cells, cells containing residual bodies, which have recently shown to belong to the cells of the sensory line (MEINIEL, 1979), and nerve cells. From a functional point of view it is relevant to note that although the photoreceptor cells are basically similar to those in the pineal organ, they are relatively rare and show signs of regression at the outer segments and the synaptic contacts. In view of the situation in mammals it is interesting that, although synaptic ribbons are present, synaptic contacts between photoreceptor cells and dendrites are very rare; instead the ribbons face either adjacent interstitial cells or basal processes of photoreceptor cells. Despite the fact that there are signs of regression of the photorecpetor cells and their synaptic connections, there is a well developed ganglion lying beneath the parapineal organ which, according to MEINIEL and COLLIN (1971), shows nervous connection with both the parapineal organ and the left habenular ganglion. It has been suggested that the ganglion represents the rostral part of the left habenular ganglion and that it may contain a parapineal tract. From a comparative point of view the nervous connections appear to be afferent, but an efferent nature cannot be excluded. Some neurosecretory fibres are present.

A detailed description of the parapineal organ in *teleost* fishes is available for *Salmo gairdneri* only. The most interesting feature is the very rare occurrence of photoreceptor cells, though supporting cells and apparently unipolar ganglion cells are abundant (RÜDEBERG, 1969b).

Very little of a definitive nature can be said about the *function* of the parapineal organ. A regular feature appears to be that in both cyclostomes (see above) and teleosts a parapineal tract courses from the parapineal organ to the *left* habenular ganglion (*Anguilla anguilla*: RÜDEBERG, 1970, 1971; *Salmo gairdneri*: KORF, 1974), which suggests that the parapineal organ, like the pineal organ, conveys messages to the brain. The observation by HAFEEZ and ZERIHUN (1974) that fibres of the *pineal* tract terminate in, or pass through, the parapineal organ invites speculation that the two organs are functionally linked. The presence of serotonin in the parapineal organ (MEINIEL, 1978) shows that the organ is apparently basically similar to the pineal organ proper (p. 331), and it is probably safe to conclude that the parapineal organ is also involved in melatonin formation.

E. The Pineal Complex in Amphibians

I. Gross Anatomy and General Histology

In the two main subclasses of amphibians, the anurans and urodeles, the pineal complex shows striking differences: in anurans, it may consist of two components, namely the intracranial pineal organ proper (epiphysis cerebri) and the extracranial frontal organ (*Stirnorgan*); urodeles and also caecilians possess a pineal organ proper and lack a frontal organ. Since detailed descriptions of the pineal complex of amphibians have been given (GLADSTONE and WAKELEY, 1960; BARGMANN, 1943), it suffices to give a brief survey.

1. The Pineal Organ Proper

The pineal organ proper of anurans and urodeles, extending from the region between the posterior and habenular commissures of the diencephalic roof (Fig. 150), is very variable in size, shape and internal structure. In anurans the pineal organ proper is basically a hollow, sac-like structure, a typical example of which is present in *Rana esculenta* (OSCHKE, 1955). In the latter species very few projections extend from the wall of the pineal organ into the lumen of the organ. In *Bufo vulgaris*, the pineal wall usually forms prominent septa (Fig. 151), which completely subdivide the lumen of the organ into a number

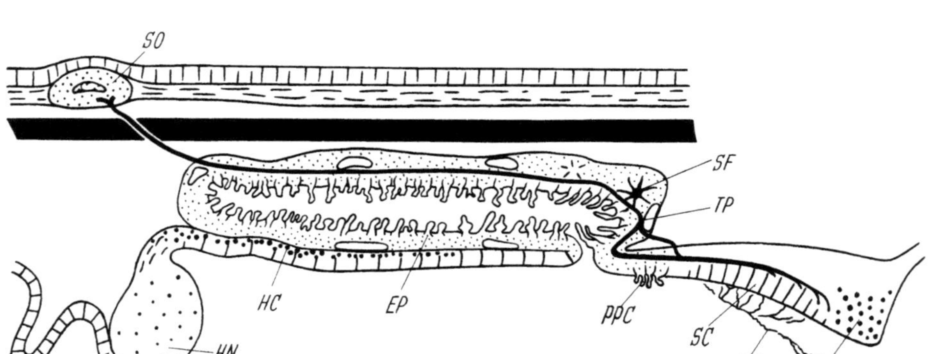

Fig. 150. Diagrammatic representation of a sagittal section of the organs of the diencephalic roof in anurans. *EP*, epiphysis cerebri; *HC*, habenular commissure; *HN*, habenular nucleus; *PC*, posterior commissure; *PPC*, posterior pineal cluster; *RF*, Reissner's fibre; *SC*, subcommissural organ; *SF*, secretory follicle; *SO*, *stirnorgan*; *TP*, pineal tract. KELLY and VAN DE KAMER, 1960, after STUDNIČKA

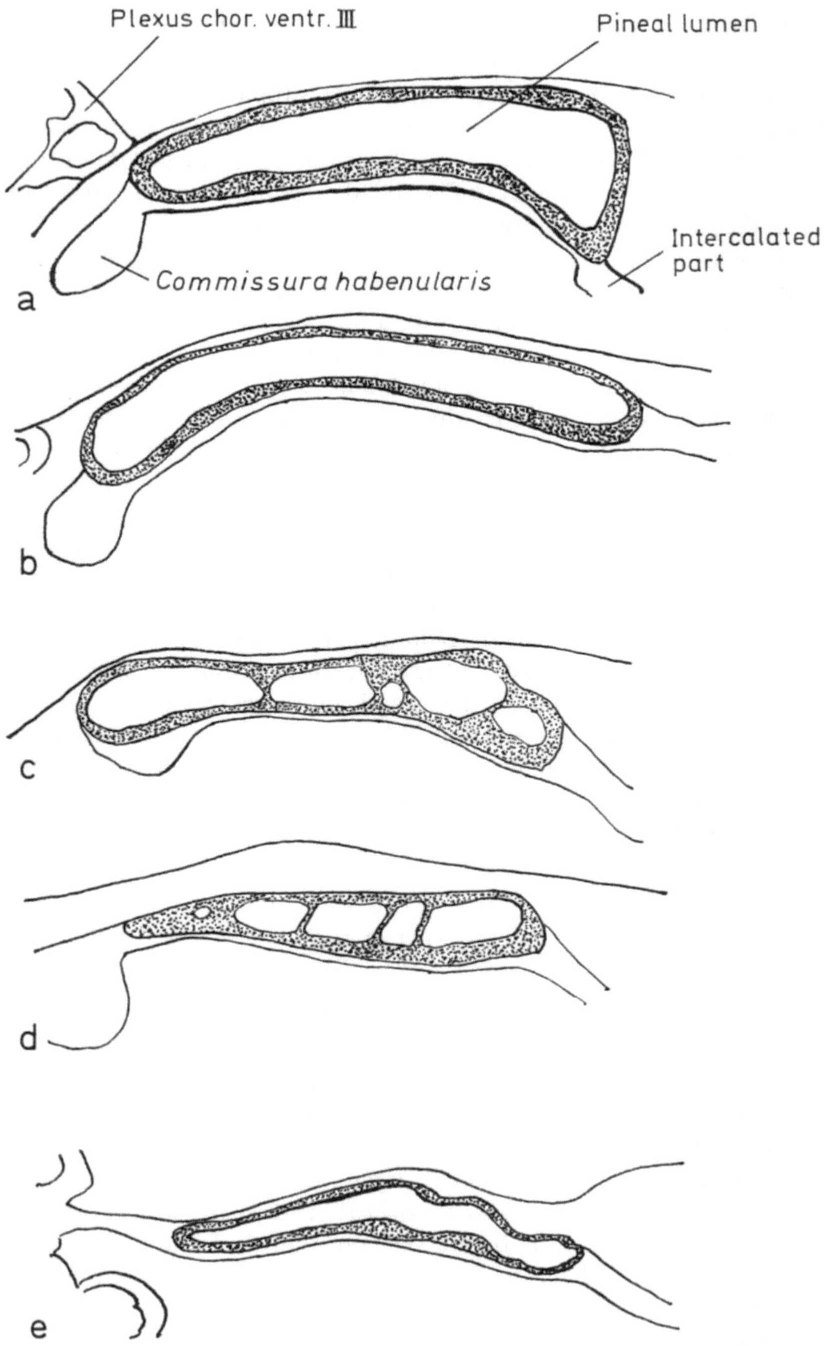

Fig. 151 a–e. Parasagittal sections through the epiphysis cerebri of anurans. **a** *Rana esculenta.* **b** *Rana temporaria* (Esculenta-type pineal body). **c** *Rana temporaria* (Bufo-type pineal body). **d** *Bufo vulgaris.* **e** *Xenopus laevis.* OKSCHE, 1955

of chambers. In *Rana temporaria* interindividual differences are pronounced (Fig. 151). In some instances the pineal organ resembles that seen in *Rana esculenta* ("Esculenta-type"), whereas in others it appears more like that in *Bufo* ("Bufo-type"). In *Xenopus laevis* the pineal is tubular rather than saccular, and a subdivision of the pineal lumen into chambers is usually not conspicuous (OSCHKE, 1955). Additional pineal tissue ("posterior pineal clusters") lying posterior to the so-called epiphyseal orifice (i.e. the region where the pineal vesicle joins the diencephalic roof) has been found in *Rana esculenta* (KELLY and VAN DE KAMER, 1960).

In urodeles, the pineal organ exhibits a wide spectrum of appearances, ranging from a hollow, sac-like structure with a few secondary lumina (*Ambystoma tigrinum*: KORF, 1976) via specimens in which the cavity of the pineal vesicle is irregularly subdivided by partitions into lobuli (*Triton*: GLADSTONE and WAKELEY, 1940) to fairly solid organs in which there are only a few scattered luminal remnants (*Notophthalmus viridescens viridescens*: FLIGHT, 1968; *Taricha torosa*: KELLY, 1963) or not even a trace of a lumen (*Desmognathus*: GLADSTONE and WAKELEY, 1940).

Size measurements of the pineal organ proper have been carried out in a few species only. In *Xenopus laevis*, the pineal organ proper measures 450 μm in length, the width being 20 μm anteriorly and 40 μm posteriorly (CHARLTON, 1968). The ovoid pineal organ of *Bufo arenarum*, the lumen of which does not communicate with the IIIrd ventricle, measures 0.7 mm in length (VITALE-CALPE, 1966). In *Taricha torosa*, the length of the pineal organ varies from 0.5–1.5 mm; its width is approximately 225 μm and the height 50–100 μm (KELLY, 1963). In the axolotl, *Ambystoma mexicanum*, the pineal organ measures up to 552 μm in length, 438 μm in width and 120 μm in height (COMBESCOT et al., 1964). Slightly larger values have been reported by SLONIMSKI (1947).

2. Frontal Organ

The second component of the pineal complex in anurans, the frontal organ, has an extracranial position (Fig. 150) and lies between the lateral eyes in the median plane beneath the epidermis. In the living animal the position of the frontal organ may be indicated by specializations that differ in the various anurans (OKSCHE, 1955): in *Rana temporaria* a very variable small, pale area devoid of pigment overlies the frontal organ, and in *Rana esculenta* the pale area has the form of a stripe. In *Xenopus laevis* and *Rana esculenta* the frontal organ and the overlying parts of the skin protrude to form a small tubercle. In adult *Bufo vulgaris*, there are no visible indications for the localization of the frontal organ.

The size of the frontal organ is very variable (OKSCHE, 1955). In *Rana esculenta* and *Xenopus laevis* it is large. In *Rana temporaria* and *Bufo vulgaris* it varies from large to extremely small, and in some specimens it is even lacking.

II. Development

The pineal complex of amphibians develops from a finger-like evagination of the diencephalic roof, in an area which later comes to lie between the habenular and posterior commissures (see BARGMANN, 1943). How the frontal organ develops was not precisely known when BARGMANN wrote his monograph.

Detailed studies carried out mainly in *Xenopus laevis* (VON HAFFNER, 1951, 1952) appear to have greatly clarified the issue. In *Xenopus*, the first clearly visible sign of a pineal anlage occurs in embryos measuring 2.4 mm in length, when the diencephalic roof shows a slight outward and caudad protrusion. At 2.7 mm, the protrusion has become more prominent, developing into what looks like a vesicle having a broad open connection with the brain ventricle at 3.0 mm. At this stage, the anterior and posterior regions of the proximal part of the anlage show a slight thickening, due to an increase in the number of cells present. At 3.6 mm, the proximal and distal parts of the anlage show distinct differences: proximally, the number of cells has further increased leading to a narrowing of the originally broad connection of the pineal vesicle with the ventricular system of the brain; distally, there is now only a single layer of flat cells, some of them showing mitoses. At 4.4 mm, the distal and the proximal walls of the anlage have narrowed to such an extent that only traces of a lumen persist. From now onwards, the pineal anlage has the shape of a mushroom; its distal cap-like part gives rise to the frontal organ, whereas the stalk region forms the pineal organ proper. The anlage of the frontal organ has reached its final size and dividing cells are no longer detectable. The anlage of the pineal organ proper, by contrast, keeps on growing, new cells being added from the tissue of the diencephalic roof lying in the immediate vicinity of the area where the anlage of the pineal organ proper is attached to the diencephalon. At the 8-mm stage, the cells of the proximal part of the anlage of the frontal organ have begun to differentiate into sensory cells, supporting cells and nerve cells, whereas the distal part retains its original epithelial character. The anlage of the pineal organ proper has increased in size, showing forward growth and lying horizontally on a part of the diencephalic roof which has been termed *Zirbelpolster* (pineal cushion). At the 10-mm stage, there is a first sign of a separation of the frontal organ and the anlage of the pineal organ proper, namely a fairly deep, wedge-shaped furrow. At the 14-mm stage, the structural elements of the frontal organ are fully differentiated and a short myelinated nerve (pineal nerve) extending from the frontal organ in the direction of the pineal organ proper is seen. The further development is characterized by further growth of the pineal organ proper and complete separation from the frontal organ, which moves anteriad, the two components of the pineal complex being linked by the pineal nerve. During metamorphosis, within a short time, the frontal organ is displaced in an anterior direction, after an elongation of the pineal nerve and the accompanying artery. In the *Xenopus laevis* larvae studied by VON HAFFNER, the frontal organ showed signs of regression after metamorphosis: pigment aggregates in the frontal organ and the sensory cells degenerate, followed by accumulation of 'lipid droplets' in the parenchyma and degeneration of both the pineal nerve and the accompanying artery. It should be noted, however, that on the basis of our present knowledge lipid droplets do not indicate degenerative processes per se. Accumulations of mitochondria and lamellae of the outer segments, due to the lipid content of their cytomembranes, may also give rise to lipid droplets in Sudan-stained preparations.

In the urodele, *Taricha torosa*, the early developmental stages of the pineal are similar to those observed in *Xenopus*. An important difference is that in *Taricha* an anlage of the frontal organ does not develop. As in *Xenopus*, the developing pineal vesicle of *Taricha* receives its supply of new cells from that part of the diencephalic roof that lies adjacent to the area where the pineal vesicle is attached to the diencephalon (the so-called orifice) (KELLY, 1963). This area is also the source of new cells in regenerating pineal vesicles (CINTRON and KELLY, 1961). By the time the larva emerges from its egg capsule, the pineal anlage has the form of an elongated vesicle with a wide lumen (KELLY, 1963), a narrow passageway connecting the lumen with the third ventricle (KELLY, 1958). The pineal vesicle then increases in length, becomes flattened against the cranial roof and the lumen is progressively obliterated. As can be expected from the constant supply of new cells from a circumscribed area, cell migration is a characteristic feature of the developing pineal. In embryos, the dominant direction taken by the new cells is toward the posterior part of the organ, mainly into the angle where the floor and the roof of the organ meet and to a lesser extent into the floor. After hatching, cell movement is almost entirely into the anterior part of the pineal organ, mainly into the floor and less frequently into the roof (HENDRICKSON and KELLY, 1969). Cell counts have revealed that the pineal organ shows a continuous increase in the total number of cells from about 79 at 2 weeks before hatching to an average of 680 in the adult, approximately 5 years after hatching. At all stages investigated the sensory cells amounted to 14%–18% of the total cell population (HENDRICKSON and KELLY, 1968). As regards the differentiation of the sensory cells, see HENDRICKSON and KELLY (1971). Pineal development in *Taricha* has been reviewed by KELLY (1962, 1965, 1971).

In the axolotl, the pineal organ reaches a constant volume of 7,500–8,500 μm^3 at 6 months of age; at this stage the sagittal axis measures 0.5–1.0 mm, the transverse axis, 0.35–0.55 mm and the dorsoventral axis, 0.08–0.15 mm (SLONIMSKI, 1947).

III. Ultrastructure and Histochemistry

Electron microscopic studies have clearly established that both the frontal organ and the pineal organ proper exhibit the typical features of a photosensory organ. According to OKSCHE and VAUPEL-VON HARNACK (1963), the frontal organ and pineal organ proper have a basically similar structural organization.

Therefore, both parts of the pineal complex are dealt with jointly. In addition to photoreceptor cells, the parenchyma of the pineal complex has been shown to consist of supporting cells and nerve cells, corroborating earlier light microscopic findings (VAN DE KAMER, 1949, 1956, 1958; OKSCHE, 1952, 1954, 1955).

1. Photoreceptor Cells

Electron microscopic studies have shown that the photoreceptor cells of the amphibian pineal complex exhibit a spectrum of images, which are not always easy to interpret. First there are typical photoreceptor cells which conform to the general pattern of vertebrate photoreceptor cells, as described above (p. 334). Then there are photoreceptor cells whose outer segments are reduced (Fig. 190). According to scanning electron microscopic studies carried out in *Rana temporaria*, 40%–50% of the inner segments lack well organized outer segments (HARTWIG and KORF, 1978). It is an open question whether these forms represent rudimentary photoreceptor cells or whether they are photoreceptors whose outer segments are in a process of renewal (see below). Finally there are photoreceptor cells whose discs of the outer segments look atypical (see below).

The *typical* pineal photoreceptor cells of both anurans and urodeles consist of outer segments, inner segments, perikarya, basal processes and finally terminal swellings which are the sites of synaptic contacts (frontal organ: *Rana temporaria, R. esculenta:* OKSCHE and VON HARNACK, 1962, 1963, 1964; *R. pipiens:* KELLY, 1962; KELLY and SMITH, 1964; *Hyla regilla:* EAKIN and WESTFALL, 1961; EAKIN, 1968; pineal organ proper: anurans: *Rana esculenta:* OKSCHE and VAUPEL-VON HARNACK, 1963, 1965b; *R. pipiens:* KELLY, 1962; KELLY and SMITH, 1964; *Hyla regilla:* EAKIN et al., 1963; *Xenopus laevis:* UECK, 1968a; CHARLTON, 1968; *Hymenochirus boettgeri, Bombina bombina:* UECK, 1968a; urodeles: *Taricha torosa:* KELLY, 1965; *Notophthalmus viridescens viridescens:* FLIGHT, 1968, 1973; *Ambystoma tigrinum:* KORF, 1976). The main morphological features of the typical photoreceptor cells are shown diagrammatically in Fig. 152. Not shown in this figure is the interesting finding obtained by scanning electron microscopy that in both *Rana temporaria* and *Ambystoma tigrinum* the majority of the pineal outer segments are arranged parallel to the roof of the skull, which may have a bearing on the perception of the *e*-vector direction of environmental polarized light (HARTWIG and KORF, 1978; and p. 394). In contrast to fishes, relatively little detailed knowledge is available regarding the number of discs present in the outer segments. A comparison of the data given in Tables 29 and 40 reveals that the number of discs is usually higher in amphibians than in fishes. In both frontal organ and pineal organ proper the outer segments have been shown to contain photopigment, those of the frontal organ (*R. catesbeiana*) a photolabile substance with an absorption maximum between 560 and 580 nm, and those of the pineal organ proper (*R. temporaria*) a rhodopsin-like pigment 502_1 (HARTWIG and BAUMANN, 1974; HARTWIG, 1975; see also OKSCHE and HARTWIG, 1979).

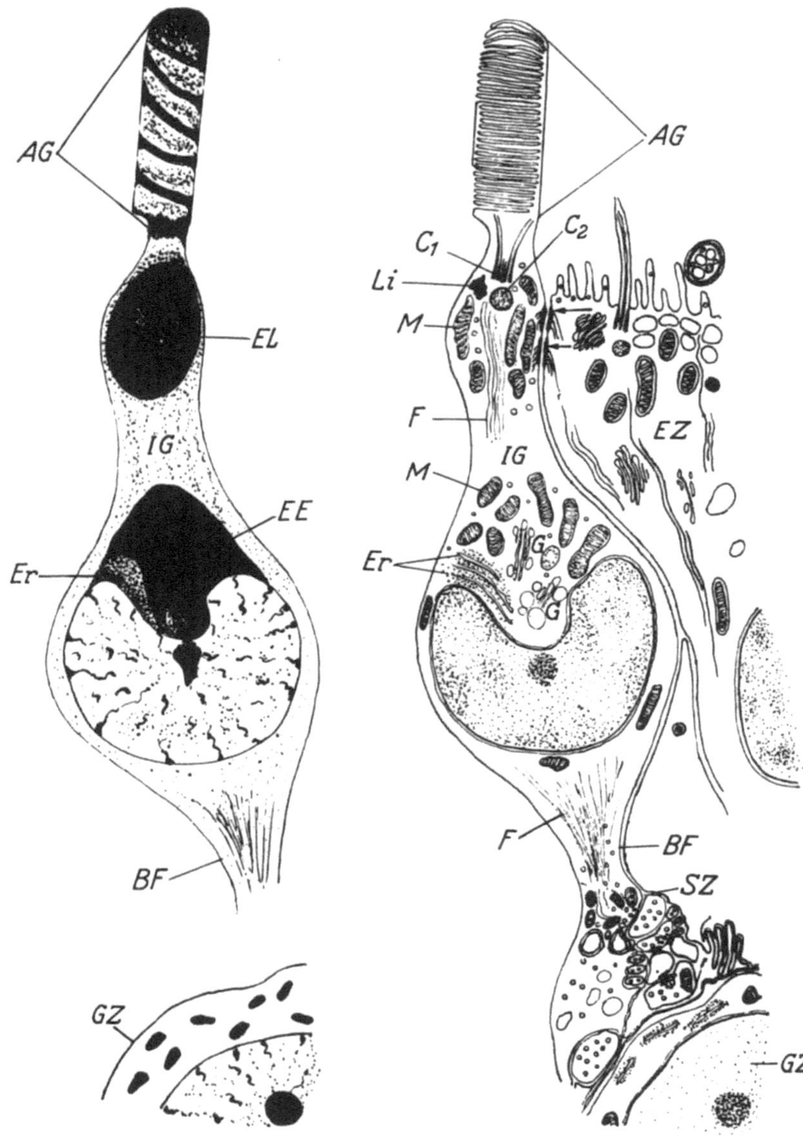

Fig. 152. Schematic representation of pineal photoreceptor cells in anurans, based on **a** light and **b** electron microscopic studies. *AG*, outer segment; *IG*, inner segment; *BF*, basal process. In **a**: *EL*, ellipsoid; *EE*, *Ersatzellipsoid*; *Er*, ergastoplasma-like, basophilic zone; *Gz*, ganglion cell. In **b**: C_1, basal body with cilium-like fibrils; C_2, centriole; *Li*, lipid droplets; *F*, filaments; *M*, mitochondria; *Er*, rough endoplasmic reticulum; *G*, Golgi apparatus; *Sz*, zone of synapses; *EZ*, ependyma-like cell. OKSCHE and VON HARNACK, 1963

Table 40. Outer segments of photoreceptor cells in the pineal complex of amphibians

Species	Pineal organ outer segments			Author(s)
	Height (µm)	No. of discs		
Rana esculenta	2–10.4	120		OKSCHE and VAUPEL-VON HARNACK (1963)
Ambystoma tigrinum		100		KORF (1976)
	Frontal organ outer segments			
	Height (µm)	Width (µm)	No. of discs	
Rana temporaria	2.5–3.1	1.0–2.6	60–110	OKSCHE and VON HARNACK (1963)
Rana esculenta	2.5–3.1	1.0–2.6	60–110	OKSCHE and VON HARNACK (1963)
Hyla regilla	4–9	3.5	100–300	EAKIN (1961), EAKIN and WESTFALL (1961)

In most of the papers cited above, the apparent breakdown and renewal of outer segments has been mentioned. This concept has been strengthened by showing that tritiated leucine incorporated into proteins is transported from the myoid region through the ellipsoid region into outer segments (BUNT and KELLY, 1971). Shedding of the outer segments occurring either apically or basally (UECK, 1968a) makes it virtually impossible to decide whether reduced-looking outer segments are rudimentary photoreceptors or cells having shed their outer segments, followed by disintegration of the discs. Systematic studies are required to elucidate whether outer segments having an appearance as shown in Fig. 153 belong to a special type of photoreceptor cells, are fixation artefacts or represent outer segments in the process of development (OKSCHE and VAUPEL-VON HARNACK, 1963).

A great number of histochemical tests have shown that the outer segments contain lipids, cystine- or cysteine-rich proteins, glycoproteins and acid mucopolysaccharides (KELLY and VAN DE KAMER, 1960; OKSCHE, 1962; OKSCHE and VON HARNACK, 1963), the latter two possibly lying on the outer surface of the plasmalemma infoldings. Whether they are an integral part of the surface coat or are involved in secretion is open to discussion. The outer segments are also moderately basophilic to thionine, weakly to moderately positive for side-chain carboxyl groups, combined sulfhydryl and disulphide and tryptophan, and are weakly positive for tyrosine and side-chain amino groups (KELLY and SMITH, 1964). An intense blackening of the outer segments after OsO_4 treatment (2%, 3 days, 40° C) has been attributed to the presence of photopigments (UECK, 1970b, 1971a).

As regards the morphological features of the other parts of the photoreceptor cells, Fig. 152 should be consulted. Points of special interest include an abundance of microvilli extending from the inner segment and surrounding the outer segment (e.g. FLIGHT, 1973). In *Ambystoma tigrinum* the microvilli have been

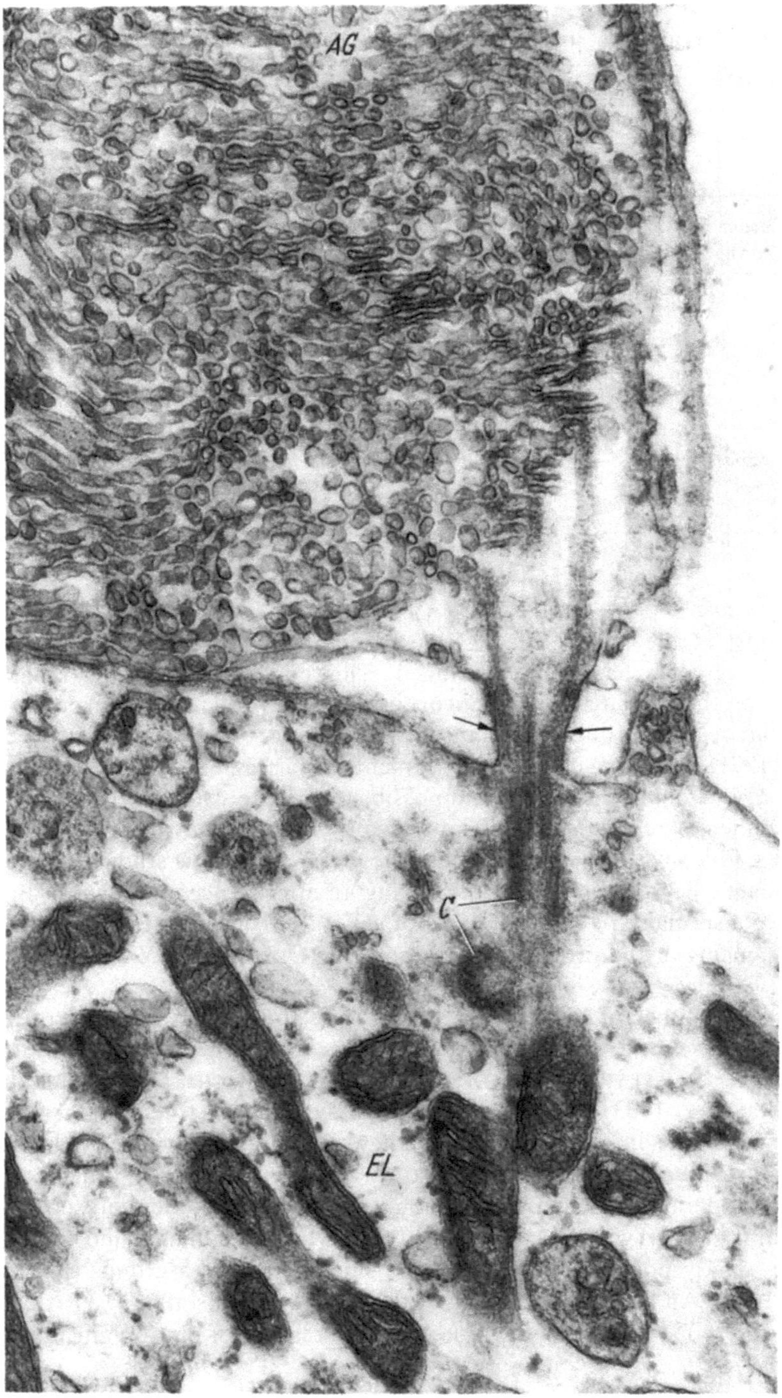

Fig. 153. Apical part of a pineal photoreceptor cell of *Rana esculenta*. The outer segment (*AG*) contains many atypical structures, instead of the usually occurring lamellar discs. *C*, centriole; *EL*, ellipsoid. OKSCHE and VAUPEL-VON HARNACK, 1963; ×43,900

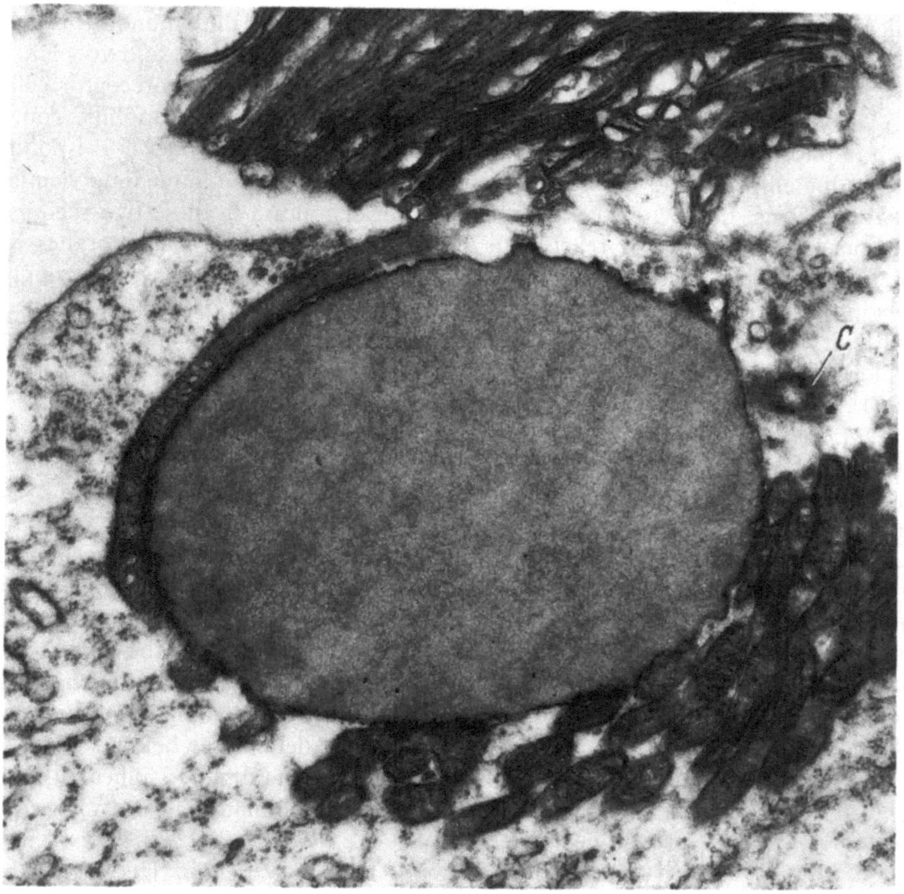

Fig. 154. Photoreceptor cell in the pineal of *Rana esculenta* showing a large lipid droplet in the inner segment. *C*, centriole. OKSCHE and VAUPEL-VON HARNACK, 1963; ×23,760

thought to be involved in the stabilization of the spatial orientation of the outer segments which are arranged parallel to the roof of the skull in order to facilitate detection of the *e*-vector direction of polarized light (HARTWIG and KORF, 1978). In some of the photoreceptor cells a large fat droplet (Fig. 154) has been noted in the ellipsoid region of the inner segment (OKSCHE and VAUPEL-VON HARNACK, 1963). Whether these fat droplets correspond to the oil droplets present in other vertebrate photoreceptor cells which may have a bearing on the reflectance of light is an open question. The presence of fat droplets in photoreceptor cells has also been demonstrated by histochemical means (KELLY and VAN DE KAMER, 1960). Frequently the presence of glycogen has been mentioned (OKSCHE, 1962; KELLY and VAN DE KAMER, 1960; EAKIN et al., 1963; KELLY and SMITH, 1964), but occasionally also its absence (CHARLTON, 1968).

The detection of high concentrations of RNA and protein-lipid complexes forming cap-like structures in the immediate vicinity of the nuclei of photoreceptor cells (OKSCHE, 1954, 1955, 1956, 1962) is in good accordance with electron microscopic findings pointing to the presence of abundant mitochondria, rough endoplasmic reticulum and dictyosomes (*Ersatzellipsoid*) (OKSCHE and VON HARNACK, 1963; OKSCHE and VAUPEL-VON HARNACK, 1963). In photoreceptor cells of *Xenopus laevis* (CHARLTON, 1968) there are two intensely acidophilic zones rich in phospholipids (Baker's acid hematein test), one of the zones corresponding to the ellipsoid and the other lying supranuclearly in the immediate vicinity of the nucleus. In view of the abundance of mitochondria in these two regions, the positive phospholipid reaction is probably due to the cytomembranes of the mitochondria. Between the mitochondria-rich acidophilic zones a strongly basophilic region has been detected. Tests for side-chain carboxyl and combined sulfhydryl and disulphide groups were distinctly positive in the ellipsoid region which stained slightly more strongly than the outer segments (KELLY and SMITH, 1964).

Enzyme histochemical studies at the light microscopic level failed to demonstrate acid and alkaline phosphatases (CHARLTON, 1968, *Xenopus*). At the electron microscopic level acid phosphatase was demonstrable in subspherical vesicles of the photoreceptor ellipsoid (*Rana pipiens:* BUNT and KELLY, 1971). Light microscopically, the cells bordering the lumen of the epiphyseal sac showed an intensely positive reaction for succinate dehydrogenase and β-hydroxybutyrate dehydrogenase, the reaction for the latter enzyme and lactate dehydrogenase demonstrating material in and droplets bordering the lumen of the epiphyseal sac (PAUL, 1968). Acetylcholinesterase activity is not demonstrable in photoreceptor and supporting cells, but is a typical feature of intrapineal nerve cells (UECK, 1971a; WAKE et al., 1974; KORF, 1976) (for details, see below).

In view of the presence of indoleamines in the mammalian pineal it was logical to search for similar substances in the amphibian pineal. While the first such attempts proved negative (EAKIN et al., 1963; ITURRIZA, 1967; KELLY and SMITH, 1964; CHARLTON, 1968), OWMAN et al. (1970) and UECK (1970b) clearly demonstrated yellow serotonin fluorescence (FALCK-HILLARP technique) in both photoreceptors and supporting cells, provided that the animals had been pretreated with the MAO inhibitor, nialamide. This leaves, however, the site of indoleamine synthesis open to discussion, especially as the indoleamines may be taken up by adjacent cells which do not synthetize them. Using an improved microspectrofluorimetric technique, HARTWIG (unpublished results) was able to show 5-HT fluorophore in pineal receptor cells of previously *untreated* frogs. To date it is impossible to say whether these receptors represent the fully differentiated or the rudimentary photoreceptor cells. Morphological correlates of secretory products are usually not demonstrable in amphibian pineal photoreceptor cells (OKSCHE and VON HARNACK, 1963). A few dense-core granules in basal processes of photoreceptor cells have been demonstrated by UECK (1968a).

A point of special interest is the *synaptic connections* of photoreceptor cells. This important aspect of pineal structure and function has been studied in detail in three species only; thus it is not known how representative the findings are for amphibians in general. Electron microscopic studies have shown that three types of synapses are present, namely ribbon synapses, conventional synapses and synapses characterized by subsurface cisterns.

The most thoroughly investigated synapses are the ribbon synapses, which are typical of vertebrate photoreceptors, as mentioned above (p. 335). Following initial studies pointing to the presence of synaptic ribbons in photoreceptor cells of *Rana* (OKSCHE and VON HARNACK, 1963; OKSCHE and VAUPEL-VON HARNACK, 1963; KELLY and SMITH, 1964; KELLY, 1965) and *Hyla regilla* (EAKIN et al., 1963), subsequent studies have revealed the high complexity of ribbon synapses. In the anuran, *Bombina variegata*, ribbon synapses have been observed opposite dendrites (axodendritic synapses) and nerve cell perikarya (axosomatic synapses). As shown in Fig. 155, BAYRHUBER (1972) subdivided the axodendritic synapses into two main types, depending on the number of ribbons present in a given contact and the number of postsynaptic dendrites. In type I synapses two to five dendrites lie opposite a single synaptic ribbon, whereas in type II synapses a varying number of ribbons (1–6) oppose a single dendrite. The dendrites are usually not invaginated into photoreceptor cells. Further details are depicted in Fig. 155. Not shown in this figure is the observation that in one instance a ribbon has been found in the cell *body* of a photoreceptor cell.

The ribbon synapses of the pineal photoreceptor cells of the urodele, *Ambystoma tigrinum*, are diagrammatically shown in Fig. 156, which is self-explanatory. In this species the interesting observation has been made that the ribbon synapses are not restricted to the basal processes of photoreceptor cells, but occur also on the cell body (KORF, 1976). In the urodele, *Diemictylus viridescens viridescens*, the situation of the ribbon synapses is virtually identical to that described in *Ambystoma*, with the additional finding of ribbon synapses on nerve cell perikarya (FLIGHT, 1973).

Conventional synapses, i.e. synapses lacking ribbons, have been observed as well, but less frequently than ribbons synapses. They are diagrammatically shown in Figs. 155 and 156. Their functional significance is obscure, but it has been suggested that they may be linked to interneurons (BAYRHUBER, 1972; KORF, 1976).

Synapses characterized by the presence of subsurface cisterns have been found in *Rana pipiens* and are thought to be involved in efferent transmission or feedback mechanisms (KELLY and SMITH, 1964).

2. Supporting Cells

In addition to photoreceptor cells, both the frontal organ and the pineal organ proper comprise supporting cells. When compared with photoreceptor cells, the supporting cells have been little investigated. It is therefore not known whether or not they represent a homogeneous population of cells. In early electron microscopic studies by OKSCHE and VON HARNACK (1963) and OKSCHE and VAUPEL-VON HARNACK (1963) ependymal-like cells (*ependymale Begleitzellen*) or ependymal cells, glial cells and epithelial cells were distinguished, the latter of which were thought to represent rudimentary photoreceptor cells. In other studies only one type of cell has been mentioned, usually referred to as supporting cell (EAKIN et al., 1963; UECK, 1968a). In view of this discrepancy

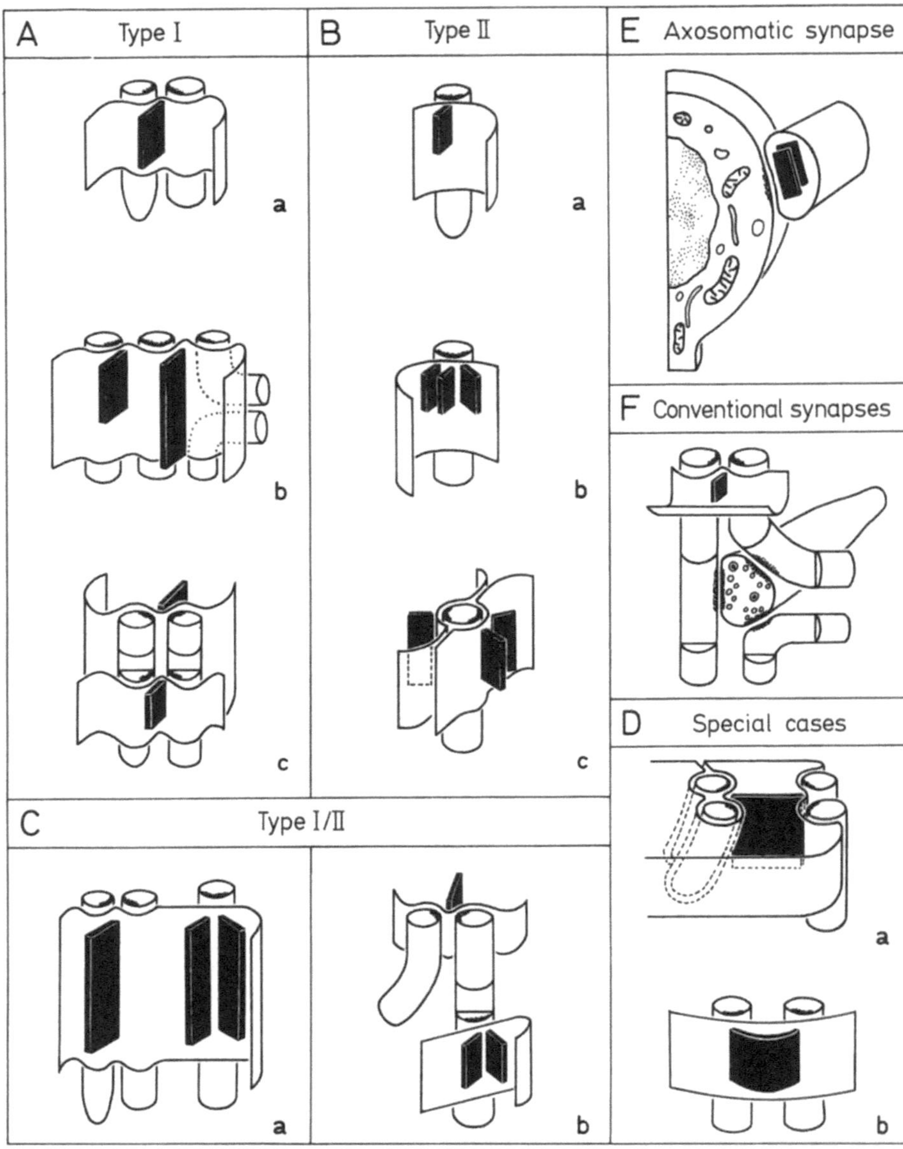

Fig. 155 A–F. Schematic representation of different types of synapses in the pineal organ of *Bombina variegata*. **A** Type I synapses, in which two (*a*, *c*) or more (*b*) dendrites are associated with a single ribbon. **B** Type II synapses characterized by a number of ribbons associated with a single dendrite. **C** Joint presence of type I and type II synapses. **D** special cases. **E** axosomatic synapse. **F** conventional synapse. BAYRHUBER, 1972

only a cursory description of supporting cells will be given, based on the papers in which cells have been explicitly termed supporting cells.

Apart from the usual cell organelles, repeated mention has been made of the presence of smooth endoplasmic reticulum and myeloid bodies (Fig. 157), which

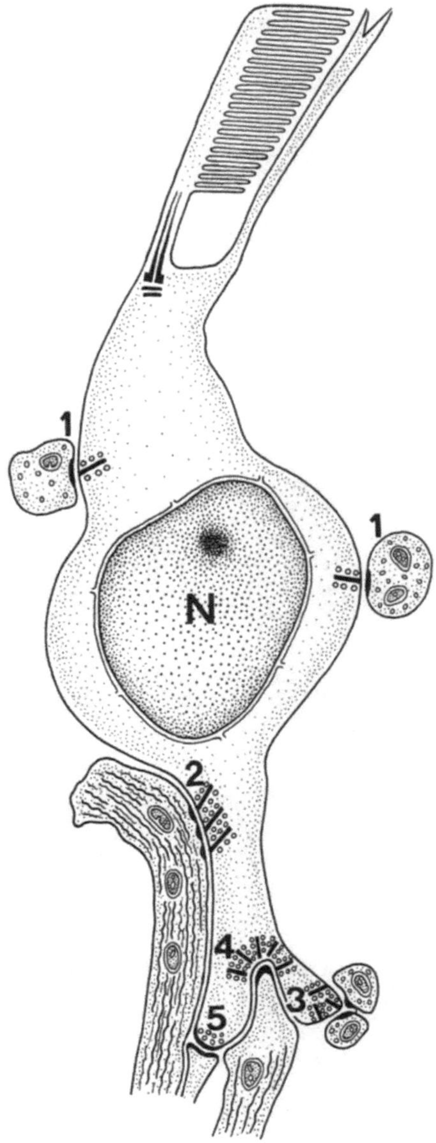

Fig. 156. Diagram to show the possible types of synapses between a photoreceptor cell and dendritic elements in the pineal body of *Ambystoma tigrinum*. *1*, smooth-surfaced ribbon contacts on the perikaryon of the photoreceptor; *2*, tangential synapse; *3*, dyad complex; *4*, ribbon synapse on an invaginated dendritic spine; *5*, conventional synapse. KORF, 1976

appear to be specializations of the endoplasmic reticulum (KELLY and SMITH, 1964; UECK, 1968a; FLIGHT, 1973). Moreover, microvilli, pigment granules, many (pinocytotic?) vesicles (EAKIN et al., 1963), lipid droplets (UECK, 1968a) and glycogen (FLIGHT, 1973) have been mentioned.

As regards the function of supporting cells, very little of a definitive nature is known. Histochemical studies by OKSCHE (1954, 1955, 1956, 1962) have shown that the cells are very active, comprising large amounts of RNA and protein-lipid complexes in the immediate vicinity of the cell nucleus. A high metabolic

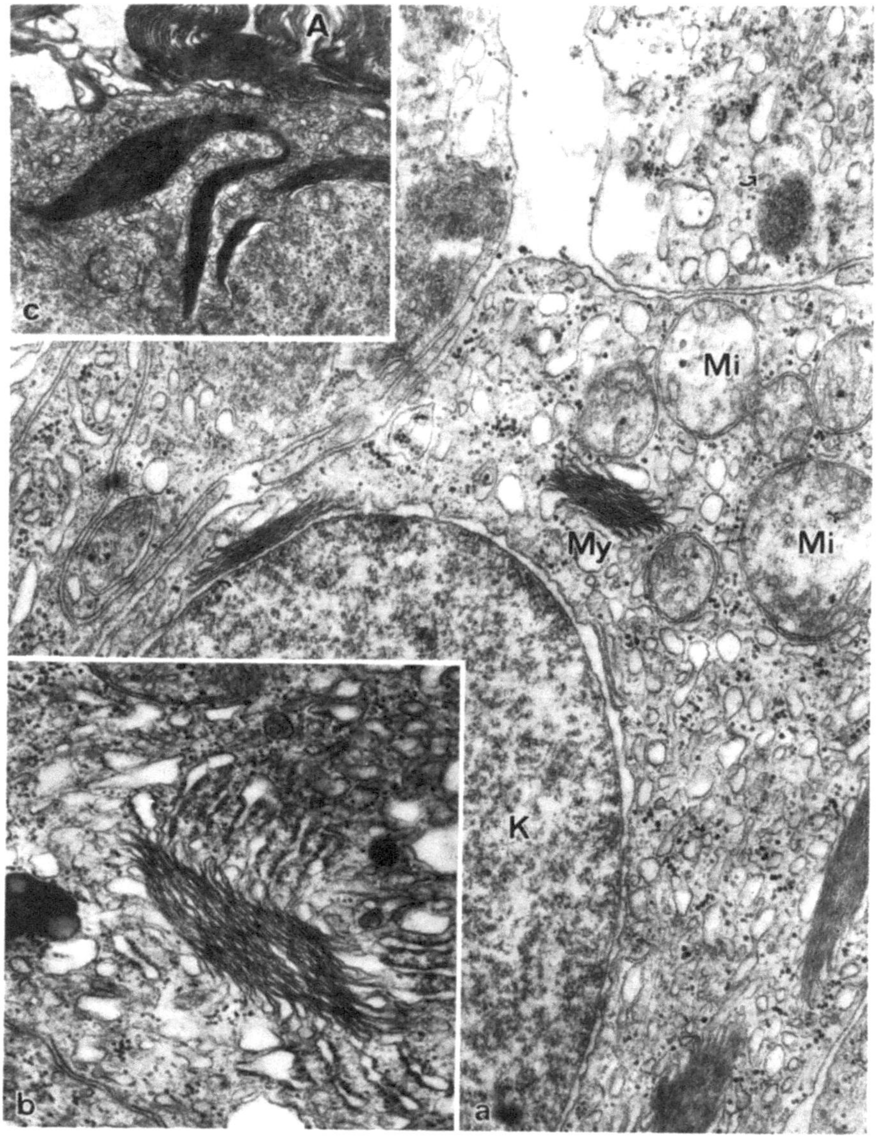

Fig. 157 a–c. Myeloid bodies (*My*) in supporting cells. **a** *Xenopus laevis*; *Mi*, mitochondria; *K*, nucleus. × 25,000. **b** *Xenopus laevis*; myeloid body showing transition to cisternae of rough endoplasmic reticulum. × 25,500. **c** *Bombina bombina*; U-shaped myeloid body. × 10,700. UECK, 1968a

activity of the supporting cells is also suggested by the fact that they actively incorporate ^{3}H-vitamin A, the cells showing the highest label of all pineal structures, the most strongly labelled components being the myeloid bodies mentioned above (FLIGHT and VAN DONSELAAR, 1975). According to these authors, supporting cells may be involved in photopigment metabolism and/or

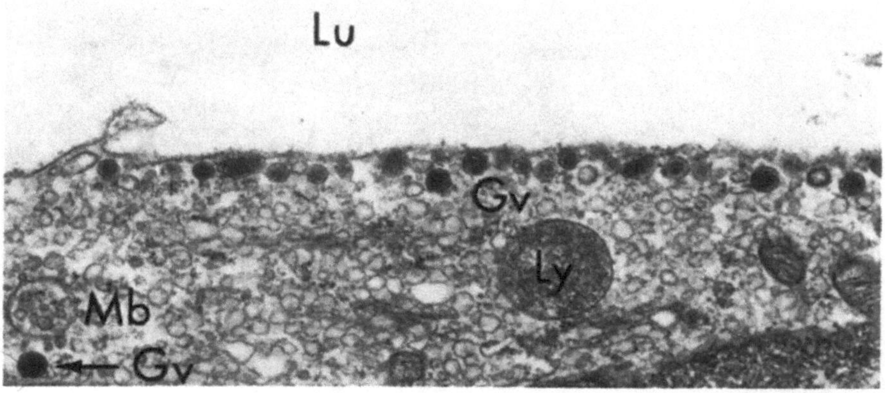

Fig. 158. Supporting cell in the pineal organ of *Diemictylus viridescens viridescens* showing granular vesicles (*Gv*) predominantly arranged near the apical plasmalemma. *Lu*, pineal lumen; *Ly*, lysosomal body; *Mb*, multivesicular body. FLIGHT, 1973; ×20,000

may represent a second type of photoreceptive element in the epiphysis cerebri. There is clear evidence for an involvement of the cells in phagocytosis, as shown by the observation that both macrophages and supporting cells contain lamellar bodies resembling degenerated discs of outer segments (UECK, 1968a; FLIGHT, 1973).

A secretory function of supporting cells has been discussed as well. OKSCHE and VON HARNACK (1963, 1964) concluded that among all the different cell types present in the frontal organ it is the *ependymale Begleitzellen* which may be involved in secretory processes, because they comprise apically a fairly large number of vacuoles which may release substances into the lumen of the organ. As mentioned above (p. 378), the supporting cells exhibit yellow serotonin fluorescence after appropriate pretreatment (OWMAN et al., 1970; UECK, 1970b). It is not known though whether the cells actively synthesize serotonin which is then converted to melatonin or whether serotonin is released from neighbouring photoreceptor cells and subsequently taken up by supporting cells. Very convincing pictures of what could well be secretory granules or monoamine storage granules in supporting cells have been published by FLIGHT (1973) (Fig. 158). In this context it is interesting to note that by using a modified OsO_4 technique (2% OsO_4 for 3 days at 40° C) UECK (1970b, 1971a) demonstrated the presence of intensely reacting granules measuring 80–250 nm in diameter (Fig. 159). An interesting, but functionally poorly understood finding, is that ependymal-like supporting cells may be synaptically innervated at their apical pole by axons running freely through the pineal lumen and belonging to the system of CSF contacting neurons (*Pleurodeles waltlii:* VIGH-TEICHMANN et al., 1973).

3. Intrapineal Nerve Cells

In addition to photoreceptor cells and supporting cells, both the frontal organ and the pineal organ proper comprise nerve cells (OKSCHE, 1952, 1954,

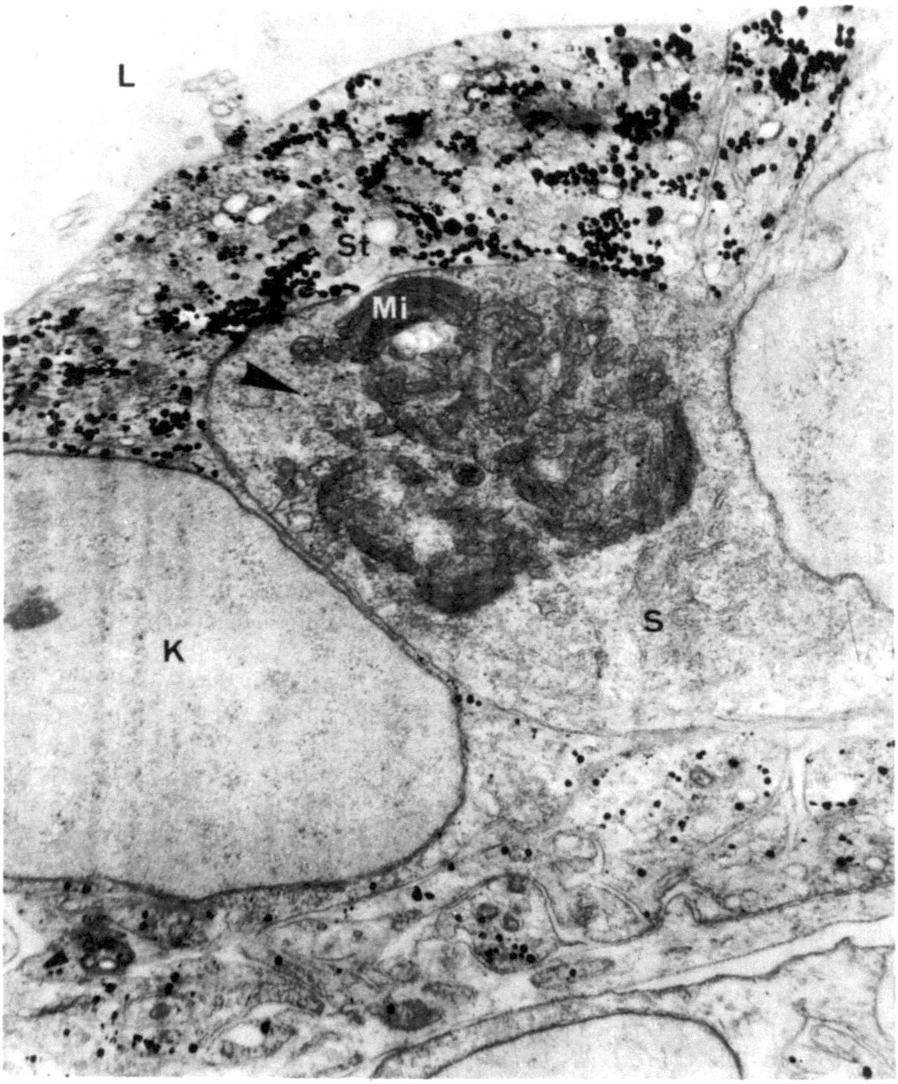

Fig. 159. Supporting cell (*St*) adjacent to the pineal lumen (*L*) and overlying a sensory cell (*S*). Note the abundance of specifically stained granules in the supporting cells and the occasional presence of smaller granules in the sensory cell (*arrowhead*). Prolonged osmication. *K*, nucleus of a supporting cell; *Mi*, mitochondrion. UECK, 1971a; ×9,240

1955). They have already been mentioned when the synaptic connections of photoreceptor cells were discussed (p. 379). The intrapineal nerve cells have attracted special attention in attempts to elucidate the neuro-anatomical basis for the electrophysiological findings that direct illumination of the frontal organ is followed by chromatic and achromatic responses, whereas the pineal organ proper usually responds only achromatically (for details see p. 394). In comparison to the neuronal wiring diagram of the retina of the lateral eyes, it was especially relevant to examine whether the neuronal chain of the pineal complex consisted of two elements only, namely photoreceptor cells (first order neurons) and a nerve cell (second order neuron), or whether bipolar nerve cells, horizontal cells and amacrine cells were present (PAUL et al., 1971). Moreover, it was relevant to define the nerve cells which give rise to the pineal nerve and pineal tract.

Although supravital methylene blue staining, silver impregnation techniques (PAUL et al., 1971) and histochemical studies demonstrating acetylcholinesterase (WAKE et al., 1974) have clearly shown the presence of different types of neurons in the epiphyseal complex of anurans, their precise spatial and functional interrelationships are difficult to disentangle, one of the difficulties being that, unlike the retina, the wall of the pineal complex does not show a clear lamination as a result of a lack of a precise arrangement of the various cell types present. In a first study (PAUL et al., 1971), multipolar nerve cells, bipolar nerve cells and amacrine-like cells were distinguished and it was thought that it was mainly the large multipolar nerve cells and the smaller scarcely ramified nerve cells which gave rise to the afferent pineal tract, a conclusion which was later revised (see below). Moreover, the view was expressed that interneurons may be present, which differ in appearance from the classic bipolar, horizontal and amacrine cells of the amphibian retina.

The best results to date were obtained by using the acetylcholinesterase reaction after Karnovsky and Roots (WAKE et al., 1974; UECK, 1971a). With this technique the pineal complex of *Rana ridibunda* and *R. esculenta* was shown to comprise large and small multipolar nerve cells, pseudo-unipolar nerve cells and a few bipolar nerve cells (Figs. 160, 161), the distributions of which are shown in Fig. 162 (WAKE et al., 1974). It can be seen that the nerve cells of the dorsal wall of the pineal organ outnumber those of the ventral wall (ratio 3:1). The ratios of multipolar to pseudo-unipolar nerve cells are 1:4 in the frontal organ and 3:5 in the pineal organ proper. As regards the function of these two main types of nerve cell, it has been suggested, in partial contrast to studies using methylene blue staining and silver impregnation (PAUL et al., 1971), that the pseudo-unipolar nerve cells give rise to the pineal tract, whereas the multipolar nerve cells have been interpreted as interneurons, arranged in a manner as shown diagrammatically in Fig. 163. Further studies will have to show whether a similar neuronal wiring is present in other species and especially whether this arrangement accounts for the differences in chromatic and achromatic responses observed between frontal and pineal organs. Also the functional significance of the bipolar nerve cells remains to be established. The same applies to two groups of unipolar nerve cells, each consisting of 30–50 elements. One group lies in juxtaposition of the pineal tract, while one lies along the base of the subcommissural organ; they send projections to the mesencephalon and the pineal organ respectively (WAKE et al., 1974).

4. Axons Running Through the Lumen of the Pineal Organ

In addition to the intrapineal neurons described above, there is a system of axons running through the lumen of the pineal body (VIGH-TEICHMANN

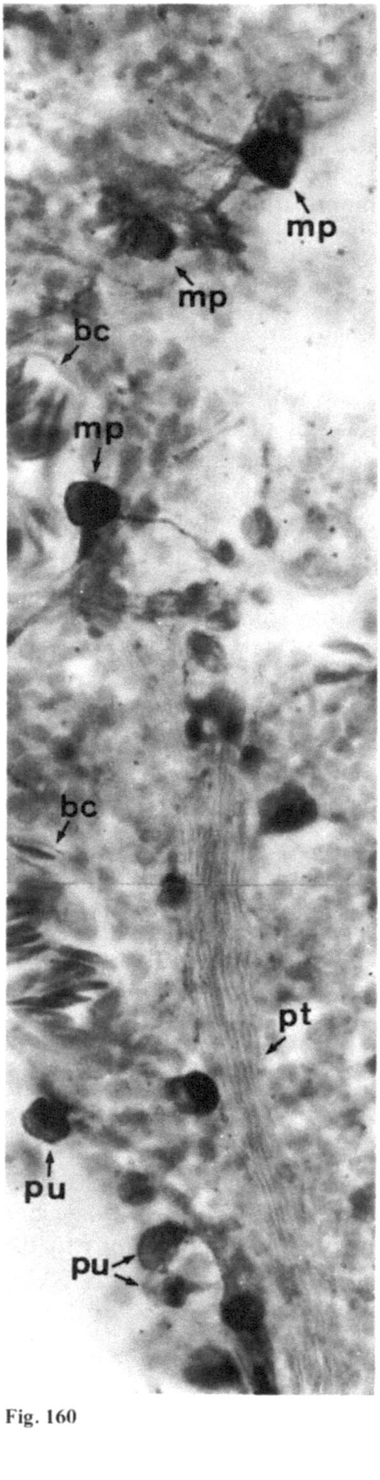

Fig. 160

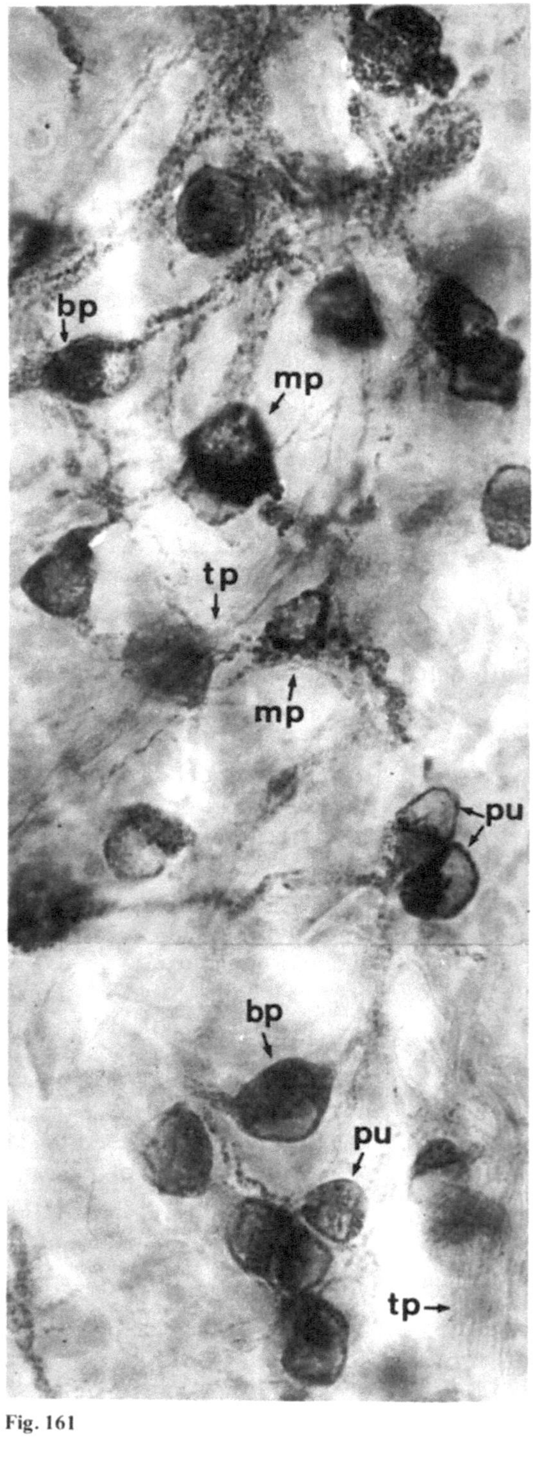

Fig. 161

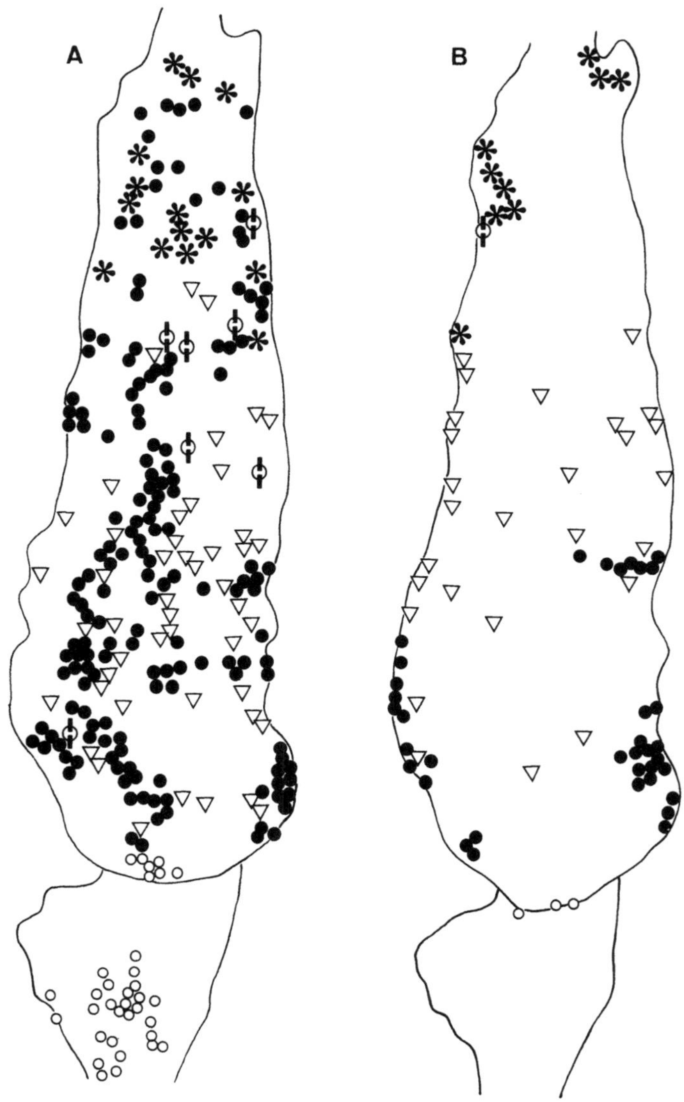

Fig. 162 A, B. Distribution of various types of AChE-positive neurons in **A** the dorsal and **B** the ventral walls of the pineal organ of *Rana ridibunda*. ∗, large multipolar nerve cells exceeding 2,000 μm^3; ▽, small multipolar nerve cells measuring less than 2,000 μm^3; ⏀, bipolar nerve cells; ●, pseudo-unipolar nerve cells; ○, nerve cells in the area of the subcommissural organ. WAKE et al., 1974

Figs. 160, 161. Types of AChE-positive neurons in the pineal organ of *Rana esculenta* (**Fig. 160**) and *R. ridibunda* (**Fig. 161**). Note the presence of multipolar neurons (*mp*), bipolar neurons (*bp*) and pseudo-unipolar neurons (*pu*). *bc*, blood capillary; *pt*, *tp*, pineal tract. **Fig. 160**: section; × 500. **Fig. 161**: whole mount preparation; × 640. WAKE et al., 1974

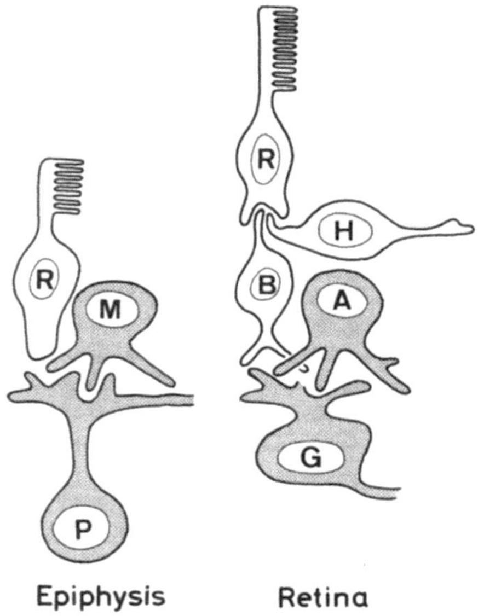

Fig. 163. Diagram depicting the neuronal organization of the frog pineal organ compared with that of the retina. *A*, amacrine cell; *B*, bipolar cell; *G*, ganglion cell; *H*, horizontal cell; *M*, multipolar cell; *P*, pseudo-unipolar cell; *R*, receptor cell. The dark elements are AChE-positive. WAKE et al., 1974

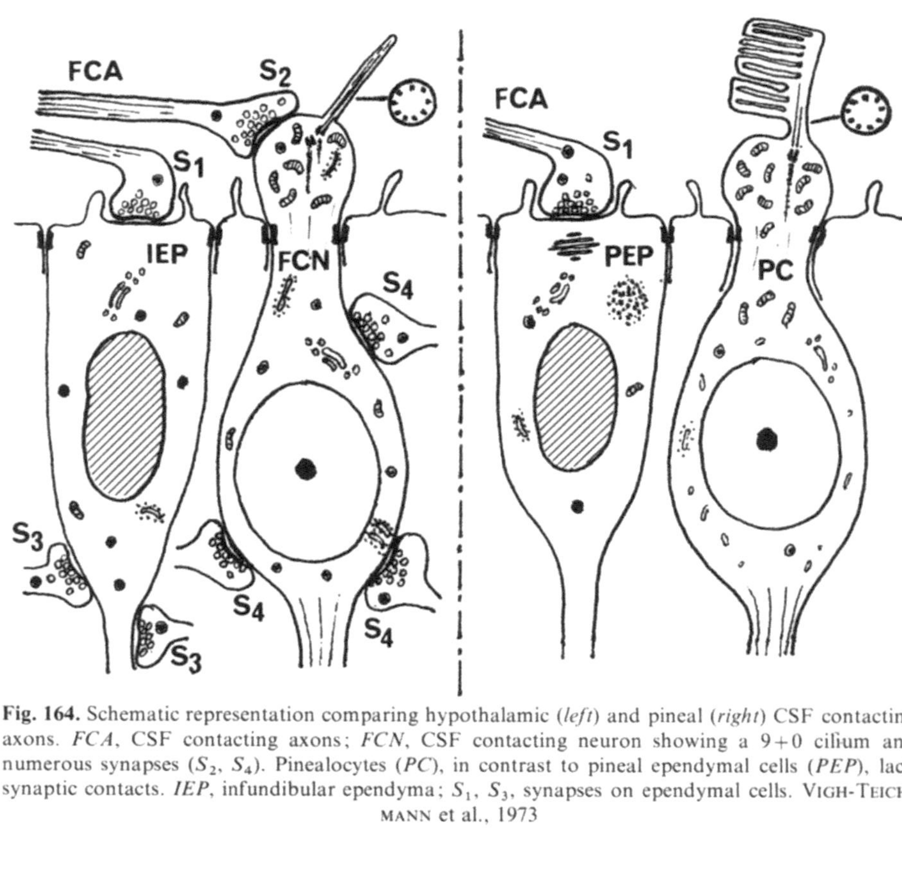

Fig. 164. Schematic representation comparing hypothalamic (*left*) and pineal (*right*) CSF contacting axons. *FCA*, CSF contacting axons; *FCN*, CSF contacting neuron showing a 9+0 cilium and numerous synapses (S_2, S_4). Pinealocytes (*PC*), in contrast to pineal ependymal cells (*PEP*), lack synaptic contacts. *IEP*, infundibular ependyma; S_1, S_3, synapses on ependymal cells. VIGH-TEICHMANN et al., 1973

et al., 1973). As shown in Fig. 164, these fibres form synapses on the apical plasmalemma of ependymal-like supporting cells and bear some resemblance with CSF-contacting axons of the other parts of the central nervous system. The non-myelinated naked axons run singly or in bundles, the latter comprising some 20–25 nerve fibres. The diameter of these axons measures 200–400 nm, but occasionally up to 700 nm. Their most conspicuous component is dense-core vesicles measuring 50–150 nm in diameter. Both the origin and the function of these fibres are obscure.

IV. Nervous Connections of the Pineal Complex

As can be expected from the foregoing sections, both the frontal organ and pineal organ proper have prominent nervous connections with the brain.

The nomenclature of these nervous connections has been the matter of some dispute (see the discussion after the paper by OKSCHE and VON HARNACK, 1964). According to OKSCHE, in comparative studies the term pineal nerve has been used for the fibre tract leaving the frontal organ, whereas the term pineal tract (tractus epiphyseos) is applied to the nerve bundle emanating from the pineal organ proper. The distinction between a *nerve* and a *tract* has been made in analogy to the situation of the optic system, in which a fibre bundle lying outside the brain matter is called a nerve whereas a fibre bundle inside the brain is termed a tract. In the discussion mentioned above, ARIËNS KAPPERS has proposed to use the term parietal nerve for the nerve fibres leaving the frontal organ and to use the term pineal nerve for the nervous connections of the epiphysis or pineal organ proper, because otherwise some confusion would be inevitable. OKSCHE, however, criticized the use of the term *parietal* nerve, because this term is usually used for the nervous connection of the parietal eye of lizards. In a subsequent paper KAPPERS (1965a) has discussed the terminological problems in detail, from which it transpires that it is mainly the nerve bundle leaving the frontal organ which poses a problem. KAPPERS avoided the term frontal nerve because a nerve with an identical name exists, viz. a branch of the ramus ophthalmicus of the trigeminal nerve, and instead proposed the term frontal tract. The term tract was also used by EAKIN (1973) because developmentally the frontal organ is part of the central nervous system. Electron microscopic studies, however, have shown that the nerve bundle in question has the structure of a peripheral nerve rather than a central nervous tract (BÖTTGER and BÖTTGER, 1973). Therefore in a review ADLER (1976) decided to use the term frontal nerve. In the present review this term will be used, too, because if the terms pineal nerve and pineal tract are used the origins of the two nerve bundles may easily be mistaken.

1. The Frontal Nerve

As discussed above, the frontal nerve is the nerve fibre bundle leaving the frontal organ. It should be recalled (p. 370) that size and differentiation of the frontal organ are variable. It is therefore not surprising that the frontal nerve also shows great variation. According to OKSCHE (1955), the frontal nerve is well developed in *Rana esculenta*, shows great interindividual variation in *R. temporaria* and is poorly developed in *Bufo vulgaris*. Without giving a reference, DODT (1973) mentions the absence of this nerve in *Hyla*.

In some species the frontal nerve has been thoroughly investigated because of electrophysiological studies showing that the chromatic and achromatic responses of the pineal complex are somehow related to the composition of the afferent nerve fibres. DODT's (1964b) findings suggest that the achromatic response involves myelinated nerve fibres whereas the chromatic response is linked to both myelinated and non-myelinated nerve fibres. Before considering the composition of the nerve, a few general points should be made.

As mentioned above, the fibres of the frontal nerve originate from pseudounipolar nerve cells lying in the wall of the frontal organ (WAKE et al., 1974; *Rana ridibunda, R. esculenta*). The fibres leave the frontal organ at its base (OKSCHE, 1955). As depicted in Fig. 150, the nerve runs through the dorsal lymph sac where it is easily accessible, e.g. for electrophysiological recordings, before it penetrates the skull and reaches the dorsal wall of the pineal organ proper where it merges with the fibres of the pineal tract. That fibres of the frontal nerve are indeed present in the pineal tract has been clearly demonstrated by lesioning of the frontal nerve, which was followed by the occurrence of degenerated fibres in the pineal tract (*R. temporaria, R. esculenta*: PAUL et al., 1971). Some fibres do not reach the pineal organ (UECK et al., 1971). The extracranial portion of the nerve has a length of 3–5 mm (DODT and HEERD, 1962). A characteristic feature is that the nerve decreases in width along its course. The main reason for this is that the nerve fibres decrease in number. In *Rana catesbeiana*, UECK et al. (1971) counted 20 myelinated and 95 non-myelinated nerve fibres in the extracranial portion of the nerve and only 9 myelinated and 30 non-myelinated fibres in its intracranial portion. Another factor for the decrease in width of the nerve is that the thickness of the myelin sheath is reduced. The decrease in number of the nerve fibres is not due to the fact that fibres leave the bundle, as ruled out by silver staining. The average width of the frontal nerve within the dorsal lymph sac after fixation is 33 μm in *Rana catesbeiana*, 28 μm in *R. esculenta* and 26 μm in *R. temporaria* (UECK et al., 1971). Supravital phase-contrast examination has yielded values between 100 and 120 μm (*R. esculenta:* OKSCHE and VAUPEL-VON HARNACK, 1965a). DODT and HEERDT (1962) measured a width of 100 μm (*R. temporaria*).

The overall structure of the frontal nerve has been studied in detail by BÖTTGER and BÖTTGER (1973). They point out that the arrangement of connective tissue within the nerve, the way in which the nerve degenerates after severance and the structural arrangement of wrapping cells is that of a peripheral rather than a central tract.

Studies on the composition of the frontal nerve have shown that the nerve comprises myelinated and non-myelinated axons, as assessed by electrophysiology (DODT, 1964b) and electron microscopy (OKSCHE and VON HARNACK, 1963; OKSCHE and VAUPEL-VON HARNACK, 1965a; UECK et al., 1971; BÖTTGER and BÖTTGER, 1973). In addition to the above-mentioned results obtained by UECK et al. (1971) in *Rana catesbeiana*, quantitative studies on *Rana esculenta* (OKSCHE and VAUPEL-VON HARNACK, 1965a) point to great interindividual variation in the number of myelinated (1–22) and non-myelinated nerve fibres (35–146). In materials studied both electrophysiologically and electron microscopically, no correlation existed between the amount of the two types of fibre in a given nerve and the chromatic and achromatic responses of frontal organs (OKSCHE and VAUPEL-VON HARNACK, 1965a). The diameter of the myelinated nerve fibres lies between 2 and 7 µm, 0.2–0.6 µm being the width of the myelin sheath.

In addition to the afferent fibres just described, the frontal nerve comprises a certain number of efferent fibres (UECK et al., 1971; PAUL, 1972; BÖTTGER and BÖTTGER, 1973). Using horseradish peroxidase, ZILLES and NICKELEIT (1979) demonstrated that in *R. temporaria* these fibres originate in the medial part of the amygdala, the pre-optic area, the nucleus rotundus, the pretectal area and the lateral parts of the mid-brain central griseum.

As mentioned above, the fibres of the frontal nerve join the pineal tract, where they amount to 2%–4% of the total fibre population (UECK, 1970a; UECK et al., 1971). They apparently do not form synapses in the wall of the pineal organ (OKSCHE, 1955; UECK et al., 1971). The central connections of the frontal nerve are described below (p. 392).

2. The Pineal Tract

The pineal tract originates from intramural pseudo-unipolar nerve cells of the epiphysis cerebri and about 2%–4% of its fibre population stem from the frontal nerve (see above). It is a rather prominent fibre bundle running in the dorsal wall of the epiphyseal sac (OKSCHE, 1955; OKSCHE and VAUPEL-VON HARNACK, 1965a; UECK et al., 1971). The pineal tract has an average diameter of 17–50 µm and increases in width in a rostrocaudal direction. Although the diameter of the pineal tract is in the same order of magnitude as the frontal nerve, the pineal tract consists of 60–80 (frontal nerve 22) myelinated fibres and up to 200 (35–146) non-myelinated fibres. The diameter of the myelinated fibres measures 1–6 µm, the thickness of the myelin sheath being 0.1–0.4 µm. The width of the non-myelinated fibres varies between 0.1 and 1.0 µm (OKSCHE and VAUPEL-VON HARNACK, 1965a). In contrast to the intramural nerve cells, the fibres of the pineal tract are only slightly, if at all, stained by the histochemical technique for acetylcholinesterase (WAKE et al., 1974).

In addition to the undoubtedly afferent fibres, the pineal tract contains some aminergic fibres with granulated vesicles measuring 85–95 nm in diameter (OKSCHE and VAUPEL-VON HARNACK, 1965a). UECK (1968b) found such fibres in the peripheral regions of the epiphysis and suspected that they surrounded the organ in a net-like fashion. A mainly leptomeningeal localization of green fluorescent fibres (Falck-Hillarp technique) was confirmed by OWMAN et al. (1970). While according to UECK (1968b) fluorescent fibres are confined to blood vessels, ITURRIZA (1967) found that fluorescent fibres were particularly abundant in the ventral region of the epiphysis (see also OWMAN et al., 1970) and that they showed no special relationship to blood vessels. An efferent nature of these fibres is indicated by the fact that transection of the posterior commissure was followed by a complete disappearance of fluorescence in the epiphysis cerebri and an accumulation of fluorescent material in the proximal stump of the pathway. The presence of granulated nerve fibres was also described by KORF (1976, *Ambystoma tigrinum*). Indications for an efferent innervation of the epiphysis cerebri were also obtained by KELLY and VAN DE KAMER (1960), KELLY and SMITH (1960) and PAUL (1972).

Central connections: In adequately stained histological sections the pineal tract can be easily followed (MAUTNER, 1965). It leaves the epiphysis cerebri near its caudal border and comes into close contact with the base of the subcommissural organ (SCO), where it gives off fibres to the region of the SCO. The majority of fibres continue to run caudad, where they penetrate the posterior commissure. Along its path the pineal tract becomes progressively thinner. Measurements carried out in fixed (unfixed) materials showed that in the rostral region of the SCO the tract exhibited a width of 25 μm (54 μm); in the middle region it measured 19 μm (36 μm) and in the caudal region, 14 μm (12 μm). Lesioning experiments (MAUTNER, 1965; PAUL et al., 1971; PAUL, 1972; ZIMMERMANN and PAUL, 1972) reveal that the fibres coming from the frontal organ innervate the region of the SCO, while those from the pineal organ reach the posterior commissure, pretectal region and periventricular grey. Indirectly, the pineal complex influences the habenular ganglion and the dorsal thalamus.

3. The Pineal Complex After Severing Its Nervous Connections

This aspect has been thoroughly studied by ZIMMERMANN and PAUL (1972) in *Rana temporaria*. These authors have shown that cutting the frontal nerve diminishes the nuclear volume of frontal organ cells more strongly than cutting the pineal tract. In the pineal organ proper the tentatively identified photoreceptor cells exhibited differential effects depending on the density of the nuclear volume. Normally two classes of cells are present, exhibiting values of 192 and 288×10^3 mm^{-3}. After lesioning the pineal tract, the epiphyseal cells with the small nuclei predominate. Severance of the frontal nerve results in an increase in number of the second cell class and the occurrence of a new class measuring 336×10^3 mm^{-3}.

V. Blood Supply

The blood supply of the epiphysis cerebri of anurans has been carefully studied by MAUTNER (1964, 1965) using supravital microscopy and Indian ink injections. Based on observations in *Rana temporaria, R. esculenta* and *Hyla arborea*, the epiphysis has been shown to possess a rich capillary network which forms part of an extensive periventricular network. The arterial supply of the capillary network comes from arteries lying at the base of the brain. The venous drainage shows differences, three types having been distinguished. About 75% of the animals investigated show a type I drainage, characterized by a blood flow in a lateral direction into the venae diencephali posteriores. In the rostral region, the blood flows in a rostral direction into anastomoses linked to blood vessels of the choroid plexus. One or two vessels in which the blood flows towards the epiphysis accompany the pineal tract. Instead of supplying the epiphyseal parenchyma they receive blood coming from the parenchyma and drain into the veins mentioned above. In 17% of the animals investigated, a type II drainage was observed. Here one of the venae diencephali posteriores was lacking, so that the blood from the corresponding half of the epiphysis cerebri drained directly into the vena longitudinalis mesencephali. Type III was characterized by an inverted blood flow in the vessels accompanying the pineal tract and the vena longitudinalis mesencephali. A system of portal blood vessels similar to that of the pituitary was not found to exist between the epiphysis cerebri and the subcommissural organ.

In the frontal organ the capillaries lie mainly in the connective tissue capsule, having little contact with the parenchyma. The frontal nerve is accompanied by a large vessel, apparently a vein.

In *Xenopus laevis* (Von Haffner, 1952), a single *artery* runs from the pineal organ proper to the frontal organ, either on the right hand side or the left-hand side. A vein was supposed to be absent.

VI. Functional Aspects

1. Light Perception

It has been clearly established by electrophysiological studies that the pineal complex of amphibians is directly photosensitive. Since this aspect has been thoroughly reviewed (Dodt, 1973; Adler, 1976; Hamasaki and Eder, 1977), it suffices to give a brief survey. In view of the fact that the frontal organ and the frontal nerve are more easily accessible than the pineal organ proper and its nerve, the pineal tract, the frontal organ has been more widely studied than the pineal organ proper.

The *frontal organ* of frogs exhibits two types of electrical activity, namely spike activity of single neurons (Dodt and Heerd, 1962) and slow (graded) potentials passively conducted from the frontal organ into the frontal nerve (Baumann, 1962; Dodt, 1964b; Dodt, 1973). Recordings from the frontal nerve reveal that spontaneous activity in the absence of any intentional stimulus is a characteristic feature of these neurons. The spontaneous activity is clearly affected by exposure to light, whereas increasing and decreasing temperature gradients, mechanical movement with a wisp of cotton and chemical agents such as acetic acid of various concentrations are ineffective in changing the spontaneous activity (Dodt and Heerd, 1962). Although light has been clearly shown to affect the electrical activity of frontal nerve neurons, it should be noted that in *Rana temporaria* only three-quarters of the animals recorded showed a functioning frontal organ, as revealed by the impulse activity of the frontal nerve (DODT and HEERD, 1962).

The frontal organ shows two different types of response to exposure to light, i.e. chromatic and achromatic. A chromatic response means that the frontal organ reacts differently to light of different wavelengths, the electrical activity being inhibited by light of short wavelengths (maximum response at 355 nm) and stimulated by light of longer wavelengths (maximum response at 515 nm). The achromatic response, by contrast, always involves inhibition, independently of the wavelength applied (DODT and HEERD, 1962). According to DODT (1964b) the achromatic response is transmitted along fast-conducting myelinated axons, whereas the chromatic response involves both myelinated and non-myelinated fibres.

The *pineal organ proper* of most amphibian species studied shows an achromatic response only (DODT and HEERD, 1962; MORITA, 1965b), an exception being *Rana catesbeiana* in which 21 out of the 65 neurons of the pineal tract recorded showed a chromatic response (MORITA, 1969). Further details may be obtained from the following papers: HEERD and DODT (1961), DODT (1963b, 1964a, 1968), DODT and JACOBSON (1963), DODT and MORITA (1964, 1967), MORITA and DODT (1965), MORITA (1965b, 1971, 1975), HAMASAKI (1970), BAUMANN (1964), DONLEY (1975), DONLEY and MEISSL (1978) and ELDRED and NOLTE (1978).

In addition to the electrophysiological studies reported, there are a number of other findings supporting the photoreceptive nature of the pineal complex in amphibians, namely the presence of photopigments as reported by HARTWIG and BAUMANN (1974) and HARTWIG (1975) (see p. 373) and the apparent involvement of the pineal region in phototaxis (MROSOVSKY and TRESS, 1966; ADLER, 1970, 1976). Most interesting is the observation that in the tiger salamander the pineal organ plays a role in compass orientation: tiger salamanders trained to orient in a particular compass direction under the sun fail to orient in the trained direction if they are eyeless and simultaneously have the brain covered with opaque plastic or are eyeless and pinealectomized; salamanders with either the eyes or the pineal intact and unobstructed continue to orient in the trained direction (TAYLOR and ADLER, 1978; for further details, see ADLER and TAYLOR, 1973; ADLER, 1976).

2. Secretion

In view of the fact that the amphibian pineal organ contains serotonin (*Rana esculenta, R. pipiens:* OWMAN et al., 1970) melatonin (*Xenopus laevis:* VAN DE VEERDONK, 1967b) and HIOMT, one of the melatonin-forming enzymes (AXELROD et al., 1965b; *Xenopus laevis* larvae, adult *Rana pipiens*; QUAY, 1965d; *R. pipiens, Bufo marinus, Necturus maculosus*), it is probably safe to conclude that the amphibian pineal organ is involved in melatonin synthesis and secretion. The mere presence of melatonin in the pineal organ should not be taken as evidence for the occurrence of secretory processes, as there are other sources of melatonin in the body, e.g. retina (QUAY, 1965d), and as the pineal may incorporate melatonin produced elsewhere in the body. There are indications that the pineal organ rapidly accumulates radioactively labelled melatonin precursors (CHARLTON, 1964, 1966a). The *development* of HIOMT activity has been studied by BAKER et al. (1965) and that of melatonin levels by BAKER (1969).

In addition, there is a large body of indirect evidence pointing to secretory processes, to be mentioned subsequently. A completely unexplored field is whether the *frontal organ* is secretory in nature. In frogs, the frontal organ does not contain microscopically detectable indoleamine fluorophores (OKSCHE, personal communication).

3. Target Organs of the Pineal Complex

Relatively little is known about the target organs of pineal action. In view of the fact that the pineal complex of amphibians has a prominent afferent innervation, the brain may be regarded as an important target of the pineal complex, especially those areas of the brain where the fibres terminate (see p. 392). However, evidence to support this notion is scarce. In some studies locomotor activity has been studied after removal of the frontal organ. In tadpoles of the burrowing frog, *Pyxicephalus delalandi*, this operation had no statistically significant effect (STEBBINS et al., 1960) whereas in blinded frogs having their frontal organs removed the locomotor rhythms were no longer entrained by the light cycle (ADLER, 1971). In developing *Xenopus laevis* there is a fivefold increase in melatonin levels, occurring at the time of active larval swimming, though it should be noted that the source of melatonin may be the retina (BAKER, 1969). Interactions with the SCO are indicated by the fact that after transection of the frontal nerve and the pineal tract the secretory material of the SCO decreased in amount (ZIMMERMANN and PAUL, 1972). DIEDEREN (1975) suspects that it is the function of the pineal complex to prevent a direct influence of light on the secretory activity of the SCO. An increased incorporation of ^{35}S-cysteine into the preoptic nucleus of frogs after removal of the frontal organ alone or the complete pineal complex has been interpreted as meaning that the pineal complex exerts an inhibitory effect on the neurosecretory activity of this nucleus (VULLINGS, 1973).

Considerable effort has been devoted to studying the role of melatonin and the pineal complex in regulating colour change. This is understandable when one recalls that substances of the mammalian pineal organ are highly effective in causing melanin concentration in amphibian melanophores (MCCORD and ALLEN, 1917; BORS and RALSTON, 1951) and that the substance causing blanching is identical with melatonin (LERNER et al., 1958). Subsequent studies have amply confirmed that when melatonin is applied to *isolated* frog skin (LERNER and CASE, 1959; LERNER and WRIGHT, 1960; MORI and LERNER, 1960; NOVALES, 1963; NOVALES and NOVALES, 1965) it is highly effective in causing melanin concentration. According to MCGUIRE and MÖLLER (1965) it is the dermal, but not epidermal melanocytes that respond to melatonin, whereas both respond to MSH (*Rana pipiens*). Iridophores are not affected by melatonin (BAGNARA, 1964; BAGNARA and HADLEY, 1969; BAGNARA et al., 1969).

Regulation of colour change in amphibians has been thoroughly reviewed by BAGNARA and HADLEY (1970). Hence it suffices to give a brief survey. While there can be no doubt that the pineal complex is involved in colour change mechanisms, it is not precisely known which role the frontal organ, in contrast to the pineal organ proper, plays. EAKIN (1961) noted in blinded *Hyla regilla* tadpoles that the blanching reaction was weaker in stirnorganectomized animals than in sham-operated controls, when the tadpoles were placed in the dark, but on the whole the differences observed were slight. In adults and tadpoles of *Xenopus laevis* removal of the frontal organ had no effect on colour change (CHARLTON, 1966b). In *Pyxicephalus delalandi* there was likewise no effect on pigmentation (STEBBINS et al., 1960). It would therefore appear that the frontal organ has little, if any, effect on colour change.

After removal of the pineal proper, by contrast, blanching is no longer demonstrable when the animals are placed in darkness (SIMONNET et al., 1952a; BAGNARA, 1960, 1961b, 1963, 1964, 1965; BRICK, 1962; KELLY, 1962, 1963; CHARLTON, 1966b). Variable results have been obtained with regard to the background response in which normal animals resting on a white background become pale, while animals on a black background darken. This response was found to be abolished in one study (CHARLTON, 1966b), whereas in another it was not (BOGENSCHÜTZ, 1967). An involvement of the pineal in colour change is also indicated by much faster pigment aggregation in the area of the pineal than in adjacent areas when larvae are placed in the dark (BAGNARA, 1961a). Injections of either extracts from beef pineal into *Rana pipiens* (LAZO-WASEM and GRAHAM, 1960) or melatonin (SUPNIEWSKI et al., 1960, *Rana temporaria*) likewise lead to pigment aggregation. Immersion of young *Xenopus laevis* embryos in melatonin solutions has the same effect (BAGNARA, 1961b), though this effect is short-lived, lasting for ca. 20 min (BURGERS and VAN OORDT, 1962; CHARLTON, 1966b). In developing *Xenopus*, melatonin concentrations in the body are lower when the animals are reared on a light background than when reared on a dark background; serotonin concentrations, by contrast, were not affected (BAKER and HOFF, 1971). When developing *Xenopus* are raised in melatonin solution from stage 21 to stage 48, they have fewer melanophores on the head than control animals not exposed to melatonin (BAKER et al., 1978a).

It is still a matter of dispute whether melatonin acts directly on the melanophores as proposed by BAGNARA and HADLEY (1970) or indirectly via MSH (CHARLTON, 1966b; DIERST-DAVIES et al., 1971). An interesting observation was that by OSHIMA and GORBMAN (1969a, b), that in the pituitary pars intermedia of *Rana pipiens* there are light-sensitive units for which the pineal organ appears to be photoreceptor. According to ABE et al. (1969) the inhibitory effect of melatonin on MSH-induced darkening of the frog skin is due to an inhibition of cAMP accumulation. Melatonin apparently does not interfere with the darkening effect of exogenous cAMP and has no effect on skin phosphodiesterase activity.

The action of melatonin on single amphibian pigment cells has been thoroughly investigated in tissue culture (MESSENGER and WARNER, 1977). In this test system melatonin (10^{-15}–10^{-13} *M*) was 10,000 times more potent in inducing pigment granule condensation than any other compound tested. Blockage of β-adrenoreceptors and serotonin receptors of pigment cells did not affect the response to melatonin, whereas phentolamine, an α-adrenoreceptor blocker, did block the effect of melatonin. Since after removal of Na^+ or Ca^{2+} from the bathing medium and after blockage of the Ca^{2+} entry into the cells the melatonin-induced pigment condensation is inhibited, it was concluded that the action of melatonin is accompanied by depolarization and the entry of calcium ions. HEWARD and HADLEY (1975) found that melatonin consistently produced significant lightening of the skin at concentrations as low as 5×10^{-11} *M*. Tests involving a number of related compounds revealed that the intrinsic activity of indolic substances on the melatonin receptor is determined primarily by the moiety substituted on the fifth carbon atom, whereas the affinity for the receptor binding site depends primarily by that on the third carbon atom.

Melatonin has also been shown to affect the retina. KRAUS-RUPPERT and LEMBECK (1965) noted that melatonin causes contraction of the retinal pigment in light-adapted *isolated* eyes of frogs. In dark-adapted eyes, by contrast, melatonin had little effect on the retracted pigments cells. Injections of melatonin into the lymph sac of light-adapted animals were also effective in inducing pigment retraction. Moreover, there is evidence that melatonin mediates light-induced cone contraction in larval *Xenopus* (QUAY, 1968g; QUAY and MCLEOD, 1968).

The effects of the amphibian pineal complex on endocrine organs have been little investigated. Thyroid stimulation after epiphysectomy has been noted by DISCLOS (1964) and REMY and DISCLOS (1970, *Alytes obstetricans*). According to JACKSON et al. (1977), the frog pineal is a source of thyrotrophin-releasing hormone (TRH), the amount of which is particularly high in autumn, and higher in dark-exposed animals than in light-exposed animals (*Rana pipiens*). Regarding gonadal function, it was observed that administration of extracts of female, but not male, bovine pineals depressed frog spermatogenesis enhanced by chorion gonadotrophin (JUSZKIEWICZ and RAKALSKA, 1963) and that melatonin has no effect under these experimental conditions (JUSZKIEWICZ and RAKALSKA, 1965). BISWAS et al. (1978) found that pinealectomy decreases spermatogenesis accompanied by an increase of Leydig cell activity (*Bufo melanostictus*), whereas melatonin administration decreases both spermatogenesis and Leydig cell activity. In *Hyla cinerea* maintained on a long photoperiod in December melatonin inhibited gonadal enlargement (DE VLAMING et al., 1974a). In *Alytes* (DISCLOS, 1964) pinealectomy was without effect on the gonads. A decrease of nuclear size of pineal cells following administration of oestradiol or testosterone (PFLUGFELDER, 1957) points to possible feedback mechanisms between the gonads and the pineal.

Metamorphosis may be accelerated by pinealectomy (DISCLOS, 1964), but this effect is not always demonstrable (KELLY, 1958; EICHLER, 1968). This may be partly due to the fact that when the pineal is removed in embryonic stages regeneration of both the pineal organ proper and the frontal organ (*Rana pipiens*) occurs in more than half of the animals examined (EICHLER, 1968). Melatonin decreases the excitability of the frog's sciatic nerve in vitro (HERTZ-

ESHEL and RAHAMIMOFF, 1965, quoting unpublished results by RAHAMIMOFF). Oxygen consumption decreased after pinealectomy in *Rana temporaria* (KASBOHM, 1967).

The role of the pineal and other epithalamic organs with regard to their possible relation to vertebrate emergence on dry land has been discussed by STEYN (1961). Sodium and potassium metabolism after pinealectomy in the crested newt has been studied by FERRERI et al. (1964). *Xenopus laevis* embryos reared in melatonin solutions exhibited defaecation and gulping, and MAO activity was elevated in the whole embryo and gut (BAKER et al., 1978b).

F. The Pineal Complex in Reptiles

I. Gross Anatomy

As in other groups of vertebrates, the pineal complex in reptiles is very variable. It consists basically of the fairly regularly occurring pineal organ proper, which is absent in Crocodilia, and the parietal or third eye, which is a typical feature of some lizards. According to GUNDY and WURST (1976a) approximately 60% of all lizard genera possess a third eye. These authors (GUNDY and WURST, 1976b) point out that regressed parietal eyes are common in certain arboreal lizards, primarily from the family Chameleontidae, whereas the parietal eye is often retained in burrowing lizards, presumably because the parietal eye is a more suitable photoreceptor for a burrower than are lateral eyes.

In the reviews by GLADSTONE and WAKELEY (1940) and BARGMANN (1943) it has been shown that the pineal organ proper consists basically of three parts, namely an expanded proximal part, a thin intermediate part, and an end-vesicle, which is mostly lacking. The two most detailed publications on the gross anatomy of the pineal organ proper and related structures are those by KUMMER-TROST (1956) and GUNDY and WURST (1976b), which should be consulted for detailed information. Figure 165 shows a diagrammatic representation of the parietal region of *Phrynocephalus undulatus* which is fairly representative of a "complete" pineal complex. An interesting point is that the morphology of the pineal organ proper is apparently not independent of the presence or absence of the parietal eye. Both KUMMER-TROST (1956) and GUNDY and WURST (1976b) noted that when the parietal eye is absent the wall of the pineal organ shows more prominent convolutions and vice versa. In a fair number of species a fingerlike extension projects craniad from the pineal; if a parietal eye is not present an extension is rarely found, and it has been suggested that pineal convolutions seem to take its place and may serve a similar function (GUNDY and WURST, 1976b). Additional information may be obtained from the publications by GLADSTONE and WAKELEY (1940), BARGMANN (1943), OKSCHE (1965a), VAN DE KAMER (1965) and LIERSE (1965a). For data regarding the variable parietal foramen, see EDINGER (1955), TROST (1956) and GUNDY and WURST (1976b).

II. Development

Considerable effort has been devoted to studying the development of the pineal complex, especially the origin of the parietal eye, as there are uncertainties as to whether the two parts of the pineal complex develop from a single anlage

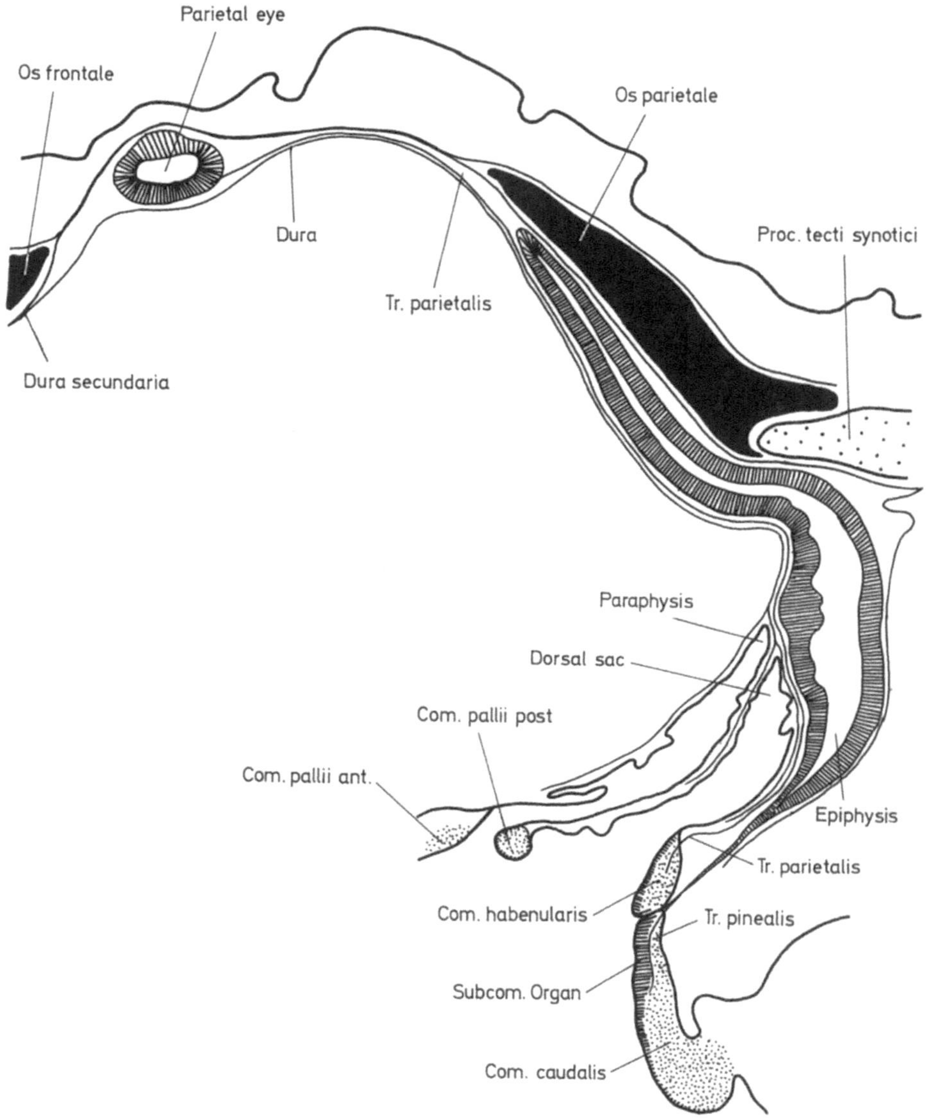

Fig. 165. Pineal complex of *Phrynocephalus undulatus* consisting of parietal eye and epiphysis cerebri. KUMMER-TROST, 1956

or whether they are formed from two distinct diencephalic evaginations. The complexity of the issue has been clearly outlined by GLADSTONE and WAKELEY (1940), who mentioned eight different views. Very briefly, these are as follows:

1. According to some authors, the pineal complex develops as a single median evagination, its tip being cut off to form the pineal eye, while the basal part, which retains its connection with the brain, forms the epiphysis cerebri. The two parts are linked by a pineal stalk, which is sometimes represented by the pineal nerve.

2. There are two separate and morphologically distinct evaginations, situated in the median plane one behind the other, of which the anterior one gives rise to the parietal eye and the posterior part to the epiphysis.
3. A single median evagination, which is subsequently divided into two: (a) an anterior vesicle giving rise to the parietal eye, which has an independent nerve, the nervus parietalis, not connected to the epiphysis but joining the habenular commissure; (b) a posterior part forming the epiphysis cerebri.
4. A single diverticulum, the distal part of which becomes divided by an incomplete transverse septum into an anterior and posterior part in such a way that the lumina of the two end-vesicles communicate with the lumen of a common stalk.
5. The parietal eye arises not as an independent structure but as an outgrowth from the anterior wall of the primary diverticulum.
6. Two completely independent outgrowths soon unite with each other.
7. The rudiments of two pairs of serially arranged sense organs become fused in the median plane, the anterior giving rise to the parietal organ, the posterior to the epiphysis.
8. The parietal eye is phylogenetically bilateral, as indicated ontogenetically in certain reptiles, e.g. *Sphenodon*.

As might be expected from the controversy about the origin of the parietal eye, most of the recent developmental studies have been concerned with lizards. From the results published it would appear that from the eight views mentioned above the true issue is a combination of possibilities 1–4 and perhaps 5. With the benefit of hindsight it seems that the confusion is partly due to the fact that (a) an insufficient number of developmental stages have been investigated, (b) there is some interspecies variation and (c) different opinions exist as to the developmental stage at which the anlagen of the parietal eye and the pineal organ proper can be distinguished. In lizards the studies by TROST (1952), VON HAFFNER (1953) and PETIT (1967) have shown (see Fig. 166) that the development of the pineal complex starts, in the prospective area of the intercommissural region of the diencephalic roof, by a slight thickening and/or outward bulging of the diencephalic roof. Close inspection reveals that in the "pineal area" there may in actual fact be two thickenings lying one behind another in an anteroposterior direction, the anterior one giving rise to the parietal eye and the posterior one forming the pineal organ proper. Normally, the development of the parietal anlage precedes that of the pineal organ proper, the former forming a fairly prominent evagination which subsequently gives rise to a vesicle, as depicted in Fig. 166. The pineal organ proper forms from the posterior part of the "pineal anlage". It is first represented by a relatively short and slightly thickened area of the diencephalic roof. It then elongates, becoming U-shaped with the vertex of the U pointing outward and the interior of the U communicating with the third ventricle. As mentioned above under No. 4, this process may well give the impression that a single diverticulum becomes divided by a transverse septum.

Usually, there is a complete separation of what is to become the parietal eye and the pineal organ proper. However, as shown by EAKIN (1964a) in the lizard *Sceloporus occidentalis*, occasionally there may be an incomplete subdivision of the parietal and pineal subdivisions of the epiphyseal diverticulum, the parietal eye lying either at its site of origin at the diencephalic level and inside the meninges and bony cranium or at its normal site, the parietal eye being connected to the tip of the pineal organ proper, which in the course

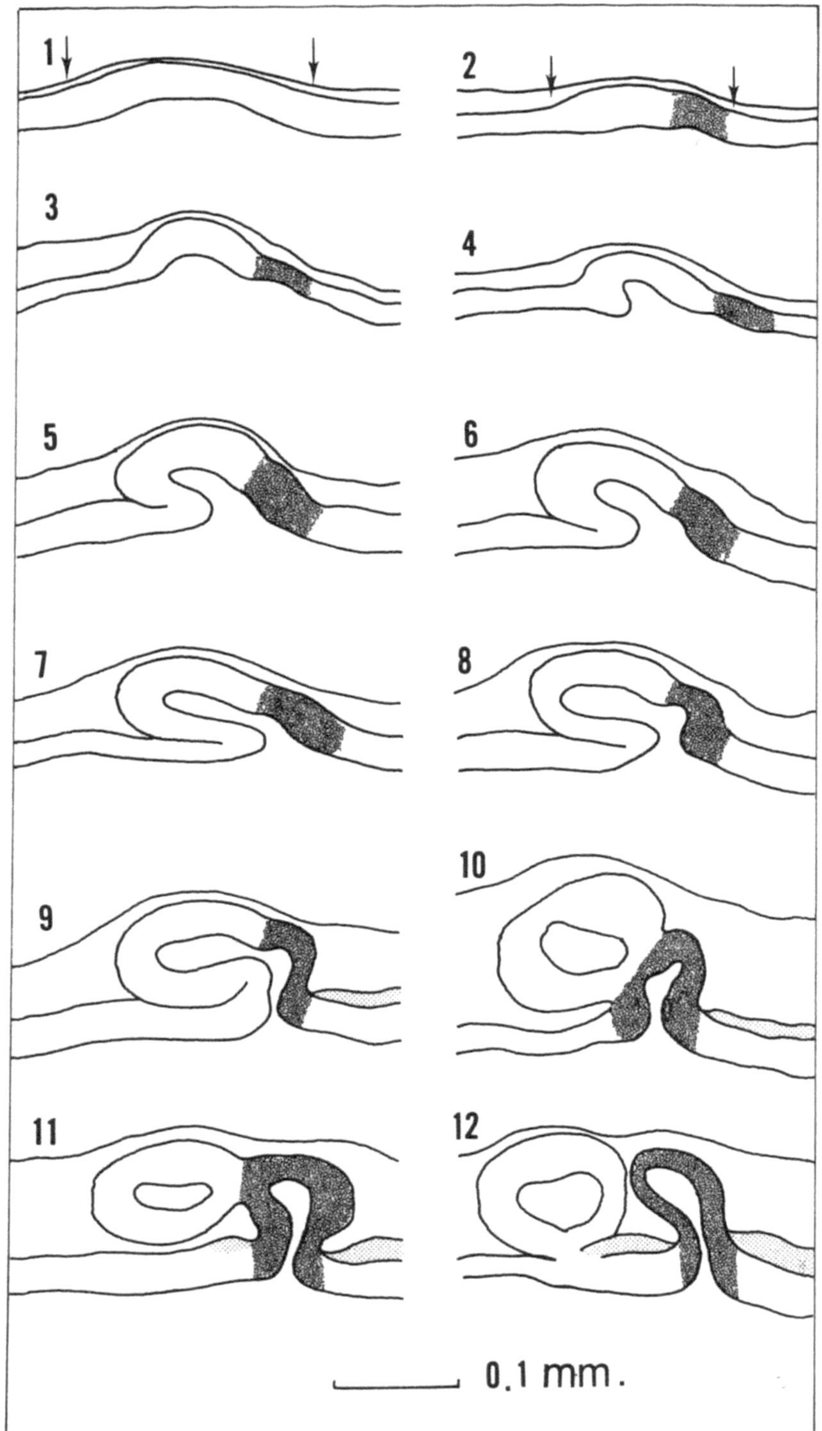

Fig. 166. Development of the pineal complex in *Anguis fragilis*. *1–12*, Various developmental stages. The *stippled areas* indicate the epiphyseal region, in front of which (*to the left*) lies the anlage of the parietal eye. In *1* and *2* the *arrows* indicate the prospective pineal region. PETIT, 1967

of development had become exceedingly elongated. While in lizards the development of the parietal eye normally precedes that of the pineal organ proper, TROST (1952) has described one case out of 43 (*Anguis fragilis*) that the pineal organ proper developed first, the anlage of the parietal eye forming a thickening of the rostral wall of the large pineal vesicle.

The other papers dealing with the development of the pineal complex in lizards contribute virtually nothing concerning the *early* development of the pineal complex. EAKIN'S (1964a) study begins at the time of constriction of the primary epiphyseal vesicle into the prospective parietal eye and the presumptive pineal organ proper. STEYN'S (1957) study on *Cordylus polyzonus* and *Mabuya sulcata* starts at a developmental phase corresponding to stages 5–8 of Fig. 166; in these two species an interesting point is the presence of a fairly prominent transverse groove indicating the borderline between the presumptive parietal eye and the presumptive pineal organ proper.

While the results mentioned above suggest the presence of a single *median* pineal anlage (which, as pointed out above, may be subdivided into a single anlage of the parietal eye and a single anlage of the pineal organ proper), a *lateral* or *bilateral* origin has been discussed as well. EAKIN (1964a) described an embryo of *Sceloporus occidentalis* in which the parietal vesicle lay slightly to the left of the midline and concluded that it "budded" off from the left side of the epiphyseal vesicle. However, PETIT (1967) is of the opinion that there is no indication for an origin from the left side, but that this is merely suggested by a secondary displacement to the left which affects both the parietal vesicle and the pineal vesicle. Although VON HAFFNER (1953) also noted an only temporary displacement of the parietal organ to the left, he proposes an originally left-hand origin of the parietal eye as its nerve, the parietal nerve, courses to the left habenular ganglion, as noticed by other authors as well (p. 434).

As regards a possible bilateral origin of the pineal complex, very little of a definitive nature has been contributed by recent publications. STEYN (1957) noted that in *Mabuya sulcata* the pineal organ proper has a short anteriorly directed limb which terminates in two vesicles, both of which resemble the parietal organ and which are situated to the left and to the right of the median plane; from the discussion in STEYN'S paper it may appear that the author regards these terminal vesicles as being related to an originally bilateral origin of the pineal complex. PETIT (1967) discusses a bilateral origin on account of the fact that occasionally a paired parietal nerve is found. MEINIEL (1977a), studying the morphogenesis of the pineal complex in *Lacerta vivipara*, observed that in the earliest developmental stages studied (stage 27 of Dufaure and Hubert) each of the two diverticula is composed of two territories, which she regarded as suggesting "that both organs originate by fusion of ancestrally paired and symmetrical primordia."

Cell proliferation in the parietal and pineal vesicles has been studied autoradiographically in *Lacerta vivipara* (MEINIEL, 1977a) showing that clear differences exist between the two organs: in the parietal vesicle cell proliferation, as assessed by ^{3}H-thymidine injection, is evenly distributed throughout the parenchyma, whereas in the pineal vesicle proliferation is found mainly in the

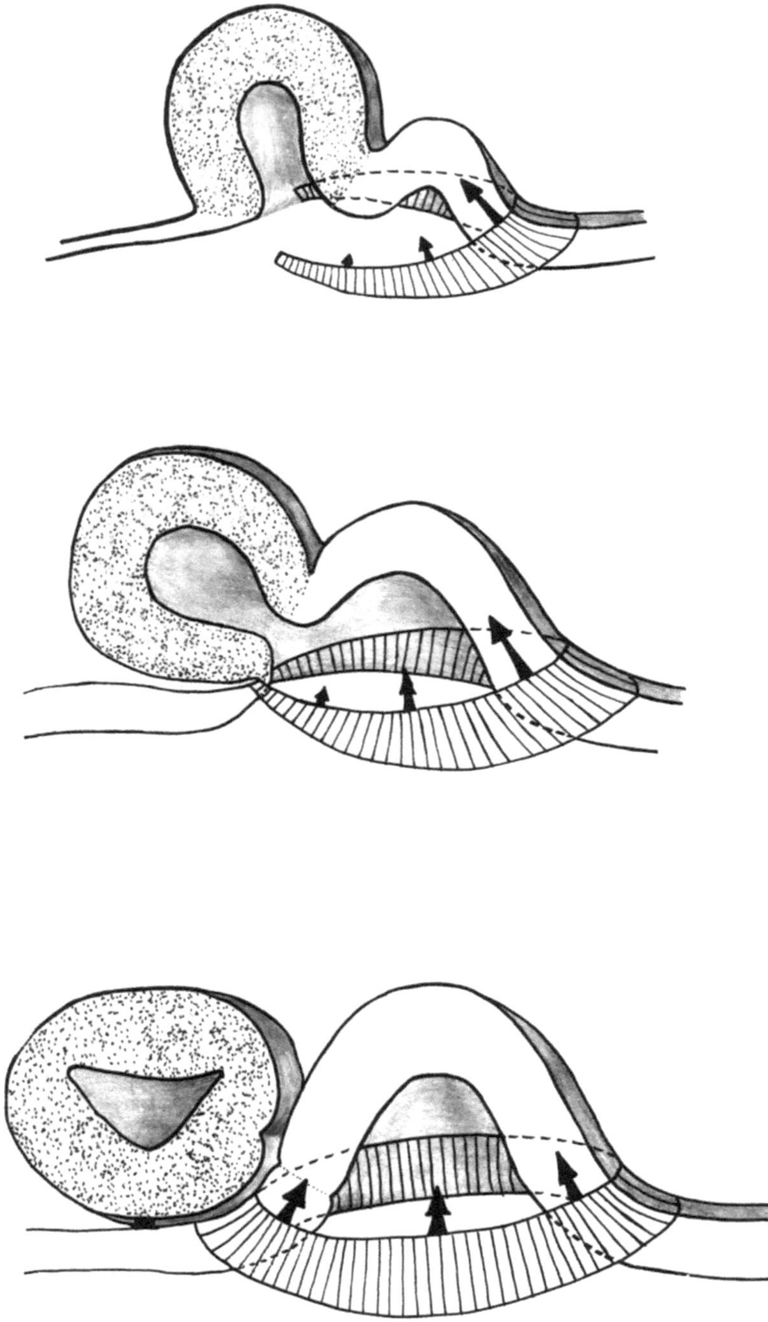

Fig. 167. Zones of cell proliferation, as studied by ^{3}H-thymidine autoradiography, in the pineal region of *Lacerta vivipara*. The *hatched area* is the proliferation zone for the pineal organ proper and the *stippled area* for the parietal eye. MEINIEL, 1977a

neuro-ectoderm of the diencephalic roof lying in the immediate vicinity of where the pineal vesicle is attached to the diencephalon (Fig. 167).

Recent developmental studies in non-lacertilian reptiles are rare. In *Tropidonotus natrix* (TROST, 1952; PETIT, 1969b), which lacks a parietal eye, the anlage of the pineal is represented by a shallow evagination of the diencephalic roof, which shows two slight thickenings, anterior and posterior, which correspond to the primordia of the parietal eye and the pineal organ proper. During the further course of development the primordium of the parietal eye is incorporated into that of the pineal organ proper, resulting in the formation of a single hollow evagination.

In *Testudo graeca* there are likewise two primordia which eventually form the single pineal evagination (VIVIEN-ROELS and PETIT, 1973; VIVIEN-ROELS, 1976a). The same applies to *Emys orbicularis* (VIVIEN-ROELS, 1976a).

The question of a communication of the lumen of the pineal organ proper with the third ventricle will be discussed below.

The histogenesis and the ultrastructural and biochemical differentiation of the pineal organ proper and the organogenesis of the parietal eye will be dealt with in separate sections (pp. 408, 409, 430).

III. Structure, Differentiation and Function of the Pineal Organ Proper

1. General Histological Pattern

As pointed out in previous publications (see BARGMANN, 1943), the general histological pattern of the pineal organ proper varies considerably, ranging from tubular or hollow to solid.

In *lizards* both TROST (1954) and GUNDY and WURST (1976b) noted that, depending on the presence or absence of a parietal eye, the basically tubular or hollow pineal organ shows varying degrees of thickening or folding of its wall. GUNDY and WURST (1976b), studying or reviewing the morphology of the pineal complex of 85 different lizards species, distinguished thick-walled, thin-walled and highly convoluted pineals, the first two types being either small or large. The connections of the pineal organ with the diencephalic roof have not been investigated in detail. In *Lacerta vivipara* (VON HAFFNER, 1953) and *Anguis fragilis* (TROST, 1952) the open communication of the pineal lumen with the third ventricle disappears during development. An opening of the pineal lumen to the ventricle is also lacking in *Chelonia mydas* and *Caretta caretta* (OWENS and RALPH, 1978). *In Cordylus polyzonus* and *Mabuya sulcata* (STEYN, 1957), the pineal organ connects to the brain by means of the pineal tract.

The pineal organ of *Chelonia* has been most thoroughly studied by VIVIEN-ROELS (1976a). As can be seen from Figs. 168 and 169, there is both interspecies

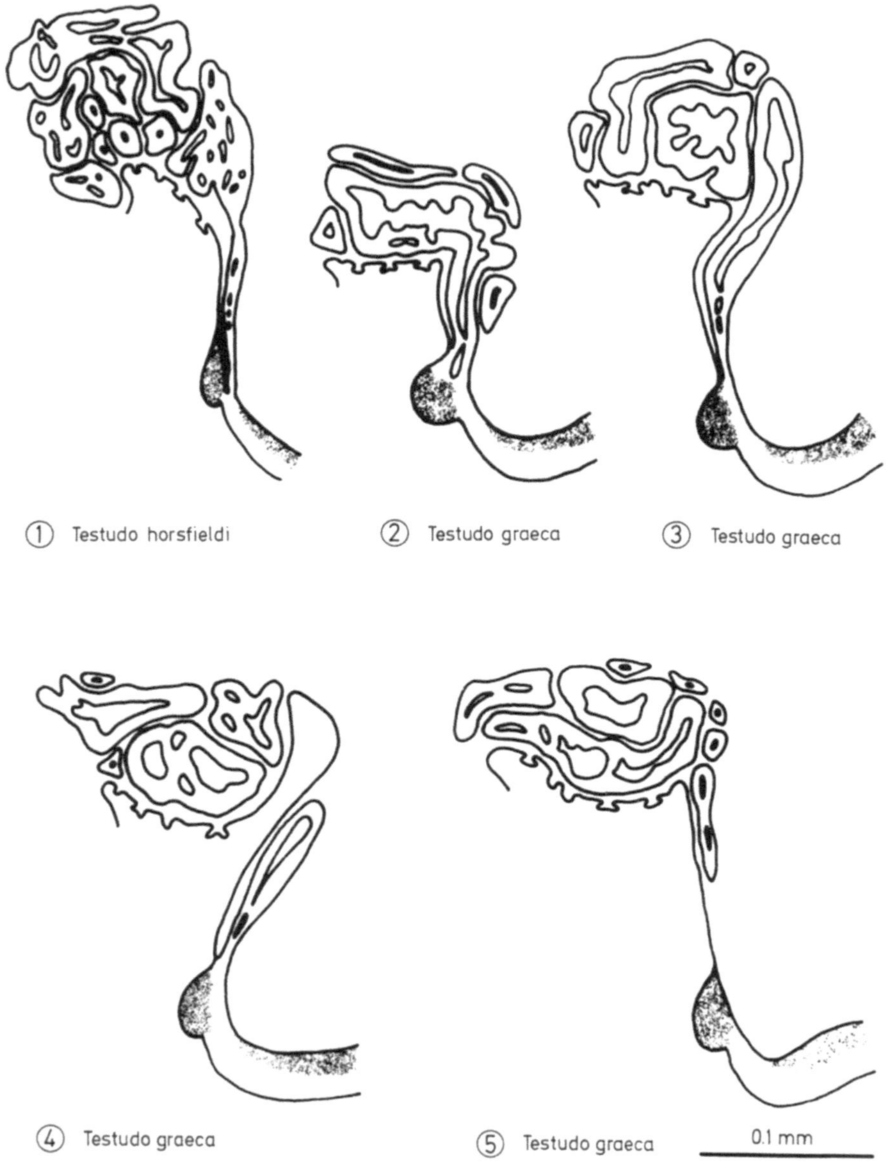

Figs. 168, 169. The epiphysis cerebri in a number of chelonian species. VIVIEN-ROELS, 1976a

and interindividual variation of the general histological pattern. As in lizards, some pineal organs are tubular, with more or less convoluted walls, while others have a follicular appearance, e.g. *Testudo hermanni*, as reported previously by MEHRING (1972). A pineal consisting of vesicles has been described in *Testudo mauritanica* (GRIGNON and GRIGNON, 1963a, b). VIVIEN-ROELS (1976a) reports the interesting observation that in some pineal specimens the histological appear-

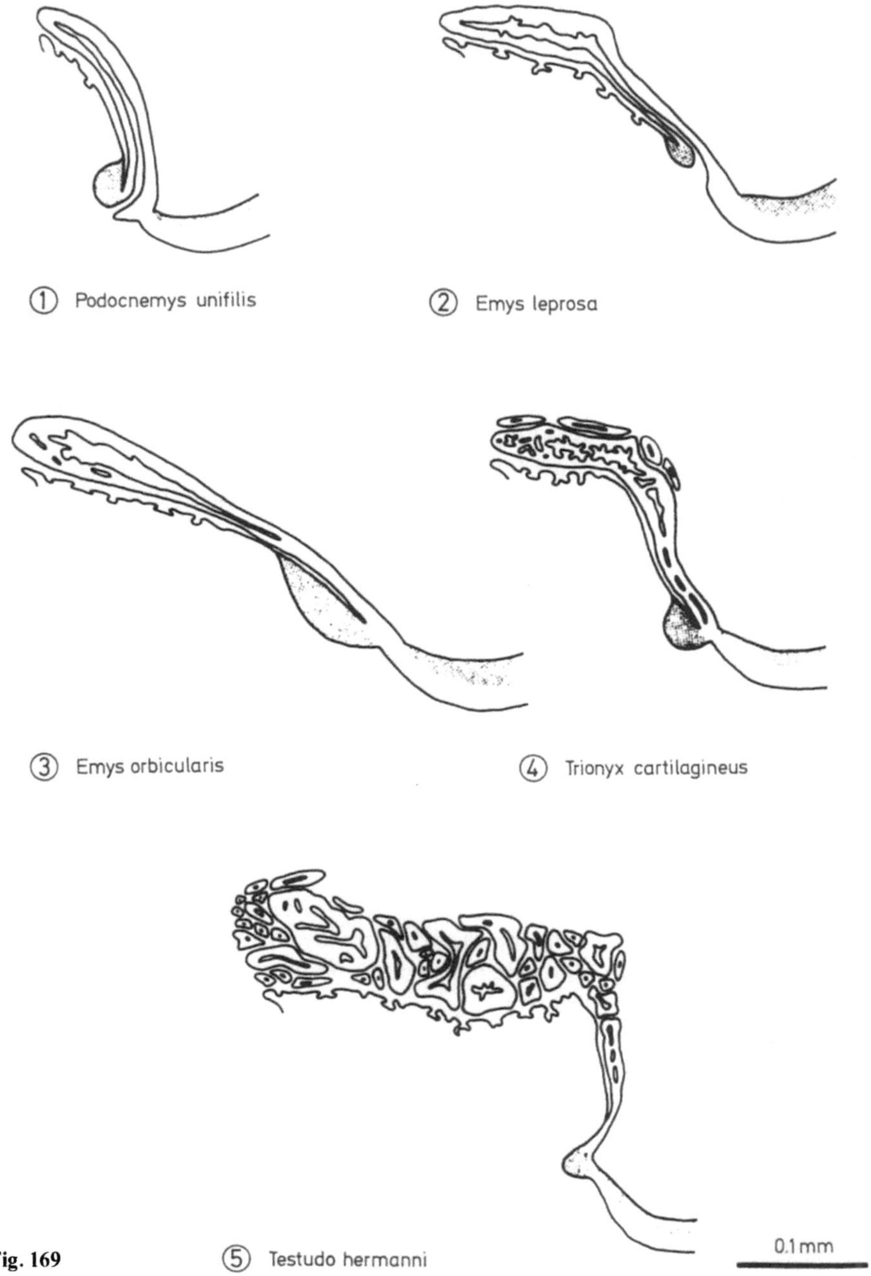

Fig. 169

ance of the dorsal wall resembles that of the lens of the parietal eye, which is not entirely surprising when one recalls that the anlage of the parietal eye is incorporated into that of the pineal organ proper (see p. 405). The variable connections of the pineal organ with the diencephalon can be deduced from Figs. 168 and 169.

The structure of the pineal organ in *Ophidia* has been little investigated. Previous observations pointing to the presence of solid pineal organs in this group of reptiles (see BARGMANN, 1943) have been corroborated by QUAY et al. (1968a, *Natrix natrix*).

2. Histogenesis

The histogenesis of the lizard pineal complex has been thoroughly investigated by TROST (1953a, b), STEYN (1959a) and MEINIEL (1976c). Differentiation of the cellular elements of the parietal vesicle precedes that of the pineal vesicle and already starts prior to the separation of the two vesicles. In the pineal vesicle cellular differentiation proceeds at a slower pace than in the parietal vesicle. Moreover, there are indications for the presence of regional differences in the pineal vesicle, insofar as cellular differentiation in the posterior epiphyseal fold precedes that of the anterior fold.

According to TROST (1953b), STEYN (1959a) and MEINIEL (1976c) sensory cells and supporting cells can be distinguished in early stages of histogenesis, the sensory cells being much more slender and more strongly staining than the supporting cells. The sensory cells develop outer and inner segments as well as a basal process. According to STEYN (1959a), who distinguishes two types of immature sensory cells, only one type shows a basal process. Whether this condition is due to a reduction of an existent process or due to a failure of such a process to develop could not be determined with certainty. For further details regarding this problem, see p. 415. Corroborating earlier findings, TROST (1953b) noted in *Anguis fragilis* and *Chalcides ocellatus* that regional differences exist in the distribution of sensory cells, which points to a similarity with the parietal eye: sensory cells are much less abundant in the dorsal region of the pineal vesicle than in the ventral and lateral areas.

Very little of a definitive nature can be said about the differentiation of the supporting cells, apart from the fact that in the distal areas of the pineal organ proper they develop pigment, like the "supporting" cells of the parietal vesicle (TROST, 1953b).

Due to the formation of basal processes of sensory cells an outer fibrous network develops; this is much later followed by the formation of an inner nervous network, which consists of a few fibres only.

Histogenesis of the pineal organ in *Chelonia* has been studied by VIVIEN-ROELS (1976a, *Testudo graeca, T. hermanni*). In these species, two cell types can be distinguished at the earliest stages of development studied. However, in contrast to lizards, there are no tinctorial differences of the cytoplasm, but rather differences in nuclear shape and staining. Prospective sensory cells can be identified as such only when they develop outer segments.

3. Ultrastructural Differentiation

The ultrastructural differentiation of the intrinsic epiphyseal cells has been thoroughly studied in both lizards (MEINIEL, 1976a) and turtles (VIVIEN-ROELS,

1976a). As mentioned previously, the epiphyseal parenchyma consists of both sensory cells and supporting cells. Electron microscopic studies have shown that at least in lizards the sensory cells are not a homogeneous population: besides typical photoreceptor cells there are secretory rudimentary photoreceptors and intermediate cell types (p. 416).

The differentiation of the cells of the sensory line, as seen in *Lacerta vivipara*, proceeds as depicted in Fig. 170 (MEINIEL, 1976a). Additional points of interest are an increase in both the mean diameter of dense-core vesicles from 96–170 nm and the number of dense-core vesicles from 20–1,154 per 100 μm^2 of basal processes. In the publication by MEINIEL (1976a) data on the development of synaptic contacts and their regression are also given.

In turtles (VIVIEN-ROELS, 1976a, *Testudo hermanni*), the differentiation of the cells of the sensory line starts with the formation of a 9+0 cilium, but typical outer segments do not develop. Often pictures are seen pointing to an hypertrophy of the ciliary shaft, giving the impression of small bulbous extensions which sometimes contain a few lamellae which are not directly comparable to discs of outer segments. In contrast to the outer segments, inner segments develop regularly, some giving rise to cytoplasmic bulbous expansions poor in cell organelles, which in light microscopic studies may have been misinterpreted as outer segments. Supporting cells apically develop microvilli.

4. Development of Indoleamine Metabolism

In view of the fact that monoamines play an important role in the function of the pineal complex, attempts have been made to study the onset of indoleamine synthesis during ontogenesis. By means of the Falck-Hillarp technique and microspectrofluorometry it was shown in *Lacerta vivipara* that serotonin was first demonstrable in embryos of stage 36 according to Dufaure and Hubert (MEINIEL and COLLIN, 1970; MEINIEL et al., 1973). At this stage yellow fluorescence is restricted to the distal regions of the epiphysis cerebri, where it is seen in the basal areas of the secretory epithelium. In the following developmental stages specific fluorescence increases in intensity, extends to the apical area of the epithelium and spreads to the proximal regions of the epiphysis. Quantitative autoradiographic studies have shown that ^{3}H-hydroxytryptophan incorporation increases strikingly from stage 35 to 3 days after birth (Table 41) (MEINIEL et al., 1975). The parietal eye, by contrast, does not show any fluorescence. Adrenergic nerve fibres exhibiting green fluorescence appear at stage 40 in the periphery of the epiphysis. At 30 h after birth, these fibres still have an extra-epithelial localization.

More detailed information on the sites and formation of serotonin during ontogenesis in *Lacerta* resulted from light microscopic autoradiography using ^{3}H-hydroxytryptophan (MEINIEL, 1975; MEINIEL et al., 1975). In accordance with the fluorescence microscopical studies it was found that specific labelling occurred at stage 36 and was restricted to sensory cells. Again, the parietal eye showed no labelling. Electron microscopic autoradiography (MEINIEL, 1976b) revealed that it was the secretory rudimentary photoreceptors, and

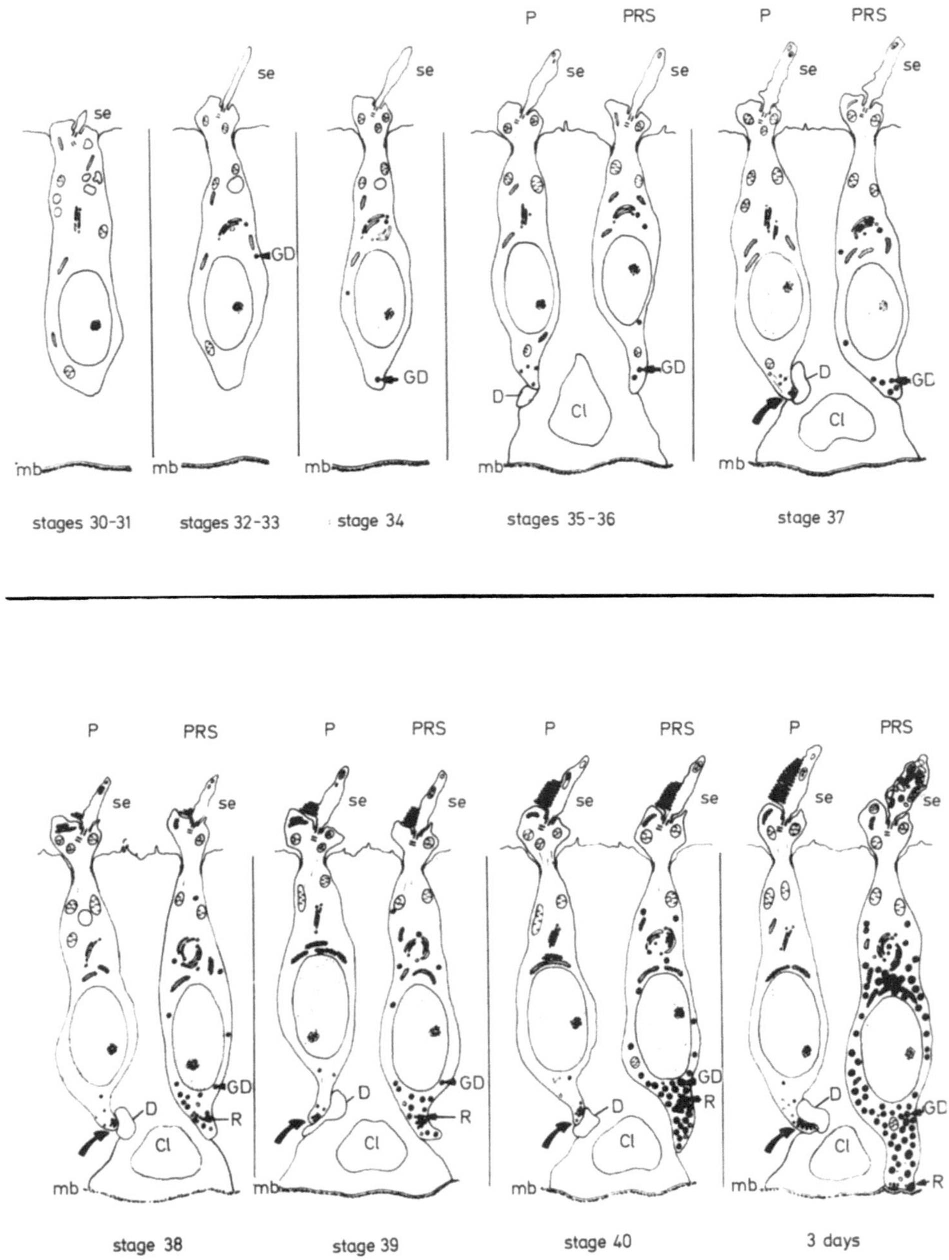

Fig. 170. Ultrastructural differentiation of the cells of the sensory line in the epiphysis cerebri of *Lacerta vivipara*. *CI*, interstitial cell; *D*, dendrite; *GD*, granular vesicle; *mb*, basal lamina; *P*, photoreceptor cell; *PRS*, secretory rudimentary photoreceptor; *R*, vesicle-crowned rodlet = synaptic ribbon; *se*, outer segment; *arrow*, synaptic contact. MEINIEL, 1976a

Table 41. Radioactivity of the epiphysis cerebri of *Lacerta vivipara* at different stages of development (MEINIEL et al., 1975)

Developmental stage[a]	Radioactivity[b]	
	Epiphysis	Brain
35	5.75 ± 1.85	3.85 ± 1.43
36	10.48 ± 2.73	8.00 ± 2.78
37	11.26 ± 4.77	4.29 ± 1.63
38	20.30 ± 9.19	5.03 ± 2.20
39	24.90 ± 8.83	3.87 ± 1.42
40	59.16 ± 28.56	9.13 ± 5.99
3 days	109.73 ± 17.88	1.20 ± 0.76

[a] According to Dufaure and Hubert.
[b] As assessed photometrically.

especially their dense granules, that were labelled. Attempts were also made to study, during ontogenesis, the presence of enzymes involved in the formation of indoleamines, namely 5-hydroxytryptophan decarboxylase (5-HTPD) and MAO (MEINIEL, 1977b). Using 5-HTPD and MAO inhibitors combined with autoradiography after injection of 5-hydroxytryptophan it was concluded that the enzymes in question were present as early as stage 35. Histochemical tests for MAO showed that the enzyme was present already at stage 34, but only in the distal parts of the pineal organ. At stage 35, a striking increase in the amount of reaction product was visible, especially in the immediate vicinity of the pineal lumen and occasionally near the basement membrane. At stages 37 and 38 the test for MAO was strongly positive in the distal and intermediate areas of the pineal organ and weakly positive in the proximal part; by stages 39 and 40 the proximal part also showed an intense reaction.

5. Histochemistry of the Adult Pineal Organ

The presence of serotonin in the pineal organ of *adult* reptiles (QUAY and WILHOFT, 1964) has been amply confirmed by histochemical techniques. Using the argentaffin reaction after Masson, the DMAB reaction after Adams and the safranin-diazo technique after Lillie, Burtner and Henson, COLLIN (1967a) obtained positive staining reactions in the secretory rudimentary photoreceptors, especially the supranuclear cytoplasm of the inner segment, the cell bodies and the basal cell processes, in *Lacerta viridis* and *L. muralis*, but not in *Anguis*, *Tarentola* and *Chalcides*. The Falck-Hillarp technique revealed an intense yellow fluorescence in *Lacerta viridis* (Fig. 171), *L. muralis* (QUAY et al., 1967; COLLIN, 1968a; WARTENBERG and BAUMGARTEN, 1969a). *L. vivipara* (MEINIEL et al., 1973), *Testudo graeca*, *T. hermanni*, *Tropidonotus natrix* (COLLIN and MEINIEL, 1972a), and *Natrix natrix* (QUAY et al., 1968). No specific fluorescence was demonstrable in *Anguis*, *Tarentola* and *Vipera* (COLLIN, 1968a; COLLIN and MEIN-

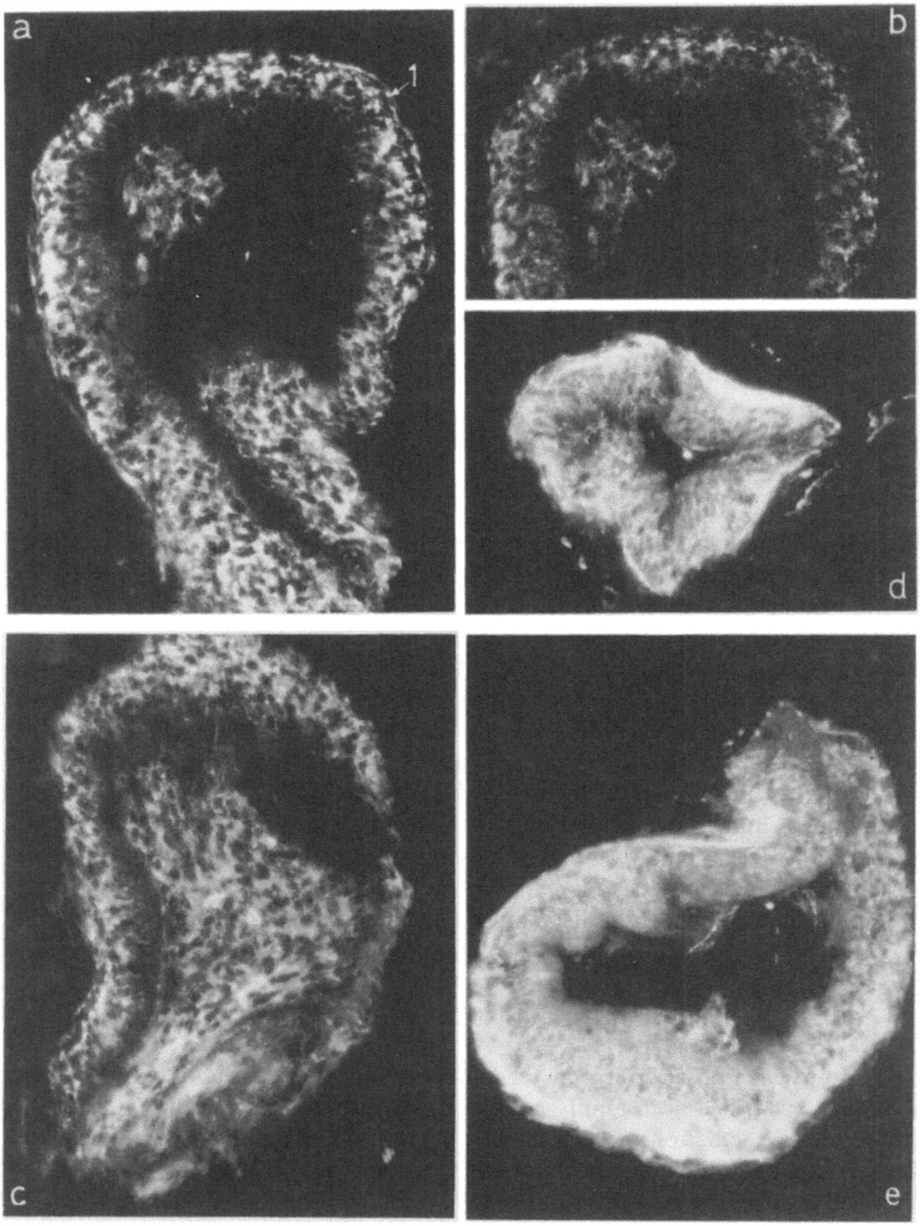

Fig. 171 a–e. Histochemical demonstration of serotonin (Falck-Hillarp technique) in the pineal organ of *Lacerta viridis*. Intense yellow fluorescence of parenchyma; *1*, green fluorescent adrenergic nerve fibres. **b** Same specimen showing fading of fluorescence; what remains is autofluorescent lipid droplets. **c** After pretreatment with 10 mg/kg reserpine. **d** After pretreatment with 300 mg/kg nialamide. **e** After pretreatment with 300 mg/kg nialamide and 30 mg/kg 5-hydroxytryptophan. WARTENBERG and BAUMGARTEN, 1969a; ×210

Table 42. Serotonin content in the reptilian epiphysis cerebri

Species	Contents		Author(s)
Terrapene carolina	6.98 ± 0.74 µg/g		QUAY and WILHOFT (1964)
Gopherus berlandieri	9.84 ± 1.55		
Sauromalus varius	0.87		
Gerrhonotus multicarinatus	41.00		
Natrix sipedon	10.00		
Pituophis catenifer	3.80		
Lacerta muralis L.	39 [a]	390 [b]	PETIT et al. (1971)
Tropidonotus natrix L.	0.4	13.3	
Testudo hermanni L.	62	21	
Pseudemys scripta elegans	33	8.4	

[a] Expressed as nanograms serotonin/epiphysis cerebri.
[b] Expressed as nanograms serotonin/mg epiphyseal tissue.

IEL, 1972a). According to MEHRING (1972) the fluorescence of the pineal organ of *Testudo hermanni* is very weak, even after pretreatment with iproniazid (up to 300 mg/kg body weight). That the yellow fluorescence is due to the presence of serotonin has been confirmed microspectrofluorometrically (WARTENBERG and BAUMGARTEN, 1969a; MEINIEL et al., 1973). Light and electron microscopic autoradiography after ^{3}H-5-hydroxytryptophan injections demonstrated a distinct labelling of the secretory rudimentary photoreceptors (COLLIN and MEINIEL, 1973a, b) in *Lacerta vivipara* and a weak labelling of the pinealocytes in *Tropidonotus natrix* (PETIT, 1971), which agrees well with the serotonin content reported (Table 42). The glutaraldehyde-silver hydroxyde technique after Cannata et al. gave a weakly positive reaction in *Lacerta* (WARTENBERG and BAUMGARTEN, 1969a) and negative results in *Tropidonotus* (PETIT, 1971). Further ultracytochemical results will be mentioned on p. 419.

Monoamine oxidase activity has been demonstrated in the apical regions of the pineal epithelium of *Lacerta vivipara* (COLLIN and MEINIEL, 1973b). Glycoproteins, polysaccharides, acid mucopolysaccharides, SH groups and different types of lipids and phospholipids have been demonstrated histochemically in what has been termed pseudosensory cells (*Pseudemys scripta elegans;* VIVIEN-ROELS, 1969). The lipopigments present in large amounts in *Testudo hermanni* have been shown to contain high concentrations of calcium associated with smaller amounts of S, K, P, Si, Al and Mg (VIVIEN-ROELS and HUMBERT, 1977).

6. Ultrastructure of the Intrinsic Cells of the Adult Pineal Organ

Electron microscopic studies in reptiles have greatly contributed to the concept that phylogenetically the epiphysis cerebri develops from a sensory organ into a secretory organ by showing that the outer segments of sensory cells tend to degenerate and that regressed photoreceptor cells contain secretory products (Fig. 190). As in some instances the outer segments show an apparently normal structure whereas in others they exhibit signs of regression, COLLIN (1969b, 1971) speaks of cells of the sensory line, in which he includes typical photoreceptor cells, secretory rudimentary photoreceptors and pinealocytes as well as probable intermediate types. Phylogenetically the photoreceptors were thought to dedifferentiate into rudimentary photoreceptors which in turn give rise to pinealocytes.

It should be noted that in some studies (e.g. PETIT, 1969a) the term pinealocyte has been applied to both sensory cells and rudimentary sensory cells. For the sake of clarity and in view of the situation in mammals, it is felt that the term pinealocyte should be reserved for cells of the sensory line which do not show typical or atypical outer segments. In reptiles typical photoreceptor cells are relatively rare; they have been found in lizards and some turtles, but are lacking in snakes. In turtles, differences exist between aquatic and terrestrial species: photoreceptor cells are restricted to aquatic turtles, where they are found in the proximal part of the pineal organ; they are absent in the pineal organ of terrestrial turtles and in the distal part of the pineal organ in aquatic turtles (VIVIEN-ROELS, 1976b). Rudimentary photoreceptors are very common in lizards, turtles and embryonic snakes. Pinealocytes are the characteristic cell type in adult snakes (see COLLIN, 1971).

a) Typical Photoreceptor Cells

As pointed out above, the photoreceptor cells undergo progressive rudimentation (Fig. 190). Since in reptiles this process is very pronounced, it is difficult to define what a typical sensory cells looks like, especially since the outer segments fall into two categories: regular and irregular. Another difficulty is that different points of reference have been used for comparisons: some authors (e.g. OKSCHE and KIRSCHSTEIN, 1968) have compared the photoreceptor cells of the pineal organ with those of the parietal eye, whereas others (e.g. COLLIN, 1969b) have compared them with those present in the pineal complex of cyclostomes. Moreover it should be borne in mind that in view of the presence of long basal processes which also undergo regression it may be difficult to characterize a complete cell and to decide whether or not it shows the typical features both apically and basally. Frequently one has to extrapolate and to assume that a cell with typical outer segments has the typical synaptic apparatus basally and vice versa.

Typical photoreceptor cells of reptiles are similar to the cones of the lateral eyes and consist of an outer segment, an inner segment (showing the usual subdivision into ellipsoid, isthmus and myoid), a cell body and a basal process (COLLIN, 1968b, 1969b, 1971; MEINIEL, 1976a). Great variation is found in the outer segments. COLLIN (1969b) studying *Lacerta muralis* found a sensory cell with an outer segment measuring 5.5 µm in length and comprising some 120 regularly arranged discs. Roughly the same appearance was observed in *Anguis fragilis* (PETIT, 1969). OKSCHE and KIRSCHSTEIN (1968, five species of lizards) found that the outer segments did not exceed 2 µm in length and contained some 50–60 discs (Fig. 172); by contrast, the outer segments in the

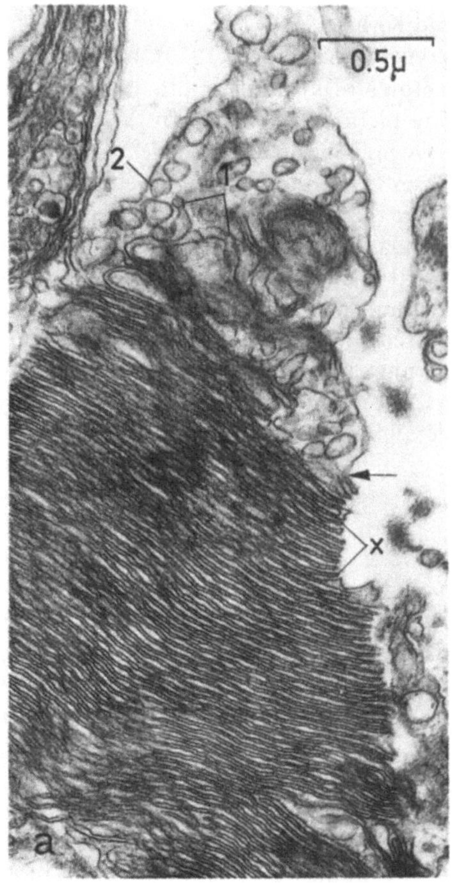

Fig. 172. Shape and ultrastructure of the outer segments of photoreceptor cells in the epiphysis cerebri of *Lacerta sicula campestris*. Note the regular lamellation of outer segment showing tubular (*1*) and vesicular (*2*) structures. *X*, cone-type invagination of the plasmalemma. Compare with Fig. 178. OKSCHE and KIRSCHSTEIN, 1968; ×31,200

parietal eye measured 15 μm in height and contained about 400 discs. In young lizards, in contrast to adult animals, outer segments with regularly arranged discs were more frequently found (*Anguis fragilis:* OKSCHE and KIRSCHSTEIN, 1968). Some 50 discs were counted in 3-day-old specimens of *Lacerta vivipara* (MEINIEL, 1976a). Very well developed outer segments have been observed in the turtle *Pseudemys scripta elegans* in which they measured up to 12.5 μm in length and contained up to 215 discs (VIVIEN-ROELS, 1970). In what has been termed an irregular type of outer segment (Fig. 178), tubules, large amounts of vesicles and some more or less regularly arranged discs were found (OKSCHE and KIRSCHSTEIN, 1968; COLLIN, 1969b). The inner segment, the cell body and basal processes apparently do not show any unusual features. According to COLLIN (1969b, 1971) the photoreceptor cells lack secretory granules.

A crucial point is whether or not basal processes of typical photoreceptor cells in reptiles form synaptic contacts with afferent neurons. As will be recalled, the synaptic contacts in fishes and amphibians are characterized by the presence of synaptic ribbons. Ribbons have been clearly demonstrated (*Lacerta muralis:*

COLLIN, 1968b; WARTENBERG and BAUMGARTEN, 1968; *Lacerta vivipara:* COLLIN and KAPPERS, 1971; *Anguis:* PETIT, 1969a; *Pseudemys:* VIVIEN and ROELS, 1968; VIVIEN-ROELS, 1969, 1970), but they are altogether rare and very few convincing pictures of synaptic contacts between sensory cells and afferent neurons exist. The pictures published by VIVIEN and ROELS (1968) and VIVIEN-ROELS (1969) showing dense-core vesicles in the post-"synaptic" neuron suggest that the fibres are efferent sympathetic fibres and do not belong to afferent neurons. More recently, very convincing pictures of synaptic contacts have been published (COLLIN and KAPPERS, 1971; PETIT and VIVIEN-ROELS, 1977a). As shown in Fig. 173, the basal process of a photoreceptor cell may form synapses with a number of dendrites, one synaptic ribbon being related to two dendrites; moreover, pictures were observed where one dendrite was related to a number of photoreceptor cells. In view of the presence of these synaptic contacts it is conceivable that photic messages may be transmitted to the brain (for further details see p. 426).

b) Secretory Rudimentary Photoreceptors

In the reptilian pineal organ, secretory rudimentary photoreceptor cells (SRP) are the predominant cell type of the sensory line (COLLIN, 1969b, 1971, 1976). As the term implies, these cells show distinct signs of rudimentation and secretion. They have been termed pseudosensory cells by VIVIEN-ROELS (1969) and pinealocytes by WARTENBERG and BAUMGARTEN (1968). As can be seen on Fig. 174, the rudimentation concerns mainly the outer segments. There are various degrees of rudimentation. In the *lizards Lacerta viridis* and *L. muralis* (WARTENBERG and BAUMGARTEN, 1968) outer segments may be completely lacking (Fig. 175), the cytoplasmic protrusions shown belonging to the inner segment which shows a prominent constriction. The apical parts of the inner segments may contain vacuoles comprising a moderately electron-dense material (Fig. 176). Although these cells lack outer segments, the inner segments still retain their abundance of densely packed mitochondria (Fig. 177). In other species small outer segments are clearly visible (Fig. 178), containing vesicles of variable sizes, vacuoles, tubules, multivesicular and multilamellar bodies and occasionally a few typical discs (*Iguana iguana, Anguis fragilis, Lacerta vivipara, L. agilis, L. sicula campestris:* OKSCHE and KIRSCHSTEIN, 1966b, 1968). In *Anguis fragilis*, PETIT (1969a) found modified outer segments in which 5–20 discs lay parallel to the longitudinal axis of the cilium connecting the inner and outer segments. Other papers dealing with outer segments in lizards are those by STEYN (1960a, *Cordylus polyzonus*), COLLIN (1967b, 1968b, 1969b, *Lacerta viridis*) and LIERSE (1965b, *Anolis carolinensis*).

In *turtles* the outer segments of the secretory rudimentary photoreceptors are basically similar to those in lizards (*Pseudemys scripta elegans:* VIVIEN-ROELS, 1969, 1970; *Pseudemys callirostris, Clemmys leprosa:* OKSCHE and KIRSCHSTEIN, 1966a, b; *Testudo hermanni:* LUTZ and COLLIN, 1968; MEHRING, 1972). The intrinsic pineal cells of *snakes* lack the structural characteristics of rudimentary photoreceptor cells (OKSCHE and KIRSCHSTEIN, 1966b, *Natrix natrix*); instead there are pinealocytes which will be described in the following section (p. 424).

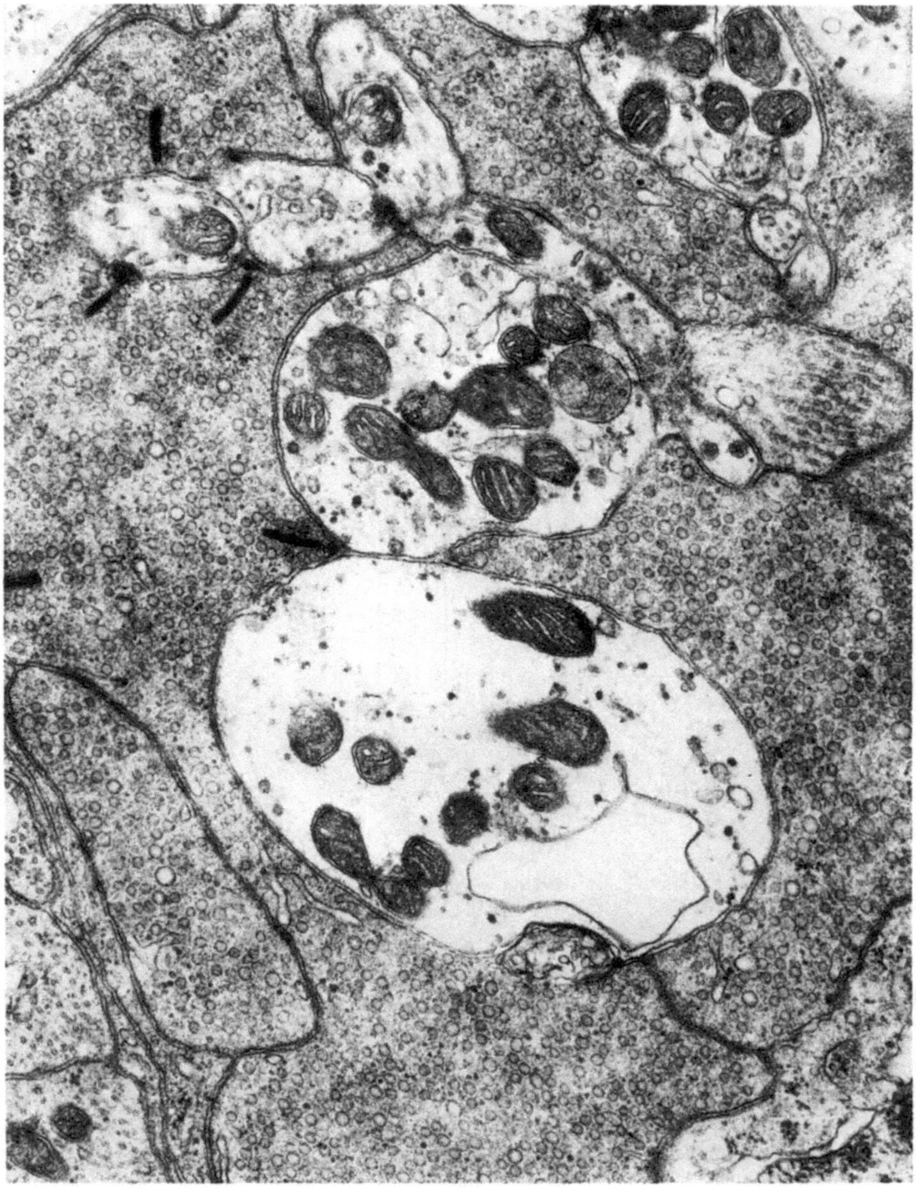

Fig. 173. Basal processes of pineal photoreceptor cells showing synapses (synaptic ribbons) with dendrites of nerve cells. *Lacerta muralis*. PETIT and VIVIEN-ROELS, 1977a; ×35,000 (reduced to 70%)

The secretory rudimentary photoreceptors of the reptilian pineal organ differ from their counterparts in other vertebrate classes by reason of the fact that they contain an abundance of secretory granules (Fig. 179), measuring between 50 and 340 nm (Table 43). The granular inclusions are apparently not a homoge-

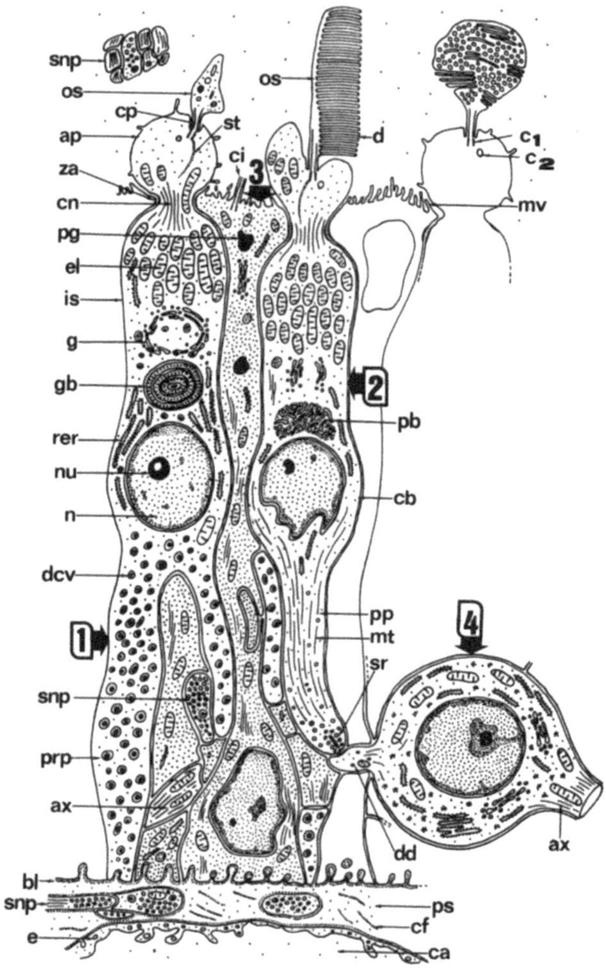

Fig. 174. Pineal organ of an adult wall lizard (*Lacerta muralis*). Diagram illustrating the sensory epithelium. *1*, secretory rudimentary photoreceptor cell; *2*, photoreceptor cell; *3*, interstitial (= supportive) cell; *4*, pineal neuron. *Abbreviations: ap*, apical process; *ax*, axon; *bl*, basal lamina; c_1, distal centriole; c_2, proximal centriole; *ca*, capillary; *cb*, cell body; *cf*, collagen fibrils in the perivascular spaces; *ci*, cilium; *cn*, cellular neck; *cp*, connecting piece; *d*, disc; *dsv*, dense-cored vesicle; *dd*, dendrite of a pineal neuron; *e*, endothelial cell; *el*, mitochondria of the ellipsoid; *g*, Golgi complex; *gb*, glycogen body (or glycogen *nebenkern*); *is*, inner segment; *mt*, microtubules; *mv*, microvilli; *n*, nucleus; *nu*, nucleolus; *os*, outer segment; *pb*, paraboloid; *pg*, pigment granule; *pp*, photoreceptor cell process; *prp*, process of a rudimentary photoreceptor cell; *ps*, perivascular space; *rer*, rough endoplasmic reticulum; *sr*, synaptic ribbon; *st*, striated ciliary rootlet; *snp*, sympathetic nerve process; *za*, zonula adherens. Collin, 1976

neous population, also including pigment granules and lysosomes (Oksche and Kirschstein, 1968). Granules are especially abundant in the perikaryon and in the basal processes of the rudimentary photoreceptors. As shown by many authors (see Table 43), the granules take their origin from the Golgi apparatus lying in the vicinity of the cell nucleus, the so-called myoid.

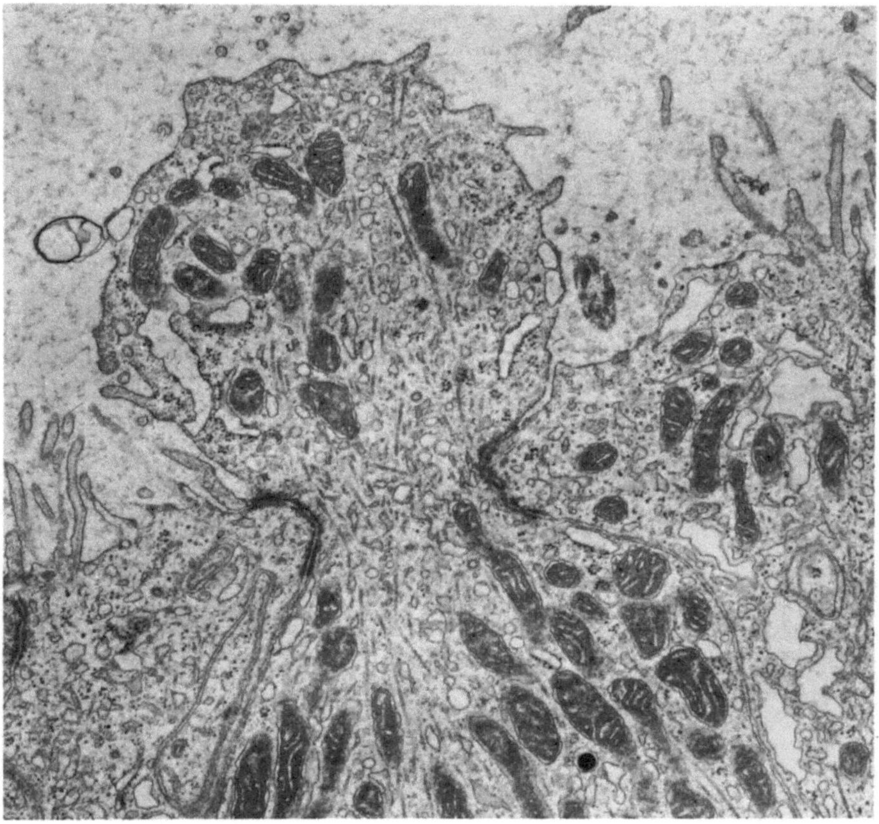

Fig. 175. Apical part of a secretory rudimentary photoreceptor of *Lacerta viridis* showing a club-like protrusion of the inner segment. An outer segment is lacking. WARTENBERG and BAUMGARTEN, 1968; × 18,000

Interspecies differences have been observed insofar as in *Tarentola, Anguis* and *Chalcides* the Golgi apparatus was less actively engaged in granule formation than in *Lacerta* (COLLIN, 1967c, 1969b). There can be no doubt that the secretory rudimentary photoreceptors are actively engaged in serotonin metabolism as shown by light and electron microscopic autoradiography using ^{3}H-hydroxytryptophan (MEINIEL et al., 1972; COLLIN and MEINIEL, 1973a, b; MEINIEL, 1976b). Opinions vary concerning the question as to whether or not the granules contain serotonin. As the granules showed little, if any, change after treatment with reserpine or the MAO inhibitor, nialamide (WARTENBERG and BAUMGARTEN, 1969a; WARTENBERG, 1969a, b), it was concluded that the granules do not contain serotonin or that two pools of serotonin existed: (a) an intravesicular pool which is fairly resistent to reserpine and tightly bound to a protein matrix and (b) an extravesicular pool which is easily available and increases after nialamide or 5-hydroxytryptophan treatment. Evidence for the presence of proteins in the granules was obtained by pronase, pepsin or papain treatment of the sections (COLLIN and MEINIEL, 1972b). In contrast to WARTENBERG and

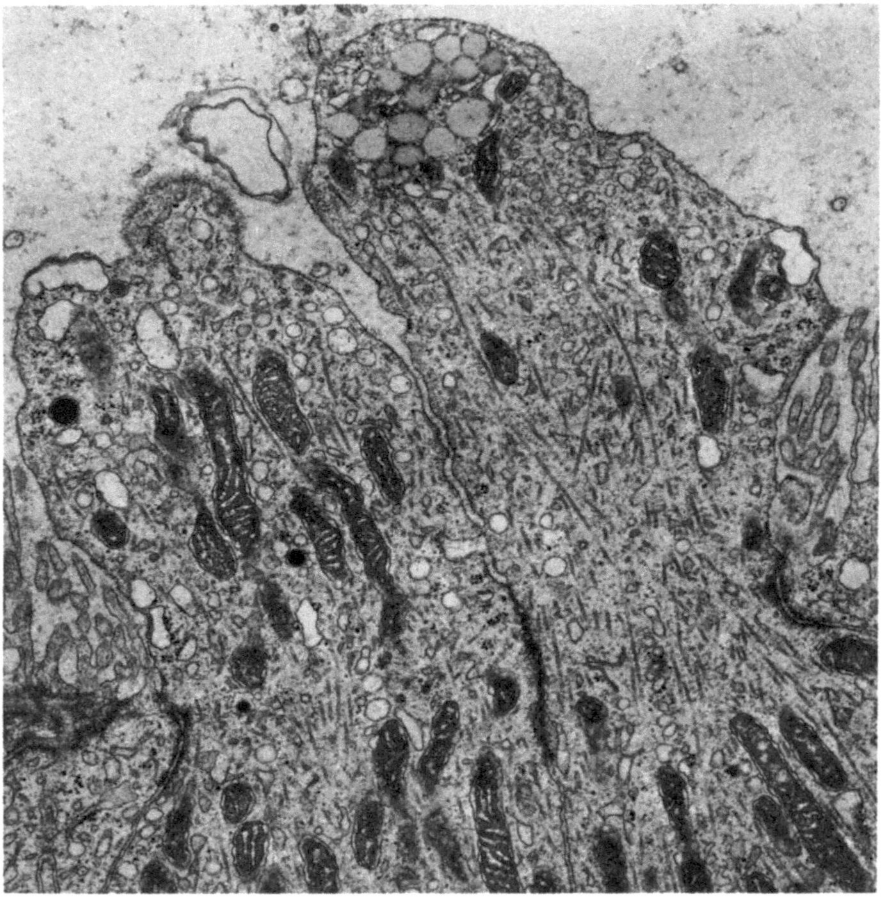

Fig. 176

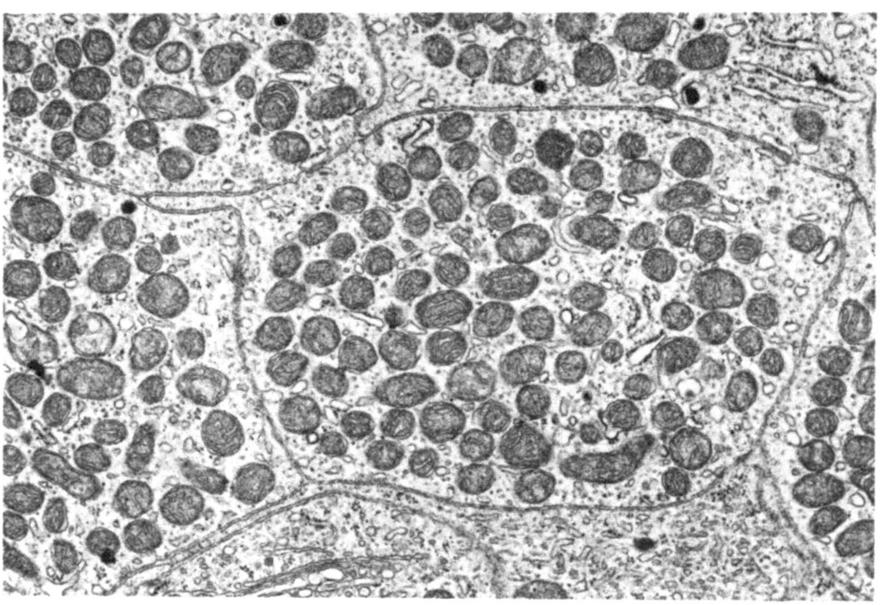

Fig. 177

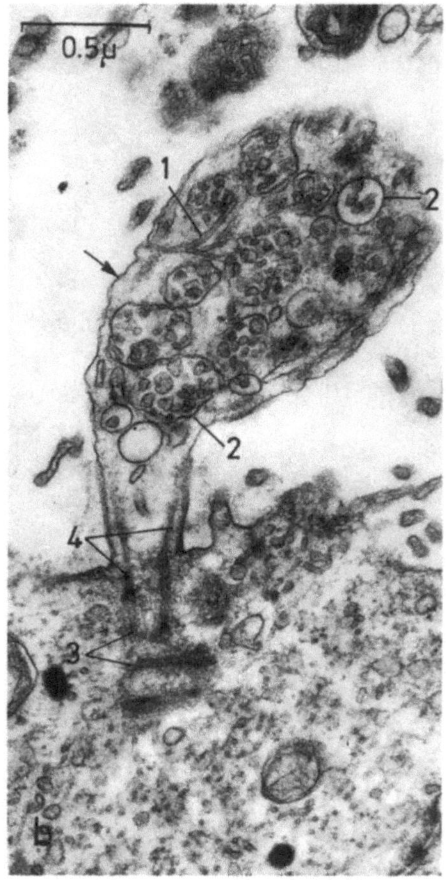

Fig. 178. Outer segment of a secretory rudimentary photoreceptor cell exhibiting an irregular ultrastructure. Note the tubules (*1*) and vesicle-containing vacuoles (*2*) of varying sizes. *3*, centrioles; *4*, cilium. OKSCHE and KIRSCHSTEIN, 1968; × 34,200

BAUMGARTEN (1969a), who were unable to demonstrate the presence of serotonin in the granules by means of a silver technique, using argentaffin and chromaffin reactions COLLIN et al. (1977) clearly demonstrated a reactivity of the granules which disappeared after *p*-CPA or reserpine administration and increased after nialamide injection. Whether the serotonin present in the secretory rudimentary photoreceptors represents the secretory product of these cells is not known. In view of the fact that the reptilian pineal organ contains HIOMT (QUAY, 1965d) and in view of the situation in other vertebrates, it is more likely that serotonin functions as a precursor of melatonin.

Fig. 176. Inner segment of a secretory rudimentary photoreceptor containing apically vacuoles filled with a moderately dense material. By courtesy of Prof. H. WARTENBERG, Department of Anatomy, University of Bonn

Fig. 177. Transverse section through an inner segment of a rudimentary secretory photoreceptor of *Lacerta viridis*, showing an accumulation of mitochondria (ellipsoid). WARTENBERG and BAUMGARTEN, 1968; × 18,000

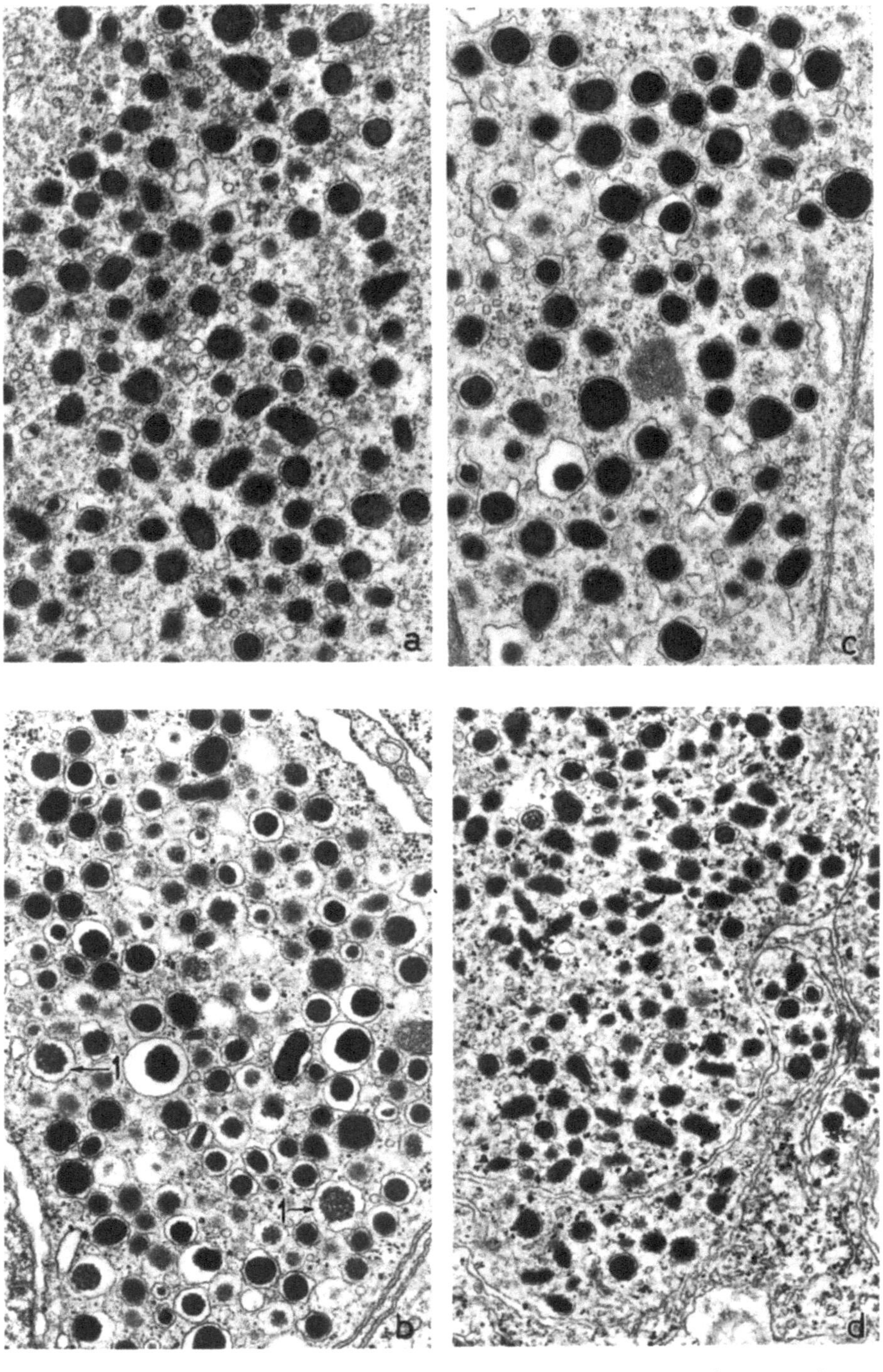

Fig. 179a–d. Profiles of pineal parenchymal cells showing large numbers of granulated vesicles. *Lacerta viridis*. **a** Untreated control animal. ×25,700. **b** After pretreatment with 10 mg/kg reserpine. Some granules show a decrease of electron density. ×25,700. **c** After pretreatment with 300 mg/kg nialamide. ×25,700. **d** Histochemical demonstration of monoamines by means of ammoniacal silver nitrate solution. The granules do not show a positive reaction. However, silver granules lie freely in the cytoplasm. ×18,000. WARTENBERG and BAUMGARTEN, 1969a

Table 43. Presence and size of secretory granules in intrinsic epiphyseal cells of reptiles

Species	Size (nm) or presence	Author(s)
Lizards		
Lacerta vivipara	70–150	OKSCHE and KIRSCHSTEIN (1968)
	50–340	COLLIN and MEINIEL (1973a, b)
Lacerta agilis	70–150	OKSCHE and KIRSCHSTEIN (1968)
Lacerta viridis	60–340	COLLIN and KAPPERS (1968), COLLIN (1969b)
	50–200	COLLIN (1967b)
	60–200	WARTENBERG and BAUMGARTEN (1968)
Lacerta muralis	133.7 ± 4.2	WARTENBERG and BAUMGARTEN (1969a)
	60–340	COLLIN and KAPPERS (1968), COLLIN et al. (1977)
Anguis fragilis	90–120	PETIT (1968)
Turtles		
Pseudemys scripta elegans	Present	VIVIEN (1964b), VIVIEN and ROELS (1967), VIVIEN-ROELS (1969)
	90–130	VIVIEN-ROELS (1970)
Testudo hermanni	Absent	LUTZ and COLLIN (1968)
	70–125	MEHRING (1972)
Snakes		
Tropidonotus natrix	Present	VIVIEN (1964a)
	60	PETIT (1971)

In addition to the outer segments and the dense-core vesicles of the secretory rudimentary photoreceptors their basal processes are of interest, especially with regard to the question as to whether they show rudimentation or not. According to COLLIN (1969b) they show distinct signs of regression, one indication being the lack of synaptic ribbons which are typically present in basal processes of non-rudimentary photoreceptors. As shown diagrammatically in Fig. 174, the basal processes of the SRP are thought to lack afferent connections. However, recently a synaptic connection between a SRP and a nerve cell soma has been found in *Lacerta muralis*, which on morphological grounds appears to belong to a pinealofugal arrangement of units. This contact lacked a synaptic ribbon; presynaptically it showed an accumulation of both synaptic vesicle-like vesicles and dense-core vesicles, while postsynaptically, an electron-dense, flattened cisterna lay parallel to the synaptic cleft and extended over the entire region of the synaptic contact (PETIT and VIVIEN-ROELS, 1977a). The functional significance of this contact is obscure. The presence of very long and thin basal processes, showing bulbous endings distally, has been described in *T. hermanni* (MEHRING, 1972).

Other points worth mentioning are the presence of tight junctions between SRP (Fig. 180) (WARTENBERG and BAUMGARTEN, 1968) and a *nebenkern* consisting of endoplasmic reticulum and glycogen (VIVIEN and ROELS, 1967; VIVIEN-ROELS, 1969, 1970; COLLIN, 1968b, 1969b). Lipopigments with high con-

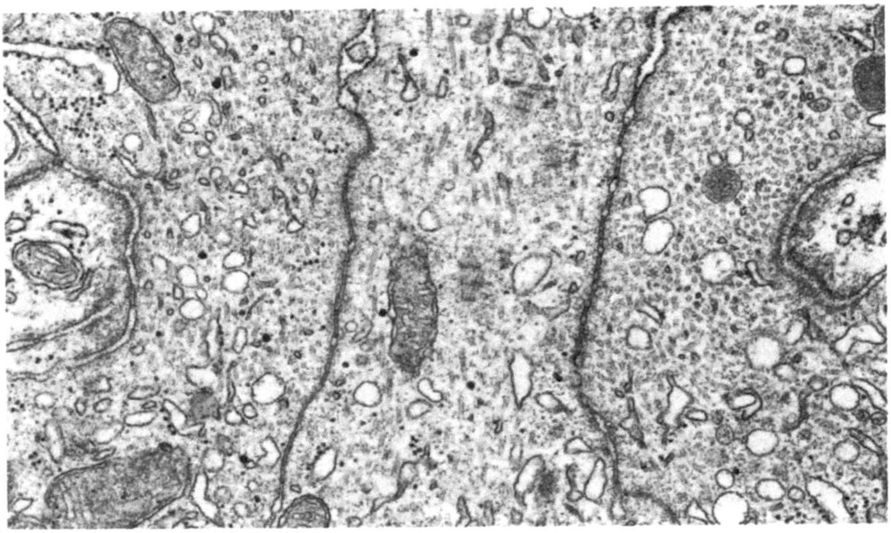

Fig. 180. Electron micrograph showing tight junctions between intrinsic pineal cells of *Lacerta viridis*. WARTENBERG and BAUMGARTEN, 1968; × 30,000

centrations of calcium have been found in *Testudo* (VIVIEN-ROELS and HUMBERT, 1977).

c) Pinealocytes

In pineal organs of snakes, the cells of the sensory line differ from photoreceptor cells and secretory rudimentary photoreceptor cells to such an extent that a different term, namely pinealocyte, has been applied to these cells (COLLIN, 1969b, 1971). The pinealocytes of adult animals lack outer and inner segments (COLLIN, 1971). Their relationship to photoreceptor cells is indicated by (a) the fact that in young embryos cells reminiscent of sensory cells are still present, which disappear quickly during ontogenesis (PETIT, 1969b; COLLIN, 1969b, 1971); (b) the presence of 9+0 cilia projecting into intercellular spaces (VIVIEN, 1964a) and (c) the presence of synaptic ribbons (PETIT, 1971). In common with secretory rudimentary photoreceptors, they possess basal processes and an abundance of dense-core vesicles (VIVIEN, 1964a; PETIT, 1969a).

The precise chemical nature of the dense-core vesicles is not known. Although the epiphyseal parenchyma exhibits the characteristic yellow serotonin fluorescence after formaldehyde treatment (QUAY et al., 1968a), PETIT (1971) failed to demonstrate catecholamines or indoleamines by means of the glutaraldehyde-potassium bichromate technique after ETCHEVERRY and ZIEHER. The osmium tetroxide-zinc iodine technique was negative as well. Administration of the false transmitter 5-hydroxydopamine had no visible effects on the dense-core vesicles, agranular vesicles and the vesicles forming part of the synaptic ribbons. Since the granules did not react to reserpine (PETIT, 1971), which had been shown to reduce epiphyseal serotonin levels (PETIT et al., 1971), and exhibited no distinct autoradiographic labelling after ^{3}H-5-hydroxytryptophan, PETIT (1971) concluded that serotonin is not stored in the dense-core vesicles, but may be present in a diffuse form. After chemical sympathectomy with 6-hydroxydopamine, the dense-core vesicles decreased in amount, became lighter and coalesced, which shows that the vesicles may be functionally related to the sympathetic nerve fibres present (p. 426).

d) Supporting Cells

In addition to the cells of the sensory line the pineal organ comprises ependymal supporting cells. So far they have been clearly identified in lizards and turtles only.

The major papers dealing with the ultrastructure of the supporting cells in the reptilian epiphysis cerebri are those by WARTENBERG and BAUMGARTEN (1968), PETIT (1969a), VIVIEN-ROELS (1969) and MEHRING (1972). Electron microscopic studies have confirmed the light microscopic findings showing that supporting cells are fairly tall, extending from the basal lamina of the epiphyseal epithelium right to the epiphyseal lumen and enwrapping the cells of the sensory line. A characteristic feature is that, as shown for *Lacerta viridis* and *L. muralis* (WARTENBERG and BAUMGARTEN, 1968), the cells possess two thickenings, apical and basal, which are linked by a thin cytoplasmic bridge sometimes measuring as little as 50–100 μm in width. The apical thickening, which at its free surface is studded with microvilli and occasionally shows some cilia, bulges laterally into the compressed neck region of the inner segment of the rudimentary photoreceptors where they are attached to one another by zonulae adherentes and occludentes. The basal thickening houses the nucleus, and the basal cell surface is more or less deeply indented, the basal lamina following the contours of the cell. In addition to the usual cell organelles, pigment granules and filaments resembling glial filaments may be present. In contrast to the cells of the sensory line they lack secretory granules. Apart from a suggested supportive role little is known about their function. They are definitely not involved in serotonin formation, as shown by the finding that they show little, if any, labelling after administration of ^{3}H-5-hydroxytryptophan (COLLIN and MEINIEL, 1973a).

7. The Nervous Apparatus

The nervous apparatus of the reptilian pineal organ consists of afferent and efferent components. As can be expected from the progressive rudimentation of the photosensory cells, the afferent innervation has a tendency to regress. In contrast to fishes and amphibians, an efferent (sympathetic) innervation is prominent, as in mammals.

a) Afferent Fibres

In view of the scarcity of fully differentiated photoreceptor cells in the reptilian pineal organ, it is not surprising that neurons linking these photoreceptors with brain centres are likewise rare (KAPPERS, 1966, 1967; WARTENBERG and BAUMGARTEN, 1968; COLLIN, 1969b; PETIT, 1969), an exception being *Pseudemys*, which possesses a high number of nerve cells in the posterior part of the pineal organ (VIVIEN and ROELS, 1968; VIVIEN-ROELS, 1970). According to KAPPERS (1966, 1967a) and COLLIN and KAPPERS (1968) the afferent fibres are basically myelinated ones which are surrounded by a myelin sheath either inside the pineal parenchyma or when they leave the pineal epithelium. These fibres form the pineal tract which courses to the habenular and posterior commissures and further to the subcommissural organ and the periventricular nerve fibre layer (KAPPERS, 1967a; OKSCHE and KIRSCHSTEIN, 1968). The synaptic connection of the photoreceptor cells to these neurons has been described above (p. 415).

b) Efferent (Sympathetic) Innervation

Efferent monoaminergic nerve fibres are regularly found in the reptilian epiphysis (Table 44). The fibres are non-myelinated and comprise dense-core vesicles measuring between 30 and 170 nm in diameter (Table 44). As in mammals, the vesicles may be subdivided into small vesicles and large vesicles, the small vesicles greatly exceeding the large ones in number. The chemical nature of the neurotransmitter present in these granules has been clearly established. In *Natrix natrix* (QUAY et al., 1968a), *Lacerta* (COLLIN, 1968a; WARTENBERG and BAUMGARTEN, 1969a), *Anguis fragilis* and *Tarentola mauritanica* (COLLIN, 1968a) and *Testudo hermanni* (MEHRING, 1972) the fibres exhibit a green (noradrenaline) fluorescence after treatment with formaldehyde vapour. Electron microscopically, the vesicles show an increased electron density after administration of the false sympathetic transmitters 5-hydroxydopamine and 5-hydroxydopa (*Lacerta muralis*: WARTENBERG and BAUMGARTEN, 1969b; *Tropidonotus na-*

Table 44. Size of dense-core vesicles in efferent neurons of the reptilian epiphysis cerebri

Species	Size (nm)	Author(s)
Lizards		
Lacerta sicula campestris	80	OKSCHE and KIRSCHSTEIN (1968)
Lacerta viridis	35–120	WARTENBERG and BAUMGARTEN (1968)
	30–167	COLLIN and KAPPERS (1968)
Lacerta muralis	35–120	WARTENBERG and BAUMGARTEN (1968)
	30–167	COLLIN and KAPPERS (1968)
Tarentola mauritanica	30–167	COLLIN and KAPPERS (1968)
Anguis fragilis	100–150	PETIT (1969a)
Turtles		
Pseudemys scripta elegans	130–170	VIVIEN-ROELS (1970)
Snakes		
Tropidonotus natrix	50–160	PETIT (1971)

trix: PETIT, 1971). Additional evidence for the aminergic nature of the fibres was obtained by application of the osmium tetroxide-zinc iodine technique, the administration of ^{3}H-hydroxytryptamine followed by electron microscopic autoradiography and the application of a number of pharmacological tests (PETIT, 1971), all of which point to the presence of both noradrenaline and serotonin in the fibres in question. It thus appears that the situation in reptiles is identical to that in mammals: noradrenaline is the basic neurotransmitter, but due to the presence of large amounts of serotonin in the parenchyma, the autonomic fibres take up serotonin.

The precise origin of the efferent fibres has not been established. In *Natrix natrix* QUAY et al. (1968a) found that the fibres derived largely from a dorsal and posterior meningeal bundle reminiscent of the nervus conarii in mammals. It is therefore possible that the fibres have the same origin as in mammals, namely from the superior cervical ganglia of the sympathetic trunk.

Studies dealing with the distribution of the fibres within the pineal organ have shown that the fluorescent fibres are localized primarily in the stromal channels accompanying the blood vessels and that some fibres innervate parenchymal lobules (QUAY et al., 1968a). Electron microscopically it was seen that the fibres apparently do not form synaptic contacts with the intrinsic cells (WARTENBERG and BAUMGARTEN, 1969b). In addition there are noradrenergic nerve fibres which lie freely in the pineal lumen, sometimes in close association with outer segments of rudimentary photoreceptors (WARTENBERG and BAUMGARTEN, 1969b).

The functional significance of the sympathetic fibres in reptiles remains to be established. It should be recalled that in mammals these fibres regulate the formation of melatonin, which shows prominent circadian changes.

8. The Blood Supply

The blood supply of the reptilian pineal organ has been studied by STEYN (1958), LIERSE (1965a) and SWAIN (1968), whose papers should be consulted for detailed information. Electron microscopy (LIERSE, 1965b) has shown that the intrapineal sinuses are lined by a very flat endothelium, measuring as little as 10–20 nm in width; occasionally pores have been noted. The fairly wide basal lamina underlying the endothelium is regularly penetrated by cytoplasmic processes of endothelial cells which come into contact with pericytes.

9. Functional Aspects

Very little is currently known about the function of the reptilian pineal organ. Although most of the pineal photoreceptor cells are rudimentary and the afferent innervation shows clear signs of regression, the epiphysis of lizards still responds to light as shown by electrophysiology (HAMASAKI and DODT, 1969; DODT, 1973). As might be expected from the scarcity of typical photoreceptor cells in the pineal, a relatively high intensity of light has to be given before

Table 45. HIOMT activity in the pineal organ of reptiles (QUAY, 1965d)

Species	HIOMT activity[a]	No. of samples
Pseudemys scripta	432 ± 32	3
Cnemidophorus lemniscatus	6,651	1
Iguana iguana	1,237 ± 309	5
Natrix vallida	2,903 ± 1,526	3

[a] Expressed as nmoles melatonin formed per hour per gram; means ±SEM.

a threshold response can be obtained. When compared with the parietal eye a light about ten times as intense is required. In contrast to the parietal eye, which shows both chromatic and achromatic responses, the pineal organ reacts achromatically only.

In addition to a direct light sensitivity of the pineal organ there are indications for melatonin formation. There is abundant evidence that the pineal organ synthesizes the melatonin precursor, serotonin (p. 411). As in mammals, pineal serotonin concentrations show characteristic changes over a period of 24 h, maximum serotonin levels being present at the beginning of the dark phase and minimum levels at the beginning of the light phase (VIVIEN-ROELS et al., 1971). The pineal possesses HIOMT activity which is required for melatonin formation (Table 45) (QUAY, 1965d, 1971d; QUAY et al., 1971), and recently melatonin formation has been demonstrated in the pineal of *Testudo hermanni* by means of a radio-immunoassay: Melatonin formation from May to October is low during the day (< 100 pg/mg pineal tissue) and high during the night (3,500 pg/mg), but these day–night differences are not demonstrable from November to March, when pineal melatonin levels stay low (~ 100 pg/mg) (VIVIEN-ROELS and ARENDT, 1979). In view of these findings it is not surprising that the pineal *serotonin* levels also show a circannual rhythm (PETIT and VIVIEN-ROELS, 1977b; VIVIEN-ROELS and PETIT, 1978).

Very little is currently known about target organs of pineal function. Most recently it has been shown that the pineal organ plays an important role in the circadian organization in the lizard *Sceloporus olivaceus*. After removal of the complete pineal complex (removal of the parietal organ alone was ineffective) the circadian activity rhythm of animals kept under continuous illumination exhibits (a) a splitting into two components, (b) marked changes in the period of the rhythm or (c) arrhythmicity. This led UNDERWOOD (1977) to the hypothesis that the pineal organ is a coupling device between circadian oscillators in a multi-oscillator system, or that the pineal organ is itself a master oscillator, which entrains other circadian oscillators and imposes its frequency upon them.

In free-living western fence lizards, *Sceloporus occidentalis*, pinealectomy resulted in a lower incidence of animals being seen on the surface of the ground than in controls. Locomotor activity increased and the animals were more widely dispersed than controls. The pinealectomized animals usually showed a marked fright reaction and the tendency to take cover was pronounced. Body temperature was slightly lower (1.1° C) when compared with sham-operated controls. Occasionally kyphoscoliosis developed after pinealectomy. Growth rates and body colouration were not changed (STEBBINS, 1960). In the same species QUAY et al. (1971) noted that pineal HIOMT activity was significantly lower on the coldest day of the experimental period, when all animals were inactive and concealed.

There is some evidence that the pineal affects endocrine organs. In the turtle, *Emys leprosa*, pinealectomy was followed by exophthalmus, thyroid hyperactivity and atrophy of the testes (ARON et al., 1960b, c; COMBESCOT and DEMARET, 1963). Testicular regression after administration of melatonin has been reported in *Callisaurus draconoides* (PACKARD and PACKARD, 1977). In *Calotes versicolor* pinealectomy stimulated early testicular recrudescence and delayed testicular regression (HALDAR and THAPLIYAL, 1977), but in *Cnemidophorus uniparens* there was very little demonstrable effect on reproduction: in 2 out of 13 pinealectomized animals vitellogenesis started 1 month earlier (CUELLAR, 1978). VIVIEN (1965), studying the pineal organ of *Tropidonotus natrix* by electron microscopy, noted that the pinealocytes showed diminished secretory activity in the prehibernation period and that after administration of chorion gonadotrophin the pinealocytes showed an increase in the amount of rough endoplasmic reticulum and Golgi complexes accompanied by a discharge of secretory granules. In *Anolis occidentalis* kept under 6L:18D destruction of the pineal organ by thermocautery in November and January caused significant follicular development in the ovaries within 14 days, which was blocked by melatonin administration; in September pinealectomy had no effect on the ovaries, pointing to seasonal differences in pineal function (LEVEY, 1973).

In vitro studies on pars intermedia tissue of the pituitary of *Anolis carolinensis* have shown that melatonin significantly inhibited MSH release by about 40% during the first hour of administration, whereas there was no significant inhibitory effect during later stages; serotonin, by contrast, stimulated MSH release (THORNTON and GESCHWIND, 1975). An involvement of the pineal organ in the regulation of water metabolism is suggested by the findings that after pinealectomy the amount of neurosecretory material in the paraventricular nuclei showed a striking increase (ARON et al., 1960a, *Emys leprosa*) and that after saline injections the height of the epithelial cells in the pineal showed an increase, as did the number of secretory granules (GRIGNON and GRIGNON, 1963a).

IV. The Parietal or Third Eye

The structure and function of the parietal or third eye has been dealt with in detail in the book by EAKIN (1973). It therefore suffices to give a cursory description of this second component of the reptilian pineal complex. GUNDY and WURST (1976a) assume that a parietal eye is present in about 60% of lizards. In their own material investigated (85 different species), 72% had a well developed parietal eye, 6% had a regressed parietal eye and 22% lacked a parietal eye (GUNDY and WURST, 1976b). Zoogeographical studies suggest that lizards without parietal eyes tend to be restricted to low latitudes, whereas lizards with parietal eyes are present at higher latitudes (GUNDY et al., 1975). Subterranean habitat of lizards may result in regression of the parietal complex (GUNDY and RALPH, 1971). Later it was concluded that regressive parietal eyes are common in certain arboreal lizards and that the parietal eye is often retained in burrowing lizards, probably because for burrowers the parietal eye is a more suitable photoreceptor than are lateral eyes (GUNDY and WURST, 1976b).

1. Gross Anatomy

The parietal eye is basically a vesicle-like organ, showing various degrees of differentiation. Highly differentiated parietal eyes consist of a cornea, lens, retina and a parietal nerve (TROST, 1953a, 1954; GUNDY and WURST, 1976b). At the other end of the spectrum lie parietal eyes consisting of vesicles with no apparent regional differentiation into lens and retina (GUNDY and WURST, 1976b). According to TROST (1954) the regression of the parietal eye follows a certain pattern: first the cornea regresses, followed by regression of the lens and the nerve fibres leaving the parietal eye.

2. Organogenesis

Since the early developmental stages of the parietal eye have already been described (p. 399), only the later developmental changes have to be considered.

Following the excellent and detailed description by TROST (1953a), in *Anguis fragilis* and *Chalcides ocellatus* the histological differentiation of the parietal vesicle begins prior to its separation from the pineal vesicle. Sensory cells develop from precursor cells characterized by dark-appearing nuclei, measuring 5.6 μm in length and 3.7 μm in width, with strong affinity to azocarmin, and a cytoplasm which, with a number of stains, stains slightly darker than the so-called light cells. The light cells give rise to ependymal cells which differentiate into either supporting cells or pigment cells. A third type of cell, intermediate in appearance, may represent precursors of ependymal cells destined to develop into sensory cells. The precursors of sensory cells soon show an uneven distribution in the parietal vesicle, being scarce in its dorsal part, which gives rise to the lens. Signs of what has been interpreted as secretory processes are prominent in both the ventral (retinal) and the dorsal (lenticular) parts of the parietal vesicle. In the retina, apical extrusion is observed in precursors of pigment cells, whereas in the lenticular area, the so-called lens cells appear particularly active in secretion. Precursors of sensory cells retract from the basal lamina, giving rise to a slender basal process in both the retina and the lens; nerve cells develop in the basal pole of the retina and give rise to the parietal nerve. The further steps of differentiation take place after the separation of the parietal vesicle from the pineal vesicle.

In the retina the not yet fully differentiated precursors of sensory cells develop into typical sensory cells exhibiting outer and inner segments. A nervous network is formed by the basal processes of sensory cells and the processes of the above-mentioned nerve cells. The nervous network lies first outside the parietal eye; later it is incorporated into the retina. The sensory cells develop a longitudinally arranged striation in both the supranuclear cytoplasm and the inner segment. The outer segments, filamentous or vesicular in appearance, show either a hyaline content or a strong granulation apically. The differentiation of the pigment cells and the formation of pigment occurs after the incorporation of the nervous network into the retina, which is brought about by an outward migration of the nucleus-bearing part of the pigment cell.

The precursors of the cells forming the *lens* have not been unequivocally identified. TROST (1953a) claims that the lens cells are mainly modified pigment cells, whereas STEYN (1959a) studying *Cordylus polyzonus* and *Mabuya sulcata* is of the opinion that the lens cells are of mixed origin, stemming from sensory and pigment cell precursors. Relatively early the lens shows a biconvex shape (TROST, 1953a; STEYN, 1959a), due to a more rapid growth of the centrally lying lens cells (TROST, 1953a). Later the biconvexity disappears. A typical feature is that cells in the lens area become dislocated and escape into the lumen of the parietal vesicle, often accompanied by epithelial disruption. Subsequently the cells disintegrate, the nuclei becoming pyknotic. The lens cells develop filamentous processes protruding into the lumen and show signs of active secretion (TROST, 1953a), which is important for the development of the vitreous body or vitreous humour, at least in the early stages. Later, the sensory cells of the parietal retina take over and show distinct signs of apocrine secretion, as a result of which coagulated-appearing fibres (fibrous reticulum) occur which link retina and lens (TROST, 1953a; STEYN, 1959a).

The cornea of the parietal eye develops on the basis of a hyaline zone lying between the epidermis and a connective tissue sheath covering the lens. The sclera is formed from loose connective tissue lying in the immediate vicinity of the parietal eye (TROST, 1953a).

3. Ultrastructure and Histochemistry

a) Retina

The retina-like part of the parietal eye has been studied in a relatively large number of species so that we have a fairly clear picture of its structure. Two cell types have been regularly found, namely photoreceptor cells and supporting or pigment cells. Nerve cells have been infrequently noted.

Photoreceptor cells show the usual structure, consisting of outer and inner segments, cell body and basal processes. The outer segments show regularly arranged discs, similar to those of cones. Quantitative data are available for a few species only. In *Lacerta sicula campestris* (OKSCHE and KIRSCHSTEIN, 1968), outer segments measured up to 15 µm in length and comprised some 400 discs. In *Anguis fragilis* (PETIT, 1968) they were 10–17 µm long, containing 250–435 discs. Between 100 and 200 discs were found in *Sceloporus occidentalis* (EAKIN and WESTFALL, 1959, 1960). Whether or not all photoreceptor cells show such long outer segments is unknown. It is clear, though, that when compared with reptilian pineal photoreceptors, those of the parietal eye show a distinctly higher degree of sensory differentiation (*Lacerta sicula campestris, Iguana iguana, Anguis fragilis, Lacerta vivipara, Lacerta agilis:* OKSCHE and KIRSCHSTEIN, 1968). Outer segments with regularly arranged discs were found also in *Xantusia henshawi* (EAKIN and WESTFALL, 1960) and *Cordylus polyzonus* (STEYN, 1959b, 1960b). Shedding of discs and subsequent uptake by luminal macrophages has been described by PETIT (1968, *Anguis fragilis*). The discs of most photoreceptors are transformed into vesicles as a result of vitamin A deficiency (EAKIN, 1964b, *Sceloporus occidentalis*).

The other parts of the photoreceptor cells do not present any unusual features. An important point is the lack of dense-core vesicles or secretory granules, which are typically present in the secretory rudimentary photoreceptors of the reptilian pineal organ (p. 417). The basal processes of photoreceptor cells are tortuous and project into the fibrous layer of the retina where they form ribbon-bearing synapses with dendrites of nerve cells which ultrastructurally present no peculiarities (PETIT, 1968). Interestingly, ribbon-bearing synapses have been found between adjacent photoreceptor cells (PETIT, 1968). Occasionally the presence of pigment has been described in photoreceptor cells (EAKIN and WESTFALL, 1960).

The ultrastructure of the second retinal cell type, the *pigment cells,* has been best described in *Sceloporus* (EAKIN and WESTFALL, 1959, 1960), *Anguis fragilis* (PETIT, 1968) and *Cordylus polyzonus* (STEYN, 1960b). As the name implies, these slender cells extending from the base of the retina to the lumen of the parietal eye are heavily laden with pigment granules. The nucleus lies in the broad basal region, and the apical plasmalemma is studded with microvilli.

Histochemical studies do not point to the presence of secretory processes in the parietal eye. It does not contain histochemically demonstrable indoleamines and catecholamines (MEINIEL et al., 1973) and does not incorporate ^{3}H-5-hydroxytryptophan (MEINIEL et al., 1975). PAS-positive substances in photoreceptor cells (EAKIN et al., 1961, *Sceloporus occidentalis*) may be glycogen, as suggested by electron microscopy (PETIT, 1968, *Anguis fragilis*). RNA has been demonstrated in the paraboloid and perinuclear regions of photoreceptor, the RNA-positive regions showing metachromasia to azure B and toluidine blue (EAKIN et al., 1961). That the chrome alum haematoxylin-positive granules found in photoreceptor cells of *Cordylus* are secretory granules (STEYN, 1959a) requires confirmation by electron microscopy. Histochemical changes occurring after exposure of the animals to different conditions of illumination are described below (p. 434).

b) Lens

Little is known about the ultrastructure of the lens of the parietal eye. According to EAKIN and WESTFALL (1960), "the cells of the clear lens are elongate elements extending in parallel array from the lumen of the eye to the basement membrane of the cornea. The "... large ovoid nuclei usually lie in the basal halves of the cells. The basal (luminal) ends of the cells exhibit a labyrinth of very irregular microvilli which project into the cavity of the eye. In the midst of the labyrinth associated with each cell is a cilium..." Near the luminal surface of the lens indentations and many vesicles are seen. Small mitochondria are abundant between the vesicular border and the nucleus. Histochemical tests (EAKIN et al., 1961) revealed the presence of PAS-positive material in the lens which decreased in amount in over 70% of the light-adapted eyes. RNA was demonstrable in the luminal part of the lens cytoplasm which also showed metachromasia to azure B and toluidine blue. Immunohistochemical studies (McDEVITT, 1972) reveal the presence of lateral eye lens crystallins in the lens of the third eye.

c) Luminal Humour

Histochemical techniques (EAKIN et al., 1961) revealed the presence of glycogen in the lumen of the eye, particularly near the surface of the lens. In light-adapted eyes, PAS-positive material decreases in amount.

d) Capsule

Electron microscopic studies (EAKIN and WESTFALL, 1960) showed that the capsule of the parietal eye usually contains flat iridocytes covering the ventral and lateral surfaces of the eye. In the cytoplasm of the iridocytes rows of rectangular blocks, assumed to be guanine, may be seen. By means of oil red O and oil blue N, pyridine-soluble lipids could be demonstrated in the chromatophores of the eye capsule (EAKIN et al., 1961).

e) Cornea

The cornea is a transparent specialized region of the integument, characterized by absence of melanophores, close packing of connective tissue fibres of the innermost layer, absence of thick fibres in the middle spongy layer and a tendency of cells and filaments within them to be oriented vertically. Above the cornea lies the thin transparent interparietal scale which is secreted by both the cornea and the surrounding skin (EAKIN, 1973).

4. Nervous Connections

The very divergent results obtained with regard to the nervous connections of the parietal eye have been carefully reviewed by KAPPERS (1965a, b; 1967a). Uncertainties existed about whether a parietal nerve is present at all and whether it is related to the right or left habenular nuclei. In view of the regression of the parietal eye in certain lizards it is not suprising that a parietal nerve cannot be found in every lizard species (STEYN, 1957). However, the presence of parietal nerve has been clearly established in *Anolis carolinensis* (ORTMAN,

1960), *Sceloporus occidentalis* (EAKIN and STEBBINS, 1959), *Lacerta vivipara* (VON HAFFNER, 1953), *L. viridis* and *L. muralis hillfordi* (KAPPERS, 1967a). Electron microscopy revealed that in *Sceloporus occidentalis* the nerve consisted of about 250–300 non-myelinated nerve fibres surrounded by a thick perineurium (EAKIN and WESTFALL, 1960). OKSCHE and KIRSCHSTEIN (1968) counted roughly the same number of nerve fibres in *Lacerta sicula campestris* (in two nerves, 279 and 308), *Iguana iguana* (300) and *Lacerta vivipara* (two bundles of 200 and 100 fibres each, measuring 0.1–0.8 μm in thickness). By contrast in *Lacerta sicula campestris* the parietal nerve comprised 39 and 12 myelinated nerve fibres in two measurements. Two different groups of authors counted about 200 (OKSCHE and KIRSCHSTEIN, 1968) and up to 371 (PETIT, 1968) non-myelinated fibres in *Anguis fragilis*. Some of the axons comprised dense-core vesicles.

According to KAPPERS (1967a) the parietal nerve runs from the caudal part of the basal surface of the parietal eye in a ventral direction between the caudal wall of the dorsal sac and rostral wall of the epiphysis cerebri. It then turns gradually to the left, pierces the dorsal part of the habenular commissure somewhat to the left to the midplane and sends most of its fibres to the left lateral habenular nucleus. The course to the left habenular nucleus was also noted by ORTMAN (1960, *Anolis carolinensis*), VON HAFFNER (1953, *Lacerta vivipara*) and TROST (1954, *Hatteria*). In *Phrynocephalus undulatus* the parietal nerve branched, reaching both the right and the left habenular nuclei (KUMMER-TROST, 1956). Connections with the *right* habenular ganglion were thought to be typical of the lacertilian material studied by TROST (1954). VON HAFFNER (1953) described that part of the fibres of the parietal nerve penetrated the distal part of the pineal organ of *Lacerta vivipara*, where they disappeared among nerve cells.

5. The Function of the Parietal Eye

As can be expected, the presence of a third eye has prompted many speculations regarding its functional significance. STEYN (1961) hypothesized that a third eye may have had survival value in the first terrestrial vertebrates by alerting them to desiccation. From zoogeographical studies GUNDY et al. (1975) concluded that it facilitates survival at higher latitudes through reproductive or thermoregulatory modulation, thus making possible the exploitation of a wider variety of habitats. The formation and transport of antirachitic substances has been discussed by STEYN (1959b).

Electrophysiologically, it has been clearly established that, as the term parietal eye implies, the organ functions as an eye which discriminates between light of different wavelengths (MILLER and WOLBARSHT, 1962; SCHERER and DODT, 1967; DODT and SCHERER, 1968; HAMASAKI, 1968, 1969a, b; DODT, 1973; HAMASAKI and EDER, 1977). Combined pharmacological and electrophysiological studies suggest that the photosensitivity of the parietal eye depends on pinealofugal fibres innervating the parietal eye (ENGBRETSON and LENT, 1976).

There is also *morphological* evidence for an effect of light on the parietal eye. Histochemically, the retinas of illuminated eyes show reduced amounts of PAS-positive material and in the retinal melanocytes melanin pigment shifts lumenward after exposure to light for 48 h and retracts basally during dark adaptation (EAKIN et al., 1961). Electron microscopic studies carried out in *Anguis fragilis* (PETIT, 1968) did not reveal any clear-cut changes in pigment localization after exposure to continuous darkness, but the outer segments of photoreceptors show a marked vesiculation of the discs, which is absent after exposure to continuous illumination. In the end-feet of the basal processes the synaptic vesicles decrease in size after exposure to darkness and glycogen diminishes in amount.

A photoreceptive role of the parietal eye is also suggested by behavioural studies showing that, after removal or shielding of the parietal eye, lizards show an increased exposure to light (STEBBINS and EAKIN, 1958; STEBBINS and WILHOFT, 1966; STEBBINS, 1970; according to PACKARD and PACKARD, 1972, this effect is restricted to males). These findings pose the important question whether it is light or temperature that matters. When lizards have a choice of selecting areas of higher or lower temperatures, parietalectomized animals select areas of higher environmental temperature (HUTCHISON and KOSH, 1974; ROTH and RALPH, 1976). As both parietalectomized and sham-parietalectomized *Anolis carolinensis* are photophobic if the light source does not provide heat (ROTH and RALPH, 1977), it would appear that it is temperature that matters most. For the thermal tolerance of parietalectomized *Anolis* acclimatized at different temperatures and photoperiods, see KOSH and HUTCHISON (1972). FIRTH and HEATWOLE (1976) noted that in *Amphibolurus muricatus* parietalectomy abolished the day–night rhythm of panting and that it lowers the panting threshold under heating with a radiant source, whereas it elevated the threshold after heating with a non-radiant source; administration of melatonin lowered the threshold.

Locomotor activity was found to be increased after shielding of the parietal eye in *Xanthusia vigilis vigilis* (GLASER, 1958) and was unaffected in *Anolis carolinensis* after removal of the parietal eye (HUTCHISON and KOSH, 1974).

In contrast to other vertebrates, the effects of the parietal eye on endocrine organs and gonads have been little investigated. This is not surprising as there is no morphological and histochemical evidence for the occurrence of secretory processes in the parietal eye (p. 432). The interesting observation by QUAY (1965d) that the parietal eye of *Iguana iguana* contains large amounts of HIOMT activity has apparently not been followed up in other studies. The *thyroid* showed an increase in height of the follicular cells in some lizards species after parietalectomy (STEBBINS and EAKIN, 1958; EAKIN et al., 1959; STEBBINS and COHEN, 1973), but in another species, *Tropidurus albemarlensis*, there was neither hypertrophy of follicular cells nor an increase in radioiodine uptake (LOWENSTEIN and STEBBINS, 1969).

As regards the *gonads* it appears that they are not greatly affected by the parietal eye. Accelerated spermatogenesis after removal of the parietal eye has been reported for *Xanthusia vigilis* (STEBBINS, 1970), but in *Anolis* (LICHT and PEARSON, 1970) and *Sceloporus* (STEBBINS and COHEN, 1973) no such effect was noted. Perhaps part of the negative results are due to seasonal differences as shown by RIDGWAY and KENT (1971) who found that parietalectomy did not affect testis weight and the structure of the seminiferous tubules from December to March, but that an inhibitory effect was demonstrable from March to mid-June. Sex differences are indicated by the fact that in *Sceloporus* the sexual organs of males are unaffected by parietalectomy, whereas females had heavier ovaries and higher ova counts than controls (STEBBINS and COHEN, 1973).

A decrease of acetylserotonin methyltransferase activity in the brain after parietalectomy has been reported by WALKER and WILHOFT (1971).

It is apparent from these results that we are far from being able to formulate a unifying concept of parietal eye function. For the most thoroughly investigated species, *Sceloporus*, EAKIN (1970) has concluded "that the parietal eye ... exerts an inhibitory influence on the activity of the lizard in relation to daylight. It serves as a kind of brake, and together with other neural and hormonal mechanisms as a fine adjustment of the animal's circadian rhythm."

G. The Avian Pineal Organ

I. Gross Anatomy

In birds, the pineal system consists of the pineal organ proper only, in contrast to fishes, amphibians and reptiles. The usage of the term pineal *system*, as opposed to pineal *organ*, is justified in view of the fact that the pineal appears to be basically a bipartite structure consisting of the primary pineal organ and, in some 20% of the animals investigated, of secondary or accessory pineal tissue (QUAY and RENZONI, 1967).

1. Primary Pineal Organ

The primary pineal organ is very variable in size, shape and localization. The size ranges from atrophic pineals to well developed pineals extending from the intercommissural region to the roof of the skull. Species with atrophic pineals have been observed in two orders of birds only, viz. Procellariiformes and Strigiformes (RENZONI, 1968a; QUAY, 1968e, 1972c), which, according to QUAY (1972c), are the groups that are believed to have been primitively primarily nocturnal. Interestingly, within each of these groups species with the largest pineal are more diurnal and those with the most atrophic pineal are more nocturnal in habits. Detailed studies on pineal size in relation to body weight are apparently lacking. Some remarks on this aspect have been made in the reviews by QUAY (1965a) and RALPH (1970). The largest pineal measuring 17 mm in length and weighing 0.1 g after formalin fixation has been found in an emu, *Dromaeus novaehollandiae* (COBB and EDINGER, 1962). Especially large pineal organs have been found in *Struthio camelus*, *Rhea americana* and *Dinornis ingens* (STARCK, 1955). In the house sparrow, *Passer domesticus*, the interesting observation has been made that young, immature sparrows of less than 180 days of age usually have larger pineals than mature ones (RALPH and LANE, 1969).

Shape and localization of the primary pineal tissue are diagrammatically shown in Fig. 181. Detailed studies by QUAY and RENZONI (1967) using serial microscopic sections of the entire median region of the brain with the pineal in situ of 71 specimens representing 54 species, 31 families and 11 orders have clearly demonstrated the great variability of the avian pineal organ, somewhat reminiscent of the situation in rodents. As can be seen from Table 46, the most frequently observed type of pineal organ is one in which the organ consists of a prominent superficially localized distal part and a tapering, sometimes slender stalk which is proximally strongly reduced, consisting of scattered paren-

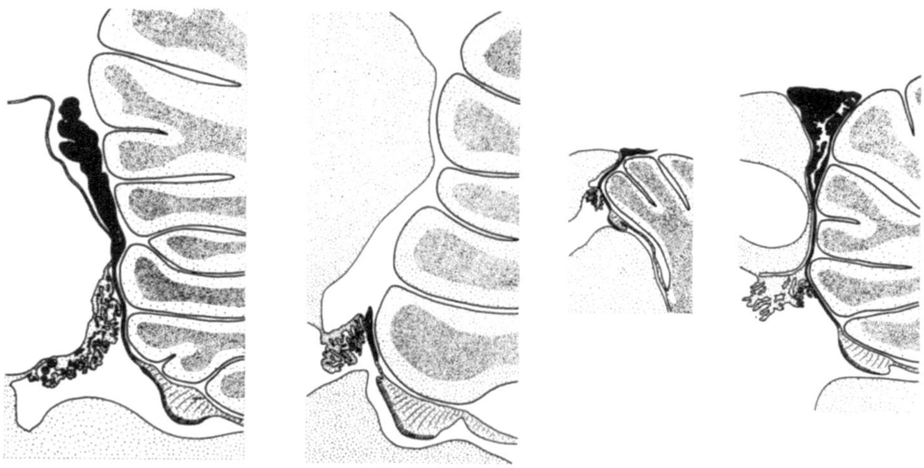

Fig. 181. Shape and localization of primary pineal tissue in a number of avian species. BREUCKER, 1967

Table 46. Forms of pineal organs in birds[a].

Type I: Prominent distal part superficially localized. Tapering parenchymal pineal stalk which is proximally reduced. Attachment of the stalk to the intercommissural region is questionable	
Anser albifrons	QUAY and RENZONI (1967)
Aythya americana	
Melanitta deglandi	
Calypte costae	
Cathartes aura	
Charadrius vociferus	
Capella gallinago	
Catoptrophorus semipalmatus	
Aphelocoma coerulescens	
Pica nuttallii	
Psaltriparus minimus	
Sitta canadensis	
Sialia mexicana	
Carpodacus mexicanus	
Spinus lawrencei	
Zonotrichia leucophrys	
Zonotrichia atricapilla	
Melospiza melodia	
Regulus calendula	
Agelaius phoeniceus	
Agelaius tricolor	
Euphagus cyanocephalus	
Pipilo erythrophthalmus	
Passerculus sandwichensis	
Junco oreganus	

Table 46 (continued)

Type II: As type I pineals, but stalk definitely not attached to intercommissural region	
Phalacrocorax penicillatus *Phalacrocorax pelagicus* *Buteo swainsonii* *Ereunetes mauri* *Salpinctes obsoletus* *Myadestes townsendi*	QUAY and RENZONI (1967)
Gallus domesticus *Meleagris gallopavo* *Anser domesticus* *Anas domestica* *Columbia domestica* *Turtur turtur* *Larus ridibundus* *Accipiter nisus* *Anser albifrons* *Ardea cinerea*	STAMMER (1961) (Remark: The material studied by Stammer shows pineals whose stalk is attached to the choroid plexus of third ventricle)
Type III: As type I, but stalk definitely attached to intercommissural region	
Gallus domesticus *Squatarola squatarola* *Erolia alpina* *Limosa fedoa* *Larus occidentalis* *Aethia cristatella* *Taeniopygia castanotis* *Padda oryzivora* *Spinus pinus* *Parus inornatus* *Dendroica auduboni* *Columba livia* *Tachycineta thalassina* *Certhia familiaris* *Anthus spinoletta* *Lanius ludovicianus* *Sturnella neglecta*	QUAY and RENZONI (1967)
Aptenodytes patagonica *Anthracothorax migricollis* *Turdus merula*	BREUCKER (1967)
Larus canus	WETZIG (1961)
Type IV: Rod-like pineal organ extending from the intercommissural region to the roof of the skull	
Podilymbus podiceps *Pelecanus occidentalis* *Falco sparverius* *Vireo huttoni*	QUAY and RENZONI (1967)
Type V: Pineal organ consists of two separate bodies, proximal and distal	
Oceanodroma homochroa	QUAY and RENZONI (1967)
Strix aluco	BREUCKER (1967)

[a] It should be noted that this classification is tentative, as the spatial relationships of the proximal part of the pineal stalk to the intercommissural region have not always been described in detail. Another complicating feature is that in some species the commissures are not seen (QUAY and RENZONI, 1967).

chymal cells or nerve fibres so that it is difficult to follow the stalk to its site of origin, the intercommissural region (IR). According to QUAY and RENZONI (1967) this type of pineal organ is typical of highly developed avian species. In a fair number of species the stalk is definitely not attached to the IR (type II), whereas in others this attachment is clearly demonstrable (type III). In some species, the pineal organ is more or less rod-like, extending from the IR to the roof of the skull (type IV). Very rarely, the pineal consists of two separate parts, proximal and distal (type V), or has an atrophic distal part (type VI).

2. Secondary or Accessory Pineal Tissue

A peculiarity of the avian pineal organ is the presence of accessory pineal tissue. According to QUAY and RENZONI (1967), it was found in about 20% of their material studied. In contrast to primary pineal tissue, attachment of accessory pineal tissue is *anterior* to the habenular commissure. Its position to the mid-sagittal plane of the brain varies: in 10 of the species possessing accessory tissue this lay on the left side, in six it followed the course of the primary pineal stalk and in one it lay on the right. The shape of the accessory tissue is variable, ranging from relatively small nodules, follicles or masses of probably little functional significance to larger masses containing active cells indistinguishable from those of the pineal organ proper. Accessory tissue has been thought to stem from what might be considered to be part of the dorsal sac or a region of the diencephalic roof anterior to the habenular commissure (QUAY and RENZONI, 1967).

3. Pinealectomy

Papers dealing with the technique of pinealectomy in birds have been published by STALSBERG (1965, 6-day-old chicken embryos) and KOBAYASHI (1968, small passerine birds).

II. Development

1. Organogenesis

Several studies have shown that the pineal organ of birds usually develops from a single small outpocketing of the area of the diencephalic roof that eventually comes to lie between the habenular and posterior commissures (domestic fowl: VIDMAR, 1953; CALVO and BOYA, 1978; *Phalacrocorax carbo:* CHARVÁT, 1954; *Spheniscus, Ardea, Larus, Phasianus, Pica, Parus, Hirundo, Charadrius:*

KRABBE, 1955; domestic fowl: SPIROFF, 1958; *Larus canus:* WETZIG, 1961; for older studies, see the publications by GLADSTONE and WAKELEY, 1935; BARGMANN, 1943; and RALPH, 1970). In the domestic fowl the first sign of a formation of a pineal out-pocketing was seen at 3 days of incubation (CALVO and BOYA, 1978). In Columbidae (RENZONI, 1970) several accessory pineal evaginations are regularly present, lying anterior to the primary evagination. In all species studied, the further development is characterized by elongation of the primary evagination. Fairly early, follicular structures appear in the distal part of the pineal tubule, either within the wall of the tubule or budding off from the pineal wall. SPIROFF (1958) mentions the development of primary, secondary and tertiary vesicles, whereas RENZONI (1970) refers to primary and secondary vesicles only. Opinions vary as to whether or not the central cavities of the follicular structures are in continuity with the primary pineal lumen. According to CALVO and BOYA (1978) there is no communication. Since the further development depends on the final histological pattern of the adult pineal, this aspect will be dealt with in the section on the histogenesis (see below). Here it suffices to add that due to the formation of the follicular structures in the distal part of the organ the club-like shape of the pineal may soon become conspicuous. The accessory evaginations found in Columbidae, especially their distal parts (RENZONI, 1970), appear to be incorporated either into the stalk or the distal part of the pineal organ. The proximal parts of the accessory evaginations remain separate and the middle parts disappear. In the swan, *Cygnus olor*, KRABBE (1955) described a detachment of pineal tissue from the distal part of the pineal organ and hypothesized that this "parietal corpuscle" corresponds to the parietal eye. A bilateral origin of the avian pineal organ has been suspected in earlier studies (see RALPH, 1970), but the recent publications do not support this notion. LANE et al. (1969) mentioned that they observed bilobated pineal organs in Japanese quail in a pilot study, but that later this finding could not be verified.

Changes in organ size occurring during development have been reported for a few species only. For *Phalacrocorax* (CHARVÁT, 1954) and *Larus canus* (WETZIG, 1961) data are available for embryos only, which show that the pineal increases steadily in size. In the domestic fowl stages of development have been observed when the increase in size is interrupted resulting in a decrease relative to previous stages (at 8, 14, 15 and 18 days of incubation; 1, 4 and 5 days post-hatching: SPIROFF, 1958; 18 days of incubation, 10 days after hatching: MOSZKOWSKA, 1958c). In *Passer domesticus*, planimetric studies have shown that the pineal bodies from birds 15–35 days of age are significantly larger than those from birds 165–185 days of age (RALPH and LANE, 1969).

2. Histogenesis

As mentioned above, the developing tubular pineal organ acquires a follicular or vesicular appearance both in vivo (SPIROFF, 1958; WETZIG, 1961) and in vitro (VIDMAR, 1953). Detailed information on the further histogenesis is available for the domestic fowl only. In this species, primary, secondary and tertiary vesicles develop, giving rise to a rather compact organ by the 16th day of incubation (SPIROFF, 1958; CAMPBELL and GIBSON, 1970). A thin connective tissue capsule becomes visible around the pineal body at 7 days of incubation; at 10 days of incubation connective tissue elements surround each vesicle and by 16 days denser strands of connective tissue delimit larger lobules (SPIROFF, 1958). According to CAMPBELL and GIBSON (1970), the percentage of vesicle area decreases between the 7th and 11th day of incubation, due to the occurrence of connective tissue strands; a rapid increase of vesicle area occurs thereafter, reaching a peak at the 17th day, followed by a sudden decrease between the 17th and 18th day and a slow decrease up to 7 days after hatching.

According to NISHIYAMA and MIKAMI (1974) pinealocytes and supporting cells are first distinguishable on the 8th day and glial cells on the 14th day of incubation. After hatching, the cavities of the follicles decrease in size, resulting in narrow clefts (BOYA and CALVO, 1978). Before and after hatching a distinct architecture of the lobules comprising a central follicle may be noted. SPIROFF (1958) described (a) radially arranged cells bordering the lumen of the follicle, which were termed ependymocytes and (b) peripheral to the ependymocytes, polygonal and loosely packed cells which were termed hypendymocytes. BOYA and CALVO (1978) and CALVO and BOYA (1978) termed these elements follicular and parafollicular cells, the parafollicular cells often forming inconspicuous rosettes. According to DIDIO et al. (1978), who studied the pineal of 5-day-old chicks, both the pseudostratified follicular layer (or subluminal layer, as these authors call it) and the diffuse basal layer consist of pinealocytes and some neuroglial cells. During further development the follicular arrangement becomes less conspicuous. It has been noted to disappear at 3–4 months after hatching (SHELLABARGER, 1953; SPIROFF, 1958), but in one study it was still evident after 4 months (BOYA and CALVO, 1978).

The histochemistry of the developing pineal gland has been studied by CAMPBELL and GIBSON (1970). Notable features are that from 15–17 days of incubation the lumina of the follicles contain an abundance of what has been termed secretory material which gives a positive reaction with the alcian blue, aldehyde fuchsin and PAS methods and stains metachromatically with toluidine blue. At the same time of development granules with similar staining properties (yet with negative reaction with alcian blue) are seen in the apical portions of pinealocytes. Furthermore, glycogen is present. Lipids are abundant on the 15th day of incubation; they rapidly decline thereafter. Both the chrome alum haematoxylin technique and the Gomori method for alkaline phosphatase failed to give a positive reaction. Pyroninophilia is moderate in intensity during the early stages of development and evenly distributed throughout the cytoplasm. From the 15th–17th day of incubation it is much stronger in the apical than the basal portions of the cells and on the 18th day and subsequent stages only the apical cytoplasm exhibits an appreciable reaction. Apically, the pyroninophilia is not appreciably influenced by ribonuclease treatment, pointing to the presence of pyroninophilic substances other than ribonucleic acid.

Sympathetic nerve fibres reach the pineal organ immediately before hatching, when they are seen in the capsule of the organ, from which they course through large and subsequently through smaller trabeculae of connective tissue.

At 3 days after hatching the axons begin to surround individual rosettes and follicles, and at 10 days the adult stage of innervation is accomplished (WIGHT, 1971; CALVO and BOYA, 1979b). According to CALVO and BOYA (1979b) the nerve fibres eventually innervating the pineal proceed from the perivascular fascicles which follow the posterior meningeal artery in its ascending course towards the pineal. After reaching the pineal stalk, most of the nerve fibres lie on the anterior surface of the pineal, subsequently reaching the superior surface. A smaller number leave the main path at the stalk level and head to the posterior surface.

Accumulations of *lymphocytes*, a typical feature of the avian pineal organ, were seen to develop as early as 4 days after hatching (SPIROFF, 1958), whereas in another study they were first seen between 10 and 15 days after hatching (BOYA and CALVO, 1978). According to the latter authors, the accumulations reach a maximum between 15 and 30 days after hatching, decreasing thereafter and still being present in 34-month-old chickens. CAMPBELL and GIBSON (1970) noted the occurrence of a lymph node within the connective tissue capsule along the posterior part of the gland at 1 month after hatching; this node consisted of a medulla and a cortical region with germinal centres and was highly vascularized; it regressed to a small patch of diffuse lymphatic tissue by 4 months after hatching and by 1 year it had disappeared completely.

A network of *capillaries* begins to enmesh the pineal tissue at 96 h of incubation. By the 6th day capillaries are seen between the developing pineal vesicles, and by the 12th day primary and secondary vesicles are encircled by capillaries (SPIROFF, 1958).

3. Ultrastructural Differentiation of Intrinsic Cells

The ultrastructural differentiation of the intrinsic cells has been carefully studied, one of the reasons being to establish whether or not the pinealocytes develop typical outer segments which later degenerate. According to OMURA (1977), distinction between pinealocytes (which have been termed secretory rudimentary photoreceptors by others, see p. 453) and supporting cells is possible in 10-day-old chicken embryos, when some of the pinealocytes develop apical swelling reminiscent of inner segments of photoreceptor cells. Mitochondria increase in number and accumulate in the supranuclear region of the cell. Occasionally, a ciliary protrusion extends from the apical swelling into the lumen of the follicles.

Already at this early developmental stage, dense-core vesicles measuring 80–120 nm in diameter (see also NISHIYAMA and MIKAMI, 1974) lie scattered near a well-developed Golgi apparatus, a finding which is of special interest in view of the alleged secretory nature of the avian pinealocytes. At 14 days of incubation, mitochondria are abundant in the apical swelling, which now resembles an inner segment more closely than before. Moreover, a basal process develops. Between 14 and 21 days of incubation (OKSCHE and VAUPEL-VON HARNACK, 1965c; OKSCHE, 1968; NISHIYAMA and MIKAMI, 1974; OMURA, 1977) some of the pinealocytes develop rudimentary outer segments with a few discs, but this process soon comes to an end. Other authors have likewise failed to demonstrate typical cone-like outer segments in avian embryos of different species (COLLIN, 1966a, b, c; BOYA and ZAMORANO, 1975). OMURA (1977) suspects that it is a lack of smooth endoplasmic reticulum in embryonic chicken pinealocytes and the disappearance of mitochondria from the inner segment after the 14th day of incubation that account for the failure of typical outer segments to develop. Other pinealocytes show microvilli or irregular cytoplasmic protrusions apically (OMURA, 1977). In 5-day-old chicks the basal processes of pinealocytes possess bulbous endings which lie in connective tissue around blood vessels; especially apically, the pinealocytes are equipped with zonulae occludentes, zonulae adherentes and maculae adherentes (DIDIO et al., 1978).

The developing supporting cells do not present any unusual features. According to OMURA (1977) they are rich in polysomes and rough endoplasmic reticulum, and possess a well-developed Golgi apparatus. In 5-day-old chicks, the supporting or neuroglial cells were noted to be richer in lysosomes than the pinealocytes (DIDIO et al., 1978). Degenerating pineal cells have been described by CALVO and BOYA (1979a).

III. Light Microscopy

Although the histological structure of the avian pineal organ is very variable, it is generally felt that three basic patterns exist, as described by STUDNIČKA (1905), namely (1) saccular pineal organs, (2) tubulofollicular pineal organs and (3) solid pineal organs, but with the understanding that intermediate forms are also present (QUAY, 1965a; OKSCHE, 1965a; RALPH, 1970; MENAKER and OKSCHE, 1974). When classifying avian pineal organs it should be borne in mind that the histological structure of the pineal organ changes with age, typical examples being the domestic fowl (SHELLABARGER, 1953; SPIROFF, 1958; BOYA and CALVO, 1978) and *Passer domesticus* (RALPH and LANE, 1969). Another complicating feature is that the histological structure may differ in a proximo-distal direction of the organ. In *Phalacrocorax pelagicus*, the distal part of the parenchyma is rather solid, whereas follicles and tubules are evident in the proximal part (QUAY, 1965a). In *Turdus merula* the proximal part of the pineal organ is saccular and the distal part follicular (BREUCKER, 1967).

Saccular pineal organs, typical of passeriform birds, have been regarded as primitive types (OKSCHE, 1965a; MENAKER and OKSCHE, 1974). *Tubulofollicular* organs are typical of pigeon and duck, whereas nearly *solid* pineal organs are seen in the adult domestic fowl (OKSCHE, 1965a; OKSCHE and MENAKER, 1974). For further details on the gross histological structure of adult avian pineal organs, see the publications by QUAY and RENZONI (1963, 1967), OKSCHE and VAUPEL-VON HARNACK (1966), RENZONI (1963, 1964a, b, 1965a, b, 1968a), BREUCKER (1967), STAMMER (1961), CHARVÁT (1954) and PIEZZI (1973), which also provide detailed information on the question as to whether or not the lumen of the pineal organ, if at all present, communicates with the third cerebral ventricle. Suffice it to mention that in a fair number of species the lumen of the pineal stalk opens more or less broadly into the third ventricle. In view of the frequent and variable closure of the lumen of the pineal stalk, QUAY and RENZONI (1967) think it unlikely that a release of secretory substances into the ventricle is an important pathway. They feel, rather, that the rich vasculature of the pineal body provides the most likely route for a secretory efflux.

Conventional histological staining reactions reveal few structural details of the intrinsic cells of the pineal organ which lend themselves to a meaningful functional interpretation. The most conspicuous cells are those forming the walls of pineal follicles. These cells may possess prominent apical protrusions or filamentous structures projecting into the lumen of the follicles (OKSCHE and VAUPEL-VON HARNACK, 1966). In addition to these large follicles, there are often small parenchymal areas in which cells are arranged in rosettes, with or without a tiny central lumen. These structures have been described under the term *cor secretorium* by BEATTIE and GLENNY (1966). Chrome alum haematoxylin staining reveals filamentous material in follicular cavities (pigeon: OKSCHE and VAUPEL-VON HARNACK, 1966), granule-containing parenchymal cells in the pineal stalk of *Passer domesticus* (RENZONI and QUAY, 1963) and cells exhibiting differences in stainability in the western sandpiper, *Ereuntes mauri* (QUAY, 1965a). In other species, the presence of ciliated cells (QUAY and RENZONI, 1967) and pigmented cells (WETZIG, 1961; CHARVÁT, 1954; QUAY and RENZONI, 1967) has been described. As regards the presence of nerve cells, see p. 458.

A typical feature of the avian pineal organ is the presence of lymphoid tissue. Its occurrence during ontogenesis has been described above (p. 443). Accumulations of lymphocytes, with or without a germinal centre, have been observed in domestic fowl (ROMIEU and JULLIEN, 1942; BARGMANN, 1943; QUAY, 1965a), western sandpipers, house sparrows (QUAY, 1965a) and *Turdus merula* (BREUCKER, 1967), in all of which they lie closely related to the capsule of the pineal organ or within connective tissue stroma. According to SPIROFF (1958) pineal lymphoid tissue shows a tendency for reduction beginning at the 4th month. QUAY (1965a) noted lymphoid tissue even in 1-year-old chickens.

IV. Histochemistry

The functionally most important histochemical findings in the avian pineal organ pertain to the demonstration of serotonin in parenchymal cells. Following biochemical studies (QUAY, 1965e, 1966b; HEDLUND and RALPH, 1967; HEDLUND et al., 1971), OKSCHE et al. (1969), UECK (1973), FUXE and LJUNGGREN (1965), JUILLARD and COLLIN (1975) and JUILLARD et al. (1977) using the Falck-Hillarp method and WIGHT and MACKENZIE (1970) applying a modification

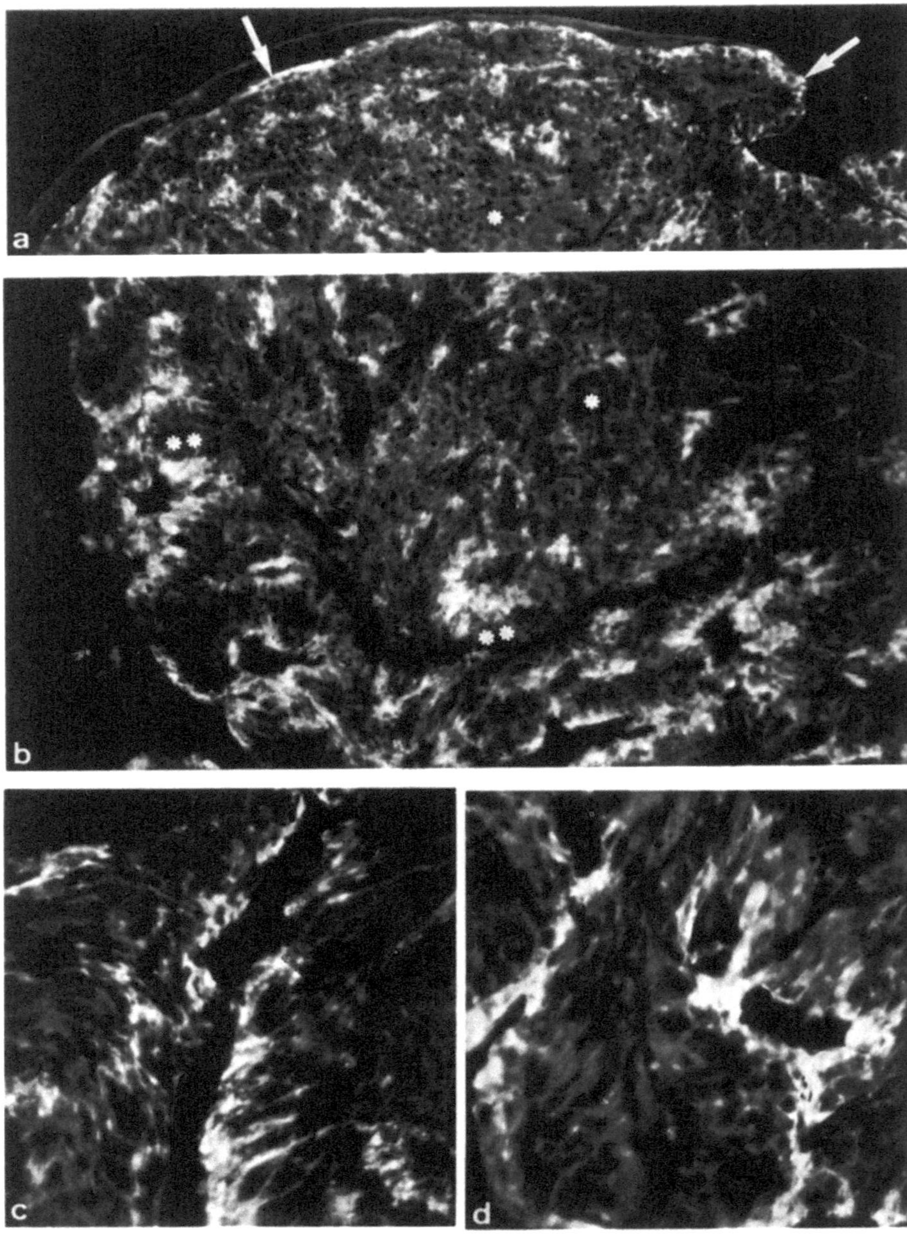

Fig. 182a–d. Histochemical demonstration of serotonin (Falck-Hillarp technique) in the pineal organ of *Columbia livia*. *, Diffuse yellow fluorescence of the parenchyma; **, intensely fluorescent, apical or basal parts of intrinsic cells; ↑, strong fluorescence of the surface of the organ. **a, b** × 112; **c, d** × 280. Oksche et al., 1969

of this technique demonstrated the presence of yellow serotonin fluorescence of varying intensities in all avian species investigated. In *Columbia livia* the strongest fluorescence intensity was noted in cells arranged in follicles, the interfollicular cells exhibiting a distinctly weaker reaction (OKSCHE et al., 1969) (Fig. 182). In the strongly fluorescent pinealocytes, the highest intensity was noted apically or basally (OKSCHE et al., 1969), the basal region showing fluorescing droplets measuring up to 10 μm in diameter (UECK, 1973). In *Melopsittacus undulatus*, on the other hand, the strongly fluorescent cells lacked specific activity in the apical regions (JUILLARD and COLLIN, 1975; JUILLARD et al., 1977). Both fluorescence (JUILLARD et al., 1977) and autoradiographic studies, the latter at the electron microscopic level (COLLIN et al., 1976), reveal that the supporting or interstitial cells are not involved in serotonin synthesis (Fig. 183). Using the ninhydrin reaction and a diazotization reaction the presence of biogenic amines was noted near sinuses and sometimes in the wall of smaller blood vessels (BEATTIE and GLENNY, 1966).

Lipids (THILLARD, 1968, *Gallus*) have been demonstrated in parenchymal cells arranged in follicles or tubules, where they lay close to the lumen. Between 5 weeks and 2 months of age the chemical properties of the stored lipid appear to change, as revealed by differences in staining properties. According to WIGHT and MACKENZIE (1971), both triglycerides and phospholipids are present in the pineal of the domestic fowl. The occurrence of cholesterol has been demonstrated by WIGHT and MACKENZIE (1971) and WELLS (1971).

PAS-positive, amylase-resistant granules lying in scattered or clustered cells near lumina or blood capillaries have been described in a number of avian species and are thought to represent possible candidates for secretory material (RENZONI and QUAY, 1963; QUAY, 1965a).

Metachromatically reacting substances were not demonstrable in the pineal of *Gallus* (BEATTIE and GLENNY, 1966).

Histochemical enzyme tests have only rarely been carried out. UECK and KOBAYASHI (1972a, b), applying an histochemical test for acetylcholinesterase activity, noted the presence of reaction product in cisternae of the endoplasmic reticulum, in the perinuclear space and over the plasmalemma of pinealocytes. The histochemistry of the *developing* pineal organ has been mentioned above (p. 442).

V. Ultrastructure of Intrinsic Cells

The parenchyma of the avian pineal organ comprises a number of cell types for which there is no standardized terminology (RALPH, 1970). Exactly how many different cell types there are has not been studied in detail. Apart from the infrequently occurring nerve cells (p. 458), perusal of the literature reveals that at least four structurally different cell types can be encountered in avian pineal organs, the first three of which may belong to the same cell

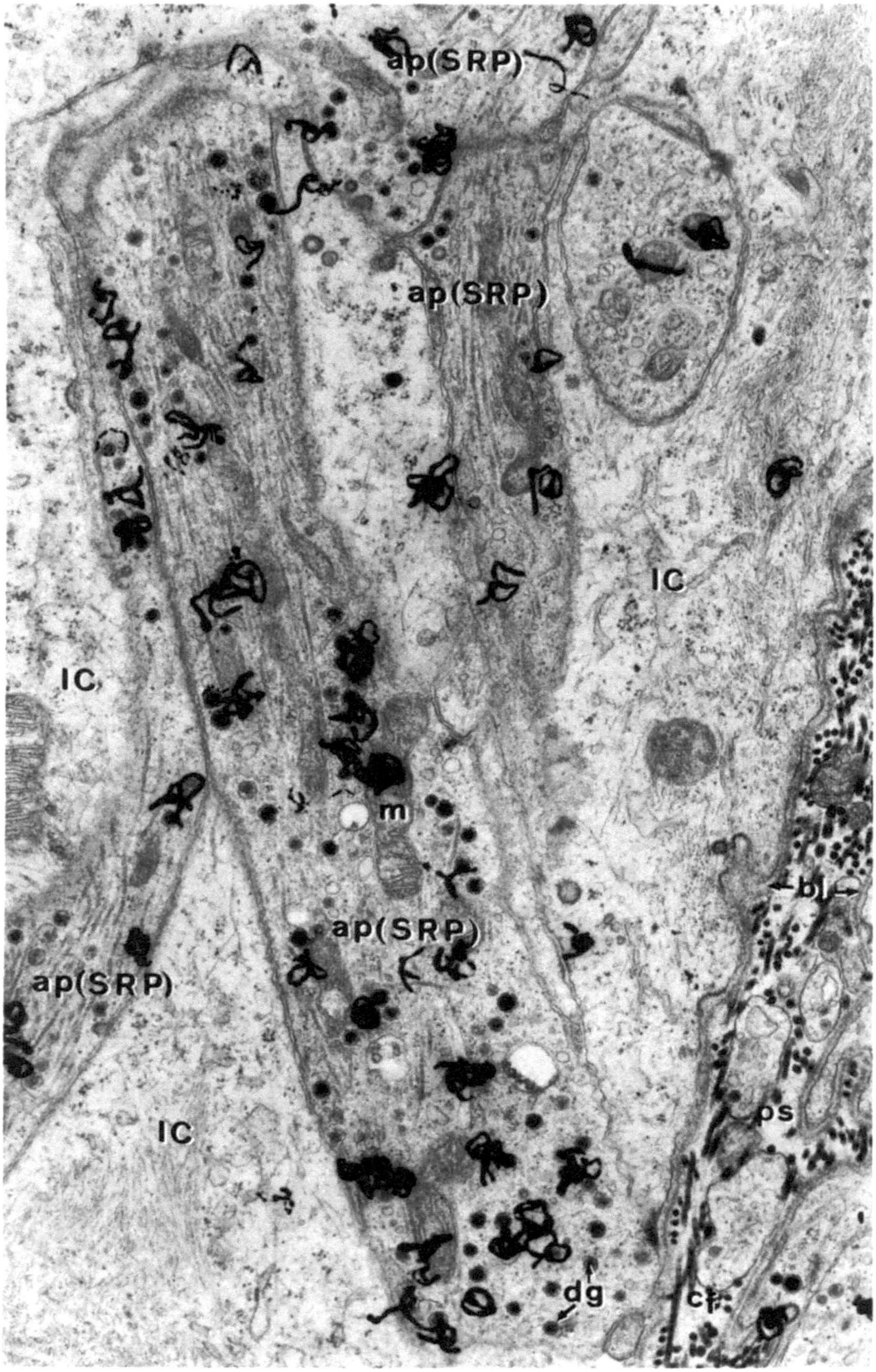

Fig. 183

line. It should be noted that not all of these cell types are necessarily present in every species.

1. Pinealocytes of the "Receptor" Type

Pinealocytes of the receptor type (see MENAKER and OKSCHE, 1974) are those intrinsic cells of the avian pineal organ which bear the greatest resemblance to pineal photoreceptor cells of lower vertebrates (Fig. 190). This type of pinealocyte appears to be restricted to saccular and follicular pineal organs (p. 444), where it lies adjacent to the lumina present. The main feature distinguishing it from the other intrinsic cells is the presence of a short outer segment comprising lamellar whorls (Fig. 184). The species in which these cells have been noted are listed in Table 47. Scanning electron microscopic studies in *Uroloncha domestica* (UECK et al., 1977) have revealed varying shapes of outer segments: bulbous, cup-like and elongate (Fig. 185). The internal structure of the outer segments and their relationships to the inner segments are not always easy to elucidate. The lamellar whorls may have a chaotic appearance or may lose their cytoplasmic contacts and degenerate (MENAKER and OKSCHE, 1974).

Following prolonged osmication the lamellar whorls show an intense staining reaction similar to that seen in photoreceptors of the lateral eyes (UECK, 1971b, 1973). The other parts of these slender pinealocytes of the receptor type appear to conform to the general pattern of photoreceptor cells (see literature mentioned in Table 47), as described above (p. 334). In the basal processes synaptic ribbons are clearly demonstrable (BISCHOFF, 1969; UECK, 1970c, d). The question of afferent nervous connections of these cells is discussed below (p. 458).

The well-established secretory function of the avian pineal organ (p. 463) and the presence in some intrinsic cells of secretory granules (see below) pose the question as to whether or not the pinealocytes of the receptor type are involved in secretory processes. This question cannot be easily answered as it is difficult to scan a complete cell right from the outer segment to the basal process and as the granules may not be present in every part of the cell. Data reported in the literature are at variance. In their review article MENAKER and OKSCHE (1974) do mention the presence of dense-core vesicles with diameters of 80–120 nm, but the further details reported suggest that these authors consider a type of secretory pinealocyte belonging to the category of pinealocytes of the receptor type. The presence of secretory granules has also been mentioned by other authors (FUJIE, 1968; QUAY et al., 1968b), but as they do not make a distinction between pinealocytes of the receptor type and other types of pinealocytes, it is not known whether the granules observed actually lie in cells of the receptor line. BISCHOFF (1969), on the other hand, makes a distinction

---

Fig. 183. Electron microscopic autoradiograph (^{3}H-5-hydroxytryptophan) showing distinct labelling of the asynaptic (*ap*) pedicles of what has been termed secretory rudimentary photoreceptor (*SRP*). Most of the silver grains are seen over mitochondria (*m*) and secretory granules (*dg*). Interstitial cells (*IC*) are virtually unlabelled. *bl*, basal lamina; *cf*, collagen fibres; *ps*, perivascular space. COLLIN et al., 1976; ×20,200

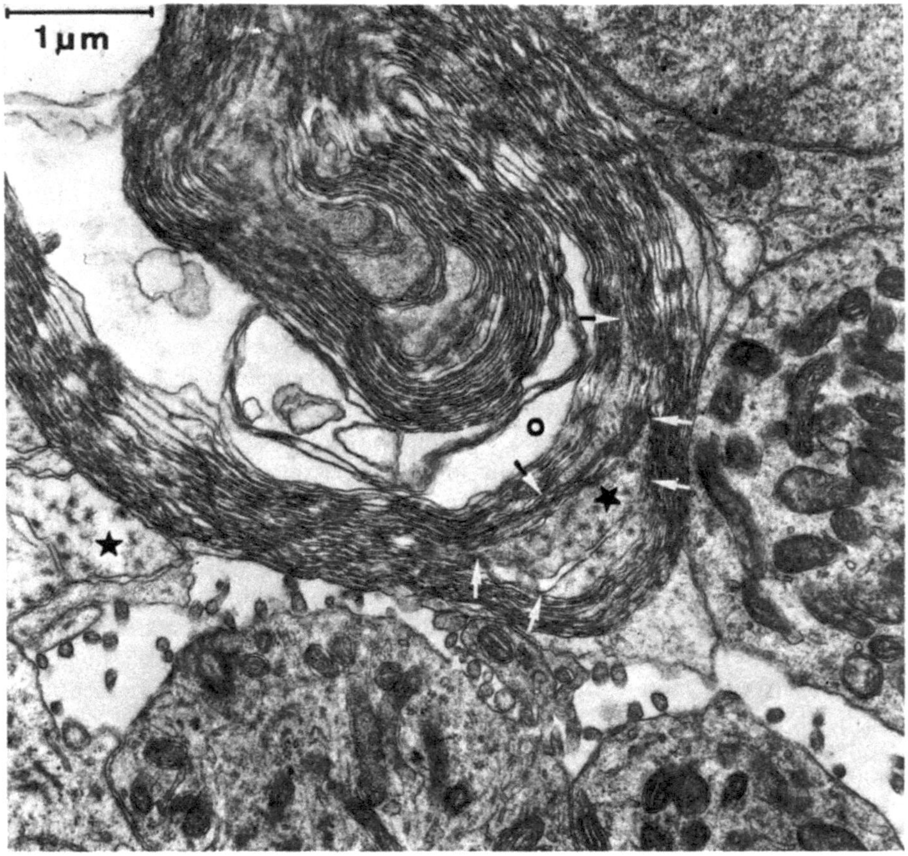

Fig. 184. Pinealocyte of the "receptor" type showing lamellar bodies (○) and bulbous cilia (*asterisk*). *Arrows* indicate sites where the marginal areas of the cilium merge with the lamellae. *Passer domesticus.* OKSCHE and KIRSCHSTEIN, 1969; ×19,600

between sensory cells and secretory cells and states that the category of secretory cells contains secretory granules. In *Columbia livia* (Fig. 186) dense-core vesicles have been seen in the immediate vicinity of lamellar whorls of outer segments (OKSCHE et al., 1969), which strongly suggests that the pinealocytes of the receptor type are involved in secretory processes.

Fig. 185a–d. Different types of outer segments in the pineal organ of *Uroloncha domestica*. **a** Bulbous structure. ×15,300. **b** Cup-shaped outer segment; *i*, invagination of cilium into outer segment; *t*, thickening of the rim of the outer segment; *m*, microvillus-like protrusion; *is*, inner segment. ×11,700. **c** Elongated outer segment showing signs of bifurcation (b_1, b_2); *r*, suture-like structure; *p*, finger-like process. ×12,600. **d** The same outer segment as shown in **c** after rotation and tilting of the specimen. ×15,300. UECK et al., 1977

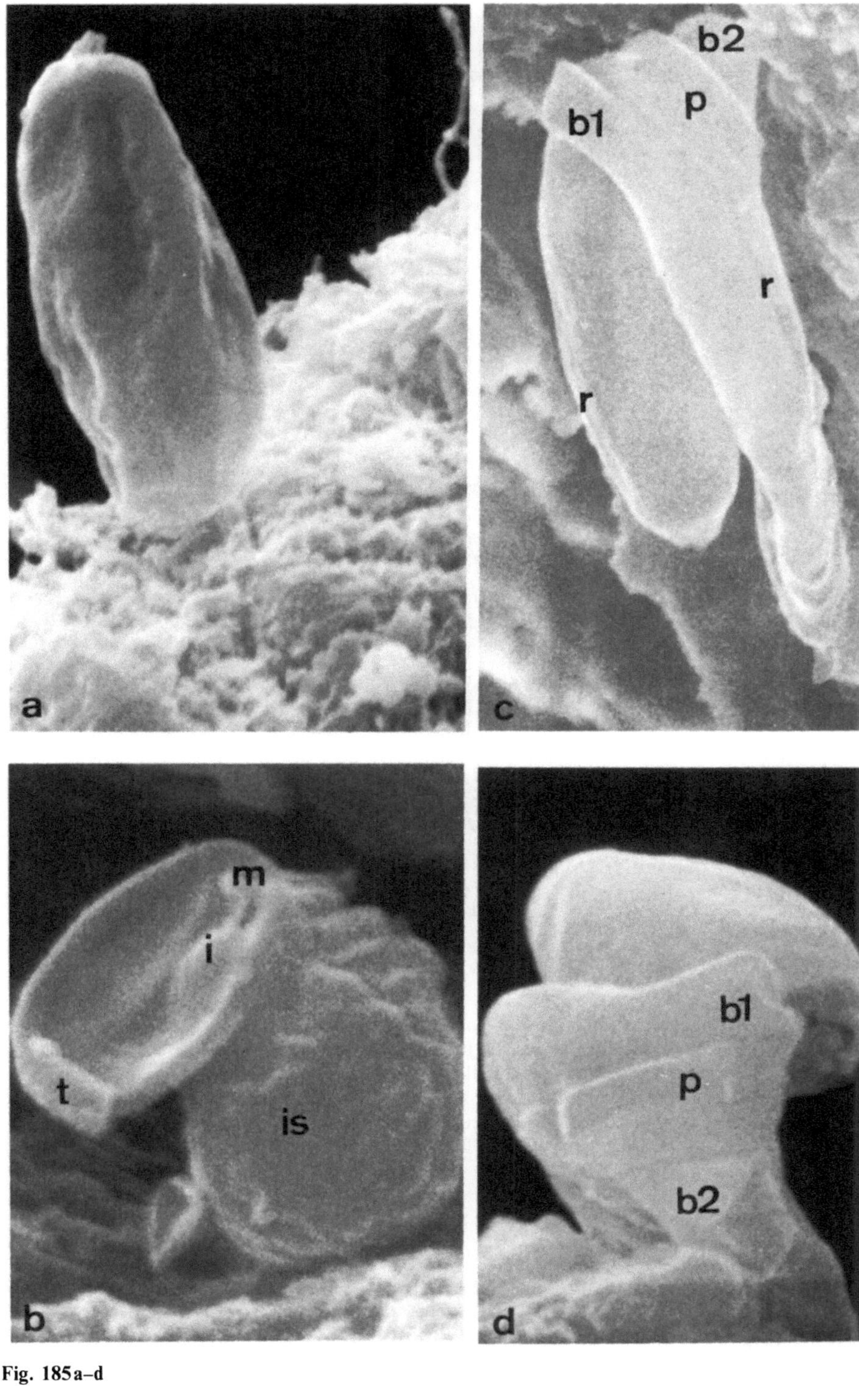

Fig. 185a–d

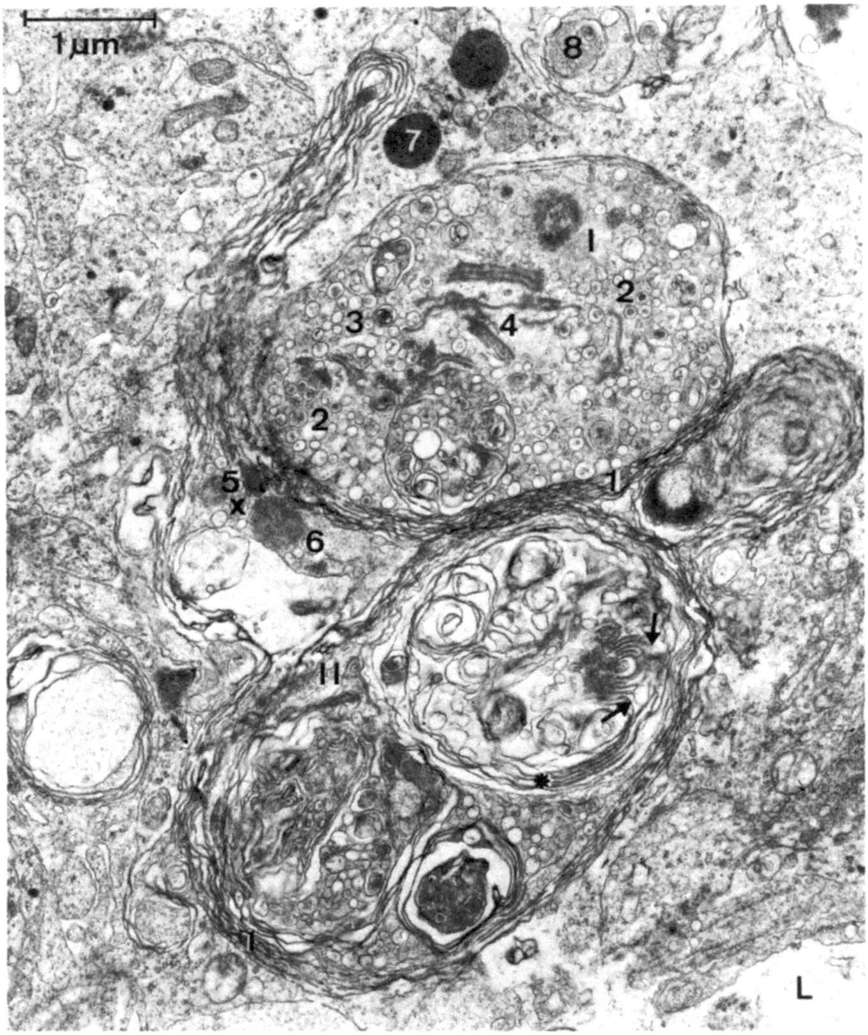

Fig. 186. Lamellar bodies (*I* and *II*) showing polymorphous internal structure. *Columba livia*. Lamellar body I: dense-core vesicle-like profiles (*2*); tubule cut obliquely (*3*); delicate tubules or fragments of lamellae, filled with electron-dense material (*4*). Lamellar body II: *asterisk*, groups of lamellae with granulated contents, aligned in the direction of the centre of spherical membrane complex (*arrow*). ×, Vicinity of a tubule filled with electron-dense material (*5*) and a cytosome (*6*). *7*, "dark" and *8*, "light" cytosomes. *L*, lumen. OKSCHE et al., 1969; × 17,000

2. Secretory Pinealocytes of the Receptor Type Lacking Membranous Whorls

This type of pinealocyte is very common in saccular and follicular pineal organs and has a localization similar to that of the aforementioned type, namely

Table 47. Avian species possessing pinealocytes of the receptor type[a]

Passer domesticus	OKSCHE and VAUPEL-VON HARNACK (1966), OKSCHE (1968), OKSCHE and KIRSCHSTEIN (1969)
Anas platyrhynchos	OKSCHE and VAUPEL-VON HARNACK (1966), OKSCHE (1968)
Columba livia	OKSCHE and VAUPEL-VON HARNACK (1966), OKSCHE (1968), OKSCHE et al. (1969), UECK (1973)
Gallus gallus[b]	GONZALES and HIDALGO (1966), FUJIE (1968), BISCHOFF (1969)
Coturnix coturnix japonica	BISCHOFF (1969)
Zonotrichia leucophrys gambelii[b]	OKSCHE (1968)
Melopsittacus undulatus	QUAY et al. (1968b)

[a] It should be noted that the cells of the sensory line of the avian species studied by COLLIN (1966a, b, c, 1967c, 1968d, 1969d, 1971) do not possess membranous whorls and hence, by definition, are not included in this category. These cells are classified as "pinealocytes of the receptor type lacking membranous whorls", in the present review, and are termed secretory rudimentary photoreceptors by COLLIN (1971, review).

[b] Absence of lamellar whorls in the above species reported for *Gallus gallus* by BOYA and ZAMORANO (1975), OKSCHE and VAUPEL-VON HARNACK (1966) and for *Zonotrichia leucophrys gambelii* by OKSCHE et al. (1972).

bordering pineal lumina. It is the cell that has been termed secretory rudimentary photoreceptor by COLLIN (1969b, 1971) and pinealocyte of the "rudimentary-receptor" type by OKSCHE et al. (1972). It has many features in common with typical photoreceptor cells. The outer segment, which by definition lacks membranous whorls, may consist of polymorphic cytoplasmic protrusions attached to the inner segment by a 9+0 cilium. Many profiles of these cells do not even show these cilia but just an accumulation of mitochondria in what corresponds to an inner segment (for literature, see Table 47 and COLLIN, 1966b, 1967c, 1969b; BISCHOFF, 1967; BISCHOFF and RICHTER, 1966). As the term secretory pinealocyte implies, this cell comprises what looks like secretory granules measuring between 50 and 350 nm in diameter (Table 48). The granules originate in the Golgi apparatus (COLLIN, 1967c; BISCHOFF, 1967, 1969; FUJIE, 1968; OKSCHE et al., 1969, 1972; JUILLARD and COLLIN, 1978) and have been shown to accumulate at different sites. They aggregate in both the cell apex and the basal processes in *Gallus* (BISCHOFF, 1969), *Columbia livia* (OKSCHE et al., 1969) and *Passer domesticus* (UECK, 1970d), whereas they restrict themselves to the cell body and the basal processes in *Melopsittacus* (QUAY et al., 1968b; COLLIN et al., 1976) and *Anas* (COLLIN et al., 1976). The amount of secretory granules ranges from small to abundant, at least in certain regions of the pinealocytes (Fig. 187). In *Columbia livia* (UECK, 1973), *Pica pica* and *Melopsittacus undulatus* (JUILLARD and COLLIN, 1978) relatively large numbers of secretory granules are demonstrable. Histochemical studies at the electron microscopic level reveal that the granules contain serotonin (perhaps also serotonin derivatives such as melatonin) and a core consisting of protein but lacking glycoproteins (JUILLARD and COLLIN, 1978; see also COLLIN et al., 1976).

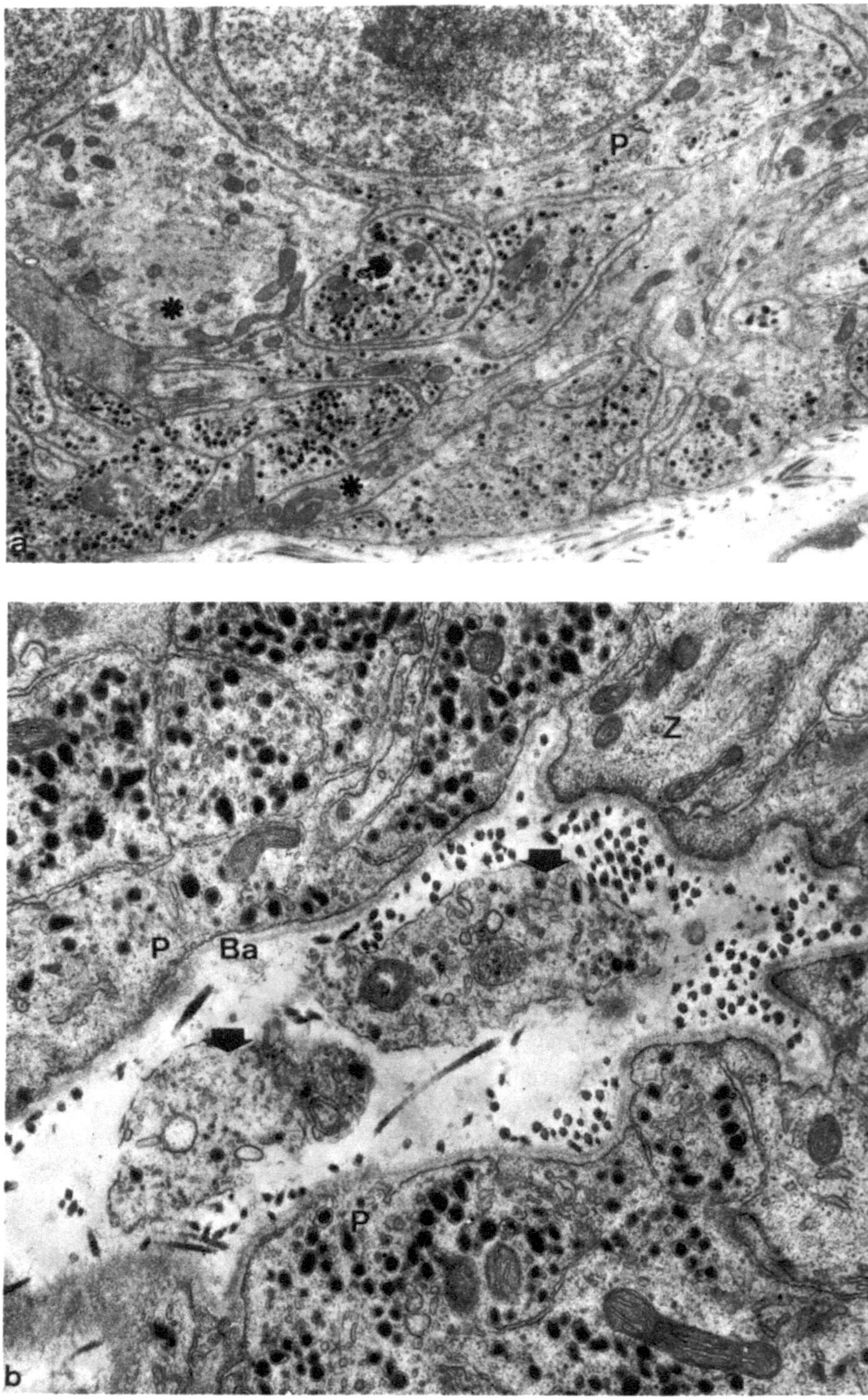

Fig. 187 a, b

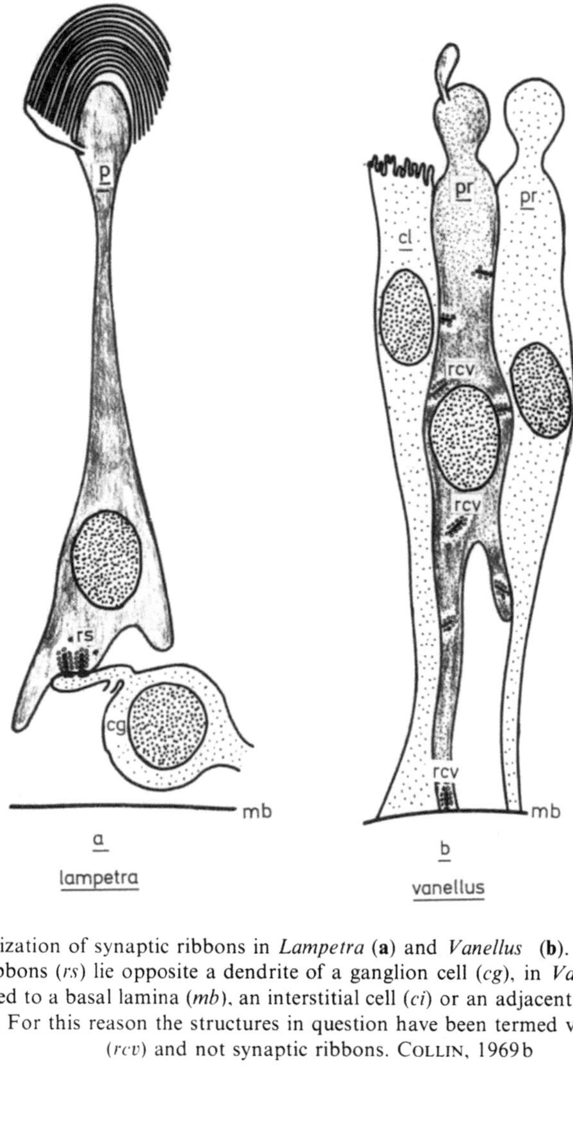

Fig. 188. Localization of synaptic ribbons in *Lampetra* (**a**) and *Vanellus* (**b**). Whereas in *Lampetra* the synaptic ribbons (*rs*) lie opposite a dendrite of a ganglion cell (*cg*), in *Vanellus* these structures lie closely related to a basal lamina (*mb*), an interstitial cell (*ci*) or an adjacent rudimentary photoreceptor cell (*pr*). For this reason the structures in question have been termed vesicle-crowned rodlets (*rcv*) and not synaptic ribbons. COLLIN, 1969b

---

Fig. 187a, b. Basal region of an epiphyseal follicle (*Columba livia*) showing a great number of processes containing dense-core vesicles of 100 nm in diameter. Note the presence of vesicle-free processes (*asterisk*). **a** *P*, Pinealocyte with dense-core vesicles in the perikaryon. ×9,200. **b** Some of the processes appear to pierce the basal lamina (*Ba*) and to lie freely in a perilobular space (*arrows*). *P*, pinealocyte processes broadly attached to the basal lamina; *Z*, process devoid of vesicles. ×23,000. UECK, 1973

Table 48. Presence and size of secretory granules in intrinsic cells of avian pineal organ

Species	Size (nm) or presence	Author(s)
Columba livia	90–120 100	OKSCHE et al. (1969) UECK (1973)
Passer domesticus	80–120	OKSCHE and KIRSCHSTEIN (1969), UECK (1970d)
Melopsittacus undulatus	50–120	QUAY et al. (1968b), COLLIN et al. (1976)
Vanellus vanellus	Present	COLLIN (1967c, 1968d, 1969b)
Sturnus vulgaris	Present	COLLIN (1969b)
Meleagris gallopavo	Present	COLLIN (1967c, 1969b)
Coturnix coturnix	Present	COLLIN (1967c, 1969b)
Pica pica	50–200	COLLIN (1969b), JUILLARD and COLLIN (1978)
Gallus gallus	200–350 200–300 70	BISCHOFF (1967) BISCHOFF (1969) FUJIE (1968)
Anas platyrhynchos	ca. 100	COLLIN et al. (1976)
Pygoscelis papua	300–400 and 50–80	PIEZZI (1973), PIEZZI and GUTIÉRREZ (1975)
Zonotrichia leucophrys gambelii	80–140	OKSCHE et al. (1972)

This type of pinealocyte has been clearly shown to comprise the so-called synaptic ribbons, e.g. in *Vanellus vanellus*, in which they are abundant, and *Coturnix coturnix japonica* (COLLIN, 1968d, 1969b). They are apparently absent in *Pica pica* (COLLIN 1969b, d). The localization of these structures in *Vanellus vanellus* is shown in Fig. 188, which led COLLIN (1969b) to suspect that the ribbons are not involved in processes of transmission, especially as an afferent nervous apparatus is apparently absent.

3. Pinealocytes Lacking Associations with a Lumen

Relatively little attention has been paid to pinealocytes which apparently do not have connections with a lumen of the pineal organ. They occur in the thicker portions of pineal organs of, for example, passerine birds (MENAKER and OKSCHE, 1974), and correspond to the parafollicular cells described by BOYA and ZAMORANO (1975) and by CALVO and BOYA (1979a). Unfortunately, no detailed description of these cells has been given, except in chickens between 2 and 5 days old (BOYA and ZAMORANO, 1975) and in embryos (CALVO and BOYA, 1979a). According to MENAKER and OKSCHE (1974) these cells produce dense-core vesicles in their Golgi zones and, citing unpublished observations by UECK, ZIMMERMANN and KOBAYASHI, appear under the picture of light and dark cells; whether these cells represent two different cell types or two different functional stages of one basic cell type could not be decided by means of morphometric studies after different pharmacological treatments. Apparently the intrinsic secretory cells of the penguin pineal also fall into this category of pinealocyte (PIEZZI, 1973; PIEZZI and GUTIÉRREZ, 1975). In this species, two types of secretory cells have been noted, the most frequent type containing secretory granules measuring 300–400 nm in diameter and the rare type comprising distinctly smaller granules with diameters of 50–80 nm. In addition, they comprise the usual cell organelles, characteristic features of the cell with the large granules being an abundance of cytolysosomes and phagolysosomes and the occasional presence of 9+2 cilia.

4. Supporting Cells

Supporting cells may be found in two sites, namely near pineal lumina and peripherally. The first type, ependyma-like, has been most thoroughly investigated. It is a slender cell, the apical pole of which is studded with microvilli (OKSCHE and VAUPEL-VON HARNACK, 1966; FUJIE, 1968; QUAY et al., 1968b; OKSCHE and KIRSCHSTEIN, 1969; PIEZZI and GUTIÉRREZ, 1975). Microvilli are few in number in Japanese quail and White Leghorn chickens, which instead possess numerous cilia of the 9+2 configuration (BISCHOFF, 1969). In adult penguins both microvilli and cilia have been reported as abundant (PIEZZI and GUTIÉRREZ, 1975). Mitochondria are common and irregularly shaped; vesicles of smooth and rough endoplasmic reticulum, polyribosomes and Golgi complexes are present in moderate amounts, whereas lysosomes are usually rare (QUAY et al., 1968b; BISCHOFF, 1969; FUJIE, 1968; BOYA and ZAMORANO, 1975). Microtubules and microfilaments may be numerous (see also OKSCHE et al., 1969). Clustering of mitochondria, which is typical of the pinealocytes, has not been observed. Crystalloid cytoplasmic inclusions are present in *Melopsittacus* (QUAY et al., 1968b). An interesting though functionally obscure observation is that after extended osmication the endoplasmic reticulum cisternae, perinuclear space and mitochondria show a strong staining reaction which is absent in these organelles in pinealocytes (UECK, 1973).

Whether the ependyma-like supporting cells have other than supportive functions is not known. QUAY et al. (1968b) thought that due to the frequent abundance of ribosomes and mitochondria and the large size of the nucleolus the cells are synthetically active and are not merely ependymal or supportive cells. The presence of an elaborate system of cisternae containing material of unspecified nature in the cell processes bordering small blood vessels was taken as a sign of a possible secretory activity. The cells do not appear to be involved in indoleamine synthesis, as shown by the fact after administration of ^{3}H-5-hydroxytryptophan or ^{3}H-5-serotonin and application of electron microscopic autoradiography the cells were not labelled (COLLIN et al., 1976).

Very little attention has been given to supporting or glial cells which do not abut pineal lumina. FUJIE (1968), studying the domestic chicken, identified glial cells in the peripheral part of pineal lobules, amounting to about 3%–5% of all cells of the pineal body. They are small in size, lie singly or in groups and often the cytoplasm contains a large number of filaments, especially in the periphery.

5. Intrinsic Cells In Vitro

Pineal organs from adult *Passer domesticus* cultured in vitro show intrinsic cells which are basically similar to those in vivo (MÖLLER, 1973, 1975, 1978).

VI. The Nervous Apparatus

It can be concluded from BARGMANN's (1943) monograph that earlier studies have contributed virtually nothing to knowledge of the innervation of the avian pineal. Since then, the presence of neuronal elements has been clearly established. Three components have to be considered: afferent fibres, efferent fibres and neurosecretory fibres stainable with chrome alum haematoxylin.

1. Afferent Nervous Connections

The trend towards a reduction of the afferent nervous connections of the pineal organ, as observed in reptiles, continues in birds. This reduction is understandable in view of the regression of photoreceptive elements in the avian pineal. Nevertheless, there is clear evidence for the presence of what can be regarded as afferent neurons. Conventional light and electron microscopic preparations do not provide a clear picture as to the presence or absence of nerve cells in the avian pineal organ.

It is therefore not surprising that in some studies (e.g. STAMMER, 1961; WETZIG, 1968) a lack of nerve cells has been reported. However, soon thereafter several papers were published in which the presence of nerve cells was mentioned (QUAY and RENZONI, 1963; RENZONI, 1965a; QUAY et al., 1968b; OKSCHE and KIRSCHSTEIN, 1969; UECK, 1970d). These findings were beautifully corroborated by applying Karnovsky's histochemical technique for acetylcholinesterase (UECK and KOBAYASHI, 1972a, b), which revealed great interspecies variation of the neuronal components. When interpreting the results obtained, it has to be taken into consideration that in addition to the afferent nervous elements, efferent fibres may also be stained.

According to UECK and KOBAYASHI (1972a, b), nerve cell bodies are apparently absent in *Coturnix coturnix japonica*. They are rare in *Columbia livia*, in which species the axons do not stain. Nerve cell bodies are more abundant in *Gallus domesticus* and *Excalfactoria chinensis*, and in these species enzyme

---

Fig. 189a–e. Acetylcholinesterase reaction in the pineal organ of *Passer domesticus*. **a** Longitudinal section through the pineal. Nerve cells and nerve fibres are scarce in the distal part of the organ; they show an increase in the intermediate part; in the stalk region a fairly compact fibre bundle together with some delicate fibres can be seen. *L*, pineal lumen. ×80. **b** Parasagittal section. In the proximal and intermediate parts, the course of the nerve fibres is predominantly directed

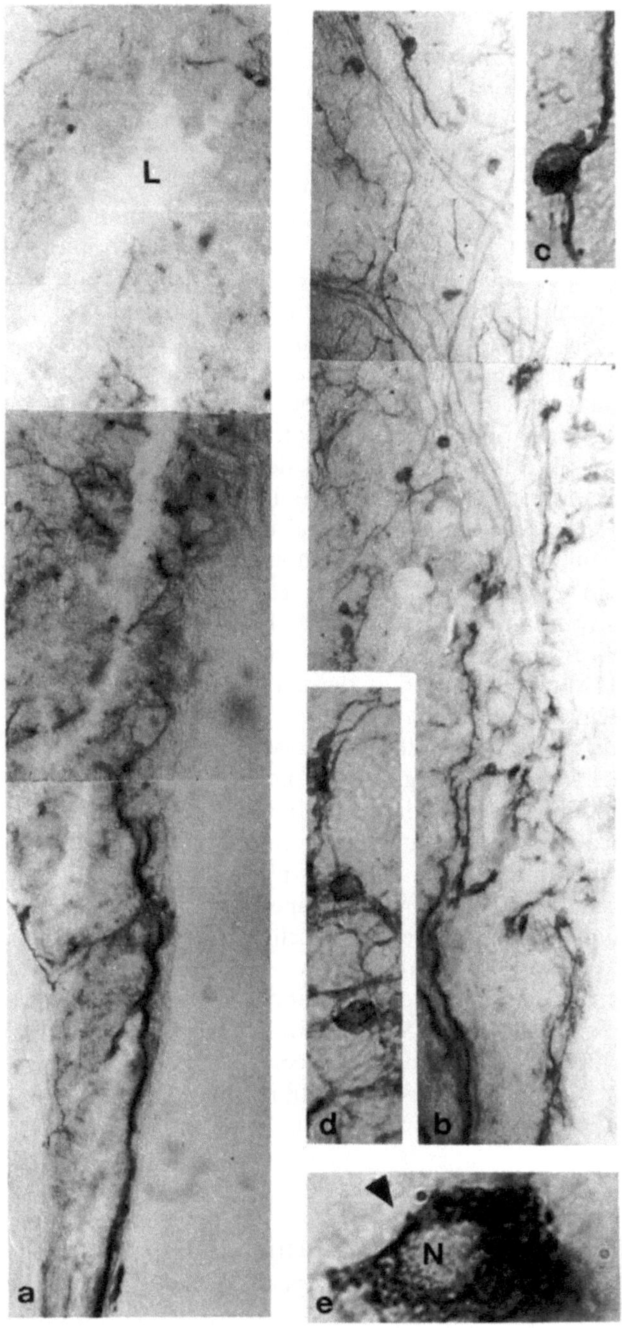

towards the brain. ×130. **c** Higher magnification of an area of **b**, showing bipolar nerve cell. ×280. **d** At the base of epiphyseal follicles cholinergic nerve cells and fibre networks are present. ×270. **e** High power light micrograph of a cholinergic nerve cell exhibiting granular reaction product. *N*, nucleus. ×1,120. UECK and KOBAYASHI, 1972a activity is not demonstrable in axons. Nerve cell bodies plus processes together with other nerve fibres are stained in *Passer domesticus, Amadina faciata alexander* and *Serinus canaria*. Nerve cell bodies without processes can be demonstrated together with stainable nerve fibres in the pineal stalk of *Melopsittacus undulatus*. In the Passeriformes investigated, the perikarya were restricted to the proximal part of the organ. Regional differences were also noted in *Melopsittacus undulatus* (QUAY et al., 1968b) in which large nerve cells and fibres were more abundant in the proximal than in the distal regions of the epiphysis.

The afferent innervation of the avian pineal organ has been most thoroughly investigated in *Passer domesticus*. In this species nerve cells are rare in the distal part of the pineal, but numerous in the middle part of the organ; the shape of these nerve cells cannot always be clearly identified, but often bipolar elements and occasionally pseudo-unipolar cells have been observed (Fig. 189) (UECK and KOBAYASHI, 1972a). The axons of these cells form bundles which course along the stalk of the pineal (UECK and KOBAYASHI, 1972a), in which prominent nerve fibre bundles had been demonstrated previously (QUAY and RENZONI, 1963; OKSCHE, 1968; OKSCHE and KIRSCHSTEIN, 1969; UECK, 1970d). Fibre bundles corresponding to the pineal tract in lower vertebrates and lying in the pineal stalk have also been demonstrated in *Columba livia* (OKSCHE et al., 1969) and *Zonotrichia leucophrys gambelii* (OKSCHE et al., 1972). The fibres are mostly non-myelinated (UECK, 1970d; OKSCHE et al., 1972). That most of the fibres are afferent has been clearly established by showing with the FALCK-HILLARP technique that the nerve fibre bundle in the pineal stalk is non-fluorescent (UECK, 1970d).

Although there can be no doubt that afferent neurons may be principally present in the avian pineal organ, it is not known whether this pathway is indeed a functioning one. As indicated above, there is little direct evidence for the presence of functioning synaptic contacts between the pinealocytes of the receptor type and afferent neurons (p. 449). In view of the fact that in *Passer domesticus* several hundreds of pinealocytes, though not necessarily of the receptor type, face a single nerve cell (UECK, 1970d), the afferent type of innervation, on morphological grounds, does not appear to play an important role (see also p. 462). The precise destination of the afferent fibres is not known. In *Pipilo erythrophthalmus* and *Sialia mexicana* most of the fibres can be traced into the anterior end of the *posterior* commissure, where they become lost in the laterally directed bundles; some fibres may fan off more dorsally into the *habenular* commissure (QUAY and RENZONI, 1963). A similar distribution with additional fibres disappearing in the intercommissural region has been noted in *Columba livia* (RENZONI, 1970). In *Regulus calendula*, by contrast, it appeared that most if not all of the tract fibres disappeared laterally at the level of the habenular commissure (QUAY and RENZONI, 1963). In *Passer domesticus* (UECK, 1971d) the fibres appear to merge directly with those of the habenular commissure. These variable results underline that more precise information on the central connections is very much needed (MENAKER and OKSCHE, 1974).

2. Efferent Autonomic Innervation

In contrast to the rather sparse afferent neurons, the avian pineal organ is richly innervated by autonomic fibres. While there can be no doubt that a large number of postganglionic *sympathetic* fibres are present, the presence of cholinergic fibres has posed a problem. According to WIGHT and MACKENZIE (1970) there is a dual innervation of the pineal body in *Gallus domesticus*, one fibre type being adrenergic and the other cholinergic. However, UECK and KOBAYASHI (1972a) suspect that, as in some mammals (p. 193), certain sympathetic fibres may display both the adrenergic and cholinergic type of reaction.

With the FALCK-HILLARP technique the presence and localization of the sympathetic fibres have been clearly established in a number of avian species. In several species the typical green fluorescence indicative of noradrenaline has been demonstrated in pineal nerve fibres (HEDLUND and NALBANDOV, 1969; HEDLUND, 1970; UECK, 1970c, 1971b, 1973; WIGHT and MACKENZIE, 1970). In some species, the autonomic fibres show a yellow *serotonin* fluorescence (e.g. *Carduelis chloris*: UECK, 1973). Microspectrofluorometrically, the presence of noradrenaline in the nerve fibres has been confirmed (UECK, 1973).

The *origin* of the sympathetic fibres from the superior cervical ganglia has been studied by HEDLUND and NALBANDOV (1969), who showed that the green fluorescent fibres disappear after ganglionectomy.

The *number* of intrapineal sympathetic fibres varies considerably. In chicken and jungle fowl, the pineal is densely innervated; fibres are rare in quail (HEDLUND and NALBANDOV, 1969; HEDLUND, 1970) and *Columba livia* (OKSCHE et al., 1969; UECK, 1973).

The intrapineal *localization* of the fibres is also very variable. An entirely perivascular localization is present in *Passer domesticus, Carduelis chloris, Pica pica, Corvus frugilegus, Gallus domesticus* (UECK, 1973) and *Melopsittacus undulatus* (JUILLARD et al., 1977). *Anas platyrhynchos* and *Excalfactoria chinensis* exhibit both a perivascular and parenchymal innervation (UECK, 1973). An entirely parenchymal innervation has been noted in *Columba livia* (UECK, 1973).

Ultrastructural studies have confirmed the presence of aminergic fibres containing the typical dense-core vesicles and electron-lucent vesicles (GONZALES and HIDALGO, 1966; FUJIE, 1968; BISCHOFF, 1969; COLLIN et al., 1976) measuring 80–120 nm and 50 nm in diameter respectively (UECK, 1970d). Occasionally a lack of vesicles has been reported (OKSCHE and KIRSCHSTEIN, 1969; OKSCHE et al., 1972). There are apparently no indications that the nerve fibres come into direct contact with parenchymal cells (UECK, 1970d; MENAKER and OKSCHE, 1974). The unmyelinated axons measure 0.1–0.35 μm in diameter in *Passer domesticus* (OKSCHE and KIRSCHSTEIN, 1969) and 0.4–1.0 μm in *Zonotrichia leucophrys gambelii* (OKSCHE et al., 1972). The functional significance of the sympathetic innervation will be discussed on p. 468.

3. Neurosecretory Neurons?

In the vicinity of the posterior and habenular commissures of a great number of avian species cells stainable with chrome alum haematoxylin-phloxine are embedded in pineal parenchyma; they were interpreted as neurosecretory cells (QUAY and RENZONI, 1963, 1966b). Cells stainable with aldehyde-fuchsin which are probably identical with the previously described cells have been mentioned by OKSCHE et al. (1972), who point out that the selectively staining material in the cell bodies often forms a loop, which is seen also in silver preparations and which appears to correspond to filament bundles that envelop the cell nucleus. The neurosecretory nature of the cells in question thus remains questionable.

VII. Vascularization

Blood supply and venous drainage of the avian pineal organ have been studied by STAMMER (1961), BAUMEL (1962) and BEATTIE and GLENNY (1966). The most detailed study is that by the latter group of authors who describe the presence of a rete pineale, which represents a complex network of vascular spaces and finer vessels which tend to divide the organ into lobes of varying size. Indian ink-injected pineal organs of the house sparrow have revealed a capillary network which is denser than in adjacent brain areas (OKSCHE et al., unpublished observations; see MENAKER and OKSCHE, 1974).

VIII. Functional Aspects

While there can be no doubt that the avian pineal organ is actively engaged in secretory processes, it is an open question whether or not the organ is directly photoreceptive. This latter aspect will be considered first.

1. Direct Photoreception?

As will be recalled, as yet there is insufficient morphological evidence available for the assumption that the avian pineal organ is *generally* directly photoreceptive: cells with outer segments are not uncommon, but the outer segments are rather rudimentary when compared to those of the lateral eyes and the pineal complex of lower vertebrates. Were these rudimentary cells indeed photoreceptive, then the question of their functional relevance would depend on the number of these cells present and, if a possible direct intrapineal action of the light perceived by these cells is excluded, on the presence of functioning afferent nervous connections. The rather fragmentary knowledge on both the synaptic contacts of the photoreceptor-like pinealocytes and the presence of afferent nervous connections at best allows one to suggest that on morphological grounds a direct photoreception by the avian pineal organ cannot be ruled out. Unfortunately the application of other techniques has not provided clear-cut answers either.

Electrophysiological studies which have so greatly contributed to the concept of a direct photosensitivity of the pineal complex in lower vertebrates (see above) suggest that the avian pineal organ is not directly light sensitive. In the first such studies (MORITA, 1966b; RALPH and DAWSON, 1968; OKSCHE et al., 1969) light had no recordable effect on the electrical activity of the pineal organ. In a more recent study (HERBUTÉ and BAYLÉ, 1974) light flashes were shown to have an effect (i.e. a marked inhibition of electrical discharges), but only when the optic nerves were intact, an observation which clearly speaks against a direct photoreception by the pineal organ, at least in the species investigated (male quail).

A direct effect of light on *explanted* pineal organs has been demonstrated in the duck by means of biochemical methods which show (a) that HIOMT activity is twice as high in pineal cultured for 24 h in continuous light than in darkness (ROSNER et al., 1971) and (b) that light stimulates the incorporation of ^{3}H-uridine into pineal RNA (ROSNER et al., 1972). An effect of light has also been demonstrated in cultured chick pineals: in this test system exposure to darkness for 6 h had a clear stimulatory effect on NAT (WAINWRIGHT and WAINWRIGHT, 1978).

Moreover, a number of different approaches have been used to study whether or not the avian pineal organ is directly light sensitive, viz. covering of the pineal area (MUNNS, 1970), blinding (LAUBER et al., 1968; BINKLEY et al., 1975), blinding combined with direct illumination of the pineal region (CUELLO et al., 1972; HISANO et al., 1972) and evaluation of photodependent processes after blinding and pinealectomy (GASTON and MENAKER, 1968; HARRISON and BECKER, 1969; MENAKER, 1971, 1972a, b; GASTON, 1971; HOMMA and SAKAKIBARA, 1971; MENAKER and UNDERWOOD, 1976), but the results of all of these studies cannot be taken as proof or disproof of a direct photoreception of the avian pineal organ, because there is convincing indirect evidence for the presence of extraretinal and extrapineal photoreception (see MENAKER and OKSCHE, 1974). Hence, direct photoreception by the avian pineal organ still awaits confirmation.

2. Secretion

The avian pineal organ, like that of other vertebrates, actively synthesizes indoleamines. Three substances in particular have been studied: melatonin, serotonin and 5-methoxytryptophol. In addition, 5-hydroxytryptophol (WURTMAN et al., 1968a), 5-hydroxyindole-3-acetic acid (QUAY, 1966b; BALEMANS et al., 1970) and 5-methoxyindole-3-acetic acid (BALEMANS et al., 1970) are present.

a) Basic Biochemistry of Pineal Indoleamines

The biosynthesis of melatonin has been described above (p. 5). The first substance in the chain of events, 5-hydroxytryptophan, could not be measured in the pigeon pineal organ, presumably due to the low levels present and its rapid decarboxylation to serotonin (QUAY, 1966b). The presence of serotonin has been demonstrated by a number of different techniques, viz. spectrofluorometry (QUAY, 1965e, 1966b; MEYER et al., 1973), bio-assay (HEDLUND and RALPH, 1967, 1968; HEDLUND et al., 1971), fluorescence histochemistry (FUXE and LJUNGGREN, 1965; OKSCHE et al., 1969; UECK, 1973; JUILLARD and COLLIN, 1975; JUILLARD et al., 1977) and microspectrography (UECK, 1973; JUILLARD et al., 1977). *N*-acetyltransferase (NAT) converts serotonin to *N*-acetylserotonin. This compound has been measured by means of radio-immunoassay in the chicken pineal, in which the daytime levels are slightly higher than those during night-time (PANG et al., 1977). Hydroxyindole-*O*-methyltransferase (HIOMT) which catalyses the last step of melatonin formation poses a problem in the avian pineal organ insofar as it O-methylates both serotonin and *N*-acetylserotonin, showing clear interspecies differences (AXELROD and LAUBER, 1968). It is therefore not surprising that HIOMT measurements have yielded equivocal results (see below). Melatonin has been demonstrated in both the pineal (VAN DE VEERDONK, 1967a; QUAY, 1966b; BALEMANS et al., 1970; and many others, see below) and blood (PELHAM et al., 1971, 1972; PANG et al., 1977), and 5-methoxytryptophol in the pineal by WURTMAN et al. (1968a) and BALEMANS et al. (1970).

The avian pineal organ also contains high activities of histamine-*N*-methyltransferase, catechol-*O*-methyltransferase and phenyl-ethanolamine-*N*-methyltransferase (AXELROD and LAUBER, 1968).

b) Melatonin-Forming Capacity of the Avian Pineal Organ Compared to that in Other Vertebrates

The avian pineal organ appears to have the highest melatonin-forming capacity of all vertebrates when assessed by HIOMT (AXELROD et al., 1964; QUAY, 1965d; BINKLEY et al., 1973c; BINKLEY, 1976a) and NAT measurements (BINKLEY, 1976b). While HIOMT activities are up to several hundred times higher than in other species, the actual melatonin content of the pineal is only slightly higher (LYNCH, 1971; PANG et al., 1977) or even lower, depending on the species studied and the time of sampling (LYNCH, 1971; RALPH, 1976). In the chicken serum melatonin levels during the night amounted to 0.34 ng/ml (PANG et al., 1977).

c) Extrapineal Sites of Melatonin Formation

As in other vertebrates, the pineal organ is not the only site of melatonin formation. HIOMT activity is also demonstrable in the retina (QUAY, 1965d), as are melatonin and *N*-acetylserotonin (PANG et al., 1977). When compared with the melatonin-forming activity of the pineal organ, the capacity of the retina is relatively low. After pinealectomy, melatonin was no longer demonstrable in blood (PELHAM et al., 1972) and brain tissue (PANG et al., 1974), but the very sensitive melatonin RIA's have yet to be applied to pinealectomized birds.

d) Mode of Secretion of Pineal Melatonin

The problem as to whether the pineal organ releases melatonin into the CSF or into the systemic circulation has been studied by PANG and RALPH (1975; see also RALPH, 1976). These authors conclude that melatonin is not secreted into the CSF, because when the pineal organ is transplanted into the anterior chamber of the eye, the brain-serum ratios of melatonin are identical to those of controls having the pineal in situ.

e) Circadian Changes of Indoleamines

As in mammals, the synthesis of indoleamines undergoes striking day–night changes in the avian pineal organ. In contrast to mammals, some techniques (e.g. HIOMT measurements) do not accurately reflect the melatonin-forming capacity of the pineal organ in birds. Hence the results obtained with this technique will be mentioned last.

It is now well established that melatonin formation is up to ten fold lower during the day than during the night (RALPH et al., 1967, 1974, 1975; LYNCH and RALPH, 1970; PELHAM and RALPH, 1972b; PELHAM et al., 1972; BINKLEY et al., 1973c, 1975; PANG et al., 1974, 1977; PELHAM, 1975; RALPH, 1976), except in Antarctic penguins (BENELBAZ et al., 1976).

These data agree well with NAT measurements (BÄCKSTRÖM et al., 1972; BINKLEY, 1976a, b; BINKLEY et al., 1973c, 1974, 1977a; RALPH et al., 1975; PRESLOCK, 1975, 1976) yielding values which are 27-fold higher during the night than during the day (BINKLEY et al., 1973c). For further details on pineal NAT, see p. 469.

For reasons discussed below (p. 469), HIOMT activity is not suited to detect regularly day–night differences in pineal function. In several studies no statistically significant differences were noted between day and night HIOMT levels (LYNCH and RALPH, 1970; BÄCKSTRÖM et al., 1972; BINKLEY et al., 1973c; BINKLEY and GELLER, 1975a; BENELBAZ et al., 1976), whereas in others an inverse relationship to NAT and melatonin was found (AXELROD et al., 1964; WINGET et al., 1967). Occasionally, the values obtained were higher during the night than during the day (SAYLOR and WOLFSON, 1969, quail, 46 and 48 days of age; PELHAM and RALPH, 1972a).

Pineal *serotonin* usually shows peak levels early in the morning (QUAY, 1966b; HEDLUND and RALPH, 1967, 1968; HEDLUND et al., 1971; STURKIE and MEYER, 1972; MEYER et al., 1973).

Pineal ^{32}P uptake shows a maximum in the light phase and a minimum in the dark (SACKMAN, 1977).

f) Seasonal Changes

Seasonal changes of pineal activity have been little investigated. In one such study, free-living male house sparrows showed a striking increase of HIOMT activity between August and September, high activities up to February/March and low values from April to July (BARFUSS and ELLIS, 1971).

g) Sex Differences

In the present review no attempt has been made to scan the literature with respect to sex differences of pineal secretory activity. BÄCKSTRÖM et al. (1972) found that NAT and HIOMT activities of the female quail were always higher than in males.

h) Onset of Secretion During Ontogenesis

The actual onset of melatonin secretion during ontogenesis has not been assessed. However, enzyme determinations have been carried out which give some insight into this problem. The development of HIOMT activity has been studied by WAINWRIGHT (1974), who demonstrated traces of enzyme activity as early as the 10th day of incubation; levels of activity rose rapidly as development continued and from the graph published it would appear that enzyme levels typical of adults are reached between 3 and 4 weeks after hatching. Exposure to continuous illumination speeds up the increase of HIOMT activity.

A comparison between HIOMT and NAT development in chicks has been carried out by BINKLEY and GELLER (1975a), who basically confirmed the HIOMT findings mentioned above and showed that throughout development no statistically significant day–night differences of HIOMT activity existed. NAT activity, by contrast, showed higher night values as early as 2 days after hatching. Moreover, the interesting observation was made that daytime levels of NAT activity did not increase between hatching and 56 days of age, whereas dark-time activity did increase, reaching adult levels at 18 days after hatching.

During phases of sexual maturation, HIOMT activity involved in melatonin formation decreases (ALEXANDER et al., 1970b; PRESLOCK, 1974), whereas HIOMT activity giving rise to 5-methoxytryptamine does not change significantly. NAT activity, by contrast, has been shown to *increase* during sexual maturation (PRESLOCK, 1974).

3. Factors Influencing Pineal Structure and Function

a) The Effects of Light and Darkness

Exposure to continuous illumination or continuous darkness as well as short-term changes in environmental lighting conditions have frequently been used to study the function of the avian pineal organ. Morphological studies, by contrast, have infrequently been carried out. Exposure to continuous illumination has been reported to decrease the size of the intrinsic pineal cells and to diminish the number of mitochondria (S. MILCOU and POSTELNICOU, 1964). THILLARD (1968) observed an increase in the amount of sudanophilic material in cells forming follicles in constant light and no change in constant darkness. Pineal weight has been found to increase under continuous light and to decrease under constant darkness (AXELROD et al., 1964). In light-deprived chickens (MCFARLAND et al., 1969) the pineal cells decreased in volume, had fewer cytoplasmic lipid granules and showed a complete loss of the interlobular sudanophilic material.

The effects of long-term exposure to constant light and dark on *biochemical* parameters are listed in Table 49. It can be seen that under constant light the circadian activity cycles are usually suppressed, HIOMT activity showing an increase of activity, whereas under constant dark the circadian changes persist, however with a lowering of the amplitude.

Short-term changes in environmental lighting conditions may be very effective in altering the secretory activity of the avian pineal organ, provided that the changes occur at the right moment of the circadian activity cycle. During daytime, exposure to darkness has no effect on pineal melatonin levels and NAT activity (BINKLEY et al., 1975). However, continuation of light into the evening inhibits the usual nocturnal decline of serotonin (QUAY, 1965e, 1966b) and the striking increase of NAT activity does not occur (BINKLEY et al., 1975). By contrast, when chickens are transferred from dark to light during the night, a rapid decrease of both melatonin content and NAT activity is seen (BINKLEY et al., 1975). This decrease is more pronounced during the first half of the night than during the second (BINKLEY et al., 1977a). Shortening of the dark period from 16 to 8 h leads to a disappearance of ultradian sub-peaks of NAT activity occurring during night-time (BINKLEY et al., 1977a). Since cultured chick pineal organs dissected out at different time points of the night all show a decline of NAT levels to daytime levels at precisely the same time, this has been thought to indicate a timekeeping ability of the avian pineal organ (BINKLEY et al., 1977b). Moreover, it has been shown that when chick pineal glands are explanted at the mid-light phase of a diurnal cycle of illumination and incubated in the dark, they develop marked increases in NAT activity, whereas after exposure to continuous illumination this increase does not occur (WAINWRIGHT and WAINWRIGHT, 1978).

Which wavelengths are most effective in influencing the secretory activity of the avian pineal organ has apparently not been studied in great detail. In a short abstract, MACHEMER et al. (1971a) reported that in the cockerel pineal HIOMT activity was lowest in blue light, intermediate in red light and highest in green and white lights.

b) Other External Factors

Noise does not depress NAT activity and melatonin levels during darkness (BINKLEY et al., 1975). Swimming stress, cold stress and immobilization also have no effect, but alcohol introduced into the alimentary canal inhibits the

Table 49. Effects of constant light and constant darkness on biochemical parameters of avian pineal organ

Parameter studied	Effect noted	Author(s)
Constant light:		
^{32}P uptake	Circadian rhythm disappears	SACKMAN (1977a)
Melatonin	Circadian cycle suppressed	RALPH et al. (1971)
HIOMT	Increase of activity	OISHI and LAUBER (1973), AXELROD et al. (1964), MUNNS (1970)
	No change	ALEXANDER et al. (1970c)
Monoamine oxidase activity	Not affected	AXELROD et al. (1964)
Constant darkness:		
^{32}P uptake	Lowering of amplitude of daily variations	SACKMAN (1977a)
Melatonin	Persistence of daily changes with lowering of amplitude	RALPH et al. (1971, 1974, 1975)
HIOMT	No change and decrease respectively	RALPH et al. (1975)
	Lower activity	OISHI and LAUBER (1973), AXELROD et al. (1964), ALEXANDER et al. (1970c)
NAT	Persistence of daily changes	RALPH et al. (1975), BINKLEY and GELLER (1975b)
5-Hydroxytryptophan decarboxylase	Increase of activity	ALEXANDER et al. (1970c)
Monoamine oxidase activity	Not affected	AXELROD et al. (1964)

dark-time rise in NAT activity (BINKLEY et al., 1978a). The effects of other chemical substances will be mentioned below.

4. Nervous Pathways and Chemical Substances Regulating Pineal Function in Birds as Compared with Mammals

As will be recalled, in the mammalian pineal organ melatonin formation is regulated by sympathetic nerve fibres which interact with β-adrenergic receptors. In the avian pineal organ also, sympathetic nerve fibres coming from the superior cervical ganglia are present (p. 461), but their role in regulating melatonin formation is enigmatic.

In contrast to mammals, sympathetic denervation by means of removal of the superior cervical ganglia does not abolish the effects of light and darkness on melatonin formation (LAUBER et al., 1968; MACBRIDE et al., 1973; RALPH et al., 1975). Likewise, electrophysiological studies have shown that ganglionectomy does not suppress the inhibitory influence of flashlight stimulations upon pineal rate of multi-unit discharges (HERBUTÉ and BAYLÉ, 1976). However, as pineal multi-unit response to flashlight is suppressed by habenular lesions (HERBUTÉ and BAYLÉ, 1977), it would appear that the pineal tract characterized as afferent (p. 458) may contain efferent fibres. In this context it is interesting to note that granulated nerve fibres of unknown origin have occasionally been noted in the intramural portion of the pineal stalk (MENAKER and OKSCHE, 1974).

Unlike in mammals, melatonin formation in the avian pineal organ is not regulated by β-adrenergic mechanisms, as shown by the fact that noradrenaline and isoproterenol do not increase pineal NAT and melatonin levels (BINKLEY et al., 1975, 1978a; BINKLEY, 1976a, b; WAINWRIGHT and WAINWRIGHT, 1978) and that the β-adrenergic blocking agent propranolol *stimulates* NAT activity (BINKLEY et al., 1978a). Administration of dibutyryl cAMP or cAMP has little or no effect on pineal NAT activity (WAINWRIGHT and WAINWRIGHT, 1978). According to BINKLEY et al. (1978a) administration of cycloheximide is inhibitory to NAT activity throughout the dark-time; dark-time rise of NAT activity is inhibited by isoproterenol, adrenaline, L-dopa, ethanol, pargyline, serotonin or histamine; without effect are melatonin, caffeine, theophylline, Lorazepam, phenoxybenzamine, Benadryl, PMS, progesterone, testosterone, oestrogen, reserpine, arginine vasotocin, dopamine, Equithesin, mersalyl acid, carbachol, pilocarpine, cysteamine, taurine, penicillamine and atropin sulphate. The question of an effect of melatonin administration on pineal melatonin formation will be discussed below (p. 470).

Careful comparisons of pineal NAT and HIOMT activities have indicated that the two enzymes appear to be regulated independently. In this context it should be recalled that pineal HIOMT does not appear to be a single enzyme, but one which acts on different substrates to yield different products, namely melatonin and 5-methoxytryptophol (AXELROD and LAUBER, 1968; ALEXANDER et al., 1970b). As will be recalled, pineal NAT activity is greatly influenced by changes in environmental lighting conditions (p. 466; PRESLOCK, 1975, 1976). HIOMT activity, the measurements of which have yielded inconsistent results, by contrast, appears to be regulated by gonadal steroids. The first hint in this direction was that during sexual maturation HIOMT activity decreases (SAYLER and WOLFSON, 1969; PRESLOCK, 1974), whereas NAT activity continues to increase (PRESLOCK, 1974). After castration, pineal HIOMT activity is depressed and can be restored to normal values by administration of gonadal steroids, whereas NAT activity is not affected (PRESLOCK, 1975, 1976).

From recent results it appears that the avian pineal organ has a high degree of independence, e.g. organ-cultured pineals retain their typical circadian rhythm of NAT activity for several days (BINKLEY et al., 1978b; KASAL et al., 1979; DEGUCHI, 1979). This has led to the view that the pineal gland of chicks contains or is a self-sustained circadian oscillator (KASAL et al., 1979), is a biological clock in vitro (BINKLEY et al., 1978b) and that NAT represents a timekeeping enzyme (BINKLEY, 1979).

5. The Effects of Pineal Substances on Avian Pineal Structure and Function

As melatonin injections are frequently carried out to unravel the biological significance of the pineal organ, it is relevant to note that melatonin administration may influence avian pineal structure and function. SINGH and TURNER (1967) noted in chickens that low dosages of melatonin increased pineal weight in both males and females; after higher dosages pineal weight increased in males and decreased in females. Pineal HIOMT and NAT activities were both significantly decreased by implants of 10 or 1 μg melatonin (female quail: WRIGHT and PRESLOCK, 1975; PRESLOCK, 1976). However, in chickens both melatonin and arginine vasotocin have been found to be without effect on pineal NAT activity (BINKLEY et al., 1978a).

6. Possible Target Organs of Pineal Substances

a) Brain

As in mammals, the avian brain appears to be a major target organ of pineal substances. The presence of melatonin in the brain has been clearly demonstrated in chickens. In this species melatonin shows identical day–night rhythms in both pineal and brain, with low levels during the day and high levels during the night. Melatonin concentrations are highest in the hypothalamus and lowest in the telencephalon; in the mid-brain, cerebellum, pons and medulla they are roughly twice as high as in the telencephalon, but are still low when compared with hypothalamic levels. The fact that the melatonin content is five times higher than in the serum appears to suggest a concentrating mechanism in brain tissue. That the pineal is the source of the brain melatonin has been demonstrated by the fact that pinealectomy renders melatonin in the brain undetectable (PANG et al., 1974).

α) *Regulation of Circadian Locomotor Rhythms*

There is strong evidence that the pineal plays an important role in the circadian organization of birds (MENAKER and ZIMMERMAN, 1976). Parameters influenced are locomotor activity (or perch-hopping activity in passerine birds), *Zugunruhe*, body temperature and uric acid excretion (MENAKER, 1972a). The most widely studied parameter is motor activity, which shows a more or less close dependence on the pineal organ, depending on the species investigated. Very close interrelationships have been shown to exist in passerine birds. The key observation was that by GASTON and MENAKER (1968) showing that in house sparrows, *Passer domesticus*, kept in constant darkness the circadian rhythm of locomotor activity is abolished by pinealectomy, while the daily activity totals are unaffected (BINKLEY et al., 1972). In white-throated sparrows, *Zonotrichia albicollis*, in which arrhythmicity develops within 3 days after pineal removal, the rhythms of spring migratory restlessness, summer daytime locomotor activity and autumnal *Zugunruhe* have been shown to be affected (McMillan, 1972).

The importance of the pineal for the rhythmicity in the house sparrow is indicated by the fact that consistent and prolonged aperiodicity occurs only in birds which are completely free of pineal tissue as judged by histological inspection. Even small remnants of pineal tissue suffice to support free-running circadian rhythms (Menaker and Zimmerman, 1976). An intact innervation of the pineal is apparently not required, as chemical sympathectomy with 6-hydroxydopamine does not abolish the circadian rhythm. Surgical interruption of the neural output is likewise without effect. Of great importance is the observation that rhythmic locomotor activity can be restored when pineal tissue is transplanted to the anterior chamber of the eye, a prerequisite being that several small cuts have been made in the gland before inserting it in the eye. Rhythmicity supported by the transplanted organ is usually long-lasting (up to 9 months). When the eye bearing the transplanted pineal is removed, the bird becomes aperiodic again (Zimmerman and Menaker, 1975; Menaker and Zimmerman, 1976; see also Zimmerman Headrick, 1976). In house sparrows, continuous administration of low levels of melatonin via Silastic capsules shortens the free-running period of locomotor activity in some animals and induces continuous activity in others (Turek et al., 1976c). In house sparrows maintained under LD 3:21 or LD 8:16, continuous administration of melatonin clearly reduces the total amount of perch-hopping activity (Hendel and Turek, 1978).

The results obtained in European starlings (*Sturnus vulgaris*) differ from those in house sparrows. In one study (Rutledge and Angle, 1977) surgical removal of the pineal organ did not alter circadian perch-hopping activity rhythms in constant darkness. More recently, Gwinner (1978) confirmed that a circadian rhythmicity of locomotor activity persists in most pinealectomized birds maintained under constant environmental conditions, but found that the period of the rhythm shortens and that the activity time lengthens. Moreover, the clear separation between activity and rest time is lost. The role of melatonin is underlined by the observation that daily administration of melatonin for several weeks results in synchronization of locomotor activity with the 24-h injection in almost all the animals so treated, whereas only one out of ten control animals synchronized with the daily control injections of sesame oil (Gwinner and Benzinger, 1978). Gwinner (1978), modifying a model proposed by Menaker and Zimmerman (1976), assumes that

> ... the avian pineal is the seat of a self-sustaining circadian pacemaker that acts on a population of secondary self-sustaining oscillators. These secondary oscillators, located outside the pineal, are only weakly coupled to each other but strongly dependent on the pineal driver. Their net circadian period is shorter than that of the pineal driver but, like the pineal driver, they can be synchronized by light.

The circadian rhythm of body temperature will be dealt with below.

β) Other Brain Functions

In addition to locomotor activity rhythms the pineal may influence other aspects of brain function. According to Hishikawa et al. (1969) melatonin (0.01–

0.06 mg/g, chicken) has a powerful sedative and hypnotic effect which appears 1–2 min after injection and lasts for 30–60 min. Melatonin-induced sleep is characterized by slow EEG activity of high voltage similar to that in normal slow wave sleep. Moreover, melatonin delays the appearance of the first paradoxical sleep and decreases total PS time. In young chickens (PANG et al., 1976) melatonin injections lower the frequency and increase the amplitude of the EEG within two minutes, while pinealectomy leads to lower frequency of cycles over 50 μV.

Most interesting findings have been obtained with respect to the hallucinatory-cataleptic agents, ketamine and gamma-hydroxybutyrate. The response to these drugs, i.e. catatonia, shows a clear diurnal rhythm, the duration of catatonia being longer during the night than during the day. Pinealectomy abolishes the dark-induced increase in response to both drugs and pre-treatment of intact chicks with serotonin, *N*-acetylserotonin or melatonin augments the response (GIEDT and WINTERS, 1976; GIEDT et al., 1978). Further studies involving both intact and pinealectomized chicks indicate that serotonin and *N*-acetylserotonin augment the ketamine response only when the pineal is present, whereas for the action of melatonin an intact pineal is not required (LAKIN, 1978).

There is evidence that the pineal is involved in *thermoregulation*. BINKLEY et al. (1971), studying the circadian rhythm of body temperature in constant darkness, noted that pinealectomy abolishes this circadian rhythm and reduces the amplitudes of the temperature fluctuations. In pinealectomized sparrows kept under LD 12:12 the daily body temperature rise occurs earlier than in intact controls. Pinealectomized *chickens* (COGBURN et al., 1976) exposed to high environmental temperature (42° C) for 60 min show a higher body temperature during scotophase, but not photophase, than intact controls. Pinealectomized *pigeons* (JOHN et al., 1978) show higher body temperature in the photophase as well as scotophase than intact or sham-operated controls when examined 2 or 3 weeks after post-surgical acclimatization to 25° or 3° C. Subcutaneous implantation of melatonin pellets into pinealectomized pigeons nullifies or reverses the hyperthermic effect of pinealectomy. After short-term exposure to −18° C for 280 min pinealectomized pigeons show a higher body temperature during photophase, but not scotophase, when compared to intact or sham-operated controls.

The effect of melatonin administration on punished and non-punished operant behaviour in the pigeon has been studied by SCHOENFELD (1971).

The target organs directly or indirectly under *hypothalamic* control will be considered below.

b) Pituitary

Relatively few studies have dealt with the influence of the pineal organ on pituitary parameters, such as weight, cellular composition and hormone content. The results obtained have been variable, probably due to the fact that the experimental animals used differed with respect to age and sex and that interspecies differences exist. Pinealectomy had no striking effect on pituitary *weight* in adult female ducks (CUELLO et al., 1972) and in chicks when the operation had been carried out at 6 days of incubation (STALSBERG, 1965) or at 3 (MACHEMER et al., 1971b) or 4 days after hatching (SHELLABARGER and BRENEMAN, 1950). Sex differences become apparent in the Japanese quail, the females of which show a decrease in pituitary weight after pinealectomy at 43 days of age only, while the males show no change at all (SAYLER and WOLFSON, 1967, 1968a). Pituitary weight elevated by exposure of animals to continuous illumination is not lowered by pinealectomy (OISHI and LAUBER, 1974a).

Administration of melatonin had no effect on pituitary weight in male chickens, but in females an increase of 56% was observed (SINGH and TURNER, 1967). Adenohypophyseal weight was not affected by melatonin in juvenile and adult Japanese quail kept under a stimulatory photoperiod (16L:8D), but under a non-stimulatory photoperiod (8L:16D) juvenile animals responded with an increase in adenohypophyseal weight (SAYLER and WOLFSON, 1968a).

The cellular composition of the adenohypophysis did not change after pinealectomy (STALSBERG, 1965) or administration of crude pineal homogenate (PFLUGFELDER, 1957). However, MIHAIL and MOLNÁR (1965) reported that the adenohypophysis was totally acidophilic after pinealectomy (pigeon).

Hormonal parameters have been little investigated. Pituitary LH levels increased from 387 to 686 ng after pinealectomy in *Fringilla montifringilla* (HAASE and FOLLET, 1974). Melatonin increased plasma STH levels when the indoleamine was given in low dosages during the photophase; higher dosages were without effect, as was melatonin administration during scotophase (MCKEOWN et al., 1975, pigeon). In the immature domestic fowl, melatonin inhibits the LHRH-induced increase in ^{32}P uptake by the testes (SACKMAN, 1977b). In vitro studies have shown that embryonic chicken pineals are capable of abolishing the gonad-stimulating activity of the embryonic adenohypophysis (MOSZKOWSKA, 1956).

Moreover, the effects of hypophysectomy or administration of pituitary hormones on the pineal have been investigated. OISHI and LAUBER (1974b) noted that in the Japanese quail hypophysectomy was not followed by changes in pineal weight or pineal HIOMT activity.

Prolactin treatment decreased pineal HIOMT activity at 33 days of age, but not at 43 and 53 days, and only in females (ALEXANDER and WOLFSON, 1968). LH administration decreased pineal ^{32}P uptake in the 13-day-old cockerel, but not in 3- or 10-day-old animals (SACKMAN, 1975).

c) Gonads

As the effects of the pineal organ on the gonads have recently been reviewed (RALPH, 1978), it suffices to give a brief survey. As regards the older literature on this topic, the reader is referred to another publication by RALPH (1970).

The present state of affairs is that we are far from being able to formulate a unifying concept of pineal-gonadal interrelationships. What we do know is that the pineal may be progonadotropic or antigonadotropic or may have no demonstrable effect.

A progonadotropic effect is usually exerted during early post-hatching stages of development, as demonstrated by pinealectomy as well as by administration of melatonin or 5-methoxytryptophol (SHELLABARGER and BRENEMAN, 1950; SHELLABARGER, 1952, 1953; THILLARD, 1964; HARRISON, 1968; SAYLER and WOLFSON, 1967, 1968a, b; ZADURA et al., 1969; BALEMANS, 1972, 1973a; BALEMANS and VAN DE KAMER, 1972; BALEMANS et al., 1977a, b); no effect could be visualized by RENZONI (1967b, 1968b) and SAYLER and WOLFSON (1968a), whereas an antigonadotropic effect is suggested by the findings of DARRE and HARRISON (1977). In *maturing* birds, the pineal appears to exert an inhibitory effect on the gonads (SHELLABARGAR and BRENEMAN, 1952; BALEMANS et al., 1977a). In *adults*, depending inter alia on the environmental lighting conditions, the pineal may be antigonadotropic (PATAY and DU CHALLARD, 1952; PFLUGFELDER, 1957; HOMMA et al., 1967; SINGH and TURNER, 1967; BALASUBRAMANIAN and SAXENA, 1973; HAASE and FOLLET, 1974; BALEMANS, 1973b; BALEMANS et al., 1977a), without effect (RENZONI, 1967a; SAYLER and WOLFSON, 1968a; OISHI and KATO, 1968; DONHAM and WILSON, 1969; HARRISON and BECKER, 1969; OKSCHE et al., 1972; OISHI and LAUBER, 1974b) or occasionally progonadotropic (CARDINALI et al., 1971; HISANO et al., 1972; CUELLO et al., 1972; PRESLOCK, 1976). Photorefractoriness as revealed by the occurrence of gonadal regression in males after long periods of photostimulation is apparently not under the direct influence of the pineal organ (DONHAM and WILSON, 1970; HAMNER and BARFIELD, 1970; OKSCHE et al., 1972; STOREY and NICHOLLS, 1978; TUREK and WOLFSON, 1978).

Finally, the effects of gonadal hormones on the pineal have also been investigated. Morphological studies are rare. In one such study administration of diethylstilboestrol did not markedly alter nuclear size of pinealocytes (VON FABER, 1954). Castration decreased pineal HIOMT activity in both male and female quail, whereas NAT activity was unaffected. Conversely, administration of diethylstilboestrol to females and testosterone proprionate to males increased HIOMT activities in both intact and castrated quail. In castrated quail oestradiol benzoate, progesterone and androstenedione also elevated HIOMT activities but not NAT activities, which together with other findings was taken as evidence that HIOMT activity was regulated by gonadal steroids and NAT activity mainly by photoperiod (PRESLOCK, 1975, 1976). In reproductively inactive house sparrows, testosterone treatment did not change pineal HIOMT activity (BARFUSS and ELLIS, 1971).

d) Thyroid

Studies concerning possible interactions between the pineal and the thyroid have yielded equivocal results. Pinealectomy was found to increase thyroid *weight* in ducks (CUELLO et al., 1972), but it was without effect in young chickens pinealectomized at 6 days of incubation (STALSBERG, 1965) or 4 days after hatching (SHELLABARGER and BRENEMAN, 1950). MIHAIL and MOLNÁR (1965) noted thyroid activation after pinealectomy in pigeons. Melatonin treatment had no effect on thyroid weight in male and female chickens (SINGH and TURNER, 1967). PFLUGFELDER (1957) reported that the cell nuclei in the thyroid decreased in size after administration of crude pineal homogenate.

e) Adrenals

In chickens pinealectomized at 6 days of incubation (STALSBERG, 1965) or 4 days after hatching (SHELLABARGER and BRENEMAN, 1950) the adrenals were not markedly affected, but melatonin administration has been claimed to decrease adrenal weight (SINGH and TURNER, 1967).

f) Miscellaneous Observations

Spleen and thymus weights as well as the histological appearance of these organs were not affected by pinealectomy carried out at 6 days of incubation (STALSBERG, 1965). Thymus weight was, however, depressed after melatonin administration (SINGH and TURNER, 1967).

The width of the third ventricle, mean nuclear volume of the subcommissural organ, feather colour, body weight and toe length were not affected by pinealectomy before hatching (STALSBERG, 1965). When pinealectomy was performed after hatching, body weight increased (PATAY and DU CHALARD, 1952) or was unaffected (SHELLABARGER and BRENEMAN, 1950). Administration of crude pineal homogenate led to a decrease of body weight by 20% (PFLUGFELDER, 1957). Plasma glucose levels increased after melatonin administration (MCKEOWN et al., 1975, pigeon) and MIHAIL and MOLNÁR (1965) described B cell hyperplasia in islets of Langerhans after pinealectomy in pigeons. Haematocrit values were not affected by melatonin administration (MCKEOWN et al., 1975).

H. Epilogue

Although the "pineal odyssey" (Reiter, 1970e) has not yet come to an end, the "era of chaos" (Kelly, 1962) characterized by "an extraordinarily high ratio of hypotheses to data" (Kitay, 1967) is over. In many respects, we are on safe ground. We know precisely the basic structure, ultrastructure, histochemistry and biochemistry of the pineal complex in mammals and submammalian vertebrates. In some respects the pineal gland–superior cervical ganglion complex has become "an easy-to-manipulate model for the study of basic neuroendocrine mechanisms at the subcellular level" (Cardinali and Vacas, 1978b).

This review has shown that, at least in mammals, a great many parameters of bodily functions appear to be affected by the pineal. Therefore it may be difficult to see the "new horizons of pineal research" (Reiter et al., 1976d) and a few should be mentioned.

The vertebrate pineal has undergone profound changes during phylogeny (Fig. 190). While it was previously held that the pineal is a sensory organ in lower vertebrates and a secretory organ in higher vertebrates (see Kappers, 1968; Collin, 1971), it is now clear that the pineal organ is basically sensory *and* endocrine and that it "gradually loses its sensory activity and retains only its endocrine function" (Collin, 1971; see also Oksche et al., 1971; Oksche, 1965b, 1971).

As regards the melatonin "story" of the pineal, a word of caution is necessary. Although there can be no doubt that melatonin exerts extrapineal actions, the synthesis of melatonin in the retina (p. 237) and the close structural resemblance between retinal and pineal photoreceptor cells makes one wonder whether melatonin produced in the pineal is really meant for export. Perhaps melatonin is primarily required for photoreceptor processes and, in pineal organs which have lost their direct photosensitivity, for processes other than photoreception. Quay (1974a) has suggested that pineal indoleamines "could possibly function as chemical messengers within pineal tissue" (see also Quay, 1965f, 1969a), and it may well be that melatonin is somehow involved in the synthesis and release of pineal peptides (e.g. Reiter et al., 1976d; Benson, 1977). Further progress in pineal research is likely to be made by studying the role of melatonin in the retina and in enterochromaffin cells, and by applying the results obtained to the pineal. Now that suitable animal models for studying the effects of melatonin administration, especially on the hypothalamus–pituitary–gonad axis, are available, it appears important to check whether melatonin is effective also when given in *physiological* doses. Pineal peptides still await their breakthrough, as do pineal indoleamines other than melatonin. The potential importance of pineal peptides is highlighted by observations in the early days of pineal research

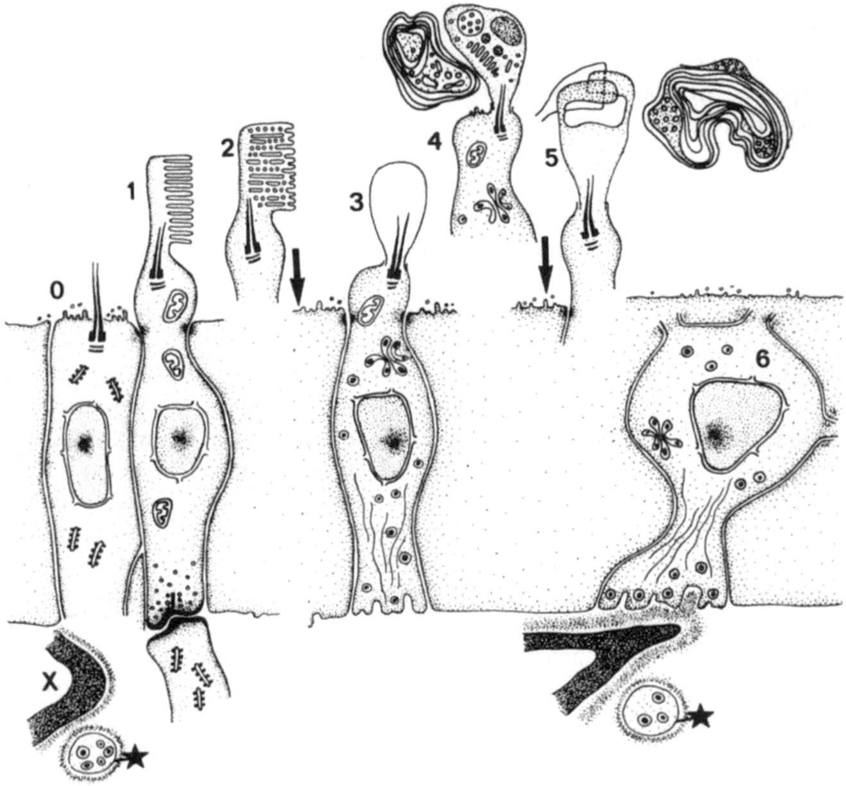

Fig. 190. Schematic representation of different types of pinealocytes, according to OKSCHE (1971). *0*, supporting cell; *1*, regular cone-like outer segment of sensory pinealocyte; *2*, outer segment showing tubular and vesicular structures (sensory pinealocyte); *3–5*, rudimentary sensory pinealocyte: *3*, bulbous cilium (9+0 type); *4*, degenerated outer segment showing fragments of lamellae, vesicles and segregated membrane whorls (v); *5*, bulbous cilium (9+0 type) with ectopic lamellar complexes (vv); *6*, secretory mammalian pinealocyte. ×, blood vessels; *asterisk*, autonomic nerve fibres. Outer segments of types 1 and 2 have been observed in the pineal organs of fish and anurans. Type 1 shows variations in length and form. Long outer segments of type 1 are predominant in the parietal eye of lizards; the lacertilian pineal organ shows the outer segment types 1, 2, 3 and 4. Type 5 is the characteristic vestigial outer segment structure of the avian pineal organ. For further details, see text. OKSCHE, 1971

when it was found that the effects of crude pineal extracts or melatonin-free pineal extracts are often opposite to those of melatonin. An almost completely neglected field is to what extent the pineal influences circadian rhythms in other organs.

What we are left with is that in mammals the pineal appears to be a potentially important "centre of neurovegetative regulation" (KAPPERS, 1967b). In mammals, especially hamsters, it is the gonads which are most clearly affected by the pineal; this has led REITER (1975a) to propose that "one of the most important functions of the pineal may be to synchronize the breeding activity in seasonal breeders with the appropriate time of the year". The importance which is attributed to the pineal with respect to reproductive processes is highlighted by the fact that recently two publications dealing solely with this aspect have appeared in book form (REITER, 1978; MESS et al., 1978), and that the pineal has als found its place in symposia on reproduction (e.g. ANAND KUMAR, 1976).

While in mammals studies on the hypothalamus–anterior pituitary axis have turned out to be most fruitful, in submammalian vertebrates it is the locomotor activity of birds which is greatly controlled by the pineal. The original concept that the pineal organ is a biological clock (WURTMAN and AXELROD, 1965) has recently gained support by studies showing that circadian rhythmicity persists in vitro (BINKLEY, 1979; BRAMMER, 1979; DEGUCHI, 1979). In this context it is interesting to note that BINKLEY (1979) speaks of a "time-keeping enzyme" (NAT). How important this time-keeping ability is for an individual remains to be established. Although the precise role of the pineal in most species is not yet known, several authors have expressed its function in general terms. Just to mention a few: According to ARVY (1966), the pineal is a "regulating centre of biogenic amines". GUSEK (1968) points out that by influencing pituitary function the pineal belongs to the most important endocrine regulatory systems. It is therefore not surprising that "a functional interplay between the pineal gland and practically all the endocrine organs has been demonstrated" (MESS, 1969). "In general, the pineal gland appears to inhibit the function of other organs or systems" (KAPPERS, 1976), and ROMIJN (1978), characterizing the pineal as a "tranquillizing organ", is of the opinion that "the pineal gland acts as a general synchronizing, stabilizing and moderating organ on behalf of several physiological processes". One reason for the sympathetic innervation of the pineal gland may reside in the fact that "its function consists in influencing the activity of the sympathetic part of the autonomous nervous system", a conclusion derived from pupillographic studies (GRUNEWALD LOWENSTEIN, 1952), which in my view represent a highly original approach to studying pineal function. CSABA (1977), who distinguishes morphogenetic and adaptive hormonal systems, claims that the pineal controls the adaptive system. All these facets show clearly that research into the pineal calls for "a modern interdisciplinary approach" (KAPPERS, 1976).

Finally, attempts have been made to work out similarities between pinealocytes and other cell types. This approach has led to a classification of the pinealocyte as a paraneuron (FUJITA, 1976; UECK and WAKE, 1977), as a type of CSF-contacting neuron (VIGH and VIGH-TEICHMANN, 1975; VIGH et al., 1975) and as APUD cell (PEARSE, 1969; LEONG and MATTHEWS, 1979). For the time being I am inclined not to classify the mammalian pinealocyte into any category of cell for the simple reason that this might stop one from posing pointed questions concerning the pinealocyte, which are even now badly required to unravel its true nature and functional significance.

Now I know in part;
then I shall understand fully ...
1 Corinthians 13, 12 (RSV)

References

Abe, K., Robison, G.A., Liddle, G.W., Butcher, R.W., Nicholson, W.E., Baird, C.E.: Role of cyclic AMP in mediating the effects of MSH, norepinephrine, and melatonin on frog skin color. Endocrinology **85**, 674–682 (1969)

Abe, T., Yamada, Y., Hashimoto, P.H., Shimuzu, N.: Histochemical study of glucose-6-phosphate dehydrogenase in the brain of normal adult rat. Med. J. Osaka Univ. **14**, 67–98 (1963)

Adam, H.: Discussion remark. Progr. Brain Res. **10**, 28–29 (1965)

Adams, W.C., Wan, L., Sohler, A.: Effect of melatonin on anterior pituitary luteinizing hormone. J. Endocr. **31**, 295–296 (1965)

Adler, K.: The role of extraoptic photoreceptors in amphibian rhythms and orientation: a review. J. Herpetol. **4**, 99–112 (1970)

Adler, K.: Pineal end organ: role in extraoptic entrainment of circadian locomotor rhythm in frogs. In: Menaker, M. (ed.): Biochronometry, pp. 342–350. Washington, D.C., National Academy of Sciences 1971

Adler, K.: Extraocular photoreception in amphibians. Photochem. Photobiol. **23**, 275–298 (1976)

Adler, K., Taylor, D.H.: Extraocular perception of polarized light by orienting salamanders. J. comp. Physiol. **87**, 203–212 (1973)

Aguado, L.I., Benelbaz, G.A., Gutierrez, L.S., Rodríguez, E.M.: Ultrastructure of the rat pineal gland grafted under the kidney capsule. Cell Tiss. Res. **176**, 131–142 (1977)

Airaksinen, M., Idänpään-Heikkilä, J., McIsaak, W.M.: Effect of p-chlorophenylalanine, oxypertine and 6-methoxytetrahydro-β-carboline on pineal and pituitary gland and sex organs in young male rats. Scand. J. clin. Lab. Invest. **25**, Suppl. 113, 43 (1970)

Airaksinen, N.M., Karppanen, H.: Effect of pinealectomy and oxypertine on hypothalamic FSH-releasing factor and circadian rhythm in rats. Scand. J. clin. Lab. Invest. **25**, (Suppl. 113), 43 (1970)

Albertazzi, E., Barbanti-Silva, C., Trentini, G.P., Botticelli, A.: Influence de l'épiphysectomie et du traitement avec la 5-hydroxytryptamine sur le cycle oestradial de la ratte albino. Ann. Endocr. (Paris) **27**, 93–100 (1966)

Alcozer, G., Giordano, G., Masciocco, D.: Studi sull epifisi: influenza dell'estratto acquoso di pineale su alcuni aspetti del metabolismo glicidico in soggetti sani ed eucrinici. Arch. E. Maragliano Pat. Clin. **12**, 1105–1113 (1956)

Alexander, B., Wolfson, A.: Effect of prolactin on gonadal growth and pineal hydroxyindole-O-methyltransferase (HIOMT) during sexual maturation in the Japanese quail, *Coturnix coturnix japonica*. Excerpta Med. Int. Congr. Series **157**, 180 (1968)

Alexander, B., Dowd, A.J., Wolfson, A.: Effects of prepuberal hypophysectomy and ovariectomy on hydroxyindole-O-methyltransferase activity in the female rat. Endocrinology **86**, 1166–1168 (1970a)

Alexander, B., Dowd, A.J., Wolfson, A.: Pineal hydroxyindole-O-methyltransferase (HIOMT) activity in female Japanese quail. Neuroendocrinology **6**, 236–246 (1970b)

Alexander, B., Dowd, A.J., Wolfson, A.: Effects of continuous light and darkness on hydroxyindole-O-methyltransferase and 5-hydroxytryptophan decarboxylase activities in the Japanese quail. Endocrinology **86**, 1441–1443 (1970c)

Alleva, J.J., Waleski, M.V., Alleva, F.R.: The Zeitgeber for ovulation in rats: non-participation of the pineal gland. Life Sci. **9**, Part I, 241–246 (1970)

Alliot, J., Dale, A., Thiéblot, L., Boucher, D.: Innervation sympathique de la glande pinéale et excrétion d'un principe urinaire antigonadotrope chez le rat soumis à différentes durées d'éclairement. J. Physiol. (Paris) **70**, 22B (1975)

Alonso, R., Prieto, L., Hernandez, C., Mas, M.: Antiandrogenic effects of pineal gland and melatonin in castrated and intact prepubertal male rats. J. Endocr. **79**, 77–83 (1978)
Alphs, L., Heller, A.: A circadian rhythm in dark induction of rat pineal serotonin: coenzyme A: N-acetyltransferase activity. Brain Res. **139**, 374–377 (1978)
Altieri, A., Sorrentino, F.: Wirkung der Epiphysen-Extrakte im Experiment auf den menschlichen Hoden. Urol. int. (Basel), **2**, 350–360 (1956).
Altner, H.: Histologische und histochemische Untersuchungen an der Epiphyse von Haien. Progr. Brain Res. **10**, 154–171 (1965)
Altschule, M.D.: Some effects of aqueous extracts of acetone-dried beef-pineal substance in chronic schizophrenia. New Engl. J. Med. **257**, 919–922 (1957)
Altschule, M.D. (ed.): Frontiers of pineal physiology. Cambridge, Mass.-London: MIT Press 1975a
Altschule, M.D.: Effect of a pineal extract on carbohydrate metabolism in schizophrenic patients. In: Frontiers of pineal physiology (M.D. Altschule, ed.), pp. 197–203. Cambridge, Mass.-London: MIT Press 1975b
Altschule, M.D., Giancola, J.N.: Pineal-like effects of central nervous system tissue extracts. Proc. Soc. exp. Biol. (N.Y.) **104**, 399–401 (1960)
Altschule, M.D., Siegel, E.P., Goncz, R.M., Murnane, J.P.: Effect of pineal extracts on blood glutathione level in psychotic patients. Arch. Neurol. Psychiat. (Chic.) **71**, 615–618 (1954)
Anand Kumar, T.C.: Cellular and humoral pathways in the neuroendocrine regulation of reproductive function. J. Reprod. Fertil., Suppl. **20**, 11–25 (1973)
Anand Kumar, T.C. (ed.): Neuroendocrine Regulation of Fertility. A potential new approach to human fertility regulation by interfering with neuroendocrine pathways. Proc. Int. Symp. Neuroendocrine Regulation of Fertility, Simla, October 28–31, 1974. Basel-München-Paris-London-New York-Sydney: S. Karger 1976
Anand Kumar, T.C., David, F.F.X., Kumar, K.: Circumventricular structures in the neuroendocrine regulation of sexual cycles. Proc. Indian natn. Sci. Acad. **39**, 249–267 (1973)
Anderson, E.: Some cytological observations on the fine structure of a mammalian pineal organ. Anat. Rec. **136**, 328–329 (1960)
Anderson, E.: Some fine structural features of the pineal body. Amer. Zool. **2**, 386 (1962)
Anderson, E.: The anatomy of bovine and ovine pineals. Light and electron microscopic studies. J. Ultrastruct. Res., Suppl. **8**, 1–80 (1965)
Andre, J.S., Parrish, J.: Inhibition of estrous cyclicity in Golden hamsters by melatonin administration on the day of proestrus. J. exp. Zool. **207**, 161–166 (1979)
Angervall, L., Berger, S., Röckert, H.: A microradiographic and x-ray crystallographic study of calcium in the pineal body and in intracranial tumors. Acta path. microbiol. scand. **44**, 113–119 (1958)
Anton-Tay, F.: Pineal-brain relationships. In: Wolstenholme, G.E.W. and Knight, J. (eds.), The pineal gland. A Ciba Foundation symposium, pp. 213–227. Edinburgh and London: Churchill Livingstone 1971
Anton-Tay, F., Wurtman, R.J.: Stimulation of hydroxyindole-O-methyltransferase activity in hamster pineal gland by blinding or continuous darkness. Endocrinology **82**, 1245–1246 (1968)
Anton-Tay, F., Wurtman, R.J.: Regional uptake of ^{3}H-melatonin from blood or cerebrospinal fluid by rat brain. Nature (Lond.) **221**, 474–475 (1969)
Anton-Tay, F., Escobar, A., Anton, S.M.: Estudios sobre la glandula pineal II. Inhibición de la hipertrofia suprarrenal compensadora por la administración de extract pineal. Bol. Inst. Estud. méd. biol. (Mex.) **20**, 281–286 (1962)
Anton-Tay, F., Anton-Tay, S., Zamora, E.T.: Inhibición de la function tiroidea producida por la administración de extracto pineal. Rev. Invest. clin. **15**, 367–375 (1963)
Anton-Tay, F., Chou, C., Anton, S., Wurtman, R.J.: Brain serotonin concentration: Elevation following intraperitoneal administration of melatonin. Science **162**, 277–278 (1968)
Anton-Tay, F., Sepulveda, J., Gonzales, S.: Increase of brain pyridoxal phosphokinase activity following melatonin administration. Life Sci. **9**, Part II, 1283–1288 (1970)
Anton-Tay, F., Diaz, J.L., Fernandez-Guardiola, A.: On the effect of melatonin upon human brain. Its possible therapeutic implication. Life Sci. **10**, Part I, 841–850 (1971)
Arendt, J.: Melatonin as a tumour marker in a patient with pineal tumour. Brit. med. J. **1978 II**, 635–636

Arendt, J., Paunier, L., Sizonenko, P.C.: Melatonin radioimmunoassay. J. clin. Endocr. **40**, 347–350 (1975)

Arendt, J., Wetterberg, L., Heyden, T., Sizonenko, P. C., Paunier, L.: Radioimmunoassay of melatonin: human serum and cerebrospinal fluid. Hormone Res. **8**, 65–75 (1977)

Arendt, J., Wirz-Justice, A., Bradtke, J.: Annual rhythm of serum melatonin in man. Neurosci. Letters **7**, 327–330 (1978)

Arieti, S.: The pineal gland in old age. J. Neuropath. exp. Neurol. **13**, 482–491 (1954)

Arnold, J.D., Lalli, F., Martin, D.C.: Augmentation of growth and division synchrony and of vascular sequestration of *Plasmodium berghei* by the photoperiodic rhythm. J. Parasit. **55**, 597–608 (1969a)

Arnold, J.D., Berger, A., Martin, D.C.: The role of the pineal in mediating photoperiodic control of growth and division synchrony and capillary sequestration of *Plasmodium berghei* in mice. J. Parasit. **55**, 609–616 (1969b)

Arnold, J.D., Berger, A.E., Martin, D.C.: Chemical agents effective in mediating control of growth and division synchrony of *Plasmodium berghei* in pinealectomized mice. J. Parasit. **55**, 617–625 (1969c)

Aro, S., Karppanen, H., Klinge, E.: Acetylcholinesterase activity is decreased after sexual maturation in bovine pineal gland. Acta physiol. scand. **87**, 575–576 (1973)

Aron, E., Combescot, C., Demaret, J., Guyon, L.: Neurosécrétion chez la Tortue d'eau *Emys leprosa* Schw. Après destruction de la région épiphysaire. C.R. Acad. Sci. (Paris) **251**, 1914–1916 (1960a)

Aron, E., Combescot, C., Demaret, J.: Thyréostimulation et repos sexuel après destruction de la région épiphysaire chez *Emys leprosa* Schw. C.R.Soc. Biol. (Paris) **154**, 1856–1859 (1960b)

Aron, E., Combescot, C., Demaret, J.: Exophthalmie de la Tortue d'eau *Emys leprosa* Schw. produite par la destruction de la région épiphysaire. C.R. Acad. Sci. (Paris) **250**, 3386–3387 (1960c)

Aron, E., Combescot, C., Demaret, J., Guyon, L., Mauvernay, R.Y.: Modifications de la neurosécrétion observées dans l'encéphale du Rat après injection d'un extrait épiphysaire. C.R. Soc. Biol. (Paris) **155**, 593–595 (1961)

Arstila, A.U.: Electron microscopic studies on the structure and histochemistry of the pineal gland of the rat. Neuroendocrinology **2**, Suppl. 1–101 (1967)

Arstila, A.U., Hopsu, V.K.: Studies in the rat pineal gland. I. Ultrastructure. Ann. Acad. Sci. fenn. A II **113**, 3–21 (1964)

Arstila, A.U., Hopsu-Havu, V.K.: Nuclear and cytoplasmic microfilaments in the pineal chief cells of the rat. Z. Zellforsch. **80**, 22–28 (1967)

Arstila, A.U., Rinne, U.K.: Electron microscopic studies on the perivascular secretory processes in the pineal gland of the rat. Acta neurol. scand. **43**, Suppl. 31, 211 (1967)

Arstila, A.U., Kalimo, H.O., Hyyppä, M.: Secretory organelles of the rat pineal gland: electron microscopic and histochemical studies in vivo and in vitro. In: The pineal gland (G.E.W. Wolstenholme and J. Knight, eds.), pp. 147–195. Edinburgh and London: Churchill Livingstone 1971

Arutyunyan, G.S., Mashkovskii, M.D., Roshchina, L.F.: Pharmacological properties of melatonin. Fed. Proc. **23**, T 1330-T 1332 (1964)

Arvy, L.: Contribution à la connaissance de la glande pinéale de *Bos taurus* L., d'*Ovis aries* L. et de *Sus scrofa* L. C.R. Acad. Sci. (Paris) **253**, 1361–1363 (1961)

Arvy, L.: Histo-enzymologie des glandes endocrines. Paris: Gauthier-Villars 1963

Arvy, L.: Activités enzymatiques histochimiquement décelables dans la glande pinéale, chez quelques artiodactyles. Progr. Brain Res. **10**, 473–475 (1965)

Arvy, L.: La glande pinéale, centre régulateur des amines biogènes. Ann. Biol. anim. **5**, 565–594 (1966)

Asagoe, Y., Hamamoto, A.: Biological and biochemical study of pinealectomized rats. Yonago Acta med. **3**, 192–198 (1959)

Atkins, T.W., Bailey, C.Y., Matty, A.J.: The effect of melatonin on insulin secretion in rat and mouse. J. Endocr. **58**, XVII–XVIII (1973)

Aubert, C., Prade, M., Bomuon, C.: Effect de la pinéalectomie sur les tumeurs mélaniques du hamster doré induites par l'administration (per os) d'une seule dose de 9,10-diméthylbenzanthracène. C.R. Acad. Sci. (Paris) **271**, 2465–2468 (1970)

Axelrod, J., Lauber, J.K.: Hydroxyindole-O-methyltransferase in several avian species. Biochem. Pharmacol. **17**, 828–830 (1968)

Axelrod, J., Saavedra, J.M.: Octopamine. Nature (Lond.) **265**, 501–504 (1977)

Axelrod, J., Weissbach, H.: Enzymatic O-methylation of N-acetyl-serotonin to melatonin. Science **131**, 1312 (1960)

Axelrod, J., Weissbach, H.: Purification and properties of hydroxyindole-O-methyltransferase. J. biol. Chem. **236**, 211–213 (1961)

Axelrod, J., Wurtman, R.J.: Photic and neural control of indoleamine metabolism in the rat pineal gland. Advanc. Pharmacol. **6**, Part A, 157–166 (1968)

Axelrod, J., Wurtman, R.J., Winget, C.M.: Melatonin synthesis in the hen pineal gland and its control by light. Nature (Lond.) **201**, 1134 (1964)

Axelrod, J., Wurtman, R.J., Snyder, S.H.: Control of hydroxyindole-O-methyltransferase activity in the rat pineal gland by environmental lighting. J. biol. Chem. **240**, 949–954 (1965a)

Axelrod, J., Quay, W.B., Baker, P.C.: Enzymatic synthesis of the skin-lightening agent, melatonin, in amphibians. Nature (Lond.) **208**, 386 (1965b)

Axelrod, J., Snyder, S.H., Heller, A., Moore, R.Y.: Light-induced changes in pineal hydroxyindole-O-methyltransferase: abolition by lateral hypothalamic lesions. Science **154**, 898–899 (1966)

Axelrod, J., Shein, H.M., Wurtman, R.J.: Stimulation of C^{14}-melatonin synthesis from C^{14}-tryptophan by noradrenaline in rat pineal in organ culture. Proc. nat. Acad. Sci. (Wash.) **62**, 544–549 (1969)

Azzali, G., Carreras, M., Dalla Rosa, V.: Primi resultati sulle modificazioni istofisiologiche della costellazioni endocrina e del sistema ipotalamoneuroipofisario dopo distruzione della epifisi e dell'organo subcommissurale. Monit. Zool. Ital., Suppl. **71**, 16–17 (1963)

Azzali, G., Carreras, M., Dalla Rosa, V.: Ricerche sperimentali sui rapporti epifisi-nuclei abenulari – organo subcommissurale e centri neurosecretori ipotalamici – ghiandole endocrine. Monit. Zool. Ital. Suppl. **72**, 24–25 (1964)

Bade, P., Rommelspacher, H., Strauss, S.: N-acetyltransferase activity in pineal gland of rats treated with pargyline. Naunyn-Schmiedeberg's Arch. Pharmacol. **297**, 143–147 (1977)

Bäckström, M., Wetterberg, L.: Catechol-O-methyltransferase, histamine-N-methyltransferase and methanol forming enzyme in the rat pineal gland. Life Sci. **11**, Part I, 293–299 (1972)

Bäckström, M., Wetterberg, L.: Beta-adrenergic regulation of N-acetyltransferase activity in rat pineal gland organ culture. Acta physiol. Scand. **87**, 27A–28A (1973a)

Bäckström, M., Wetterberg, L.: Increased N-acetylserotonin and melatonin formation induced by d-amphetamine in rat pineal gland organ culture via a β-adrenergic receptor mechanism. Acta physiol. scand. **87**, 113–120 (1973b)

Bäckström, M., Wetterberg, L.: Melatonin formation in rat pineal gland organ culture: induction by an adrenergic β_2-stimulating drug. Yale J. Biol. Med. **46**, 630 (1973c)

Bäckström, M., Wetterberg, L.: Influence of psoralen on melatonin formation in rat pineal gland in organ culture. Acta endocr. (Kbh.) **86**, 659–666 (1977)

Bäckström, M., Hetta, J., Wahlström, G., Wetterberg, L.: Enzyme regulation of melatonin synthesis in the pineal gland of Japanese quail. Life Sci. **11** (Part II) 493–498 (1972)

Bagchi, S.P., Zarycki, E.P.: Occurrence of phenylalanine hydroxylation in the pineal gland *in vivo*: possible role of tyrosin and tryptophan hydroxylases. Res. Commun. Chem. Pathol. Pharmacol. **2**, 370–381 (1971a)

Bagchi, S.P., Zarycki, E.P.: The effect of p-chlorophenylalanine on the hydroxylation of phenylalanine in the pineal gland and brainstem. Fed. Proc. **30**, 381 (1971b)

Bagnara, J.T.: Pineal regulation of the body lightening reaction in amphibian larvae. Science **132**, 1481–1483 (1960)

Bagnara, J.T.: The pineal gland and body lightening in *Xenopus* larvae. Anat. Rec. **139**, 204 (1961a)

Bagnara, J.T.: Onset of pineal and hypophyseal regulation of melanophores in *Xenopus*. Amer. Zoologist **1**, 339–340 (1961b)

Bagnara, J.T.: The pineal and the body lightening reaction of larval amphibians. Gen. comp. Endocr. **3**, 86–100 (1963)

Bagnara, J. T.: Independent actions of pineal and hypophysis in the regulation of chromatophores of anuran larvae. Gen. comp. Endocr. **4**, 299–303 (1964)

Bagnara, J.T.: Pineal regulation of body blanching in amphibian larvae. Progr. Brain Res. **10**, 489–506 (1965)

Bagnara, J.T., Hadley, M.E.: The control of bright colored cells of fishes and amphibians. Amer. Zoologist **9**, 465–478 (1969)
Bagnara, J.T., Hadley, M.E.: Endocrinology of the amphibian pineal. Amer. Zoologist **10**, 201–216 (1970)
Bagnara, J.T., Hadley, M.E., Taylor, J.D.: Regulation of bright-colored pigmentation of amphibians. Gen. comp. Endocr., Suppl. **2**, 425–438 (1969)
Bailey, C.J., Atkins, T.W., Matty, A.J.: Melatonin inhibition of insulin secretion in the rat and mouse. Hormone Res. **5**, 21–28 (1974)
Bak, I.J., Kim, J.H., Hassler, R.: Electron microscopic autoradiography for demonstration of pineal serotonin in rat. Z. Zellforsch. **105**, 167–175 (1970)
Baker, P.C.: Melatonin levels in developing *Xenopus laevis*. Comp. Biochem. Physiol. **28**, 1387–1393 (1969)
Baker, P.C., Hoff, K.M.: The effect of light and dark background upon melatonin levels in larval *Xenopus*. Comp. Gen. Pharmacol. **2**, 59–62 (1971)
Baker, P.C., Quay, W.B., Axelrod, J.: Development of hydroxyindole-O-methyltransferase activity in eye and brain of the amphibian, *Xenopus laevis*. Life Sci. **4**, 1981–1987 (1965)
Baker, P.C., Hoff, K.M., Buda, R.E.: Exogenous melatonin and melanophore development in *Xenopus*. Experientia (Basel) **34**, 1521–1522 (1978 a)
Baker, P.C., Hoff, K.M., Buda, R.E.: The effects of exogenous melatonin upon the gut of larval *Xenopus laevis*. Gen. Pharmacol. **9**, 431–432 (1978 b)
Balasubramanian, K.S., Saxena, R.N.: Effect of pinealectomy and photoperiodism in the reproduction of Indian weaver birds, *Ploceus philippinus*. J. exp. Zool. **185**, 333–340 (1973)
Baldessarini, R.J., Kopin, I.J.: S-Adenosylmethionine in brain and other tissues. J. Neurochem. **13**, 769–777 (1966)
Balemans, M.G.M.: Age-dependent effects of 5-methoxytryptophol and melatonin on testes and comb growth of the white leghorn (*Gallus domesticus* L.). J. Neural Transm. **33**, 179–194 (1972)
Balemans, M.G.M.: The stimulatory effects of several concentrations of 5-methoxytryptophol on testicular growth in the white leghorn (*Gallus domesticus* L.). J. Neural Transm. **34**, 49–60 (1973 a)
Balemans, M.G.M.: The inhibitory effect of 5-methoxytryptophol on ovarian weight, follicular growth and egg production of adult white leghorns hens (*Gallus domesticus* L.). J. Neural Transm. **34**, 159–169 (1973 b)
Balemans, M.G.M., Van de Kamer, J.C.: The influence of melatonin and 5-methoxytryptophol on gonadal maturation in *Gallus gallus*. Gen. comp. Endocr. **18**, 574 (1972)
Balemans, M.G.M., Ebels, I., Vonk-Visser, D.M.A.: Separation of pineal extracts on Sephadex G-10. I. A spectrofluorimetric study of indoles in a cockerel pineal extract. J. neur.-visc. Rel. **32**, 65–73 (1970)
Balemans, M.G.M., Van de Veerdonk, F.C.G., Van de Kamer, J.C.: The influence of 5-methoxytryptophol, a pineal compound, on comb growth, ovarian weight, follicular growth and egg production of juvenile, maturing and adult white Leghorn hens (*Gallus domesticus* L.). J. Neural Transm. **41**, 37–46 (1977 a)
Balemans, M.G.M., Van de Veerdonk, F.C.G., Van de Kamer, J.C.: Effect of injecting 5-methoxy indoles, pineal compound, on testicular weight of white Leghorn cockerels (*Gallus domesticus* L.). J. Neural Transm. **41**, 47–57 (1977 b)
Balemans, M.G.M., Bary, F.A.M., Legerstee, W.C., Noordegraaf, E.M.: Circadian rhythmicity in the methylation of 5-hydroxyindoles and norepinephrine in the pineal gland of 10 day old rats. J. Neural Transm. **43**, 167–175 (1978 a)
Balemans, M.G.M., Noordegraaf, E.M., Bary, F.A.M.: Estimation of the methylating capacity of the pineal gland. With special reference to indole metabolism. Experientia (Basel) **34**, 887–888 (1978 b)
Balemans, M.G.M., Bary, F.A.M., Legerstee, W.C., Van Benthem, J.: Estimation of the methylating capacity in the pineal gland of the rat with special reference to the methylation of N-acetylserotonin and 5-hydroxytryptophol separately. Experientia (Basel) **34**, 1434–1435 (1978 c)
Balestreri, R., Foppiani, E., Jacopino, E., Giordano, G.: Ghiandola pineale, melatonina e steroidogenesi testicolare. Arch. E. Maragliano Pat. Clin. **25**, 119–125 (1969)
Banerjee, S.E., Margulis, L.: Mitotic arrest by melatonin. Exp. Cell Res. **78**, 314–318 (1973)

Banerji, T.K., Quay, W.B.: Adrenal dopamine-β-hydroxylase activity: 24-hour rhythmicity and evidence for pineal control. Experientia (Basel) **32**, 253–254 (1976)

Banerji, T.K., Quay, W.B.: Modification of plasma dopamine-β-hydroxylase activity by adrenal and pineal extirpations, and time of day dependency of changes. Chronobiologia **5**, 379–385 (1978)

Bara, D., Skaliczki, J., Ormos, J.: Comparative study of histoenzymological properties of the supraoptic and paraventricular nuclei and of other nuclear regions of the hypothalamus, the epithalamus-epiphysis region and the subfornical organ. Hormones (Lond.) **2**, 164–183 (1971)

Baráth, P., Csaba, G.: Effect of pinealectomy on the tritiated serotonin uptake of the adrenal medulla. Horm. Metab. Res. **5**, 466–470 (1973)

Barbarossa, C., De Martino, C., Peruzy, A.D., Torlonia, G.: Attività gonadotropa ipofisaria nel ratto privato di pineale. Folia endocr. (Roma) **12**, 535–541 (1959)

Barbarossa, C., De Martino, C., Peruzy, D., Torlonia, G.: The pituitary-pineal antagonism. Acta endocr. (Kbh.), Suppl. **51**, 79–80 (1960)

Barbarossa, C., Peruzy, A.D., De Martino, C., Capone, M.: Comportamento della diuresi nel ratto privato di pineal e trattato con estratto epifisario. Folia endocr. (Kbh.) **14**, 895–900 (1961)

Barbarossa, C., Berni, A., Micarelli, R., Torlonia, G.: Contributo isto-morfologico alle correlazioni epifiso-tiroidee. G. Geront. **10**, 1333–1335 (1962a)

Barbarossa, C., Carnovale, M., Martina, G.D., Orofino, C.: Captazione tiroidea dello I^{131} nel ratto geronte. (Suoi rapporti con l'ipotalamo e l'epifisi.) G. Geront. **10**, 1329–1331 (1962b)

Barbarossa, C., Carnovale, M., Martina, G.D., Orofino, G.: Captazione tiroidea dello I^{131} nel vecchio trattato con estratto di epifisi. G. Geront. **11**, 939–941 (1963)

Barber, S.G.: Melatonin as a marker for pineal tumours. Brit. med. J. **1978 II**, 896

Barber, S.G., Smith, J.A., Hughes, R.C.: Melatonin as a tumour marker in a patient with pineal tumour. Brit. med. J. **1978 II**, 328

Barbolini, G., Barbansi, C., Trentini, G.P.: The possible relationship between the pineal gland and red cells destruction by the spleen. Virchows Arch. path. Anat. **341**, 255–258 (1966)

Barchas, J.D., Lerner, A.B.: Localization of melatonin in the nervous system. J. Neurochem. (Oxford) **11**, 489–491 (1964)

Barchas, J., Da Costa, F., Spector, S.: Acute pharmacology of melatonin. Nature (Lond.) **214**, 919–920 (1967)

Barchas, J., Conner, R., Levine, S., Vernikos-Danellis, J.: Effect of chronic melatonin and saline injections on pituitary adrenal secretion. Experientia (Basel) **25**, 413–414 (1969)

Bardasano Rubio, J.L.: La glandula pineal. Madrid: H. Blume Ediciones 1978

Barfuss, D.W., Ellis, L.C.: Seasonal cycles in melatonin synthesis by the pineal gland as related to testicular function in the House Sparrow (*Passer domesticus*). Gen. comp. Endocr. **17**, 183–193 (1971)

Barfuss, D.W., Tait, G.R., Ellis, L.C.: Melatonin synthesis in rat pineal glands inhibited by x-irradiation. Proc. Soc. exp. Biol. (N.Y.) **132**, 35–36 (1969)

Bargmann, W.: Die Epiphysis cerebri. In: Hdb. mikrosk. Anat. Mensch., Bd. VI, 4, pp. 309–502. Hrsg. W.v. Möllendorff. Berlin: Springer 1943

Bargmann, W.: Neurosekretion und hypothalamisch-hypophysäres System. Verh. anat. Ges. **51**, 30–45 (1954)

Bargmann, W.: Epiphysis cerebri. In: Pathologie der Laboratoriumstiere (P. Cohrs, R. Jaffé, H. Meessen, Hrsg.), S. 459–466. Berlin-Göttingen-Heidelberg: Springer 1958

Barone, R.M., Das Gupta, T.K.: Role of pinealectomy on Walker 256 carcinoma in rats. J. Surg. Oncol. **2**, 313–322 (1970)

Barrell, G.K., Lapwood, K.R.: Effects of pinealectomy of rams on secretory profiles of luteinizing hormone, testosterone, prolactin and cortisol. Neuroendocrinology **27**, 216–227 (1978)

Barrett, R.E., Neff, N.H., Costa, E., Duffy, P.: Compartmentation of rat pineal monoamines as revealed by histochemical, electronmicroscopic and pharmacological studies. J. Neuropath. exp. Neurol. **28**, 166 (1969)

Barry, J.: Etude histophysiologique des monoamines centrales chez le Lérot (*Eliomys quercinus*) en phase d'estivation et de préhibernation. C.R. Soc. Biol. (Paris) **165**, 853–858 (1971)

Bartlett, F.K.: A case of acromegaly and polyglandular syndrome, with special reference to the pineal gland. Arch. intern. Med. **12**, 201–213 (1913)

Baschieri, L., De Luca, F., Cramarossa, L., De Martino, C., Olieverio, A., Negri, M.: Modifications of thyroid activity by melatonin. Experientia (Basel) **19**, 15–17 (1963)

Basinska, J., Sastry, P.S., Stancer, H.C.: Lipid composition of human, bovine and sheep pineal glands. J. Neurochem. **16**, 707–714 (1969a)

Basinska, J., Sastry, P.S., Stancer, H.C.: ^{32}P uptake into phospholipids of pineal and other tissues. Fed. Proc. **28**, 835 (1969b)

Basinska, J., Sastry, P.S., Stancer, H.C.: Incorporation of ^{32}P orthophosphate into phospholipids of calf pineal slices in the presence and absence of neurotransmitters. Endocrinology **92**, 1588–1595 (1973)

Baum, M.J.: Pineal gland: influence on development of copulation in male rats. Science **162**, 586–587 (1968)

Baum, M.J.: Light-synchronization of rat feeding rhythms following sympathectomy and pinealectomy. Physiol. Behav. **5**, 325–329 (1970)

Baumann, Ch.: Lichtabhängige langsame Potentiale aus dem Stirnorgan des Frosches. Pflügers Arch. ges. Physiolog. **276**, 56–65 (1962)

Baumann, Ch.: Lichtaktivierte opponierende Prozesse im Stirnorgan (nach Untersuchungen langsamer Potentiale vom Pinealnerven des Frosches). Progr. Brain Res. **5**, 206–208 (1964)

Baumel, J.J.: Asymmetry of encephalic arteries in the pigeon (*Columba livia*). Anat. Anz. **111**, 91–102 (1962)

Bayer, A., Bayerová, G.: Elektronenmikroskopischer Beitrag zur Morphogenese der Konkremente in der menschlichen Epiphyse. Endokrinologie **52**, 1–3 (1967)

Bayerová, G., Bayer, A.: Beitrag zur cytochemischen Charakteristik einiger Zellarten in der menschlichen Epiphyse. Acta histochem. (Jena) **10**, 276–285 (1960)

Bayerová, G., Bayer, A.: Beitrag zur Fermenthistochemie der menschlichen Epiphyse. Acta histochem. (Jena) **28**, 169–173 (1967)

Bayerová, G., Malinský, J.: Quantitative ultrastructural study of rat pinealocytes during postnatal development. Folia morph. (Praha) **20**, 56–59 (1972)

Bayerová, G., Bayer, A., Obručník, M.: Zur Frage der fluoreszenz- und polarisations-mikroskopischen Untersuchungen an der menschlichen Epiphyse. Acta histochem. (Jena) **14**, 276–283 (1962)

Bayerová, G., Bayer, A., Malinský, D., Zapletal, B.: On functional significance of epiphysial pigment (electron-optic study). [In Russian.] Arch. anat. (Moskva) **48**, 18–21 (1965)

Bayrhuber, H.: Über die Synapsenformen und das Vorkommen von Acetylcholinesterase in der Epiphyse von *Bombina variegata* (L.), (Anura). Z. Zellforsch. **126**, 278–296 (1972)

Beattie, C.W., Glenny, F.H.: Some aspects of the vascularization and chemical histology of the pineal gland in *Gallus*. Anat. Anz. **118**, 396–404 (1966)

Beauvallet, M., Goodfeoy, F., Fugazza, J.: Influence d'une surcharge du régime en chlorure de sodium sur la teneur au 5-hydroxytryptamine du cerveau et de l'épiphyse chez le Rat. C.R. Soc. Biol. (Paris) **160**, 1218–1220 (1966)

Belsare, D.K.: Morphology of the pineal organ in some carps. Zool. Beitr. N.F. **20**, 47–54 (1974a)

Belsare, D.K.: Studies on the development of endocrine glands in fishes. II. Development of the pineal organ in *Clarias batrachus* L. Anat. Anz. **136**, 268–273 (1974b)

Benelbaz, G.A., Piezzi, R.S., Lynch, H.J.: Hydroxyindole-O-methyltransferase (HIOMT) and melatonin in the pineal gland of the antartic penguins (*Pygoscelis adeliae* and *P. papua*). Gen. comp. Endocr. **30**, 43–46 (1976)

Bensinger, R., Vaughan, M., Klein, D.C.: Isolation of a non-melatonin lipophylic antigonadotrophic factor from the bovine pineal gland. Fed. Proc. **32**, 252 (1973)

Benson, B.: Current status of pineal peptides. Neuoendocrinology **24**, 241–258 (1977)

Benson, B., Krasovich, M.: Circadian rhythm in the number of granulated vesicles in the pinealocytes of mice. Effects of sympathectomy and melatonin treatment. Cell Tiss. Res. **184**, 499–506 (1977)

Benson, B., Matthews, M.J.: Studies on non-melatonin pineal antigonadotropin. Endocrinology **88** (Suppl.) A-106 (1971)

Benson, B., Orts, R.J.: Regulation of ovarian growth by the pineal gland. In: Regulation of organ and tissue growth (R.J. Gross, ed.), pp. 315–335. New York: Academic Press 1972

Benson, B., Satterfield, V.: Ultrastructural characteristics of mouse pinealocytes following optic enucleation or continuous illumination. Anat. Rec. **181**, 312 (1975)

Benson, B., Matthews, M.J., Rodin, A.E.: A melatonin-free extract of bovine pineal with antigonadotropic activity. Life Sci. (I) **10**, 607–612 (1971)

Benson, B., Matthews, M.J., Rodin, A.E.: Studies on a non-melatonin pineal antigonadotrophin. Acta endocr. (Kbh.) **69**, 257–266 (1972a)

Benson, B., Matthews, M.J., Orts, R.J.: Presence of an antigonadotropic substance in rat pineal incubation media. Life Sci. (I) **11**, 669–677 (1972b)

Benson, B., Matthews, M.J., Hadley, M.E., Powers, S., Hruby, V.J.: Differential localization of antogonadotropic and vasotocic activities in bovine and rat pineal. Life Sci. **19**, 747–754 (1976a)

Benson, B., Matthews, M.J., Hruby, V.J.: Characterization and effects of a bovine pineal antigonadotropic peptide. Amer. Zoologist **16**, 17–24 (1976b)

Bererhi, A., Abbas-Terki, M.: Structure fine de l'épiphyse du magot d'Algérie (*Macacus sylvanus* L.). Bull. Ass. Anat. **148**, 285–294 (1970)

Berg, G.R., Klein, D.C.: Pineal gland melatonin production: site of action of norepinephrine and dibutyryl cyclic adenosine monophosphate. Fed. Proc. **29**, 615 (1970)

Berg, G.R., Klein, D.C.: Pineal gland in organ culture. II. Role of adenosine 3', 5'-monophosphate in the regulation of radiolabeled melatonin production. Endocrinology **89**, 453–464 (1971)

Berg, G.R., Klein, D.C., Glinsmann, W.H.: Pineal gland in organ culture: effect of norepinephrine and dibutyryl cyclic AMP on ^{32}P incorporation into phospholipids. Fed. Proc. **30**, 364 (1971)

Bergmann, G.: Elektronenmikroskopische Untersuchungen am Pinealorgan von *Pterophyllum scalare* Cuv. et Val. (Cichlidae, Teleostei). Z. Zellforsch. **119**, 257–288 (1971)

Bergmann, W., Engel, P.: Über den Einfluß von Zirbelextrakten auf Tumoren bei weißen Mäusen und bei Menschen. Wien. klin. Wschr. **62**, 79–82 (1950)

Bernád, I., Csaba, G.: Localization of biogenic monoamines in tissue cultures of rat pineal. Acta biol. Acad. Sci. hung. **21**, 235–241 (1970)

Bertler, Å., Falck, B., Owman, C.: Cellular localization of 5-hydroxytryptamine in the rat pineal gland. Kgl. Fysiogr. Sällsk. Lund, Förhdl. **33**, 13–16 (1963)

Bertler, Å., Falck, B., Owman, Ch.: Studies on 5-hydroxytryptamine stores in pineal gland of the rat. Acta physiol. scand. **63**, Suppl. 239, 1–18 (1964)

Bertolini, B., Mangia, F.: Osservazioni sulla ultrastruttura dell'occhio pineale della lampreda. Rend. Acc. Naz. Lincei (Roma) **41**, 147–153 (1966)

Bestle, J.: Extragonadal endodermal sinus tumors originating in the region of the pineal gland. Acta path. microbiol. scand. **74**, 214–222 (1968)

Bhargava, H.N.: The pineal organ of the minnow, *Phoxinus phoxinus* L. and a note on the effects of the temperature, light, background responses, hypophysectomy and gonadectomy. Zool. Jb. Anat. (Abt. Anat. Ontogen. Tier) **91**, 478–499 (1973)

Bianchini, P., Osima, B.: Studi su di un probabile ormone della pineale: l'anestrina. Boll. Soc. ital. Biol. sper. **36**, 1674–1677 (1960)

Bianchini, P., Osima, B.: Azione della pineale sulla ipertrofia gonadica compensatoria. Boll. Soc. ital. Biol. sper. **38**, 1547–1549 (1962)

Bick, R.L., Stuart, R.R., Giolli, R.A.: The effects of pinealectomy and environmental lighting on the gonads and thyroid gland of female Long-Evans rats. Anat. Rec. **160**, 316 (1968)

Bick, R.L., Giolli, R.A., Dearden, L.C., Stuart, R.R.: The effect of pinealectomy and environmental lighting on the gonadal, thyroid, and total body weight of female Long-Evans rats. Experientia (Basel) **25**, 531–532 (1969)

Bigelow, L.B.: Some effects of aqueous pineal extract administration on schizophrenia symptoms. In: Frontiers in pineal physiology (M.D. Altschule, ed.), pp. 225–263. Cambridge, Mass.-London, England: MIT Press 1975

Bigelow, L.B., Vaughan, M.K., Vaughan, G.M., Reiter, R.: Gonadal influence of aqueous pineal extracts: effects of pineal substances on compensatory ovarian hypertrophy in the mouse. Fed. Proc. **31**, 222 (1972)

Bindoni, M., Cambria, A.: Incorporazione *in vivo* del P^{32} nel DNA e frequenza delle mitosi in alcuni organi dell ratto pinealectomizzato. Boll. Soc. ital. Biol. sper. **42**, 1985–1987 (1966)

Bindoni, M., Cambria, A.: Effects of pinealectomy on the in vivo and in vitro biosynthesis of nucleic acids and on mitotic rate in some organs of the rat. Arch. Sci. biol. (Bologna) **52**, 271–283 (1968)

Bindoni, M., Raffaele, R.: Mitotic activity in the adenohypophysis of rats after pinealectomy. J. Endocr. **41**, 451–452 (1968)

Bindoni, M., Rizzo, R.: Effeti di lesioni elettrolitiche della ghiandola pineale sulla attivitá elettrica di alcune strutture encefaliche del coniglio. Boll. Soc. ital. Biol. sper. **40**, 2010–2013 (1964)

Bindoni, M., Rizzo, R.: Hippocampal evoked potentials and convulsive activity after electrolytic lesions of the pineal body, in chronic experiments on rabbits. Arch. Sci. biol. (Bologna) **49**, 223–233 (1965)

Binkley, S.: Comparative biochemistry of the pineal glands of birds and mammals. Amer. Zoologist. **16**, 57–65 (1976a)

Binkley, S.: Pineal gland biorhythms: N-acetyltransferase in chickens and rats. Fed. Proc. **35**, 2347–2352 (1976b)

Binkley, S.: A timekeeping enzyme in the pineal gland. Sci. Amer. **240**, 50–55 (1979)

Binkley, S., Geller, E.B.: Pineal enzymes in chickens: development of daily rhythmicity. Gen. Comp. Endocr. **27**, 424–429 (1975a)

Binkley, S., Geller, E.: Pineal N-acetyltransferase in chickens: J. comp. Physiol. **99**, 67–70 (1975b)

Binkley, S., Kluth, E., Menaker, M.: Pineal function in sparrows: circadian rhythms and body temperature. Science **174**, 311–314 (1971)

Binkley, S., Kluth, E., Menaker, M.: Pineal and locomotor activity: Levels and arrhythmia in sparrows. J. comp. Physiol. **77**, 163–169 (1972)

Binkley, S., Weller, J.L., Klein, D.C.: Protection of induced pineal N-acetyltransferase activity. Fed. Proc. **32**, 251 (1973a)

Binkley, S.A., Klein, D.C., Weller, J.L.: Dark induced increase in pineal serotonin N-acetyltransferase activity: a refractory period. Experientia (Basel) **29**, 1339–1340 (1973b)

Binkley, S., MacBride, S.E., Klein, D.C., Ralph, C.L.: Pineal enzymes: Regulation of avian melatonin synthesis. Science **181**, 273–275 (1973c)

Binkley, S., MacBride, S.E., Klein, D.C., Ralph, C.L.: Regulation of pineal rhythms in chickens: Refractory period and nonvisual light perception. Endocrinology **96**, 848–853 (1975)

Binkley, S., Klein, D.C., Weller, J.L.: Pineal serotonin N-acetyltransferase activity: protection of stimulated activity by acetyl-CoA and related compounds. J. Neurochem. **26**, 51–55 (1976)

Binkley, S., Stephens, J.L., Riebman, J.B., Reilly, K.B.: Regulation of pineal rhythms in chickens: photoperiod and dark-time sensitivity. Gen. comp. Endocr. **32**, 411–416 (1977a)

Binkley, S., Riebman, J.B., Reilly, K.B.: Timekeeping by the pineal gland. Science **197**, 1181–1183 (1977b)

Binkley, S., Riebman, J.B., Reilly, K.: Regulation of pineal rhythms in chickens: inhibition of dark-time N-acetyltransferase activity. Comp. Biochem. Physiol. **59C**, 165–171 (1978a)

Binkley, S.A., Riebman, J.B., Reilly, K.B.: The pineal gland: a biological clock in vitro. Science **202**, 1198–1201 (1978b)

Bischoff, M.B.: Ultrastructural evidence for secretory and photoreceptor functions in the avian pineal organ. J. Cell Biol. **35**, 13A (1967)

Bischoff, M.B.: Photoreceptoral secretory structures in the avian pineal organ. J. Ultrastruct. Res. **28**, 16–26 (1969)

Bischoff, M.B., Richter, W.R.: Some ultrastructural features of the pineal organ in Japanese quail (*Coturnix coturnix japonica*). 6th Int. Congr. Electron Microscopy, 1966. Vol. II, p. 523. Tokyo: Maruzen 1966

Biswas, N.M., Chakraborty, J., Chanda, S., Sanylas, S.: A basic experimental approach in perspective of pineal and melatonin involvement in photoperiod-induced alteration of spermatogenesis in toad (*Bufo melanostictus*). Endokrinologie **71**, 143–148 (1978)

Bittman, E.L.: Melatonin prevents refractoriness to short days in male hamsters. Proc. Soc. exp. Biol. (N.Y.) **158**, 359–362 (1978a)

Bittman, E.L.: Hamster refractoriness: the role of insensitivity of pineal target tissues. Science **202**, 648–650 (1978b)

Björklund, A., Owman, Ch., West, K.A.: Peripheral sympathetic innervation and serotonin cells in the habenular region of the rat brain. Z. Zellforsch. **127**, 570–579 (1972)

Blair, I.A., Seaborn, C.J.: The synthesis of melatonin antigens. Aust. J. Chem. **32**, 399–403 (1979)

Blair-West, J.R., Coghlan, J.P., Denton, D.A., Munro, J.A., Wintour, M., Wright, R.D.: The effect of bilateral nephrectomy and midcollicular decerebration with pinealectomy and hypophysectomy on the cortico-steroid secretion of sodium-deficient sheep. J. clin. Invest. **43**, 1576–1595 (1964)

Blake, C.A.: Effects of pinealectomy on the rat oestrous cycle and pituitary gonadotrophin release. J. Endocr. **69**, 67–75 (1976)

Blask, D.E., Reiter, R.J.: The pineal gland of the blind-anosmic female rat: Its influence on medial basal hypothalamic LRH, PIF and/or PRF activity in vivo. Neuroendocrinology **17**, 362–374 (1975a)

Blask, D.E., Reiter, R.J.: Pituitary and plasma LH and prolactin levels in female rats rendered blind and anosmic. Influence of the pineal gland. Biol. Reprod. **12**, 329–334 (1975b)

Blask, D.E., Reiter, R.J.: Pineal removal or denervation: effects on hypothalamic PRF activity in rat. Mol. Cell. Endocr. **11**, 243–248 (1978)

Blask, D.E., Vaughan, G.M.: PRF and PIF activity in the bovine, rat and human pineal gland. Anat. Rec. **184**, 361 (1976)

Blask, D.E., Vaughan, M.K., Reiter, R.J., Johnson, L.Y., Vaughan, G.M.: Prolactin-releasing and release-inhibiting factor activities in bovine, rat, and human pineal gland: *in vitro* and *in vivo* studies. Endocrinology **99**, 152–162 (1976)

Blask, D.E., Reiter, R.J., Johnson, L.Y.: Pineal-induced alterations in reproductive function and pituitary prolactin in the female rat: the effects of bilateral superior cervical ganglionectomy and nervi conarii transection. J. Neurosci. Res. **3**, 127–133 (1977)

Blask, D.E., Reiter, R.J., Vaughan, M.K., Johnson, L.Y.: Differential effects of the pineal gland on LH-RH and FSH-RH activity in the medial basal hypothalamus of the male Golden hamster. Neuroendocrinology **28**, 36–43 (1979)

Blin, P.C., Maurin, C.: Anatomie macroscopique de l'épiphyse des mammifères domestiques. Rec. Méd. vét. **132**, 36–53 (1956)

Bliss, D.K., Bates, P.L.: Choice intake of saline and water in pinealectomized rats under diurnal or constant lighting conditions. Physiol. Behav. **9**, 429–433 (1972)

Bliss, D.K., Bates, P.L.: A rapid and reliable technique for pinealectomizing rats. Physiol. Beh. **11**, 111–112 (1973)

Bloom, F.E., Giarman, N.J.: Fine structure of granular vesicles in pineal autonomic nerve endings after serotonin depletion. Anat. Rec. **157**, 351 (1967)

Bloom, F.E., Giarman, N.J.: The effects of p-Cl-phenylalanine on the content and cellular distribution of 5-HT in the rat pineal gland: combined biochemical and electron microscopic analysis. Biochem. Pharmacol. **19**, 1213–1219 (1970)

Blum, K., Reiter, R.J., Merritt, H.J., Wallace, J.E.: Ethanol preference as a function of the pineal gland. Fed. Proc. **32**, 730 (1973a)

Blum, K., Merritt, J.H., Reiter, R.J., Wallace, J.E.: A possible relationship between the pineal gland and ethanol preference in the rat. Cur. Ther. Res. **15**, 25–30 (1973b)

Blumfield, M.G., Tapp, E.: Measurements of pineal parenchymal cells and their nuclei in the albino rat at different ages. Acta morph. neerl. scand. **8**, 1–8 (1970)

Boccabella, A.V., Salgado, E.D., Alger, E.A.: Testicular function and histology following serotonin administration. Endocrinology **71**, 827–837 (1962)

Boeckmann, D.: Morphological investigation of the deep pineal of the rat. Cell Tiss. Res. **210**, 283–294 (1980)

Böttger, W.V., Böttger, E.M.: Degenerationsstudien am Nervus pinealis von *Rana esculenta* L. nach stirnorgannaher und -ferner Durchtrennung. Z. Zellforsch. **136**, 365–391 (1973)

Bogenschütz, H.: Über den Farbwechsel von *Rana esculenta* nach Epiphysektomie. Experientia (Basel) **23**, 967–968 (1967)

Bondareff, W.: Submicroscopic morphology of granular vesicles in sympathetic nerves of rat pineal body. Z. Zellforsch. **67**, 211–218 (1965a)

Bondareff, W.: Electron microscope study of the pineal body in aged rats. J. Geront. **20**, 321–327 (1965b)

Bondareff, W.: Localization of α-methylnorepinephrine in sympathetic nerve fibers of the pineal body. Exp. Neurol. **16**, 131–135 (1966a)

Bondareff, W.: Submicroscopic localization of norepinephrine in sympathetic nerves of rat pineal. Anat. Rec. **154**, 498–499 (1966b)

Bondareff, W., Gordon, B.: Submicroscopic localization of norepinephrine in sympathetic nerves of rat pineal. J. Pharmacol. exp. Ther. **153**, 42–47 (1966)

Borchardt, R.T., Wu, Y.S., Wu, B.S.: Mechanism of inhibition of pineal hydroxyindole-O-methyltransferase by pyridoxal 5′-phosphate. Biochem. Pharmacol. **27**, 120–122 (1978)

Borell, U., Örström, A.: Metabolism in different parts of the brain, especially in the epiphysis, measured with radioactive phosphorus. Acta physiol. scand. **10**, 231–242 (1945)

Borell, U., Örström, A.: The turnover of phosphate in the pineal body compared with that in other parts of the brain. Biochem. J. **41**, 398–403 (1947)

Bors, O., Ralston, W.C.: A simple assay of mammalian pineal extracts. Proc. Soc. exp. Biol. (N.Y.) **77**, 807–808 (1951)

Bostelmann, W.: Beitrag zur Kenntnis der Epiphysis cerebri unter Berücksichtigung ihrer Zytochemie. Wiss. Z. Univ. Rostock **12**, 437–453 (1963)

Bostelmann, W.: Beitrag zur submikroskopischen Zytologie der Epiphysis cerebri und zur experimentellen Beeinflussung ihrer Zellelemente. Zbl. allg. Path. path. Anat. **107**, 430–440 (1965)

Bostelmann, W.: Das ultrastrukturelle und enzymhistochemische Verhalten der Rattenzirbeldrüse nach Funktionsphasenwechsel durch Dauerbeleuchtung und ständige Dunkelheit. Endokrinologie **53**, 365–384 (1968a)

Bostelmann, W.: Enzymhistochemische und submikroskopische Untersuchungen zur Reparation der Rattenzirbeldrüse nach experimenteller Parenchymnekrose. Acta neuropath. (Berl.) **11**, 140–156 (1968b)

Bostelmann, W.: Submikroskopische und enzymhistochemische Veränderungen der Rattenzirbeldrüse während der Pubertät. Acta biol. med. germ. **22**, 151–157 (1969a)

Bostelmann, W.: Elektronenoptische und enzymhistochemische Untersuchungsergebnisse an der experimentell dysregulierten Ratten-Zirbeldrüse. In: Zirkumventrikuläre Organe und Liquor (Hrsg. G. Sterba), p. 107–111. Jena: Gustav Fischer 1969b

Bostelmann, W.: Der Einfluß der bilateralen Kastration auf die Feinstruktur der Zirbeldrüse. Endokrinologie **54**, 56–65 (1969c)

Bostelmann, W., Bienengräber, A.: Enzymhistochemie der Epiphysis cerebri des Kaninchens. Acta neuropath. (Berl.) **2**, 461–469 (1963)

Botticelli, A., Trentini, G.P., Barbolini, G., De Gaetani, C.F.: Histoenzymology of the "circumventricular organs" of the rat. V. The oxidoreductases, the transferases and the hydroxylases of the pineal gland. Acta histochem. (Jena) **44**, 288–299 (1972)

Bourges, M., Boucher, D.: Évolution infrastructurale de la glande pinéale de rat jusqu'à la naissance. J. Physiol. (Paris) **70**, 7B (1975)

Bowie, E.P., Herbert, D.C.: Immunocytochemical evidence for the presence of arginine vasotocin in the rat pineal gland. Nature (Lond.) **261**, 66 (1976)

Boya, J., Calvo, J.: Post-hatching evolution of the pineal gland of the chicken. Acta anat. (Basel) **101**, 1–9 (1978)

Boya, J., Zamorano, L.: Ultrastructural study of the pineal gland of the chicken (*Gallus gallus*). Acta anat. (Basel) **92**, 202–226 (1975)

Brack, E.: Morpho- und histogenetische Untersuchungen der Epiphysis cerebri vom Rind. Zbl. Veterinärmed. **9**, 905–924 (1962)

Brackmann, M.: Melatonin delays puberty in the Djungarian hamster. Naturwissenschaften **64**, 642–643 (1977)

Brackmann, M., Hoffmann, K.: Pinealectomy and photoperiod influence testicular development in the Djungarian hamster. Naturwissenschaften **64**, 341 (1977)

Brammer, G.L., Morley, J.E., Geller, E., Yuwiler, A., Hershman, J.M.: Hypothalamus-pituitary-thyroid axis interactions with pineal gland in the rat. Amer. J. Physiol. **236**, E 416-E 420 (1979)

Brammer, M.: Daily serotonin changes persist in rat pineals in organ culture. Life Sci. **24**, 967–972 (1979)

Braun, R.K., Gideon, L.A., Ribold, T.W.: Surgical approach for pinealectomy in the calf. Amer. J. vet. Res. **38**, 1973–1976 (1977)

Breder, C.M., Rasquin, P.: A preliminary report on the role of the pineal organ in the control of pigment cells and light reactions in recent teleost fishes. Science **111**, 10–12 (1950)

Bret, A.-J., Legros, R.: Essai d'utilisation clinique d'un extrait épiphysaire à action anti-FSH. Presse méd. **67**, 13–14 (1959)

Breucker, H.: Vergleichende histologische Studien an der Zirbel der Vögel. Verh. Anat. Ges., Anat. Anz., Suppl. **120**, 177–183 (1967)

Breucker, H., Horstmann, E.: Elektronenmikroskopische Untersuchungen am Pinealorgan der Regenbogenforelle (*Salmo irideus*). Progr. Brain Res. **10**, 259–269 (1965)
Brewer, G.F., Quay, W.B.: Pineal and hypophysial phosphate uptake after castration and administration of sex hormones. Proc. Soc. exp. Biol. (N.Y.) **98**, 361–364 (1958)
Brick, I.: Relationship of the pineal to the pituitary-melanophore effector system in *Ambystoma opacum*. Anat. Rec. **142**, 299 (1962)
Brooks, C. McC., Ishikawa, T., Koizumi, K.: Autonomic system control of the pineal gland and the role of this complex in the integration of body function. Brain Res. **87**, 181–190 (1975)
Brown, G.M., Basinska, J., Bubenik, G., Sibony, D., Grota, L.J., Stancer, H.C.: Gonadal effects of pinealectomy and immunization against N-acetylindolealkylamines in the hamster. Neuroendocrinology **22**, 289–297 (1976)
Brown, R.D., Cowan, R.L., Kavanaugh, J.F.: Effect of pinealectomy on seasonal androgen titers, antler growth and feed intake in white-tailed deer. J. Animal Sci. **47**, 435–440 (1978)
Brown, W., Jones, R., Forbes, J.M.: Effect of pinealectomy on prolactin, growth hormone and growth, in lambs. J. Endocr. **72**, 35 (1977)
Brown-Grant, K., Östberg, A.J.C.: Lack of effect of pineal denervation on the responses of the female albino rat to exposure to constant light. J. Endocr. **62**, 45–50 (1974)
Brownstein, M.J.: The pineal gland. Life Sci. **16**, 1363–1374 (1975)
Brownstein, M., Axelrod, J.: Pineal gland: 24-hour rhythm in norepinephrine turnover. Science **184**, 163–165 (1974)
Brownstein, M.J., Heller, A.: Hydroxyindole-O-methyltransferase activity: Effect of sympathetic nerve stimulation. Science **162**, 367–368 (1968)
Brownstein, M.J., Saavedra, J., Holz, R., Axelrod, J.: The control of indole metabolism in the rat pineal gland by a β-adrenergic receptor. Fed. Proc. **32**, 695 (1973a)
Brownstein, M.J., Saavedra, J.M., Axelrod, J.: Control of pineal N-acetylserotonin by a beta adrenergic receptor. Molec. Pharmacol. **9**, 605–611 (1973b)
Brownstein, M., Holz, R., Axelrod, J.: The regulation of pineal serotonin by a beta adrenergic receptor. J. Pharmacol. exp. Ther. **186**, 109–113 (1973c)
Brownstein, M.J., Palkovits, M., Saavedra, J.M., Bassiri, R.M., Utiger, R.D.: Thyrotropin-releasing hormone in specific nuclei of rat brain. Science **185**, 267–269 (1974)
Bruinvels, J., Houten, J.V. van, Noordwijk, J. van: Influence of pinealectomy and hypophysectomy on the renin content of rat kidneys. Quart. J. exp. Physiol. **49**, 95–102 (1964)
Bubenik, G.A., Brown, G.M., Uhlir, I., Grota, L.J.: Immunohistological localization of N-acetylindolealkylamines in pineal gland, retina and cerebellum. Brain Res. **81**, 233–242 (1974)
Bubenik, G.A., Brown, G.M., Grota, L.J.: Immunohistological investigations of N-acetylserotonin in the rat cerebellum after parachlorophenylalanine. Experientia (Basel) **32**, 579–581 (1976a)
Bubenik, G.A., Brown, G.M., Grota, L.J.: Differential localization of melatonin and N-acetylserotonin in CNS and the Harderian gland. Anat. Rec. **184**, 365–366 (1976b)
Bubenik, G.A., Brown, G.M., Grota, L.J.: Immunohistochemical localization of melatonin in the rat Harderian gland. J. Histochem. Cytochem. **24**, 1173–1177 (1976c)
Bubenik, G.A., Brown, G.M., Grota, L.J.: Immunohistological localization of melatonin in rat digestive system. Experientia (Basel) **33**, 662–663 (1977)
Bubenik, A., Purtill, R.A., Brown, G.M., Grota, L.J.: Melatonin in the retina and the Harderian gland. Ontogeny, diurnal variations and melatonin treatment. Exp. Eye Res. **27**, 323–333 (1978)
Bucana, C.D., Nadakavukaren, M.J., Frehn, J.L.: Annulate lamellae in hamster pineal gland. Tissue Cell **3**, 405–412 (1971)
Bucana, C.D., Nadakavukaren, M.J., Frehn, J.L.: Ultrastructural features of the pineal gland from cold-exposed golden hamsters. J. Neurocyt. **2**, 237–247 (1973)
Bucana, C.D., Nadakavukaren, M.J., Frehn, J.L.: Novel features of hamster pinealocyte ultrastructure. Tissue Cell **6**, 85–93 (1974)
Buda, M., Klein, D.C.: Suspension culture of pinealocytes: Regulation of N-acetyltransferase activity. Endocrinology **103**, 1483–1493 (1978)
Budd, G.C., Salpeter, M.M.: ^{3}H-Norepinephrine-labeled axons in electron microscope radioautograms of pineal body and adrenal capsule. J. Cell Biol. **39**, 19a (1968)
Budd, G.C., Salpeter, M.M.: The distribution of labeled norepinephrine within sympathetic nerve terminals studied with electron microscope radioautography. J. Cell Biol. **41**, 21–32 (1969)

Bugnon, C., Moreau, N.: Modifications cytologiques préhypophysaires chez le rat blanc épiphysectomisé ou soumis à l'action d'extraits acétoniques d'épiphyse de bœuf. Ann. Sci. Univ. Besançon, Méd. Sér. **2**, Fasc. **5**, 37–46 (1961 a)

Bugnon, C., Moreau, N.: Les formations neurosecretoires diencéphaliques du rat blanc, leur réponse à l'épiphysectomie ou à l'administration d'extraits acétonique d'épiphyse du bœuf. Ann. Sci. Univ. Besançon, Méd. Sér. **2**, Fasc. **5**, 47–52 (1961 b)

Bugnon, C., Moreau, N.: Recherches préliminaires sur d'éventuelles relations épiphyso-surrénaliennes chez le rat blanc. Bull. Assoc. Anat. Trimest. **111**, 197–206 (1962)

Bugnon, C., Moreau, N.: Recherches sur l'antagonisme épiphyso-hypophysaire et la glande thyroide chez le rat blanc. C.R. Ass. Anat. Bull. Ass. Anat. **117**, 377–387 (1963)

Bugnon, C., Moreau, N.: Etude histologique des corrélations entre l'épiphyse et l'adénohypophyse. Ann. Sci. Univ. Besançon, Méd. Sér. **2**, Fasc. **9**, 79–93 (1964)

Bugnon, C., Moreau, N., Lenys, R.: Recherches histophysiologiques sur les relations entre l'épiphyse et la zone glomérulaire du cortex surrénalien chez le rat blanc. Ann. Endocr. (Paris) **24**, 348–355 (1963)

Bunt, A.H.: Enzymatic digestion of synaptic ribbons in amphibian retinal photoreceptors. Brain Res. **25**, 571–577 (1971)

Bunt, A.H., Kelly, D.E.: Frog pineal photoreceptor renewal: preliminary observations. Anat. Rec. **171**, 99–116 (1971)

Burgers, A.C.J., Oordt, G.J. van: Regulation of pigment migration in the amphibian melanophore. Gen. comp. Endocr., Suppl. **1**, 99–109 (1962)

Bush, F.M., Seibel, H.R.: Blinding and pinealectomy in male hamsters: effect of pineal indoles and p-chlorophenylalanine on testes, seminal vesicles, pituitary and adrenal glands. Acta endocr. (Kbh.) **84**, 419–428 (1977)

Buss, H.: Vergleichende licht- und elektronenmikroskopische Untersuchungen zur orthologischen und experimentell variierten Feinstruktur der Epiphysis cerebri der Ratte. Thesis, Hamburg 1965

Buss, H., Gusek, W.: Zur Ultrastruktur der Epiphysis cerebri der Ratte unter dem Einfluß von Noradrenalin und Niamid. Endokrinologie **48**, 76–94 (1965)

Byrne, J.F.: Locomotor activity responses in juvenile sockeye salmon, *Oncorhynchus nerka*, to melatonin and serotonin. Canad. J. Zool. **48**, 1425–1427 (1970)

Cabassa, N., Sciascia, R.: L'influenza della melatonina sulla captaziona eritrocitaria della T_3-I^{131} nel ratto. Radioter. Radiobiol. Fis. med. **20**, 17–27 (1965)

Cady, P., Dillman, R.O.: Influence of catechol and indole amines upon pineal uptake of thyroxine. Neuroendocrinology **8**, 228–234 (1971)

Cady, P., Dillman, R.O., Rupnik, J.K.: Transport mechanisms and thyroxine uptake in pineal slices. Neuroendocrinology **8**, 86–93 (1971)

Cagnoni, M., Tarquini, B., Panerai, A., Toccafondi, R., Fantini, F.: Il ritmo circadiano della serotonina nella pineale del ratto albino trattato con corticosteroidi subito dopo la nascita. Rass. Neurol. veg. **20**, 422–426 (1966)

Calvo, J., Boya, J.: Embryonic development of the pineal gland of the chicken (*Gallus gallus*). Acta anat. (Basel) **101**, 289–303 (1978)

Calvo, J., Boya, J.: Ultrastructural study of the embryonic development of the pineal gland of the chicken (*Gallus gallus*). Acta anat. (Basel) **103**, 39–73 (1979 a)

Calvo, J., Boya, J.: Development of the innervation in the chicken pineal gland (*Gallus gallus*). Acta anat. (Basel) **103**, 212–225 (1979 b)

Cambria, A., Bindoni, M.: Incorporazione in vivo del P^{32} nell'RNA di alcuni organi del ratto pinealectomizzato. Boll. Soc. ital. Biol. sper. **42**, 1982–1985 (1966)

Camins, M.B., Schlesinger, E.B.: Treatment of tumours of the posterior part of the third ventricle and the pineal region: A long term follow-up. Acta neurochir. (Wien) **40**, 131–143 (1978)

Cammermeyer, J.: Mast cells and postnatal topographic anomalies in mammalian subfornical body and supraoptic crest. Z. Anat. Entwickl.-Gesch. **140**, 245–269 (1973)

Campbell, E., Gibson, M.A.: A histological and histochemical study of the development of the pineal gland in the chick, *Gallus domesticus*. Canad. J. Zool. **48**, 1321–1328 (1970)

Canepa, G.: Contributi all conoscenza delle relazioni tra organo subcommessurale ed epifisi in alcuni mammiferi. Arg. Anat. Antropol. (Lisboa) **26**, 513–528 (1949)

Cardinali, D.P.: Melatonin and the endocrine role of the pineal organ. In: Current topics in experimental endocrinology (V.H.T. James and L. Martini, eds.), Vol. 2, pp. 107–127. New York: Academic Press 1974

Cardinali, D.P.: Changes in hypothalamic neurotransmitter uptake following pinealectomy, superior cervical ganglionectomy or melatonin administration to rats. Neuroendocrinology **19**, 91–95 (1975)

Cardinali, D.P.: Nuclear receptor estrogen complex in the pineal gland. Modulation by sympathetic nerves. Neuroendocrinology **24**, 333–346 (1977)

Cardinali, D.P., Freire, F.: Melatonin effects on brain: interaction with microtubule protein, inhibition of fast axoplasmic flow and induction of crystalloid and tubular formations in the hypothalamus. Mol. Cell. Endocr. **2**, 317–330 (1975)

Cardinali, D.P., Rosner, J.M.: Retinal localization of the hydroxyindole-O-methyltransferase in the rat. Endocrinology **89**, 301–303 (1971)

Cardinali, D.P., Vacas, M.I.: Progesterone-induced decrease of pineal protein synthesis in rats. Possible participation in estrous-related changes of pineal function. J. Neural Transm. **42**, 193–205 (1978a)

Cardinali, D.P., Vacas, M.I.: Mechanisms underlying hormone effects on pineal function: a model for the study of integrative neuroendocrine processes. J. Endocr. Invest. **1**, 89–96 (1978b)

Cardinali, D.P., Wurtman, R.J.: Hydroxyindole-O-methyltransferases in rat pineal, retina and Harderian gland. Endocrinology **91**, 247–252 (1972)

Cardinali, D.P., Wurtman, R.J.: Methods for assessing the biological activity of the mammalian pineal organ. Meth. Enzymol. **39**, 376–397 (1975)

Cardinali, D.P., Cuello, A.E., Tramezzani, J.H., Rosner, J.M.: Effects of pinealectomy on the testicular function of the adult male duck. Endocrinology **89**, 1082–1093 (1971)

Cardinali, D.C., Larin, F., Wurtman, R.J.: Action spectra for effects of light on hydroxyindole-O-methyltransferases in rat pineal, retina and Harderian gland. Endocrinology **91**, 877–886 (1972a)

Cardinali, D.P., Lynch, H.J., Wurtman, R.J.: Binding of melatonin to human and rat plasma proteins. Endocrinology **91**, 1213–1218 (1972b)

Cardinali, D.P., Larin, F., Wurtman, R.J.: Control of the rat pineal gland by light spectra. Proc. nat. Acad. Sci. (Wash.) **69**, 2003–2005 (1972c)

Cardinali, D.P., Nagle, C.A., Denari, J.H.: Lack of effects of light on the rat pineal in organ culture. Gen. comp. Endocr. **21**, 573–577 (1973a)

Cardinali, D.P., Hyyppä, M.T., Wurtman, R.J.: Fate of intracisternally injected melatonin in the rat brain. Neuroendocrinology **12**, 30–40 (1973b)

Cardinali, D.P., Nagle, C.A., Rosner, J.M.: Uptake and effects of melatonin on the synthesis of proteins by the rat cerebral cortex. Life Sci. **13**, 823–833 (1973c)

Cardinali, D.P., Nagle, C.A., Rosner, J.M.: Incorporation of ^{3}H-leucine into proteins in the hypothalamus and the anterior hypophysis. Effects of pinealectomy, superior cervical ganglionectomy and continuous exposure to light or darkness. Neuroendocrinology **16**, 74–83 (1974a)

Cardinali, D.P., Nagle, C.A., Rosner, J.M.: Metabolic fate of androgens in the pineal organ: uptake, binding to cytoplasmic proteins and conversion of testosterone into 5 α-reduced metabolites. Endocrinology **95**, 179–187 (1974b)

Cardinali, D.P., Nagle, C.A., Rosner, J.M.: Effects of estradiol on melatonin and protein synthesis in the rat pineal organ. Hormone Res. **5**, 304–310 (1974c)

Cardinali, D.P., Nagle, C.A., Rosner, J.M.: Aromatization of androgens to estrogens by the pineal gland. Experientia (Basel) **30**, 1222–1223 (1974d)

Cardinali, D.P., Nagle, C.A., Rosner, J.M.: Control of estrogen and androgen receptors in the rat pineal gland by catecholamine transmitter. Life Sci. **16**, 93–106 (1975a)

Cardinali, D.P., Nagle, C.A., Freire, F., Rosner, J.M.: Effects of melatonin on neurotransmitter uptake and release by synaptosome-rich homogenates of the rat hypothalamus. Neuroendocrinology **18**, 72–85 (1975b)

Cardinali, D.P., Nagle, C.A., Gomez, E., Rosner, J.M.: Norepinephrine turnover in the rat pineal gland. Acceleration by estradiol and testosterone. Life Sci. **16**, 1717–1724 (1975c)

Cardinali, D.P., Nagle, C.A., Rosner, J.M.: Gonadal steroids as modulators of the function of the pineal gland. Gen. Comp. Endocr. **26**, 50–58 (1975d)

Cardinali, D.P., Freire, F., Nagle, C.A., Rosner, J.M.: Effects of environmental lighting, superior cervical ganglionectomy and adrenergic drugs on microtubule protein levels of the rat hypothalamus. Neuroendocrinology **19**, 44–53 (1975e)

Cardinali, D.P., Nagle, C.A., Rosner, J.M.: Gonadotrophin- and prolactin-induced increase of rat pineal hydroxyindole-O-methyltransferase: involvement of the sympathetic nervous system. J. Endocr. **68**, 341–342 (1976a)

Cardinali, D.P., Gomez, E., Rosner, J.M.: Changes in (^{3}H)leucine incorporation into pineal proteins following estradiol or testosterone administration: involvement of the sympathetic superior cervical ganglion. Endocrinology **98**, 849–858 (1976b)

Cardinali, D.P., Vacas, M.I., Boyer, E.E.: High affinity of melatonin in bovine medial basal hypothalamus. IRCS Med. Sci. **6**, 357 (1978)

Carnicelli, A., Saba, P., Cella, P., Marescotti, V.: Effects of epiphysectomy on karyometry of hypothalamic nuclei in rats. Folia endocr. **16**, 229–234 (1963)

Carson, R.S., Matthews, C.D., Findlay, J.K., Symons, R.G., Burger, H.G.: Biological and immunological luteinising hormone-releasing hormone (LH-RH) activity of the ovine pineal. Neuroendocrinology **24**, 221–225 (1977)

Cassano, C., Torsoli, A., Peruzy, A.D., De Martino, C.: Studi su l'epifisi. Indagini nell'animale da esperimento e nell'uomo. Folia endocr. (Pisa) **14**, 755–790 (1961)

Cattabeni, F., Koslow, S.H., Costa, E.: Gas chromatographic-mass spectrometric assay of four indole alkylamines of rat pineal. Science **178**, 166–168 (1972)

Chang, P.L.: Light and electron microscopic studies on the interstitial fibres in the rat pineal gland. Mem. Coll. Med. Taiwan Univ. **14**, 1–14 (1969)

Charlton, H.M.: Uptake of labelled precursors of melatonin by the epiphysis of *Xenopus laevis*. Nature (Lond.) **204**, 1093–1094 (1964)

Charlton, H.M.: The uptake of ^{14}C 5-hydroxytryptamine creatinine sulphate and ^{14}C methylmethionine by the epiphysis of *Xenopus laevis* Daudin. Comp. Biochem. Physiol. **17**, 777–784 (1966a)

Charlton, H.M.: The pineal gland and color change in *Xenopus laevis* Daudin. Gen. comp. Endocr. **7**, 384–397 (1966b)

Charlton, H.M.: The pineal gland of *Xenopus laevis*, Daudin: a histological, histochemical, and electron microscopic study. Gen. comp. Endocr. **11**, 465–480 (1968)

Charlton, H.M., Grocock, C.A., Ostberg, A.: The effects of pinealectomy and superior cervical ganglionectomy on the testis of the vole, *Microtus agrestis*. J. Reprod. Fert. **48**, 377–379 (1976)

Charvát, Z.: Vývoj epifysy u kormorána (*Phalacrocorax carbo* L.). Čs. Morfol. **2**, 156–164 (1954)

Cheesman, D.W.: Structure elucidation of a gonadotropin inhibiting substance from the bovine pineal gland. Biochim. biophys. Acta (Amst.) **207**, 247–253 (1970)

Cheesman, D.W., Fariss, B.L.: Isolation and characterization of a gonadotropin-inhibiting substance from the bovine pineal gland. Proc. Soc. exp. Biol. (N.Y.) **133**, 1254–1256 (1970)

Cheesman, D.W., Forsham, P.H.: Inhibition of induced ovulation by a highly purified extract of the bovine pineal gland. Proc. Soc. exp. Biol. (N.Y.) **146**, 722–724 (1974a)

Cheesman, D.W., Forsham, P.H.: Action of a gonadotropin inhibitor from the bovine pineal gland. Proc. Soc. exp. Biol. (N.Y.) **147**, 438–440 (1974b)

Cheesman, D.W., Osland, R.B., Forsham, P.H.: Effects of 8-arginine vasotocin on plasma prolactin and follicle-stimulating hormone surges in the proestrous rat. Proc. Soc. exp. Biol. (N.Y.) **156**, 369–372 (1977a)

Cheesman, D.W., Osland, R.B., Forsham, P.H.: Suppression of the preovulatory surge of luteinizing hormone and subsequent ovulation in rat by arginine vasotocin. Endocrinology **101**, 1194–1202 (1977b)

Chen, C.L., Besch, E.L.: Effect of altered photoperiod on estrous cycle, pregnancy and onset of puberty in the rat. Biol. Reprod. **12**, 396–399 (1975)

Cherdyntsev, S.G.: On the functional connection of the epiphysis with the thyroid gland. Probl. Éndokr. Gormonoter. (Mosk.) **10**, 83–86 (1967)

Chèze, G.: Contribution á l'étude morphologique de l'encéphale de *Pleuronectes platessa* L., étude particulière de l'épiphyse. Bull. Soc. Zool. Fr. **91**, 593–604 (1966)

Chèze, G.: Étude morphologique, histologique et expérimentale de l'épiphyse de *Symphodus melops* (Poisson, Labridé). Bull. Soc. Zool. Fr. **94**, 697–704 (1969)

Chèze, G., Lahaye, J.: Étude morphologique de la région épiphysaire de *Gambusia affinis holbrooki* G. Incidences histologiques de certains facteurs externes sur 'e toit diencéphalique. Ann. Endocr. (Paris) **30**, 45–53 (1969)

Chu, E.W., Wurtman, R.J., Axelrod, J.: An inhibitory effect of melatonin on the estrous phase of the estrous cycle of the rodent. Endocrinology **75**, 238–242 (1964)

Cintron, C., Kelly, D.E.:An analysis of pineal regeneration in developing newts. Amer. Zool. **1**, 441 (1961)

Citharel, A., Ebels, I., L'Héritier, A., Moszkowska, A.: Epiphyseal-hypothalamic interaction. An in vitro study with some sheep pineal fractions. Experientia (Basel) **29**, 718–719 (1972)

Clabough, J.W.: Ultrastructural features of pineal organs from hamsters with pineal-induced gonadal atrophy. Anat. Rec. **163**, 169 (1969)

Clabough, J.W.: Ultrastructural study of pineal cytogenesis in fetal rats and hamsters. Anat. Rec. **166**, 291 (1970)

Clabough, J.W.: Ultrastructural features of the pineal gland in normal and light deprived golden hamsters. Z. Zellforsch. **114**, 151–164 (1971)

Clabough, J.W.: Cytological aspects of pineal development in rats and hamsters. Amer. J. Anat. **137**, 215–230 (1973)

Clabough, J.W., Norvell, J.E.: Effects of castration, blinding, and the pineal gland on the Harderian glands of the male golden hamster. Neuroendocrinology **12**, 344–353 (1973)

Clabough, J.W., Norvell, J.E.: Pineal influence on the Harderian glands of female golden hamsters. Anat. Rec. **178**, 119–125 (1974)

Clabough, J.W., Seibel, H.R.: Studies of the pineal glands of rats and hamsters during ontogenesis and under various experimental conditions. Anat. Rec. **160**, 463 (1968)

Clara, M.: Beiträge zur Histotopochemie des Vitamin C im Nervensystem des Menschen. Z. mikr.-anat. Forsch. **52**, 359–392 (1942)

Clarke, J.R., Farrar, G.: The effect of pinealectomy and chemical sympathectomy upon gonadal function in the vole, *Microtus agrestis*. Acta endocr. (Kbh.), Suppl. **199**, 248 (1975)

Clementi, F., Fraschini, F., Müller, E., Ornesi, A., Zanoboni, A.: Sui rapporti tra epifisi e secrezione aldosteronica: e sul meccanismo della inibizione della diuresi da reserpina. Atti Acad. med. lombarda **17**, 209–221 (1962)

Clementi, F., Müller, E., Zanoboni, A.: Pineal function and modifications of its ultrastructural aspects. Proc. 2nd Int. Congr. Endocrinology. Excerpta Med. Int. Congr. Series, No. **83**, 364–374 (1965a)

Clementi, F., Fraschini, F., Müller, E., Zanoboni, A.: The pineal gland and the control of electrolyte balance and of gonadotropic secretion. Functional and morphological observations. Progr. Brain Res. **10**, 585–603 (1965b)

Clementi, F., De Virgiliis, G. Fraschini, F., Mess, B.: Modifications of pituitary morphology following pinealectomy and the implantation of the pineal body in different areas of the brain. Proc. 6th Int. Congr. Electron. Microscopy, Vol. 2, pp. 539–540. Tokyo: Maruzen 1966

Clementi, F., De Virgiliis, G., Mess, B.: Influence of pineal gland principles on gonadotrophin-producing cells of the rat anterior pituitary gland: an electron microscopic study. J. Endocr. **44**, 241–246 (1969)

Clive, D., Snell, R.S.: Effect of melatonin on mammalian hair color. J. invest. Derm. **53**, 159–162 (1969)

Cobb, S., Edinger, T.: The brain of the emu (*Dromaeus novaehollandiae*, Lath) I. Gross anatomy of the brain and pineal body. Breviora (Cambr.) **170**, 1–18 (1962)

Cogburn, L.A., Harrison, P.C., Brown, D.E.: Scotophase-dependent thermoregulatory dysfunction in pinealectomized chickens. Proc. Soc. exp. Biol. (N.Y.) **153**, 197–201 (1976)

Cohen, M., Roselle, D., Chabner, B., Schmidt, T.J., Lippman, M.: Evidence for a cytoplasmic melatonin receptor. Nature (Lond.) **274**, 894–895 (1978)

Cole, E.R., Crank, G.: Tryptamines I. The chromatography of melatonin. J. Chromatogr. **61**, 225–230 (1971)

Collier, R.: Über den Feinbau der Epiphysis cerebri von Nagetieren und die Frage seiner funktionellen Veränderungen. Z. Zellforsch. **33**, 51–67 (1943)

Collin, J.-P.: Contribution a l'étude des follicules de l'épiphyse embryonnaire d'Oiseau. C.R. Acad. Sci. (Paris) D **262**, 2263–2266 (1966a)

Collin, J.-P.: Étude préliminaire des photorécepteurs rudimentaires de l'épiphyse de *Pica pica* L. pendant la vie embryonnaire et postembryonnaire, C.R. Acad. Sci. (Paris) D **263**, 660–663 (1966b)

Collin, J.-P.: Sur l'évolution des photorécepteurs rudimentaires épiphysaires chez la Pie (*Pica pica* L.). C.R. Soc. Biol. (Paris) **160**, 1876–1880 (1966c)

Collin, J.-P.: Recherches préliminaires sur les propriétés histochimiques de l'épiphyse de quelques Lacertiliens. C.R. Acad. Sci. (Paris) D **265**, 1827–1830 (1967a)
Collin, J.-P.: Structure, nature sécrétoire, dégénéréscence partielle des photorécepteurs rudimentaires chez *Lacerta viridis* (Laurenti). C.R. Acad. Sci. (Paris) D **264**, 647–650 (1967b)
Collin, J.-P.: Nouvelles remarques sur l'épiphyse de quelques Lacertiliens et Oiseaux. C.R. Acad. Sci. (Paris) D **265**, 1725–1728 (1967c)
Collin, J.-P.: L'épiphyse des Lacertiliens: relations entre les données microélectroniques et celles de l'histochimie (en fluorescence ultraviolette) pour la détection des indole- et catécholamines. C.R. Soc. Biol. (Paris) **162**, 1785–1789 (1968a)
Collin, J.-P.: Pluralité des "photorécepteurs" dans l'épiphyse de *Lacerta*. C.R. Acad. Sci. (Paris) D **267**, 1047–1050 (1968b)
Collin, J.-P.: L'épithélium sensoriel de l'organe pinéal de la larve âgée et de l'adulte de *Lampetra planeri*. C.R. Acad. Sci. (Paris) D **267**, 1768–1771 (1968c)
Collin, J.P.: Rubans circonscrits par des vésicules dans les photorécepteurs rudimentaires épiphysaires de l'Oiseau: *Vanellus vanellus* (L.) et nouvelles considérations phylogénétiques relatives aux pinéalocytes (ou cellules principales) des Mammifères. C.R. Acad. Sci. (Paris) D **267**, 758–761 (1968d)
Collin, J.-P.: La cupule sensorielle de l'organe pinéal de la Lamproie de Planer. L'ultrastructure des cellules sensorielles et ses implications fonctionelles. Arch. Anat. micr. Morph. exp. **58**, 145–182 (1969a)
Collin, J.-P.: Contribution à l'étude de l'organe pinéal. De l'épiphyse sensorielle à la glande pinéale: modalités de transformation et implications fonctionelles. Ann. Stat. Biol. Besse-en-Chandesse, Suppl. **1**, 1–357 (1969b)
Collin, J.-P.: Cellules ganglionnaires et tractus de l'organe pinéal de *Lampetra planeri*. J. neur.-visc. Rel. **31**, 308–333 (1969c)
Collin, J.-P.: Distinction et rapports entré les pédicules basaux des photorécepteurs rudimentaires sécrétoires et les afférences nerveuses monoaminergiques de l'épiphyse d'Oiseau. Recherches chez le poussin de Passereau (*Pica pica* L.). C.R. Soc. Biol. (Paris) **163**, 1137–1142 (1969d)
Collin, J.-P.: Differentiation and regression of the cells of the sensory line in the epiphysis cerebri. In: The pineal gland. (eds. Wolstenholme, G.E.W., Knight, J.), pp. 79–125. Edinburgh and London: Churchill Livingstone 1971
Collin, J.-P.: La rudimentation des photorecepteurs dans l'organe pinéal des vertébrés. In: Colloques internationaux C.N.R.S. N°-266: Mécanismes de la rudimentation des organes chez les embryons de vertébrés. Toulouse 19.–26.9.1976, pp. 393–408 (1976)
Collin, J.-P., Kappers, J.A.: Electron microscopic study of pineal innervation in lacertilians. Brain Res. **11**, 85–106 (1968)
Collin, J.-P., Kappers, J.A.: Synapses of the ribbon type in the pineal organ of *Lacerta vivipara* (Reptiles, Lacertilians). Experientia (Basel) **27**, 1456–1457 (1971)
Collin, J.-P., Meiniel, A.: Les synapses de l'organe pinéal de l'ammocète. C.R. Acad. Sci. (Paris) D **266**, 1293–1295 (1968a)
Collin, J.-P., Meiniel, A.: Contribution à la connaissance des structures synaptiques du type ruban dans l'organe pinéal des vertébrés. Étude particulière en microscopie électronique des connexions de l'innervation efférente chez l'ammocète de Lamproie de Planer. Arch. Anat. micr. Morph. exp. **57**, 275–296 (1968b)
Collin, J.-P., Meiniel, A.: Amines biogènes de l'organe pinéal des reptiles. Gen. comp. Endocr. **18**, 583–584 (1972a)
Collin, J.-P., Meiniel, A.: L'organe pinéal de genre *Lacerta* (Reptile, Lacertilien): action d'enzymes proteolytiques sur les grains denses (500–3400 Å) des photorécepteurs rudimentaires. C.R. Soc. Biol. (Paris) **166**, 370–374 (1972b)
Collin, J.-P., Meiniel, A.: Métabolisme des indolamines dans l'organe pinéal de *Lacerta* (Reptiles, Lacertiliens). I. Intégration sélective de 5-HTP-^{3}H (5-hydroxytryptophane-^{3}H) et rétention de ses dérivés dans les photorécepteurs rudimentaires sécrétoires. Z. Zellforsch. **142**, 549–570 (1973a)
Collin, J.-P., Meiniel, A.: Métabolisme des indolamines dans l'organe pinéal de *Lacerta* (Reptiles, Lacertiliens). II. L'activité MAO et l'incorporation de 5-HTP-^{3}H et de 5-HT-^{3}H, dans les conditions normales et expérimentales. Z. Zellforsch. **145**, 331–361 (1973b)
Collin, J.-P., Calas, A., Juillard, M.T.: The avian pineal organ. Distribution of exogenous indolea-

mines: a qualitative study of the rudimentary photoreceptor cells by electron microscopic radioautography. Exp. Brain Res. **25**, 15–33 (1976)
Collin, J.-P., Juillard, M.T., Falcon, J.: Localization of 5-hydroxytryptamine and protein(s) in secretion granules of rudimentary photoreceptor cells in pineal of *Lacerta*. J. Neurocyt. **6**, 541–554 (1977)
Collu, R., Fraschini, F.: The pineal gland – a neuroendocrine transducer. In: Advances in metabolic disorders, Vol. 6 (R. Levine and R. Luft, eds.), pp. 161–175. New York: Academic Press 1972
Collu, R., Fraschini, F., Martini, L.: The effect of pineal methoxyindoles on rat vaginal opening time. J. Endocr. **50**, 679–683 (1971 a)
Collu, R., Fraschini, F., Martini, L.: Blockade of ovulation by melatonin. Experientia (Basel) **27**, 844 (1971 b)
Collu, R., Fraschini, F., Visconti, P., Martini, L.: Adrenergic and serotoninergic control of growth hormone secretion in adult male rats. Endocrinology **90**, 1231–1237 (1972)
Combescot, C., Demaret, J.: Histophysiologie de l'épiphyse chez la Tortue d'eau *Emys leprosa* (Schw.). Ann. Endocr. (Paris) **24**, 204–214 (1963)
Combescot, C., Demaret, J., Reynouard-Brault, F., Guyon, L.: L'épiphyse de l'axolotl. Arch. Anat. micr. Morph. exp. **53**, 261–270 (1964)
Cooper, E.R.A.: Further studies of arachnoid granulations in man. Acta anat. (Basel) **42**, 88–104 (1960)
Corbin, A.L., Schottelius, A.: Hypothalamic neurohormonal agents and sexual maturation of immature female rats. Amer. J. Physiol. **201**, 1176–1180 (1961)
Cordonnier, J.L., Roux, M.: Étude en histofluorescence des monoamines épiphysaires chez le Lérot (*Eliomys quercinus* L.) dans diverses conditions expérimentales au cours de la période hivernale. Ann. Endocr. (Paris) **39**, 403–410 (1978)
Cosnier, J.: Premiers résultats de greffes intratesticulaires d'épiphyse chez le Rat albinos. C.R. Soc. Biol. (Paris) **154**, 651–652 (1960)
Cotzias, G.C., Papavasiliou, P.S., Ginos, J., Steck, A., Düby, S.: Metabolic modification of Parkinson's disease and of chronic manganese poisoning. Ann. Rev. Med. **22**, 305–326 (1971 a)
Cotzias, G.C., Tang, L.C., Miller, S.T., Ginos, J.Z.: Melatonin and abnormal movements induced by l-dopa in mice. Science **173**, 450–452 (1971 b)
Cramer, H., Rudolph, J., Consbruch, U., Kendel, K.: On the effects of melatonin on sleep and behaviour in man. Advanc. Biochem. Psychopharmacol. **11**, 187–191 (1974)
Cravioto, H., Dart, D.: The ultrastructure of 'pinealoma' (Seminoma-like tumor of the pineal region). J. Neuropath. exp. Neurol. **32**, 522–565 (1973)
Cremer-Bartels, G.: Age dependence of melatonin biosynthesis in rat retina. Interdiscipl. Topics Geront. **13**, 46–50 (1978)
Csaba, G.: Hormonal regulation: morphogenetic and adaptive systems. Biol. Rev. **52**, 295–303 (1977)
Csaba, G., Baráth, P.: Are Langerhans's islets influenced by the pineal body? Experientia (Basel) **27**, 962 (1971)
Csaba, G., Baráth, P.: The effect of pinealectomy on the parafollicular cells of the rat thyroid gland. Acta anat. (Basel) **88**, 137–146 (1974 a)
Csaba, G., Baráth, P.: Effect of pinealectomy on the ^{3}H-^{5}HT uptake of mast cells in the thyroid gland of the rat. Acta anat. (Basel) **89**, 442–451 (1974 b)
Csaba, G., Baráth, P.: Morphological changes of thymus and the thyroid gland after postnatal exstirpation of pineal body. Endocrinol. exp. **9**, 59–67 (1975)
Csaba, G., Bernád, I.: In vitro model of feedback mechanism between different endocrine glands and rat pineal in tissue culture. Acta biol. Acad. Sci. hung. **23**, 91–93 (1972)
Csaba, G., Bókay, J.: The effect of melatonin and corpus pineale extract on serum electrolytes in the rat. Acta Biol. **28**, 143–144 (1977)
Csaba, G., Nagy, S.U.: The regulatory role of the pineal gland on the thyroid gland, adrenal medulla and the islets of Langerhans. Acta biol. med. germ. **31**, 617–619 (1973)
Csaba, G., Nagy, Z.U.: Effect of TSH and melatonin on thyroid activity in the rat. Acta physiol. **48**, 101–103 (1976)
Csaba, G., Bodoky, M., Törö, I.: Hormonal relationships of mastocytogenesis in lymphatic organs. II. Effect of epiphysectomy on the genesis of mast cells. Acta anat. (Basel) **61**, 289–296 (1965)

Csaba, G., Bodoky, M., Fischer, J., Ács, T.: The effect of pinealectomy and thymectomy on the immune capacity of the rat. Experientia (Basel) **22**, 168–169 (1966)

Csaba, G., Kiss, J., Bodoky, M.: The effect of pinealectomy on the Ca and SO_4 turnover of the lymphatic organs. Experientia (Basel) **23**, 148–149 (1967)

Csaba, G., Kiss, J., Bodoky, M.: Uptake of radioactive iodine by the thyroid after pinealectomy. Acta biol. Acad. Sci. hung. **19**, 35–41 (1968a)

Csaba, G., Dunay, C., Fischer, J., Bodoky, M.: Hormonal relationships of mastocytogenesis in lymphatic organs. III. Effect of the pineal body-thyroid-thymus system on mast cell production. Acta anat. (Basel) **71**, 565–580 (1968b)

Csaba, G., Reti, I., Fischer, J.: Effect of pineal body on thyroid-thymus correlations. ^{131}I uptake by rat thymus and salivary gland. Acta med. Acad. Sci. hung. **27**, 183–189 (1970)

Csillik, B.: Monoamines in pinealocytes. J. Cell. Biol. **19**, 17A (1963)

Cuellar, H.S.: Continuance of circannual reproductive refractoriness in pinealectomized parthogenetic whiptails (*Cnemidophorus uniparens*). J. exp. Zool. **206**, 207–214 (1978)

Cuello, A.C.: Ultrastructural characteristics and innervation of the pineal organ of the Antartic seal *Leptonychotes weddelli*. J. Morph. **141**, 217–226 (1973)

Cuello, A.C., Tramezzani, J.H.: The epiphysis cerebri of the Weddell seal: its remarkable size and glandular pattern. Gen. comp. Endocr. **12**, 153–164 (1969a)

Cuello, A.C., Tramezzani, J.H.: La epifisis cerebri de la foca de Weddell, su notable tamano y organizacion glandular. Contr. Inst. Antartico Argent. N° **121**, 1–20 (1969b)

Cuello, A.C., Hisano, N., Tramezzani, J.H.: The pineal gland and the photosexual reflex in female ducks. Gen. comp. Endocr. **18**, 162–168 (1972)

Curzio, A., Rendace, P., De Arcangelis, A.: Comportamento della metionina libera epatica in conigli trattori con estratto di pineale. Boll. Soc. ital. Biol. sper. **43**, 528–530 (1967)

Czarnocki, J., Sastry, P.S., Stancer, H.C.: The lipids of human pineal gland. In: Richter, D. (ed.), Biochemical factors concerned in the functional activity of the nervous system, p. 47. London: Pergamon Press 1969

Czyba, J.-C., Girod, C., Durand, M.: Sur l'antagonisme épiphyso-hypophysaire et les variations saisonnières de la spermatogénèse chez le Hamster doré (*Mesocricetus auratus*). C.R. Soc. Biol. (Paris) **158**, 742–745 (1964)

Czyba, J.-C., Girod, C., Curé, M., Durand, N.: Sur les corrélations épiphyse-testiculaires chez le hamster doré (*Mesocricetus auratus* Waterh.) C.R. Soc. Anat. **131**, 324–333 (1965)

Dafny, N.: Electrophysiological evidence of photic, acoustic, and central input to the pineal body and hypothalamus. Exp. Neurol. **55**, 449–457 (1977)

Dafny, N., Burks, T.F.: Opiate and endocrine interaction: morphine effects on hypothalamus and pineal body. Neuroendocrinology **22**, 72–88 (1977)

Dafny, N., McClung, R.: Pineal body: Neural recording. Experientia (Basel) **31**, 321–322 (1975)

Dafny, N., McClung, R., Strada, S.J.: Neurophysiological properties of the pineal body. 1. Field potentials. Life Sci. **16**, 611–620 (1975)

Damian, E.: Actiunea pinealei asupra functiei sexuale si asupra metabolismului lipidic. Stud. Cercet. Endocr. **23**, 389–396 (1972)

Damian, E.: Actiunea antigonadotropa a extractelor pineale. (Antigonadotropic effects of pineal extracts). Stud. Cercet. Endocr. **24**, 31–38 (1973)

Damian, E., Ianăs, O.: Controlul Prezentei Melatoninei in Extractele epifizare. Stud. Cercet. Endocr. **22**, 449–453 (1971)

Damian, E., Ionescu, M., Teodoru, V.: Actiunea extractelor din epifize de boi, vaci, tauri si juninci asupra respiratiei testiculare *in vitro*. Stud. Cercet. Endocr. **13**, 373–375 (1962)

Damian, E., Ianăs, O., Bădescu, I.: Actiunea Melatoninei Asupra Colesterolului Si a 17-Cetosteroizilor la sobolan. Stud. Cercet. Endocr. **24**, 105–108 (1973a)

Damian, E., Ianăs, O., Bădescu, I.: Actiunea Polipeptidelor Pineale in Ateroscleroza Experimentala. Stud. Cercet. Endocr. **24**, 393–403 (1973b)

Damian, E., Ianăs, O., Bădesco, I.: Réduction des lipides sériques et de l'aorte, après l'administration des polypeptides pinéau dans l'athéromatose expérimentale chez le Lapin. Rev. roum. Endocr. **11**, 201–207 (1974)

Damian, E., Ianăs, O., Bădescu, I., Oprescu, M.: Anti-LH and FSH activity of melatonin-free pineal extract. Neuroendocrinology **26**, 325–332 (1978)

Darre, M.J., Harrison, P.C.: Influence of gonadotropic releasing hormone, luteinizing hormone and pineal extracts on testes of immature cockerels. Poultry Sci. **56**, 2016–2021 (1977)

Dartnall, H.J.A. (ed.): Photochemistry of vision. Berlin-Heidelberg-New York: Springer 1972

Das Gupta, T.K.: Cellular hypertrophy in rat pineal after castration. J. Endocr. **41**, 607–608 (1968 a)

Das Gupta, T.K.: The anatomy of the pineal organ in the Syrian hamster. Acta anat. (Basel) **71**, 28–35 (1968 b)

Das Gupta, T.K.: Suppression of development of pineal gland by multiple pineal transplants in prepubertal rats. Acta anat. (Basel) **73**, 142–147 (1969)

Das Gupta, T.K., Terz, J.: Influence of pineal gland on the growth and spread of melanoma in the hamster. Cancer Res. **27**, 1306–1311 (1967 a)

Das Gupta, T.K., Terz, J.: Influence of pineal body on melanoma of hamsters. Nature (Lond.) **213**, 1038–1040 (1967 b)

David, G.F.X., Anand Kumar, T.C.: Histochemical localization of cholinesterase in neural tissue of pineal in rhesus monkey. Experientia (Basel) **34**, 1067–1068 (1978)

David, G.F.X., Herbert, J.: Experimental evidence for a synaptic connection between habenula and pineal ganglion in the ferret. Brain Res. **64**, 327–343 (1973)

David, G.F.X., Herbert, J., Wright, G.D.S.: The ultrastructure of the pineal ganglion in the ferret. J. Anat. (Lond.) **115**, 79–97 (1973)

David, G.F.X., Umberkoman, B., Kumar, K., Anand Kumar, T.C.: Neuroendocrine significance of pineal. In: Brain-endocrine interaction II. The ventricular system in neuroendocrine mechanisms. Proceeding of the Second Intern. Symposium on Brain-Endocrine Function held in Shizuola, Japan, October 16–18, 1974. Edit. by K.M. Knigge, D.E. Scott, H. Kobayashi, S. Ishii, pp. 365–375. Basel: Karger 1975

David, M., Bernard-Weil, E., Dilenge, D.: Les tumeurs de la glande pinéale. Ann. Endocr. (Paris) **24**, 287–330 (1963)

Davis, G.A.: Response of adenosine 3′,5′-monophosphate to norepinephrine in hypothalamus and pineal organ of female rats in proestrus or diestrus. Endocrinology **103**, 1048–1053 (1978)

Davis, J.O.: Mechanism regulating the secretion and metabolism of aldosterone in experimental secondary hyperaldosteronism. Recent Progr. Hormone Res. **17**, 293–330 (1961)

Dayan, A.D., Marshall, A.H.E., Miller, A.A., Pick, F.J., Pankin, N.E.: Atypical teratomas of the pineal and hypothalamus. J. Path. Bact. **92**, 1–28 (1966)

Debeljuk, L.: Effect of melatonin on the gonadotrophic function of the male rat under constant illumination. Endocrinology **84**, 937–939 (1969)

Debeljuk, L., Feder, V.M., Paulucci, O.A.: Effect of treatment with melatonin on the pituitary-testicular axis of the male rat. J. Reprod. Fertil. **21**, 363–364 (1970 a)

Debeljuk, L., Feder, V.M., Paulucci, O.A.: Effect of melatonin on changes induced by castration and testosterone in sexual structures of male rats. Endocrinology **87**, 1358–1363 (1970 b)

Debeljuk, L., Vilchez, J.A., Schnitman, M.A., Paulucci, O.A., Feder, V.M.: Further evidence for a peripheral action of melatonin. Endocrinology **89**, 1117–1119 (1971)

De Fronzo, R.A., Roth, W.D.: Lack of effect of pinealectomy on ovarian function: Evidence for a time-dependent action. Proc. Soc. exp. Biol. (N.Y.) **138**, 1058–1061 (1971)

De Fronzo, R.A., Roth, W.D.: Evidence for the existence of a pineal-adrenal and a pineal-thyroid axis. Acta endocr. (Kbh.) **70**, 35–42 (1972)

De Gallardo, M.G.P., Piezzi, R.S.: Serotonin content in the pineal gland of the Antarctic penguin (*Pygoscelis papua*). Gen. comp. Endocr. **21**, 468–471 (1973)

De Gallardo, M.G.P., Tramezzani, J.H.: Hydroxyindole-O-methyl-transferase activity in the pineal gland of the rabbit. J. Neural Transm. **36**, 51–57 (1975)

Degen, P.H., Doamaral, J.R., Barchas, J.D.: A gas-liquid chromatographic assay of melatonin and indoleamines using heptafluorobutyryl derivatives. Analyt. Biochem. **45**, 634–644 (1972)

Deguchi, T.: Role of the beta adrenergic receptor in the elevation of adenosine cyclic 3′,5′-monophosphate and induction of serotonin N-acetyltransferase in rat pineal glands. Molec. Pharmacol. **9**, 184–190 (1973)

Deguchi, T.: Shift of circadian rhythm of serotonin: acetyl coenzyme A N-acetyltransferase activity in pineal gland of rat in continuous darkness or in the blinded rat. J. Neurochem. **25**, 91–93 (1975 a)

Deguchi, T.: Characteristics of serotonin-acetyl-coenzyme A N-acetyltransferase in pineal gland of rat. J. Neurochem. **24**, 1083–1085 (1975b)

Deguchi, T.: Ontogenesis of a biological clock for serotonin acetyl coenzyme N-acetyltransferase in pineal gland of the rat. Proc. nat. Acad. Sci. (Wash.) **72**, 2814–2818 (1975c)

Deguchi, T.: Tryptophan hydroxylase in pineal gland of rat: postsynaptic localization and absence of circadian change. J. Neurochem. **28**, 667–668 (1977a)

Deguchi, T.: Circadian rhythms of enzyme and running activity under ultradian lighting schedule. Amer. J. Physiol. **232**, E 375-E 381 (1977b)

Deguchi, T.: Ontogenesis of circadian rhythm of melatonin synthesis in pineal gland of rat. J. Neural Transm., Suppl. **13**, 115–128 (1978)

Deguchi, T.: Circadian rhythm of serotonin N-acetyltransferase activity in organ culture of chicken pineal gland. Science **203**, 1245–1247 (1979)

Deguchi, T., Axelrod, J.: Control of circadian change of serotonin N-acetyltransferase activity in the pineal organ by the beta adrenergic receptor. Proc. nat. Acad. Sci. (Wash.) **69**, 2547–2550 (1972a)

Deguchi, T., Axelrod, J.: Induction and superinduction of serotonin N-acetyltransferase by adrenergic drugs and denervation in rat pineal organ. Proc. nat. Acad. Sci. (Wash.) **69**, 2208–2211 (1972b)

Deguchi, T., Axelrod, J.: Sensitive assay for serotonin N-acetyltransferase activity in rat pineal gland. Analyt. Biochem. **50**, 174–179 (1972c)

Deguchi, T., Axelrod, J.: Supersensitivity and subsensitivity of the β-adrenergic receptor in pineal gland regulated by catecholamine transmitter. Proc. nat. Acad. Sci. (Wash.) **70**, 2411–2414 (1973a)

Deguchi, T., Axelrod, J.: Superinduction of serotonin N-acetyltransferase and supersensitivity of adenyl cyclase to catecholamines in denervated pineal gland. Molec. Pharmacol. **9**, 612–618 (1973b)

Deguchi, T., Barchas, J.: Effect of p-chlorophenylalanine on tryptophan hydroxylase in rat pineal. Nature (Lond.) New Biol. **235**, 92–93 (1972a)

Deguchi, T., Barchas, J.: Effect of p-chlorophenylalanine on hydroxylation of tryptophan in pineal and brain of rats. Molec. Pharmacol. **8**, 770–779 (1972b)

Delahunty, G., Bauer, Prack, M., De Vlaming, V.: Effects of pinealectomy and melatonin treatment on liver and plasma metabolites in the goldfish, *Carassius auratus*. Gen. comp. Endocr. **35**, 99–109 (1978)

De la Motte, I.: Untersuchungen zur vergleichenden Physiologie der Lichtempfindlichkeit geblendeter Fische. Naturwissenschaften **50**, 363 (1963)

De la Motte, I.: Untersuchungen zur vergleichenden Physiologie der Lichtempfindlichkeit geblendeter Fische. Z. vergl. Physiol. **49**, 58–90 (1964)

De Luca, F., Cramarossa, L., Peruzy, A.D., Oliverio, A.: Variazioni della funzione tiroidea in corso di trattamento con estratto di pineale. Rass. Fisiopat. clin. ter. **33**, 396–405 (1961)

De Martino, C., Pavoni, P., Peruzy, A.D.: Modificazioni istologiche e captazione del ^{32}P della ghiandola pineale nel ratto maschio di varie etá. G. Geront. **9**, 275–276 (1961)

De Martino, C., Peruzy, A.D., Pavoni, P., Capone, M.: Captazione del ^{32}P ed aspetti istologici della ghiandola pineale nelle varie etá. Folia endocr. (Roma) **15**, 363–369 (1962)

De Martino, C., De Luca, F., Paluello, F.M., Tonietti, G., Orci, L.: The osmiophilic granules of the pineal body in rats. Experientia (Basel) **19**, 639–641 (1963)

De Martino, C., Tonietti, G., Accinni, L.: Electron microscopic study of impuberal and adult rats pineal body. Experientia (Basel) **20**, 556–557 (1964)

Demmel, U., Höck, A., Kasperek, K., Feinendegen, L.E.: Trace element concentrations in the human pineal body. In: Nuclear activation techniques in the life sciences 1978, pp. 193–203. Vienna: Int. Atomic Energy Agency 1979

Demoulin, A., Hudson, B., Legros, J.J., Franchimont, P.: Influence d'un extrait de glandes pinéales ovines sur la libération de prolactine *in vitro*. C.R. Soc. Biol. (Paris) **171**, 1134–1140 (1977a)

Demoulin, A., Hudson, B., Franchimont, P., Legros, J.J.: Arginine-vasotocin does not affect gonadotrophin secretion *in vitro*. J. Endocr. **72**, 105–106 (1977b)

Dengler, H.J., Wilson, C.W.M., Spiegel, H.E., Titus, E.: Uptake of norepinephrine by isolated pineal bodies. Biochem. Pharmacol. **11**, 795–801 (1962)

De Prospo, N., Hurley, J.: Effects of injecting melatonin and its precursors into the lateral cerebral ventricles on selected organs in rats. J. Endocr. **49**, 545–546 (1971a)

De Prospo, M.D., Hurley, J.: A comparison of intracerebral and intraperitoneal injections of melatonin and its precursors on ^{131}I uptake by the thyroid glands of rats. Agents Actions **2**, 14–17 (1971b)

De Prospo, N., Melgar, M.: Effects of p-chlorophenylalanine on pineal serotonin and ^{131}I uptake by the thyroid glands of rats. J. Endocr. **66**, 295–296 (1975)

De Prospo, N.D., De Martino, L.J., McGuinness, E.T.: Melatonin's effect on ^{131}I uptake by the thyroid glands in normal and ovariectomized rats. Life Sci. **7**, 183–188 (1968)

De Prospo, N.D., Safinsky, R.J., De Martino, L.J., McGuinness, E.T.: Melatonin and its precursors' effects on ^{131}I uptake by the thyroid gland under different photic conditions. Life Sci. **8**, 837–842 (1969)

Derenbach, K.: Über die Häufigkeit des Vorkommens quergestreifter Muskelfasern in der Epiphyse des Rindes. Morph. Jb. **91**, 266–272 (1951)

De Robertis, E., Pellegrino de Iraldi, A.: Plurivesicular secretory processes and nervey endings in the pineal gland of the rat. J. biophys. biochem. Cytol. **10**, 361–373 (1961a)

De Robertis, E., Pellegrino de Iraldi, A.: A plurivesicular component in adrenergic nerve endings. Anat. Rec. **139**, 299 (1961b)

Des Gouttes, M.-N.: Etude de quelques effets de la castration pratiquée à la naissance chez le Rat mâle. C.R. Soc. Biol. (Paris) **158**, 1260–1267 (1964)

De Smet, J., Van de Velde, J.: Influence d'un extrait d'épiphyse sur les dépôts de graisse du Rat. C.R. Soc. Biol. (Paris) **148**, 1907–1909 (1954)

Deussen-Schmitter, M., Garweg, G., Schwabedal, P.E., Wartenberg, H.: Simultaneous changes of the perivascular contact area and HIOMT activity in the pineal organ after bilateral adrenalectomy in the rat. Anat. Embryol. **149**, 297–305 (1976)

Devečerski, V.: Contribution à l'étude de l'effect de l'épiphysectomie sur l'histophysiologie du thymus. Acta anat. (Basel) **54**, 352–353 (1963)

Devečerski, V.: Contribution histoenzymologique à l'étude du metabolisme de la glande pinéale chez la ratte gravide. C.R. Soc. Biol. (Paris) **166**, 731–732 (1972a)

Devečerski, V.: Etude histoenzymologique de la glande pinéale du Rat en adaptation au froid. C.R. Soc. Biol. (Paris) **166**, 970–972 (1972b)

De Vlaming, V.L.: Effects of pinealectomy on gonadal activity in the cyprinid teleost, *Notemigonus crysoleucas*. Gen. comp. Endocr. **26**, 36 (1975)

De Vlaming, V.L., Vodicnic, M.J.: Effects of pinealectomy on pituitary gonadotrophs, pituitary gonadotropin potency and hypothalamic gonadotropin releasing activity in *Notemigonus crysoleucas*. J. Fish. Biol. **10**, 73–86 (1977)

De Vlaming, V., Vodicnik, M.J.: Seasonal effects of pinealectomy on gonadal activity in the goldfish, *Carassius auratus*. Biol. Repr. **19**, 57–63 (1978)

De Vlaming, V.L., Sage, M., Charlton, C.B.: The effects of melatonin treatment on gonosomatic index in the teleost, *Fundulus similis*, and the tree frog, *Hyla cinerea*. Gen. comp. Endocr. **22**, 433–438 (1974a)

De Vlaming, V.L., Sage, M., Charlton, C.B., Tiegs, B.: The effects of melatonin on lipid deposition in cyprinodontid fishes and on pituitary prolactin activity in *Fundulus similis*. J. comp. Physiol. **94**, 309–319 (1974b)

De Vries, R.A.C.: Some aspects of the functional relationship of the rat pineal gland to the hypothalamic magnocellular neurosecretory nuclei under different experimental conditions. Thesis, Purmerend, The Netherlands, Novy's Drukkerij (1972a)

De Vries, R.A.C.: Influence of pinealectomy on hypothalamic magnocellular neurosecretory activity in the female rat during normal light conditions, light-induced persistent oestrus, and after gonadectomy. Neuroendocrinology **9**, 244–249 (1972b)

De Vries, R.A.C.: Abolition of the effect of pinealectomy on hypothalamic magnocellular neurosecretory activity in male rats by hypothalamic pineal implants. Neuroendocrinology **9**, 358 (1972c)

De Vries, R.A.C., Kappers, J.A.: Influence of the pineal gland on neurosecretory activity of the supraoptic hypothalamic nucleus in the male rat. Neuroendocrinology **8**, 359–366 (1971)

Diaz-Flores, L., Aguilar, D., Linares, J., Garcia del Moral, R., Arjona, V.: Ultrastructura del germinoma pineal. Presencia de sistemas de laminillas anuladas y puffs (Cuerpos de Balbiani). Morfol. norm. y pat. Sec. B. **2**, 177–196 (1978)

Dickson, K.L., Hasty, D.L.: Effects of the pineal gland in unilaterally adrenalectomized rats. Acta endocr. (Kbh.) **70**, 438–444 (1972)

Dickson, K., Benson, B., Tate, G.: The effect of blinding and pinealectomy in unilaterally ovariectomized rats. Acta endocr. (Kbh.) **66**, 177–182 (1971)

DiDio, L.J.A., Allen, D.J., Litke, L.L., Yeasting, R.L.: Light and electron microscopic study of the pineal gland in five-day-old chicks. J. submicr. Cytol. **10**, 265–279 (1978)

Diederen, J.H.B.: A possible functional relationship between the subcommissural organ and the pineal complex and lateral eyes in *Rana esculenta* and *Rana temporaria*. Cell Tiss. Res. **158**, 37–60 (1975)

Diehl, B.J.M.: Occurrence and regional distribution of calcareous concretions in the rat pineal gland. Cell Tiss. Res. **195**, 359–366 (1978a)

Diehl, B.J.M.: Occurrence and regional distribution of striated muscle fibers in the rat pineal gland. Cell Tiss. Res. **190**, 349–355 (1978b)

Dierst-Davies, K., Landgrebe, F.W., Mitchell, G.M.: Melanophore-stimulating hormone-melatonin antagonism in relation to colour change in *Xenopus laevis*. J. Endocr. **49**, 573–580 (1971)

Dill, R.E.: The distribution of striated muscle in the epiphysis cerebri of the rat. Acta anat. (Basel) **54**, 310–316 (1963)

Dill, R.E., Walker, B.E.: Pineal cell proliferation in the mouse. Proc. Soc. exp. Biol. Med. (N.Y.) **121**, 911–912 (1966)

Dillman, R.O., Cady, P.: Thyroxine metabolism by pineal slices. Neuroendocrinology **8**, 198–204 (1971)

Disclos, P.: Épiphysectomie chez le têtard d'Alytes. C.R. Acad. Sci. (Paris) **258**, 3101–3103 (1964)

Dobrovolskaia-Zavadskaia, N., Zéphiroff, P.: Effect des produits d'origine épiphysaire et d'origine hépatique sur la croissance des tumeurs chez la souris. C.R. Soc. Biol. (Paris) **134**, 60–63 (1940)

Dodt, E.: Photosensitivity of the pineal organ in the teleost, *Salmo irideus* (Gibbons). Experientia (Basel) **19**, 642–643 (1963a)

Dodt, E.: Reversible Umsteuerung lichtempfindlicher Systeme bei Pflanzen und Tieren (Beispiel: Anuren). Experientia (Basel) **19**, 53–56 (1963b)

Dodt, E.: Physiologie des Pinealorgans anurer Amphibien. Vision Res. **4**, 23–31 (1964a)

Dodt, E.: Aktivierung markhaltiger und markloser Fasern im Pinealnerven bei Belichtung des Stirnorgans. Progr. Brain Res. **5**, 201–205 (1964b)

Dodt, E.: Das "dritte" Auge beim Grasfrosch. Umschau in Wiss. Technik **68**, 54 (1968)

Dodt, E.: The parietal eye (pineal and parietal organs) of lower vertebrates. In: Handbook of sensory physiology, Vol. VII/3B (E. Jung, ed.), pp. 113–140. Berlin-Heidelberg-New York: Springer 1973

Dodt, E., Heerd, E.: Mode of action of pineal nerve fibres in frogs. J. Neurophysiol. **25**, 405–429 (1962)

Dodt, E., Jacobson, M.: Photosensitivity of a localized region of the frog diencephalon. J. Neurophysiol. **26**, 752–758 (1963)

Dodt, E., Morita, Y.: Purkinje-Verschiebung, absolute Schwelle und adaptives Verhalten einzelner Elemente der intrakranialen Anuren-Epiphyse. Vision Res. **4**, 413–421 (1964)

Dodt, E., Morita, Y.: Conduction of nerve impulses within the pineal system of frog. Pflügers Arch. ges. Physiol. **293**, 184–192 (1967)

Dodt, E., Scherer, E.: Photic responses from the parietal eye of the lizard *Lacerta sicula campestris* (De Betta). Vision Res. **8**, 61–72 (1968)

Domański, E., Przekop, F., Skubiszewski, B., Wolińska, E.: The effect and site of action of indoleamines on the hypothalamic centers involved in the control of LH release and ovulation in sheep. Neuroendocrinology **17**, 265–273 (1975)

Donat, J.F., Okazaki, H., Gomez, M.R., Reagan, T.J., Baker, H.L., Laws, E.R.: Pineal tumors – 53-year experience. Arch. Neurol. (Chic.) **35**, 736–740 (1978)

Donham, R.S., Wilson, F.E.: Pinealectomy in Harris' sparrow. Auk. **86**, 553–555 (1969)

Donham, R.S., Wilson, F.E.: Photorefractoriness in pinealectomized Harris' sparrows. Condor **72**, 101–102 (1970)

Donley, C.S.: Color opponent slow potential interactions in the frontal organ of the frog: *Rana pipiens*. Vision Res. **15**, 245–251 (1975)

Donley, C.S., Meissl, H.: Sensitive change of the chromatic response of the pineal organ of the frog, *Rana esculenta*, by light adaptation. J. Physiol. (Lond.) **284**, 115P (1978)

Donofrio, R.J., Reiter, R.J.: Depressed pituitary prolactin levels in blinded anosmic female rats: role of the pineal gland. J. Reprod. Fertil. **31**, 159–162 (1972)

Donovick, P.J., Bliss, D.K., Burright, R.G., Wertheim, L.M.: Effect of pinealectomy or septal lesions on intake of unpalatable fluids by rats given sodium deplete or replete diets. Physiol. Behav. **10**, 1095–1099 (1973)

Dorn, E.: Über den Feinbau der Paraphyse von *Protopterus annectens*. Z. Zellforsch. **46**, 115–120 (1957)

Dowling, J.E., Boycott, B.B.: Organization of the primate retina: electron microscopy. Proc. Soc. B **166**, 80–111 (1966)

Dretzki, J.: Licht- und elektronenmikroskopische Untersuchungen zum Problem der Blut-Hirn-Schranke zirkumventrikulärer Organe der Ratte nach Behandlung mit Myofer. Z. Anat. Entwickl.-Gesch. **134**, 278–297 (1971)

Drexler, J.: The calcified pineal body and mammary carcinoma. Cancer (Philad.) **16**, 1616–1617 (1963)

Drexler, J., Meaney, T.F., McCormack, L.J.: Calcified pineal body and carcinoma. Cleveland Clin. Quart. **24**, 242–247 (1957)

Dubé, D., Leclerc, R., Pelletier, G., Arimura, A., Schally, A.V.: Immunohistochemical detection of growth hormone-release inhibiting hormone (somatostatin) in the guinea-pig brain. Cell Tiss. Res. **161**, 385–392 (1975)

Duchesne, P.Y.: Mise en évidence histochimique de catécholamines ou d'indolamines dans la glande pinéale du Rat et tentative d'identification. C.R. Soc. Biol. (Paris) **158**, 1174–1176 (1964)

Duffy, P.E., Markesbery, W.R.: Granulated vesicles in sympathetic nerve endings in the pineal gland: observations of the effects of pharmacologic agents by electron microscopy. Amer. J. Anat. **128**, 97–116 (1970)

Dunaway, J.E.: Effects of pinealectomy on timing of PMS-induced ovulation. Anat. Rec. **160**, 342 (1968)

Dunaway, J.E.: Alterations in the timing of PMS-induced ovulation following pinealectomy. Neuroendocrinology **5**, 281–289 (1969)

Dunaway, J.E., O'Steen, W.K.: Stimulatory effects of pinealectomy on gonadotrophin-induced ovulation in immature rats. Tex. Rep. Biol. Med. **25**, 525–529 (1967)

Duncan, D., Micheletti, G.: Notes on the fine structure of the pineal organ of cats. Tex. Rep. Biol. Med. **24**, 576–587 (1966)

Dupont, A., Labrie, F., Pelletier, G., Puviani, R., Coy, D.H., Coy, E.J., Schally, A.V.: Organ distribution of radioactivity and disappearance of radioactivity from plasma after administration of [^{3}H] luteinizing hormone-releasing hormone to mice and rats. Neuroendocrinology **16**, 65–73 (1974)

Eakin, R.M.: Photoreceptors in the amphibian frontal organ. Proc. nat. Acad. Sci. (Wash.) **47**, 1084–1088 (1961)

Eakin, R.M.: Lines of evolution of photoreceptors. In: General physiology of cell specialization (Mazia, D., and Tyler, A., eds.), pp. 393–425. New York: McGraw-Hill 1963

Eakin, R.M.: Development of the third eye in the lizard *Sceloporus occidentalis*. Rev. suisse Zool. **71**, 267–285 (1964a)

Eakin, R.M.: The effect of vitamin A deficiency on photoreceptors in the lizard *Sceloporus occidentalis*. Vision Res. **4**, 17–22 (1964b)

Eakin, R.M.: Evolution of photoreceptors. In: Evolutionary biology (Dobzhansky, T., Hecht, M.K., Steere, W.C., eds.), Vol. 2, pp. 194–242. Amsterdam: North Holland Publishing Company 1968

Eakin, R.M.: A third eye. Amer. Sci. **58**, 73–79 (1970)

Eakin, R.M.: The third eye. Berkeley: University of California Press 1973

Eakin, R.M., Stebbins, R.C.: Parietal eye nerve in the fence lizard. Science **130**, 1573–1574 (1959)

Eakin, R.M., Westfall, J.A.: Fine structure of the retina in the reptilian third eye. J. biophys. biochem. Cytol. **6**, 133–134 (1959)

Eakin, R.M., Westfall, J.A.: Further observations on the fine structure of the parietal eye of lizards. J. biophys. biochem. Cytol. **8**, 483–499 (1960)

Eakin, R.M., Westfall, J.A.: The development of photoreceptors in the stirnorgan of the treefrog, *Hyla regilla*. Embryologia (Nagoya) **6**, 84–98 (1961)

Eakin, R.M., Stebbins, R.C., Wilhoft, D.C.: Effects of parietalectomy and sustained temperature on thyroid of lizard, *Sceloporus occidentalis*. Proc. Soc. exp. Biol. (N.Y.) **101**, 162–164 (1959)

Eakin, R.M., Quay, W.B., Westfall, J.A.: Cytochemical and cytological studies of the parietal eye of lizard, *Sceloporus occidentalis*. Z. Zellforsch. **53**, 449–470 (1961)

Eakin, R.M., Quay, W.B., Westfall, J.A.: Cytological and cytochemical studies on the frontal and pineal organs of the tree frog, *Hyla regilla*. Z. Zellforsch. **59**, 663–683 (1963)

Earle, K.M.: X-ray diffraction and other studies on the calcareous deposits in human pineal glands. J. Neuropath. exp. Neurol. **24**, 108–118 (1965)

Ebadi, M.S., Weiss, B., Costa, E.: A sensitive and specific method for assaying cyclic 3′,5′-AMP in rat pineal gland. Fed. Proc. **29**, 263 (1970a)

Ebadi, M.S., Weiss, B., Costa, E.: Adenosine 3′,5′-monophosphate in rat pineal gland: Increase induced by light. Science **170**, 188–190 (1970b)

Ebadi, M.S., Weiss, B., Costa, E.: Diurnal periodicity of cyclic 3′,5′-adenosine monophosphate in rat pineal gland. Pharmacologist **12**, 489 (1970c)

Ebels, I.: Étude chimique des extraits épiphysaires fractionnés. Biol. méd. (Paris) **56**, 395–402 (1967)

Ebels, I.: Pineal factors other than melatonin. Gen. comp. Endocr. **25**, 189–198 (1975)

Ebels, I.: Isolation of avian and mammalian pineal indoles and antigonadotropic factors. Amer. Zoologist **16**, 5–15 (1976)

Ebels, I., Horwitz-Bresser, A.E.M.: Separation of pineal extracts by gelfiltration. IV. Isolation, location and identification from sheep pineals of three indoles, identical with 5-hydroxytryptophol, 5-methoxytryptophol and melatonin. J. Neural Transm. **38**, 31–41 (1976)

Ebels, I., Prop, N.: A study of the effect of melatonin on the gonads, the oestrus cycle, and the pineal organ of the rat. Acta endocr. (Kbh.) **49**, 567–577 (1965)

Ebels, L., Moszkowska, A., Scemana, A.: Étude in vitro des extraits épiphysaires fractionnés. Résultats préliminaires. C.R. Acad. Sci. (Paris) **260**, 5126–5129 (1965a)

Ebels, I., Versteeg, D.H.G., Vliegenthart, J.F.G.: An attempt to isolate arginine vasotocin from sheep and bovine pineal body. Proc. kon. ned. Akad. Wet. **63**, 27–30 (1965b)

Ebels, I., Moszkowska, A., Scemama, A.: An attempt to separate a sheep pineal extract fraction showing antigonadotropic activity. J. neuro-visc. Rel. **32**, 1–10 (1970)

Ebels, I., Balemans, M.G.M., Tommel, D.K.J.: Separation of pineal extracts on Sephadex G-10. III. Isolation and comparison of extracted and synthetic melatonin. Analyt. Biochem. **50**, 234–244 (1972a)

Ebels, I., Balemans, M.G.M., Verkley, A.H.: Separation of pineal extracts on Sephadex G-10. II. A spectrofluorimetric and thinlayer chromatographic study of indoles in a sheep pineal extract. J. neuro-visc. Rel. **32**, 270–282 (1972b)

Ebels, I., Benson, B., Matthews, M.J.: Localization of a sheep pineal antigonadotropin. Analyt. Biochem. **56**, 546–565 (1973)

Ebels, I., Citharel, A., Moszkowska, A.: Separation of pineal extracts by gel-filtration. III. Sheep pineal factors acting either on the hypothalamus, or on the anterior hypophysis of mice and rats in *in vitro* experiments. J. Neural Transm. **36**, 281–302 (1975)

Ebels, I., Benson, B., Bria, C.F., McDonnell, D., Chang, S.Y., Hruby, V.J.: Location by paper chromatography of compensatory ovarian hypertrophy (COH) inhibiting activity in acetic acid extracts from bovine pineals. J. Neural Transm. **42**, 275–292 (1978)

Ebels, I., De Morée, A., Hus-Citharel, A., Moszkowska, A.: A survey of some active pineal fractions and a discussion on the possible significance of pteridines in those fractions in *in vitro* and *in vivo* assays. J. Neural Transm. **44**, 97–116 (1979)

Eckhardt, W.: Über den Verlauf der Gliafasern in der Epiphysis cerebri des Rhesusaffen (*Macaca mulatta*). Acta anat. (Basel) **81**, 237–252 (1972)

Eddy, J.M.P.: Metamorphosis and the pineal complex in the brook lamprey, *Lampetra planeri*. J. Endocr. **44**, 451–452 (1969)

Eddy, J.M.P., Strahan, R.: The role of the pineal complex in the pigmentary effector system of the lampreys, *Mordacia mordax* (Richardson) and *Geotria australis* Gray. Gen. comp. Endocr. **11**, 528–534 (1968)

Eddy, J.M.P., Strahan, R.: The structure of the epiphyseal complex of *Mordacia mordax* and *Geotria australis* (Petromyzonidae). Acta zool. (Stockh.) **51**, 67–84 (1970)

Edinger, T.: The size of parietal foramen and organ in reptiles. A rectification. Bull. Mus. Comp. Zool. Harvard Univ. **114**, 1–34 (1955)

Eichberg, J., Schwartz, M., Shein, H., Hauser, G.: Stimulation of ^{32}P incorporation into phosphatidylglycerol (PG) and phosphatidylinositol (PI) by catecholamines and propranolol in rat pineal organ cultures. Fed. Proc. **31**, 453 (1972)

Eichler, V.B.: Pineal regeneration in the frog, *Rana pipiens*, following embryonic extirpation. J. Morph. **125**, 253–258 (1968)

Eichler, V.B., Moore, R.Y.: Alterations of gonadal and pineal responses to light by cervical sympathectomy in the male hamster. Amer. Zool. **10**, 300 (1970)

Eichler, V.B., Moore, R.Y.: Pineal hydroxyindole-O-methyltransferase and gonadal responses to blinding or continuous darkness blocked by pineal denervation in the male hamster. Neuroendocrinology **8**, 81–85 (1971)

Elden, C.A., Keyes, M.C., Marshall, C.E.: Pineal body of the northern fur seal (*Callorhinus ursinus*): A model for studying the probable function of the mammalian pineal body. Amer. J. vet. Res. **32**, 639–647 (1971)

Eldred, S.H., Bell, N.W., Sherman, L.J.: A pilot study comparing the effects of pineal extract and a placebo in patients with chronic schizophrenia. New Engl. J. Med. **263**, 1330–1335 (1960)

Eldred, W.D., Nolte, J.: Pineal photoreceptors: evidence for a vertebrate visual pigment with two physiologically active states. Vision Res. **18**, 29–32 (1978)

Ellis, L.C.: The direct action of melatonin and serotonin on testicular androgen production in vitro. J. Reprod. Fertil. **18**, 159 (1969)

Ellis, L.C.: Some interrelationships between experimental lighting, melatonin synthesis, and x-irradiation induced changes in the germinal epithelium of rat testes. Fed. Proc. **29**, 452 Abstr. (1970)

Ellis, L.C.: Inhibition of rat testicular androgen synthesis *in vitro* by melatonin and serotonin. Endocrinology **90**, 17–28 (1972)

Ellis, L.C.: Introduction to the symposium: The endocrine role of the pineal. Amer. Zool. **16**, 3–4 (1976a)

Ellis, L.C.: Peripheral and CNS effects of the pineal gland: Target enzymes common of tissues and species. Amer. Zool. **16**, 67–78 (1976b)

Ellis, L.C., Balph, D.F.: Age and seasonal differences in the synthesis and metabolism of testosterone by testicular tissue and pineal HIOMT activity of Uinta ground squirrels (*Spermophilus armatus*). Gen. comp. Endocr. **28**, 42–51 (1976)

Ellis, L.C., Buhrley, L.E.: Inhibitory effects of melatonin, prostaglandin E_1, cyclic AMP, dibutyryl-cyclic AMP and theophylline on rat seminiferous tubular contractility in vitro. Biol. Reprod. **19**, 217–222 (1978)

Ellis, L.C., Barfuss, D.W., Van Kampen, K.R.: Modification of rat testicular response to x-irradiation by various lighting regimes: 1. The pineal gland and melatonin synthesis as a possible mediator of the response. Life Sci. **9**, Part I, 1011–1020 (1970)

Ellis, L.C., Jaussi, A.W., Tait, G.R., Urry, R.L.: In vivo and in vitro effects of x-irradiation and histamine-PO_4 on rat and bovine pineal HIOMT activity and melatonin synthesis. Life Sci. **13**, 835–845 (1973)

Ellison, N., Weller, J.L., Klein, D.C.: Development of a circadian rhythm in the activity of pineal serotonin N-acetyltransferase. J. Neurochem. **19**, 1335–1341 (1972)

Endersby, C.A., Robson, J.M., Sullivan, F.M., Wilson, C.: The effect of 5-hydroxytryptamine on ovulation in rats. J. Endocrinol. **48**, 63–64 (1970)

Engbretson, G., Lent, C.M.: Parietal eye of the lizard: neuronal photoresponses and feedback from the pineal gland. Proc. nat. Acad. Sci. (Wash.) **73**, 654–657 (1976)

Engel, P., Bergmann, W.: Die physiologische Funktion der Zirbeldrüse und ihre therapeutische Anwendung. Z. Vitamin-, Hormon- u. Fermforsch. **4**, 564–594 (1951/52)

Eränkö, O., Eränkö, L.: Loss of histochemically demonstrable catecholamines and acetylcholinesterase from sympathetic nerve fibres of the pineal body of the rat after chemical sympathectomy with 6-hydroxydopamine. Histochem. J. **3**, 357–363 (1971)

Eränkö, O., Rechardt, L., Eränkö, L., Cunningham, A.: Light and electron microscopic histochemical observations on cholinesterase-containing sympathetic nerve fibres in the pineal body of the rat. Histochem. J. **2**, 479–489 (1970a)

Eränkö, O., Rechardt, L., Eränkö, L., Cunningham, A.: Acetylcholinesterase (AChE) activity in the sympathetic nerve fibres of the pineal body (PB) of the rat. Scand. J. clin. Lab. Invest. **25**, (Suppl. 113) 82 (1970b)

Erdinç, F.: Concrement formation encountered in rat pineal gland. Experientia (Basel) **33**, 514 (1977)

Ewig, J.E., Wickersham, E.W.: Effects of melatonin on steroidgenesis by bovine luteal tissue in vitro. Proc. Pa. Acad. Sci. **42**, 214–216 (1968)

Fabre, L.F., Jr., Banks, R.C., McIsaac, W.M., Farrell, G.: Effect of ubiquinone and related substances on secretion of aldosterone and cortisol. Amer. J. Physiol. **208**, 1275–1280 (1965)

Fajer, A., Hoxter, G., Fraga, E.: Electrophoretic pattern of the soluble proteins of a pineal body extract. Nature (Lond.) **196**, 274 (1962)

Falck, B., Hillarp, N.-Å., Thieme, G., Torp, A.: Fluorescence of catecholamines and related compounds condensed with formaldehyde. J. Histochem. Cytochem. **10**, 348–354 (1962)

Falck, B., Owman, C., Rosengren, E.: Changes in rat pineal stores of 5-hydroxytryptamine after inhibition of its synthesis or break-down. Acta physiol. scand. **67**, 300–305 (1966)

Falcon, J.: Pluralité et sites d'élaboration des messages de l'organe pinéal. Étude chez un vertébré inferieur: le brochet (*Esox lucius* L.) Thesis, Université de Poitiers, 1978

Falcon, J.: L'organe pinéal du Brochet (*Esox lucius* L.). I. Étude anatomique et cytologique. Ann. Biol. anim. **19**, 445–465 (1979a)

Falcon, J.: L'organe pinéal du Brochet (*Esox lucius* L.). II. Étude en microscopie électronique de la différenciation et de la rudimentation partielle des photorécepteurs; conséquences possibles sur l'élaboration des messages photosensoriels. Ann. Biol. anim. **19**, 661–688 (1979b)

Fantini, F., Tarquini, B., Panerai, A., Toccafondi, R., Gagnoni, M.: Effeti del trattamento precoce con androgeni sul ritmo circadiano della serotonina epifisaria nel ratto albino. Rass. Neurol. veg. **20**, 427–431 (1966)

Fariello, R.G., Bubenik, G.A.: Melatonin-induced changes in the sensory activation of acute epileptic foci. Neurosci. Letters **3**, 151–155 (1976)

Fariss, B.L., Botz, L.D.: Comparison of hydroxyindole-O-methyltransferase levels in pineal glands of normal, blinded, and congenitally anophthalmic guinea-pigs. Endocrinology **96**, 1595–1596 (1975)

Farmer, R.: A possible feedback of dehydroepiandrosterone upon the pineal in mice. Life Sci. **14**, 297–301 (1974)

Farrell, G.: Regulation of aldosterone secretion. Physiol. Rev. **38**, 709–728 (1958)

Farrell, G.: Glomerulotropic activity of an acetone extract of pineal tissue. Endocrinology **65**, 239–241 (1959a)

Farrell, G.: The physiological factors which influence the secretion of aldosterone. Recent Progr. Hormone Res. **15**, 275–310 (1959b)

Farrell, G.: Epiphysis cerebri in the control of steroid secretion. Fed. Proc. **19**, 601–604 (1960a)

Farrell, G.: Adrenoglomerulotropin. Circulation **21**, 1009–1015 (1960b)

Farrell, G.L., McIsaac, M.W.: Adrenoglomerulotropin. Arch. Biochem. Biophys. **94**, 543–544 (1961)

Farrell, G., Powers, D., Otani, T.: Inhibition of ovulation in the rabbit: seasonal variation and the effects of indoles. Endocrinology **83**, 599–603 (1968)

Fasano, A.V.: Histochemical and gravimetric analyses of the pineal gland of the albino rat during the estrous cycle, pregnancy and pseudopregnancy. Ph. D. Thesis. Loyola Univ. Chicago. Diss/1–238 (1972)

Fasano, A.V., Kasprow, B.A., Kovarik, F.A., Velardo, J.T.: Enzyme study of the pineal gland of the albino rat during the estrous cycle. Physiologist **14**, 141 (1971a)

Fasano, A.V., Kasprow, B.A., Velardo, J.T.: Gravimetric analyses of the pineal gland, endocrine organs and uterus of the albino rat during the estrous cycle. Amer. Zool. **11**, 649 (1971b)

Fasano, A.V., Kasprow, B.A., Velardo, J.T.: Dehydrogenase activity of the pineal gland during the estrous cycle. Endocrine Res. Comm. **4**, 257–266 (1977)

Fassbender, E.: Topographie und mikroskopisch-anatomischer Feinbau der Epiphysis cerebri des Pferdes. Gegenbaurs morph. Jb. **103**, 457–483 (1962)

Faure, J., Vincent, D., Bensch, C.: Couple habenulo-épiphysaire et régulation hypothalamique de la fonction gonadotrope chez le Lapin femelle adulte. C.R. Soc. Biol. (Paris) **160**, 1557–1560 (1966)

Fazio, B., Giordano, G.: Studi sull'epifisi. Comportamento delle frazioni colesteroliche surrenali in cavie adulte trattate con estratto acquoso epifisario. Arch. E. Maragliano Pat. Clin. **8**, 365–370 (1953)

Feldberg, W., Fleischhauer, K.: Penetration of bromophenol blue from the perfused cerebral ventricles into the brain tissue. J. Physiol. (Lond.) **150**, 451–462 (1960)

Feldman, J.M., Lebovitz, H.E.: Structural determinants of indole amine action on *in vitro* insulin release. Endocrinology **91**, 809–816 (1972)

Feldstein, A., Williamson, O.: Serotonin metabolism in pineal homogenates. Advanc. Pharmacol. Chemother. **6**A, 91–96 (1968)

Fenwick, J.C.: The functions of the fish pineal organ. Ph. D. Thesis, University of British Columbia, Vancouver, Canada (1969)

Fenwick, J.C.: The pineal organ. In: Fish physiology (ed. by W.S. Hoar and D.J. Randall), Vol. IV. The nervous system, circulation and respiration, pp. 91–108. New York and London: Academic Press 1970a

Fenwick, J.C.: Demonstration and effect of melatonin in fish. Gen. comp. Endocr. **14**, 86–97 (1970b)

Fenwick, J.C.: Effects of pinealectomy and bilateral enucleation on the phototactic response and on the conditioned response to light of the goldfish *Carassius auratus* L. Canad. J. Zool. **48**, 175–182 (1970c)

Fenwick, J.C.: The pineal organ: photoperiod and reproductive cycles in the goldfish, *Carassius auratus* L. J. Endocr. **46**, 101–111 (1970d)

Fenwick, J.C.: Brain serotonin and swimming activity in the goldfish, *Carassius auratus*. Comp. Biochem. Physiol. **32**, 803–806 (1970e)

Ferreri, E., Peyrot, A., Mazzi, V., Socino, M.: Il metabolismo del sodio e del potassio nel tritone crestato. III) Effetti dell'ablazione dell'epifisi e del corpo sottocommissurale e dell'interrenalectomia subtotale. Boll. Soc. ital. Biol. sperm. **40**, 222–226 (1964)

Fevre, M., Segel, T., Marks, J.F., Boyar, R.M.: LH and melatonin secretion patterns in pubertal boys. J. clin. Endocr. **47**, 1383–1386 (1978)

Fideleff, H., Aparicio, N.J., Guitelman, A., Debeljuk, L., Mancini, A., Cramer, C.: Effect of melatonin on the basal and stimulated gonadotropin levels in normal men and postmenopausal women. J. clin. Endocr. **42**, 1014–1017 (1976)

Finkelstein, J.S., Baum, F.R., Campbell, C.S.: Entrainment of the female hamster to reverse photoperiod: role of the pineal. Physiol. Behav. **21**, 105–111 (1978)

Fioretti, M.C., Riccardi, C., Menconi, E., Martini, L.: Control of the circadian rhythm of the body temperature in the rat. Life Sci. **14**, 2111–2119 (1974)

Firth, B.T., Heatwole, H.: Panting thresholds of lizards: the role of the pineal complex in panting responses in an agamid, *Amphibolurus muricatus*. Gen. comp. Endocr. **29**, 388–401 (1976)

Fiske, V.M.: Serotonin rhythms in the pineal organ: control by the sympathetic nervous system. Science **146**, 253–254 (1964)

Fiske, V.M., Huppert, L.C.: Melatonin action on pineal varies with photoperiod. Science **162**, 279 (1968)

Fiske, V.M., MacDonald, G.J.: Melatonin and reproduction. In: Endocrinology, ed. by R.O. Scow. Excerpta Med. Sec. III/Int. Congr. Series **273**, 880–885 (1973)

Fiske, V.M., Bryant, G.K., Putnam, J.: Effect of light on the weight of the pineal in the rat. Endocrinology **66**, 489–491 (1960)

Fiske, V.M., Pound, J., Putnam, J.: Effect of light on the weight of the pineal organ in hypophysectomized, gonadectomized, adrenalectomized or thiouracil-fed rats. Endocrinology **71**, 130–133 (1962)

Fiske, V.M., MacDonald, G.J., Bigelow, L.B., Altschule, M.D.: Estrogen and progesterone secretion in monkeys following melatonin or Altschule's pineal extract. Fertil. and Steril. **26**, 613–616 (1975)

Fitzgerald, T.J., Veal, A.: Melatonin antagonizes colchicine-induced mitotic arrest. Experientia (Basel) **32**, 372–373 (1976)

Flaugh, M.E., Crowell, T.A., Clemens, J.A., Sawyer, B.D.: Synthesis and evaluation of the antiovulatory activity of a variety of melatonin analogues. J. med. Chem. **22**, 63–69 (1979)

Fleischhauer, K.: Regional differences in the structure of the ependyma and subependymal layers of the cerebral ventricles of the cat. In: Kety, S.S., Elkes, J. (eds.) Regional neurochemistry, pp. 279–283. Oxford: Pergamon 1961

Fleischhauer, K.: Fluorescenzmikroskopische Untersuchungen über den Stofftransport zwischen Ventrikelliquor und Gehirn. Z. Zellforsch. **62**, 639–654 (1964)

Flight, W.F.G.: Some observations on pineal ultrastructure in the newt, *Notophthalmus (Diemictylus) viridescens viridescens*. Proc. kon. ned. Akad. Wet., Ser. C **71**, 525–528 (1968)

Flight, W.F.G.: Observations on the pineal ultrastructure of the urodele, *Diemictylus viridescens viridescens*. I, II. Proc. kon. ned. Akad. Wet., Ser. C **76**, 425–448 (1973)

Flight, W.F.G., Van Donselaar, E.: Ultrastructural aspects of the incorporation of ^{3}H-vitamin A in the pineal organ of the urodele, *Diemictylus viridescens viridescens*. Proc. kon. ned. Akad. Wet., Ser. C **78**, 130–142 (1975)

Folk, G.E., Hagelstein, K.A., Ringens, P.J.: Cold acclimatization and the pineal gland of the lemming. Fed. Proc. **36**, 419 Abs (1977)

Fontana, J.A., Lovenberg, W.: A cyclic AMP-dependent protein kinase of the bovine pineal gland. Proc. nat. Acad. Sci. (Wash.) **68**, 2787–2790 (1971)

Fontana, J.A., Lovenberg, W.: Pineal protein kinase and the regulation of hydroxyindole metabolism. Fed. Proc. **31**, 440 (1972)

Fontana, J.A., Lovenberg, W.: Pineal protein kinase: effect of enzyme phosphorylation on actinomycin D binding by, and template activity of, chromatin. Proc. nat. Acad. Sci. (Wash.) **70**, 755–758 (1973)

Fontana, P., Jr., Alves de Azevedo, B.: Effect of sound stimulation on pineal glands of female albino rats. An. Acad. bras. Ciêcias **50**, 116 (1978)

Ford, D.H.: Uptake of ^{131}I-labeled triiodothyronine in the pineal body as compared with the cerebral grey and other tissues of the rat. Progr. Brain Res. **10**, 530–539 (1965)

Ford, D.H., Gross, J.: The metabolism of ^{131}I-labelled thyroid hormones in the hypophysis and brain of the rabbits. Endocrinology **62**, 416–436 (1958)

Fore, H., Morton, R.A.: Manganese in rabbit tissues. Biochem. J. **51**, 600–603 (1952)

Franchimont, P.: Action de la mélatonine et de la glomérulotrophine sur les fonctions génitale et surrénalienne du rat. Arch. int. Physiol. Biochem. **72**, 735–740 (1964)

Fraschini, F.: The pineal gland and the control of LH and FSH secretion. In: C. Gual (ed.), Progress in endocrinology, pp. 637–644. Amsterdam: Excerpta Med. Fdn. 1969

Fraschini, F., Martini, L.: Rhythmic phenomena and pineal principles. In: Martini, L., Motta, M., Fraschini, F. (eds.), The hypothalamus, pp. 529–549. New York-London: Academic Press 1970

Fraschini, F., Mess, B., Martini, L.: Pineal gland, melatonin, and the control of luteinizing hormone secretion. Endocrinology **82**, 919–924 (1968)

Fraschini, E., Collu, R., Martini, L.: Mechanisms of inhibitory action of pineal principles on gonadotropin secretion. In: The pineal gland (G.E.W. Wolstenholme, J. Knight, eds.), pp. 259–278. Edinburgh and London: Churchill Livingstone 1971

Frauchiger, E.: Vergleichendes zum Fragenkreis über die Epiphyse. Psychiat. Neurol. Neurochir. **64**, 188–191 (1961)

Frauchiger, E.: Altes und Neueres über die Zirbeldrüse (Epiphysis cerebri). Schweiz. Arch. Tierheilk. **105**, 183–194 (1963)

Frauchiger, E., Sellei, K.: Dünnschichtchromatographie der Lipoide in der Glandula pinealis. Schweiz. Arch. Neurol. Neurochir. Psychiat. **98**, 240–243 (1968)

Frehn, J.L., Urry, R.L., Balph, D.F., Ellis, L.C.: Photoperiod and crowding effects on testicular serotonin metabolism and lack of effects on melatonin synthesis in Uinta ground squirrels (*Spermophilus armatus*) J. exp. Zool. **183**, 139–144 (1973)

Frehn, J.L., Urry, R.L., Ellis, L.C.: Effect of melatonin and short photoperiod on Δ^4-reductase activity in liver and hypothalamus of the hamster and the rat. J. Endocr. **60**, 507–515 (1974)

Freire, F., Cardinali, D.P.: Effects of melatonin treatment and environmental lighting on the ultrastructural appearance, melatonin synthesis, norepinephrine turnover and microtubule protein content of the rat pineal gland. J. Neural Transm. **37**, 237–257 (1975)

Freund, D., Arendt, J., Vollrath, L.: Tentative immunohistochemical demonstration of melatonin in the rat pineal gland. Cell Tiss. Res. **181**, 239–244 (1977)

Friedman, N.B.: Germinoma of the pineal. Its identity with germinoma (seminoma) of the testis. Cancer Res. **7**, 363–368 (1947)

Frink, R., Krupp, P.P., Young, R.A.: Seasonal ultrastructural variations in pinealocytes of the Woodchuck, *Marmota monax*. J. Morph. **158**, 91–108 (1978)

Fujie, E.: Ultrastructure of the pineal body of the domestic chicken, with special reference to the changes induced by altered photoperiods. Arch. Histol. Jap. **29**, 271–303 (1968)

Fujii, R.: Demonstration of the adrenergic nature of transmission at the junction between melanophore-concentrating nerve and melanophore in bony fish. J. Fac. Sci. Univ. Tokyo, IV **9**, 171–196 (1961)

Fujita, T.: The gastro-enteric endocrine cell and its paraneuronic nature, pp. 191–208. In: R.E. Coupland and T. Fujita (eds.), Chromaffin, enterochromaffin and related cells. Amsterdam: Elsevier 1976

Fujiwara, M., Shibata, M., Watanabe, Y., Nukiwa, T., Hirata, F., Mizuno, N., Hayaishi, O.: Indoleamine 2,3-dioxygenase. Formation of L-kynurenine from L-tryptophan in cultured rabbit pineal gland. J. biol. Chem. **253**, 6081–6085 (1978)

Fuller, R.W., Perry, K.W.: Effect of p-chloramphetamine on serotonin content of the rat pineal gland. Res. Comm. Chem. Path. Pharmacol. **18**, 769–772 (1977a)

Fuller, R.W., Perry, K.W.: Increase of pineal noradrenaline concentration in rats by desipramine but not fluoxetine: implications concerning the specificity of these inhibitors. J. Pharmacol. **29**, 710–711 (1977b)

Fuortes, M.G.F.: Physiology of photoreceptor organs. Berlin-Heidelberg-New York: Springer 1972

Fuxe, K., Ljunggren, L.: Cellular localization of monoamines in the upper brain stem of the pigeon. J. comp. Neurol. **125**, 355–382 (1965)

Garbarg, M., Julien, C., Schwartz, J.-C.: Circadian rhythm of histamine in the pineal gland. Life Sci. **14**, 539–543 (1974)

Gardner, J.H.: Development of the pineal body in the hooded rat. Anat. Rec. **103**, 538–539 (1949)

Gardner, J.H.: Innervation of pineal gland in hooded rat. J. comp. Neurol. **99**, 319–330 (1953)

Garweg, G., Gollmer, D., Kortmann, H.: Zur Wirkung des Lichtentzuges auf Melatoninsynthese und Hodenaktivität beim Goldhamster. Verh. anat. Ges. **70**, 77–83 (1976)

Gaston, S.: The influence of the pineal organ on the circadian activity rhythm in birds. In: Biochronometry (M. Menaker, ed.), pp. 541–548. Washington, D.C.: National Academy of Sciences 1971

Gaston, S., Menaker, M.: Pineal function: The biological clock in the sparrow? Science **160**, 1125–1127 (1968)

Geller, I.: Ethanol preference in the rat as a function of photoperiod. Science **173**, 456–459 (1971)

Gerebtzoff, M.A., Devečerski, V.: Recherches histoenzymologiques sur les pinéalocytes et sur les astrocytes épiphysaires. C.R. Soc. Biol. (Paris) **166**, 713–715 (1972)

Gern, W.A., Ralph, C.L.: Melatonin synthesis by the retina. Science **204**, 183–184 (1979)

Gern, W.A., Owens, D.W., Ralph, Ch.L.: Plasma melatonin in the trout: Day-night change demonstrated by radioimmunoassay. Gen. comp. Endocr. **34**, 453–458 (1978a)

Gern, W.A., Owens, D.W., Ralph, Ch.L.: The synthesis of melatonin by the trout retina. J. exp. Zool. **206**, 263–270 (1978b)

Gern, W.A., Owens, D.W., Ralph, Ch.L.: Persistence of the nycthemeral rhythm of melatonin secretion in pinealectomized or optic tract-sectioned trout (*Salmo gairdneri*). J. exp. Zool. **205**, 371–376 (1978c)

Giacomelli, F.: Über Veränderungen der Nebennierenrinde nach Pinealektomie. Endokrinologie **42**, 144–150 (1962)

Giarman, N.J., Day, M.: Presence of biogenic amines in the bovine pineal body. Biochem. Pharmacol. **1**, 235 (1958/59)

Giarman, N.J., Day, S.M., Pepeu, G.: Presence of neuro-humors in bovine pineal glands. Fed. Proc. **18**, 394 (1959)

Giarman, N.J., Freedman, D.X., Picard-Ami, L.: Serotonin content of the pineal glands of man and monkey. Nature (Lond.) **186**, 480–481 (1960)

Giedt, W.R., Jr., Winters, D.: Pharmacological stimulation of the pineal: potentiation of ketamine response in the chick. Fed. Proc. **35**, 564 Abs. (1976)

Giedt, W.R., Lakin, M.L., Winters, W.D.: Diurnal response to ketamine and gamma-hydroxybutyrate and its possible relationship to pineal indoleamines. Neuropharmacology **17**, 221–228 (1978)

Giordano, G., Balestreri, R.: La glande épiphysaire dans la régulation de la biosynthèse de l'aldostérone. Ann. Endocr. (Paris) **24**, 331–347 (1963)

Giordano, G., Balestreri, R., Jacopino, G.E., Foppiani, E., Bertolini, S.: L'action *in vitro* de la mélatonine sur l'hormono-synthèse cortico-surrénale du Rat. Ann. Endocr. (Paris) **31**, 1071–1080 (1970)

Girod, C., Curé, M., Czyba, J.C., Durand, N.: Influence de l'épiphysectomie sur les cellules gonadotropes antéhypophysaires du Hamster doré (*Mesocricetus auratus* Waterh.). C.R. Soc. Biol. (Paris) **158**, 1636–1637 (1964)

Girod, C., Durand, N., Czyba, J.C.: Influence de l'épiphysectomie sur l'ovarie du Hamster doré (*Mesocricetus auratus* Waterh.). C.R. Soc. Biol. (Paris) **161**, 1575–1576 (1967)

Giroud, A., Martinet, R.: Sur la localisation des encéphalocéles et des meningocéles dans la region épiphysaire. Bull. Ass. Anat. (Nancy) No. **125**, 696–698 (1965)

Gittes, R.F., Chu, E.W.: Reversal of the effect of pinealectomy in female rats by multiple isogeneic pineal transplants. Endocrinology **77**, 1061–1067 (1965)

Gladstone, R.J., Wakeley, C.P.G.: Development and histogenesis of the human pineal organ. J. Anat. (Lond.) **69**, 427–454 (1935)

Gladstone, R.J., Wakeley, C.P.G.: The pineal gland. London: Baillière, Tindall and Cox (1940)

Glaser, R.: Increase in locomotor activity following shielding of the parietal eye in night lizards. Science **128**, 1577–1578 (1958)

Gogan, F., Moszkowska, A., Rotsztejn, W., Scemama, A.: Action de l'épiphysectomie sur l'incorporation hypothalamique et hypophysaire d'oestradiol 17β chez la Ratte. C.R. Acad. Sci. (Paris) D **275**, 2391–2394 (1972)

Goldman, H.: The nervous control of blood flow to the pineal body. Life Sci. **6**, 2071–2077 (1967)

Goldman, H., Wurtman, R.J.: Flow of blood to the pineal body of the rat. Nature (Lond.) **203**, 87–88 (1964)

Goldman, M., Peaslee, M.H., Naber, S.P.: The iodide concentrating mechanisms of the thyroid gland in weanling Long-Evans rats injected with melatonin soon after birth. Proc. South. Dak. Acad. Sci. **48**, 33–38 (1969)

Goldstein, R., Pavel, S.: Vasotocin release into the cerebrospinal fluid of cats induced by luteinizing hormone releasing hormone, thyrotrophin releasing hormone and growth hormone release-inhibiting hormone. J. Endocr. **75**, 175–176 (1977)

Gonzales, G., Blasquez, E.: Ultrastructural evidence of a secretory process in the rat pineal gland. Experientia (Basel) **31**, 969–971 (1975)

Gonzales, G.G., Hidalgo, F.G.: Ultraestructura de la glándula pineal de las aves. Trab. Inst. Cajal Invest. biol. **58**, 55–67 (1966)

Gonzales, G.G., Alvarez-Uria, M., Fernandez Ruiz, R., Fernandez Ruiz, B.: Ultrastructura de la glándula pineal de los mamiferos. I. Imágenes sináptica en los pineócitos. Trab. Inst. Cajal Invest. biol. **61**, 41–62 (1969)

Goridis, C., Neff, N.H.: Evidence for a specific monoamine oxidase associated with sympathetic nerves. Neuropharmacology **10**, 557–564 (1971)

Goridis, C., Neff, N.H.: Evidence for specific monoamine oxidases in human sympathetic nerve and pineal gland. Proc. Soc. exp. Biol. (N.Y.) **140**, 573–574 (1972)

Gorray, K.C., Quay, W.B.: Effects of pinealectomy and of sham-pinealectomy on blood glucose levels in the alloxan-diabetic rat. Horm. Metabol. Res. **10**, 389–392 (1978)

Grachev, I.I., Usanova, R.I., Seliverstow, U.A.: The influence of the pineal extract on the growth activity of the rat pituitary gland. Sechenov Fiziol. S.S.S.R. **58**, 272 (1972) [in Russian, cited after Reiter, 1974a]

Gray, E.G., Pease, H.L.: On understanding the organization of the retinal receptor synapses. Brain Res. **35**, 1–15 (1971)

Green, A.R., Koslow, S.H., Costa, E.: Identification and quantitation of a new indolealkylamine in rat hypothalamus. Brain Res. **51**, 371–374 (1973)

Green, J.P., Day, M., Robinson, J.D., Jr.: Some acidic substances in neoplastic mast cells and in the pineal body. Biochem. Pharmacol. **11**, 957–960 (1962)

Greenberg, L.H., Weiss, B.: β-Adrenergic receptors in aged rat brain: Reduced number and capacity of pineal gland to develop supersensitivity. Science **201**, 61–63 (1978)

Greer, M., Williams, C.M.: Gas-chromatographic determination of melatonin and 6-hydroxymelatonin. Clin. chim. Acta **15**, 165–168 (1967)

Gregorek, J.C.: The ultrastructure of the pineal gland of normal and enucleated gerbils. Anat. Rec. **175**, 333 (1973)

Gregorek, J.C., Seibel, H.R., Reiter, R.J.: The pineal complex and its relationship to other epithalamic structures. Acta anat. (Basel) **99**, 425–434 (1977)

Greiner, A.C., Chan, S.C.: Melatonin content of the human pineal gland. Science **199**, 83–84 (1978)

Griffiths, E.C.: Peptidase inactivation of hypothalamic releasing hormones. Hormone Res. **7**, 179–191 (1976)

Grignon, G., Grignon, M.: Sur la structure de l'épiphyse de la tortue terrestre. Bull. Ass. Anat. (Nancy) No. **118**, 685–696 (1963a)

Grignon, M., Grignon, G.: Etude histologique et histophysiologique de l'épiphyse chez la tortue terrestre (*Testudo mauritanica* Dumer). Ann. Endocr. (Paris) **24**, 356–364 (1963b)

Grishchenko, V.I., Zubkov, G.V.: Summary of results and prospects of research into the role of the epiphysis in midwifery and gynaecology. Akush Ginek. (Moskva) **5**, 4–8 (1974)

Gromova, E.A., Kraus, M., Křeček, J.: Effect of melatonin and 5-hydroxytryptamine on aldosterone and corticosterone production by adrenal glands of normal and hypophysectomised rats. J. Endocr. **39**, 345–350 (1967)

Gross, D.S.: Distribution of gonadotropin-releasing hormone in the mouse brain as revealed by immunohistochemistry. Endocrinology **98**, 1408–1417 (1976)

Grota, L.J., Brown, G.M.: Antibodies to indolealkylamines: Serotonin and melatonin. Canad. J. Biochem. **52**, 196–202 (1974)

Grunewald-Lowenstein, M.: Studies on the function of the pineal body. Exp. Med. Surg. **10**, 135–154 (1952)

Grunewald-Lowenstein, M.: Influence of light and darkness on the pineal body in *Astyanax mexicanus* (Filippi). Zoologica **41**, 119–128 (1956)

Guchhait, R.B., Grau, J.E., Jr.: Biosynthesis of S-adenosyl-L-methionine in the rat pineal gland. J. Neurochem. **31**, 921–925 (1978)

Guérillot, C., Da Lage, C.: Effects de l'ablation du ganglion cervical supérieur sur le cycle oestrien de la ratte préalablement soumise à un eclairement continu. C.R. Soc. Biol. (Paris) **162**, 1031–1033 (1968)

Guérillot, C., Vendrely, E., Da Lage, C.: Modifications du cycle oestrien de la ratte maintenue à l'obscurité après ablation bilatérale du ganglion cervical supérieur. C.R. Soc. Biol. (Paris) **167**, 890–894 (1973)

Guerra, M.O., Silva, N.O.G., Guimarães, C.S., Souza, J.P., Andrade, A.T.L.: Pinealectomy and blindness during pregnancy in the rat. Amer. J. Obstet. Gynec. **115**, 582–583 (1973)

Gundy, G.C., Ralph, C.L.: A histological study of the third eye and related structures in scincid lizards. Herpetol. Rev. **3**, 65 (1971)

Gundy, G.C., Wurst, G.Z.: The occurrence of parietal eyes in recent Lacertilia. J. Herpetol. **10**, 113–121 (1976a)

Gundy, G.C., Wurst, G.Z.: Parietal eye-pineal morphology in lizards and its physiological implications. Anat. Rec. **185**, 419–432 (1976b)

Gundy, G.C., Ralph, C.L., Wurst, G.Z.: Parietal eyes in lizards: Zoogeographical correlates. Science **190**, 671–673 (1975)

Gusek, W.: Neue Befunde zur Morphologie und Funktion der Epiphysis cerebri. Ergebn. allg. Path. path. Anat. **50**, 104–148 (1968)

Gusek, W.: Der Einfluß von Antiandrogen (Cyproteronacetat) auf die Morphologie der Zirbeldrüse. Acta endocr. (Kbh.) Suppl. **152**, 37 (1971)

Gusek, W.: Die Feinstruktur der Rattenzirbel und ihr Verhalten unter Einfluß von Antiandrogenen und nach Kastration. Endokrinologie **67**, 129–151 (1976)

Gusek, W., Buss, H.: Morphologische und histochemische Veränderungen der Zirbeldrüse unter dem Einfluß von Prolan und nach Ovarektomie. Frankfurt. Z. Path. **75**, 172–186 (1966a)

Gusek, W., Buss, H.: Über den experimentellen Strukturwandel der Zirbeldrüse und seine Interpretation. Zbl. allg. Path. path. Anat. **109**, 452–453 (1966b)

Gusek, W., Santoro, A.: Elektronenoptische Beobachtungen zur Ultramorphologie der Pinealzellen der Ratte. Biol. Lat. **10**, 451–464 (1960)

Gusek, W., Santoro, A.: Zur Ultrastruktur der Epiphysis cerebri der Ratte. Endokrinologie **41**, 105–129 (1961)

Gusek, W., Buss, H., Wartenberg, H.: Weitere Untersuchungen zur Feinstruktur der Epiphysis cerebri normaler und vorbehandelter Ratten. Progr. Brain Res. **10**, 317–331 (1965)

Gutte, G., Grütze, I.: Histochemische Untersuchungen an den Parenchymzellen der Epiphysis cerebri von Rind (*Bos taurus domesticus*) und Schwein (*Sus scrofa domesticus*). Acta histochem. (Jena) **60**, 204–210 (1977)

Gwinner, E.: Effects of pinealectomy on circadian locomotor activity rhythms in European starlings, *Sturnus vulgaris*. J. comp. Physiol. A **126**, 123–129 (1978)
Gwinner, E., Benzinger, I.: Synchronization of a circadian rhythm in pinealectomized European starlings by daily injections of melatonin. J. comp. Physiol. A **127**, 209–213 (1978)
Haase, E., Follett, B.K.: Effects of pinealectomy on the pituitary gonadal axis of male bramblings (*Fringilla montifringilla*). Gen. comp. Endocr. **22**, 386–387 (1974)
Hafeez, M.A.: Effect of melatonin on body coloration and spontaneous swimming activity in rainbow trout, *Salmo gairdneri*. Comp. Biochem. Physiol. **36**, 639–656 (1970)
Hafeez, M.A.: Light microscopic studies on the pineal organ in teleost fishes with special regard to its function. J. Morph. **134**, 281–314 (1971)
Hafeez, M.A., Ford, P.: Histology and histochemistry of the pineal organ in the sockeye salmon, *Oncorhynchus nerka* Walbaum. Canad. J. Zool. **45**, 117–126 (1967)
Hafeez, M.A., Merhige, M.E.: Light and electron microscopic study on the pineal complex of the coelacanth, *Latimeria chalumnae* Smith. Cell Tiss. Res. **178**, 249–265 (1977)
Hafeez, M.A., Quay, W.B.: Histochemical and experimental studies of 5-hydroxytryptamine in pineal organs of teleosts (*Salmo gairneri* and *Atherinopsis californiensis*). Gen. comp. Endocr. **13**, 211–217 (1969)
Hafeez, M.A., Quay, W.B.: Pineal acetylserotonin methyltransferase activity in the teleost fishes, *Hesperoleucus symmetricus* and *Salmo gairdneri*, with evidence for lack of effect of constant light and darkness. Comp. Gen. Pharmacol. **1**, 257–262 (1970a)
Hafeez, M.A., Quay, W.B.: The role of the pineal organ in the control of phototaxis and body coloration in rainbow trout (*Salmo gairdneri* Richardson). Z. vergl. Physiol. **68**, 403–416 (1970b)
Hafeez, M.A., Zerihun, L.: Studies on central projections of the pineal nerve tract in rainbow trout, *Salmo gairdneri* Richardson, using cobalt chloride iontophoresis. Cell Tiss. Res. **154**, 485–510 (1974)
Hafeez, M.A., Zerihun, L.: Autoradiographic localization of ^{3}H-5-HTP and ^{3}H-5-HT in the pineal organ and circumventricular areas in the rainbow trout, *Salmo gairdneri* Richardson. Cell Tiss. Res. **170**, 61–76 (1976)
Hafeez, M.A., Wagner, H.H., Quay, W.B.: Mediation of light-induced changes in pineal receptor and supporting cell nuclei and nucleoli in steelhead trout (*Salmo gairdneri*). Photochem. Photobiol. **28**, 213–218 (1978)
Håkanson, R., Owman, Ch.: Effect of denervation and enzyme inhibition on DOPA decarboxylase and monoamine oxidase activities of rat pineal gland. J. Neurochem. **12**, 417–429 (1965)
Håkanson, R., Owman, Ch.: Pineal DOPA decarboxylase and monoamine oxidase activities as related to the monoamine stores. J. Neurochem. **13**, 597–605 (1966)
Håkanson, R., Lombard des Gouttes, M.N., Owman, Ch.: Activities of tryptophan hydroxylase, DOPA decarboxylase and monoamine oxidase as correlated with the appearance of monoamines in developing rat pineal gland. Life Sci. **6**, 2577–2585 (1967)
Håkanson, R., Owman, Ch., Sjölund, K.: Cytospectrofluorometric characterization of OPT-induced fluorescence in rat pinealocytes. Histochemistry **42**, 323–331 (1974)
Halaris, A.: Effect of continuous illumination on mitochondria of the rat pineal body. Acta endocr. (Kbh.), Suppl. **138**, 239 (1969)
Halaris, A., Matussek, N.: Effect of continuous illumination on mitochondria of the rat pineal body. Experientia (Basel) **25**, 486–487 (1969)
Haldar, C., Thapliyal, J.P.: Effect of pinealectomy on the annual testicular cycle of *Calotes versicolor*. Gen. comp. Endocr. **32**, 395–399 (1977)
Halpern, C.C., Martin, C.R.: Effect of melatonin and thymectomy on food and fluid intake and renal excretion rhythms of adrenalectomized rats. Fed. Proc. **31**, 327 (1972)
Hamasaki, D.I.: Properties of the parietal eye of the green iguana. Vision Res. **8**, 591–599 (1968)
Hamasaki, D.I.: Spectral sensitivity of the parietal eye of the green iguana. Vision Res. **9**, 515–523 (1969a)
Hamasaki, D.I.: Interaction of the slow responses of the parietal eye. Vision Res. **9**, 1453–1459 (1969b)
Hamasaki, D.I.: Interaction of excitation and inhibition in the stirnorgan of the frog. Vision Res. **10**, 307–316 (1970)
Hamasaki, D.I., Dodt, E.: Light sensitivity of the lizard's epiphysis cerebri. Pflügers Arch. **313**, 19–29 (1969)

Hamasaki, D.I., Eder, D.J.: Adaptive radiation of the pineal system. p. 497–548. In: Handbook of sensory physiology, Vol. VII/5. The visual system in vertebrates (F. Crescitelli, ed.). Berlin-Heidelberg-New York: Springer 1977
Hamasaki, D.I., Streck, P.: Properties of the epiphysis cerebri of the small-spotted dogfish shark, *Scyliorhinus caniculus* L. Vision Res. **11**, 189–198 (1971)
Hamner, W.M., Barfield, R.J.: Ineffectiveness of pineal lesions on the testis cycle of a finch. Condor **72**, 99–101 (1970)
Hansen, G.: Karcinommetastase in der Epiphyse. Z. allg. Path. path. Anat. **93**, 357–360 (1955)
Hanyu, I.: Light sense of fish. Bull. Japan. Soc. Sci. Fish. **33**, 255–261 (1967)
Hanyu, I., Niwa, H.: Pineal photosensitivity in three teleosts, *Salmo irideus, Plecoglossus altivelus* and *Mugil cephalus*. Rev. Canad. Biol. **29**, 133–140 (1970)
Hanyu, I., Niwa, H., Tamura, T.: A slow potential from the epiphysis cerebri of fishes. Vision Res. **9**, 621–623 (1969)
Hanyu, I., Niwa, H., Tamura, T.: Salient features in photosensory function of teleostean pineal organ. Comp. Biochem. Physiol. **61 A**, 49–54 (1978)
Harrison, P.C.: Reproductive development of pinealectomized chicks. Poul. Sci. **47**, 1679 (1968)
Harrison, P.C., Becker, W.C.: Extraretinal photocontrol of oviposition in pinealectomized domestic fowl. Proc. Soc. exp. Biol. (N.Y.) **132**, 161–164 (1969)
Hartmann, F.: Über die Innervation der Epiphysis cerebri einiger Säugetiere. Z. Zellforsch. **46**, 416–429 (1957)
Hartwig, H.G.: Photolabile Substanzen im Pinealkomplex von Anuren. Verh. anat. Ges. **69**, 439–441 (1975)
Hartwig, H.G., Baumann, Ch.: Evidence for photosensitive pigments in the pineal complex of the frog. Vision Res. **14**, 597–598 (1974)
Hartwig, H.G., Korf, H.W.: The epiphysis cerebri of poikilothermic vertebrates: A photosensitive neuroendocrine circumventricular organ. In: R.P. Becker, O. Johari (eds.), Scanning Electron microscopy, Vol. II. pp. 163–168. Chicago: III Research Institute 1978.
Hartwig, H.G., Pfautsch, M.: Rasterelektronenmikroskopische Beobachtungen an pinealen Sinneszellen der Forelle, *Salmo gairdneri* (Teleostei). Z. Zellforsch. **138**, 585–589 (1973)
Hassler, R., Bak, I.J.: Effects of amine-depleting and amine-storing substances on the axon terminals of the pineal gland. In: Proc. 6th Int. Cong. Electron Micr. (R. Uyeda, ed.) **2**, 521–522. Tokyo 1966
Hata, T., Kita, T.: Newly designed method for removal of pineal body, and depression of convulsions and enhancement of exploratory movements by pinealectomy in mice. Endocr. jap. **25**, 407–413 (1978)
Hattori, S.: Electron microscopic study on the postnatal development of the hamster pineal body. Acta anat. Nippon. **46**, 111–124 (1971)
Haug, H.: Die Epiphyse und die circumventrikulären Strukturen des Epithalamus im Gehirn des Elefanten (*Loxodonta africana*). Z. Zellforsch. **129**, 533–547 (1972)
Haulica, I., Branisteanu, D.D., Rosca, V., Stratone, A., Berbeleu, V., Balan, G., Ionescu, L.: A renin-like activity in pineal gland and hypophysis. Endocrinology **96**, 530–532 (1975)
Hauser, G., Eichberg, J., Shein, H.M.: Effects of norepinephrine (NE) and propranolol (PR) on labeling patterns of CDP-diglyceride derived phospholipids in rat pineal. Fed. Proc. **32**, 602 (1973)
Hauser, G., Shein, H.M., Eichberg, J.: Relationship of α-adrenergic receptors in rat pineal gland to drug-induced stimulation of phospholipid metabolism. Nature (London) **252**, 482–483 (1974)
Hayes, M.M.M., Knight, B.K., Symington, R.B.: A histochemical study of the pineal body of the vervet monkey (*Cercopithecus aethiops*). S. Afr. J. med. Sci. **39**, 143–148 (1970)
Hayes, M.M.M., Knight, B.K., Warton, C.M.R.: Preliminary studies on the functional relationships between the pineal, hypothalamus and adenohypophysis using bovine tissues in organ culture. Cent. Afr. J. Med. **19**, 193–194 (1973)
Healey, E.G., Ross, D.M.: The effects of drugs on the background colour response of the minnow *Phoxinus phoxinus* L. Comp. Biochem. Physiol. **19**, 545–580 (1966)
Hearse, D.J., Steinberg, M.S., Weber, W.W.: Individual variation of N-acetyltransferase activity in rabbit pineal bodies. Fed. Proc. **28**, 580 (1969)
Hedlund, L.: Sympathetic innervation of the avian pineal body. Anat. Rec. **166**, 406 (1970)
Hedlund, L., Nalbandov, A.V.: Innervation of the avian pineal body. Amer. Zoologist **9**, 1090 (1969)

Hedlund, L., Ralph, C.L.: Daily variation of pineal serotonin in Japanese quail and Sprague-Dawley rats. Amer. Zoologist **7**, 712 (1967)

Hedlund, L., Ralph, C.L.: Effect of photoperiod phase and ganglionectomy on pineal serotonin in Japanese quail. Amer. Zoologist **8**, 756 (1968)

Hedlund, L., Ralph, C.L., Chepko, J., Lynch, H.J.: A diurnal serotonin cycle in the pineal body of Japanese quail: photoperiod phasing and the effect of superior cervical ganglionectomy. Gen. comp. Endocr. **16**, 52–58 (1971)

Hedlund, L., Lischko, M.M., Rollag, M.D., Niswender, G.N.: Melatonin: Daily cycle in plasma and cerebrospinal fluid of calves. Science **195**, 686–687 (1977)

Heerd, E., Dodt, E.: Wellenlängen-Diskriminatoren im Pinealorgan von *Rana temporaria*. Pflügers Arch. ges. Physiol. **274**, 33–34 (1961)

Heidel, G.: Die Häufigkeit des Vorkommens von Kalkkonkrementen im Corpus pineale des Kindes. Anat. Anz. **116**, 139–154 (1965)

Heinecke, H.: Kernkugeln in den Parenchymzellen der Schweineepiphyse. Z. mikr.-anat. Forsch. **65**, 282–288 (1959)

Heiniger, H.J.: Histologie der Epiphyse des Schweines hinsichtlich Geschlecht und Alter. Zeiss Mitt. Fortschr. techn. Opt. **3**, 301–353 (1965a)

Heiniger, H.J.: Histochemische Untersuchungen über den Ribonucleinsäure-Gehalt der Pineocyten der Glandula pinealis bei Schwein und Mensch. Acta neuropath. (Berl.) **4**, 340–344 (1965b)

Heldmaier, G., Hoffmann, K.: Melatonin stimulates growth of brown adipose tissue. Nature (Lond.) **247**, 224–225 (1974)

Heller, A.: Neuronal control of brain serotonin. Fed. Proc. **31**, 81–90 (1972)

Hellman, B., Larsson, S.: Utilization of uniformly labelled ^{14}C-glucose in the pineal body of goats. Acta endocr. (Kbh.) **38**, 353–360 (1961)

Hendel, R.C., Turek, F.W.: Suppression of locomotor activity in sparrows by treatment with melatonin. Physiol. Behav. **21**, 275–278 (1978)

Hendrickson, A.E., Kelly, D.E.: Development of the amphibian pineal organ; cell proliferation and migration. Anat. Rec. **165**, 211–228 (1969)

Hendrickson, A.E., Kelly, D.E.: Development of the amphibian pineal organ; fine structure during maturation. Anat. Rec. **170**, 129–142 (1971)

Henzl, M.R., Spaur, C.L., Magoun, R.E., Kincl, F.A.: A note of endocrine functions of neonatally pinealectomized rats. Endocrinol. Exp. (Bratisl.) **4**, 77–81 (1970)

Herbert, J.: Prevention by pinealectomy of light-accelerated oestrus in the ferret. J. Endocr. **41**, XX (1968)

Herbert, J.: The effect of removal or sympathetic denervation of the pineal upon light-induced oestrus in the ferret. Acta endocr. (Kbh.) **119**, (Supplement) 46 (1967)

Herbert, J.: The pineal gland and light-induced oestrus in ferrets. J. Endocr. **43**, 625–636 (1969)

Herbert, J.: The role of the pineal gland in the control by light of the reproductive cycle of the ferret. In: The pineal gland (G.E.W. Wolstenholme and J. Knight, eds.), pp. 303–327. Edinburgh and London: Churchill Livingstone 1971

Herbert, J.: Initial observations of pinealectomized ferrets kept for long periods in either daylight or artificial illumination. J. Endocr. **55**, 591–597 (1972)

Herbert, J., Vincent, D.S.: Light and the breeding season in mammals. Excerpta med. (Amst.) **273**, 875–879 (1973)

Herbuté, S., Baylé, J.D.: Multiple-unit activity in the pineal gland of the Japanese quail: spontaneous firing and responses to photic stimulations. Neuroendocrinology **16**, 52–64 (1974)

Herbuté, S., Baylé, J.D.: Pineal multiunit activity in conscious quail: effects of light, blinding, ganglionectomy. Amer. J. Physiol. **231**, 136–140 (1976)

Herbuté, S., Baylé, J.D.: Suppression of pineal multiunit response to flash after habenular lesion in quail. Amer. J. Physiol. **233**, E 293-E 297 (1977)

Hernandez, J., Illnerová, H.: 5-Hydroxytryptophan decarboxylase activity in the rat pineal gland during the first 20 days of postnatal life: effect of light. Neuroendocrinology **6**, 343–348 (1970)

Hertz-Eshel, M., Rahamimoff, R.: Effect of melatonin on uterine contractility. Life Sci. **4**, 1367–1372 (1965)

Herwig, H.J.: Comparative ultrastructural investigations of the pineal organ of the blind cave fish, *Anoptichthys jordani*, and its ancestor, the eyed river fish, *Astyanax mexicanus*. Cell Tiss. Res. **167**, 297–324 (1976)

Heward, Ch.B., Hadley, M.E.: Structure activity relationships of melatonin and related indoleamines. Life Sci. **17**, 1167–1178 (1975)

Hewing, M.: Die postnatale Entwicklung der Epiphysis cerebri beim Goldhamster. Verh. anat. Ges. **70**, 85–92 (1976)

Hewing, M.: A liquor contacting area in the pineal recess of the golden hamster (*Mesocricetus auratus*). Anat. Embryol. **153**, 295–304 (1978)

Hewing, M.: Synaptic ribbons during postnatal development of the pineal gland in the golden hamster (*Mesocricetus auratus*). Cell Tiss. Res. **199**, 473–482 (1979)

Hipkin, L.J.: Effect of 5-methoxytryptophol and melatonin on uterine weight responses to human chorionic gonadotrophin. J. Endocr. **48**, 287–288 (1970a)

Hipkin, L.J.: Gonadotropin inhibitory properties of pineal extracts. Nature (Lond.) **228**, 1202 (1970b)

Hirata, F., Hayaishi, O., Tokuyama, T., Senoh, S.: *In vitro* and *in vivo* formation of two new metabolites of melatonin. J. biol. Chem. **249**, 1311–1313 (1974)

Hisano, N., Cardinali, D.P., Rosner, J.M., Nagle, C.A., Tramezzani, J.H.: Pineal role in the duck extraretinal photoreception. Endocrinology **91**, 1318–1322 (1972)

Hishikawa, Y., Cramer, H., Kuhlo, W.: Natural and melatonin-induced sleep in young chickens. A behavioral and electrographic study. Exp. Brain Res. **7**, 84–94 (1969)

Hoar, W.S.: Phototactic and pigmentary responses of the sockeye salmon smolts following injury to the pineal organ. J. Fisheries Res. Board Can. **12**, 178–185 (1955)

Hofer, H.O., Merker, G., Oksche, A.: Atypische Formen des Pinealorgans der Säugetiere. Verh. anat. Ges. **70**, 97–102 (1976)

Hoffman, R.A.: The epiphyseal complex in fish and reptiles. Amer. Zoologist **10**, 191–199 (1970)

Hoffman, R.A., Reiter, R.J.: Rapid pinealectomy in hamsters and other small rodents. Anat. Rec. **153**, 19–21 (1965a)

Hoffman, R.A., Reiter, R.J.: Pineal gland: influence on gonads of male hamsters. Science **148**, 1609–1611 (1965b)

Hoffman, R.A., Reiter, R.J.: Influence of compensatory mechanisms and the pineal upon light-induced gonadal atrophy in male hamsters. Nature (Lond.) **207**, 658–659 (1965c)

Hoffman, R.A., Reiter, R.J.: Responses of some endocrine organs of female hamsters to pinealectomy and light. Life Sci. **5**, 1147–1151 (1966)

Hoffmann, J.C., Cullin, A.M.: Effects of pinealectomy and constant light on the estrous cycles of female rats. Neuroendocrinology **17**, 167–174 (1975)

Hoffmann, J.C., Pomerantz, D.K.: Effect of pinealectomy on photoperiodic control of estrous cycle length in the rat. Fed. Proc. **30**, 363 Abs. (1971)

Hoffmann, K.: Melatonin inhibits photoperiodically induced testis development in a dwarf hamster. Naturwissenschaften **59**, 218–219 (1972)

Hoffmann, K.: The influence of photoperiod and melatonin on testis size, body weight, and pelage colour in the Djungarian hamster (*Phodopus sungorus*). J. comp. Physiol. **85**, 267–282 (1973)

Hoffmann, K.: Testicular involution in short photoperiods inhibited by melatonin. Naturwissenschaften **61**, 364–365 (1974)

Hoffmann, K.: Die Funktion des Pineals bei der Jahresperiodik der Säuger. Nova Acta Leopoldina N.F. **46**, 217–229 (1977)

Hoffmann, K., Küderling, I.: Pinealectomy inhibits stimulation of testicular development by long photoperiods in a hamster (*Phodopus sungorus*). Experientia (Basel) **31**, 122–123 (1975)

Hoffmann, K., Küderling, I.: Antigonadal effects of melatonin in pinealectomized Djungarian hamsters. Naturwissenschaften **64**, 339–340 (1977)

Hofstätter, R.: Beitrag zur therapeutischen Verwendung von Zirbelextrakten. Wien. klin. Wschr. **62**, 338–339 (1950)

Hofstätter, R.: Versuche der postoperativen Krebsbehandlung mit Zirbelstoffen. Krebsarzt **14**, 307–316 (1959)

Holban, R., Sahleanu, V., Voiculet, N.: Evolutia agresiunii formolice la şobolanul alb pinealectomizat. Fiziol. norm. si pat. **11**, 55–59 (1965)

Hollmann, P.: Über Herkunft und Bedeutung der gliösen Elemente in der Epiphysis cerebri. Untersuchungen an Haussäugetieren. Zbl. Vet.-Med. (Ser. A) **10**, 203–226 (1963)

Holmes, R.L.: Effect of pinealectomy on the rat pituitary. Nature (Lond.) **177**, 791 (1956)

Holmes, R.L.: Iris reactions to homo- and auto-grafts of pineal tissue in rats and rabbits. Anat. Rec. **127**, 173–185 (1957)

Holmes, S.W., Sugden, D.: The effect of melatonin on pinealectomy-induced hypertension in the rat. Brit. J. Pharmacol. **56**, 360P–361P (1976)

Holmgren, U.: Secretory material in the pineal body as shown by aldehyde fuchsin following performic acid oxidation. Stain Technol. **33**, 148–149 (1958a)

Holmgren, U.: On the pineal organ of the tuna, *Thynnus thynnus*. Breviora Mus. comp. Zool. (Cambridge, Mass.) **100**, 1–5 (1958b)

Holmgren, U.: On the structure of the pineal area of teleost fishes with special reference to a few deep sea fishes. Göteborgs Kungl. Vetensk. Vitterhets-Samh. Handl., Ser. B **8**, 1–66 (1959a)

Holmgren, U.: On the pineal area and adjacent structures of the brain of the dipnoan fish, *Protopterus annectens* (OWEN). Breviora **109**, 1–7 (1959b)

Holmgren, U.: On the ontogeny of the pineal and parapineal organs in teleost fishes. Progr. Brain Res. **10**, 172–182 (1965)

Holmgren, U., Altschule, M.D., Wurtman, R.J.: Effects of injection of bovine pineal extract on the nuclei of rat pineal parenchymal cells. Nature (Lond.) **186**, 393–394 (1960)

Holz, R.W., Deguchi, T., Axelrod, J.: Stimulation of serotonin N-acetyltransferase in pineal organ culture by drugs. J. Neurochem. **22**, 205–209 (1974)

Homma, K., Sakakibara, Y.: Encephalic photoreceptors and their significance in photoperiodic control of sexual activity in Japanese quail. In: Menaker, M. (ed.): Biochronometry, pp. 333–341. Washington, D.C.: National Academy of Sciences 1971

Homma, K., McFarland, L.Z., Wilson, W.O.: Response of the reproductive organs of the Japanese quail to pinealectomy and melatonin injections. Poult. Sci. **46**, 314–319 (1967)

Hooper, J., Leone, R.M., Silman, R.E., Edwards, R., Smith, I., Mullen, P.E., Linsell, C., Savage, M.: Preliminary studies in man on the pineal hormone 5-methoxytryptophol. J. Endocr. **80**, 42P–43P (1979)

Hopsu, V.K., Arstila, A.V.: An apparent somato-somatic structure in the pineal gland of the rat. Exp. Cell Res. **37**, 484–487 (1965)

Horányi, B.: Das Corpus pineale im Senium. Wien. Z. Nervenheilk. **17**, 129–139 (1960)

Hori, S.: Effects of sulfhydryl agents on the activation of tryptophan-5-monooxygenase from bovine pineal glands. Biochim. biophys. Acta (Amst.) **384**, 58–68 (1975)

Hori, S., Ohotani, S.: Solubilization of tryptophan-5-monooxygenase from the pineal glands and existence of an activating substance in the tissue extract. J. Neurochem. **31**, 663–671 (1978)

Hori, S., Kuroda, Y., Saito, K., Ohotani, S.: Subcellular localization of tryptophan-5-mono-oxygenase in bovine pineal glands and raphe nuclei. J. Neurochem. **27**, 911–914 (1976)

Horita, N., Ishii, T., Moroji, T.: Effects of long-term administration of chlorpromazine on the pineal gland of rats. Acta neuropath. (Berl.) **42**, 49–52 (1978)

Hosaka, T., Nakai, A., Kushima, S.: Innervation of the pineal body and the supracommissural and subcommissural organs of the Japanese monkey. Bull. Tokyo Med. Dent. Univ. **4**, 365–378 (1957)

Houssay, A.B., Barcelo, A.C.: Effects of estrogens and progesterone upon the biosynthesis of melatonin by the pineal gland. Experientia (Basel) **28**, 478–479 (1972a)

Houssay, A.B., Barcelo, A.C.: Effects of testosterone upon the biosynthesis of melatonin by the pineal gland. Acta physiol. lat.-amer. **22**, 274–275 (1972b)

Houssay, A.B., Pazo, J.H.: Accion de la melatonina sobre el crecimento del pelo en ratones C_3H/EP, p. 147. 3rd Int. Pharmacological Congr., São Paulo 1966

Houssay, A.B., Pazo, J.H.: Role of the pituitary in the thyroid hypertrophy of pinealectomized rats. Experientia (Basel) **24**, 813–814 (1968)

Houssay, A.B., Pazo, J.H., Epper, C.E.: Effects of the pineal gland upon the hair cycles in mice. Acta physiol. lat.-amer. **16**, 202–205 (1966a)

Houssay, A.B., Pazo, J.H., Epper, C.E.: Effect of the pineal gland upon the hair cycles in mice. J. invest. Derm. **47**, 230–234 (1966b)

Howd, R.A., Seo, K.S., Wurtman, R.J.: Rat liver N-acetyltransferase: inhibition by melatonin. Biochem. Pharmacol. **25**, 977–978 (1976)

Hsieh, K.S., Ota, M.: Improved procedure of pinealectomy in rat. Endocr. jap. **16**, 477–478 (1969)

Huang, C.Y., Everitt, A.V.: The effect of pregnancy on pineal weight in the rat. J. Endocr. **32**, 261–262 (1965)

Hülsemann, M.: Vergleichende histologische Untersuchungen über das Vorkommen von Gliafasern in der Epiphysis cerebri von Säugetieren. Acta anat. (Basel) **66**, 249–278 (1967)

Hülsemann, M.: Development of the innervation in the human pineal organ. Light and electron microscopic investigations. Z. Zellforsch. **115**, 396–415 (1971)

Hungerford, G.F., Panagiotis, N.M.: Response of pineal lipid to hormone imbalances. Endocrinology **71**, 936–942 (1962)

Hungerford, G.F., Pomerat, C.M.: Rat pineal parenchymal cells as observed in tissue culture. Z. Zellforsch. **57**, 809–817 (1962)

Hungerford, G.F., Pomerat, C.M.: Observations on the rat pineal in tissue culture. Progr. Brain Res. **10**, 465–472 (1965)

Hus-Citharel, A., Roseau, S., Zurburg, W.: Effects of precocious pinealectomy and hemicastration on pituitary and plasma LH levels in immature male rats. J. Neural Transm. **40**, 33–46 (1977)

Hutchison, V.H., Kosh, R.J.: Thermoregulatory function of the parietal eye in the lizard *Anolis carolinensis*. Oecologia (Berl.) **16**, 173–177 (1974)

Huxley, M., Tapp, E.: The histochemical measurements of the lipid content of the pineal gland in rats suffering from malignancy. Brain Res. **37**, 123–125 (1972)

Hyyppä, M.: Hypothalamic monoamines and pineal dopamine during the sexual differentiation of the rat brain. Experientia (Basel) **27**, 336–337 (1971)

Hyyppä, M., Lehtinen, P., Rinne, U.K.: Effect of L-dopa on the hypothalamic, pineal and striatal monoamines and on the sexual behaviour of the rat. Brain Res. **30**, 265–272 (1971)

Hyyppä, M.T., Cardinali, D.P., Wurtman, R.J.: Sex-dependent increase in pineal hydroxyindole-O-methyltransferase activity after a single intraventricular injection of 6-hydroxydopamine to newborn rats. Neuroendocrinology **11**, 274–283 (1973)

Ifft, J.D.: Effects of pinealectomy, a pineal extract and pineal grafts on light-induced prolonged estrus in rats. Endocrinology **71**, 181–182 (1962)

Illnerová, H.: 5-Hydroxytryptophan decarboxylase activity in the rat pineal body during ontogenesis. Physiol. bohemoslov. **18**, 119–123 (1969)

Illnerová, H.: Effect of light on the serotonin content of the pineal gland. Life Sci. **10**, Part I, 955–960 (1971a)

Illnerová, H.: Effect of environmental lighting on serotonin rhythm in rat pineal gland during postnatal development. Life Sci. **10**, Part I, 583–590 (1971b)

Illnerová, H.: The effect of light on the development of a diurnal rhythm in serotonin content and on the development of hydroxyindole-O-methyl transferase activity in the rat pineal gland. Activ. nerv. sup. (Praha) **14**, 130–131 (1972)

Illnerová, H.: The effect of constant light and chemical sympathectomy on the development of serotonin N-acetyltransferase and monoamine oxidase activities in the rat epiphysis. Physiol. bohemoslov. **22**, 399–400 (1973a)

Illnerová, H.: The effect of pargyline and darkness on the activity of N-acetyltransferase in the rat pineal gland. Physiol. bohemoslov. **22**, 399 (1973b)

Illnerová, H.: The effect of constant light from birth on the appearance of a diurnal rhythm in serotonin content in the rat epiphysis. Develop. Biol. **46**, 418–421 (1975a)

Illnerová, H.: Effect of estradiol on the activity of serotonin N-acetyltransferase in the rat epiphysis. Endocr. exp. (Prag) **9**, 141–148 (1975b)

Illnerová, H., Albrecht, I.: Isoprotenerol induction of pineal serotonin N-acetyltransferase in normotensive and spontaneously hypertensive rats. Experientia (Basel) **31**, 95–96 (1975)

Illnerová, H., Škopková, J.: Regulation of the diurnal rhythm in rat pineal serotonin-N-acetyltransferase activity and serotonin content during ontogenesis. J. Neurochem. **26**, 1051–1052 (1976)

Illnerová, H., Tolar, M.: Enzymes of serotonin metabolism in the developing rat pineal gland. Physiol. bohemoslov. **22**, 69 (1973)

Illnerová, H., Bäckström, M., Sääf, L., Wetterberg, L., Vango, B.: Melatonin in rat pineal gland and serum: rapid parallel decline after light exposure at night. Neurosci. Letters **9**, 189–193 (1978)

Illnerová, H., Vaněček, J., Křeček, J., Wetterberg, L., Sääf, J.: Effect of one minute exposure to light at night on rat pineal serotonin N-acetyltransferase and melatonin. J. Neurochem. **32**, 673–675 (1979)

Ishibashi, T., Hahn, D.W., Srivastava, L., Kumaresan, P., Turner, C.W.: Effect of pinealectomy and melatonin on feed consumption and thyroid hormone secretion rate. Proc. Soc. exp. Biol. (N.Y.) **122**, 644–647 (1966)

Ito, T., Matsushima, S.: A quantitative morphological study of the postnatal development of the pineal body of the mouse. Anat. Rec. **159**, 447–452 (1967)

Ito, T., Matsushima, S.: Electron microscopic observations on the mouse pineal, with particular emphasis on its secretory nature. Arch. Histol. Jap. **30**, 1–15 (1968a)

Ito, T., Matsushima, S.: Some morphological aspects of the pineal body. Hokkaido Univ. Med. Libr. Ser. **1**, 1–20 (1968b)

Ito, T., Matsushima, S.: Effects of gonadectomy and hypophysectomy on the pineal body in the mouse: a quantitative morphological study. Anat. Rec. **162**, 479–482 (1968c)

Iturriza, F.C.: Histochemical demonstration of biogenic monoamines in the pineal gland of the toad, *Bufo arenarum*. J. Histochem. Cytochem. **15**, 301–303 (1967)

Iwata, K., Nakajima, N., Nakajima, M., Sato, N.: Report of a case of pineal tumor with extensive metastasis into lungs. Brain Nerve **14**, 815–818 (1962)

Jackson, I.M.D., Saperstein, R., Reichlin, S.: Thyrotropin releasing hormone (TRH) in pineal and hypothalamus of frog: effect of season and illumination. Endocrinology **100**, 97–100 (1977)

Jacobs, J.: The effect of pinealectomy on adrenocortical rhythm in the blinded rat. Amer. J. Anat. **139**, 437–441 (1974)

Jacobson, C.D., Mann, D.R.: Influences of pineal and adrenal glands on the proestrous release of hormones in rats maintained in a reduced photoperiod. Biol. Reprod. **18**, 712–718 (1978)

Jaim-Etcheverry, G., Zieher, L.M.: Cytochemistry of 5-hydroxytryptamine at the electron microscope level. II. Localization in the autonomic nerves of the rat pineal gland. Z. Zellforsch. **86**, 393–400 (1968)

Jaim-Etcheverry, G., Zieher, L.M.: Selective demonstration of a type of synaptic vesicle by phosphotungstic acid staining. J. Cell Biol. **42**, 855–860 (1969)

Jaim-Etcheverry, G., Zieher, L.M.: Ultrastructural cytochemistry and pharmacology of 5-hydroxytryptamine in adrenergic nerve endings. III. Selective increase of norepinephrine in the rat pineal gland consecutive to depletion of neuronal 5-hydroxytryptamine. J. Pharmacol. exp. Ther. **178**, 42–48 (1971)

Jaim-Etcheverry, G., Zieher, L.M.: Octopamine probably coexists with noradrenaline and serotonin in vesicles of pineal adrenergic nerves. J. Neurochem. **25**, 915–917 (1975)

Japha, J.L., Eder, T.J., Goldsmith, E.D.: Morphological and histochemical features of the gerbil pineal system. Anat. Rec. **178**, 381–382 (1974)

Japha, J.L., Eder, T.J., Goldsmith, E.D.: Calcified inclusions in the superficial pineal gland of the mongolian gerbil, *Meriones unguiculatus*. Acta anat. (Basel) **94**, 533–544 (1976)

Japha, J.L., Eder, T.J., Goldsmith, E.D.: A histochemical study of aldehyde fuchsin-positive material and "high-esterase cells" in the pineal gland of the mongolian gerbil. Amer. J. Anat. **149**, 23–38 (1977)

Jelínek, J., Křeček, J.: The effect of age and pinealectomy on the hypertension produced by adrenal regeneration. Experientia (Basel) **24**, 912–913 (1968)

Jellinger, K.: Primary intracranial germ cell tumors. Acta neuropath. (Berl.) **25**, 291–306 (1973)

Jellinger, K., Minauf, M., Kraus, H., Sunderplasmann, M.: Embryonales Carcinom der Epiphysenregion. Acta neuropath. (Berl.) **15**, 176–182 (1970)

Jenkin, R.D.T., Simpson, W.J.K., Keen, C.W.: Pineal and suprasellar germinomas. Results of radiation treatment. J. Neurosurg. **48**, 99–107 (1978)

Jequier, E., Robinson, D.S., Lovenberg, W., Sjoerdsma, A.: Further studies on tryptophan hydroxylase in rat brain-stem and beef pineal. Biochem. Pharmacol. **18**, 1071–1081 (1969)

Jimerson, D.C., Lynch, H.J., Post, R.M., Wurtman, R.J., Bunney, W.E.: Urinary melatonin rhythms during sleep deprivation in depressed patients and normals. Life Sci. **20**, 1501–1508 (1977)

Jöchle, W.: Über die Wirkung eines Epiphysenextraktes (Glanepin) auf Sexualentwicklung und Sexualcyclus junger weiblicher Ratten unter normalen Haltungsbedingungen und die Dauerbeleuchtung. Endokrinologie **33**, 287–295 (1956)

John, T.M., Itoh, S., George, J.C.: On the role of the pineal in thermoregulation in the pigeon. Hormone Res. **9**, 41–56 (1978)

Johnson, J.I., Jr.: Central Nervous System of Marsupials. In: The biology of marsupials (D. Hunsaker II, ed.), pp. 157–278. New York-San Francisco-London: Academic Press 1977

Johnson, J.R., Meyer, P.A.R., Westaby, D., Herbert, J.: An experimental study of the autonomic innervation of the ferret's pineal gland. J. Anat. (Lond.) **113**, 278–279 (1972)

Jones, R.L., McGeer, P.L., Greiner, A.C.: Metabolism of exogenous melatonin in schizophrenic and non-schizophrenic volunteers. Clin. chim. Acta. **26**, 281–285 (1969)

Jóo, I., Kaháw, A.: The porphyrin content of Harderian glands in rats and the melatonin-melanocyte stimulating hormone-system. Endokrinologie **65**, 308–312 (1975)

Joss, J.M.P.: Pineal-gonad relationships in the lamprey *Lampetra fluviatilis*. Gen. comp. Endocr. **21**, 118–122 (1973a)

Joss, J.M.P.: The pineal complex, melatonin, and color change in the lamprey *Lampetra*. Gen. comp. Endocr. **21**, 188–195 (1973b)

Joss, J.M.P.: Hydroxyindole-O-methyltransferase (HIOMT) activity and the uptake of ^{3}H-melatonin in the lamprey, *Geotria australis* Gray. Gen. comp. Endocr. **31**, 270–275 (1977)

Jouan, P.: Action de l'adrénoglomérulotrophine sur la corticoidogénèse in vitro chez le Rat. C.R. Soc. Biol. (Paris) **156**, 1567–1569 (1962)

Jouan, P.: Epiphyse, 5-hydroxytryptamine et corticoidogenèse *in vitro*. Ann. endocr. (Paris) **24**, 365–370 (1963)

Jouan, P., Mendell, Ph.: Sur la libération in vitro des corticostéroides par les surrénales du rat sous l'influence des extraits épiphysaires et de la 5-hydroxytryptamine (sérotonine). In: Major problems in neuroendocrinology (E. Bajusz and G. Jasmin, eds.), pp. 409–418. Basel-New York: Karger 1964

Jouan, P., Rocaboy, J.-C.: Etude de quelques activités enzymatiques de l'épiphyse du porc. C.R. Soc. Biol. (Paris) **159**, 591–594 (1965)

Jouan, P., Rocaboy, J.-C.: Étude de l'activité peptidasique de la glande pinéale du porc. C.R. Soc. Biol. (Paris) **160**, 859–862 (1966)

Jouan, P., Samperez, S.: Recherches sur la spécificité d'action de la 5-hydroxytryptamine vis-à-vis de la sécrétion in vitro de l'aldostérone. Ann. endocr. (Paris) **25**, 70–75 (1964)

Jouan, P., Samperez, S.: Étude de la sécrétion des corticosteroides et de l'hormone adrénocorticotrope hypophysaire chez le rat épiphysectomisé. Progr. Brain Res. **10**, 604–611 (1965a)

Jouan, P., Samperez, S.: Recherches sur l'action de la N-acétyl 5-méthoxytryptamine (mélatonine) vis à vis de la sécrétion adrénocorticotrope hypophysaire chez le Rat. C.R. Soc. Biol. (Paris) **159**, 1270–1273 (1965b)

Jouan, P., Dai, T.V., Cormier, M.: Sur la nature des phospholipides de la glande pinéale (épiphyse) du porc. Bull. Soc. Chim. biol. (Paris) **46**, 1121–1129 (1964)

Jouan, P., Garreau, A., Samperez, S.: Extraction et séparation des peptides de la glande pinéale du mouton. Ann. endocr. (Paris) **26**, 535–543 (1965)

Jouan, P., Samperez, S., Garreau, A.: Isolement de la 5-hydroxytryptamine (sérotonine) des épiphyses de rats par filtration sur gel de sephadex G25. Ann. endocr. (Paris) **27**, 561–569 (1966)

Jouan, P., Boisseau, J., More, E.: Isolement des acides aminés et des peptides de la glande pinéale de mouton. Rev. roum. Endocr. **5**, 117–124 (1968)

Joyner, J.E.: Metastatic gastric adenocarcinoma to the pineal body. A case report. Acta neuropath. (Berl.) **1**, 416–419 (1962)

Juillard, M.T., Collin, J.P.: Rôle des photorécepteurs rudimentaires sécrétoires de l'organe pinéal de la Perruche (*Melopsittacus undulatus*). J. Physiol. (Paris) **70**, 5B–6B (1975)

Juillard, M.-Th., Collin, J.-P.: The avian pineal organ: Evidence for a proteinaceous component in the secretion granules of the rudimentary photoreceptor cells. An ultracytochemical and pharmacological study in the parakeet. Biol. Cellulaire **31**, 51–58 (1978)

Juillard, M.T., Hartwig, H.G., Collin, J.P.: The avian pineal organ. Distribution of endogenous monoamines; a fluorescence microscopic, microspectrofluorimetric and pharmacological study in the parakeet. J. Neural Transm. **40**, 269–287 (1977)

Juszkiewicz, T., Rakalska, Z.: Antioestrogenic effects of bovine pineal glands. Nature (Lond.) **200**, 1329–1330 (1963)

Juszkiewicz, T., Rakalska, Z.: Lack of the effect of melatonin on the frog spermatogenic reaction. J. Pharm. Pharmacol. **17**, 189–190 (1965)

Kachi, T., Ito, T.: Neural control of glycogen content and its diurnal rhythm in mouse pineal cell. Amer. J. Physiol. **232**, E 584–E 589 (1977)

Kachi, T., Matsushima, S., Ito, T.: Effects of continuous lighting on glycogen in the pineal cells of the mouse: a quantitative histochemical study. Z. Zellforsch. **118**, 214–220 (1971a)

Kachi, T., Matsushima, S., Ito, T.: Diurnal changes in glycogen content in the pineal cells of the male mouse. A quantitative histochemical study. Z. Zellforsch. **118**, 310–314 (1971b)

Kachi, T., Matsushima, S., Ito, T.: Diurnal variations in pineal glycogen content during the estrous cycle in female mice. Arch. Histol. Jap. **35**, 153–159 (1973)

Kachi, T., Matsushima, S., Ito, T.: Effect of continuous darkness on diurnal rhythm in glycogen content in pineal cells of the mouse: a semiquantitative histochemical study. Anat. Rec. **197**, 405–410 (1974)

Kachi, T., Matsushima, S., Ito, T.: Postnatal observations on the diurnal rhythm and the light-responsiveness in the pineal glycogen content in mice. Anat. Rec. **183**, 39–46 (1975)

Kachi, T., Banerji, T.K., Quay, W.B.: Daily rhythmic changes in synaptic vesicle contents of nerve endings on adrenomedullary adrenaline cells, and their modification by pinealectomy and sham operations. Neuroendocrinology **28**, 201–211 (1979)

Kageyama, N., Belsky, R.: Ectopic pinealoma in the chiasma region. Neurology (Minneap.) **11**, 318–327 (1961)

Kalm, H., Magun, R.: Beitrag zur Klinik und patholog. Anatomie der Pinealome. Dtsch. Z. Nervenheilk. **164**, 453–468 (1950)

Kalsow, C.M., Wacker, W.B.: Localization of a uveitogenic soluble retinal antigen in the normal guinea-pig eye by an indirect fluorescent antibody technique. Int. Arch. Allergy **44**, 11–20 (1973)

Kalsow, C.W., Wacker, W.B.: Pineal reactivity of anti-retina sera. Invest. Ophthalmol. Visual Sci. **16**, 181–184 (1977)

Kalsow, C.M., Wacker, W.B.: Pineal gland involvement in retina-induced experimental allergic uveitis. Invest. Ophthal. Vis. Sci. **17**, 774–783 (1978)

Kamberi, I.A.: The role of brain monoamines and pineal indoles in the secretion of gonadotrophins and gonadotrophin-releasing factors. Progr. Brain Res. **39**, 261–280 (1973)

Kamberi, I., Mical, R.S., Porter, J.C.: Effect of anterior pituitary perfusion and intraventricular injection of catecholamines and indoleamines on LH release. Endocrinology **87**, 1–12 (1970)

Kamberi, I.A., Mical, R.S., Porter, J.C.: Effects of melatonin and serotonin on the release of FSH and prolactin. Endocrinology **88**, 1288–1293 (1971)

Kanazawa, I., Iversen, L.L., Kelly, J.S.: Glutamate decarboxylase activity in the rat posterior pituitary, pineal gland, dorsal root ganglion and superior cervical ganglion. J. Neurochem. **27**, 1267–1269 (1976)

Kantarjian, A.H.: Functional aspects of the mammalian pineal gland during pregnancy. Thesis, University of London 1976

Kappers, J.A.: The development, topographical relations and innervation of the epiphysis cerebri in the albino rat. Z. Zellforsch. **52**, 163–215 (1960a)

Kappers, J.A.: Innervation of the epiphysis cerebri in the albino rat. Anat. Rec. **136**, 220–221 (1960b)

Kappers, J.A.: Die Innervation der Epiphyse cerebri der Albinoratte. Acta neuroveg. (Wien) **23**, 111–114 (1961)

Kappers, J.A.: Melatonin, a pineal compound. Preliminary investigations on its function in the rat. Gen. comp. Endocr. **2**, 610–611 (1962)

Kappers, J.A.: Survey of the innervation of the pineal organ in vertebrates. Amer. Zool. **4**, 47–51 (1964)

Kappers, J.A.: Survey of the innervation of the epiphysis cerebri and the accessory pineal organs of vertebrates. Progr. Brain Res. **10**, 87–153 (1965a)

Kappers, J.A.: On the innervation of the pineal organs and on their function in relation to their innervation. In: Bargmann, W. (ed.). Aus der Werkstatt der Anatomen, pp. 93–117. Stuttgart: Georg Thieme 1965b

Kappers, J.A.: Note préliminaire sur l'innervation de l'épiphyse du lizard *Lacerta viridis*. Bull. Ass. Anat. (Paris) **134**, 111–116 (1966)

Kappers, J.A.: The sensory innervation of the pineal organ in the lizard, *Lacerta viridis*, with remarks on its position in the trend of pineal phylogenetic structural and functional evolution. Z. Zellforsch. **81**, 581–618 (1967a)

Kappers, J.A.: The mammalian epiphysis cerebri as a center of neurovegetative regulation. Acta neuroveg. (Wien) **30**, 190–200 (1967b)

Kappers, J.A.: The morphological and functional evolution of the pineal organ during its phylogenetic development. Excerpta Med. Int. Congr. Series **184**, 619–626 (1968)

Kappers, J.A.: The mammalian pineal organ. J. neuro-visc. Rel., Suppl. **9**, 140–184 (1969)

Kappers, J.A.: The pineal organ: an introduction. In: The pineal gland (G.E.W. Wolstenholme and J. Knight, eds.), pp. 3–34. Edinburgh-London: Churchill Livingstone 1971a

Kappers, J.A.: Discussion remark. In: The pineal gland (G.E.W. Wolstenholme and J. Knight, eds.), p. 121. Edinburgh-London: Churchill Livingstone 1971b

Kappers, J.A.: The mammalian pineal gland, a survey. Acta neurochir. (Wien) **34**, 109–149 (1976)

Kappers, J.A., Schadé, J.P. (eds.): Structure and function of the epiphysis cerebri. Progr. Brain Res. **10** (1965)

Kappers, J.A., Prop, N., Braams, W.G.: Histochemical and electron microscopical investigation of the "Golgi complex" in the parenchymal cells of the pineal gland in the rat. Acta morph. neerl.-scand. **4**, 290 (1962)

Kappers, J.A., Prop, N., Zweens, J.: Qualitative evaluation of pineal fats in the albino rat by histochemical methods and paper chromatography and the changes in pineal fat contents under physiological and experimental conditions. Progr. Brain. Res. **5**, 191–200 (1964)

Kappers, J.A., Smith, A.R., De Vries, R.A.C.: The mammalian pineal gland and its control of hypothalamic activity. Progr. Brain. Res. **41**, 149–174 (1974)

Karahasanoğlu, A.M., Özand, P.T.: The purification and properties of bovine pineal S-adenosylmethionine: N-acetylserotonin O-methyltransferase. J. Neurochem. **19**, 411–430 (1972)

Karasek, M.: A study of the nature of so-called nuclear spheres ("Kernkugeln"). Folia morph. (Warszawa) **26**, 307–312 (1967)

Karasek, M.: The cilia in the white rat pineal gland. J. Microsc. **9**, 1103–1104 (1970)

Karasek, M.: The influence of hypophysectomy on the ultrastructure of the pineal gland in white rats. Preliminary investigations. Acta med. pol. **12**, 153–156 (1971a)

Karasek, M.: Ultrastructure of the pineal body in white rats in normal conditions and after pituitary gland extraction. Endokryn. Pol. **22**, 15 (1971b)

Karasek, M.: Ultrastructure of rat pineal gland in organ culture; influence of norepinephrine, dibutyryl cyclic adenosine 3',5'-monophosphate and adenohypophysis. Endokrinologie **64**, 106–114 (1974)

Karasek, M.: Szyszynka. Warszawa: Paústwowy Zaktad Wydawnictw Lekarskich 1976a

Karasek, M.: Quantitative changes in number of "synaptic" ribbons in rat pinealocytes after orchidectomy and in organ culture. J. Neural Transm. **38**, 149–157 (1976b)

Karasek, M., Marek, K.: Influence of gonadotropic hormones on the ultrastructure of rat pinealocytes. Cell Tiss. Res. **188**, 133–141 (1978)

Karasek, M., Oginski, M., Surma, M., Pawlikowski, M.: The influence of time of day on ^{32}P uptake by the pineal body, superior cervical ganglia and hypophysis in rats. Pol. Endocr. **23**, 199–202 (1972)

Karasek, M., Pawlikowski, M., Kappers, J.A., Stępién, H.: Influence of castration followed by administration of LH-RH on the ultrastructure of rat pinealocytes. Cell Tiss. Res. **167**, 325–339 (1976a)

Karasek, M., Pawlikowski, M., Pévet, P., Stępién, H.: Ultrastructural and fluorescence histochemical studies of the rat pineal gland after castration. Annals Med. Sect. Pol. Acad. Sci. **21**, 57–58 (1976b)

Karasek, M., Karasek, E., Stępién, H.: Effect of castration on the concentration of adenosine 3',5'-monophosphate in the rat pineal organ. J. Neural Transm. **42**, 145–149 (1978a)

Karasek, M., Marek, K., Kunert-Radek, J.: Ultrastructure of rat pinealocytes in vitro: influence of gonadotropic hormones and LH-RH. Cell Tiss. Res. **195**, 547–556 (1978b)

Karler, A.: Report summarizing work on the Bergmann pineal preparations. Z. Vitamin-, Hormon- u. Fermentforsch. **12**, 350–354 (1962/63)

Karnovsky, M.J., Roots, L.: A "direct-coloring" thiocholine method for cholinesterase. J. Histochem. Cytochem. **12**, 219–221 (1964)

Karppanen, H.: Effect of propranolol on the blood pressure of normotensive and pinealectomized hypertensive rats. Naunyn-Schmiedeberg's Arch. Pharmacol. **281**, 1–12 (1974)

Karppanen, H., Vapaatalo, H.: Effects of an aldosterone antagonist, spironolactone on pinealectomized rats. Pharmacology **6**, 257–264 (1971)

Karppanen, H., Vapataalo, H., Lahovaara, S., Paasonen, M.K.: Studies with pinealectomized rats. Pharmacology **3**, 76–84 (1970a)

Karppanen, H., Vapaatalo, H., Lahovaara, S., Paasonen, M.K.: Effects of spironolactone on the pinealectomized rats. Scand. J. clin. Lab. Invest. **25**, Suppl. **113**, 42 (1970b)

Karppanen, H., Airaksinen, M.M., Särkimäki, I.: Effects in rats of pinealectomy and oxypertine on spontaneous locomotor activity and blood pressure during various light schedules. Ann. Med. exp. Fenn. **51**, 93–103 (1973)

Karppanen, H., Lahovaara, S., Männiströ, P.: Plasma renin activity and *in vitro* synthesis of aldosterone by the adrenal glands of rats with spontaneous, renal, or pinealectomy-induced hypertension. Acta physiol. scand. **94**, 184–188 (1975)

Kasal, C.A., Menaker, M., Perez-Polo, J.R.: Circadian clock in culture: N-acetyltransferase activity of chick pineal glands oscillates in vitro. Science **203**, 656–658 (1979)

Kasbohm, P.: Der Einfluß des Lichts auf die Temperaturadaptation bei *Rana temporaria*. Helgoländer wiss. Meeresunters. **16**, 157–178 (1967)

Kastin, A.J., Schally, A.V.: Autoregulation of release of melanocyte stimulating hormone from the rat pituitary. Nature (Lond.) **213**, 1238–1240 (1967)

Kastin, A.J., Redding, T.W., Schally, A.V.: MSH activity in rat pituitaries after pinealectomy. Proc. Soc. exp. Biol. (N.Y.) **124**, 1275–1277 (1967a)

Kastin, A.J., Redding, R., Schally, A.V.: Elevation of rat pituitary MSH levels after pinealectomy. Clin. Res. **15**, 32 (1967b)

Kastin, A.J., Schally, A.V., Viosca, S., Barrett, L., Redding, T.W.: MSH activity in the pituitaries of rats exposed to constant illumination. Neuroendocrinology **2**, 257–262 (1967c)

Kastin, A.J., Schally, A.V., Viosca, S., Miller, M.C.: MSH activity in plasma and pituitaries of rats after various treatments. Endocrinology **84**, 20–27 (1969)

Kastin, A.J., Viosca, S., Nair, R.M.G., Schally, A.V., Miller, M.C.: Interactions between pineal, hypothalamus and pituitary involving melatonin, MSH release-inhibiting factor and MSH. Endocrinology **91**, 1323–1328 (1972)

Kastin, A.J., Plotnikoff, N.P., Sandman, C.A., Sprites, M.A., Kostrzewa, R.M., Paul, S.M., Stratton, L.O., Miller, L.H., Labrie, F., Schally, A.V., Goldman, H.: The effects of MSH and MIF on the brain. In: Anatomical neuroendocrinology (W.E. Stumpf and L.D. Grant, eds.), pp. 290–297. Basel: S. Karger 1975

Kawashima, T., Takeuchi, K., Nakamura, M., Ogata, T.: The relationship between histochemical enzyme activities of brain tumors and clinical features of the patients. Acta Med. Okayama **19**, 293–306 (1965)

Kebabian, J.W., Zatz, M., Romero, J.A., Axelrod, J.: Rapid changes in pineal β-adrenergic receptor: alteration in l-[^{3}H] alprenolol binding and adenylate cyclase. Proc. nat. Acad. Sci. (Wash.) **72**, 3735–3739 (1975)

Kehl, R., Czyba, J.C., Durand, M.: Sur l'action lutéinique de l'épiphysiectomie chez le Rat. C.R. Soc. Biol. (Paris) **156**, 98–100 (1962)

Kelly, D.E.: Embryonic and larval epiphysectomy in the salamander, *Taricha torosa*, and observations on scoliosis. J. Morph. **103**, 503–538 (1958)

Kelly, D.E.: Pineal organs: photoreception, secretion, and development. Amer. Sci. **50**, 597–625 (1962)

Kelly, D.E.: The pineal organ of the newt. A developmental study. Z. Zellforsch. **58**, 693–713 (1963)

Kelly, D.E.: Ultrastructure and development of amphibian pineal organs. Progr. Brain Res. **10**, 270–287 (1965)

Kelly, D.E.: Developmental aspects of amphibian pineal systems. In: The pineal gland (G.E.W. Wolstenholme and J. Knight eds.), pp. 53–77. Edinburgh-London: Churchill Livingstone 1971

Kelly, D.E., Smith, S.W.: Fine structure of the pineal organs of the adult frog, *Rana pipiens*. J. Cell Biol. **22**, 565–587 (1964)

Kelly, D.E., Van de Kamer, J.C.: Cytological and histochemical investigations on the pineal organ of the adult frog (*Rana esculenta*). Z. Zellforsch. **52**, 618–639 (1960)

Kennaway, D.J., Seamark, R.F.: The occurrence of hydroxyindole-O-methyltransferase activity in fetal sheep pineal tissue and its relationship to preparturient endocrine changes. J. Reprod. Fertil. **45**, 529–531 (1975)

Kennaway, D.J., Frith, R.G., Phillipou, G., Matthews, C.D., Seamark, R.F.: A specific radioimmunoassay for melatonin in biological tissue and fluids and its validation by gas chromatography-mass spectrometry. Endocrinology **101**, 119–127 (1977)

Kennaway, D.J., Porter, K.J., Seamark, R.F.: Changes in plasma tryptophan and melatonin content in penned sheep. Aust. J. Biol. Sci. **31**, 49–52 (1978)

Kenny, G.C.T.: The "nervus conarii" of the monkey. (An experimental study). J. Neuropath. exp. Neurol. **20**, 563–570 (1961)

Kenny, G.C.T.: The innervation of the mammalian pineal body (A comparative study). Proc. Aust. Assoc. Neurol. **3**, 133–140 (1965a)

Kenny, G.C.T.: Transversely striated muscle fibres in the pineal region of mammals. J. Anat. (Lond.) **99**, 945 (1965b)
Kenny, G.C.T., Scheelings, F.T.: Observations of the pineal region of non-eutherian mammals. Cell. Tiss. Res. **198**, 309–324 (1979)
Kenny, G.C.T., Louis, C.J., Bradley, K.C., Anderson, R.McD., Walden, N.B.: Effect of optic chiasmotomy upon serotonin concentration in the pineal gland of the monkey. Amer. J. Phys. Anthropol. **38**, 399–402 (1973)
Kerényi, N.A., Sarkar, K.: The postnatal transformation of the pineal gland. Acta morph. Acad. Sci. hung. **16**, 223–236 (1968)
Kérenyi, N.A., Von Westarp, C.: The effect of blindness on the postnatal transformation of the pineal gland. Excerpta med. **157**, 180 (1968)
Kerényi, N.A., Von Westarp, C.: Post-natal transformation of the pineal gland: effect of constant darkness. Endocrinology **88**, 1077–1079 (1971)
Kerényi, N.A., Lucis, O.J., Borkow, J., Lucis, R.: The effect of constant darkness on the pineal gland and gonads. Gen. comp. Endocr. **18**, 600 (1972)
Kerényi, N.A., Sótonyi, P., Somogyi, E.: Localizing acetyl-serotonin transferase by electron microscopy. Histochemistry **46**, 77–80 (1975a)
Kerényi, N.A., Somogyi, E., Sotonyi, P.: Electron microscopic studies on the aryl-sulphatase activity of the pineal gland. Acta histochem. (Jena) **53**, 192–197 (1975b)
Kevorkian, J., Wessel, W.: So-called "nuclear pellets" ("Kernkugeln") of pineocytes. Arch. Path. **68**, 513–524 (1959)
Keyes, M.C., Wurtman, R.J., Elden, C.A., Chou, C., McGuire, R.: Seasonal fluctations of hydroxyindole levels in the pineal gland of the Northern fur seal. J. Wildl. Dis. **7**, 235 (abst.) 1971
Khelimsky, A.M.: On the origin of cerebral sand from the colloid of epiphysis. Prob. Éndokr. Gormonoter. **4**, 86 (1958), cited after Quay (1974a)
Khelimsky, A.M.: Histophysiology of the colloid and brain sand of the pineal gland (histochemical studies). Probl. Endocr. Mosk. **15**, 50–54 (1969)
Kincl, F.A., Benagiano, G.: The failure of the pineal gland removal in neonatal animals to influence reproduction. Acta endocr. (Kbh.) **54**, 189–192 (1967)
Kincl, F.A., Chang, C.C., Zbuzkova, V.: Observation of the influence of changing photoperiod on spontaneous wheelrunning activity of neonatally pinealectomized rats. Endocrinology **87**, 38–42 (1970)
Kinson, G.A., Liu, C.C.: Effects of blinding and pinealectomy on diurnal variations in plasma testosterone. Experientia (Basel) **29**, 1415–1416 (1973a)
Kinson, G.A., Liu, C.-C.: Testicular responses to melatonin and serotonin implanted peripherally in immature rats. Life Sci. **12**, Part I, 173–184 (1973b)
Kinson, G.A., Peat, F.: The influences of illumination, melatonin and pinealectomy on testicular function in the rat. Life Sci. **10**, Part I, 259–269 (1971)
Kinson, G.A., Robinson, S.: Gonadal function of immature male rats subjected to light restriction, melatonin administration and removal of the pineal gland. J. Endocr. **47**, 391–392 (1970)
Kinson, G.A., Singer, B.: The pineal gland and the adrenal response to sodium deficiency in the rat. Neuroendocrinology **2**, 283–293 (1967)
Kinson, G.A., Singer, B., Grant, L.: Adrenocortical hormone secretion at various time intervals after pinealectomy in the rat. Gen. comp. Endocr. **10**, 447–449 (1968a)
Kinson, G.A., Singer, B., Grant, L.: Tissue weights and adrenocortical function at various time intervals after pinealectomy in the rat. Excerpta med. **157**, 179 (1968b)
Kinson, G., Wahid, A.K., Singer, B.: Effect of chronic pinealectomy on adrenocortical hormone secretion rates in normal and hypertensive rats. Gen. comp. Endocr. **8**, 445–454 (1967)
Kiss, J., Bánhegyi, D., Csaba, G.: Endocrine regulation of blood calcium level. II. Relationship between the pineal body and the parathyroid glands. Acta med. Acad. Sci. hung. **26**, 363–370 (1969)
Kitay, J.I.: Effects of pinealectomy on ovarian weight in immature rats. Endocrinology **54**, 114–116 (1954a)
Kitay, J.I.: Pineal lesions and precocious puberty: A review. J. clin. Endocr. **14**, 622–625 (1954b)
Kitay, J.-I.: Relations entre la glande pinéale et la fonction hypophysocortico-surrénalienne. Ann. endocr. (Paris) **24**, 227 (1963)
Kitay, J.I.: Possible functions of the pineal gland. In: Neuroendocrinology (L. Martini, W.F. Ganong, eds.), Vol. 2, pp. 641–664. New York and London: Academic Press 1967

Kitay, J.I., Altschule, M.D.: The pineal gland. Cambridge: Harvard Univ. Press 1954a

Kitay, J.I., Altschule, M.D.: Effects of pineal extract administration on ovary weight in rats. Endocrinology **55**, 782–784 (1954b)

Kizer, J.S., Zivin, J.A., Jacobowitz, D.M., Kopin, I.J.: The nyctohemeral rhythm of plasma prolactin: Effects of ganglionectomy, pinealectomy, constant light, constant darkness or 6-OH-dopamine administration. Endocrinology **96**, 1230–1240 (1975)

Klatzo, I., Miquel, J., Ferris, P.J., Prokop, J.D., Smith, D.E.: Observations on the passage of the fluorescein labelled serum proteins (FLSP) from the cerebrospinal fluid. J. Neuropath. exp. Neurol. **23**, 18–35 (1964)

Klein, D.C.: Pineal gland metabolism. The relationship between hydroxyindole-O-methyl transferase, melatonin production and secretion as stimulated by norepinephrine. Fed. Proc. **28**, 734 Abs. (1969)

Klein, D.C.: Evidence for the placental transfer of ^{3}H-acetyl-melatonin. Nature (Lond.) New Biol. **237**, 117–118 (1972)

Klein, D.C., Berg, G.R.: Pineal gland: Stimulation of melatonin production by norepinephrine involves cyclic AMP-mediated stimulation of N-acetyltransferase. In: Role of cyclic AMP in cell function (P. Greengard, E. Costa, eds.). Advanc. Biochem. Psychopharmacol. **3**, 241–263 (1970)

Klein, D.C., Lines, S.V.: Pineal hydroxyindole-O-methyltransferase activity in the growing rat. Endocrinology **84**, 1523–1525 (1969)

Klein, D.C., Notides, A.: Thin-layer chromatographic separation of pineal gland derivatives of serotonin-^{14}C. Analyt. Biochem. **31**, 480–483 (1969)

Klein, D.C., Rowe, J.: Pineal gland in organ culture. I. Inhibition by harmine of serotonin-^{14}C oxidation, accompanied by stimulation of melatonin-^{14}C production. Molec. Pharmacol. **6**, 164–171 (1970)

Klein, D.C., Weller, J.: Input and output signals in a model neural system: the regulation of melatonin production in the pineal gland. In Vitro **6**, 197–204 (1970a)

Klein, D.C., Weller, J.: Pineal gland in culture: serotonin N-acetyltransferase activity is stimulated by norepinephrine and dibutyryl cyclic adenosine monophosphate. Fed. Proc. **29**, 615 (1970b)

Klein, D.C., Weller, J.L.: Indole metabolism in the pineal gland: a circadian rhythm in N-acetyltransferase. Science **169**, 1093–1095 (1970c)

Klein, D.C., Weller, J.: Rapid light-induced decrease in pineal serotonin N-acetyltransferase activity. Science **177**, 532–533 (1972a)

Klein, D.C., Weller, J.L.: Pineal N-acetylserotonin. Fed. Proc. **31**, 222 (1972b)

Klein, D.C., Weller, J.L.: Adrenergic-adenosine 3′,5′-monophosphate regulation of serotonin N-acetyltransferase activity and the temporal relationship of serotonin N-acetyltransferase activity to synthesis of ^{3}H-N-acetylserotonin and ^{3}H-melatonin in the cultured rat pineal gland. J. Pharmacol. exp. Ther. **186**, 516–527 (1973)

Klein, D.C., Berg, G.R., Weller, J.: Melatonin synthesis: adenosine 3′,5′-monophosphate and norepinephrine stimulate N-acetyltransferase. Science **168**, 979–980 (1970a)

Klein, D.C., Berg, G.R., Weller, J., Glinsmann, W.: Pineal gland: dibutyryl cyclic adenosine monophosphate stimulation of labeled melatonin production. Science **167**, 1738–1740 (1970b)

Klein, D.C., Weller, J.L., Moore, R.Y.: Melatonin metabolism: Neural regulation of pineal serotonin: acetyl coenzyme A N-acetyltransferase activity. Proc. nat. Acad. Sci. (Wash.) **68**, 3107–3110 (1971a)

Klein, D.C., Ellison, N., Weller, J.L., Klinsmann, W.: Development and regulation of the diurnal rhythm in pineal serotonin-N-acetyltransferase. Fed. Proc. **30**, 364 Abs. (1971b)

Klein, D.C., Reiter, R.J., Weller, J.L.: Pineal N-acetyltransferase activity in blinded and anosmic male rats. Endocrinology **89**, 1020–1023 (1971c)

Klein, D.C., Juwiler, A., Weller, J.L., Plotkins, S.: Postsynaptic adrenergic-cyclic AMP control of the serotonin content of cultured rat pineal gland. J. Neurochem. **21**, 1261–1271 (1973a)

Klein, D.C., Juwiler, A., Weller, J.L., Plotkins, S.: Postsynaptic adrenergic-cyclic AMP regulation of pineal serotonin. Fed. Proc. **32**, 775 (1973b)

Klein, D.C., Buda, M.J., Kapoor, C.L., Krishna, G.: Pineal serotonin N-acetyltransferase activity: abrupt decrease in adenosine 3′,5′-monophosphate may be signal for "turnoff". Science **199**, 309–311 (1978)

Knigge, K.M., Sheridan, M.N.: Pineal function in hamsters bearing melatonin antibodies. Life Sci. **19**, 1235–1238 (1976)

Knowles, F.G.W.: Photomechanical changes in the pineal of lampreys. J. exp. Biol. **16**, 524–529 (1939)
Kobayashi, H.: A technique of pinealectomy in small birds. Tori **18**, 247–251 (1968)
König, A., Meyer, A.: Tagesperiodische Schwankungen einer antidiuretischen Aktivität aus der Epiphysis cerebri ausgewachsener männlicher Ratten. Naturwissenschaften **54**, 93 (1967)
König, A., Meyer, A.: The effect of continuous illumination on the circadian rhythm of the antidiuretic activity of the rat pineal. J. interdiscipl. Cycle Res. **2**, 255–262 (1971)
König, A., Meyer, A., Thieme, U.: Die akute antidiuretische Aktivität der Epiphysis cerebri von Wistar-Ratten. Endokrinologie **55**, 353–358 (1970)
Komatsu, K., Hiratsuka, H., Inaba, Y.: Pathological study of four cases with tumor in the region of the pineal body. Brain Nerve **23**, 917–926 (1971)
Kopin, I.J., Pare, C.M.B., Axelrod, J., Weissbach, H.: 6-Hydroxylation, the major metabolic pathway for melatonin. Biochim. biophys. Acta (Amst.) **40**, 377–378 (1960)
Kopin, I.J., Pare, C.M.B., Axelrod, J., Weissbach, H.: The fate of melatonin in animals. J. biol. Chem. **236**, 3072–3075 (1961)
Kopsch, F.: Lehrbuch und Atlas der Anatomie des Menschen. Bd. III. Nervensystem-Sinnesorgane. 18. Aufl. Leipzig: Georg Thieme 1953
Korf, H.-W.: Acetylcholinesterase-positive neurons in the pineal and parapineal organs of the rainbow trout, *Salmo gairdneri* (with special reference to the pineal tract). Cell Tiss. Res. **155**, 475–489 (1974)
Korf, H.-W.: Histological, histochemical and electron microscopical studies on the nervous apparatus of the pineal organ in the tiger salamander, *Ambystoma tigrinum*. Cell Tiss. Res. **174**, 475–497 (1976)
Kosh, R., Hutchison, V.: Thermal tolerances of parietalectomized *Anolis carolinensis* acclimatized at different temperatures and photoperiods. Herpetologia **28**, 183–191 (1972)
Koslow, S.H.: 5-Methoxytryptamine: A possible central nervous system transmitter. Advanc. Biochem. Psychopharmacol. **11**, 95–100 (1974)
Kowalzig, H.G., Gusek, W.: Die Feinstruktur der Rattenzirbel nach Kastration und unter dem Einfluß von Antiandrogen. Verh. dtsch. Ges. Path. **55**, 677–681 (1971)
Krabbe, K.H.: Development of the pineal organ and a rudimentary parietal eye in some birds. J. comp. Neurol. **103**, 139–149 (1955)
Krapp, Ch.: Der Einfluß der Epiphyse auf die Lokomotionsaktivität bei Ratten. Experientia (Basel) **33**, 731–732 (1977)
Krapp, Ch.: The ependyma on the pineal of the guinea pig (*Cavia cobaya*). A scanning electron microscopic investigation. Anat. Embryol. **152**, 217–222 (1978)
Krass, M.E., LaBella, F.S.: Biochemical evidence for a secretory role of the pineal body. Oxidation of [1-^{14}C]- and [6-^{14}C] glucose *in vitro* by pineal body, pituitary and brain from young and adult animals. J. Neurochem. (Oxford) **13**, 1157–1162 (1966)
Krass, M.-E., LaBella, F.S.: Hexosemonophosphate shunt in endocrine tissues. Quantitative estimation of the pathway in bovine pineal body, anterior pituitary, posterior pituitary, and brain. Biochim. biophys. Acta (Amst.) **148**, 384–391 (1967)
Krass, M.E., LaBella, F.S., Shin, S.H., Minnich, J.: Biochemical features of the pineal compared with other endocrine and nervous tissues. Mem. Soc. Endo. **19**, 49–76 (1971)
Kraus-Ruppert, R., Lembeck, F.: Die Wirkung von Melatonin auf die Pigmentzellen der Retina von Fröschen. Pflügers Arch. ges. Physiol. **284**, 160–168 (1965)
Křeček, J., Palatý, V.: The effect of epiphysectomy on androgenic activity in normally and prematurely weaned rats. Gen. comp. Endocr. **9**, 466–467 (1967)
Křeček, J., Páner, M., Šalátová, J., Zicha, P.: The pineal gland and the effect of neonatal administration of androgen upon the development of spontaneous salt and water intake in female rats. Neuroendocrinology **18**, 137–143 (1975)
Krstić, R.: Veränderungen der Epithelkörperchen der Ratte nach Epiphysektomie. Naturwissenschaften **52**, 40–41 (1965a)
Krstić, R.: Über die Wirkung von Epiphysenextrakt auf die Epithelkörperchen der Ratte. Naturwissenschaften **52**, 164 (1965b)
Krstić, R.: Die Treffermethode in der Messung der Parathyreozyten-Aktivität nach der Epiphysektomie. Experientia (Basel) **22**, 336–337 (1966)
Krstić, R.: Über Veränderungen der Epithelkörperchen nach Epiphysektomie. Z. Zellforsch. **77**, 8–24 (1967)

Krstić, R.: Quantitative investigations of the dark and light parafollicular cells in the rat. Z. Anat. Entwickl.-Gesch. **129**, 353–358 (1969)

Krstić, R.: Die Einwirkung von Kälte auf mit Zinkjodid-Osmiumtetroxyd reagierende synaptische Bläschen in den Nervenendigungen im Corpus pineale der Ratte. Z. Anat. Entwickl.-Gesch. **135**, 301–306 (1972a)

Krstić, R.: Elektronenmikroskopische Untersuchung der quergestreiften Muskelfasern im Corpus pineale von Wistar-Ratten. Z. Zellforsch. **128**, 227–240 (1972b)

Krstić, R.: Influence du froid sur la morphodynamique de la glande pinéale du rat. Acta anat. (Basel) **86**, 320–321 (1973)

Krstić, R.: Ultrastructure of rat pineal gland after preparation by freeze-etching technique. Cell Tiss. Res. **148**, 371–379 (1974)

Krstić, R.: Perjodatreaktive Stellen in der Glandula pinealis der Ratte. Acta anat. (Basel) **93**, 316 (1975a)

Krstić, R.: Scanning electron microscope observations in the canaliculi in the rat pineal gland. Experientia (Basel) **31**, 1072–1073 (1975b)

Krstić, R.: Contribution to the study of the ultracytochemistry of the rat pineal body. Fol. Anat. Iugosl. **4**, 9–14 (1975c)

Krstić, R.: Beitrag zur Kenntnis der "synaptic ribbons" in der Glandula pinealis der Ratte (Vorläufige Mitteilung). Verh. anat. Ges. **69**, 443–446 (1975d)

Krstić, R.: Ultracytochemistry of the synaptic ribbons in the rat pineal organ. Cell Tiss. Res. **166**, 135–143 (1976a)

Krstić, R.: A combined scanning and transmission electron microscopic study and electron probe microanalysis of human pineal acervuli. Cell Tiss. Res. **174**, 129–137 (1976b)

Krstić, R.: Ultrastruktur des Corpus pineale der Ratte nach Autotransplantation in die vordere Augenkammer. Verh. anat. Ges. **70**, 93–96 (1976c)

Krstić, R.V.: Ultrastruktur der Säugetierzelle. Ein Atlas zum Studium für Mediziner und Biologen. Berlin-Heidelberg-New York: Springer 1976d

Krstić, R.: Ultracytochemical localisation and comparison of adenyl cyclase activities in pineal bodies of Wistar rats and Mongolian gerbils. Histochemistry **53**, 249–255 (1977a)

Krstić, R.: Glande pinéale de Rat. Analyse morphométrique aux microscopes photonique et électronique. Labirint, Proceedings of the First Yugoslav Symposium on the Labyrinth, Sarajevo 1977. Edited by Dept of Anatomy Sarajevo. pp. 121–124. Sarajevo: Association of the Anatomists of Bosnia and Herzegovina 1977b

Krstić, R., Golaz, J.: Ultrastructural and X-ray microprobe comparison of gerbil and human pineal acervuli. Experientia (Basel) **33**, 507–508 (1977)

Krstić, R., Tarsoly, E.: Enzymhistochemische Untersuchungen der Calcitonineinwirkung auf das Corpus pineale bei der Ratte. Experientia (Basel) **28**, 1234–1235 (1972)

Kummer-Trost, E.: Die Bildungen des Zwischenhirndaches der Agamidae, nebst Bemerkungen über die Lagebeziehungen des Vorderhirns. Morph. Jb. **97**, 143–192 (1956)

Kunkel, A.: The influence of melatonin on the male gonad in Sprague-Dawley rats. Pol. Endocr. **20**, 32–35 (1969)

Kurosumi, K., Kawabata, I.: Electron microscopic studies on the fine structure of pineal glands in normal and experimental rats. In: Uyeda, R. (ed.): Proc. 6th Int. Congr. Electron Microscopy, 1966, Vol. 2, pp. 519–520. Tokyo: Maruzen 1966

Kurumado, K., Mori, W.: A morphological study of the circadian cycle of the pineal gland of the rat. Cell Tiss. Res. **182**, 565–568 (1977)

Kusche, P.: Über Ependym und Gliafasern in der Epiphyse der erwachsenen Katze. Z. Zellforsch. **71**, 405–414 (1966)

Kuszak, J., Rodin, M.: A new technique of pinealectomy for adult rats. Experientia (Basel) **33**, 283–284 (1977)

Kuwano, R., Takahashi, Y.: A simple method for the preparation of the immunoglobulins to hydroxyindole O-methyltransferase. J. Neurochem. **31**, 809–814 (1978)

Kuwano, R., Yoshida, Y., Takahashi, Y.: Purification of bovine pineal hydroxyindole O-methyltransferase by immunoadsorption chromatography. J. Neurochem. **31**, 815–824 (1978)

Kveder, S., McIsaac, W.M.: The metabolism of melatonin (N-acetyl-5-methoxytryptamine) and 5-methoxytryptamine. J. biol. Chem. **236**, 3214–3220 (1961)

LaBella, F.S., Krass, M.E.: Oxidation of ^{14}C-1- and ^{14}C-6-glucose in vitro by bovine pineal body (PB), anterior pituitary (AP), posterior pituitary (PP), and brain (B). Fed. Proc. **25**, 254 (1966)

LaBella, F.S., Shin, S.: Estimation of cholinesterase and choline acetyltransferase in bovine anterior pituitary, posterior pituitary, and pineal body. J. Neurochem. **15**, 335–342 (1968)

LaBella, F., Vivian, S., Queen, G.: Abundance of cystathionine in the pineal body. Free amino acids and related compounds of the bovine pineal, anterior and posterior pituitary, and brain. Biochem. biophys. Acta (Amst.) **158**, 286–288 (1968)

Lakin, M.L., Giedt, W.R., Winters, W.D.: Effect of pineal indolamines on response to ketamine in pinealectomized chick. Proc. West. pharmacol. Soc. **21**, 41–45 (1978)

Landolt, R., Hess, H.H.: Regional distribution of some chemical structural components of the human nervous system. II. Cerebrosides, proteolipid proteins and residue proteins. J. Neurochem. **13**, 1453–1459 (1966)

Landolt, R., Hess, H.H., Thacheimer, C.: Regional distribution of some chemical structural components of the human nervous system. I. DNA, RNA and ganglioside sialic acid. J. Neurochem. **13**, 1441–1452 (1966)

Lane, K.B., Ralph, C.L., Gilbert, S.: Lack of correlation between sexual maturation and pineal cytology in Japanese quail. Anat. Rec. **163**, 215 (1969)

Lang, K.: Anatomische und histologische Untersuchungen der Epiphysis cerebri von Rind und Schaf. Thesis, München 1959

Lapin, V.: Pineal gland and malignancy. Öst. Z. Onkol. **3**, 51–60 (1976)

Lapin, V.: Effects of reserpine on the incidence of 9,10-dimethyl-1,2-benzanthracene-induced tumors in pinealectomised and thymectomised rats. Oncology **35**, 132–135 (1978)

Lauber, J.K., Boyd, J.E., Axelrod, J.: Enzymatic synthesis of melatonin in avian pineal body: Extraretinal response to light. Science **161**, 489–490 (1968)

Lazo-Wasem, E.A., Graham, C.E.: Quantitative in vivo assay of pineal melanophore contracting principle (melatonin). Fed. Proc. **19**, 150 (1960)

Leduc, E.H., Wislocki, G.B.: The histochemical localization of acid and alkaline phosphatases, non-specific esterase and succinic dehydrogenase in the structures comprising the hematoencephalic barrier of the rat. J. comp. Neurol. **97**, 241–280 (1952)

Legait, H., Legait, E.: Contribution à l'étude de la glande pinéale humaine. Étude faite à l'aide de 747 glandes. Bull. Ass. Anat. (Nancy) **61**, 107–121 (1977)

Legait, H., Oboussier, H.: Étude interspécifique des correlations statistiques existant entre le volume ou le poids de la glande pinéale et les poids encéphalique et somatique dans deux groupes de mammifères. Étude intraspécifique des correlations entre glande pinéale et encephals chez le Rat et chez l'Homme. Bull. Ass. Anat. (Nancy) **61**, 123–132 (1977)

Legait, H., Roux, M., Sirjean, D., Contet-Audonneau, J.L., Burlet, C.: Etude volumétrique de l'hypothalamus, de l'épiphyse et de l'organe subfornical chez divers Rongeurs. C.R. Soc. Biol. (Paris) **167**, 1436–1439 (1973)

Legait, H., Legait, E., Contet-Audonneau, J.L., Mur, J.M., Roux, M., Richoux, J.P., Sirjean, D., Dussart, G.: Étude des corrélations liant les volumes de l'hypophyse, de l'épiphyse et de l'OSH au poids somatique et au volume de l'hypothalamus chez les Rongeurs. Bull. Ass. Anat. (Nancy) **59**, No. 164, 185–201 (1975a)

Legait, H., Roux, M., Dussart, G., Richoux, J.P., Contet-Audonneau, J.L.: Données morphométriques sur la glande pinéale du Loir (*Glis glis*) et du Lérot (*Eliomys quercinus*) au cours du cycle annuel. C.R. Soc. Biol. (Paris) **169**, 132–136 (1975b)

Legait, H., Bauchot, R., Stephan, H., Contet-Audonneau, J.L.: Etude des corrélations liant le volume de l'épiphyse aux poids somatique et encéphalique chez les rongeurs, les insectivores, les chiroptères, les prosimiens et les simiens. Mammalia **40**, 327–337 (1976a)

Legait, H., Bauchot, R., Contet-Audonneau, J.L.: Etude des corrélations liant les volumes des lobes hypophysaires et de l'épiphyse au poids somatique et au poids encéphalique chez les Chiroptères. Bull. Ass. Anat. (Nancy) **60**, No. 168, 175–188 (1976b)

Legait, H., Legait, E., Roux, M., Sirjean, D.: Comparaison de la croissance intraspécifique des organes endocrâniens (lobes hypophysaires, glande pinéale, OSF) du Rat blanc Wistar *Rattus norvegicus* (Beckenhout) avec leur croissance interspécifique chez les Rongeurs. Bull. Ass. Anat. (Nancy) **60**, No. 168, 189–200 (1976c)

Legait, H., Legait, E., Oboussier, M.: Corrélation entre le poids de la glande pinéale, le poids somatique et le poids de l'encéphale chez les Artiodactyles. Ann. Biol. animal. **17**, 187–191 (1977)

Legait, H., Harteman, P., Cunin, B.: Mise en évidence d'une involution de la glande pinéale et de l'hypophyse chez des souris intoxiquées au tétrachlorure de carbone. C.R. Soc. Biol. (Paris) **172**, 289–292 (1978)

Legros, J.-J., Louis, F., Gröschel-Stewart, U., Franchimont, P.: Presence of immunoreactive neurophysin-like material in human target organs and pineal gland: physiological meaning. Ann. N.Y. Acad. Sci. **248**, 157–171 (1975)

Legros, J.J., Louis, F., Demoulin, A., Franchimont, P.: Immunoreactive neurophysins and vasotocin in human foetal pineal glands. J. Endocr. **69**, 289–290 (1976)

Le Gros Clark, W.E.: The nervous and vascular relations of the pineal gland. J. Anat. (Lond.) **74**, 471–492 (1939/40)

Lenti, G., Molinatti, G.M., Pizzini, A.: Sull'azione svolta da un estratto liofilizzato di ghiandola pineale sull'apparato genitale del topo e del ratto (Richerche sui rapporti epifisi-ipofisi). Folia endocr. (Roma) **10**, 11–21 (1957)

Lenys, R.: Contribution à l'étude de la structure et du rôle de l'organe sous-commissural. Ann. Sci. Univ. Besançon 3e sér. Médecine fasc. **2**, 1–147 (1966)

Leonardelli, J., Tramu, G., Hermand, E.: Mélatonine et cellules à gonadolibérine (LH-RH) de l'hypothalamus du Rat. C. R. Soc. Biol. (Paris) **172**, 481–484 (1978)

Leone, R.M., Hooper, J., Silman, R.E., Finnie, M., Carter, S., Smith, I., Mullen, P.E.: The use of gas chromatography-mass spectrometry for the specific and sensitive assay of the pineal indole 5-methoxytryptophol. J. Endocr. **80**, 42P (1979)

Leong, A.S.Y., Matthews, C.D.: The pineal gland as an APUD organ; supporting evidence and implications. Med. Hypotheses **5**, 265–273 (1979)

Leonhardt, H.: Charakteristische Anordnung von Mitochondrien und Lamellen in der Kaninchenepiphyse. Naturwissenschaften **53**, 556–557 (1966)

Leonhardt, H.: Über axonähnliche Fortsätze, Sekretbildung und Extrusion der hellen Pinealozyten des Kaninchens. Z. Zellforsch. **82**, 307–320 (1967)

Leonhardt, H.: Ependym und Circumventriculäre Organe. In: Hdb. d. mikrosk. Anatomie d. Menschen (A. Oksche u. L. Vollrath, Hrsg.), Bd. IV/10 (A. Oksche, Hrsg.) S. 177–666. Berlin-Heidelberg-New York: Springer 1980

Leopold, P.G.: Über den histotopochemischen Nachweis von Vitamin C im Zentralnervensystem (mit Berücksichtigung der Epiphysis cerebri). Zugleich ein Beitrag zur Frage der Spezifität der Vitamin C-Reaktion. Z. Zellforsch. **31**, 502–512 (1941)

Lerner, A.B., Case, J.D.: Pigment cell regulatory factors. J. invest. Derm. **32**, 211–221 (1959)

Lerner, A.B., Nordlund, J.J.: Comment. Administration of melatonin to human subjects. In: Frontiers of pineal physiology. Edit. by M.D. Altschule, pp. 42–43. Cambridge, Mass.-London: MIT Press 1975

Lerner, A.B., Wright, R.M.: In vitro frog skin assay for agents that darken and lighten melanocytes. Meth. biochem. anal. **8**, 295–307 (1960)

Lerner, A.B., Case, J.D., Takahashi, Y., Lee, Y., Mori, W.: Isolation of melatonin, the pineal gland factor that lightens melanocytes. J. Amer. chem. Soc. **80**, 2587 (1958)

Lerner, A.B., Case, J.D., Heinzelmann, R.V.: Structure of melatonin. J. Amer. chem. Soc. **81**, 6084–6085 (1959a)

Lerner, A.B., Case, J.D., Mori, W., Wright, M.R.: Melatonin in the peripheral nerve. Nature (Lond.) **27**, 1826 (1959b)

Lerner, A.B., Case, J.D., Biemann, K., Heinzelman, R.V., Szmuszkovicz, J., Anthony, W.C., Krivis, A.: Isolation of 5-methoxyindole-3-acetic acid from bovine pineal glands. J. Amer. chem. Soc. **81**, 5264 (1959c)

Lerner, A.B., Case, J.D., Takahashi, Y.: Isolation of melatonin and 5-methoxyindole-3-acetic acid from bovine pineal gland. J. biol. Chem. **235**, 1992–1997 (1960)

Letellier, M.A., Plotka, E.D., Seal, U.S., Verme, L.J., Ozoga, J.J.: A technique for pinealectomy in deer, with notes on the neuroanatomy. Amer. J. vet. Res. **39**, 1617–1620 (1978)

Levey, F.L.: Effects of pinealectomy and melatonin injections at different seasons on ovarian activity in the lizard *Anolis carolinensis*. J. exp. Zool. **185**, 169–174 (1973)

Levine, L., Riceberg, L.J.: Radioimmunoassays for melatonin. Res. Commun. Chem. Pathol. Pharmacol. **10**, 693–702 (1975)

Licht, P., Pearson, A.K.: Failure of parietalectomy to affect the testes in the lizard *Anolis carolinensis*. Copeia **1970**, 172–173 (1970)

Lierse, W.: Die Gefäßversorgung der Epiphyse und Paraphyse bei Reptilien. Progr. Brain Res. **10**, 183–192 (1965a)

Lierse, W.: Elektronenmikroskopische Untersuchungen zur Cytologie und Angiologie des Epiphysenstiels von *Anolis carolinensis*. Z. Zellforsch. **65**, 397–408 (1965b)

Lin, H.-S.: Microcylinders within mitochondrial cristae in the rat pinealocyte. J. Cell Biol. **25**, 435–441 (1965)

Lin, H.-S.: A peculiar configuration of agranular reticulum (lamellar body) in the rat pinealocyte. J. Cell Biol. **33**, 15–25 (1967)

Lin, H.-S.: A special morphological expression of centriolar formation. Anat. Rec. **163**, 313 (1969)

Lin, H.-S.: The fine structure and transformation of centrioles in the rat pinealocyte. Cytobios **2**, 128–151 (1970)

Lin, H.-S.: Transformation of centrioles in pinealocytes of adult guinea pigs. J. Neurocytol. **1**, 61–68 (1972)

Lin, H.-S., Wing, T.-Y.: The influence of activation, removal or denervation of the pineal on the fine structure of the Leydig cell and seminal vesicle epithelium in golden hamsters. Cell Tiss. Res. **191**, 367–378 (1978)

Lin, H.-S., Tseng, C.Y., Huang, B.S.: Effects of blinding and superior cervical ganglionectomy on fine structure of the hamster pineal gland and adenohypophysis. Anat. Rec. **175**, 508–509 (1973)

Lin, H.-S., Hwang, H.-B., Tseng, C.-Y.: Fine structural changes in the hamster pineal gland after blinding and superior cervical ganglionectomy. Cell Tiss. Res. **158**, 285–299 (1975)

Lingjaerde, P., Malm, O.J., Skaug, O.E.: P^{32} uptake in pinealectomized-adrenalectomized rats. Acta endocr. (Kbh.) **31**, 305–308 (1959)

Liss, L., Scharenberg, K.: The pinealomas. Lab. Invest. **14**, 556 (1965)

Liu, C.-C., Frehn, J.L.: Failure of the pineal removal to prevent some cold-induced testicular changes in golden hamsters. Experientia (Basel) **29**, 1507–1509 (1973)

Löfgren, F.O.: Vertebral angiography in the diagnosis of tumors in the pineal region. Acta radiol. (Stockh.) **50**, 108–124 (1958)

Lombard-des Gouttes, M.N.: Evolution pondérale des glandes endocrines et des glandes génitales annexes chez le rat mâle de la naissance à la puberté. Ann. Endocr. (Paris) **27**, 121–133 (1966)

Lombard-des Gouttes, M.N.: Croissance pondérale de l'épiphyse comparée à celle des glandes endocrines au cours de premiers jours de la vie chez le Rat mâle. C. R. Soc. Biol. (Paris) **160**, 1358–1362 (1967a)

Lombard-des Gouttes, M.N.: Epiphysectomie à la naissance chez la Souris mâle. C. R. Acad. Sci. (Paris) D **264**, 2141–2144 (1967b)

Lommer, D.: Hemmung der Corticosteroid-11-β-Hydroxylierung durch einen Extrakt aus Corpus pineale. Experientia (Basel) **22**, 122–123 (1966)

Longenecker, D.E., Gallo, D.G.: The inhibition of PMSG-induced ovulation in immature rats by melatonin. Proc. Soc. exp. Biol. (N.Y.) **137**, 623–625 (1971)

Losada, J.: Efectos de la pinealectomia experimental. Annales Anat. **26**, 133–153 (1977)

Losada, J., Sarrat, R.: Pinealectomia y gonada masculina. An. Anat. Numero extraordinario ad vol. **25**, 225–228 (1977)

Lott, I.T., Quarles, R.H., Klein, D.C.: Glycoproteins are synthesized and secreted by pineal glands in culture. Fed. Proc. **30**, 364 Abs. (1971)

Lott, I.T., Quarles, R.H., Klein, D.C.: Biosynthesis and release of glycoprotein by pineal gland in organ culture. Biochim. biophys. Acta (Amst.) **264**, 144–151 (1972)

Louis, C.J., Kenny, G.C., Anderson, R.McD.: Autoradiographic localization of 5-hydroxytryptamine in monkey pineal gland. Experientia (Basel) **26**, 756–757 (1970)

Lovenberg, W., Jequier, E., Sjoerdsma, A.: Tryptophan hydroxylation: Measurement in pineal gland, brain stem, and carcinoid tumor. Science **155**, 217–219 (1967)

Lowenstein, J.M., Stebbins, R.C.: Iodine metabolism and thyroid activity in parietalectomized tropical lizards. Copeia **1969**, 664–669 (1969)

Lu, K.H., Meites, J.: Effects of serotonin precursors and melatonin on serum prolactin release in rats. Endocrinology **93**, 152–155 (1973)

Lues, G.: Die Feinstruktur der Zirbeldrüse normaler, trächtiger und experimentell beeinflußter Meerschweinchen. Z. Zellforsch. **114**, 38–60 (1971)

Lukaszyk, A., Reiter, R.J.: Neurosecretion in the pineal gland of *Macaca rhesus*. Experientia (Basel) **30**, 654–655 (1974)

Lukaszyk, A., Reiter, R.J.: Histophysiological evidence for the secretion of polypeptides by the pineal gland. Amer. J. Anat. **143**, 451–464 (1975)

Lupulescu, A.: Ultrastructure of the pineal gland after hypophysectomy. Experientia (Basel) **24**, 482–484 (1968)

Luttge, W.G., Wallis, C.J.: *In vitro* accumulation and saturation of ^{3}H-progestins in selected brain regions and in the adenohypophysis, uterus and pineal of the female rat. Steroids **22**, 493–502 (1973)

Lutz, H., Collin, J.-P.: Sur la régression des cellules photoréceptrices épiphysaires chez la tortue terrestre: *Testudo hermanni* (Gmelin) et la phylogénie des photorécepteurs épiphysaires chez les vertébrés. Bull. Soc. Zool. Fr. **92**, 797–808 (1968)

Lynch, H.J.: Diurnal oscillations in pineal melatonin content. Life Sci. **10** (Part I), 791–795 (1971)

Lynch, H.J., Ralph, C.L.: Diurnal variation in pineal melatonin and its non-relationship to HIOMT activity. Amer. Zool. **10**, 300 (1970)

Lynch, H.J., Wurtman, R.J.: Pineal indole biosynthesis: Effects of changes in sympatho-adrenal function induced by factors other than environmental lighting. Fed. Proc. **32**, 251 Abs. (1973)

Lynch, H.J., Wurtman, R.J., Greenhouse, G.: Relative merits of *Rana pipiens* and *Xenopus laevis* larvae in the quantitative assay of melatonin. Am. Zool. **11**, 650 (1971)

Lynch, H.J., Eng, J.P., Wurtman, R.J.: Control of pineal indole biosynthesis by changes in sympathetic tone caused by factors other than environmental lighting. Proc. nat. Acad. Sci. (Wash.) **70**, 1704–1707 (1973a)

Lynch, H.J., Wang, P., Wurtman, R.J.: Increase in rat pineal melatonin content following L-dopa administration. Life Sci. **12**, Part I, 145–151 (1973b)

Lynch, H.J., Wurtman, R.J., Moskowitz, M.A., Archer, M.C., Ho, M.H.: Daily rhythm in human urinary melatonin. Science **187**, 169–171 (1975a)

Lynch, H.J., Ozaki, Y., Shakal, D., Wurtman, R.J.: Melatonin excretion of man and rats: effect of time of day, sleep, pinealectomy and food consumption. Int. J. Biomet. **19**, 267–279 (1975b)

Lynch, H.J., Ho, M., Wurtman, R.J.: The adrenal medulla may mediate the increase in pineal melatonin synthesis induced by stress, but not caused by exposure to darkness. J.Neural Transm. **40**, 87–97 (1977)

Lynch, H.J., Jimerson, D.C., Ozaki, Y., Post, R.M., Bunney, W.E., Wurtman, R.J.: Entrainment of rhythmic melatonin secretion in man to a 12-hour phase shift in light-dark cycle. Life Sci. **23**, 1557–1564 (1978)

MacBride, S.E., Ralph, C.L., Binkley, S., Klein, D.C.: Pineal rhythms persist in superior cervical ganglionectomized chickens. Fed. Proc. **32**, 251 (1973)

Machado, A.B.M.: Dados histológicos e embriológicos sôbre a região pineal em alguns marsupiais brasileiros. Cienc. Cult. (Sao Paulo) **17**, 249 (1965)

Machado, A.B.M.: Ultrastructure of the pineal body of the newborn rat. Anat. Rec. **154**, 381 (1966)

Machado, A.B.M.: Straight OsO_4 versus glutaraldehyde-OsO_4 in sequence as fixatives for the granular vesicles in sympathetic axons of the rat pineal body. Stain Technol. **42**, 293–300 (1967a)

Machado, A.B.M.: Ultrastructural study of the sympathetic innervation in the developing pineal body of the rat. Anat. Rec. **157**, 282 (1967b)

Machado, A.B.M.: Electron microscopy of developing sympathetic fibres in the rat pineal body. The formation of granular vesicles. Prog. Brain Res. **34**, 171–185 (1971)

Machado, A.B.M., Da Silva, C.R.: Acão de extratos de corpo pineal e órgão subcomissural na excreção de sódio no rato. Ciénc. Cult. (São Paulo) **14**, 244–245 (1962)

Machado, A.B.M., Da Silva, C.R.: Pineal body and urinary sodium excretion in the rat. Experientia (Basel) **19**, 264–265 (1963)

Machado, A.B.M., Lemos, V.P.J.: Histochemical evidence for a cholinergic sympathetic innervation of the rat pineal body. J. neuro-visc. Rel. **32**, 104–111 (1971)

Machado, A.B.M., Machado, C.R.S.: Occurrence and distribution of alkaline phosphatase in the human subcommissural organ. Anat. Rec. **151**, 463 (1965)

Machado, A.B.M., Faleiro, L.C.M., Da Silva, W.D.: Study of mast cell and histamine contents of the pineal body. Zschr. Zellforsch. **65**, 521–529 (1965)

Machado, A.B.M., Machado, C.R.S., Wragg, L.E.: Catecholamines and granular vesicles in adrenergic axons of the developing pineal body of the rat. Experientia (Basel) **24**, 464–465 (1968)

Machado, C.R.S.: Estudos histológicos e histoquímicos sôbre a regeneração do cortex da supra-renal após sua enucleação em ratos normais e pinealectomizados. Cienc. Cult (São Paulo) **17**, 250 (1965)

Machado, C.R.S.: Histoquantitative studies on the regeneration of the adrenal cortex after enucleation in normal and pinealectomized rats. Anat. Rec. **154**, 381 (1966)

Machado, C.R.S.: Histochemical study of catecholamines and 5-HT in the developing pineal body. Anat. Rec. **157**, 282 (1967)

Machado, C.R.S., Wragg, L.E., Machado, A.B.M.: A histochemical study of sympathetic innervation and 5-hydroxytryptamine in the developing pineal body of the rat. Brain Res. **8**, 310–318 (1968)

Machado, C.R.S., Wragg, L.E., Machado, A.B.M.: Circadian rhythm of serotonin in the pineal body of immunosympathectomized immature rats. Science **164**, 442–443 (1969a)

Machado, C.R.S., Machado, A.B.M., Wragg, L.E.: Circadian serotonin rhythm control: Sympathetic and nonsympathetic pathways in rat pineals of different ages. Endocrinology **85**, 846–848 (1969b)

Machemer, R.H., Foss, D.C., Carew, L.B.: The effect of selected wave-lengths of environmental light on pineal gland activity in cockerels. Poultry Sci. **50**, 1599 (1971a)

Machemer, R.H., Foss, D.C., Carew, L.B.: Effects of pinealectomy on physiological development of cockerels. Poultry Sci. **50**, 1599 (1971b)

Machová, A., Jošt, K.: Comparison of natriuretic action of carbo-analogues of deaminooxytocin and (4-Leucine, 8-arginine)-vasotocin in rats. Endocr. exp. **9**, 269–277 (1975)

MacKinnon, P.C.B., Simpson, R.A.: Preliminary observations on the uptake and incorporation of C^{14} uridine into RNA in brains of female rats from birth to senility. In: Mitchell, R.P. (ed.), Endocrinology and human behavior, pp. 330–340 London: Oxford Univ. Press 1968

MacPhee, A.A., Cole, F.E., Rice, B.F.: The effect of melatonin on steroidogenesis by the human ovary in vitro. J. clin. Endocr. **40**, 688–696 (1975)

Maeda, T., Shimizu, N., Wegmann, R.: Further investigation of histochemistry of fructose-1,6-diphosphate D-glyceraldehyde-3-phosphate-lyase in the rat brain. J. Neurochem. **15**, 519–524 (1968)

Maickel, R., Miller, F.P.: The fluorometric determination of indolealkylamines in brain and pineal gland. Advanc. Pharmacol. Chemother. **6**A, 71–77 (1968)

Maickel, R.P., Bosin, T.R., Harrison, S.D. Jr., Riddle, M.A.: Comparative physiological disposition of melatonin and its benzo(b)thiophene analog in the rat. Life Sci. **14**, 1735–1739 (1974)

Malinský, J., Bayerová, G., Bayer, A., Zapletal, V.: Vorkommen von Falten der Kernmembran in den Pinealzellen des Menschen. Acta Univ. Palock. Olomucensis **33**, 73–78 (1963)

Malm, O.J., Skaug, O.E., Lingjaerde, P.: The effect of pinealectomy on bodily growth, survival rate and P^{32} uptake in the rat. Acta endocr. (Kbh.) **30**, 22–28 (1959)

Mann, D.R., Korowitz, C.D., Barraclough, C.A.: Adrenal gland involvement in synchronizing the preovulatory release of LH in rats. Proc. Soc. exp. Biol. (N.Y.) **150**, 115–120 (1975)

Manocha, S.L.: Histochemical distribution of alkaline and acid phosphatase and adenosine triphosphatase in the brain of squirrel monkey *(Saimiri sciureus)*. Histochemie **21**, 221–235 (1970a)

Manocha, S.L.: Histochemical distribution of acetylcholinesterase and simple esterases in the brain of squirrel monkey *(Saimiri sciureus)*. Histochemie **21**, 236–248 (1970b)

Manocha, S.L., Bourne, G.H.: Histochemical mapping of monoamine oxidase and lactic dehydrogenase in the pons and mesencephalon of squirrel monkey *(Saimiri sciureus)*. J. Neurochem. **13**, 1047–1056 (1966)

Marcean, R., Nanu, L., Ionescu, V., Milcu, I.: Actiunea pinealectomiei asupra permeabilitatii tesutului conjunctiv subcutanat la sobolan. Stud. Cercet. Endocr. **19**, 289–293 (1968)

Marczynski, T.J., Yamaguchi, N., Ling, G.M., Grodzinska, L.: Sleep induced by the administration of melatonin (5-methoxy-N-acetyltryptamine) to the hypothalamus in unrestrained cats. Experientia (Basel) **20**, 435–437 (1964)

Markına, V.V.: Lipid content and lipase activity in the pineal body of adrenalectomized rats receiving diets with different sodium contents. Bjull. Eksp. Biol. Med. **69**, 34–36 (1970)

Marks, B.H., Wu, T.K., Goldman, H.: Soluble estrogen binding protein in the rat pineal gland. Res. Commun. Chem. Pathol. Pharmacol. **3**, 595–600 (1972)

Marks, N., Galoyan, A., Grynbaum, A., Lajtha, A.: Protein and peptide hydrolases of the rat hypothalamus and pituitary. J. Neurochem. **22**, 735–739 (1974)

Marshall, A.H.E., Dayan, A.D.: An immune reaction in man against seminomas, dysgerminomas, pinealomas, and the mediastinal tumours of similar histological appearance? Lancet **1964 II**, 1102–1104

Marshall, N.B., Thines, G.L.: Studies of the brain, sense organs and light sensitivity of a blind cave fish *(Typhlogarra widdowsoni)* from Iraq. Proc. zool. Soc. Lond. **131**, 441–456 (1958)

Martin, J.E., Klein, D.C.: Melatonin inhibition of the neonatal pituitary response to luteinizing hormone-releasing factor. Science **191**, 301–302 (1976)

Mas, M., Alonso Solis, R.: Effects of constant light exposure and blindness on oxidative metabolism of selected brain areas in male rats. Experientia (Basel) **33**, 1390–1391 (1977)

Mashimo, T.: Elektronenmikroskopisch-histochemische Untersuchungen an Epiphysis cerebri bei einigen Säugetieren. [Japanese; Abstract in German.] J. Kyoto Pref. Med. Univ. **74**, 411–420 (1965)

Massa, M.: Contributo spermentale alla conoscenza di possibili rapporti fra la pineale e la tiroide. G. Clin. med. **30**, 916–925 (1950)

Mata, M.M., Schrier, B.K., Klein, D.C., Weller, J.L.: On GABA function and physiology in the pineal gland. Brain Res. **118**, 383–394 (1976)

Matsushima, S., Ito, T.: Diurnal changes in sympathetic nerve endings in the mouse pineal: semi-quantitative electron microscopic observations. J. Neural Transm. **33**, 275–288 (1972)

Matsushima, S., Morisawa, Y.: Effects of acute cold exposure on the ultrastructure of the mouse pinealocyte. Cell Tiss. Res. **195**, 461–469 (1978)

Matsushima, S., Reiter, R.J.: Ultrastructural observations of pineal gland capillaries in four rodent species. Amer. J. Anat. **143**, 265–282 (1975a)

Matsushima, S., Reiter, R.J.: Comparative ultrastructural studies of the pineal gland of rodents. In: Ultrastructure of endocrine and reproductive organs (M. Hess, ed.), pp. 335–356. New York: John Wiley and Sons 1975b

Matsushima, S., Reiter, R.J.: Fine structural features of adrenergic nerve fibres and endings in the pineal gland of the rat, ground squirrel and chinchilla. Amer. J. Anat. **148**, 463–478 (1977)

Matsushima, S., Reiter, R.J.: Electron microscopic observations on neuron-like cells in the ground squirrel pineal gland. J. Neural. Transm. **42**, 223–237 (1978)

Matsushima, S., Morisawa, Y., Mukai, S.: Diurnal variation in large granulated vesicles in sympathetic nerve fibres of the mouse pineal – Quantitative electron microscopic observations. J. Neural. Transm. **45**, 63–73 (1979)

Matthews, C.D., Kennaway, D.J., Frith, R.G., Phillipou, G., LeCornu, A., Seamark, R.F.: Plasma melatonin values in man and some domestic animals: initial observations on the effects of pregnancy in man and pinealectomy in sheep. J. Endocr. **73**, 41 P–42 P (1977)

Matthews, M.J., Benson, B.: Inactivation of pineal antigonadotropin by proteolytic enzymes. J. Endocr. **56**, 339–340 (1973)

Matthews, M.J., Benson, B., Rodin, A.E.: Antigonadotropic activity in a melatonin-free extract of human pineal glands. Life Sci. **10**, Part I, 1375–1379 (1971)

Mautner, W.: Lebendbeobachtungen über den Epiphysenkreislauf von Anuren. Naturwissenschaften **51**, 68 (1964)

Mautner, W.: Studien an der Epiphysis cerebri und am Subcommissuralorgan der Frösche. Mit Lebendbeobachtung des Epiphysenkreislaufs, Totalfärbung des Subcommissuralorgans und Durchtrennung des Reissnerschen Fadens. Z. Zellforsch. **67**, 234–270 (1965)

McClung, R., Dafny, N.: Neurophysiological properties of the pineal body. II. Single unit recording. Life Sci. **16**, 621–628 (1975)

McCord, C.P., Allen, F.B.: Evidence associating pineal gland function with alterations in pigmentation. J. exp. Zool. **23**, 207–224 (1917)

McCormack, T.J., Plassche, W.M., Lin, S.-R.: Ruptured teratoid tumor in the pineal region. J. Comput. Assist. Tomogr. **2**, 499–501 (1978)

McDevitt, D.S.: Presence of lateral eye lens crystallins in the median eye of the American chameleon. Science **175**, 763–764 (1972)

McFarland, L.Z., Wilson, W.O., Winget, C.M.: Response of the chicken pineal gland, blood and reproductive organs to darkness. Poultry Sci. (Ithaka, N.Y.) **48**, 903–907 (1969)

McGeer, E.G., McGeer, P.L.: Circadian rhythm in pineal tyrosine hydroxylase. Science **153**, 73–74 (1966)
McGovern, V.J.: Tumors of the epiphysis cerebri. J. Path. Bact. **51**, 1–9 (1948)
McGuire, J., Möller, H.: Response of melanocytes of dermis and epidermis to lightening agents. Nature (Lond.) **208**, 493–494 (1965)
McIsaac, W.M., Taborsky, R.G., Farrell, G.: 5-Methoxytryptophol: effect on estrus and ovarian weight. Science **145**, 63–64 (1964)
McIsaac, W.M., Farrell, G., Taborsky, R.G., Taylor, A.N.: Indole compounds. Isolation from pineal tissue. Science **148**, 102–103 (1965)
McKenny, C.T., Timiras, P.S., Quay, W.B.: Concentrations of 5-hydroxytryptamine in rat brain and pineal after adrenalectomy and cortisol administration. Neuroendocrinology **1**, 251–256 (1966)
McKeown, B.A., John, T.M., George, J.C.: Diurnal variation in effects of melatonin on plasma growth hormone and glucose in the pigeon. Endocr. exp. **9**, 263–268 (1975)
McKinney, T.D., Vaughan, M.K., Reiter, R.J.: Pineal influence on intermale aggression in adult house mice. Physiol. Behav. **15**, 213–216 (1975)
McMillan, J.P.: Pinealectomy abolishes the circadian rhythm of migratory restlessness. J. comp. Physiol. **79**, 105–112 (1972)
McNeill, M.E.: An unusual organelle in the pineal gland of the rat. Cell Tiss. Res. **184**, 133–137 (1977)
McNeill, M.E.: Membranous structures in pinealocytes of the infertile diabetic mutant mouse (C 57 BL/Ks-db/db). J. Neural Transm. **42**, 207–221 (1978)
McNeill, M.E., Whitehead, D.S.: The synaptic ribbons of the guinea pig pineal gland in sterile, pregnant and fertile but non-pregnant females and in reproductively active males. J. Neural Transm. **45**, 149–164 (1979)
McNulty, J.A.: A comparative study of the pineal complex in the deep-sea fishes *Bathylagus wesethi* and *Nezumia liolepis*. Cell Tiss. Res. **172**, 205–225 (1976)
McNulty, J.A.: The pineal of the troglophilic fish, *Chologaster agassizi*: an ultrastructural study. J. Neural Transm. **43**, 47–71 (1978a)
McNulty, J.A.: Fine structure of modified photoreceptor cells in pineal of the goby, *Clevelandia ios* (Pisces: Gobiidae). Experientia (Basel) **34**, 507–509 (1978b)
McNulty, J.A.: Fine structure of the pineal organ in the troglobytic fish, *Typhlichthyes subterraneous* (Pisces: Amblyopsidae). Cell Tiss. Res. **195**, 535–545 (1978c)
McNulty, J.A., Nafpaktitis, B.G.: The structure and development of the pineal complex in the lantern-fish *Triphoturus mexicanus* (Family Myctophidae). J. Morph. **150**, 579–606 (1976)
McNulty, J.A., Nafpaktitis, B.G.: Morphology of the pineal complex in seven species of lanternfishes (Pisces: Myctophidae). Amer. J. Anat. **150**, 509–529 (1977)
Mead, R.: Pineal gland: its rôle in controlling delayed implantation in the spotted skunk. J. Reprod. Fertil. **30**, 147–150 (1972)
Megyeri, L.: Cystische Veränderungen des Corpus pineale. Frankfurt. Z. Path. **70**, 699–704 (1960)
Mehring, G.: Licht- und elektronenmikroskopische Untersuchung des Pinealorgans von *Testudo hermanni*. Anat. Anz. **131**, 184–203 (1972)
Meijers, H.D.K., Zweens, J.: The pineal weight as standard for the physiologic activity of the pineal gland. Acta morph. neerl.-scand. **5**, 293–294 (1962)
Meinel, A.: Lage-, Form- und Strukturentwicklung des Subkommissuralorgans der weißen Ratte. Inaugural-Dissertation, München 1967
Meiniel, A.: Étude préliminaire de l'organe parapinéal d'ammocète de *Lampetra planeri*. Arch. Anat. micr. Morph. exp. **58**, 219–237 (1969a)
Meiniel, A.: Cellules de type photorécepteur dans la rétine dorsale de l'organe parapinéal d'ammocète de *Lampetra planeri*. C. R. Acad. Sci. (Paris) D **268**, 2265–2268 (1969b)
Meiniel, A.: L'organe parapinéal de l'ammocète de *Lampetra planeri*: Études cytophysiologiques. Thesis, Clermont-Ferrand 1970
Meiniel, A.: Étude cytophysiologique de l'organe parapinéal de *Lampetra planeri*. J. neuro-visc. Rel. **32**, 157–199 (1971)
Meiniel, A.: Mise en évidence d'une relation entre les sites de stockage des indolamines et les zones de concentration des grains denses dans les photorécepteurs rudimentaires sécrétoires de l'épiphyse embryonnaire de *Lacerta vivipara* L. J. Physiol. (Paris) **70**, 4B–5B (1975)

Meiniel, A.: L'épiphyse embryonnaire de *Lacerta vivipara* J. I. Différenciation des cellules de la lignée sensorielle (CLS). Le gradient de différenciation morphologique. J. Neural Transm. **39**, 139–174 (1976a)

Meiniel, A.: L'épiphyse embryonnaire de *Lacerta vivipara* J. II. Étude en microscopie électronique de l'incorporation du 5-hydroxytryptophane-^{3}H (5-HTP-^{3}H) au niveau des photorécepteurs rudimentaires secrétoires (PRS). Le gradient de différenciation biochimique. J. Neural Transm. **39**, 231–250 (1976b)

Meiniel, A.: Contribution à l'étude du complexe pariétal embryonnaire des lacertiliens. Différenciation cellulaire de l'épiphyse de *Lacerta vivipara* (Jacquin) en rapport avec les activités sensorielle, sécrétoire et neurohumorale (biosynthèses indoliques). Thesis, Clermont-Ferrand 1976c

Meiniel, A.: Morphogenèse du complexe pariétal de *Lacerta vivipara* J. Considérations sur l'origine de 3ème oeil de lacertiliens. J. Neural Transm. **41**, 287–311 (1977a)

Meiniel, A.: Activités 5-hydroxytryptophane décarboxylasique (5-HTPDT) et monoamine oxydasique (MAO) dans l'organe pinéal embryonnaire de *Lacerta vivipara* J. Études qualitative et quantitative de l'incorporation du 5-HTP-^{3}H dans les conditions expérimentales et histoenzymologie des monoamine oxydases. J. Neural Trans. **41**, 175–208 (1977b)

Meiniel, A.: Presence of indoleamines in the pineal and parapineal organs of *Lampetra planeri* (Petromyzontidae). C. R. Acad. Sci. (Paris) **287**, 313–316 (1978)

Meiniel, A.: Les «cellules à corps résiduels» des organes pariétaux de *Lampetra planeri;* démonstration de leur appartenance aux cellules de la lignée sensorielle. C. R. Acad. Sci. (Paris) D **288**, 101–104 (1979)

Meiniel, A.: Ultrastructure of serotonin-containing cells in the pineal organ of *Lampetra planeri* (Petromyzontidae). A second sensory all line from photoreceptor cell to pinealocyte. Cell Tiss. Res. **207**, 407–427 (1980)

Meiniel, A., Collin, J.-P.: Monoamines de l'organe pinéal au cours de stades de développement intra-uterins d'un Amniote. Recherches chez *Lacerta vivipara* (Jacquin), Reptile, Lacertilien. C. R. Acad. Sci. (Paris) D **271**, 2373–2376 (1970)

Meiniel, A., Collin, J.-P.: Le complexe pinéal de l'ammocète *(Lampetra planeri,* Bl.). Identification du ganglion sous-jacent à l'organe parapinéal et relations épithalamiques des organes pinéal et parapinéal. Z. Zellforsch. **117**, 354–380 (1971)

Meiniel, A., Collin, J.-P., Calas, A.: Incorporation du 5-hydroxytryptophane (5-MTP) dans l'organe pinéal du *Lacertilien, Lacerta vivipara* (J): étude par radioautographie à haute resolution. C. R. Acad. Sci. (Paris) D **274**, 2897–2900 (1972)

Meiniel, A., Collin, J.P., Hartwig, H.G.: Pinéale et troisième œil de *Lacerta vivipara* (J.), au cours de la vie embryonnaire et postnatale. Étude cytophysiologique des monoamines en microscopie de fluorescence et en microspectrofluorimétrie. Z. Zellforsch. **144**, 89–115 (1973)

Meiniel, A., Collin, J.P., Roux, M.: Pinéale et 3ème oeil de l'embryon de *Lacerta vivipara:* Étude qualitative et quantitative, en microscopie photonique, de l'incorporation de 5-hydroxytryptophane-^{3}H, au cours de l'ontogenèse. J. Neural Transm. **36**, 249–279 (1975)

Meissl, H., Donley, C.S., Wissler, J.H.: Free amino acids and amines in the pineal organ of the rainbow trout *(Salmo gairdneri):* Influence of light and dark. Comp. Biochem. Physiol. **61 C**, 401–405 (1978)

Menaker, M.: Synchronization with the photic environment via extraretinal receptors in the avian brain. In: Menaker, M. (ed.), Biochronometry, pp. 315–332. Washington, D.C.: National Academy of Sciences 1971

Menaker, M.: The pineal organ and circadian rhythmicity. Gen. comp. Endocr. **18**, 608 (1972a)

Menaker, M.: Nonvisual light perception. Sci. Amer. **226**, No. 3, 22–29 (1972b)

Menaker, M., Oksche, A.: The avian pineal organ. In: Avian biology (eds. D.S. Farner and J.R. King), Vol. 4, pp. 79–118, New York-London: Academic Press 1974

Menaker, M., Underwood, H.: Extraretinal photoreception in birds. Photochem. Photobiol. **23**, 299–306 (1976)

Menaker, M., Zimmerman, N.: Role of the pineal in the circadian system of birds. Amer. Zool. **16**, 45–55 (1976)

Menigot, M., Gaillard, G., Thiéblot, L.: Influence de la pinéalectomie sur le métabolism calcique chez le Rat. J. Physiol. (Paris) **62**, Suppl. 1, 189 (1970)

Merker, G.: Licht- und elektronenmikroskopische Studien über die Fasergliastruktur der Epiphysen-Subcommissuralregion der Primaten. Z. Zellforsch. **92**, 232–255 (1968)

Merrit, J.H., Sulkowski, T.S.: Alterations of pineal gland biorhythms by N-methyl-3-piperidyl benzilate. J. Pharmacol. exp. Therap. **166**, 119–124 (1969)
Mess, B.: Endocrine and neurochemical aspects of pineal function. Int. Rev.Neurobiol. **11**, 171–198 (1969)
Mess, B., Tima, L., Trentini, G.P.: The role of pineal principles in ovulation. Progr. Brain Res. **39**, 251–259 (1973)
Mess, B., Trentini, G.P., Tima, L.: The role of the pineal gland in the control of ovulation. Budapest: Akademiá Kiado 1978
Messenger, E.A., Warner, A.E.: The action of melatonin on single amphibian pigment cells in tissue culture. Brit. J. Pharmacol. **61**, 607–614 (1977)
Meyburg, P.: Zur Frage nach der Herkunft der Parenchymzellen des Pinealorgans. Arch. Neur. Psych. **95**, 245–270 (1965)
Meyer, C.J., Wurtman, R.J., Altschule, M.D., Lazo Wasem, E.A.: The arrest of prolonged estrus in "middle-aged" rats by pineal gland extract. Endocrinology **68**, 795–800 (1961)
Meyer, D.C., Quay, W.B.: Effects of continuous light and darkness, and of pinealectomy, adrenalectomy and gonadectomy on uptake of ^{3}H-serotonin by the suprachiasmatic nuclear region of male rats. Neuroendocrinology **22**, 231–239 (1976)
Meyer, D.C., Sturkie, P.D., Gross, K.: Diurnal rhythm in serotonin of blood and pineals of chickens. Comp. Biochem. Physiol. **46**A, 619–623 (1973)
Michotte, Y., Lowenthal, A., Knaepen, L., Collard, M., Massart, D.L.: A morphological and chemical study of calcification of the pineal gland. J. Neurol. **215**, 209–219 (1977)
Mieno, M., Yamashita, E., Imori, M., Yamashita, K.: An inhibitory effect of melatonin on the luteinizing hormone releasing activity of luteinizing hormone releasing hormone in immature male dogs. J. Endocr. **78**, 283–284 (1978)
Mihail, N., Molnár, B.: Folgeerscheinungen nach Epiphysektomie bei der Taube. Naturwissenschaften **52**, 433 (1965)
Mikami, S.: Cytological and histochemical studies of the pineal bodies of domestic animals. Tohoku J. Agricult. Res. **2**, 41–48 (1951)
Milcou, I., Nanu, L., Marcean, R.: De l'existence d'une hormone hypoglycémiante épiphysaire synergique de l'insuline. Ann. Endocr. (Paris) **18**, 612–620 (1957)
Milcou, I., Nanu, L., Marcean, R., Ionesco, V.: L'action de l'insuline sur le metabolisme glucidique chez l'animal épiphysectomisé. Rev. roum. Endocr. **3**, 127–131 (1966a)
Milcou, I., Damian, E., Ionesco, M., Popesco, I.: L'action de l'épiphyse sur la synthèse du cholestérol hépatique. Rev. roum. Endocr. **3**, 35–39 (1966b)
Milcou, I., Nanu, L., Ionescu, V., Marcean, R.: Changes in total nitrogen and nucleic acid contents of muscle and liver in pinealectomized rats. Rev. roum. Endocr. **4**, 109–112 (1967)
Milcou, S.M., Pavel, S.: Antigonadotropic function of the pineal gland and the oxytocin of the neurosecretory hypothalamic system. Nature (Lond.) **187**, 950–951 (1960)
Milcou, S.M., Petrea, I.: Caractère secrétoire des cellules de la glande pinéale vue au microscope électronique. Ann. Endocr. (Paris) **22**, 902–911 (1961)
Milcou, S.M., Petrea, I.: Electron microscopic studies on the secretory cytodynamics of the pineal gland. Epiphysis-hypophysis-gonadic correlation. Rev. roum. Endocr. **1**, 109–114 (1964)
Milcou, S.M., Postelnicou, D.: L'influence de l'illumination prolongée sur la structure de l'épiphyse, chez le canard. Rev. roum. Endocr. **1**, 175–177 (1964)
Milcou, S.M., Vrejoiu, G., Marcean, R., Nanu, L.: L'action de l'hormone hypoglycémiante épiphysaire sur le pancréas endocrinien chez les animaux alloxanisés. Ann. Endocr. (Paris) **18**, 621–627 (1957)
Milcou, S.M., Milcou, I., Nanu, L.: Le rôle de la glande pinéale dans le métabolisme des glucides. Ann. Endocr. (Paris) **24**, 233–254 (1963)
Milcou, S.M., Holban, R., Taşcă, C., Ghinea, E., Stănescu, O.: In vitro study of thyroxine effects on enzymatic activity and cell differentiation in the pineal gland. Rev. roum. Endocr. **5**, 203–207 (1968)
Milcu, I., Damian, E., Ionescu, M.: Action of pineal hormone on the testicular respiration in vitro. [In Roumanian.] Stud. Cercet. Endocr. **11**, 647–650 (1960)
Milcu, I., Nanu, L., Marcean, R., Sitaru, S.: Probleme în legătură cu testarea biologica a funcţiei hipoglicemiante a epifizei. Stud. Cerc. Endocr. **2**, 719–729 (1961)

Milcu, I., Nanu, L., Sitaru, S., Teodoru, V.: Studiu comparativ al continutului in hormon hipoglicemiant al epifizei de bovine. Stud. Cerc. Endocr. **13**, 365–372 (1962)
Milcu, I., Nanu, L., Marcean, R., Sitaru, S.: L'action de l'extrait pinéal et de la pinéalectomie sur le glycogène hépatique et musculaire après infusion prolongée de glucose. Stud. Cerc. Endocr. **14**, 651–655 (1963)
Milcu, I., Nanu, L., Marcean, R.: Toleranta la glucide la iepure în sindriamele epifizare experimentale. Stud. Cercet. Endocr. **15**, 307–312 (1964a)
Milcu, I., Nanu, L., Marcean, R., Sitaru, S.: Cercetări asupra valorii diabetogene a epifizectomiei la miel. Stud. Cerc. Endocr. **15**, 507–513 (1964b)
Milcu, I., Nanu, L., Marcean, R., Ionescu, V.: Efectul epifizectomiei asupra metabolismului glucidici. Studii pe şobolan. Stud. Cercet. Endocr. **16**, 17–24 (1965)
Milcu, I., Marcean, R., Ionescu, V.: Controlul acţiunii melatoninei asupra glicemiei. Stud. Cercet. Endocr. **18**, 405–409 (1967)
Milcu, I., Nanu, L., Marcean, R., Ionescu, V., Milcu, S.M.: Influence of the pineal gland on nitrogen metabolism. Excerpta Med. Int. Congr. Series No. **157**, 179 (1968)
Milcu, I., Nanu-Ionescu, L., Marcean, R., Ionescu, V.: The influence of the pineal gland on nitrogen metabolism. J. Endocr. **45**, 175–181 (1969)
Milcu, S.M., Milcu, I.: Über ein hypoglykämisch wirkendes Hormon der Zirbeldrüse. Medizinische 711–715 (1958)
Milcu, S.M., Vrejoiu, G.: Structure of the pineal gland in aged subjects. Ro. Med. Rev. **3**, 13–14 (1959)
Milcu, S.M., Petrea, I., Săhleanu, V., Holban, R.: Altrastructura epifizei după castrare şi după administrarea hormonilor sexuali. Stud. Cercet. Endocr. **13**, 617–622 (1962)
Milcu, S.M., Pavel, S., Neacsu, C.: Biological and chromatographic characterization of a polypeptide with pressor and oxytocic activities isolated from bovine pineal gland. Endocrinology **72**, 563–566 (1963)
Milcu, S.M., Petrescu, R., Taşcă, C.: The effect of thyroxine and cortisol on some dehydrogenases and lysosomal enzyme activities in rat pineal cultures. Histochemie **15**, 312–317 (1968)
Milcu, S.M., Nanu-Ionescu, L., Milcu, I.: The effect of pinealectomy on plasma insulin in rats. In: The pineal gland (G.E.W. Wolstenholme and J. Knight, eds.), pp. 345–377. Edinburgh and London: Churchill Livingstone 1971
Milcu, S.M., Damian, E., Ianas, O., Badescu, I., Oprescu, M.: Decrease in serum and testicular testosterone after administration of pineal polypeptides to rats. Endocr. exp. **9**, 259–262 (1975)
Milin, J.: Promena aktiviteta supstrat-zavisne monoamin oksidaze u pinealnoj zlezdi u audutuvnom stres-u. Change in the activity of substrate-dependent monoamine oxidase in the pineal gland in auditive stress. Folia anat. iugosl. **5**, 27–36 (1976)
Milin, R.: Cytodynamics of the pineal gland in emotional stress. Proc. Int. Symp. Neuroendocrine Regulatory Mechanisms, pp. 231–241. Belgrade: Serbian Academy of Sciences and Arts 1979
Milin, R., Nešić, L.: Contribution à l'étude d'histo-physiologie de la glande pinéale. Bull. Sci. Cons. Acad. R.P.F. Yougosl. **4**, 77 (1958)
Milin, R., Šćepović, M.: La part du complexe habénulo-épiphysaire dans l'histophysiologie de la glande thyroide. Bull. Sci. Cons. Acad. R.P.F. Yougosl. **5**, 48 (1960)
Milin, R., Stern, P.: Influence de la réserpine sur la glande pinéale. Bull. Sci. Cons. Acad. R.P.F. Yougosl. **4**, 77 (1958)
Milin, R., Stern, P.: Effet de la réserpine sur la réactivité stressogêne du complexe habénulo-épiphysaire. Bull. Sci. Cons. Acad. R.P.F. Yougosl. **5**, 119 (1960)
Milin, R., Stern, P., Ciglar, M., Huković, S.: Beitrag zur Erforschung der antiandrogenen Funktion der Pinealdrüse. Naturwissenschaften **46**, 477–478 (1959)
Miline, R.: La part de l'épiphyse dans le syndrome d'adaptation. Congrès Nat. Editions de l'Académie de la Sciences Médicales Bucarest 5–11 Mai 1957, pp. 421–444 (1957a)
Miline, R.: La part de l'épiphyse dans le syndrome d'adaptation. Bull. Sci. Cons. Acad. R.P.F. Yougosl. **3**, 113 (1957b)
Miline, R.: Sur la réactivité stressogène du complexe hypothalamo-hypophysaire. Symp. biol. hung. **1**, 105–130 (1960a)
Miline, R.: Influence de l'extrait épiphysaire sur la résistance à l'irradiation. Acta anat. (Basel) **42**, 270 (1960b)

Miline, R.: La part du noyau paraventriculaire dans l'histophysiologie corrélative de la glande thyroide et de la glande pinéale. Ann. Endocr. (Paris) **24**, 255–269 (1963)

Miline, R.: Contribution à l'étude du comportement corrélatif du complexe épithalamo-épiphysaire et de la zone glomérulaire des glandes surrénales sous l'influence de l'obscurité. Progr. Brain Res. **10**, 612–626 (1965)

Miline, R.: Le comportement du noyau supraoptique sous l'influence de la mélatonine. Acta anat. (Basel) **69**, 302 (1968)

Miline, R.: Le rôle de la glande pinéale dans l'acclimation au froid. Int. J. Biometeorol. (Suppl.) **13**, 6 (1969)

Miline, R., Krstić, R.: Sur l'histophysiologie corrélative de la glande pinéale et des glandes parathyroides. Z. Zellforsch. **69**, 428–437 (1966)

Miline, R., Šćepović, M.: La part du complexe habénulo-épiphysaire dans l'histophysiologie de la glande thyroide. Ann. Endocr. (Paris) **20**, 511–518 (1959)

Miline, R., Stern, P., Huković, S.: Sur la présence de la sérotonine dans la glande pinéale. Bull. Sci. Conc. Acad. R.P.F. Yougosl. **4**, 75 (1958)

Miline, R., Werner, R., Ciglar, M., Petruševska, M., Žorić, V., Šćepović, N.: Sur la réactivité stressogène du complexe habénulo-épiphysaire. Acta anat. (Basel) **42**, 270 (1960)

Miline, R., Devečerski, V., Krstić, R.: Influence d'excitations olfactives sur le système habénulo-épiphysaire. Ann. Endocr. (Paris) **24**, 377–379 (1963a)

Miline, R., Werner, R., Šćepović, M., Devečerski, V., Krstić, R.: Contribution à l'étude de l'influence de l'irradiation sur le ganglion de l'habenula et la glande pinéale. Ann. Endocr. (Paris) **24**, 380–384 (1963b)

Miline, R., Šćepović, M., Devečerski, V.: Influence d'irradiation sur l'histophysiologie corrélative de la zone glomérulaire des surrénales, de l'épithalamus et de l'hypothalamus chez le rat. In: Major problems in neuroendocrinology, ed. E. Bajusz et G. Jasmin, pp. 393–408. Basel-New York: S. Karger 1964

Miline, R., Devečerski, V., Krstić, R.: Les modifications épiphysaires dans le stress et en particulier dans les névroses expérimentales d'effroi. Probl. Act. Endocr. Nutr. Sér. **10**, 229–256 (1966)

Miline, R., Krstić, R., Devečerski, V.: Sur le comportement de la glande pinéale dans des conditions de stress. Acta anat. (Basel) **71**, 352–402 (1968)

Miline, R., Devečerski, V., Krstić, R.: Corpus pinéale – glande de nature sensoneuroendocrine. Akad. Nauka Umjet. Bosne Herceg. **37**, 69–84 (1969a)

Miline, R., Werner, R., Šćepović, M., Devečerski, V., Miline, J.: Influence du froid sur le comportement du noyau supraoptique chez les rats épiphysectomisés. Bull. Ass. Anat. (Nancy) **145**, 289–293 (1969b)

Miline, R., Devečerski, V., Krstić, R.: Effects des stimuli auditifs sur la glande pinéale de la chauve-souris en hibernation. Acta anat. (Basel), Suppl. **56**, 193–300 (1969c)

Miline, R., Devečerski, N., Krstić, R.: Pineal gland behaviour as affected by cold. Hormones (Lond.) **1**, 321–331 (1970)

Miline, R., Devečerski, V., Miline, J.: Glande pinéale des lièvres sous l'influence de l'émotion. Gen. comp. Endocr. **18**, 609 (1972)

Miller, F.P., Maickel, R.P., Fluorimetric determination of indole derivatives. Life Sci., **9**, Part I, 747–752 (1970)

Miller, W.H., Wolbarsht, M.L.: Neural activity in the parietal eye of a lizard. Science **135**, 316–317 (1962)

Milofsky, A.: The fine structure of the pineal in the rat, with special reference to parenchyma. Anat. Rec. **127**, 435–436 (1957)

Minneman, K.P., Iversen, L.L.: Diurnal rhythm in rat pineal cyclic nucleotide phosphodiesterase activity. Nature (Lond.) **260**, 59–61 (1976)

Minneman, K.P., Lynch, H., Wurtman, R.J.: Relationship between environmental light intensity and retina-mediated suppression of rat pineal serotonin-N-acetyltransferase. Life Sci. **15**, 1791–1796 (1975)

Mira, E.: Studio quantitativo fotometrico delle modificazioni de cromatofori di *Scardinius erythrophthalmus* L. Mikroskopie **18**, 216–222 (1963)

Mishkinsky, J., Nir, I., Lajtos, Z.K., Sulman, F.G.: Mammary development in pinealectomized rats. J. Endocr. **36**, 215–216 (1966)

Misugi, K., Liss, L., Bradel, E.J.: Electron microscopic study of an ectopic pinealoma. Acta neuropath. (Berl.) **9**, 346–356 (1967)

Mitchell, J.A., Yochim, J.M.: Influence of environmental lighting on duration of pregnancy in the rat. Endocrinology **87**, 472–480 (1970)

Mitchell, M.D., Sayers, L., Keirse, M.J.N.C., Anderson, A.B.M., Turnbull, A.C.: Melatonin in amniotic fluid during human parturition. Brit. J. Obstet. Gynaec. **85**, 684–686 (1978)

Mizobe, F., Kurokawa, M.: Induction of a specific protein by oestradiol in the rat pineals in culture. FEBS Letters **87**, 45–48 (1978)

Mizuno, H., Sensui, N.: Lack of effects of melatonin administration and pinealectomy on the milk ejection response in the rat. Endocr. Jap. **17**, 417–420 (1970)

Møller, M.: The ultrastructure of the human fetal pineal gland. I. Cell types and blood vessels. Cell Tiss. Res. **152**, 13–30 (1974)

Møller, M.: The ultrastructure of the human fetal pineal gland. II. Innervation and cell junctions. Cell Tiss. Res. **169**, 7–21 (1976)

Møller, M.: Presence of a pineal nerve (nervus pinealis) in the human fetus: a light and electron microscopical study of the innervation of the pineal gland. Brain Res. **154**, 1–12 (1978)

Møller, M., Høyer, P.E.: Histochemical demonstration of a circadian rhythm of succinate dehydrogenase in rat pineal gland. Influence of coenzyme Q_{10} addition. Histochemistry **59**, 259–269 (1979)

Møller, M., Møllgård, K., Kimble, J.E.: Presence of a pineal nerve in sheep and rabbit fetuses. Cell Tiss. Res. **158**, 451–459 (1975)

Møller, M., Ingild, A., Bock, E.: Immunohistochemical demonstration of S-100 protein and GFA protein in interstitial cells of rat pineal gland. Brain Res. **140**, 1–13 (1978a)

Møller, M., Van Deurs, B., Westergaard, E.: Vascular permeability to proteins and peptides in the mouse pineal gland. Cell Tiss. Res. **195**, 1–15 (1978b)

Möller, W.: Zur Cytologie eines circadianen aktiven neuroendokrinen Organs (Pinealorgans) in vitro. Verh. anat. Ges. **67**, 375–378 (1973)

Möller, W.: Studium zur Struktur und Funktion von organartig abgegrenzten Einheiten des Zentralnervensystems (Plexus chorioidei, Glykogenkörper, Pinealorgan) in vitro. Habilitationsschrift, Giessen 1975

Möller, W.: Circumventrikuläre Organe in der Gewebekultur. Advanc. Anat. Embryol. Cell Biol. **54**, 1–95 (1978)

Møllgaard, K., Møller, M.: On the innervation of the human fetal pineal gland. Brain Res. **52**, 428–432 (1973)

Mogler, R.K.-H.: Das endokrine System des syrischen Goldhamsters (*Mesocricetus auratus auratus* Waterhouse) unter Berücksichtigung des natürlichen und experimentellen Winterschlafs. Z. Morph. Ökol. Tiere **47**, 267–308 (1958)

Moguilevsky, J.A., Scacchi, P., Deis, R., Siseles, N.O.: Effects of melatonin on the luteinizing hormone release induced by clomiphene and luteinizing hormone-releasing hormone. Proc. Soc. exp. Biol. (N.Y.) **151**, 663–666 (1976)

Molinoff, P.B., Axelrod, J.: Distribution and turnover of octopamine in tissues. J. Neurochem. **19**, 157–163 (1972)

Moore, R.Y.: Pineal response to light: mediation by the accessory optic system in the monkey. Nature (Lond.) **222**, 781–782 (1969)

Moore, R.Y.: Visual pathways and the central neural regulation of diurnal rhythms. In: F.O. Schmitt and F.G. Worden (eds.) The neurosciences – Third intensive study program, pp. 537–542. Cambridge, Mass.: MIT Press (1974)

Moore, R.Y.: Pineal transplants to the anterior chamber of the eye: evidence for functional reinnervation. Exp. Neurol. **49**, 617–621 (1975a)

Moore, R.Y.: Indoleamine metabolism in the intact and denervated pineal, pineal stalk and habenula. Neuroendocrinology **19**, 323–330 (1975b)

Moore, R.Y.: The innervation of the mammalian pineal gland. Progr. reprod. Biol. **4**, 1–29 (1978a)

Moore, R.Y.: Neural control of pineal function in mammals and birds. J. Neural Transm., Suppl. **13**, 47–58 (1978b)

Moore, R.Y., Eichler, V.B.: Central neural mechanisms in diurnal rhythm regulation and neuroendocrine responses to light. Psychoneuroendocrinology **1**, 265–279 (1976)

Moore, R.Y., Klein, D.C.: Visual pathways and the central neural control of circadian rhythm in pineal serotonin N-acetyltransferase activity. Brain Res. **71**, 17–33 (1974)

Moore, R.Y., Rapport, R.L.: Pineal and gonadal function in the rat following cervical sympathectomy. Neuroendocrinology **7**, 361–374 (1971)

Moore, R.Y., Smith, R.A.: Postnatal development of a norepinephrine response to light in the rat pineal and salivary glands. Neuropharmacology **10**, 315–323 (1971)

Moore, R.Y., Traynor, M.E.: Diurnal rhythms in pineal N-acetyltransferase and hippocampal norepinephrine: effects of water deprivation, blinding and hypothalamic lesions. Neuroendocrinology **20**, 250–259 (1976)

Moore, R.Y., Heller, A., Wurtman, R.J., Axelrod, J.: Visual pathway mediating pineal response to environmental light. Science **155**, 220–223 (1967)

Moore, R.Y., Heller, A., Bhatnager R.K., Wurtman, R.J., Axelrod, J.: Central control of the pineal gland: visual pathways. Arch. Neurol. (Chic.) **18**, 208–218 (1968)

Moore, W.T., Jr., Hampton, J.K., Jr.: Effects of para-chlorophenylalanine on pregnancy in the rat. Biol. Reprod. **11**, 280–287 (1974)

Moreau, M.H.: Radiological localization of the pineal gland. Acta radiol. Diag. **5**, 65–67 (1966)

Morgan, W.W., Hansen, J.T.: Time course of the disappearance of pineal noradrenaline following superior cervical ganglionectomy. Exp. Brain. Res. **32**, 429–434 (1978)

Morgan, W.W., Reiter, R.J.: Pineal noradrenaline levels in Mongolian gerbil and in different strains of laboratory rats over a lighting regimen. Life Sci. **21**, 555–558 (1977)

Morgan, W.W., Reiter, R.J., Pfeil, K.A.: Hamster pineal noradrenaline: Levels over a regulated lighting period and the influence of superior cervical ganglionectomy. Life Sci. **19**, 437–440 (1976)

Morgano, G., Alcozer, G., Giordano, G.: Studi sull'epifisi. Influenza di un estratto acquoso di ghiandola pineale sulla escrezione urinaria dei 17-chetosteroidi in soggetti sani ed eucrini d'ambo i sessi. Arch. E. Maragliano Pat. Clin. **12**, 231–237 (1956)

Mori, W.: Atrophy of the pineal gland in Xeroderma pigmentosum. Yale J. Biol. Med. **46**, 712 (1973)

Mori, W., Lerner, A.B.: A microscopic bioassay for melatonin. Endocrinology **67**, 443–450 (1960)

Morita, Y.: Erregung und Hemmung pinealer Neurone der Regenbogenforelle (*Salmo irideus*) bei Belichtung des Zwischenhirns. Pflügers Arch. ges. Physiol. **283**, R30 (1965a)

Morita, Y.: Extra- und intracelluläre Ableitungen einzelner Elemente des lichtempfindlichen Zwischenhirns anurer Amphibien. Pflügers Arch. ges. Physiol. **286**, 97–108 (1965b)

Morita, Y.: Entladungsmuster pinealer Neurone der Regenbogenforelle (*Salmo irideus*) bei Beleuchtung des Zwischenhirns. Pflügers Arch. ges. Physiol. **289**, 155–167 (1966a)

Morita, Y.: Absence of electrical activity of the pigeon's pineal organ in response to light. Experientia (Basel) **22**, 402 (1966b)

Morita, Y.: Wellenlängen-Diskriminatoren im intrakranialen Pinealorgan von *Rana catesbyana*. Experientia (Basel) **25**, 1277 (1969)

Morita, Y.: Post-tetanic activity changes of the frog's neurosensory pineal end vesicle (Stirnorgan). Pflügers Arch. **328**, 135–144 (1971)

Morita, Y.: Direct photosensory activity of the pineal. In: Brain endocrine interaction. II. The ventricular system (eds. K.M. Knigge, D.E. Scott, H. Kobayashi, S. Ishii), pp. 376–387. Basel: Karger 1975

Morita, Y., Bergmann, G.: Physiologische Untersuchungen und weitere Bemerkungen zur Struktur des lichtempfindlichen Pinealorgans von *Pterophyllum scalare* Cuv. et Val. (Cichlidae, Teleostei). Z. Zellforsch. **119**, 289–294 (1971)

Morita, Y., Dodt, E.: Nervous activity of the frog's epiphysis cerebri in relation to illumination. Experientia (Basel) **21**, 221–222 (1965)

Morita, Y., Dodt, E.: Slow photic responses of the isolated pineal organ of lamprey. Nova Acta Leopoldina **38**, 331–339 (1973)

Morrissey, J.J., Lovenberg, W.: Pineal RNA synthesis and protein phosphorylation during N-acetyltransferase induction. Fed. Proc. **36**, 879 Abs. (1977)

Morrissey, J.J., Lovenberg, W.: Synthesis of RNA in the pineal gland during N-acetyltransferase induction. Biochem. Pharmacol. **27**, 551–555 (1978a)

Morrissey, J.J., Lovenberg, W.: Synthesis of RNA in the pineal gland during N-acetyltransferase induction. The effects of actinomycin D, α-amanitin and cordycepin. Biochem. Pharmacol. **27**, 557–562 (1978b)

Morrissey, J.J., Lovenberg, W.: Protein synthesis in pineal gland during serotonin-N-acetyltransferase induction. Arch. Biochem. Biophys. **191**, 1–7 (1978c)

Moszkowska, A.: Différence d'activité entre l'épiphyse de quelques mammifères et l'épiphyse de poule. Ann. Endocr. (Paris) **8**, 138–140 (1947)

Moszkowska, A.: Action des extraits épiphysaires sur la femelle cobaye adulte jeune. C.R. Soc. Biol. (Paris) **145**, 845–847 (1951a)

Moszkowska, A.: Action d'extraits aqueux de glandes pinéales de moutons sur les souris mâles et femelles. Antagonisme épiphyso-hypophysaire. C.R. Soc. Biol. (Paris) **145**, 847–850 (1951b)

Moszkowska, A.: L'antagonisme épiphyso-hypophysaire. Étude *in vitro* par la méthode de E. Wolff. C.R. Acad. Sci. (Paris) **243**, 315–317 (1956)

Moszkowska, A.: Mise en évidence, par les greffes sur l'ovaire de la Ratte impubère, de l'antagonisme épiphyso-hypophysaire. C.R. Acad. Sci. (Paris) **246**, 3685–3687 (1958a)

Moszkowska, A.: Etude *in vitro* du rôle de l'épiphyse dans l'excrétion d'hormones gonadotropes hypophysaires. C.R. Acad. Sci. (Paris) **247**, 1659–1662 (1958b)

Moszkowska, A.: L'antagonisme épiphyso-hypophysaire. Étude in vivo and in vitro chez l'embryon de poulet Sussex. Annal. Endocr. (Paris) **19**, 69–79 (1958c)

Moszkowska, A.: L'antagonisme épiphyso-hypophysaire. Ann. Endocr. (Paris) **24**, 215–226 (1963)

Moszkowska, A.: Quelques arguments en faveur de la spécificité zoologique de l'activité antigonadotrope de l'épiphyse. Ann. Endocr. (Paris) **25**, Suppl., 79–85 (1964)

Moszkowska, A.: Contribution à l'étude du mécanisme de l'antagonisme épiphyso-hypophysaire. Progr. Brain Res. **10**, 564–576 (1965a)

Moszkowska, A.: Quelques donnees nouvelles sur le mécanisme de l'antagonisme épiphyso-hypophysaire. Rôle possible de la sérotonine et de la mélatonine. Rev. suisse Zool. **72**, 145–160 (1965b)

Moszkowska, A.: Données nouvelles concernant le mécanisme de l'action de l'épiphyse sur la fonction gonadotrope hypophysaire. Probl. Act. Endocr. Nutr. Série **10**, 213–228 (1966)

Moszkowska, A.: Étude des extraits épiphysaires fractionnées. Physiologie. Biol. Méd. **56**, 403–412 (1967a)

Moszkowska, A.: Quelques données nouvelles concernant les rélations épiphyso-hypophysaires et la fonction gonadotrope hypophysaire. Rev. europ. endocr. **4**, 351–372 (1967b)

Moszkowska, A., Ebels, I.: A study of the antigonadotrophic action of synthetic arginine vasotocin. Experientia (Basel) **24**, 610–611 (1968)

Moszkowska, A., Ebels, I.: The influence of the pineal body on the gonadotropic function of the hypophysis. J. neuro-visc. Rel., Suppl. X, 160–176 (1971)

Moszkowska, A., Scémama, A.: Effet de l'épiphysectomie sur la photosensibilité du rat mâle. C.R. Soc. Biol. (Paris) **162**, 636–640 (1968a)

Moszkowska, A., Scémama, A.: L'épiphysectomie et la réponse photosexuelle du rat. Arch. Anat. Histol. Embryol. Norm. Exp. **51**, 475–479 (1968b)

Moszkowska, A., Ebels, I., Scémama, A.: Étude in vitro des extraits fractionnées d'épiphyses d'agneau. C.R. Soc. Biol. (Paris) **159**, 2298–2302 (1965)

Moszkowska, A., Kordon, C., Ebels, I.: Biochemical fractions and mechanisms involved in the pineal modulation of pituitary gonadotropin release. In: G.E.W. Wolstenholme and J. Knight, The pineal gland, pp. 241–258. Edinburgh and London: Churchill Livingstone 1971

Moszkowska, A., Scémama, A., Lombard, M.N., Héry, M.: Experimental modulation of hypothalamic content of the gonadotropic releasing factors by pineal factors in the rat. J. Neural Transm. **34**, 11–22 (1973)

Moszkowska, A., Citharel, A., l'Héritier, A., Ebels, I., Laplante, E.: Some new aspects of a sheep pineal gonadotropic inhibiting activity in in vitro experiments. Experientia (Basel) **30**, 964–965 (1974)

Moszkowska, A., Hus-Citharel, A., l'Héritier, A., Zurburg, W., Ebels, I.: Separation of pineal extracts by gelfiltration. V. Location by paper chromatography of a sheep pineal principle inhibiting hypophyseal gonadotropic activity. J. Neural Transm. **38**, 239–247 (1976)

Motta, M., Fraschini, F., Martini, L.: Endocrine effects of pineal gland and melatonin. Proc. Soc. exp. Biol. (N.Y.) **126**, 431–435 (1967)

Motta, M., Schiaffini, O., Piva, E., Martini, L.: Pineal principles and the control of adrenocorticotropin secretion. In: The pineal gland (G.E.W. Wolstenholme and J. Knight, eds.), pp. 279–301. Edinburgh and London: Churchill Livingstone 1971

Mouret, J., Coindet, J., Chouvet, G.: Effet de la pinéalectomie sur les états et rhythmes de sommeil du rat mâle. Brain Res. **81**, 97–105 (1974)

Mrosovsky, N., Tress, K.H.: Plasticity of reactions to light in frogs and a possible role for the pineal eye. Nature (Lond.) **210**, 1174–1175 (1966)

Munns, T.W.: Effect of different photoperiods on melatonin synthesis in the pineal gland of the canary (*Serenus canarius*) and testicular activity. Anat. Rec. **166**, 352 (1970)

Murphy, R.C.: The structure of the pineal organ in the bluefin tuna, *Thunnus thynnus*. J. Morph. **133**, 1–16 (1971)

Murthy, G.G., Modesto, R.R.: Effects of melatonin and serotonin on ^{14}C-glucose metabolism in rabbit adipose tissue. Experientia (Basel) **31**, 383–384 (1975)

Naber, S.P., Goldman, M., Peaslee, M.H.: The effect of melatonin on the iodine-concentrating mechanism of the thyroid gland in young adult rats. Proc. Soc. Dak. Acad. Sci. **48**, 44–46 (1969)

Nagle, C.A., Cardinali, D.P., Rosner, J.M.: Light regulation of rat retinal hydroxyindole-O-methyltransferase (HIOMT) activity. Endocrinology **91**, 423–427 (1972a)

Nagle, C.A., Neuspiller, R.N., Cardinali, D.P., Rosner, J.M.: Uptake and effect of 17β-estradiol on pineal hydroxyindole-O-methyltransferase (HIOMT) activity. Life Sci. **11**, Part II, 1109–1116 (1972b)

Nagle, C.A., Cardinali, D.P., Rosner, J.M.: Retinal and pineal hydroxyindole-O-methyl transferases in the rat: changes following cervical sympathectomy, pinealectomy or blinding. Endocrinology **92**, 1560–1565 (1973a)

Nagle, C.A., Cardinali, D.P., Rosner, J.M.: Uptake of estradiol by the rat pineal organ. Effects of cervical sympathectomy, stage of the estrous cycle and estradiol treatment. Life Sci. **13**, 1089–1103 (1973b)

Nagle, C.A., Cardinali, D.P., Rosner, J.M.: Effects of castration and testosterone administration on pineal and retinal hydroxyindole-O-methyl transferase of male rats. Neuroendocrinology **14**, 14–23 (1974a)

Nagle, C.A., Cardinali, D.P., Rosner, J.M.: Diurnal rhythms in tissue radioactivity uptake after ^{3}H-estradiol and ^{3}H-testosterone administration to castrated rats. Steroid Lipids Res. **5**, 107–112 (1974b)

Nagle, C.A., Cardinali, D.P., De Laborde, N.P., Rosner, J.M.: Sex-dependent changes in rat retinal hydroxyindole-O-methyl transferase. Endocrinology **94**, 294–297 (1974c)

Nagle, C.A., Cardinali, D.P., Rosner, J.M.: Testosterone effects on protein synthesis in the rat pineal gland. Modulation by the sympathetic nervous system. Life Sci. **16**, 81–91 (1975)

Namboodiri, M.A.A., Namboodiri, M.A., Weller, J.L., Klein, D.C.: A comparative study of pineal acetyl-CoA hydrolase and N-acetyltransferase activities. J. Neurochem. **32**, 31–33 (1979)

Nanu-Ionescu, L., Ionescu, V.: Glanda pineal˘ şi metabolismul insulinei. I. Nivelul plasmatic al activităţii insulinice total si al insulinei "libre" si "legate" in apinealismul experimental. Stud. Cercet. Endocr. **20**, 237–243 (1969)

Nanu-Ionescu, L., Marcean, R., Ionescu, V.: Modèle expérimental pour l'étude du métabolisme azoté chez la souris. Application du contrôle de l'activité d'un extrait pinéal. Rev. roum. Endocr. **6**, 293–297 (1969)

Narang, G.D., Singh, D.V., Turner, C.W.: Effect of melatonin on thyroid hormone secretion rate and feed consumption of female rats. Proc. Soc. exp. Biol. (N.Y.) **125**, 184–188 (1967)

Nasr, H., Hamed, M.Y., Soliman, F.A.: Studies on pineal body function. I. The relation between the pineal body and some anterior pituitary hormones in male rabbits. Zbl. Vet.-Med. **8**, 192–200 (1961)

Nathanson, M.A., Binkley, S., Hilfer, S.R.: Cultivation of mammalian pineal cells: retention of organization and function in tissue culture. Ann. clin. Res. **9**, 843–848 (1977)

Neacşu, C.: Structure-activity data on a pineal peptide with oxytocic and vasopressor activities. In: Structure-activity relationships of protein and polypeptide hormones. Proc. 2nd Int. Symp., Liège, 1971, p. 275. Amsterdam: Excerpta medica 1972a

Neacşu, C.: The mechanism of antigonadotropic action of a polypeptide extracted from a bovine pineal gland. Rev. roum. Physiol. **9**, 161–169 (1972b)

Neacşu, C., Gheorghiu, C.: Rélage endocrinien du métabolisme minéral. I. Modifications du métabolisme du magnésium sous l'influence de l'épiphyse. C.R. Soc. Biol. (Paris) **160**, 2047–2051 (1966)
Neff, N.H., Barrett, R.E., Costa, E.: Compartmentation of serotonin (5HT) in rat pineal. Fed. Proc. **27**, 288 Abs. (1968)
Neff, N.H., Barrett, R.E., Costa, E.: Kinetic and fluorescent histochemical analysis of the serotonin compartments in rat pineal gland. Europ. J. Pharmacol. **5**, 348–356 (1969)
Negoescu, I., Constantinescu, A., Marcean, R.P., Dumitrescu, C.: Influenta pinealectomiei asupra tiroxinei libere se legate la sobulanul alb Wistar. Stud. Cercet. Endocr. **19**, 125–129 (1968)
Negus, N.C., Berger, P.J.: Pineal weight response to a dietary variable in *Microtus montanus*. Experientia (Basel) **27**, 215–216 (1971)
Nešić, Lj.: Effet de l'extrait épiphysaire sur l'histo-physiologie du thymus. Acta anat. (Basel) **46**, 175–176 (1961)
Nešić, Lj.: Contribution à l'étude du rhythme saisonnier de la glande pinéale de brebis. Acta anat. (Basel) **49**, 376–377 (1962)
Neuvonen, P., Karppanen, H.: Effects of some saluretic diuretics on the pinealectomized rats. Scand. J. clin. Lab. Invest. **25**, Suppl. **113**, 42 (1970)
Newman Taylor, A.: Effect of midbrain and pineal body lesions on aldosterone secretion. Acta endocr. (Kbh.), Suppl. **51**, 73–74 (1960)
Newman Taylor, A., Farrell, G.L.: Effects of brain stem lesions on aldosterone and cortisol secretion. Endocrinology **70**, 556–565 (1962)
Newman Taylor, A., Farrell, G.: Facteur glomérulotrope. Ann. Endocr. (Paris) **24**, 228–232 (1963)
Newman Taylor, A., Wilson, R.: Electrophysiological evidence for the action of environmental lighting on the pineal gland and the pathways involved. Anat. Rec. **163**, 327 (1969)
Newman Taylor, A., McIsaac, W.M., Farrell, G.: Steroidogenic properties of 1-methyl-6-methoxytetrahydrocarboline. Fed. Proc. **21**, 186 (1962)
Nielsen, J.T., Møller, M.: Nervous connections between the brain and the pineal gland in the cat (*Felis catus*) and the monkey (*Cercopithecus aethiops*). Cell Tiss. Res. **161**, 293–301 (1975)
Nielsen, J.T., Møller, M.: Innervation of the pineal gland in the Mongolian gerbil (*Meriones unguiculatus*). A fluoresence microscopical study. Cell. Tiss. Res. **187**, 235–250 (1978)
Nielsen, S.L., Wilson, C.B.: Ultrastructure of a "pineocytoma". J. Neuropath. exp. Neurol. **34**, 148–158 (1975)
Niemi, M., Ikonen, M.: Histochemical evidence of amino-peptidase activity in rat pineal gland. Nature (Lond.) **185**, 928 (1960)
Niles, L.P., Brown, G.M., Grota, L.J.: Effects of neutralization of circulating melatonin and N-acetylserotonin on plasma prolactin levels. Neuroendocrinology **23**, 14–22 (1977a)
Niles, L.P., Brown, G.M., Grota, L.J., Wilson, D.A., Pang, S.F.: Antigonadal effects of blinding in presence of antibody against circulating melatonin. Life Sci. **21**, 613–618 (1977b)
Nir, I., Hirschmann, N.: Pineal N-acetyltransferase depression in rats exposed to heat. Experientia (Basel) **34**, 1645–1646 (1978a)
Nir, I., Hirschmann, N.: The effect of thyroid hormones on rat pineal indoleamine metabolism in vitro. J. Neural Transm. **42**, 117–126 (1978b)
Nir, I., Hirschmann, N., Mishkinsky, J., Sulman, F.G.: The effect of light and darkness on nucleic acids and protein metabolism of the pineal gland. Life Sci. **8**, Part II, 279–287 (1969a)
Nir, I., Behroozi, K., Assael, M., Ivriani, I., Sulman, F.G.: Changes in the electrical activity of the brain following pinealectomy. Neuroendocrinology **4**, 122–127 (1969b)
Nir, I., Kaiser, N., Hirschmann, N., Sulman, F.G.: The effect of 17β-estradiol on pineal metabolism. Life Sci **9**, Part I, 851–858 (1970a)
Nir, I., Hirschmann, N., Sulman, F.G.: The effect of light and darkness on lactic acid content of the pineal gland. Proc. Soc. exp. Biol. (N.Y.) **133**, 452–455 (1970b)
Nir, I., Hirschmann, N., Sulman, F.G.: Diurnal rhythms of pineal nucleic acids and protein. Neuroendocrinology **7**, 271–277 (1971a)
Nir, I., Schmidt, U., Hirschmann, N., Sulman, F.G.: The effect of pinealectomy on rat plasma corticosterone levels under various conditions of light. Life Sci. **10**, Part I, 317–324 (1971b)
Nir, I., Hirschmann, N., Sulman, F.G.: Influence of the oestrous cycle on the nucleic acid and protein content of the rat pineal gland. Experientia (Basel) **28**, 88–89 (1972a)

Nir, I., Hirschmann, N., Sulman, F.G.: Pineal gland changes of rats exposed to heat. Experientia (Basel) **28**, 701–702 (1972b)

Nir, I., Briel, G., Dames, W., Neuhoff, V.: Rat pineal free amino acids diurnal rhythm and effect of light. Arch. int. Physiol. Biochim. **81**, 617–627 (1973a)

Nir, I., Dames, W., Neuhoff, V.: Micro-disc electrophoresis of rat pineal proteins. Diurnal rhythm and effect of light. Arch. int. Physiol. Biochim. **81**, 607–616 (1973b)

Nir, I., Briel, G., Dames, W., Neuhoff, V.: Pineal proteins and free amino acids during ontogenesis in rats. Neuroendocrinology **14**, 34–43 (1974a)

Nir, I., Hirschmann, N., Kremer, N., Sulman, F.G.: Inversion of pineal N-acetyltransferase rhythm by reversed environmental lighting. Neuroendocrinology **15**, 231–235 (1974b)

Nir, I., Hirschmann, N., Sulman, F.G.: The effect of heat on rat pineal hydroxyindole-O-metyl transferase activity. Experientia (Basel) **31**, 867–868 (1975)

Nir, I., Reiter, R.J., Wurtman, R.J.: The pineal gland. Proc. Int. Symposium, Jerusalem, November 1977. J. Neural Transm., Suppl. **13** (1978). Wien: Springer 1978a

Nir, I., Hirschmann, N., Puder, M., Petrank, J.: Changes following treatment with pineal indolic compounds. Arch. int. Physiol. Biochim. **86**, 353–362 (1978b)

Nishii, K.: A pathological study of pineal tumors. Arch. jap. Chir. **32**, 548–567 (1963)

Nishino, H.: Suprachiasmatic nuclei and circadian rhythms. The role of suprachiasmatic nuclei on rhythmic activity of neurons in the lateral hypothalamic area, ventromedian nuclei and pineal gland. Fol. pharmacol. jap. **72**, 941–954 (1976)

Nishiyama, H., Mikami, S.: Cytodifferentiation of the pineal body of the domestic fowl. J. Fac. Agr. Iwate Univ. **12**, 107–131 (1974)

Nomokonova, L.M., Shapiro, B.I.: On epiphysis and subcommisural organ in some vertebrates. (In Russian). Arkh. Anat. Gistol. Embriol. **48**, 36–44 (1965)

Norris, J.T.: Effects of ovarian subcapsular and intraperitoneal injections of melatonin on compensatory ovarian hypertrophy. Anat. Rec. **166**, 355 (1970)

Norris, J.T.: The ovary as a potential target organ for pineal principles. Anat. Rec. **169**, 389 (1971)

Notario, A.: Compartamento istologico di alcuni organi endocrini della cavia adulta dopo prolungata somministrazione di un estratto epifisario. Arch. De Vecchi Anat. Patol. **24**, 305–318 (1956)

Nováková, V., Sandritter, W., Křeček, J., Zelenková, M.: Veränderungen des Nucleinsäuregehaltes von Zellen der Epiphyse während der postnatalen Entwicklung bei Ratten. Virchows Arch. Abt. B **2**, 292–296 (1969)

Nováková, V., Sterc, J., Sandritter, W., Křeček, J.: The day-night differences in total RNA content of parenchymal cells of rat pineal gland. Beitr. Path. **144**, 211–215 (1971)

Novales, R.R.: Reponses of cultured melanophores to the synthetic hormones α-MSH, melatonin and epinephrine. Ann. N.Y. Acad. Sci. **100**, 1035–1047 (1963)

Novales, R.R., Novales, B.J.: Analysis of antagonisms between pineal melatonin and other agents which act on the amphibian melanophore. Progr. Brain Res. **10**, 507–512 (1965)

Novotná, B., Ulvrová, L., Hromada, J.: Some observations on the pineal body of macaques. Folia morph. (Praha) **14**, 1–6 (1966)

Nuzzi, R., De Martino, C., Pavoni, P., Ciampalini, L.: Variazioni del contenuto di corticotropina nella ipofisi del ratto trattato con estratti epifisari o privato de pineale. Sperimentale **112**, 216–221 (1962)

Obručník, M., Bayerová, G.: Možnostifluorescenčně a polarisačně mikroskopických vyšetření v epifyse člověka. Ćs. Morfol. **10**, 329–335 (1962)

O'Dea, R.F., Zatz, M.: Catecholamine-stimulated cyclic GMP accumulation in rat pineal: Apparent presynaptic site of action. Proc. nat. Acad. Sci. (Wash.) **73**, 3398–3402 (1976)

O'Dea, R.F., Gagnon, C., Zatz, M.: Regulation of guanosine 3′,5′ cyclic monophosphate in the rat pineal and posterior pituitary glands. J. Neurochem. **31**, 733–738 (1978)

Ogle, T.F., Kitay, J.I.: Effects of pinealectomy on adrenal function in vivo and in vitro in female rats. Endocrinology **98**, 20–24 (1976)

Ogle, T.F., Kitay, J.I.: *In vitro* effects of melatonin and serotonin on adrenal steroidogenesis. Proc. Soc. exp. Biol. (N.Y.) **157**, 103–105 (1978)

Oguri, M., Omura, Y., Hibiya, T.: Uptake of ^{14}C-labelled 5-hydroxytryptophan into the pineal organ of rainbow trout. Bull. Japan. Soc. Fish **34**, 687–690 (1968)

Oishi, T., Kato, M.: Pineal organ as a possible photoreceptor in photoperiodic testicular response in Japanese quail. Mem. Fac. Sci. Kyoto Univ., Ser. B **2**, 12–18 (1968)

Oishi, T., Lauber, J.K.: Effect of light and darkness on pineal hydroxyindole-O-methyl transferase (HIOMT) in Japanese quail. Life Sci. **13**, 1105–1116 (1973)

Oishi, T., Lauber, J.K.: Pineal control of photo-endocrine responses in growing Japanese quail. Endocrinology **94**, 1731–1734 (1974a)

Oishi, T., Lauber, J.K.: Pituitary-pineal-endocrine relationships in Japanese quail. Endocrinology **94**, 1735–1738 (1974b)

Oksche, A.: Der Feinbau des Organon frontale bei *Rana temporaria* und seine funktionelle Bedeutung. Morph. Jb. **92**, 123–167 (1952)

Oksche, A.: Über die Art und Bedeutung sekretorischer Zelltätigkeit in der Zirbel und im Subkommissuralorgan. Anat. Anz., Suppl. **101**, 88–96 (1954)

Oksche, A.: Untersuchungen über die Nervenzellen und Nervenverbindungen des Stirnorgans, der Epiphyse und des Subkommissuralorgans bei anuren Amphibien. Morph. Jb. **95**, 393–425 (1955)

Oksche, A.: Funktionelle histologische Untersuchungen über die Organe des Zwischenhirndaches der Chordaten. Anat. Anz. **102**, 404–419 (1956)

Oksche, A.: Histologische, histochemische und experimentelle Studien am Subkommissuralorgan von Anuren (mit Hinweisen auf den Epiphysenkomplex). Z. Zellforsch. **57**, 240–326 (1962)

Oksche, A.: Survey of the development and comparative morphology of the pineal organ. Progr. Brain Res. **10**, 3–29 (1965a)

Oksche, A.: Neue Erkenntnisse über die Ultrastruktur und Funktion des Pinealorgans. (Photorezeptoren und ihr Strukturwandel). 8th Int. Congr. of Anatomists, Wiesbaden, 1965, pp. 88–89. Stuttgart: Thieme 1965b

Oksche, A.: Zur Frage extraretinaler Photorezeptoren im Pinealorgan der Vögel. Arch. Anat. (Strasbourg) **51**, 497–507 (1968)

Oksche, A.: Sensory and glandular elements of the pineal organ. In: The pineal gland (G.E.W. Wolstenholme and J. Knight, eds.), pp. 127–146, Edinburgh and London: Churchill Livingstone 1971

Oksche, A., Hartwig, H.-G.: Pineal sense organs – Components of photoneuroendocrine systems. Progr. Brain Res. **52**, 113–130 (1979)

Oksche, A., Kirschstein, H.: Elektronenmikroskopische Feinstruktur der Sinneszellen im Pinealorgan von *Phoxinus laevis* L. (Pisces, Teleostei, Cyprinidae). (Mit vergleichenden Bemerkungen.) Naturwissenschaften **53**, 591 (1966a)

Oksche, A., Kirschstein, H.: Zur Frage der Sinneszellen im Pinealorgan der Reptilien. Naturwissenschaften **53**, 46 (1966b)

Oksche, A., Kirschstein, H.: Die Ultrastruktur der Sinneszellen im Pinealorgan von *Phoxinus laevis*. Z. Zellforsch. **78**, 151–166 (1967)

Oksche, A., Kirschstein, H.: Unterschiedlicher elektronenmikroskopischer Feinbau der Sinneszellen im Parietalauge und im Pinealorgan (Epiphysis cerebri) der Lacertilia. Ein Beitrag zum Epiphysenproblem. Z. Zellforsch. **87**, 159–192 (1968)

Oksche, A., Kirschstein, H.: Elektronenmikroskopische Untersuchungen am Pinealorgan von *Passer domesticus*. Z. Zellforsch. **102**, 214–241 (1969)

Oksche, A., Kirschstein, H.: Weitere elektronenmikroskopische Untersuchungen am Pinealorgan von *Phoxinus laevis* (Teleostei, Cyprinidae). Z. Zellforsch. **112**, 572–588 (1971)

Oksche, A., Vaupel-von Harnack, M.: Elektronenmikroskopische Untersuchungen an der Epiphysis cerebri von *Rana esculenta*. Z. Zellforsch. **59**, 582–614 (1963)

Oksche, A., Vaupel-von Harnack, M.: Elektronenmikroskopische Untersuchungen an den Nervenbahnen des Pinealkomplexes von *Rana esculenta* L. Z. Zellforsch. **68**, 389–426 (1965a)

Oksche, A., Vaupel-von Harnack, M.: Vergleichende elektronenmikroskopische Studien am Pinealorgan. Progr. Brain Res. **10**, 237–258 (1965b)

Oksche, A., Vaupel-von Harnack, M.: Über rudimentäre Sinneszellstrukturen im Pinealorgan des Hühnchens. Naturwissenschaften **52**, 662–663 (1965c)

Oksche, A., Vaupel-von Harnack, M.: Elektronenmikroskopische Untersuchungen zur Frage der Sinneszellen im Pinealorgan der Vögel. Z. Zellforsch. **69**, 41–60 (1966)

Oksche, A., Von Harnack, M.: Elektronenmikroskopische Untersuchungen am Stirnorgan (Epiphysenendblase, Frontalorgan) von *Rana temporaria* und *Rana esculenta*. Naturwissenschaften **49**, 429–430 (1962)

Oksche, A., Von Harnack, M.: Elektronenmikroskopische Untersuchungen am Stirnorgan von Anuren (Zur Frage der Lichtrezeptoren). Z. Zellforsch. **59**, 239–288 (1963)

Oksche, A., Von Harnack, M.: Die elektronenmikroskopische Feinstruktur des Stirnorgans (Epiphysenendblase) der Anuren. Progr. Brain Res. **5**, 209–222 (1964)

Oksche, A., Morita, Y., Vaupel-von Harnack, M.: Zur Feinstruktur und Funktion des Pinealorgans der Taube (*Columba livia*). Z. Zellforsch. **102**, 1–30 (1969)

Oksche, A., Ueck, M., Rüdeberg, C.: Comparative ultrastructural studies of sensory and secretory elements in pineal organs. Mem. Soc. Endocr. **19**, 7–25 (1971)

Oksche, A., Kirschstein, H., Kobayashi, H., Farner, D.S.: Electron microscopic and experimental studies of the pineal organ in the white-crowned sparrow, *Zonotrichia leucophrys gambelii*. Z. Zellforsch. **124**, 247–274 (1972)

Olcese, I., De Vlaming, V.: Possible pineal organ involvement in the control of hypothalamic monoamine oxidase activity in the goldfish. Amer. Zoologist **17**, 100 Abs. (1977)

Olson, L., Boréus, L.O., Seiger, Å.: Histochemical demonstration and mapping of 5-hydroxytryptamine- and catecholamine-containing neuron systems in the human fetal brain. Z. Anat. Entwickl.-Gesch. **139**, 259–282 (1973)

Olsson, R.: Studies on the subcommissural organ. Acta zool. (Stockh.) **39**, 71–102 (1958)

Olsson, R.: Subcommissural ependyma and pineal organ development in human fetuses. Gen. comp. Endocr. **1**, 117–123 (1961)

Omura, Y.: Influence of light and darkness on the ultrastructure of the pineal organ in the blind cave fish, *Astyanax mexicanus*. Cell Tiss. Res. **160**, 99–112 (1975)

Omura, Y.: Ultrastructural study of embryonic and post-hatching development in the pineal organ of the chicken (Brown Leghorn, *Gallus domesticus*). Cell Tiss. Res. **183**, 255–271 (1977)

Omura, Y., Oguri, M.: Histological studies on the pineal organ of 15 species of teleosts. Bull. Jap. Soc. Sci. Fish. **35**, 991–1000 (1969)

Omura, Y., Kitoh, J., Oguri, M.: The photoreceptor cell of the pineal organ of Ayu, *Plecoglossus altivelis*. Bull. Jap. Soc. Sci. Fish. **35**, 1067–1071 (1969)

Onoyama, Y., Kouji, O., Toshibumi, N., Masahiro, H., Mitsuyuki, A.: Radiation therapy of pineal tumors. Radiology **130**, 757–760 (1979)

O'Rahilly, R.: The development of the epiphysis cerebri and the subcommissural complex in staged human embryos. Anat. Rec. **160**, 488–489 (1968)

Orme, S.K., Wells, S.A., Rabson, A.A., Wurtman, R.J.: In vitro neoplastic transformation of hamster pineal cells by three oncogenic DNA viruses. Cancer (Philad.) **21**, 477–482 (1968)

Orofino, G.: Ultrastruttura della pineale al microscopico elettronico. Policlinico **70**, 179–192 (1964)

Orsi, L., Denari, J.H., Nagle, C.A., Cardinali, D.P., Rosner, J.M.: Effects of melatonin on the synthesis of proteins by the rat hypothalamus, hypophysis and pineal organ. J. Endocr. **58**, 131–132 (1973)

Ortega, P., Malamud, N., Shimkin, M.G.: Metastasis to the pineal body. Arch. Path. **52**, 518–528 (1951)

Ortman, R.: Parietal eye and nerve in *Anolis carolinensis*. Anat. Rec. **137**, 386 (1960)

Orts, R.J.: Reduction of serum LH and testosterone in male rats by a partially purified bovine pineal extract. Biol. Reprod. **16**, 249–254 (1977)

Orts, R.J., Benson, B.: Inhibitory effects on serum and pituitary LH by a melatonin-free extract of bovine pineal glands. Life Sci. **12**, 513–519 (1973)

Orts, R.J., Benson, B., Cook, B.F.: Some antigonadotrophic effects of melatonin-free bovine pineal extracts in rats. Acta endocr. (Kbh.) **76**, 438–448 (1974a)

Orts, R., Benson, B., Cook, B.F.: LH inhibitory properties of aqueous extracts of rat pineal glands. Life Sci. **14**, 1501–1510 (1974b)

Orts, R.J., Kocan, K.N., Wilson, I.B.: Inhibitory action of melatonin on a pineal antigonadotropin. Life Sci. **17**, 845–850 (1975)

Orts, R.J., Kocan, K.M., Johnson, R.P.: Antifertility properties of bovine pineal extracts: reduction of ovulation and pre-ovulatory luteinizing hormone in the rat. Acta endocr. (Kbh.) **85**, 225–234 (1977)

Osborne, M.P., Thornhill, R.A.: The effect of monoamine depleting drugs upon the synaptic bars in the inner ear of the bullfrog (*Rana catesbeiana*). Z. Zellforsch. **127**, 347–355 (1972)

Oshima, K., Gorbman, A.: Evidence for a doubly innervated secretory unit in the anuran pars intermedia. I. Electrophysiologic studies. Gen. comp. Endocr. **13**, 98–107 (1969a)

Oshima, K., Gorbman, A.: Pars intermedia: unitary electrical activity regulated by light. Science **163**, 195–197 (1969b)

Osland, R.B., Cheesman, D.W., Forsham, P.H.: Studies on the mechanism of the suppression of the preovulatory surge of luteinizing hormone in the rat by arginine vasotocin. Endocrinology **101**, 1203–1209 (1977)

Osman, P., Welschen, R.W., Moll, J.: Anti-gonadotropic and anti-growth effects of the pineal gland in immature female rats. Neuroendocrinology **10**, 121–128 (1972)

O'Steen, W.K.: Suppression of ovarian activity in immature rats with serotonin. Endocrinology **77**, 937–939 (1965)

O'Steen, W.K., Dill, R.E.: Intracellular monoamine oxidase and lipids in diffusion chamber cultures of the pineal organ. J. Histochem. Cytochem. **12**, 615–620 (1964)

Ota, M., Hsieh, K.S.: Failure of melatonin to inhibit ovulation induced with pregnant mare serum and human chorionic gonadotropin in rats. J. Endocr. **41**, 601–602 (1968)

Ota, M., Hsieh, K.S., Obara, K.: Absence of gonadotropin-inhibiting substance in the urine of pinealectomized rats. Endocrinology **88**, 816–820 (1971)

Ota, M., Horiuchi, S., Obara, K.: Inhibition of ovulation induced with PMS and HCG by a melatonin-free extract of bovine pineal powder. Neuroendocrinology **18**, 311–321 (1975)

Ota, M., Shimizu, E., Obara, K.: Separation of non-melatonin antiovulatory substances from bovine pineal powder by ultrafiltration. Molec. Cell. Endocr. **9**, 21–31 (1977)

Otani, T., Györkey, F., Farrell, G.: Enzymes of the human pineal body. J. clin. Endocr. **28**, 349–354 (1968 a)

Otani, T., Farrell, G.L., Creaven, P.J.: Studies on the formation of 5-methoxytryptophol in pineal. Fed. Proc. **27**, 288 Abs. (1968 b)

Otani, T., Creaven, P.J., Farrell, G., McIsaac, W.M.: Studies on the biosynthesis of 5-methoxytryptophol in the pineal. Biochim. biophys. Acta (Amst.) **184**, 184–190 (1969)

Ouyang, R., Rozdilsky, B.: Metastasis of carcinoma to pineal body. Report of two cases. Arch. Neurol. (Chic.) **15**, 399–403 (1966)

Owens, D.W., Ralph, Ch.L.: The pineal-paraphyseal complex of sea turtles. I. Light microscopic description. J. Morph. **158**, 169–180 (1978)

Owens, D.W., Gern, W.A., Ralph, Ch.L., Boardman, T.J.: Nonrelationship between plasma melatonin and background adaptation in the rainbow trout (*Salmo gairdneri*). Gen. comp. Endocr. **34**, 459–467 (1978)

Owman, Ch.: Secretory activity of the fetal pineal gland of the rat. Acta morph. neerl.-scand. **3**, 367–394 (1960/61)

Owman, Ch.: On the effects of pinealectomy in the fetal rat. A preliminary report. Communications from Dept. of Anat. Univ. of Lund (Sweden) **2**, 1–8 (1962)

Owman, Ch.: Pinealectomy of the rat foetus in utero. A method for producing localized intracerebral lesions during the last third of gestation. Quart. J. exp. Physiol. **48**, 402–407 (1963 a)

Owman, Ch.: Prenatal changes in epithelium of small intestine of rat foetus pinealectomized in utero. Quart. J. exp. Physiol. **48**, 408–422 (1963 b)

Owman, Ch.: On the presence of 5-hydroxytryptamine in pineal nerves. Biochem. Pharmacol. **12**, (Suppl.) 112 (1963 c)

Owman, Ch.: New aspects of the mammalian pineal gland. Acta physiol. scand. **63**, Suppl. **240**, 1–40 (1964 a)

Owman, Ch.: Sympathetic nerves probably storing two types of monoamines in the rat pineal gland. Int. J. Neuropharmacol. **3**, 105–112 (1964 b)

Owman, Ch.: Études morphologiques et éxpérimentales sur la fonction de la glande pinéale foetale. Ann. Endocr. (Paris) **25**, Suppl., 86–92 (1964 c)

Owman, Ch.: Further studies on prenatal functional relations between rat pineal gland and epithelium of lower ileum. Acta endocr. (Kbh.) **47**, 500–516 (1964 d)

Owman, Ch.: Localization of neuronal and parenchymal monoamines under normal and experimental conditions in the mammalian pineal gland. Progr. Brain Res. **10**, 423–453 (1965)

Owman, Ch.: On the significance of the 5-hydroxytryptamine stores in the pineal gland. Advanc. Pharmacol. Chemother. **6** A, 167–169 (1968)

Owman, Ch., Rüdeberg, C.: Light, fluorescence, and electron microscopic studies on the pineal organ of the pike, *Esox lucius* L., with special regard to 5-hydroxytryptamine. Z. Zellforsch. **107**, 522–550 (1970)

Owman, Ch., Rüdeberg, C., Ueck, M.: Fluoreszenzmikroskopischer Nachweis biogener Monoamine in der Epiphysis cerebri von *Rana esculenta* und *Rana pipiens*. Z. Zellforsch. **111**, 550–558 (1970)

Ozaki, Y., Lynch, H.J.: Presence of melatonin in plasma and urine of pinealectomized rats. Endocrinology **99**, 641–644 (1976)

Ozaki, Y., Lynch, H.J., Wurtman, R.J.: Melatonin in rat pineal, plasma, and urine: 24-hour rhythmicity and effect of chloropromazine. Endocrinology **98**, 1418–1424 (1976)

Ozaki, Y., Wurtman, R., Alonso, R., Lynch, H.J.: Melatonin secretion decreases during the proestrous stage of the rat estrous cycle. Proc. nat. Acad. Sci. (Wash.) **75**, 531–534 (1978)

Packard, G.C., Packard, M.J.: Photic exposure of the lizard *Callisaurus draconoides* following shielding of the parietal eye. Copeia **1972**, 695–701 (1972)

Packard, M.J., Packard, G.C.: Antigonadotrophic effect of melatonin in male lizards (*Callisaurus draconoides*). Experientia (Basel) **33**, 1665 (1977)

Palkovits, M.: Participation of the epithalamo-epiphyseal system in the regulation of water and electrolyte metabolism. Progr. Brain Res. **10**, 627–634 (1965)

Palkovits, M.: Karyometrische Untersuchungen zur Klärung der osmo- bzw. volumenregulatorischen Rolle des Subcommissuralorganes und seiner funktionellen Verbindung mit der Nebennierenrinde. Z. Zellforsch. **84**, 59–71 (1968)

Palkovits, M., Földvari, P.I.: Effect of the subcommissural organ and the pineal body on the adrenal cortex. Endocrinology **72**, 28–32 (1963)

Palkovits, M., Inke, G., Lukacs, G.: Die Bedeutung der topographischen Beziehungen des Subcommissuralorgans und des Corpus pineale bei der Beurteilung der Wirkung des Epiphysen-Organextraktes. Z. Naturforsch. B **17**, 207 (1962a)

Palkovits, M., Inke, G., Lukacs, G.: Der Subcommissuralkomplex (Subcommissuralorgan mit seinen Komponenten und Nachbargebilden) des Menschen während der postnatalen Lebensperioden. Z. mikr.-anat. Forsch. **69**, 88–108 (1962b)

Palkovits, M., Inke, G., Lukacs, G.: Topographische Beziehungen des menschlichen Subcommissuralorgans zur Epiphyse und ihre funktionelle Bedeutung. Endokrinologie **42**, 194–202 (1962c)

Palladini, G., Alfei, L., Appiciutoli, L.: Osservazioni istochimiche sui corpora arenacae dell'epifisi umana. Arch. ital. Anat. Embriol. **70**, 253–270 (1965)

Panagiotis, N.M., Hungerford, G.F.: Response of the pineal and adrenal glands to sodium restriction. Endocrinology **69**, 217–224 (1961a)

Panagiotis, N.M., Hungerford, G.F.: Response of the pineal and adrenal glands to sodium restriction. Anat. Rec. **139**, 262 (1961b)

Panagiotis, N.M., Hungerford, G.F.: Response of the pineal gland to hypophysectomy, hormone administration and dietary sodium restriction. Anat. Rec. **142**, 264–265 (1962)

Panagiotis, N.M., Hungerford, G.F.: Response of pineal sympathetic nerve processes and endings to angiotensin. Nature (Lond.) **211**, 374–376 (1966)

Panda, J.N., Turner, C.W.: The role of melatonin in the regulation of thyrotrophin secretion. Acta endocr. (Kbh.) **57**, 363–373 (1968)

Pang, P.K.T.: Light sensitivity of the pineal gland in blinded *Fundulus heteroclitus*. Amer. Zool. **5**, 682 (1965)

Pang, P.K.T.: The effect of pinealectomy on adult male killifish, *Fundulus heteroclitus*. Amer. Zool. **7**, 715 (1967)

Pang, S.F., Ralph, C.L.: Mole of secretion of pineal melatonin in the chicken. Gen. comp. Endocr. **27**, 125–128 (1975)

Pang, S.F., Yew, D.T.: Pigment aggregation by melatonin in the retinal pigment epithelium and choroid of guinea-pigs, *Cavia porcellus*. Experientia (Basel) **35**, 231–232 (1979)

Pang, S.F., Ralph, C.L., Reilly, D.P.: Melatonin in the chicken brain: its origin, diurnal variation and regional distribution. Gen. comp. Endocr. **22**, 499–506 (1974)

Pang, S.F., Ralph, C.L., Petrozza, J.A.: Effects of melatonin administration and pinealectomy on the electroencephalogram of the chicken (*Gallus domesticus*) brain. Life Sci. **18**, 961–966 (1976)

Pang, S.F., Brown, G.M., Grota, L.J., Chambers, J.W., Rodman, R.L.: Determination of N-acetylserotonin and melatonin activities in the pineal gland, retina, Harderian gland, brain and serum of rats and chickens. Neuroendocrinology **23**, 1–13 (1977)

Panke, E.S., Rollag, M.D., Reiter, R.J.: Pineal melatonin concentrations in the Syrian hamster. Endocrinology **104**, 194–197 (1979)

Papavasiliou, P.S., Cotzias, G.C., Düby, S.E., Steck, A.J., Bell, M., Lawrence, W.H.: Melatonin and Parkinsonism. J. Amer. med. Ass. **221**, 88–89 (1972)

Parfitt, A.G., Klein, D.C.: Sympathetic nerve endings in pineal gland protect against acute stress-induced increase in N-acetyltransferase (EC 2.3.1.5) activity. Endocrinology **99**, 840–851 (1976)

Parfitt, A., Klein, D.C.: Increase caused by desmethylimipramine in the production of (^{3}H) melatonin by isolated pineal glands. Biochem. Pharmacol. **26**, 904–905 (1977)

Parfitt, A., Weller, J.L., Klein, D.C., Sakai, K.K., Marks, B.H.: Blockade by ouabain or elevated potassium ion concentration of the adrenergic and adenosine cyclic 3′,5′-monophosphate-induced stimulation of pineal serotonin N-acetyltransferase activity. Molec. Pharmacol. **11**, 241–255 (1975)

Parfitt, A., Weller, J.L., Klein, D.C.: Beta-adrenergic-blockers decrease adrenergically stimulated N-acetyltransferase activity in pineal glands in organ culture. Neuropharmacology **15**, 353–358 (1976)

Parhon, C.I., Potop, I., Felix, E., Boeru, V.: Influenta epifisectomiei si a administrării de extract epifisar asupra unor date metabolice privind mineralele (Ca, K, P, Şi, Mg), lipidele, protidele si glucidele, la sobolanul alb adult. Stud. Cercet. Endocr. **3**, 321–329 (1952)

Patay, R., Du Chalard, A.: A propos de l'épiphysectomie chez *Gallus domesticus* L. C.R. Acad. Sci. (Paris) **235**, 498–500 (1952)

Paterson, A.T., Rickerby, J., Simpson, J.: Pineal facilitation of territorial aggression. J. Physiol. (Lond.) **284**, 173P–174P (1978a)

Paterson, A.T., Rickerby, J., Simpson, J., Vickers, C.: Melanocyte-stimulating hormone and the pineal in the control of territorial aggression. J. Physiol. (Lond.) **285**, 45p (1978b)

Paul, E.: Histochemische Studien an den Plexus chorioidei, an der Paraphyse und am Ependym von *Rana temporaria* L. Z. Zellforsch. **91**, 519–546 (1968)

Paul, E.: Innervation und zentralnervöse Verbindungen des Frontalorgans von *Rana temporaria* und *Rana esculenta*. Faserdegeneration nach operativer Unterbrechung des Nervus pinealis. Z. Zellforsch. **128**, 504–511 (1972)

Paul, E., Hartwig, H.G., Oksche, A.: Neurone und zentralnervöse Verbindungen des Pinealorgans der Anuren. Z. Zellforsch. **112**, 466–493 (1971)

Paul, S.M., Hsu, L.L., Mandell, A.J.: Extra-pineal N-acetyltransferase activity in rat brain. Life Sci. **15**, 2135–2143 (1974)

Pavel, S.: Investigations on a new pineal hormone with a peptidic structure. Stud. Cercet. Endocr. **14**, 665 (1963)

Pavel, S.: Evidence for the presence of lysine vasotocin in the pig pineal gland. Endocrinology **77**, 812–817 (1965)

Pavel, S.: Endocrine functions of arginine vasotocin from mammalian pineal gland. Gen. comp. Endocr. **9**, 481 (1967)

Pavel, S.: Effects of pineal and synthetic arginine vasotocin on the gonads of prepuberal male mice. 5th Conf. Europ. Comparative Endocrinologists, Utrecht, Abstract 121 (1969)

Pavel, S.: Tentative identification of arginine vasotocin in human cerebrospinal fluid. J. clin. Endocr. **31**, 369–371 (1970)

Pavel, S.: Evidence for the ependymal origin of arginine vasotocin in the bovine pineal gland. Endocrinology **89**, 613–614 (1971)

Pavel, S.: Arginine vasotocin release into cerebrospinal fluid of cats induced by melatonin. Nature (Lond.) New Biol. **246**, 183–184 (1973)

Pavel, S.: Opposite effects of vasotocin injected intrapituitarly and intraventricularly on corticotropin release in mice. Experientia (Basel) **31**, 1469–1470 (1975)

Pavel, S., Coculescu, M.: Arginine vasotocin-like activity of cerebrospinal fluid induced by injection of hypertonic saline into the third cerebral ventricle of cats. Endocrinology **91**, 825–827 (1972)

Pavel, S., Petrescu, S.: Inhibition of gonadotrophin by a highly purified pineal peptide and by synthetic arginine vasotocin. Nature (Lond.) **212**, 1054 (1966)

Pavel, S., Dorcescu, M., Petrescu-Holban, R., Ghinea, E.: Biosynthesis of a vasotocin-like peptide in cell cultures from pineal glands of human fetuses. Science **181**, 1252–1253 (1973a)

Pavel, S., Petrescu, M., Vicoleanu, N.: Evidence of central gonadotropin inhibiting activity of arginine vasotocin in the female mouse. Neuroendocrinology **11**, 370–374 (1973b)

Pavel, S., Matrescu, L., Petrescu, M.: Central corticotropin inhibition by arginine vasotocin in the mouse. Neuroendocrinology **12**, 371–375 (1973c)

Pavel, S., Dumitru, I., Klepsh, I., Dorcescu, M.: A gonadotropin inhibiting principle in the pineal of human fetuses. Evidence for its identity with arginine vasotocin. Neuroendocrinology **13**, 41–46 (1973/74)

Pavel, S., Gheorghiu, C., Calb, M., Petrescu, M.: Reversal by vasotocin of pinealectomy and constant light effects on the pituitary melanocyte-stimulating hormone (MSH) content in the mouse. Endocrinology **97**, 674–676 (1975a)

Pavel, S., Goldstein, R., Calb, M.: Vasotocin content in the pineal gland of foetal, newborn and adult male rats. J. Endocr. **66**, 283–284 (1975b)
Pavel, S., Calb, M., Georgescu, M.: Reversal of the effects of pinealectomy on the pituitary prolactin content in mice by very low concentrations of vasotocin injected into the third cerebral ventricle. J. Endocr. **66**, 289–290 (1975c)
Pavel, S., Goldstein, R., Gheorghiu, C., Calb, M.: Pineal vasotocin: Release into cat cerebrospinal fluid by melanocyte-stimulating hormone release-inhibiting factor. Science **197**, 179–180 (1977a)
Pavel, S., Psatta, D., Goldstein, R.: Slow-wave sleep induced in cats by extremely small amounts of synthetic and pineal vasotocin injected into the third ventricle of the brain. Brain Res. Bull. **2**, 251–254 (1977b)
Pavel, S., Cristoveanu, A., Goldstein, R., Calb, M.: Inhibition of release of corticotropin releasing hormone in cats by extremely small amounts of vasotocin injected into 3rd ventricle of brain. Evidence for involvement of 5-hydroxytryptamine-containing neurons. Endocrinology **101**, 672–678 (1977c)
Pavel, S., Goldstein, R., Ghinea, E., Calb, M.: Chromatographic evidence for vasotocin biosynthesis by cultured pineal ependymal cells from rat fetuses. Endocrinology **100**, 205–208 (1977d)
Pavel, S., Ghinea, E., Goldstein, R., Matulevicius, V.: Vasotocin biosynthesis by cultured pineal glands from adult male rats. J. Endocr. **77**, 147–148 (1978)
Pawlikowski, M., Karasek, M.: On the presence of "multilamellar bodies" in neurohypophysis, in pineal gland and in upper cervical ganglion of white rat. Endokr. pol. **18**, 153–156 (1967)
Pazo, J.H., Houssay, A.B., Davidson, T.A., Chait, R.J.: On the mechanism of the thyroid hypertrophy in pinealectomized rats. Acta physiol. lat.-amer. **18**, 332–340 (1968)
Pearse, A.G.E.: The cytochemistry and ultrastructure of polypeptide hormone-producing cells of the APUD series and the embryologic, physiologic and pathologic implications of the concept. J. Histochem. Cytochem. **17**, 303–313 (1969)
Pearse, A.G.E.: Histochemistry. Theoretical and applied. 3rd ed., Vol. 2. Edinburgh and London: Churchill Livingstone 1972
Peaslee, M.H., Goldman, M., Naber, S.P.: Influence of melatonin upon melanocyte-stimulating hormone (MSH) activity in weanling and young adult rats. Proc. South Dakota Acad. Sci. **48**, 39–43 (1969)
Pecker, J., Scarabin, J.M., Brucher, J.M., Vallée, B.: Apport des techniques stéréotaxiques au diagnostic et au traitment des tumeurs de la région pinéale. Rev. neurol. **134**, 287–299 (1978)
Pedroza García, E., Cardinali, D.P., Laborde, N.P., Garcia Bienere, W., Nagle, C.A., Rosner, J.M.: Binding of estradiol to macromolecules in the rat neurohypophysis: Effects of time of the day, pinealectomy or melatonin administration. Neuroendocrinology **14**, 174–186 (1974)
Pelayo, F., Dubocovich, M.L., Langer, S.Z.: Regulation of noradrenaline release in the rat pineal through a negative feedback mechanism mediated by presynaptic α-adrenoceptors. Europ. J. Pharmacol. **45**, 317–318 (1977)
Pelayo, F., Dubocovich, M.L., Langer, S.Z.: Possible role of cyclic nucleotides in regulation of noradrenaline release from rat pineal through presynaptic adrenoceptors. Nature (Lond.) **274**, 76–78 (1978)
Pelham, R.: A serum melatonin-rhythm in chickens and its abolition by pinealectomy. Endocrinology **96**, 543–546 (1975)
Pelham, R.W., Ralph, C.L.: Pineal hydroxyindole-O-methyltransferase (HIOMT) in the chicken: effect of diurnal lighting and substrate concentration. Life Sci. **11**, Part II, 51–59 (1972a)
Pelham, R., Ralph, C.L.: Diurnal rhythm of serum melatonin in chicken: abolition by pinealectomy. Physiologist **15**, 236 (1972b)
Pelham, R.W., Mull, D.R., Ralph, C.L.: A melatonin-like principle in chicken serum. Amer. Zool. **11**, 650–651 (1971)
Pelham, R.W., Ralph, C.L., Campbell, J.M.: Mass spectral identification of melatonin in blood. Biochem. biophys. Res. Commun. **46**, 1236–1241 (1972)
Pelham, R.W., Vaughan, G.M., Sandock, K.L., Vaughan, M.K.: Twenty-four-hour cycle of a melatonin-like substance in the plasma of human males. J. clin. Endocr. **37**, 341–344 (1973)
Pellegrino De Iraldi, A.: Granular vesicles in pinealocytes of the hamster. Anat. Rec. **154**, 481 (1966)
Pellegrino De Iraldi, A.: Granulated vesicles in the pineal gland of the mouse. Z. Zellforsch. **101**, 408–418 (1969)

Pellegrino De Iraldi, A., De Robertis, E.: Action of reserpine on the submicroscopic morphology of the pineal gland. Experientia (Basel) **17**, 122–124 (1961)
Pellegrino De Iraldi, A., De Robertis, E.: Action of reserpine, iproniazid and pyrogallol on nerve endings of the pineal gland. Int. J. Neuropharmacol. **2**, 231–239 (1963)
Pellegrino De Iraldi, A., De Robertis, E.: Ultrastructure and function of catecholamine containing systems. Excerpta Med. Int. Congr. Series **83**, 355–363 (1965)
Pellegrino De Iraldi, A., Gueudet, R.: Action of reserpine on the osmium tetroxide zinc iodide reactive site of synaptic vesicles in pineal nerves of the rat. Z. Zellforsch. **91**, 178–185 (1968)
Pellegrino De Iraldi, A., Gueudet, R.: Catecholamine and serotonin in granulated vesicles of nerve endings in the pineal gland of the rat. Int. J. Neuropharmacol. **8**, 9–14 (1969)
Pellegrino De Iraldi, A., Letzen, D.: Zinc iodide-osmium tetroxide reactive substances in the matrix of granulated vesicles in the rat pineal nerves. Experientia (Basel) **29**, 844–846 (1973)
Pellegrino De Iraldi, A., Rodriguez de Lores Arnaiz, G.: 5-Hydroxytryptophan decarboxylase activity in normal and denervated pineal glands of rats. Life Sci. **3**, 589–593 (1964)
Pellegrino De Iraldi, A., Rodriguez de Lores Arnaiz, G.: Neural regulation of acetylcholinesterase in the superior cervical ganglia and the pineal gland of the rat. Brain Res. **113**, 435–440 (1976)
Pellegrino De Iraldi, A., Suburo, A.M.: Electron staining of synaptic vesicles using the Champy-Maillet technique. J. Microsc. **91**, 99–103 (1970)
Pellegrino De Iraldi, A., Suburo, A.M.: Two compartments in the granulated vesicles of the pineal nerves. In: Wolstenholme, G.E.W., Knight, J. (eds.), The pineal gland, pp. 177–195. Edinburgh and London: Churchill Livingstone 1971 a
Pellegrino De Iraldi, A., Suburo, A.M.: Action of p-chlorophenylalanine on the synaptic vesicles from rat pineal nerves. Experientia (Basel) **27**, 289–290 (1971 b)
Pellegrino De Iraldi, A., Suburo, A.M.: Effect of tyramine on the compartments of the granulated vesicles in rat pineal nerve endings. Europ. J. Pharmacol. **19**, 251–259 (1972)
Pellegrino De Iraldi, A., Zieher, L.M.: Noradrenaline and dopamine content of normal, decentralized and denervated pineal gland of the rat. Life Sci. **5**, 149–154 (1966)
Pellegrino De Iraldi, A., Zieher, L.M., De Robertis, E.: The 5-hydroxytryptamine content and synthesis of normal and denervated pineal gland. Life Sci. **2**, 691–696 (1963)
Pellegrino De Iraldi, A., Zieher, L.M., De Robertis, E.: Ultrastructure and pharmacological studies of nerve endings in the pineal organ. Progr. Brain Res. **10**, 389–422 (1965)
Pellegrino De Iraldi, A., Gueudet, R., Suburo, A.M.: Differentiation between 5-hydroxytryptamine and catecholamines in synaptic vesicles. Progr. Brain Res. **34**, 161–170 (1971)
Pelletier, G.: Immunohistochemical localization of hypothalamic hormones at the electron microscope level. In: Current topics in molecular endocrinology, Vol. 3, pp. 433–450. New York: Plenum Press 1976
Pende, V.: La calcificazione della glandola pineale: correlazioni con le endocrinopatie. Studio clinico radiologico su 1121 casi. Folia endocr. (Roma) **6**, 191–200 (1953)
Penney, D.P., Reiter, R.J.: Fine structural localization of a possible acyl transferase in the pineal gland of male golden hamster. J. Histochem. Cytochem. **15**, 793 (1967)
Pennington, S.N.: A note on the chromatography of melatonin. J. Chromatogr. (Amst.) **32**, 406–407 (1968)
Peress, N.S., Murthy, G.G., Balcolm, R.J.: Pineal effects upon pituitary protein and nucleic acid synthesis in vitro. Neurosci. Letters **2**, 207–210 (1976)
Perez-Polo, J.R., Hall, K., Vaughan, M.K., Reiter, R.J.: Effect of pinealectomy, blinding, castration and melatonin on nerve growth factor levels in the submaxillary glands of mice. Neurosci. Letters **10**, 83–87 (1978)
Perrelet, A., Orci, L.: Experimentelle Untersuchung zur Ultrastruktur der Rattenepiphyse. Anat. Anz., Suppl., **125**, 429–437 (1969)
Perrelet, A., Orci, L., Rouiller, C.: Clarification of the osmiophilic granules of the rat pinealocytes by p-chlorophenylalanin. Experientia (Basel) **24**, 1047–1049 (1968)
Peschke, E.: Morphologische, klinische und statistische Untersuchungen zur Bedeutung der Epiphyse für die Schilddrüse. Morph. Jb. **122**, 480–484 (1976)
Peschke, E., Stenzel, I., Peil, J.: Kernvariationsstatistische Untersuchungen der Rattenschilddrüse nach Epiphysektomie. Zool. Jb. Anat. **92**, 579–592 (1974)
Peter, R.E.: Failure to detect an effect of pinealectomy in goldfish. Gen. comp. Endocr. **10**, 443–445 (1968)

Peter, R.E.: Hypothalamic control of thyroid gland activity and gonadal activity in the goldfish, *Carassius auratus*. Gen. comp. Endocr. **14**, 334–356 (1970)

Petit, A.: Nouvelles observations sur la morphogénèse et l'histogénèse du complex épiphysaire des Lacertiliens. Arch. Anat. (Strasbourg) **50**, 227–257 (1967)

Petit, A.: Ultrastructure de la rétine de l'œil pariétal d'un Lacertilien, *Anguis fragilis*. Z. Zellforsch. **92**, 70–93 (1968)

Petit, A.: Ultrastructure, innervation et fonction de l'épiphyse de l'orvet (*Anguis fragilis* L.). Z. Zellforsch. **96**, 437–465 (1969a)

Petit, A.: Embryogénèse de l'épiphyse et de l'organe sous-commissural de la couleuvre à collier (*Tropidonotus natrix* L.). Arch. Anat. (Strasbourg) **52**, 1–25 (1969b)

Petit, A.: L'épiphyse d'un serpent: *Tropidonotus natrix* L. II. Études cytochimique, autoradiographique et pharmacologique. Z. Zellforsch. **120**, 246–260 (1971)

Petit, A., Vivien-Roels, B.: Présence de contacts neurosensoriels et de synapses d'un type nouveau dans l'épiphyse du Lézard des murailles (*Lacerta muralis*, Laurenti). C.R. Acad. Sci. (Paris) D **284**, 1911–1913 (1977a)

Petit, A., Vivien-Roels, B.: Les rhythmes circadiens et circannuels du taux de 5-hydroxytryptamine (sérotonine) chez *Lacerta muralis* Laurenti (Reptiles-Lacertiliens). Arch. Biol. (Liège) **88**, 217–234 (1977b)

Petit, A., Vivien-Roels, B., Lichte, C.: Dosage spectrofluorimétrique de la 5-hydroxy-tryptamine (5 HT=sérotonine) épiphysaire chez les reptiles. Étude des variations du taux de 5-HT en fonction de quelques drogues. C.R. Acad. Sci. (Paris) D **272**, 1283–1285 (1971)

Petrelli, F.: Metabolismo delle proteine e degli acidi nucleici in ratti pinealectomizzati e in ratti normali trattati con estratto di pineale di bovino. Boll. Soc. ital. Biol. sper. **40**, 1907–1911 (1964)

Petronio, L.: Sul comportamento delle fosfatasi alcaline del plasma e di alcuni organi dopo prolungata somministrazione de estratto epifisario. Biol. lat. (Milano) **11**, 89–102 (1958)

Petronio, L., Dori, A.: Influenza di un estratto epifisario acquoso sulla struttura dei principali parenchimi. Biol. lat. (Milano) **10**, 174–182 (1957)

Petronio, L., Tavazza, D.: Effetti istomorfologici della prolungata somministrazione di un estratto epifisario acquoso in ratti ipofisectomizzati. Nota I: Comportamento delle ghiandole sessuali. Biol. lat. (Milano) **11**, 199–210 (1958a)

Petronio, L., Tavazza, D.: Effetti istomorfologici della prolungata somministrazione di un estratto epifisario acquoso in ratti ipofisectomizzati. II. Comportamento di alcune ghiandole endocrine (tiroide, paratiroidi, pancreas e surrene). Arch. E. Maragliano Pat. Clin. **14**, 1761–1777 (1958b)

Pévet, P.: Etude ultrastructurale de l'épiphyse du Hérisson mâle. Evolution en fonction du cycle sexuel. Thesis, Poitiers 1972

Pévet, P.: The pineal gland of the mole (*Talpa europaea* L.). The fine structure of the pinealocytes. Cell Tiss. Res. **153**, 277–292 (1974)

Pévet, P.: Vacuolated pinealocytes in the hedgehog (*Erinaceus europaeus* L.) and the mole (*Talpa europaea* L.). Cell. Tiss. Res. **159**, 303–309 (1975)

Pévet, P.: Correlations between pineal gland and sexual cycle. An electron microscopical and histochemical investigation on the pineal gland of the hedgehog, mole, mole-rat and white rat. Thesis, Amsterdam 1976

Pévet, P.: On the presence of different populations of pinealocytes in the mammalian pineal gland. J. Neural Transm. **40**, 289–304 (1977a)

Pévet, P.: The pineal gland of the mole (*Talpa europaea* L.). IV. Effect of pronase on material present in cisternae of the granular endoplasmic reticulum of pinealocytes. Cell Tiss. Res. **182**, 215–219 (1977b)

Pévet, P., Collin, J.P.: Les pinéalocytes de Mammifère: Diversité, homologies, origine. Etude chez la Taupe adulte (*Talpa europaea* L.). J. Ultrastruct. Res. **57**, 22–31 (1976)

Pévet, P., Kappers, J.A.: Secretory processes in the mammalian pineal gland. An ultrastructural identification. Acta endocr. (Kbh.) Suppl. **212**, 157 (1977a)

Pévet, P., Kappers, J.A.: Secretory process differing from that producing granular vesicles in the mammalian pinealocytes. An ultrastructural identification. XXVIIe International Congress of Physiological Sciences, Paris, July 18–23. Abstract N° 1760. 1977b

Pévet, P., Karasek, M.: Are the pineal active compounds of mammals proteinaceous in nature? An ultrastructural contribution. Acta med. pol. **18**, 351–353 (1977)

Pévet, P., Kuyper, M.A.: The ultrastructure of pinealocytes in the golden mole (*Amblysomus hottentotus*) with special reference to the granular vesicles. Cell Tiss. Res. **191**, 39–56 (1978)

Pévet, P., Saboureau, M.: Modifications ultrastructurales dans l'épiphyse du Hérisson mâle au cours du repos sexuel. Gen. comp. Endocr. **18**, 615–616 (1972)

Pévet, P., Saboureau, M.: L'épiphyse du Hérisson (*Erinaceus europaeus* L.) mâle. I. Les pinéalocytes et leur variations ultrastructurales considérées au cours du cycle sexuel. Z. Zellforsch. **143**, 367–385 (1973)

Pévet, P., Saboureau, M.: Effect of serotonin administration on the ultrastructure of pinealocytes during the period of maximal sexual activity of the male hedgehog (*Erinaceus europaeus* L.). Experientia (Basel) **30**, 1069–1070 (1974)

Pévet, P., Smith, A.R.: The pineal gland of the mole (*Talpa europaea* L.). II. Ultrastructural variations observed in the pinealocytes during different parts of the sexual cycle. J. Neural Transm. **36**, 227–248 (1975a)

Pévet, P., Smith, A.R.: The mole pinealocytes. J. Endocr. **64**, 64 (1975b)

Pévet, P., Smith, A.R., Van de Kar, L., Van Bronswijk, H.: Effect of castration on the rat pineal gland: a fluorescence histochemical and biochemical study. Experientia (Basel) **31**, 1237–1238 (1975a)

Pévet, P., Kappers, J.A., Nevo, E.: Etude ultrastructurale des pinéalocytes de deux mammifères souterrains aveugles, la Taupe (*Talpa europaea*, Insectivore) et le spalax (*Spalax ehrenbergi*, Rongeur): Importance des synthèses protéiques. J. Physiol. (Paris) **70**, 6B (1975b)

Pévet, P., Smith, A.R., Kappers, A.J.: Les pinéalocytes de la Taupe et leur variations ultrastructurales considérées au cours du cycle sexuel. J. Microsc. Biol. Cell. **22**, 30a (1975c)

Pévet, P., Juillard, M.T., Smith, A.R., Kappers, J.A.: The pineal gland of the mole (*Talpa europaea* L.). III. A fluorescence histochemical study. Cell Tiss. Res. **165**, 297–306 (1976a)

Pévet, P., Kappers, J.A., Nevo, E.: The pineal gland of the mole-rat (*Spalax ehrenbergi*, Nehring). 1. The fine structure of pinealocytes. Cell Tiss. Res. **174**, 1–24 (1976b)

Pévet, P., Kappers, J.A., Voûte, A.M.: The pineal gland of nocturnal mammals. I. The pinealocytes of the bat (*Nyctalus noctula*, Schreber). J. Neural Transm. **40**, 47–68 (1977a)

Pévet, P., Kappers, J.A., Voûte, A.M.: Morphologic evidence for differentiation of pinealocytes from photoreceptor cells in the adult noctule bat (*Nyctalus noctula*, Schreber). Cell Tiss. Res. **182**, 99–109 (1977b)

Pévet, P., Balemans, M.G.M., Bary, F.A.M., Noordegraaf, E.M.: The pineal gland of the mole (*Talpa europaea*, L.). V. Activity of hydroxyindole-O-methyltransferase (HIOMT) in the formation of melatonin/5-methoxytryptophol in the eyes and the pineal gland. Ann. Biol. anim. Bioch. Biophys. **18**, 259–264 (1978)

Pévet, P., Dogterom, J., Buijs, R.M., Reinharz, A.: Is it the vasotocin or a vasotocin-like peptide which is present in the mammalian pineal and subcommissural organ? J. Endocr. **80**, 49P (1979)

Pfister, A., Guerillot, C., Muller, J., Vendrely, E., Da Lage, C.: Existence d'une innervation d'origine centrale dans l'épiphyse du hamster et du rat. J. Physiol. (Paris) **70**, 10B–11B (1975)

Pflugfelder, O.: Wirkungen der Epiphysektomie auf die Postembryonalentwicklung von *Lebistes reticulatus* Peters. Arch. Entwickl.-Mech. Org. **146**, 115–136 (1953)

Pflugfelder, O.: Wirkungen partieller Störungen der Parietalregion von *Lebistes reticulatus*. Arch. Entwickl.-Mech. Org. **147**, 42–60 (1954)

Pflugfelder, O.: Wirkungen von Epiphysan und Thyroxin auf die Schilddrüse epiphysektomierter *Lebistes reticulatus* Peters. Arch. Entwickl.-Mech. Org. **148**, 463–473 (1956)

Pflugfelder, O.: Physiologie der Epiphyse. Zool. Anz., Suppl. **20**, 53–75 (1957)

Pflugfelder, O.: Wirkungen lokaler Hirnläsionen auf Hypophyse und Thyreoidea von *Carassius gibelio auratus* Bloch. Arch. Entwickl.-Mech. Org. **155**, 535–548 (1964)

Pflugfelder, O.: Weitere Untersuchungen über Kypho-Lordose und Skoliose nach Zerstörung der Epiphysenregion bei Fischen und Haushühnern. Arch. Entwickl.-Mech. Org. **158**, 170–187 (1967)

Philo, R., Reiter, R.J.: Characterization of pinealectomy induced convulsions in the Mongolian gerbil (*Meriones unguiculatus*). Epilepsia **19**, 485–492 (1978a)

Philo, R., Reiter, R.R.J.: Brain amines and convulsions in four strains of parathyroidectomized, pinealectomized rats. Epilepsia **19**, 133–137 (1978b)

Piechowiak, H., Schnizer, W.: On electrolyte determinations in the pituitary and pineal gland: is there a diurnal rhythm? Endokrinologie **67**, 51–59 (1976)

Piekut, D.T., Knigge, K.M.: Primary cultures of dispersed cells of rat pineal gland. I. Fine structure and indole metabolism. Cell Tiss. Res. **188**, 285–297 (1978)

Piezzi, R.S.: The pineal, biological clock. Acta physiol. lat.-amer. **23**, 445–447 (1973)

Piezzi, R.S., Gutierrez, L.S.: Electron microscopic studies on the pineal organ of the antarctic penguin (*Pygoscelis papua*). Cell Tiss. Res. **164**, 559–570 (1975)

Piezzi, R.S., Wurtman, R.J.: Pituitary serotonin content: Effects of melatonin or deprivation of water. Science **169**, 285–286 (1970)

Pilling, J.R., Hawkins, T.D.: Distribution of calcification within the pineal gland. Brit. J. Radiol. **50**, 796–798 (1977)

Pitis, M., Scherzer, M.: The action of the pineal gland on the compensatory hyperplasia in immature rats. Rev. roum. Endocr. **6**, 209–213 (1969)

Pitis, M., Nicolescu-Catargi, A., Taşcă, C.: Effects of pinealectomy on the gonads in newly born and immature rats. Rev. roum. Endocr. **7**, 37–41 (1970)

Pohl, C.R., Gibbs, F.P.: Circadian rhythms in blinded rats: correlation between pineal and activity cycles. Amer. J. Physiol. **234**, R110–R114 (1978)

Poletti, H.M., Castellano, M.A.: Role of the nuclear membrane in smooth endoplasmic reticulum formation in white rat pinealocytes. Experientia (Basel) **23**, 465–466 (1967)

Polvani, F.: Studio anatomico della glandola pineale umana. Folia neuro-biol. (Lpz.) **7**, 655 (1963)

Pomerantz, G., Reiter, R.J.: Age dependent changes in the susceptibility of parathyroidectomized, pinealectomized rats to seizures. Exp. Neurol. **40**, 254–257 (1973)

Pomerantz, G., Reiter, R.J.: Influence of intraocularly-injected pineal indoles on PMS-induced ovulation in immature rats. Int. J. Fertil. **19**, 117–120 (1974)

Pomerantz, G., Sorrentino, S., Jr.: The influence of melatonin administered subcutaneously, intravenously or intraocularly upon ovulation in the PMS-treated immature rat. Neuroendocrinology **12**, 354–361 (1973)

Popova, N.K., Kolaeva, S.G., Dianova, I.I.: State of the pineal gland during hibernation. Bull. exp. Biol. Med. **79**, 467–468 (1975)

Poppen, J.L.: The right occipital approach to a pinealoma, J. Neurosurg. **25**, 706–710 (1966)

Porter, J.R., Heiman, M.: The effect of two pineal indoles and a crude bovine pineal extract on ACTH mediated corticosterone release by isolated adrenal cells. Fed. Proc. **36**, 367 Abs. (1977a)

Porter, J.R., Heimann, H.: Effects of pineal indoles and a crude aqueous pineal extract on ACTH mediated corticosterone release by isolated adrenal cells. Life Sci. **20**, 1363–1371 (1977b)

Povlishock, J.T., Kriebel, R.M., Seibel, H.R.: A light and electron microscopic study of the pineal gland of the ground squirrel, *Citellus tridecemlineatus*. Amer. J. Anat. **143**, 465–484 (1975)

Preslock, J.P.: Pineal hydroxyindole-O-methyl transferase and N-acetyl transferase during sexual maturation of *Coturnix* quail. Life Sci. **15**, 1805–1813 (1974)

Preslock, J.P.: Photoperiodic and gonadal steroid regulation of pineal hydroxyindole-O-methyl transferase and N-acetyl transferase in *Coturnix* quail. Life Sci. **17**, 1227–1232 (1975)

Preslock, J.P.: Regulation of pineal enzymes by photoperiod, gonadal hormones and melatonin in *Coturnix* quail. Hormone Res. **7**, 108–117 (1976)

Prop, N.: A paper chromatographic study of the epiphysis cerebri of the albino rat. Acta morph. neerl.-scand. **3**, 407 (1961)

Prop, N.: Fats in the pineal gland of the rat. Acta morph. neerl.-scand. **5**, 285 (1963)

Prop, N.: Lipids in the pineal body of the rat. Progr. Brain Res. **10**, 454–464 (1965)

Prop, N., Ebels, I.: Effects of sheep and young calf pineal extracts and continuous light on the pineal gland, the gonads and the oestrous cycle of the rat. Acta endocr. (Kbh.) **57**, 585–594 (1968)

Prop, N., Kappers, J.A.: Demonstration of some compounds present in the pineal organ of the albino rat by histochemical methods and paper chromatography. Acta anat. (Basel) **45**, 90–109 (1961)

Pun, J.Y., Lombrozo, L.: Microelectrophoresis of brain and pineal proteins in polyacrylamide gel. Analyt. Biochem. **9**, 9–20 (1964)

Quastel, M.R., Rahamimoff, R.: Effect of melatonin on spontaneous contractions and response to 5-hydroxytryptamine of rat isolated duodenum. Brit. J. Pharmacol. **24**, 455–461 (1965)

Quay, W.B.: Volumetric and cytologic variation in the pineal body of *Peromyscus leucopus* (Rodentia) with respect to sex, captivity and day-length. J. Morph. **98**, 471–495 (1956a)

Quay, W.B.: The demonstration of a secretory material and cycle in the parenchymal cells of the mammalian pineal organ. Exp. Cell Res. **10**, 541–544 (1956b)

Quay, W.B.: Cytochemistry of pineal lipids in rat and man. J. Histochem. Cytochem. **5**, 145–153 (1957a)

Quay, W.B.: Cytochemistry of pineal lipids in rat and man and their changes with age. Anat. Rec. **127**, 351 (1957b)

Quay, W.B.: The pineal cells and fibres staining with paraldehyde fuchsin and the effects of pineal tractotomy in the rodent. Anat. Rec. **127**, 441–442 (1957c)

Quay, W.B.: Pineal blood content and its experimental modification. Amer. J. Physiol. **195**, 391–395 (1958a)

Quay, W.B.: Localization, changes with age and experimental augmentation of pineal succinoxidase activity. Anat. Rec. **130**, 360 (1958b)

Quay, W.B.: Experimental modifications and changes with age in pineal succinic dehydrogenase activity. Amer. J. Physiol. **196**, 951–955 (1959a)

Quay, W.B.: Striated muscle in the mammalian pineal organ. Anat. Rec. **133**, 57–64 (1959b)

Quay, W.B.: Experimental and comparative studies of succinic dehydrogenase activity in mammalian choroid plexuses, ependyma and pineal organ. Physiol. Zoology **33**, 206–212 (1960)

Quay, W.B.: Reduction of mammalian pineal weight and lipid during continuous light. Gen. comp. Endocr. **1**, 211–217 (1961a)

Quay, W.B.: Photic modification of mammalian pineal weight and composition and its anatomical basis. Anat. Rec. **139**, 265–266 (1961b)

Quay, W.B.: Experimental and cytological studies of pineal cells staining with acid hematein in the rat (*Rattus norvegicus*). Acta morph. neerl.-scand. **5**, 87–100 (1962a)

Quay, W.B.: Metabolic and cytologic evidence of pineal inhibition by continuous light. Amer. Zoologist **2**, 217 (1962b)

Quay, W.B.: The respiration of ependyma and pineal cells in vitro with diverse substrates and concentrations of inorganic ions. Amer. Zoologist **2**, 439 (1962c)

Quay, W.B.: Differential extraction for the spectrofluorometric measurement of diverse 5-hydroxy- and 5-methoxyindoles. Analyt. Biochem. **5**, 51–59 (1963a)

Quay, W.B.: Cytologic and metabolic parameters of pineal inhibition by continuous light in the rat (*Rattus norvegicus*). Z. Zellforsch. **60**, 479–490 (1963b)

Quay, W.B.: Circadian rhythm in rat pineal serotonin and its modifications by estrous cycle and photoperiod. Gen. comp. Endocr. **3**, 473–479 (1963c)

Quay, W.B.: Pineal and ependymal respiration with diverse substrates and inorganic ions. Amer. J. Physiol. **204**, 245–250 (1963d)

Quay, W.B.: Effect of dietary phenylalanine and tryptophan on pineal and hypothalamic serotonin levels. Proc. Soc. exp. Biol. (N.Y.) **114**, 718–721 (1963e)

Quay, W.B.: Circadian and estrous rhythms in pineal melatonin and 5-hydroxyindole-3-acetic acid. Proc. Soc. exp. Biol. (N.Y.) **115**, 710–713 (1964a)

Quay, W.B.: Circadian and estrous rhythms in pineal and brain serotonin. Progr. Brain Res. **8**, 61–63 (1964b)

Quay, W.B.: Histological structure and cytology of the pineal organ in birds and mammals. Progr. Brain Res. **10**, 49–86 (1965a)

Quay, W.B.: Photic relations and experimental dissociation of circadian rhythms in pineal composition and running activity in rats. Photochem. Photobiol. **4**, 425–432 (1965b)

Quay, W.B.: Experimental evidence for pineal participation in homeostasis of brain composition. Progr. Brain Res. **10**, 646–653 (1965c)

Quay, W.B.: Retinal and pineal hydroxyindole-O-methyl transferase activity in vertebrates. Life Sci. **4**, 983–991 (1965d)

Quay, W.B.: Daily rhythms and photic responses of pineal hydroxyindole derivatives in the pigeon (*Columba livia*). Amer. Zoologist **5**, 218 (1965e)

Quay, W.B.: Indole derivatives of pineal and related neural and retinal tissues. Pharmacol. Rev. **17**, 321–345 (1965f)

Quay, W.B.: 24-hour rhythms in pineal 5-hydroxytryptamine and hydroxyindole-O-methyl transferase activity in the macaque. Proc. Soc. exp. Biol. (N.Y.) **121**, 946–948 (1966a)

Quay, W.B.: Rhythmic and light-induced changes in levels of pineal 5-hydroxyindoles in the pigeon (*Columba livia*). Gen. comp. Endocr. **6**, 371–377 (1966b)

Quay, W.B.: Pineal structure and composition in red and grey kangaroos. Anat. Rec. **154**, 405 (1966c)

Quay, W.B.: The significance of darkness and monoamine oxidase in the nocturnal changes in 5-hydroxytryptamine and hydroxyindole-O-methyltransferase activity of the macaque's epiphysis cerebri. Brain Res. **3**, 277–286 (1966/67)

Quay, W.B.: Lack of day-night rhythm and effect of darkness on rat pineal content of N-acetylserotonin-O-methyltransferase. Physiologist **10**, 286 (1967a)

Quay, W.B.: Spontaneous adenomata of the hypophyseal anterior lobe in aged pinealectomized rats. Experientia (Basel) **23**, 129–130 (1967b)

Quay, W.B.: Specificity and structure-activity relationships in the *Xenopus* larval melanophore assay for melatonin. Gen. comp. Endocr. **11**, 253–254 (1968a)

Quay, W.B.: Specificity and fluorometry of 5-hydroxytryptamine by means of products with ninhydrin. J. pharm. Sci. **57**, 1568–1572 (1968b)

Quay, W.B.: Relation of pineal acetylserotonin methyltransferase activity to daily photoperiod and light intensity. Arch. Anat. Histol. Embryol. **51**, 565–571 (1968c)

Quay, W.B.: Comparative physiology of serotonin and melatonin. Advanc. Pharmacol. Chemother. **6**, Part A, 283–297 (1968d)

Quay, W.B.: Occurrences and biological significance of pineal atrophy among families of birds. Amer. Zoologist **8**, 762–763 (1968e)

Quay, W.B.: Individuation and lack of pineal effect in the rat's circadian locomotor activity. Physiol. Behav. **3**, 109–118 (1968f)

Quay, W.B.: Melatonin as a retinal neurohumor mediating light-induced cone contraction in the larval amphibian. Proc. Int. Union. Physiol. Sc. **7**, 357 (1968g)

Quay, W.B.: Catecholamines and tryptamines. I. Neurovisc. Relat., Suppl. **9**, 212 (1969a)

Quay, W.B.: Evidence for a pineal contribution in the regulation of vertebrate reproductive system. Gen. Comp. Endocr. Suppl. **2**, 101–110 (1969b)

Quay, W.B.: The role of the pineal gland in environmental adaptation, pp. 508–550. In: E. Bajusz. (ed.), Physiology and pathology of adaptation mechanisms. New York: Pergamon Press 1969c

Quay, W.B.: Comparative histology and histochemistry of the primate pineal region with particular reference to the orangutan (*Pongo pygmaeus*). Anat. Rec. **166**, 364 (1970a)

Quay, W.B.: Pineal structure and composition in the orangutan (*Pongo pygmaeus*). Anat. Rec. **168**, 93–104 (1970b)

Quay, W.B.: Pineal organ. In: Biology of bats (Wimsatt, W.A., ed.), pp. 311–318. New York: Academic Press 1970c

Quay, W.B.: Physiological significance of the pineal during adaptation to shifts in photoperiod. Physiol. Behav. **5**, 353–360 (1970d)

Quay, W.B.: Effect of pinealectomy on circadian phase shifting following changes in photoperiod. Fed. Proc. **29**, 325 (1970e)

Quay, W.B.: Diagnosis of destructive lesions of the pineal. Lancet **1970f II**, 42

Quay, W.B.: Precocious entrainment and associated characteristics of activity patterns following pinealectomy and reversal of photoperiod. Physiol. Behav. **5**, 1281–1290 (1970g)

Quay, W.B.: Effects of cutting nervi conarii and tentorium cerebelli on pineal composition and activity shifting following reversal of photoperiod. Physiol. Behav. **6**, 681–688 (1971a)

Quay, W.B.: Dissimilar functional effects of pineal stalk and cerebral meningeal interruptions on phase shifts of circadian activity rhythms. Physiol. Behav. **7**, 557–567 (1971b)

Quay, W.B.: Effects of interruption of pineal stalk and anterior dorsal meninges on phase shifting of circadian activity following reversal of photoperiod. Fed. Proc. **30**, 610 Abs. (1971c)

Quay, W.B.: Factors in the measurement of pineal acetylserotonin methyltransferase activity in the lizard *Sceloporus occidentalis*. Gen. comp. Endocr. **17**, 220–226 (1971d)

Quay, W.B.: Twenty-four-hour rhythmicity in carbonic anhydrase activities of choroid plexuses and pineal gland. Anat. Rec. **174**, 279–287 (1972a)

Quay, W.B.: Pineal vasoconstriction at daily onset of light: its physiological correlates and control. Physiologist **15**, 241 (1972b)

Quay, W.B.: Infrequency of pineal atrophy among birds and its relation to nocturnality. Condor **74**, 33–45 (1972c)

Quay, W.B.: Twenty-four-hour rhythmicity of pineal canaliculi and evidence for their intrinsic humoral regulation. Physiologist **16**, 427 (1973a)

Quay, W.B.: Retrograde perfusions of the pineal region and the question of pineal vascular routes to brain and choroid plexuses. Amer. J. Anat. **137**, 387–402 (1973b)

Quay, W.B.: Pineal chemistry. Springfield, Ill., USA: Charles C. Thomas 1974a

Quay, W.B.: Pineal canaliculi: demonstration, twenty-four-hour rhythmicity and experimental modification. Amer. J. Anat. **139**, 81–94 (1974b)

Quay, W.B.: Effects of early postnatal pinealectomy on adult brain and divisional pituitary weights in relation to type of illumination. Endocr. Res. Commun. **1**, 359–371 (1974c)

Quay, W.B.: Seasonal cycle and physiological correlates of pinealocyte nuclear and nucleolar diameters in the bats, *Myotis lucifugus* and *Myotis sodalis*. Gen. comp. Endocr. **29**, 369–375 (1976)

Quay, W.B.: Quantitative morphology and environmental responses of the pineal gland in the collared lemming (*Dicrostonyx groenlandicus*) Amer. J. Anat. **153**, 545–562 (1978)

Quay, W.B., Bagnara, J.T.: Relative potencies of indolic and related compounds in the body-lightening reaction of larval *Xenopus*. Arch. int. Pharmacodyn. **150**, 137–143 (1964)

Quay, W.B., Baker, P.C.: Form, weight, and indole content of pineal organs of red and grey kangaroos. Austr. J. Zool. **13**, 727–733 (1965)

Quay, W.B., Halevy, A.: Experimental modification of the rat pineal's content of serotonin and related indoleamines. Physiol. Zool. **35**, 1–7 (1962)

Quay, W.B., Kahn, R.H.: Pineal histology and cytochemistry in organ culture. Cellule **63**, 247–257 (1963)

Quay, W.B., Levine, B.E.: Pineal growth and mitotic activity in the rat and the effects of colchicine and sex hormones. Anat. Rec. **129**, 65–77 (1957a)

Quay, W.B., Levine, B.: Postnatal growth and mitotic activity in the rat pineal and the effects of colchicine and sex hormones. Anat. Rec. **127**, 441 (1957b)

Quay, W.B., McLeod, R.W.: Melatonin and photic stimulation of cone contraction in the retina of larval *Xenopus laevis*. Anat. Rec. **160**, 491 (1968)

Quay, W.B., Millar, R.: Organization and histology of the pineal region in the hyrax (*Procavia capensis*). Amer. J. Anat. **130**, 377–392 (1971)

Quay, W.B., Renzoni, A.: Comparative and experimental studies of pineal structure and cytology in passeriform birds. Riv. Biol. **56**, 363–407 (1963)

Quay, W.B., Renzoni, A.: Twenty-four-hour rhythms in the pineal mitotic activity and nuclear and nucleolar dimensions. Growth **30**, 315–324 (1966a)

Quay, W.B., Renzoni, A.: Studies on the "commissuro-pineal neurosecretory cells" of birds. Riv. Biol. **59**, 231–266 (1966b)

Quay, W.B., Renzoni, A.: I rapporti diencefalici e la variabilita nella duplice struttura del complesso epifisario degli uccelli. The diencephalic relations and variably bipartite structure of the avian pineal complex. [Italian and English translation] Riv. Biol. **60**, 9–75 (1967)

Quay, W.B., Smart, L.I.: Substrate specificity and post-mortem effects in mammalian pineal acetylserotonin methyltransferase activity. Arch. int. Physiol. Biochim. **75**, 197–210 (1967)

Quay, W.B., Wilhoft, D.C.: Comparative and regional differences in the serotonin content of reptilian brains. J. Neurochem. **11**, 805–811 (1964)

Quay, W.B., Jongkind, J.F., Kappers, J.A.: Localizations and experimental changes in monoamines of the reptilian pineal complex studied by fluorescence histochemistry. Anat. Rec. **157**, 304–305 (1967)

Quay, W.B., Kappers, J.A., Jongkind, J.F.: Innervation and fluorescence histochemistry of monoamines in the pineal organ of a snake (*Natrix natrix*). J. neuro-visc. Rel. **31**, 11–25 (1968a)

Quay, W.B., Renzoni, A., Eakin, R.M.: Pineal ultrastructure in *Melopsittacus undulatus* with particular regard to cell types and functions. Riv. Biol. **61**, 371–393 (1968b)

Quay, W.B., Bennett, E.L., Rosenzweig, M.R., Krech, D.: Effects of isolation and environmental complexity on brain and pineal organ. Physiol. Behav. **4**, 489–494 (1969a)

Quay, W.B., Smart, L.I., Hafeez, A.: Substrate specificity and tissue localization of acetylserotonin methyltransferase in eyes of trout (*Salmo gairdneri*). Comp. Biochem. Physiol. **28**, 947–953 (1969b)

Quay, W.B., Stebbins, R.C., Kelley, T.D., Cohen, N.W.: Effects of environmental and physiological factors on pineal acetylserotonin methyltransferase activity in the lizard, *Sceloporus occidentalis*. Physiol. Zoology **44**, 241–248 (1971)

Rahamimoff, R., Bruderman, I., Colshani, G.: Effect of melatonin on 5-hydroxytryptamine induced contraction of isolated cat trachea. Life Sci. **4**, 2281–2287 (1965)

Raikhlin, N.T., Kvetnoy, I.M.: Melatonin and enterochromaffine cells. Acta histochem. (Jena) **55**, 19–24 (1976)

Raikhlin, N.T., Kvetnoy, I.M.: Immunohistochemical detection of cells containing melatonin and N-acetylserotonin. Bull. exp. Biol. Med. **84**, 1785–1786 (1977)

Raikhlin, N.T., Kvetnoy, I.M., Tolkachev, V.N.: Melatonin may be synthesized in enterochromaffine cells. Nature (Lond.) **255**, 344–345 (1975)

Raikhlin, N.T., Kvetnoy, I.M., Kadagidze, Z.G., Sokolov, A.V.: Immunomorphological studies on synthesis of melatonin in enterochromaffin cells. Acta histochem. (Jena) **11**, 75–77 (1978)

Ralph, C.L.: Structure and alleged functions of avian pineals. Amer. Zoologist **10**, 217–235 (1970)

Ralph, C.L.: The pineal complex: A retrospective view. Amer. Zoologist **15** (Suppl. 1) 105–116 (1975)

Ralph, C.L.: Correlations of melatonin content in pineal gland, blood and brain of some birds and mammals. Amer. Zoologist **16**, 35–43 (1976)

Ralph, C.L.: Pineal control of reproduction: Nonmammalian vertebrates. Progr. Reprod. Biol. **4**, 30–50 (1978)

Ralph, C.L., Dawson, D.C.: Failure of the pineal body of two species of birds (*Coturnix coturnix japonica* and *Passer domesticus*) to show electrical responses to illumination. Experientia (Basel) **24**, 147–148 (1968)

Ralph, C.L., Lane, K.B.: Morphology of the pineal body of wild house sparrows (*Passer domesticus*) in relation to reproduction and age. Canad. J. Zool. **47**, 1205–1208 (1969)

Ralph, C.L., Lynch, H.J.: A quantitative melatonin assay. Gen. comp. Endocr. **15**, 334–338 (1970)

Ralph, C.L., Hedlund, L., Murphy, W.A.: Diurnal cycles of melatonin in bird pineal bodies. Comp. Biochem. Physiol. **22**, 591–599 (1967)

Ralph, C.L., Mull, D., Lynch, H.J.: Locomotor activity rhythms of rats under constant condition as predictors of melatonin content of their pineals. Amer. Zoologist **10**, 302 (1970)

Ralph,. C.L., Mull, D., Lynch, H.J., Hedlund, L.: A melatonin rhythm persists in rat pineals in darkness. Endocrinology **89**, 1361–1366 (1971)

Ralph, C.L., Pelham, R.W., MacBride, S.E., Reilly, D.P.: Persistent rhythms of pineal and serum melatonin in cockerels in continuous darkness. J. Endocr. **63**, 319–324 (1974)

Ralph, C.L., Binkley, S., MacBride, S.E., Klein, D.C.: Regulation of pineal rhythms in chickens: Effects of blinding, constant light, constant dark, and superior cervical ganglionectomy. Endocrinology **97**, 1373–1378 (1975)

Ramaley, J.A., Bartosik, D.: Effect of adrenalectomy on light-induced precocious puberty in rats. Biol. Reprod. **13**, 347–352 (1975)

Ramsey, H.J.: Ultrastructure of a pineal tumor. Cancer (Philad.) **18**, 1014–1025 (1966)

Rasquin, P.: Studies in the control of pigment cells and light response in recent teleost fishes. I. Morphology of the pineal organ. II. Reactions of the pigmentary system to hormonal stimulation. Bull. Amer. Mus. Natur. Hist. **115**, 1–68 (1958)

Ratliff, B., Miller, L.P., Lovenberg, W.: Ornithine decarboxylase: Apparent lack of involvement in the induction of N-acetyltransferase in rat pineal gland. Biochem. Pharmacol. **28**, 389–391 (1979)

Rausch, L.J., Long, C.J.: Habenular nuclei: A crucial link between the olfactory and motor systems. Brain Res. **29**, 146–150 (1971)

Redding, T.W., Nair, R.M.G., Kastin, A.J.: Physiological studies with ^{14}C-melanocyte stimulating hormone-inhibiting factor (MIF). Fed. Proc. **31**, 811 (1972)

Reed, B.L.: The control of circadian pigment changes in the pencil fish: A proposed role for melatonin. Life Sci. **7**, Part II, 961–973 (1968)

Reed, B.L., Finnin, B.C., Ruffin, N.E.: The effects of melatonin and epinephrine on the melanophores of freshwater teleosts. Life Sci. **8**, Part II, 113–120 (1969)

Rees, H.D., Stumpf, W.E., Sar, M.: Autoradiographic studies with ^{3}H-dexamethasone in the rat brain and pituitary. In: Anatomical neuroendocrinology (W.E. Stumpf and L.D. Grant, eds.), pp. 262–269. Basel: S. Karger 1975

Reinharz, A., Vallotton, M.B.: Presence of peptidase in bovine pineal gland. Experientia (Basel) **30**, 691 (1974)

Reinharz, A.C., Vallotton, M.B.: Presence of two neurophysins in human pineal gland. Endocrinology **100**, 994–1001 (1977)

Reinharz, A.C., Czernichow, P., Vallotton, M.B.: Neurophysin-like protein in bovine pineal gland. J. Endocr. **62**, 35–44 (1974)

Reinharz, A.C., Czernichow, P., Vallotton, M.B.: Neurophysins I and II from the bovine posterior pituitary lobe and neurophysin-like proteins from bovine pineal gland. Ann. N.Y. Acad. Sci. **248**, 172–183 (1975)

Reinharz, A.C., Pavel, S., Vallotton, M.B.: Evidence for in vitro release of neurophysin by rat pineal gland. Experientia (Basel) **34**, 1232–1233 (1978)

Reiss, M., Badrick, F.E., Halkerston, J.M.: The uptake of radioactive tracers by the pineal gland. J. Anat. (Lond.) **83**, 81 (1949a)

Reiss, M., Badrick, F.E., Halkerston, J.M.: The influence of the pituitary on phosphorus metabolism of brain. Biochem. J. **44**, 257–260 (1949b)

Reiss, M., Davis, R.H., Sideman, M.B., Plichta, E.S.: Pineal gland and spontaneous activity of rats. J. Endocr. **28**, 127–128 (1963a)

Reiss, M., Mauer, I., Sideman, M.B., Davis, R.H., Plichta, E.S.: Pituitary-pineal-brain interrelationships. J. Neurochem. **10**, 851–857 (1963b)

Reiss, M., Davis, R.H., Sideman, M.B., Mauer, I., Plichta, E.S.: Action of pineal extracts on the gonads and their function. Endocrinol. **27**, 107–118 (1963c)

Reiss, M., Sideman, M.B., Plichta, E.S.: Spontaneous activity and pineal gland cell density. J. Endocr. **37**, 475–476 (1967)

Reiter, R.J.: The effect of pineal grafts, pinealectomy and denervation of the pineal gland on the reproductive organs of male hamsters. Neuroendocrinology **2**, 138–146 (1967a)

Reiter, R.J.: Failure of the pineal gland to prevent gonadotrophin-induced ovarian stimulation in blinded hamsters. J. Endocr. **38**, 199–200 (1967b)

Reiter, R.J.: Pineal function in androgen-sterilized female rats. Amer. Zoologist **7**, 712 (1967c)

Reiter, R.J.: Morphological studies on the reproductive organs of blinded male hamsters and the effect of pinealectomy and superior cervical ganglionectomy. Anat. Rec. **160**, 13–24 (1968a)

Reiter, R.J.: The pineal gland and gonadal development in male rats and hamsters. Fertil. Steril. **19**, 1009–1017 (1968b)

Reiter, R.J.: Photic regulation of the pineal gland and its effect on reproduction in early androgen treated female rats. Fed. Proc. **27**, 319 (1968c)

Reiter, R.J.: Pineal function in long term blinded male and female golden hamsters. Gen. comp. Endocr. **12**, 460–468 (1969a)

Reiter, R.J.: Failure of constant light during the preweaning period to influence the activity of the hamster pineal gland after adulthood. Anat. Rec. **163**, 248–249 (1969b)

Reiter, R.J.: Antigonadotropic activity of the pineal gland in blinded anosmic female rats. Fed. Proc. **28**, 318 (1969c)

Reiter, R.J.: Relationships among the pineal gland, the photoperiod, olfaction and reproduction in female rats. J. Reprod. Fertil. **18**, 152 (1969d)

Reiter, R.J.: Comparative endocrinology of the pineal. Symposium for the Division of Comperative Endocrinology, American Society of Zoologists. Boston, Mass. 1969. Amer. Zoologist. **10**, 187–267 (1970a)

Reiter, R.J.: Physiological evidence for the participation of the medial forebrain bundle in pineal-mediated gonadal responses. Anat. Rec. **166**, 366 (1970b)

Reiter, R.J.: Pineal inhibition of compensatory testicular enlargement in light-deprived hamsters. Experientia (Basel) **26**, 1153–1154 (1970c)

Reiter, R.J.: Physiological effects of the pineal gland in androgen-sterilized female rats. Acta endocr. (Kbh.) **63**, 667–678 (1970d)

Reiter, R.J.: Introduction to the Symposium: A précis of the pineal odyssey. Amer. Zoologist **10**, 189–190 (1970e)

Reiter, R.J.: Evidence for refractoriness of the pituitary-gonadal axis to the pineal gland in golden hamsters and its possible implications in annual reproductive rhythms. Anat. Rec. **173**, 365–371 (1972a)

Reiter, R.J.: Compensatory growth of the ovaries, adrenal glands and kidneys in blinded, anosmic rats. Experientia (Basel) **28**, 1492–1493 (1972b)

Reiter, R.J.: Surgical procedures involving the pineal gland which prevent gonadal degeneration in adult male hamsters. Ann. Endocr. (Paris) **33**, 571–581 (1972c)

Reiter, R.J.: The effects of various surgical manipulations on pineal gland function in the golden hamster. Anat. Rec. **172**, 392 (1972d)

Reiter, R.J.: Comparative physiology: pineal gland. Ann. Rev. Physiol. **35**, 305–328 (1973a)
Reiter, R.J.: Involvement of pineal indoles and polypeptides with the neuroendocrine axis. Progr. Brain Res. **39**, 281–287 (1973b)
Reiter, R.J.: Pineal control of a seasonal reproductive rhythm in male golden hamsters exposed to natural daylight and temperature. Endocrinology **92**, 423–430 (1973c)
Reiter, R.J.: Evidence for pineal-induced seasonal changes in reproductive physiology of male hamsters kept under natural environmental conditions. Fed. Proc. **32**, 251 Abs. 219 (1973d)
Reiter, R.J.: Comparative effects of continual lighting and pinealectomy on the eyes, the Harderian glands and reproduction in pigmented and albino rats. Comp. Biochem. Physiol. A **44**, 503–509 (1973e)
Reiter, R.J.: Influence of pinealectomy on the breeding capability of hamsters maintained under natural photoperiodic and temperature conditions. Neuroendocrinology **13**, 366–370 (1973/74)
Reiter, R.J.: Pineal-anterior pituitary gland relationships. In: MTP Int'l Review of Science, Physiology Series One, Vol. 5, Endocrine physiology. Edit. by S.M. McCann, pp. 277–308. London: Butterworths 1974a
Reiter, R.J.: Pineal regulation of hypothalamico-pituitary axis: gonadotrophins. In: Handbook of physiology-endocrinology IV, Part 2 (E. Knobil and W.H. Sawyer, eds.), pp. 519–550. Washington: American Physiological Society 1974b
Reiter, R.J.: Evidence for a seasonal rhythm in pineal gland function. In: Chronobiology (L.E. Scheving, F. Halberg and J.E. Pauly, eds.), pp. 155–159. Tokyo: Igaku Shoin Ltd, 1974c
Reiter, R.J.: Circannual reproductive rhythms in mammals related to photoperiod and pineal function. A review. Chronobiologica **1**, 365–395 (1974d)
Reiter, R.J.: Endocrine rhythms associated with pineal gland function. In: Biological rhythms and endocrine function (L.W. Hedlund, J.M. Franz and A.K. Kenny, eds.), pp. 43–78. New York: Plenum Press 1975 (1975a)
Reiter, R.J.: Changes in pituitary prolactin levels of female hamsters as a function of age, photoperiod and pinealectomy. Acta endocr. (Kbh.) **79**, 43–50 (1975b)
Reiter, R.J.: Exogenous and endogenous control of the annual reproductive cycle in the male golden hamster: participation of the pineal gland. J. exp. Zool. **191**, 111–120 (1975c)
Reiter, R.J.: Pineal and associated neuroendocrine rhythms. Psychoneuroendocrinology **1**, 255–263 (1976a)
Reiter, R.J.: Regulation of pituitary gonadotropins by the mammalian pineal gland. In: Neuroendocrine regulation of fertility (T.C. Anand Kumar, ed.). Int. Symp., Simla 1974, pp. 215–226. Basel: Karger 1976b
Reiter, R.J.: The pineal 1977. Montreal: Eden Press 1977
Reiter, R.J. (ed.): Pineal and reproduction. Progr. Reprod. Biol., Vol. 4. Basel: S. Karger 1978
Reiter, R.J., Banks, A.F.: Apparent paradoxical effects of melatonin on reproduction in rodents. Anat. Rec. **184**, 511 (1976)
Reiter, R.J., Ellison, N.M.: Delayed puberty in blinded anosmic female rats: role of the pineal gland. Biol. Reprod. **2**, 216–222 (1970)
Reiter, R.J., Fraschini, F.: Endocrine aspects of the mammalian pineal gland: a review. Neuroendocrinology **5**, 219–255 (1969)
Reiter, R.J., Hedlund, L.: Peripheral sympathetic innervation of the deep pineal of the golden hamster. Experientia (Basel) **22**, 1071–1072 (1976)
Reiter, R.J., Hester, R.J.: Interrelationships of the pineal gland, the superior cervical ganglia and the photoperiod in the regulation of the endocrine systems of hamsters. Endocrinology **79**, 1168–1170 (1965)
Reiter, R.J., Hester, R.J.: Regulation of endocrine functions of black rats by light and the pineal gland. Amer. Zoologist **6**, 313 (1966)
Reiter, R.J., Johnson, L.Y.: Elevated pituitary LH and depressed pituitary prolactin levels in female hamsters with pineal-induced gonadal atrophy and the effects of chronic treatment with synthetic LRF. Neuroendocrinology **14**, 310–320 (1974a)
Reiter, R.J., Johnson, L.Y.: Pineal regulation of immunoreactive luteinizing hormone and prolactin in light-deprived female hamsters. Fertil. and Steril. **25**, 958–964 (1974b)
Reiter, R.J., Johnson, L.Y.: Depressant action of the pineal gland on pituitary luteinizing hormone and prolactin in male hamsters. Hormone Res. **5**, 311–320 (1974c)
Reiter, R.J. Klein, D.C.: Observations on the pineal gland, the Harderian glands, the retina,

and the reproductive organs of adult female rats exposed to continuous light. J. Endocr. **51**, 117–125 (1971)

Reiter, R.J., Morgan, W.W.: Attempt to characterize the convulsive response of parathyroidectomized rats to pineal gland removal. Physiol. Behav. **9**, 203–208 (1972)

Reiter, R.J., Sorrentino, S., Jr.: Factors influenced in determining the gonad-inhibiting activity of the pineal gland. In: The pineal gland (G.E.W. Wolstenholme and I. Knight, eds.), pp. 329–344. London: Churchill Livingstone 1971a

Reiter, R.J., Sorrentino, S., Jr.: Inhibition of luteinizing hormone release and ovulation in PMS-treated rats by peripherally administered melatonin. Contraception **4**, 385–392 (1971b)

Reiter, R.J., Sorrentino, S.: Prevention of pineal-mediated reproductive responses in light-deprived hamsters by partial or total isolation of the medial basal hypothalamus. J. neuro-visc. Rel. **32**, 355–367 (1972)

Reiter, R.J., Vaughan, M.K.: Pineal antigonadotrophic substances: polypeptides and indoles. Life Sci. **21**, 159–172 (1977)

Reiter, R.J., Hoffman, R.A., Hester, R.J.: Inhibition of I^{131} uptake by thyroid glands of male rats treated with melatonin and pineal extract. Amer. Zoologist **5**, 727–728 (1965)

Reiter, R.J., Hoffman, R.A., Hester, R.J.: The effects of thiourea, photoperiod and the pineal gland on the thyroid, adrenal and reproductive organs of female hamsters. J. exp. Zool. **162**, 263–268 (1966)

Reiter, R.J., Rubin, P.H., Richert, J.R.: Pineal-induced ovarian atrophy in rats treated neonatally with testosterone. Life Sci. **7**, 299–305 (1968a)

Reiter, R.J., Hoffmann, J.C., Rubin, P.H.: Pineal gland: influence on gonads of male rats treated with androgen three days after birth. Science **160**, 420–421 (1968b)

Reiter, R.J., Sorrentino, S.D., Hoffmann, J.C., Rubin, P.H.: Pineal neural and photic control of reproductive organ size in early androgen-treated male rats. Neuroendocrinology **3**, 246–255 (1968c)

Reiter, R.J., Klein, D.C., Donofrio, R.J.: Preliminary observations on the reproductive effects of the pineal gland in blinded, anosmic male rats. J. Reprod. Fertil. **19**, 563–565 (1969)

Reiter, R.J., Sorrentino, S., Jr., Jarrow, E.L.: Central and peripheral neural pathways necessary for pineal function in the adult female rat. Neuroendocrinology **8**, 321–333 (1971)

Reiter, R.J., Morgan, W.W., Talbot, J.A.: Pineal interaction with the brain: evidence from thyroparathyroidectomized rats. Fed. Proc. **31**, 221 (1972a)

Reiter, R.J., Sorrentino, S., Jr., Hoffman, R.A.: Muscular spasms and death in thyroparathyroidectomized rats subjected to pinealectomy. Life Sci. **11**, Part I, 123–133 (1972b)

Reiter, R.J., Morgan, W.W., Talbot, J.A.: Seizures in rats induced by pinealectomy: influence of diazepam, chlordiazepoxide, diphenylhydantoin and pineal substances. Arch. int. Pharmacodyn. **202**, 219–230 (1973a)

Reiter, R.J., Blask, D.E., Talbot, J.A., Barnett, M.P.: Nature and the time course of seizures associated with surgical removal of the pineal gland from parathyroidectomized rats. Exp. Neurol. **38**, 386–397 (1973b)

Reiter, R.J., Vaughan, M.K., Blask, D.E., Johnson, L.Y.: Melatonin: its inhibition of pineal antigonadotrophic activity in male hamsters. Science **185**, 1169–1171 (1974a)

Reiter, R.J., Blum, K., Wallace, J.E., Merritt, J.H.: Pineal gland: evidence for an influence on ethanol preference in male Syrian hamsters. Comp. Biochem. Physiol. **47A**, 11–16 (1974b)

Reiter, R.J., Johnson, L.Y., Bigelow, L.: Effect of drugs which inhibit tryptophan hydroxylase or dopamine-β-hydroxylase on pinealectomy-induced seizures in parathyroidectomized rats. Gen. Pharmac. **6**, 53–57 (1975a)

Reiter, R.J., Vaughan, M.K., Blask, D.E., Johnson, L.Y.: Pineal methoxyindoles: New evidence concerning their function in the control of pineal-mediated changes in the reproductive physiology of male golden hamsters. Endocrinology **96**, 206–213 (1975b)

Reiter, R.J., Vaughan, M.K., Rudeen, P.K., Vaughan, G.M., Waring, P.J.: Melatonin-pineal relationships in female golden hamsters. Proc. Soc. exp. Biol. (N.Y.) **149**, 290–293 (1975c)

Reiter, R.J., Blask, D.E., Vaughan, M.K.: A counter antigonadotrophic effect of melatonin in male rats. Neuroendocrinology **19**, 72–80 (1975d)

Reiter, R.J., Vaughan, M.K., Vaughan, G.M., Sorrentino, S., Jr., Donofrio, R.J.: The pineal gland as an organ of internal secretion. In: Frontiers of pineal physiology (M.D. Altschule, ed.), pp. 54–174. Cambridge, Mass., and London: MIT Press 1975e

Reiter, R.J., Vaughan, M.K., Blask, D.E.: Possible role of the cerebrospinal fluid in the transport of pineal hormones in mammals. In: Brain-Endocrine Interaction II. The ventricular system in neuroendocrine mechanisms (K.M. Knigge, D.E. Scott, H. Kobayashi, eds.), 2nd Int. Symposium Ventricular System, Shizouka, 1974, pp. 337–354. Basel: Karger 1975f

Reiter, R.J., Welsh, M.G., Vaughan, M.K.: Age-related changes in the intact and sympathetically denervated gerbil pineal gland. Amer. J. Anat. **146**, 427–432 (1976a)

Reiter, R.J., Blask, D.E., Johnson, L.Y., Rudeen, P.K., Vaughan, M.K., Waring, P.J.: Melatonin inhibition of reproduction in the male hamster: its dependency on time of day of administration and on an intact and sympathetically innervated pineal gland. Neuroendocrinology **22**, 107–116 (1976b)

Reiter, R.J., Vaughan, M.K., Rudeen, P.K., Philo, R.C.: Melatonin induction of testicular recrudescence in hamsters and its subsequent inhibitory action on the antigonadotrophic influence of darkness on the pituitary-gonadal axis. Amer. J. Anat. **147**, 235–241 (1976c)

Reiter, R.J., Lukaszyk, A.J., Vaughan, M.K., Blask, D.E.: New horizons of pineal research. Amer. Zoologist **16**, 93–101 (1976d)

Reiter, R.J., Rudeen, P.K., Sackman, J.W., Vaughan, M.K., Johnson, L.Y., Little, J.C.: Subcutaneous melatonin implants inhibit reproductive atrophy in male hamsters induced by daily melatonin injections. Endocr. Res. Comm. **4**, 35–44 (1977a)

Reiter, R.J., Vaughan, M.K., Waring, P.J.: Prevention by melatonin of short day induced atrophy of the reproductive systems of male and female hamsters. Acta endocr. (Kbh.) **84**, 410–418 (1977b)

Reiter, R.J., Rudeen, P.K., Philo, R.C.: Influence of chronic melatonin availability on the reproductive quiescent period in male hamsters exposed to natural photoperiods during the winter months. Curr. Stud. Hypothal. Function **2**, 175–182 (1978)

Relkin, R.: Pineal function relation to absolute darkness and sexual maturation. Amer. J. Physiol. **213**, 999–1002 (1967)

Relkin, R.: Critical time in pineal "priming". Endocrinology **82**, 1249–1250 (1968)

Relkin, R.: The influence of prepuberal and postpuberal pinealectomy on treadwheel activity. Physiol. Behav. **5**, 341–343 (1970)

Relkin, R.: Effects of pinealectomy, constant light and darkness on growth hormone levels in the pituitary and plasma of the rat. J. Endocr. **53**, 289–293 (1972a)

Relkin, R.: Rat pituitary and plasma prolactin levels after pinealectomy. J. Endocr. **53**, 179–180 (1972b)

Relkin, R.: Effects of pinealectomy and constant light and darkness on thyrotropin levels in the pituitary and plasma of the rat. Neuroendocrinology **10**, 46–52 (1972c)

Relkin, R.: Effects of variations in environmental lighting on pituitary and plasma prolactin levels in the rat. Neuroendocrinology **9**, 278–284 (1972d)

Relkin, R.: Effects of sodium deprivation and pinealectomy on pituitary and plasma prolactin in the rat. J. Endocr. **59**, 383–384 (1973)

Relkin, R.: The pineal. Montreal: Eden Press 1976

Relkin, R.: Use of melatonin and synthetic TRH to determine site of pineal inhibition of TSH secretion. Neuroendocrinology **25**, 310–318 (1978)

Relkin, R., Adachi, M.: Effects of pinealectomy, constant light and darkness on pituitary and plasma prolactin levels in the rat. Fed. Proc. **31**, 811 (1972)

Relkin, R., Schneck, L.: Effect of pinealectomy on rat brain myelin. Proc. Soc. exp. Biol. (N.Y.) **148**, 337–338 (1975)

Relkin, R., Adachi, M., Schneck, L.: Pineal function in Tay-Sachs disease. Fed. Proc. **30**, 364 (1971)

Relkin, R., Adachi, M., Kahan, S.A.: Effects of pinealectomy and constant light and darkness on prolactin levels in the pituitary and plasma and on pituitary ultrastructure of the rat. J. Endocr. **54**, 263–268 (1972)

Relkin, R., Fok, W.Y., Schneck, L.: Pinealectomy and brain myelination. Endocrinology **92**, 1426–1428 (1973)

Remane, A., Storch, V., Welsch, U.: Systematische Zoologie. Stämme des Tierreiches. Stuttgart: Gustav Fischer 1976

Remley, N.R., Bryon, D.M., Wilson, F.R.: The effects of pinealectomy on the activity levels in rats. Psycho. Sci. **15**, 175–176 (1969)

Remy, C., Disclos, P.: Influence de l'épiphysectomie sur le développement de la thyroide et des gonades chez les têtards d'*Alytes obstetricans*. C.R. Soc. Biol. (Paris) **164**, 1989–1993 (1970)

Rennels, E.G., Dill, R.E.: Effect of pinealectomy on the histology and lipid content of the rat adrenal gland. Tex. Rep. Biol. Med. **19**, 843–850 (1961)

Renzoni, A.: Assenza di epifisi nei rapaci notturni. Atti Accad. Fisiocr. Ser., 13, **12**, 864–866 (1963)

Renzoni, A.: Ulteriori vedute sull'epifisi degli ucelli. Boll. Zool. **31**, 351–356 (1964a)

Renzoni, A.: Forma e struttura dell'epifisi in uno psittacide. Monit. Zool. Ital., Suppl. **72**, 109 (1964b)

Renzoni, A.: L'epifisi nel *Melopsittacus undulatus*. Riv. Biol. **48**, 343–364 (1965a)

Renzoni, A.: Ancora sull'epifisi degli ucelli. Boll. Zool. **32**, 743–749 (1965b)

Renzoni, A.: La fisiologia dell'epifisi negli ucelli. I. Pinealectomia in *Coturnix coturnix japonica*. Soc. ital. Biol. sper. **43**, 585–588 (1967a)

Renzoni, A.: La fisiologia dell'epifisi negli ucelli. II. L'azione di estratti epifisari sullo sviluppo sessuale di *Coturnix coturnix japonica*. Soc. ital. Biol. sper. **43**, 588–590 (1967b)

Renzoni, A.: Osservazioni comparative sull'epifisi degli Strigiformi ed Ordini affini. Arch. ital. Anat. Embriol. **73**, 321–336 (1968a)

Renzoni, A.: La fisiologia dell'epifisi negli ucelli. III. L'azione della melatonina sullo sviluppo sessuale di *Coturnix coturnix japonica*. Soc. ital. Biol. sper. **44**, 212–215 (1968b)

Renzoni, A.: Developmental morphology of the pineal complex in doves and pigeons (Aves; Columbidae). Z. Zellforsch. **104**, 19–28 (1970)

Renzoni, A., Quay, W.B.: Comparative studies of pineal structure and composition in birds. Amer. Zoologist **3**, 554 (1963)

Renzoni, A., Quay, W.B.: Daily karyometric and mitotic rhythms of pineal parenchymal cells in the rat. Amer. Zoologist **4**, 416–417 (1964)

Richards, J.G., Tranzer, J.P.: Electron microscopic localization of 5-hydroxydopamine, a 'false' adrenergic neurotransmitter, in the autonomic nerve endings of the rat pineal gland. Experientia (Basel) **25**, 53–54 (1969)

Richoux, J.P., Roux, M., Legait, H.: Étude de l'appareil génital femelle et de l'épiphyse du Lérot dans diverses conditions au cours de la période hivernale. J. Physiol. (Paris) **70**, 21B–22B (1975)

Rickards, D.A.: The therapeutic effect of melatonin on canine melanosis. J. invest. Derm. **44**, 13–16 (1965)

Ridgway Jr., P.M., Kent, G.C.: Effects of parietalectomy and blinding on the testes of the lizard *Anolis carolinensis*. Amer. Zoologist **11**, 651 (1971)

Ringertz, N., Nordenstam, H., Flyger, G.: Tumors of the pineal region. J. Neuropath. exp. Neurol. **13**, 540–561 (1954)

Rivas, L.R.: The pineal apparatus of tunas and related scrombid fishes as a possible light receptor controlling phototactic movements. Bull. Marine Sci. Gulf Caribbean **3**, 168–180 (1953)

Rizkalla, W.: The morphology of the pineal organ in the teleost, *Clarias lazera*, C.V. Acta biol. Acad. Sci. hung. **21**, 25–33 (1970)

Robson, J.M., Botros, M.: The effect of 5-hydroxytryptamine and of monoamine oxidase inhibitors on sexual maturity. J. Endocr. **22**, 165–175 (1961)

Roche, J.F., Dziuk, P.J.: A technique for pinealectomy of the ewe. Amer. J. vet. Res. **30**, 2031–2035 (1969)

Roche, J.F., Foster, D.L., Karsch, F.J., Dziuk, P.J.: Effect of castration and infusion of melatonin on levels of luteinizing hormone in sera and pituitaries of ewes. Endocrinology **87**, 1205–1210 (1970a)

Roche, J.F., Karsch, F.J., Foster, D.L., Takagi, S., Dziuk, P.J.: Effect of pinealectomy on estrus, ovulation and luteinizing hormone in ewes. Biol. Reprod. **2**, 251–254 (1970b)

Rodieck, R.W.: The vertebrate retina. Principles of structure and function. San Francisco: W.H. Freeman & Co 1973

Rodin, A.E.: The growth and spread of Walker 256 carcinoma in pinealectomized rats. Cancer Res. **23**, 1545–1548 (1963)

Rodin, A.E., Overall, J.: Statistical relationships of weight of the human pineal to age and malignancy. Cancer (Philad.) **20**, 1203–1214 (1967)

Rodin, A.E., Turner, R.A.: The relationship of intravesicular granules to the innervation of the pineal gland. Lab. Invest. **14**, 1644–1652 (1965a)

Rodin, A.E., Turner, R.A.: The location and significance of catecholamine granules in the pineal gland. Lab. Invest. **14**, 556 (1965b)

Rodin, A.E., Turner, R.A.: The perivascular space of the pineal gland. Tex. Rep. Biol. Med. **24**, 153–163 (1966)

Rodriguez De Lores Arnaiz, G., Pellegrino De Iraldi, A.: Cholinesterase in cholinergic and adrenergic nerves: A study of the superior cervical ganglia and the pineal gland of the rat. Brain Res. **42**, 230–233 (1972)

Rodriguez Perez, A.P.: Contribution al conocimento de la inervacion de las glandulas endocrinas. IV. Primeros resultados experimentales en torno a la inervación de la epífisis. Trab. Inst. Cajal Invest. biol. (Madr.) **54**, 225–236 (1962)

Rohen, J.W.: Das Auge und seine Hilfsorgane. In: Hdb. der mikr. Anat. d. Menschen (W. Bargmann, Hrsg.), Bd. III, 4. Berlin-Göttingen-Heidelberg-New York: Springer 1964

Roldán, R., Antón-Tay, F.: EEG and convulsive threshold changes produced by pineal extract administration. Brain Res. **11**, 238–245 (1968)

Rollag, M.D., Niswender, G.D.: Radioimmunoassay of serum concentrations of melatonin in sheep exposed to different lighting regimens. Endocrinology **98**, 482–489 (1976)

Rollag, M.D., Morgan, R.J., Niswender, G.D.: Route of melatonin secretion in sheep. Endocrinology **102**, 1–8 (1978a)

Rollag, M.D., O'Callaghan, P.L., Niswender, G.D.: Dynamics of photo-induced alterations in pineal blood flow. J. Endocr. **76**, 547–548 (1978b)

Rollag, M.D., O'Callaghan, P.L., Niswender, G.D.: Serum melatonin concentrations during different stages of the annual reproductive cycle in ewes. Biol. Reprod. **18**, 279–285 (1978c)

Romero, J.A., Axelrod, J.: Pineal β-adrenergic receptor: Diurnal variation in sensitivity. Science **184**, 1091–1092 (1974)

Romero, J.A., Axelrod, J.: Regulation of sensitivity to beta-adrenergic stimulation in induction of pineal N-acetyltransferase. Proc. nat. Acad. Sci. (Wash.) **72**, 1661–1665 (1975)

Romero, J.A., Zatz, M., Axelrod, J.: Beta-adrenergic stimulation of pineal N-acetyltransferase: Adenosine 3′:5′-cyclic monophosphate stimulates both RNA and protein synthesis. Proc. nat. Acad. Sci. (Wash.) **72**, 2107–2111 (1975a)

Romero, J.A., Zatz, M., Kebabian, J.W., Axelrod, J.: Circadian cycles in binding of ^{3}H-alprenolol to β-adrenergic receptor sites in rat pineal. Nature (Lond.) **258**, 435–436 (1975b)

Romieu, M., Jullien, G.: Sur l'existence d'une formation lymphoide dans l'épiphyse des gallinacés. C.R. Soc. Biol. (Paris) **136**, 626–628 (1942)

Romijn, H.J.: Structure and innervation of the pineal gland of the rabbit, *Oryctolagus cuniculus* (L.), with some functional considerations. A light and electron microscopic investigation. Thesis, Amsterdam 1972

Romijn, H.J.: Structure and innervation of the pineal gland of the rabbit, *Oryctolagus cuniculus* (L.) I. A light microscopic investigation. Z. Zellforsch. **139**, 473–485 (1973a)

Romijn, H.J.: Structure and innervation of the pineal gland of the rabbit, *Oryctolagus cuniculus (L.)* II. An electron microscopic investigation of the pinealocytes. Z. Zellforsch. **141**, 545–560 (1973b)

Romijn, H.J.: Parasympathetic innervation of the rabbit pineal gland. Brain Res. **55**, 431–436 (1973c)

Romijn, H.J.: The ultrastructure of the rabbit pineal gland after sympathectomy, parasympathectomy, continuous illumination, and continuous darkness. J. Neural Transm. **36**, 183–194 (1975a)

Romijn, H.J.: Structure and innervation of the pineal gland of the rabbit, *Oryctolagus cuniculus* (L.) III. An electron microscopic investigation of the innervation. Cell Tiss. Res. **157**, 25–51 (1975b)

Romijn, H.J.: The influence of some sympatholytic, parasympatholytic and serotonin-synthesis-inhibiting agents on the ultrastructure of the rabbit pineal organ. Cell Tiss. Res. **167**, 167–177 (1976)

Romijn, H.J.: Pineal, a tranquillizing organ? Life Sci. **23**, 2257–2274 (1978)

Romijn, H.J., Gelsema, A.J.: Electron microscopy of the rabbit pineal organ in vitro. Evidence of norepinephrine-stimulated secretory activity of the Golgi apparatus. Cell Tiss. Res. **172**, 365–377 (1976)

Romijn, H.J., Mud, M.T., Wolters, P.S.: Diurnal variations in number of Golgi-dense core vesicles in light pinealocytes of the rabbit. J. Neural Transm. **38**, 231–237 (1976)

Romijn, H.J., Mud, M.T., Wolters, P.S.: Electron microscopic evidence of glycogen storage in the dark pinealocytes of the rabbit pineal gland. J. Neural Transm. **40**, 69–79 (1977a)

Romijn, H.J., Mud, M.T., Wolters, P.S.: A pharmacological and autoradiographic study on the ultrastructural localization of indoleamine synthesis in the rabbit pineal gland. Cell. Tiss. Res. **185**, 199–214 (1977b)

Rønnekleiv, O.K., McCann, S.M.: Effects of pinealectomy, anosmia and blinding on serum and pituitary prolactin in intact and castrated male rats. Neuroendocrinology **17**, 340–353 (1975a)

Rønnekleiv, O.K., McCann, S.M.: Effects of pinealectomy, anosmia and blinding alone or in combination on gonadotropin secretion and pituitary and target gland weight in intact and castrated male rats. Neuroendocrinology **19**, 97–114 (1975b)

Rønnekleiv, O.K, McCann, S.M.: Growth hormone release in conscious pinealectomized and sham-operated rats. Endocrinology **102**, 1694–1701 (1978)

Rønnekleiv, O.K., Krulich, L., McCann, S.M.: An early morning surge of prolactin in the male rat and its abolition by pinealectomy. Endocrinology **92**, 1339–1342 (1973)

Rønnekleiv, O.K., Kelly, M.J., Møller, M., Wuttke, W.: Electrophysiological and morphological evidence of direct central innervation of the pineal gland. Pflügers Arch. Supp. **373**, Abstr. 187 (1978a)

Rønnekleiv, O., Krulich, L., McCann, S.M.: The effect of pinealectomy on the pattern of prolactin secretion in conscious freely moving male rats. Neuroendocrinology **27**, 279–290 (1978b)

Rosenbloom, A.A., Fisher, D.A.: Radioimmunoassayable AVT and AVP in adult mammalian brain tissue: Comparison of normal and Brattleboro rats. Neuroendocrinology **17**, 345–361 (1975)

Rosenblum, I.Y., Benson, B., Bria, C.F., McDonnell, D., Hruby, V.J.: Localization and chemical characterization of a partially purified bovine pineal antigonadotropin. J. Neural Transm. **44**, 197–220 (1979)

Rosenbluth, J.: Subsurface cisterns and their relationship to the neuronal plasma membrane. J. Cell Biol. **13**, 405–421 (1962)

Rosengarten, J., Meller, E., Friedhoff, A.J.: *In vitro* enzymatic formation of melatonin by human erythrocytes. Res. Commun. Chem. Pathol. Pharmacol. **4**, 457–465 (1972)

Rosner, J.M., Declercq de Pérez Bedés, G., Cardinali, D.P.: Direct effect of light on duck pineal explants. Life Sci. **10**, Part II, 1065–1069 (1971)

Rosner, J.M., Denari, J.H., Nagle, C.A., Cardinali, D.P., Declercq de Pérez Bedés, G., Orsi, L.: Direct action of light on serotonin metabolism and RNA biosynthesis in duck pineal explants. Life Sci. **11**, Part II, 829–836 (1972)

Roth, C., Braun, R.: Zur Funktion des Parietalauges der Blindschleiche *Anguis fragilis* (Reptilia; Lacertilia, Anguidae). Naturwissenschaften **45**, 218–219 (1958)

Roth, J.J., Ralph, C.L.: Body temperature of the lizard (*Anolis carolinensis*): Effect of parietalectomy. J. exp. Zool. **198**, 17–28 (1976)

Roth, J.R., Ralph, C.L.: Thermal and photic preferences in intact and parietalectomized *Anolis carolinensis*. Behav. Biol. **19**, 341–348 (1977)

Roth, W.D.: Comments on J. Ariëns Kappers' review and observations on pineal activity. Amer. Zoologist **4**, 53–57 (1964)

Roth, W.D.: Metabolic and morphologic studies on the rat pineal organ during puberty. Progr. Brain Res. **10**, 552–563 (1965)

Roth, W.D., Wurtman, R.J., Altschule, M.D.: Morphologic changes in the pineal parenchyma cells of rats exposed to continuous light or darkness. Endocrinology **71**, 888–892 (1962)

Roux, M., Richoux, J.-P., Dussart, G.: Etude ultrastructurale de l'épiphyse du lérot (*Eliomys quercinus* L.). Bull. Ass. Anat. (Nancy) **58**, 1031–1042 (1974)

Roux, M., Richoux, J.P., Cordonnier, J.L.: Influence de la photopériode sur l'ultrastructure de l'épiphyse avant et pendant la phase génitale saisonnière chez la femelle du lérot (*Eliomys quercinus*). J. Neural Transm. **41**, 209–223 (1977)

Rowe, J.W., Richert, J.R., Klein, D.C., Reichlin, S.: Relation of the pineal gland and environmental lighting to thyroid function in the rat. Neuroendocrinology **6**, 247–254 (1970)

Rowe, V., Neale, E.A., Avins, L., Guroff, G., Schrier, B.K.: Pineal gland cells in culture. Morphology, biochemistry, differentiation, and co-culture with sympathetic neurons. Exp. Cell Res. **104**, 345–356 (1977)

Rubin, B.D., Traum, R.E.: The effect of melatonin on ovarian compensatory hypertrophy in the rat. J. Endocr. **50**, 179–180 (1971)

Rubin, P., Kramer, S.: Ectopic pinealoma: A radiocurable neuroendocrinologic entity. Radiology **85**, 512–523 (1965)

Rubinstein, L.J., Okazaki, H.: Gangliogliomatous differentiation in a pineocytoma. J. Path. Bact. **102**, 27–32 (1970)

Rubio, M.C., Jaim-Etcheverry, G., Zieher, L.M.: Tyrosine hydroxylase activity increases in pineal sympathetic nerves after depletion of neuronal serotonin. Naunyn-Schmiedeberg's Arch. Pharmacol. **301**, 75–78 (1977)

Rudeen, P.K., O'Steen, W.K.: The effects of the pineal gland on light-induced retinal photoreceptor damage. Exp. Eye Res. **28**, 37–44 (1979)

Rudeen, P.K., Reiter, R.J.: Effect of shortened photoperiods on pineal serotonin N-acetyltransferase activity and rhythmicity. J. Interdisc. Cycle Res. **8**, 47–54 (1977)

Rudeen, P.K., Reiter, R.J., Vaughan, M.K.: Pineal serotonin-N-acetyltransferase activity in four mammalian species. Neurosci. Letters **1**, 225–229 (1975)

Rudman, D.: Injection of melatonin into cisterna magna increases concentration of 3′, 5′ cyclic guanosine monophosphate in cerebrospinal fluid. Neuroendocrinology **20**, 235–242 (1976)

Rudman, D., Scott, J.W.: Melanotropic-lipolytic peptides of the pineal gland and other CNS regions. In: Frontiers of pineal physiology (M.D. Altschule, ed.), pp. 44–53. Cambridge, Mass.-London: MIT Press 1975

Rudman, D., Del Rio, A.E., Garcia, L.A., Barnett, J., Bixler, T., Hollins, B.: Lipolytic substances in bovine thyroid, parotid and pineal glands. Endocrinology **87**, 27–37 (1970)

Rudman, D., Del Rio, A.E., Hollins, B., Houser, D.H.: Observations on the lipolytic and melanotropic activities of the pineal gland. J. biol. Chem. **246**, 324 (1971)

Rudman, D., Del Rio, A.E., Hollins, B., Houser, D.H., Sutin, J., Mosteller, R.C.: Comparison of lipolytic and melanotropic factors in bovine choroid plexus and in bovine pineal gland. Endocrinology **90**, 1139–1146 (1972)

Rudman, E., Del Rio, A.E., Hollins, B., Houser, D.H., Keeling, M.E., Sutin, J., Scott, J.W., Sears, R.A., Rosenberg, M.Z.: Melanotropic-lipolytic peptides in various regions of bovine, simian and human brains and in simian and human cerebrospinal fluids. Endocrinology **92**, 372–379 (1973)

Rüdeberg, C.: Electron microscopical observations on the pineal organ of the teleosts *Mugil auratus* (Risso) and *Uranoscopus scaber* (Linné). Pubbl. Staz. Zool. Napoli **35**, 47–60 (1966)

Rüdeberg, C.: Receptor cells in the pineal organ of the dogfish *Scyliorhinus canicula* Linné. Z. Zellforsch. **85**, 521–526 (1968a)

Rüdeberg, C.: Structure of the pineal organ of the sardine, *Sardina pilchardus sardina* (Risso), and some further remarks on the pineal organ of *Mugil* spp. Z. Zellforsch. **84**, 219–237 (1968b)

Rüdeberg, C.: Light and electron microscopic studies on the pineal organ of the dogfish, *Scyliorhinus canicula* L. Z. Zellforsch. **96**, 548–581 (1969a)

Rüdeberg, C.: Structure of the parapineal organ of the adult rainbow trout, *Salmo gairdneri* Richardson. Z. Zellforsch. **93**, 282–304 (1969b)

Rüdeberg, C.: Light and electron microscopic investigations on the pineal and parapineal organs of fishes. Thesis, Lund (Sweden) 1970

Rüdeberg, C.: Structure of the pineal organs of *Anguilla anguilla* L. and *Lebistes reticulatus* Peters (Teleostei). Z. Zellforsch. **122**, 227–243 (1971)

Rust, C.C., Meyer, R.K.: Hair color, molt, and testis size in male, short-tailed weasels treated with melatonin. Science **165**, 921–923 (1969)

Rutledge, J.T., Angle, M.J.: Persistence of circadian activity rhythms in pinealectomized European starlings (*Sturnus vulgaris*) J. exp. Zool. **202**, 333–337 (1977)

Ruzsas, C., Trentini, G.P., Mess, B.: The role of the pineal gland in the regulation of LH-release in rats with different types of the anovulatory syndrome. Endocrinology **70**, 142–149 (1977)

Rzeszowska-Tokarska, G.: Obserwacje histochemiczne nad aktywnością enzymów hydrolitycznych w szyszynce szczurów w różnych okresach ciąży. Observations histochimiques de l'activité des enzymes hydrolytiques dans la glande pinéale chez les rats dans les diverses periodes de grossesse. Ann. Univ. Mariae Curie-Sklodowska Lublin, Sect. D **24**, 395–421 (1969)

Saavedra, J.M.: Enzymatic-isotopic method for octopamine at the picogram level. Analyt. Biochem. **59**, 628–633 (1974)

Sackman, J.W.: The effect of luteinizing hormone on chicken testis and pineal gland uptake of ^{32}P. Experientia (Basel) **31**, 1362–1363 (1975)

Sackman, J.W.: Circadian rhythm of pineal uptake of ^{32}P in domestic fowl. J. Endocr. **83**, 99–103 (1977a)

Sackman, J.W.: The effect of melatonin on LH-RH-induced changes in pituitary and gonadal function of immature domestic fowl. Biol. Reprod. **17**, 718–722 (1977b)

Sackman, J.W., Little, J.C., Rudeen, P.K., Waring, P.J., Reiter, R.J.: The effects of pineal indoles given late in the light period on reproductive organs and pituitary prolactin levels in male golden hamsters. Hormone Res. **8**, 84–92 (1977)

Sakai, K.K., Marks, B.H.: Adrenergic effects on pineal cell membrane potential. Life Sci. **11**, Part I, 285–291 (1972)

Salas, M., Schapiro, S.: Effects of light upon pineal electrical rhythms. Fed. Proc. **29**, 325 (1970)

Saliichuk, L.I.: On the formation of epiphyseal sand at different age periods in the development of man. Probl. Endokrinol. (Mosk.) **10**, 44 (1964) cited after Quay (1974a)

Sampson, P.H.: Behavior and pineal functioning. In: Frontiers of pineal physiology (M.D. Altschule, ed.), pp. 204–222. Cambridge, Mass.-London: MIT Press 1975

Sampson, P.H., Bigelow, L.B.: Pineal influence on exploratory behavior of the female rat. Physiol. Behav. **7**, 713–715 (1971)

Sander, G., Schmid, S.: Über die Wirkung von Epiphysenimplantationen und Epiphysenextrakten bei menschlichen malignen Tumoren. Wien. klin. Wschr. **64**, 505–508 (1952)

Sankaran, K., Hanbauer, I., Lovenberg, W.: Heat-stable low molecular weight form of phosphodiesterases from bovine pineal gland. Proc. nat. Acad. Sci. (Wash.) **75**, 3188–3191 (1978)

Sano, Y., Mashimo, T.: Elektronenmikroskopische Untersuchungen an der Epiphysis cerebri beim Hund. Z. Zellforsch. **69**, 129–139 (1966)

Santamarina, E.: Melanin pigmentation in bovine pineal glands and its possible correlation with gonadal function. Canad. J. Biochem. **36**, 227–235 (1958)

Santamarina, E., Meyer-Arendt, J.: Identification of melanin in the bovine pineal gland. Acta Histochem. (Jena) **3**, 1–5 (1956)

Santamarina, E., Venzke, W.G.: Physiological changes in the mammalian pineal gland correlated with the reproductive system. Amer. J. vet. Res. **14**, 555–562 (1953)

Sartin, J.L., Bruot, B.C., Orts, R.J.: Interaction of arginine vasotocin and norepinephrine upon pineal indoleamine synthesis in vitro. Mol. Cell. Endocr. **11**, 7–18 (1978)

Satake, N., Morton, B.E.: Scotophobin A causes several responses in goldfish if the pineal gland is present. Pharmacol. Biochem. Behav. **10**, 183–188 (1979)

Satodate, R., Sasaki, K., Ota, M.: The pineal gland of intact, hypophysectomized, or ovariectomized rats. Light and electron microscopic studies. Arch. Neurol. (Chic.) **23**, 278–286 (1970a)

Satodate, R., Hsieh, K.S., Ota, M.: Morphological changes in the pineal gland of the albino rat by hypophysectomy and ovariectomy. Experientia (Basel) **26**, 638–640 (1970b)

Satodate, R., Katsura, S., Ota, M.: Structural changes of the pineals of blinded rats fed long term. Experientia (Basel) **29**, 1414–1415 (1973)

Saxena, P.K., Anand, K.: A comparison of ovarian recrudescence in the catfish, *Mystus tengara* (Ham.), exposed to short photoperiods, to long photoperiods, and to melatonin. Gen. comp. Endoc. **33**, 506–511 (1977)

Sayler, A., Wolfson, A.: Avian pineal gland: progonadotropic response in the Japanese quail. Science **158**, 1478–1479 (1967)

Sayler, A., Wolfson, A.: Influence of the pineal gland on gonadal maturation in the Japanese quail. Endocrinol. **83**, 1237–1246 (1968a)

Sayler, A., Wolfson, A.: Role of the eyes and superior cervical ganglia on the effects of light on the pineal and gonads of the Japanese quail. Arch. Anat. Histol. Embryol. **51**, 615–626 (1968b)

Sayler, A., Wolfson, A.: Hydroxyindole-O-methyl transferase (HIOMT) activity in the Japanese quail in relation to sexual maturation and light. Neuroendocrinology **5**, 322–332 (1969)

Scardapane, L., Cardinali, D.P.: Effect of estradiol and testosterone on catechol-O-methyl transferase activity of rat superior cervical ganglion, pineal gland, anterior hypophysis and hypothalamus. J. Neural Transm. **40**, 81–86 (1977)

Šćepović, M.: Contribution à l'étude de l'histo-physiologie de la glande thyroïde de rats épiphysectomisés. Acta anat. (Basel) **42**, 273 (1960)

Šćepović, M.: Effet tardif de l'épiphysectomie sur l'histo-physiologie de la glande thyroïde. Acta anat. (Basel) **46**, 178–179 (1961)

Šćepović, M.: Contribution à l'étude de l'effet de l'épiphysectomie sur le cortex de la glande surrénale. Acta anat. (Basel) **54**, 364–365 (1963a)
Šćepović, M.: Contribution à l'étude histophysiologique de la glande thyroïde chez les rats épiphysectomisés. Ann. Endocr. (Paris) **24**, 371–376 (1963b)
Šćepović, M., Schippers, M.: Effets de l'épiphysectomie sur quelques activités enzymatiques de la glande thyroïde du Rat. Bull. Ass. Anat. (Nancy), No. **128**, 1583–1585 (1965)
Schapiro, S., Salas, M.: Effects of age, light and sympathetic innervation on electrical activity of the rat pineal gland. Brain Res. **28**, 47–55 (1971)
Scharenberg, K., Liss, L.: The histologic structure of the human pineal body. Progr. Brain Res. **10**, 193–217 (1965)
Scherer, E., Dodt, E.: Wellenlängenspezifische Erregungen im Parietalorgan der Wieseneidechse, *Lacerta sicula campestris* (De Betta). Pflügers Arch. ges. Physiol. **294**, R 52 (1967)
Scherzer, M.: Tehnică de epifizectomie la iepure. St. Cercet. Endocr. **15**, 313–318 (1964)
Schindler, E., Kretzschmar, K., Aulich, A., Wende, S., Kutzner, J.: Serial CT studies of a metastatic pinealoma with reference to the radiotherapeutic problems. Neuroradiology **14**, 127–132 (1977)
Schneider, H.P.G., Staemmler, H.J., Sachs, S.L., Glöckner, C.: Über die Wirkung von gonadotropininhibierendem Material aus Epiphysen- und Urinextrakten. Arch. Gynäk. **206**, 72–81 (1968)
Schoenfeld, R.I.: Melatonin: Effect on punished and nonpunished operant behavior of the pigeon. Science **171**, 1258–1260 (1971)
Schönherr, J.: Über die Abhängigkeit der Instinkthandlungen vom Vorderhirn und Zwischenhirn (Epiphyse) bei *Gasterosteus aculeatus* L. Zool. Jb. Abt. allg. Zool. Physiol. Tiere **65**, 357–386 (1955)
Schon, F., Beart, P.M., Chapman, D., Kelly, J.S.: On GABA metabolism in the gliocyte cells of the rat pineal gland. Brain Res. **85**, 479–490 (1975)
Schott, H.F., Masuoka, D.T., Vivonia, C.: In utero sensitivity of rat pineal to nerve growth factor antiserum. Life Sci. **9**, 713–720 (1970)
Schrier, B.K., Klein, D.C.: Absence of choline acetyltransferase in rat and rabbit pineal gland. Brain Res. **79**, 347–351 (1974)
Segal, S., Sadovsky, E., Palti, Z., Pfeifer, Y., Polishuk, W.Z.: Serotonin and 5-hydroxyindoleacetic acid in fertile and subfertile men. Fertil. Steril. **26**, 314–316 (1975)
Seibel, H.R., Schweisthal, M.R.: Relationship between the pineal gland, other endocrine glands and reproductive organs of single and parabiosed golden hamsters. Acta endocr. (Kbh.) **74**, 434–448 (1973)
Sellei, K., Frauchiger, E.: Melatonin und Serotonin in der Glandula pinealis bei einigen Tierarten. Schweiz. Arch. Tierheilk. **110**, 395–398 (1968)
Semm, P.: Electrophysiological and morphological aspects of the guinea-pig epiphysis cerebri. J. Neural Transm., Suppl. **13**, 394–395 (1978)
Semm, P., Vollrath, L.: Electrophysiological properties of single cells of the guinea-pig epiphysis cerebri. Pflügers Arch., Suppl. **373**, R 55 (1978)
Semm, P., Vollrath, L.: Electrophysiology of the guinea-pig pineal organ: Sympathetically influenced cells responding differently to light and darkness. Neurosci. Lett. **12**, 93–96 (1979)
Shani (Mishkinsky), J., Knaggs, G.S., Tindal, J.S.: The effect of noradrenaline, dopamine, 5-hydroxytryptamine and melatonin on milk yield and composition in the rabbit. J. Endocr. **50**, 543–544 (1971)
Shaw, K.M.: The pineal gland: a review of the biochemistry, physiology and pharmacological potential of melatonin and other pineal substances. Adv. Drug. Res. **11**, 75–96 (1977)
Shein, H.M.: Pineal gland organ culture techniques. Meth. Enzymol. **39**, 398–403 (1975)
Shein, H.M., Wurtman, R.J.: Cyclic adenosine monophosphate: Stimulation of melatonin and serotonin synthesis in cultured rat pineals. Science **166**, 519–520 (1969)
Shein, H.M., Wurtman, R.J.: Stimulation of (^{14}C) tryptophan 5-hydroxylation by norepinephrine and dibutyryl adenosine 3′, 5′ monophosphate in rat pineal organ cultures. Life Sci. **10**, Part I, 935–940 (1971)
Shein, H.M., Wurtman, R.J., Axelrod, J.: Serotonin synthesis in pineal gland in organ culture. Nature (Lond.) **213**, 730–731 (1967)
Shein, H.M., Larin, F., Wurtman, R.J.: Lack of direct effect of morphine on the synthesis of pineal ^{14}C-indoles in organ culture. Life Sci. **9**, 29–33 (1970)
Shein, H.M., Wilson, S., Larin, F., Wurtman, R.J.: Stimulation of ^{14}C-serotonin synthesis from ^{14}C-tryptophan by mescaline in rat pineal organ cultures. Life Sci. **10**, Part II, 273–282 (1971)

Shellabarger, C.J.: Pinealectomy vs. pineal injection in the young cockerel. Endocrinology **51**, 152–154 (1952)
Shellabarger, C.J.: Observations of the pineal in the White Leghorn capon and cockerel. Poultry Sci. **32**, 189–197 (1953)
Shellabarger, C.J., Breneman, W.R.: The effects of pinealectomy on young white Leghorn cockerels. Proc. Indiana Acad. Sci. **59**, 299–302 (1950)
Sheridan, M.N.: Fine structure of the hamster pineal gland. Anat. Rec. **157**, 320 (1967)
Sheridan, M.N.: Further observations of the fine structure of the hamster pineal gland. Anat. Rec. **163**, 262 (1969)
Sheridan, M.N.: Pineal gland fine structure. Dense-cored vesicles. In: Brain-Endocrine Interaction II. The ventricular system in neuroendocrine mechanisms. Ed. by K.M. Knigge, D.E. Scott, H. Kobayashi, pp. 324–336. Basel: S. Karger 1975
Sheridan, M.N.: Pineal function in golden hamsters bearing melatonin antibody. Anat. Rec. **184**, 528 (1976)
Sheridan, M.N., Keppel, J.F.: The effect of p-chlorophenylalanine (PCPA) and 6-hydroxydopamine (6-HD) on ultrastructural features of hamster pineal parenchyma. Anat. Rec. **169**, 427 (1971)
Sheridan, M.N., Reiter, R.J.: The fine structure of the hamster pineal gland. Amer. J. Anat. **122**, 357–376 (1968)
Sheridan, M.N., Reiter, R.J.: Observations on the pineal system in the hamster. I. Relations of the superficial and deep pineal to the epithalamus. J. Morph. **131**, 153–161 (1970a)
Sheridan, M.N., Reiter, R.J.: Observations on the pineal system in the hamster. II. Fine structure of the deep pineal. J. Morph. **131**, 163–177 (1970b)
Sheridan, M.N., Reiter, R.J.: The fine structure of the pineal gland in the pocket gopher, *Geomys bursarius*. Amer. J. Anat. **136**, 363–382 (1973)
Sheridan, M.N., Reiter, R.J., Jacobs, J.J.: An interesting anatomical relationship between the hamster pineal gland and the ventricular system of the brain. J. Endocr. **45**, 131–132 (1969)
Sheridan, M.N., Sladek, J.R., Jr.: Histofluorescence and ultrastructural analysis of hamster and monkey pineal. Cell Tiss. Res. **164**, 145–152 (1975)
Shibuya, H., Toru, M., Watanabe, S.: A circadian rhythm of trytophan hydroxylase in rat pineals. Brain Res. **138**, 364–368 (1978)
Shiino, M., Arimura, A., Rennels, E.G.: Effects of blinding, olfactory bulbectomy, and pinealectomy on prolactin and growth hormone cells of the rat, with special reference to ultrastructure. Amer. J. Anat. **139**, 191–207 (1974)
Shimizu, N., Morikawa, N.: Histochemical studies of succinic dehydrogenase of the brain of mice, rats, guinea-pigs and rabbits. J. Histochem. Cytochem. **5**, 334–345 (1957)
Shimizu, N., Okada, M.: Histochemical distribution of phosphorylase in rodent brain from newborn to adults. J. Histochem. Cytochem. **5**, 459–471 (1957)
Shimizu, N., Morikawa, N., Okada, M.: Histochemical studies of monoamine oxidase of the brain of rodents. Z. Zellforsch. **49**, 389–400 (1959)
Shivers, B.D.: Failure of urethane anesthetic to block induction of pineal serotonin N-acetyltransferase activity in the rat. Brain Res. Bull. **3**, 579–581 (1978)
Simionescu, N., Goldstein, M.: Autogrefa glandei epifizie in tesuturile somatice (subcutan, muscular, fascial, retroperitoneal) la şobolanul alb. Stud. Cercet. Endocr. **13**, 623–628 (1962)
Simionescu, N., Scherzer, M.: Un nou model experimental: Simbioza pinealo-glomerulară la şobolanul alb. Stud. Cercet. Endocr. **15**, 117–122 (1964a)
Simionescu, N., Scherzer, M.: Demonostrarea in situ a controlului hormonal al zonei glomerulare cu ajutorul "simbiozei pinealo-glomerulare". Stud. Cercet. Endocr. **15**, 227–231 (1964b)
Simionescu, N., Goldstein, M., Teodorescu, C.: Autogrefa epifizei in cartilajul de crestere la sobolanul alb impuber. Stud. Cercet. Endocr. **13**, 629–638 (1962)
Simonnet, H., Sternberg, J.: Le rôle endocrinien de l'épiphyse; I. L'épiphyse et le métabolisme du phosphore. Rev. canad. Biol. **9**, 407–421 (1951)
Simmonet, H., Thiéblot, L.: Action de l'épiphyse sur différentes glandes endocrines. J. Physiol. Path. gén. **42**, 726–729 (1950)
Simonnet, H., Thiéblot, L.: Recherches expérimentales sur la physiologie de la glande pinéale. Acta endocr. (Kbh.) **7**, 306–320 (1951)
Simonnet, H., Thiéblot, L.: Colloque sur la glande pinéale. Clermont-Ferrand, 29–30 May, 1962. Ann. Endocr. (Paris) **24**, 197–390 (1963)

Simonnet, H., Thiéblot, L., Melik, T.: Influence de l'épiphyse sur l'ovaire de la jeune rate. Ann. Endocr. (Paris) **12**, 202–205 (1951a)

Simonnet, H., Thiéblot, L., Ségal, V.: Action de l'épiphyse sur la teneur des surrénales en cholestérol et en acide ascorbique chez le rat blanc mâle ou femelle. Ann. Endocr. (Paris) **12**, 198–202 (1951b)

Simonnet, H., Thiéblot, L., Ségal, V.: Interrelation épiphyse-hypophysaire et effect expanso-mélanophorique. Ann. Endocr. (Paris) **13**, 340–344 (1952a)

Simonnet, H., Thiéblot, L., Melik, T., Ségal, V.: Régulation hormonale de l'activité génitale femelle; influence de l'épiphyse sur la lutéinisation. Rev. Path. comp. **52**, 144–148 (1952b)

Simonnet, H., Thiéblot, L., Melik, T., Ségal, V.: Nouvelles preuves de l'endocrinie épiphysaire. Acta endocr. (Kbh.) **17**, 402–413 (1954)

Singh, D.V., Turner, C.W.: Effect of melatonin upon thyroid hormone secretion rate and endocrine glands of chicks. Proc. Soc. exp. Biol. (N.Y.) **125**, 407–411 (1967)

Singh, D.V., Turner, C.W.: Effect of light and darkness upon thyroid secretion rate and on the endocrine glands of female rats. Proc. Soc. exp. Biol. (N.Y.) **131**, 1296–1299 (1969)

Singh, D.V., Turner, C.W.: Melatonin on endocrine DNA changes in female rats at increasing ages. Acta endocr. (Kbh.) **68**, 597–604 (1971)

Singh, D.V., Turner, C.W.: Effect of melatonin upon thyroid secretion rate in female hamsters and male rats. Acta endocr. (Kbh.) **69**, 35–40 (1972)

Singh, D.V., Narang, G.D., Turner, C.W.: Effect of melatonin and its withdrawal on thyroid hormone secretion rate of female rats. J. Endocr. **43**, 489–490 (1969)

Sitaram, B.R., Lees, G.J.: Diurnal rhythm and turnover of tryptophan hydroxylase in the pineal gland of the rat. J. Neurochem. **31**, 1021–1026 (1978)

Sjöstrand, F.S.: Ultrastructure of retinal rod synapses of the guinea-pig eye as revealed by three dimensional reconstructions from serial sections. J. Ultrastruct. Res. **2**, 122–170 (1958)

Skaug, O.E., Lingjaerde, P., Malm, O.J.: P^{32} uptake in hypophysectomized rats. Acta endocr. (Kbh.) **29**, 315 (1958)

Skaug, O.E.,, Lingjaerde, P., Malm, O.J.: P^{32} uptake in thymectomized hypophysectomized rats. Acta endocr. (Kbh.) **31**, 309 (1959)

Slama-Scemama, A.: Effect of pinealectomy on gonadotrophins in immature female rats. J. Neural Transm. **39**, 251–256 (1976)

Slama-Scemama, A., L'Héritier, A.: Seasonal diurnal variations in LH release in immature female rats and the effect of pinealectomy. J. Neural Transm. **43**, 161–166 (1978)

Slonimski, P.: L'épiphyse pendant le développement post-embryonnaire de l'Axolotl. C.R. Soc. Biol. (Paris **141**, 1107–1108 (1947)

Smith, A.R.: The topographical relations of the rabbit pineal gland to the large intracranial veins. Brain Res. **30**, 339–348 (1971)

Smith, A.R.: Over de invloed van de glandula pinealis op de hypothalamus, een histochemisch onderzoek. Ned. T. Geneesk. **118**, 1418 (1974)

Smith, A.R., Kappers, J.A.: Effect of rat pineal gland on the hypothalamo-hypophyseo-gonadal axis. In: Neurosecretion – the final neuroendocrine pathway (F. Knowles and L. Vollrath, eds.), p. 323. Berlin-Heidelberg-New York: Springer 1974

Smith, A.R., Kappers, J.A.: Effect of pinealectomy, gonadectomy, pCPA and pineal extracts on the rat parvocellular neurosecretory hypothalamic system; a fluorescence histochemical investigation. Brain Res. **86**, 353–371 (1975)

Smith, A.R., Jongkind, J.F., Kappers, J.A.: Distribution and quantification of serotonin-containing and autofluorescent cells in the rabbit pineal organ. Gen. comp. Endocr. **18**, 364–371 (1972a)

Smith, A.R., Kappers, J.A., Jongkind, J.F.: Alterations in the distribution of yellow fluorescing rabbit pinealocytes produced by p-chlorophenylalanine and different conditions of illumination. J. Neural Transm. **33**, 91–111 (1972b)

Smith, A.R., Pévet, P., Van de Kar, L., Oosterom, R.V.: Effect of gonadotropic hormones on the rat pineal gland: a fluorescence histochemical and biochemical study. J. Neural Transm. **36**, 217–226 (1975)

Smith, B.: Monoamine oxidase in the pineal gland, neurohypophysis and brain of the albino rat. J. Anat. (Lond.) **97**, 81–86 (1963)

Smith, C.A., Sjöstrand, F.S.: A synaptic structure in the hair cells of the guinea-pig cochlea. J. Ultrastruct. Res. **5**, 184–192 (1961)

Smith, I., Mullen, P.E., Silman, R.E., Snedden, W., Wilson, B.W.: Absolute identification of melatonin in human plasma and cerebrospinal fluid. Nature (Lond.) **260**, 718–719 (1976)

Smith, J.A., Padwick, D., Mee, T.J.X., Minneman, K.P., Bird, E.D.: Synchronous nyctohemeral rhythms in human blood melatonin and in human post-mortem pineal enzyme. Clin. Endocr. **6**, 219–225 (1977)

Smith, J.R., Weber, L.J.: Diurnal fluctuations in acetylserotonin methyltransferase (ASMT) activity in the pineal gland of the steelhead trout (*Salmo gairdneri*). Proc. Soc. exp. Biol. (N.Y.) **147**, 441–443 (1974)

Smith, M.L.: Effects of a purified pineal extract in unilaterally ovariectomized mice. Anat. Rec. **169**, 432–433 (1971)

Smith, M.L., Jr., Orts, R.J., Benson, B.: Effects of non-melatonin pineal factors in the PMS-stimulated immature rat. Anat. Rec. **172**, 408 (1972)

Smith, M.L., Orts, R.J., Benson, B.: Effects of pinealectomy and a pineal extract upon the ovulatory cycle in rats. Tex. Rep. Biol. Med. **31**, 600–601 (1973)

Smit-Vis, J.H.: The effect of pinealectomy and of testosterone administration on the occurrence of hibernation in adult male golden hamsters. Acta morphol. neerl.-scand. **10**, 269–282 (1972)

Smythe, G.A.: The role of serotonin and dopamine in hypothalamic-pituitary function. Clin. Endocr. **7**, 325–341 (1977)

Smythe, G.A., Lazarus, L.: Growth hormone regulation by melatonin and serotonin. Nature (Lond.) **244**, 230–231 (1973a)

Smythe, G.A., Lazarus, L.: Blockade of serotonin induced GH secretion in rats by melatonin and cyproheptadine. Horm. Metab. Res. **5**, 227 (1973b)

Smythe, G.A., Lazarus, L.: Suppression of human growth hormone secretion by melatonin and cyproheptadine. J. clin. Invest. **54**, 116–121 (1974a)

Smythe, G.A., Lazarus, L.: Growth hormone responses to melatonin in man. Science **184**, 1373–1374 (1974b)

Smythe, G.A., Stuart, M.C., Lazarus, L.: Stimulation and suppression of somatomedin activity by serotonin and melatonin. Experientia (Basel) **30**, 1356–1357 (1974)

Snell, R.S.: Effect of melatonin on mammalian epidermal melanocytes. J. invest. Derm. **44**, 273–275 (1965)

Snyder, S.H.: Development of enzyme activities and a circadian rhythm in pineal gland serotonin: evidence for a nonretinal pathway of light to the pineal gland of newborn rats. Advanc. Pharmacol. **6A**, 301–305 (1968)

Snyder, S.H., Axelrod, J.: Influence of light and the sympathetic nervous system on 5-hydroxytryptophan decarboxylase (5-HTPD) activity in the pineal gland. Fed. Proc. **23**, 206 (1964)

Snyder, S.H., Axelrod, J.: Circadian rhythm in pineal serotonin: Effect of monoamine oxidase inhibition and reserpine. Science **149**, 542–544 (1965)

Snyder, S.H., Zweig, M.: Evidence for a non-retinal pathway of light to the rat pineal gland. Fed. Proc. **25**, 353 (1966)

Snyder, S.H., Zweig, M., Axelrod, J.: Control of the circadian rhythm in serotonin content of the rat pineal gland. Life Sci. **3**, 1175–1179 (1964a)

Snyder, S.H., Axelrod, J., Fischer, J.E., Wurtman, R.J.: Neural and photic regulation of 5-hydroxytryptophan decarboxylase in the rat pineal gland. Nature (Lond.) **203**, 981–982 (1964b)

Snyder, S.H., Axelrod, J., Wurtman, R.J., Fischer, J.E.: Control of 5-hydroxytryptophan decarboxylase activity in the rat pineal gland by sympathetic nerves. J. Pharmacol. exp. Ther. **147**, 371–375 (1965a)

Snyder, S.H., Zweig, M., Axelrod, J., Fischer, J.E.: Control of the circadian rhythm in serotonin content of the rat pineal gland. Proc. nat. Acad. Sci. (Wash.) **53**, 301–305 (1965b)

Snyder, S.H., Axelrod, J., Zweig, M.: Circadian rhythm in the serotonin content of the rat pineal gland: regulating factors. J. Pharmacol. exp. Ther. **158**, 206–213 (1967a)

Snyder, S.H., Axelrod, J., Smith, O.D., Pucci, G.L.: Formation of methanol by an enzyme in an ectopic pinealoma. Nature (Lond.) **215**, 773–774 (1967b)

Soffer, L.J., Fogel, M., Rudavsky, A.Z.: The presence of a "gonadotrophin inhibiting substance" in pineal gland extracts. Acta endocr. (Kbh.) **48**, 561–564 (1965)

Solarski, A., Panke, E.S., Panke, T.W.: Craniopharyngioma in pineal gland. Arch. Pathol. Lab. Med. **102**, 490–491 (1978)

Sorrentino, S., Jr.: Antigonadotrophic effects of melatonin in intact and unilaterally ovariectomized rats. Anat. Rec. **160**, 432 (1968)

Sorrentino, S., Jr.: Ovulation in PMS-treated rats with gonadotropin releasing hormone after pentobarbital and melatonin block. Neuroendocrinology **19**, 170–176 (1975)

Sorrentino, S., Jr., Reiter, R.J.: Lack of pineal-induced gonadal regression in dark-exposed or blind hamsters after surgical isolation of the medial-basal hypothalamus. Gen. comp. Endocr. **17**, 227–231 (1971)

Sorrentino, S., Jr., Reiter, R.J., Schalch, D.S.: Interaction of pineal gland, blinding and underfeeding on reproductive organ size and radioimmunoassayable growth hormone. Neuroendocrinology **7**, 105–115 (1971 a)

Sorrentino, S., Jr., Reiter, R.J., Schalch, D.S.: Pineal regulation of growth hormone synthesis and release in blinded and blinded-anosmic rats. Neuroendocrinology **7**, 210–218 (1971 b)

Sorrentino, S., Jr., Reiter, R.J., Schalch, D.S.: Role of the pineal gland in growth restraint of adult male rats by light and smell deprivation. Neuroendocrinology **8**, 116–124 (1971 c)

Sorrentino, S., Jr., Reiter, R.J., Schalch, D.S.: Hypotrophic reproductive organs and normal growth in male rats treated with melatonin. J. Endocr. **51**, 213–214 (1971 d)

Sorrentino, S., Jr., Reiter, R.J., Schalch, D.S.: Pineal regulation of growth hormone synthesis and release in blinded and blinded-anosmic male rats. Amer. J. ment. Defic. **75**, 210 (1971 e)

Sorrentino, S., Jr., Schalch, D.S., Reiter, R.J.: Environmental control of growth hormone and growth. In: Growth and growth hormone (ed. by Pecile, A. and Müller, E.E.), pp. 330–348. Amsterdam: Excerpta Medica 1972

Soulairac, A., Soulairac, M.-L.: Rhinencéphale et structures épithalamo-épiphysaires. (Introduction a l'étude neuro-endocrinienne de la glande pinéale). Ann. Endocr. (Paris) **24**, 197–203 (1963)

Spiroff, B.E.N.: Embryonic and post-hatching development of the pineal body of the domestic fowl. Amer. J. Anat. **103**, 375–401 (1958)

Stalsberg, H.: Effects of extirpation of the epiphysis cerebri in 6-day chick embryos. Acta endocr. (Paris), Suppl. **97**, 1–119 (1965)

Stammer, A.: Untersuchungen über die Struktur und die Innervation der Epiphyse bei Vögeln. Acta Biol. (Szeged) **7**, 65–75 (1961)

Stammer, A.: Studies on the pineal organ of the great horseshoe bat (*Rhinolophus ferrum equinum*). Gen. comp. Endocr. **18**, 624 (1972)

Starck, D.: Die endokraniale Morphologie der Ratiten, besonders der Apterygidae und Dinornithidae. Morph. Jb. **96**, 14–72 (1955)

Stebbins, R.C.: Effects of pinealectomy in the Western fence lizard *Sceloporus occidentalis*. Copeia **4**, 276–283 (1960)

Stebbins, R.C.: Activity changes in the striped plateau lizard with evidence on influence of the parietal eye. Copeia **1963**, 681–691 (1963)

Stebbins, R.C.: The effect of parietalectomy on testicular activity and exposure to light in the desert night lizard (*Xantusia vigilis*). Copeia **1970**, 261–270 (1970)

Stebbins, R.C., Cohen, N.W.: The effect of parietalectomy on the thyroid and gonads in free-living western fence lizards (*Sceloporus occidentalis*). Copeia **1973**, 662–668 (1973)

Stebbins, R.C., Eakin, R.M.: The role of the "third eye" in reptilian behavior. Amer. Mus. Nov. No. **1870**, 1–40 (1958)

Stebbins, R.C., Steyn, W., Peers, C.: Results of stirnorganectomy in tadpoles of the African ranid frog, *Pyxicephalus delalandi*. Herpetologica **16**, 261–275 (1960)

Stebbins, R.C., Wilhoft, D.C.: Influence of the parietal eye on activity in lizards. In: A.S. Bowman, ed. The Galapagos. Proceedings of the Symposia of the Galapagos International Scientific Project of 1964, pp. 258–268. Berkeley: University of Calif. Press 1966

Ştefănescu-Gavăt, V.: Studiul histochimic al fosfatazelor in glanda pineală la bovidee (*Bos taurus*). Stud. Cercet. Endocr. **21**, 329–334 (1970)

Stephan, F.K., Zucker, I.: Rat drinking rhythm: central visual pathways and endocrine factors mediating responsiveness to environmental illumination. Physiol. Behav. **8**, 315–326 (1972)

Stephan, H., Spatz, H.: Vergleichend-anatomische Untersuchungen an Insektivorengehirnen. IV. Gehirne afrikanischer Insektivoren. Versuch einer Zuordnung von Hirnbau und Lebensweise. Gegenbaurs morph. Jb. **103**, 108–174 (1962)

Stephens, J.L., Binkley, S.: Daily change in pineal N-acetyltransferase activity in a diurnal mammal, the ground squirrel. Experientia (Basel) **34**, 1523–1524 (1978)

Stetson, M.H., Elliott, J.A., Menaker, M.: Photoperiodic regulation of hamster testis: Circadian sensitivity to the effect of light. Biol. Reprod. **13**, 329–339 (1975)

Steyn, W.: The morphogenesis and some functional aspects of the epiphysial complex in lizards. J. comp. Neurol. **107**, 227–252 (1957)

Steyn, W.: The pineal circulation in some lizards. S. Afr. J. med. Sci. **54**, 143–147 (1958)

Steyn, W.: Epithelial organization and histogenesis of the epiphysial complex in lizards. Acta anat. (Basel) **37**, 310–335 (1959a)

Steyn, W.: Ultrastructure of pineal eye sensory cells. Nature (Lond.) **183**, 764–765 (1959b)

Steyn, W.: Electron microscopic observations on the epiphysial sensory cell problem. Z. Zellforsch. **51**, 735–747 (1960a)

Steyn, W.: Observations on the ultrastructure of the pineal eye. J. roy. micr. Soc. **79**, 47–58 (1960b)

Steyn, W.: Some epithalamic organs, the subcommissural organ, and their possible relation to vertebrate emergence on dry land. S. Afr. J. med. Sci. **57**, 283–287 (1961)

Steyn, W., Steyn, S.: Further light and electron microscopy of the pineal eye with a note on thermoregulatory aspects. Progr. Brain Res. **10**, 288–295 (1965)

Steyn, W., Webb, M.: The pineal complex in the fish *Labeo umbratus*. Anat. Rec. **136**, 79–85 (1960)

Storey, C.R., Nicholls, T.J.: Failure of exogenous melatonin to influence the maintenance or dissipation of photorefractoriness in the canary, *Serinus canarius*. Gen. comp. Endocr. **34**, 468–470 (1978)

Strada, S.J., Klein, D.C., Weller, J., Weiss, B.: Effect of norepinephrine on the concentration of adenosine 3′,5′-monophosphate of rat pineal gland in organ culture. Endocrinology **90**, 1470–1475 (1972)

Studnička, F.K.: Die Parietalorgane. In: Lehrbuch der vergleichenden mikroskopischen Anatomie (A. Oppel, ed.), Vol. 5, pp. 1–254. Jena: Fischer 1905

Stumpf, W.E., Sar, M.: Hormone-architecture of the mouse brain with ^{3}H-estradiol. In: Anatomical neuroendocrinology. W.E. Stumpf and L.D. Grant (eds.), pp. 82–103. Basel: S. Karger 1975a

Stumpf, W.E., Sar, M.: Localization of thyroid hormone in the mature rat brain and pituitary. In: Anatomical neuroendocrinology. W.E. Stumpf and L.D. Grant (eds.), pp. 318–327. Basel: S. Karger 1975b

Sturkie, P.D., Meyer, D.: Circadian rhythm in blood and pineal levels of serotonin in chickens. Fed. Proc. **31**, 327 (1972)

Suardi, L., Casirola, G., Ricotti, V.: Modificazioni del contenuto in acidi nucleici di alcune ghiandole endocrine dopo prolungato trattamento con ormoni. Nota IV: Azione di un estratto acquoso di ghiandola pineale. Biol. Lat. **12**, 623–632 (1959)

Sundararaj, B.I., Keshavanath, P.: Effects of melatonin and prolactin treatment on the hypophysial-ovarian system in the catfish, *Heteropneustes fossilis* (Bloch). Gen. comp. Endocr. **29**, 84–96 (1976)

Sung, D.II., Harisiadis, L., Chang, Ch.H.: Midline pineal tumors and suprasellar germinomas: Highly curable by irradiation. Radiology **128**, 745–751 (1978)

Supniewski, J., Marczynski, T., Misztal, S.: Biological properties of melatonin (5-methoxy-N-acetyl tryptamine). Bull. Acad. Polon. Sci. VI, **8**, 483–487 (1960)

Suzuki, J., Iwabuchi, T.: Surgical removal of pineal tumors (pinealomas and teratomas). Experience in a series of 19 cases. J. Neurosurg. **23**, 565–571 (1965)

Suzuki, Y.: Beiträge zur Anatomie des Epithalamus, besonders der Epiphyse, bei den Primaten, Arb. anat. Inst. Sendai **21**, 45–141 (1938)

Swain, R.: The pineal vascular system in *Lacerta muralis*, with notes on the venous system of other lizards. J. Zool. **154**, 487–493 (1968)

Szabo, T., Wersäll, J.: Ultrastructure of an electroreceptor (Mormyromast) in a mormyoid fish, *Guathonemus petersii*. II. J. Ultrastruct. Res. **30**, 473–490 (1970)

Szamier, R.B.: Enzymatic digestion of presynaptic structures in electroreceptors. Amer. J. Anat. **139**, 567–574 (1974)

Szamier, R.B., Wachtel, A.W.: Special cutaneous receptor organs of fish. VI. Ampullary and tuberous organs of *Hypopomus*. J. Ultrastruct. Res. **30**, 450–471 (1970)

Tabata, M., Niwa, H., Tamura, T.: On a slow potential from the epiphysis cerebri of fishes. Bull. Jap. Soc. Sci. Fish. **37**, 487–490 (1971)

Tabata, M., Tamura, T., Niwa, H.: Origin of the slow potential in the pineal organ of the rainbow trout. Vision Res. **15**, 737–740 (1975)

Taborsky, R.G., Deloigs, P., Page, I.H.: 6-Hydroxyindole and the metabolism of melatonin. J. med. Chem. **8**, 855–861 (1965)

Tabuchi, K., Yamada, O., Nishimoto, A.: The ultrastructure of pinealomas. Acta neuropath. (Berl.) **24**, 117–127 (1973)

Tagliamonte, A., Tagliamonte, P., Gessa, G.L., Brodie, B.B.: Compulsive sexual activity induced by p-chlorophenylalanine in normal and pinealectomized male rats. Science **166**, 1433–1435 (1969)

Tait, G.R., Barfuss, D.W., Ellis, L.C.: Pineal gland, melatonin synthesis, and testicular development in the rat. Life Sci. **8**, 717–725 (1969)

Takahashi, H.: Light and electron microscopic studies on the pineal organ of the goldfish, *Carassius auratus* L. Bull. Fac. Fish. Hokkaido Univ. **20**, 143–157 (1969)

Takahashi, K., Inoue, K., Takahashi, Y.: No effect of pinealectomy on parallel shift in circadian rhythms of adrenocortinal activity and food intake in blinded rats. Endocr. jap. **23**, 417–421 (1976)

Takahashi, M., Homma, K., Susuki, Y.: Resumption of cyclicity in light induced persistent estrus by treatments with serotonin or its precursors. Endocr. jap. **20**, 271–276 (1973)

Talbot, J.A., Reiter, R.J.: Influence of melatonin, 5-methoxytryptophol and pinealectomy on pituitary and plasma gonadotropin and prolactin levels in castrated adult male rats. Neuroendocrinology **13**, 164–172 (1973/74)

Tamai, M., Teirstein, P., Goldman, A., O'Brien, P., Chader, G.: Pineal gland does not control rod outer segment shedding and phagocytosis in rat retina and pigment epithelium. Invest. Ophthal. Vis. Sci. **17**, 558–562 (1978)

Tamarkin, L., Brown, S., Goldman, B.D.: Neuroendocrine regulation of seasonal reproductive cycles in the hamster, p. 458. 5th Annual Neurosci. Mtg New York 1975

Tamarkin, L., Westrom, W.K., Hamill, A.I., Goldman, B.D.: Effect of melatonin on the reproductive systems of male and female Syrian hamsters: Diurnal rhythm in sensitivity to melatonin. Endocrinology **99**, 1534–1541 (1976)

Tamarkin, L., Lefebvre, N.G., Hollister, Ch.W., Goldman, B.D.: Effect of melatonin administered during the night on reproductive function in the Syrian hamster. Endocrinology **101**, 631–634 (1977a)

Tamarkin, L., Hollister, C.W., Lefebvre, N.G., Goldman, B.D.: Melatonin induction of gonadal quiescence in pinealectomized Syrian hamsters. Science **198**, 953–955 (1977b)

Tanner, W.D., Hungerford, G.F.: Sodium and potassium excretion in pinealectomized (and adrenalectomized) rats. Proc. Soc. exp. Biol. (N.Y.) **109**, 388–390 (1962)

Tapp, E.: Melatonin as a tumour marker in a patient with pineal tumour. Brit. med. J. **1978 II**, 636

Tapp, E., Blumfield, M.: The parenchymal cells of the rat pineal gland. Acta morph. neerl.-scand. **8**, 119–131 (1970a)

Tapp, E., Blumfield, M.: The weight of the pineal gland in malignancy. Brit. J. Cancer **24**, 67–70 (1970b)

Tapp, E., Huxley, M.: The weight and degree of calcification of the pineal gland. J. Path. Bact. **105**, 31–39 (1971)

Tapp, E., Huxley, M.: The histological appearance of the human pineal gland from puberty to old age. J. Path. Bact. **108**, 137–144 (1972)

Tapp, E., Huxley, M., Davies, S.: The enzyme histochemistry of the pineal gland in rats. Histochem. J. **5**, 49–56 (1973)

Taşcă, C., Ghinea, E., Stănescu, O.: Histoenzymologic investigations on tissue culture of rat pineal gland. Rev. roum. Endocr. **5**, 21–24 (1968)

Taxi, J.: Morphological and cytochemical studies on the synapses in the autonomic nervous system. Progr. Brain Res. **31**, 5–20 (1969)

Taxi, J., Droz, B.: Étude de l'incorporation de noradrénaline-^{3}H (NA-^{3}H) et de 5-hydroxytryptophane-^{3}H (5-HTP-^{3}H) dans l'épiphyse et le ganglion cervical supérieur. C.R. Acad. Sci. (Paris) D **263**, 1326–1329 (1966)

Taylor, A.N., Wilson, R.W.: Elelctrophysiological evidence for the action of light on the pineal gland in the rat. Experientia (Basel) **26**, 267–269 (1977)

Taylor, D.H., Adler, K.: The pineal body: Site of extraocular perception of celestial cues for orientation in the tiger salamander (*Ambystoma tigrinum*). J. comp. Physiol. **124**, 357–361 (1978)

Taylor, K.M., Gfeller, E., Snyder, S.H.: Regional localization of histamine and histidine in the brain of the rhesus monkey. Brain Res. **41**, 171–179 (1972)

Tedman, R.A.: The pineal gland of the Weddell seal and the southern elephant seal. J. Anat. (Lond.) **124**, 524 (1977)

Theron, J.J., Biagio, R., Meyer, A.C., Boekkooi, S.: Microfilaments, the smooth endoplasmic reticulum and synaptic ribbon fields in the pinealocytes of the baboon (*Papio ursinus*). Amer. J. Anat. **154**, 151–162 (1979)

Thiéblot, L., Blaise, S.: Influence de la glande pinéale sur les gonades. Ann. Endocr. (Paris) **24**, 270–286 (1963)

Thiéblot, L., Blaise, S.: Influence de la glande pinéale sur la sphère génitale. Progr. Brain Res. **10**, 577–584 (1965)

Thiéblot, L., Blaise, S.: Étude biochimique du principe pinéal antigonadotrope. Probl. Actuels Endocr. Nutr. **10**, 257–275 (1966)

Thiéblot, L., Blaise, S.: Principe anti-gonadotrope de la glande pinéale. Rev. roum. Endocr. **4**, 269–283 (1967)

Thiéblot, L., Le Bars, H.: La glande pinéale ou épiphyse (Anatomie, histologie, physiologie, clinique). Paris: Maloine 1955

Thiéblot, L., Naudascher, J., Le Bars, H.: Modifications histologiques de l'épiphyse à la suite d'excitation électrique du ganglion cervical supérieur chez le chat. Ann. Endocr. (Paris) **8**, 468–469 (1947a)

Thiéblot, L., Naudascher, J., Le Bars, H.: Modifications de la structure épiphysaire à la suite d'injections répétées d'hormones sexuelles. Ann. Endocr. (Paris) **8**, 469–472 (1947b)

Thiéblot, L., Sternberg, J., Le Bars, H.: Influence de l'épiphyse sur les variations du taux phosphore dans certains organes. Étude comparative. J. Physiol. **41**, 276A–277A (1949a)

Thiéblot, L., Naudascher, J., Le Bars, H.: Modifications histologiques de l'épiphyse à la suite d'injections d'hormones gonadotropes. Ann. Endocr. (Paris) **10**, 192–194 (1949b)

Thiéblot, L., Meyniel, G., Blaise, S., Alfieri, R., Berthelay, J.: L'épiphyse joue-t-elle un rôle sur la phospho-captation au niveau des gonades. Ann. Endocr. (Paris) **24**, 385–390 (1963)

Thiéblot, L., Bastide, P., Blaise, S., Boyer, J., Datugue, C.: De l'equipement enzymatique de la glande pinéale (Étude critique et expérimentale). Ann. Endocr. (Paris) **26**, 313–314 (1965)

Thiéblot, L., Alassimone, A., Blaise, S.: Étude chromatographique et électrophorétique du facteur antigonadotrope de la glande pinéale. Ann. Endocr. (Paris) **27**, 861–866 (1966a)

Thiéblot, L., Blaise, S., Alassimone, A.: Essai de caractérisation du principe anti-gonadotrope de la glande pinéale. C.R. Soc. Biol. (Paris) **160**, 1574–1576 (1966b)

Thiéblot, L., Bastide, P., Blaise, S., Boyer, J., Sastugue, C.: De l'équipement enzymatique de la glande pinéale. Étude critique et expérimentale. Ann. Endocr. (Paris) **27**, 1–20 (1966c)

Thiéblot, L., Berthelay, J., Blaise, S.: Effets de la mélatonine chez le Rat mâle et femelle. I. Action au niveau des gonades et des annexes. Ann. Endocr. (Paris) **27**, 65–68 (1966d)

Thiéblot, L., Berthelay, J., Blaise, S.: Effets de la mélatonine chez le Rat mâle et femelle. II. Action au niveau de la thyroide. Ann. Endocr. (Paris) **27**, 69–71 (1966e)

Thiéblot, L., Berthelay, J., Blaise, S.: Action de la mélatonine sur la sécrétion gonadotrope de Rat. C.R. Soc. Biol. (Paris) **160**, 2306–2309 (1966f)

Thiéblot, L., Blaise, S., Couquelet, J.: Recherche de dérivés indoliques dans des extraits de glande pinéale. C.R. Soc. Biol. (Paris) **161**, 295–297 (1967)

Thillard, M.-J.: La greffe simultanée intraoculaire et chorioallantoidienne de gonades males et d'épiphyse chez l'embryon de poulet. Bull. Ass. Anat. (Nancy) No. **120**, 1263–1272 (1964)

Thillard, M.J.: Mise en évidence histochimique des lipides dans l'épiphyse du Poulet. C.R. Soc. Biol. (Paris) **162**, 1074–1080 (1968)

Thompson, S.W., Huseby, R.A., Fox, M.A., Davis, C.L., Hunt, R.D.: Spontaneous tumors in the Sprague-Dawley rat. J. nat. Cancer Inst. **27**, 1037–1057 (1961)

Thoresen, Th.S.: Radioimmunoassay for melatonin in human serum. Scand. J. clin. Lab. Invest. **38**, 687–692 (1978)

Thornton, V.F., Geschwind, I.I.: Evidence that serotonin may be a melanocyte stimulating hormone releasing factor in the lizard, *Anolis carolinensis*. Gen. comp. Endocr. **26**, 346–353 (1975)

Thorpe, P.A., Herbert, J.: The role of the pineal gland in the response of ferrets to artificially restricted photoperiods. J. Endocr. **63**, 56–57 (1974)
Thorpe, P.A., Herbert, J.: Studies on the duration of the breeding season and photorefractoriness in female ferrets pinealectomized or treated with melatonin. J. Endocr. **70**, 255–262 (1976)
Tigchelaar, P.V., Nalbandov, A.V.: Comparison of various commercial preparations of melatonin in immature rats. J. Reprod. Fertil. **25**, 141–143 (1971)
Tigchelaar, P.V., Nalbandov, A.V.: Effect of the pineal gland on ovulation and pregnancy in the rat. Biol. Reprod. **13**, 461–469 (1975)
Tilders, F.J.H., Smelik, P.G.: A diurnal rhythm in melanocyte-stimulating hormone content of the rat pituitary gland and its independence from the pineal gland. Neuroendocrinology **17**, 296–308 (1975)
Tilders, F.J.H., Ploem, J.S., Smelik, P.G.: Quantitative microfluorimetric studies on formaldehyde-induced fluorescence of 5-hydroxytryptamine in the pineal gland of the rat. J. Histochem. Cytochem. **22**, 967–975 (1974)
Tilstra, B., Prop, N.: On the possible function of melatonin. Acta morph. neerl.-scand. **5**, 289–290 (1962)
Tima, L., Trentini, G.P., Mess, B.: Effects of serotonin on ovulation induced by pinealectomy in anovulatory frontal-deafferented rats. Neuroendocrinology **12**, 149–152 (1973)
Timmermans, L., Devečerski, V.: Répercussions cyto-enzymologiques de l'épiphysectomie sur le testicule du Rat. C.R. Soc. Biol. (Paris) **166**, 733–736 (1972a)
Timmermans, L., Devečerski, V.: Influence de l'épiphysectomie sur la structure et le métabolisme de la paroi vésicale du Rat. C.R. Soc. Biol. (Paris) **166**, 968–970 (1972b)
Timmermans, L., Devečerski, V.: Etude de l'influence d'hormones sexuelles et de l'épiphysectomie sur l'activité d'une oxydase d'acides α-hydroxylés dans le tissue interstitiel du testicule et dans le rein. C.R. Soc. Biol. (Paris) **166**, 715–717 (1972c)
Timones, S., Carpen, E.: Multiple pregnancies and photoperiodicity. Annal. Chir. Gynaec. Fenn. **57**, 135–138 (1968)
Tinacci, F., Fazzini, G.: Azione di un estratto di ghiandola pineale sulle cartilagini dell ratto in accrescimento. Boll. Soc. ital. Biol. sper. **40**, 282–284 (1964)
Tomatis, M.E., Orias, R.: Changes in melatonin concentration in pineal gland in rats exposed to continous light or darkness. Acta physiol. lat.-amer. **17**, 227–233 (1967)
Torack, R.M., Barrnett, R.J.: The fine structural localization of nucleoside phosphatase activity in the blood-brain barrier. J. Neuropath. exp. Neurol. **23**, 46–59 (1964)
Trakulrungsi, W.K., Yeager, V.L.: Effect of photoperiod on early changes in the neonatal rat pineal gland. Experientia (Basel) **33**, 84–85 (1977)
Tranzer, J.P., Snipes, R.L.: Fine structural localization of noradrenaline in sympathetic nerve terminals: A critical study on the influence of fixation. In: Electron microscopy, Vol. II, pp. 519–520. 4th European Regional Conference on Electron Microscopy (D.S. Bocciarelli, ed.). Roma: Tipografia Poliglotta Vaticana 1968
Trautmann, A., Silber, H.: Die Beeinflussung des Fettansatzes beim Schwein durch Epiphysan. Tierärztl. Umsch. **2**, 1–3 (1947)
Trentini, G.P., Vassanelli, P., Barbanti Silva, C., Fantoni, P.A.: Rilievi sull'attività funzionale del surrene dopo trattamento con estratti di epifisi (Studio sperimentale istochimico). Riv. Istochim. Norm. Patol. **10**, 713–724 (1964)
Trentini, G.P., Barbanti Silva, C., Vassanelli, P., Botticelli, A.: Modificazioni istoenzimatiche del corticosurrene di ratto conseguenti ad epifisectomia (studio istochemico). Boll. Soc. ital. Biol. sper. **41**, 967–970 (1965)
Trentini, G.P., Ferrari de Gaetani, G., Barbanti Silva, C., Rivasi, F.: Modificazioni istomorfologiche e microfluorescopiche pinealiche indotte dalla splenectomia nella ratta albina prepubere e adulta. Arch. De Vecchi Anat. pat. **52**, 1–15 (1968)
Trentini, G.P., Ferrari de Gaetani, C., Botticelli, A., Rivasi, F.: L'antagonismo epifisi-ipofisario. Modificazioni istomorfologiche ed istoenzimatiche epifisarie indotte dalla ipergonadotropinemia sperimentala nel ratto albino accecato. Boll. Soc. ital. Biol. sper. **45**, 631–634 (1969)
Trentini, G.P., Barbieri Palmieri, F., Cristiani, G., Barbolini, G.C.: Spleen-pineal gland correlations. Histomorphological and microfluoroscopic studies of the pineal gland in experimental methylcellulose hypersplenism. Ann. Endocr. (Paris) **32**, 361–370 (1971)
Trentini, G.P., Ferrari de Gaetani, C., Barbieri Palmieri, F.: The role of sympathetic innervation

in pineal-pituitary feed back. Blocking of pineal changes induced in the rat by experimental hypergonadotropinaemia by chronic treatment with bretylium or guanethidine. Ann. Endocr. (Paris) **34**, 261–270 (1973)

Trentini, G.P., De Gaetani, C.F., Martini, L., Mess, B.: Effect of pinealectomy and of bilateral cervical ganglionectomy on serum LH levels in constant-estrous anovulatory rats. Proc. Soc. exp. Biol. (N.Y.) **153**, 590–494 (1976)

Trost, E.: Untersuchungen über die frühe Entwicklung des Parietalauges und der Epiphyse von *Anguis fragilis, Chalcides ocellatus* und *Tropidonotus natrix*. Zool. Anz. **148**, 58–71 (1952)

Trost, E.: Die Histogenese und Histologie des Parietalauges von *Anguis fragilis* und *Chalcides ocellatus*. Z. Zellforsch. **38**, 185–217 (1953a)

Trost, E.: Die Entwicklung, Histogenese und Histologie der Epiphyse, der Paraphyse, des Velum transversum, des Dorsalsackes und des subcommissuralen Organs bei *Anguis fragilis, Chalcides ocellatus* und *Natrix natrix*. Acta anat. (Basel) **18**, 326–342 (1953b)

Trost, E.: Zur Morphologie der Parietalorgane der Eidechsen. Anat. Anz. (Erg.-H.) **100**, 368–374 (1954)

Trost, E.: Über die Lage des Foramen perietale bei rezenten Reptilien und Labyrinthodontiern. Acta anat. (Basel) **26**, 318–339 (1956)

Trowell, O.A.: The culture of mature organs in a synthetic medium. Exp. Cell Res. **16**, 118–147 (1959)

Trueman, T., Herbert, J.: The distribution of monoamines and acetylcholinesterase in the pineal gland and habenula of the ferret. J. Anat. (Lond.) **106**, 406 (1970a)

Trueman, T., Herbert, J.: Monoamines and acetylcholinesterase in the pineal gland and habenula of the ferret. Z. Zellforsch. **109**, 83–100 (1970b)

Tsang, D., Martin, J.B.: Effect of hypothalamic hormones on the concentration of adenosine 3′,5′-monophosphate in incubated rat pineal glands. Life Sci. **19**, 911–918 (1976)

Tubahara, M.: Über die Innervation der Epiphyse. Kurume med. J. **2**, 218–234 (1955)

Tulsi, R.S.: Scanning electron microscopy of the pineal recess of the adult *Trichosurus vulpecula*. J. Anat. (Lond.) **124**, 515 (1977)

Tulsi, R.S.: Some evidence in support of presence of supraependymal neural plexus in pineal-subcommissural region of *Trichosurus vulpecula*. J. Anat. (Lond.) **126**, 648 (1978a)

Tulsi, R.S.: A scanning electron microscopic study of the developmental changes in the pineal recess in *Trichosurus vulpecula*. J. Anat. (Lond.) **126**, 651 (1978b)

Turek, F.W.: Antigonadal effect of melatonin in pinealectomized and intact male hamsters. Proc. Soc. exp. Biol. (N.Y.) **155**, 31–34 (1977)

Turek, F.W., Losee, S.H.: Melatonin-induced testicular growth in golden hamsters maintained on short days. Biol. Reprod. **18**, 299–305 (1978)

Turek, F.W., Wolfson, A.: Lack of an effect of melatonin treatment via silastic capsules on photic-induced gonadal growth and the photorefractory condition in white-throated sparrows. Gen. comp. Endocr. **34**, 471–474 (1978)

Turek, F.W., Desjardins, C., Menaker, M.: Melatonin: Antigonadal and progonadal effects in male golden hamsters. Science **190**, 280–282 (1975)

Turek, F.W., Desjardins, C., Menaker, M.: Melatonin-induced inhibition of testicular function in adult golden hamsters. Proc. Soc. exp. Biol. (N.Y.) **151**, 502–506 (1976a)

Turek, F.W., Desjardins, C., Menaker, M.: Differential effects of melatonin on the testes of photoperiodic and nonphotoperiodic rodents. Biol. Reprod. **15**, 94–97 (1976b)

Turek, F.W., McMillan, J.P., Menaker, M.: Melatonin: Effects on the circadian locomotor rhythm of sparrows. Science **194**, 1441–1443 (1976c)

Ueck, M.: Ultrastruktur des pinealen Sinnesapparates bei einigen Pipidae und Discoglossidae. Z. Zellforsch. **92**, 452–476 (1968a)

Ueck, M.: Granulierte marklose Nervenfasern in der Epiphysenregion von Anuren. Z. Zellforsch. **90**, 389–402 (1968b)

Ueck, M.: Ultrastrukturbesonderheiten der pinealen Sinneszellen von *Protopterus dolloi*. Z. Zellforsch. **100**, 560–580 (1969)

Ueck, M.: Zur Ultrastruktur lichtempfindlicher pinealer Sinnesapparate. Anat. Anz. **126**, Erg.-H., 321–323 (1970a)

Ueck, M.: Weitere Ultrastrukturbeiträge zur Funktionsanalyse des pinealen Sinnesapparates der Anuren. Verh. Dtsch. Zool. Ges. Köln, 64. Jahresversammlung, 92–96 (1970b)

Ueck, M.: Zur Ultrastruktur der Epiphysis cerebri der Vögel. Verh. Zool. Ges. 1969. Zool. Anz., Suppl. **33**, 509–518 (1970c)

Ueck, M.: Weitere Untersuchungen zur Feinstruktur und Innervation des Pinealorgans von *Passer domesticus* L. Z. Zellforsch. **105**, 276–302 (1970d)

Ueck, M.: Strukturbesonderheiten der Anurenepiphyse nach prolongierter Osmierung und Anwendung der Acetylcholinesterase-Reaktion. Z. Zellforsch. **112**, 526–541 (1971a)

Ueck, M.: Vergleichende Strukturstudien (Fluoreszenz- und Elektronenmikroskopie, Histochemie) am Pinealorgan der Vögel. Verh. anat. Ges. (Jena) **66**, 235–238 (1971b)

Ueck, M.: Fluoreszenzmikroskopische und elektronenmikroskopische Untersuchungen am Pinealorgan verschiedener Vogelarten. Z. Zellforsch. **137**, 37–62 (1973)

Ueck, M.: Vergleichende Betrachtungen zur neuroendokrinen Aktivität des Pinealorgans. Fortschr. Zool. **22**, 167–203 (1974)

Ueck, M., Kobayashi, H.: Vergleichende Untersuchungen über acetylcholinesterasehaltige Neurone im Pinealorgan der Vögel. Z. Zellforsch. **129**, 140–160 (1972a)

Ueck, M., Kobayashi, H.: Secretory and neuronal elements of the avian pineal organ. Gen. comp. Endocr. **18**, 624 (1972b)

Ueck, M., Kobayashi, H.: Neue Ergebnisse zu Fragen der vergleichenden Epiphysenforschung. Verh. anat. Ges. (Jena) **73**, 961–963 (1980)

Ueck, M., Wake, K.: The pinealocyte – a paraneuron? A review. Arch. histol. jap. **40**, Suppl., 261–278 (1977)

Ueck, M., Vaupel- von Harnack, M., Morita, Y.: Weitere experimentelle und neuroanatomische Untersuchungen an den Nervenbahnen des Pinealkomplexes der Anuren. Z. Zellforsch. **116**, 250–274 (1971)

Ueck, M., Ohnishi, R., Wake, K.: Photoreceptor-like outer segments in the pineal organ of the lovebird, *Uroloncha domestica* (Aves: Passeriformes). A scanning electron microscopic study. Cell Tiss. Res. **182**, 139–143 (1977)

Ueck, M., Ohnishi, R., Wake, K.: The outer segments of photoreceptive pinealocytes in the pineal organ of the funa, *Carassius gibelio langsdorfi*. A scanning electron microscopic study. Cell Tiss. Res. **186**, 259–268 (1978)

Ulrich, R., Yuwiler, A., Wetterberg, L., Klein, D.: Effects of light and temperature on the pineal gland in suckling rats. Neuroendocrinology **13**, 255–263 (1973/74)

Underwood, H.: Circadian organization in lizards: The role of the pineal organ. Science **195**, 587–589 (1977)

Upson, R.H., Benson, B.: Effects of blinding on the ultrastructure of mouse pinealocytes with particular emphasis on the dense-cored vesicles. Cell Tiss. Res. **183**, 491–498 (1977)

Upson, R.H., Benson, B., Satterfield, V.: Quantitation of ultrastructural changes in the mouse pineal in response to continuous illumination. Anat. Rec. **184**, 311–324 (1976)

Urasaki, H.: Effect of pinealectomy on gonadal development in the Japanese killifish (Medaka) *Oryzias latipes*. Annot. Zool. Jap. **45**, 10–15 (1972a)

Urasaki, H.: Role of the pineal gland in gonadal development in the fish, *Oryzias latipes*. Annot. Zool. Jap. **45**, 152–158 (1972b)

Urasaki, H.: Effects of restricted photoperiod and melatonin administration on gonadal weight in the Japanese killifish. J. Endocr. **55**, 619–620 (1972c)

Urasaki, H.: Effect of pinealectomy and photoperiod on oviposition and gonadal development in the fish, *Oryzias latipes*. J. exp. Zool. **185**, 241–246 (1973)

Urry, R.L., Ellis, L.C.: Monoamine oxidase activity of the hypothalamus and pituitary: Alterations after pinealectomy, changes in photoperiod, or additions of melatonin in vitro. Experientia (Basel) **31**, 891–892 (1975)

Urry, R.L., Barfuss, D.W., Ellis, L.C.: Hydroxyindole-O-methyl transferase activity of male rat pineal glands following hypophysectomy and HCG treatment. Biol. Reprod. **6**, 238–243 (1972)

Urry, R.L., Dougherty, K.A., Frehn, J.L., Ellis, L.C.: Factors other than light affecting the pineal gland: Hypophysectomy, testosterone, dihydrotestosterone, estradiol, cryptorchism, and stress. Amer. Zool. **16**, 79–91 (1976)

Van de Kamer, J.C.: Over de ontwikkeling de determinatie en de betekenis van de epiphyse en de paraphyse van de amphibien. Thesis, Rijksuniversiteit te Utrecht 1949

Van de Kamer, J.C.: The pineal organ of fish and amphibia. In: Progress in neurobiology (J.A. Kappers, ed.), pp. 113–120. Amsterdam-London: Elsevier 1956

Van de Kamer, J.C.: Histologische und dytologische Untersuchungen über das Ependym und seine Abkömmlinge (insbesondere die Epiphyse und den Saccus vasculosus) bei niederen Vertebraten. Experientia (Basel) **14**, 161–166 (1958)

Van de Kamer, J.C.: Histological structure and cytology of the pineal complex in fishes, amphibians and reptiles. Progr. Brain Res. **10**, 30–48 (1965)

Van de Veerdonk, F.C.G.: Separation method for melatonin in pineal extract. Nature (Lond.) **208**, 1324–1325 (1967a)

Van de Veerdonk, F.C.G.: Demonstration of melatonin in amphibia. Curr. Mod. Biol. **1**, 175–177 (1967b)

Van de Velde, J., De Smet, J.: Action d'hormones épiphysaires sur les dépôts de graisse du Rat. Différences sexuelles. C.R. Soc. Biol. (Paris) **154**, 1517–1520 (1960)

Van der Have-Kirchberg, M.L.L., De Morée, A., Van Laar, J.F., Gerwig, G.J., Versluis, C., Ebels, I., Hus-Citharel, A., L'Héritier, A., Roseau, S., Zurburg, W., Moszkowska, A.: Separation of pineal extracts by gelfiltration. VI. Isolation and identification from sheep pineals of biopterin; Comparison of the isolated compound with some synthetic pteridines and the biological activity in *in vitro* and *in vivo* bioassays. J. Neural Transm. **40**, 205–220 (1977)

Van der Wal, B., Moll, J., De Wied, D.: Effect of pinealectomy and of lesions in the subcommissural body on the rate of aldosterone secretion by rat adrenal glands *in vitro*. Progr. Brain Res. **10**, 635–645 (1965)

Vaněček, J., Illnerová, H.: The regulation of pineal serotonin N-acetyltransferase activity in newborn rats. Develop. Biol. **68**, 287–291 (1979)

Van Veen, T., Hartwig, H.G., Müller, K.: Light-dependent motor activity and photonegative behavior in the eel (*Anguilla anguilla* L.). Evidence for extraretinal and extrapineal photoreception. J. comp. Physiol. **111**, 209–219 (1976)

Van Veen, T., Brackmann, M., Moghimzadeh, E.: Post-natal development of the pineal organ in the hamsters *Phodopus sungorus* and *Mesocricetus auratus*. Cell Tiss. Res. **189**, 241–250 (1978)

Vassanelli, P., Trentini, G.P., Barbanti Silva, G., Botticelli, A.: Aspetti istochimici del cortico surrene di ratto epifisectomizzato dopo trattamento con 5-idrossitriptamina. Boll. Soc. ital. Biol. sper. **41**, 970–972 (1965)

Vaughan, G.M., Vaughan, M.K.: Effect of melatonin and other pineal indoles on adrenal enlargement produced in male and female mice by pinealectomy, unilateral adrenalectomy, castration and cold stress. Anat. Rec. **172**, 421 (1972)

Vaughan, G.M., Pelham, R.W., Pang, S.F., Loughlin, L., Wilson, K.M., Sandock, K.L., Vaughan, M.K., Koslow, S.H., Reiter, R.J.: Nocturnal elevation of plasma melatonin and urinary 5-hydroxyindoleacetic acid in young men; attempts at modification by brief changes in environmental lighting and sleep and by autonomic drugs. J. clin. Endocr. **42**, 752–764 (1976)

Vaughan, G.M., McDonald, S.D., Jordan, R.M., Allen, J.P., Bohmfalk, G.L., Abou-Samra, M., Story, J.L.: Melatonin concentration in human blood and cerebrospinal fluid: Relationship to stress. J. clin. Endocr. **47**, 220–223 (1978a)

Vaughan, G.M., Allen, J.P., Tullis, W., Siler-Khodr, T.M., De la Pena, A., Sackman, J.W.: Overnight plasma profiles of melatonin and certain adenohypophyseal hormones in men. J. clin. Endocr. **47**, 566–571 (1978b)

Vaughan, G.M., Reiter, R.J., Siler-Khodr, T.M., Sackman, J.W., Allen, J.P., Vaughan, M.K., McGuire, W.L., Johnson, L.Y., Starr, P.: Influence of pinealectomy on serum estrogen and progesterone levels in blind-anosmic female rats. Experientia (Basel) **34**, 1378–1379 (1978c)

Vaughan, G.M., Allen, J.P., Tullis, W., Sackman, J.W., Vaughan, M.K.: Stress-induced increase of pineal N-acetyltransferase activity in intact rats. Neurosci. Letters **9**, 83–87 (1978d)

Vaughan, M.K.: Sexual differentiation of the rat hypothalamus: interaction of pineal amines and steroids. Thesis, University of Texas Medical Branch, Galveston (1970)

Vaughan, M.K.: Effect of intraventricular or intraperitoneal injections of arginine vasotocin on compensatory ovarian hypertrophy in normal or pinealectomized mice. Anat. Rec. **184**, 553–554 (1976)

Vaughan, M.K., Barchas, J.: Effects of melatonin and related compounds on the release of glycerol from rat adipose tissue *in vitro*. J. Pharmacol. exp. Ther. **152**, 298–303 (1966)

Vaughan, M.K., Reiter, R.J.: Transient hypertrophy of the ventral prostate and coagulating glands and accelerated thymic involution following pinealectomy in the mouse. Tex. Rep. Biol. Med. **29**, 579–586 (1972)

Vaughan, M.K., Vaughan, G.M.: Neonatal melatonin administration and subsequent functioning of adult gonadal and adrenal neuroendocrine axes. Endocr. exp. **8**, 261–268 (1974)

Vaughan, M.K., Benson, B., Norris, J.T.: Inhibition of compensatory ovarian hypertrophy in mice by 5-hydroxytryptamine and melatonin. J. Endocr. **47**, 397–398 (1970)

Vaughan, M.K., Reiter, R.J., Vaughan, G.M.: Inhibition of compensatory ovarian hypertrophy by Altschule's pineal extract, pineal indoles and vasopressin in mice and voles. Amer. Zool. **11**, 649–650 (1971a)

Vaughan, M.K., Benson, B., Norris, J.T., Vaughan, G.M.: Inhibition of compensatory ovarian hypertrophy in mice by melatonin, 5-hydroxytryptamine and pineal powder. J. Endocr. **50**, 171–175 (1971b)

Vaughan, M.K., Vaughan, G.M., O'Steen, W.K.: Pituitary, adrenal, pineal and renal weights in offspring of rats treated with testosterone and/or melatonin during pregnancy. J. Endocr. **51**, 211–212 (1971c)

Vaughan, M.K., Reiter, R.J., Vaughan, G.M., Bigelow, L., Altschule, M.D.: Inhibition of compensatory ovarian hypertrophy in the mouse and vole: a comparison of Altschule's pineal extract, pineal indoles, vasopressin, and oxytocin. Gen. comp. Endocr. **18**, 372–377 (1972a)

Vaughan, M.K., Vaughan, G.M., Reiter, R.J., Benson, B.: Effect of melatonin and other pineal indoles on adrenal enlargement produced in male and female mice by pinealectomy, unilateral adrenalectomy, castration and cold stress. Neuroendocrinology **10**, 139–154 (1972b)

Vaughan, M.K., Vaughan, G.M., Klein, D.C.: Effects of pinealectomy on the thymus and on the hypophysectomy-induced regression of reproductive organs in rats. Fed. Proc. **32**, 251 (1973)

Vaughan, M.K., Reiter, R.J., McKinney, T., Vaughan, G.M.: Inhibition of growth of gonadal dependent structures by arginine vasotocin and purified bovine pineal fractions in immature mice and hamsters. Int. J. Fertil. **19**, 103–106 (1974)

Vaughan, M.K., Vaughan, G.M., Reiter, R.J.: Inhibition of human chorionic gonadotrophin-induced ovarian and uterine growth in the mouse by synthetic arginine vasotocin. Experientia (Basel) **31**, 862–863 (1975)

Vaughan, M.K., Vaughan, G.M., Blask, D.E., Barnett, M.P., Reiter, R.J.: Arginine vasotocin: Structure-activity relationships and influence on gonadal growth and function. Amer. Zoologist. **16**, 25–34 (1976a)

Vaughan, M.K., Vaughan, G.M., Reiter, R.J.: Inhibition of human chorionic gonadotrophin-induced hypertrophy of the ovaries and uterus in immature mice by some pineal indoles, 6-hydroxymelatonin and arginine vasotocin. J. Endocr. **68**, 397–400 (1976b)

Vaughan, M.K., Reiter, R.J., Vaughan, G.M.: Fertility pattern in female mice following treatment with arginine vasotocin or melatonin. Int. J. Fertil. **21**, 65–68 (1976c)

Vaughan, M.K., Blask, D.E., Vaughan, G.M., Reiter, R.J.: Dose-dependent prolactin releasing activity of arginine vasotocin in intact and pinealectomized estrogen-progesterone treated adult male rats. Endocrinology **99**, 1319–1322 (1976d)

Vaughan, M.K., Vaughan, G.M., Blask, D.E., Reiter, R.J.: Influence of melatonin, constant light, or blinding on reproductive system of gerbils (*Meriones unguiculatus*). Experientia (Basel) **32**, 1341–1342 (1976e)

Vaughan, M.K., Little, J.C., Johnson, L.Y., Blask, D.E., Vaughan, G.M., Reiter, R.J.: Effects of melatonin and natural and synthetic analogues of arginine vasotocin on plasma prolactin levels in adult male rats. Hormone Res. **9**, 236–246 (1978)

Veech, R.L., Nielsen, R., Harris, R.L.: Effects of pineal and other hormones on the free [$NADP^+$]/[NADPH] ratio in rat liver. In: Frontiers of pineal physiology (M.D. Altschule, ed.), pp. 177–196. Cambridge, Mass.-London: MIT Press 1975

Vellan, E.J., Gjessing, L.R., Stalsberg, H.: Free amino acids in the pineal and pituitary glands of human brain. J. Neurochem. **17**, 699–701 (1970)

Venzke, W.G., Gilmore, J.W.: Histological observations on the epiphysis cerebri and on the choroid plexus of the third ventricle of the dog. Proc. Iowa Acad. Sci. **47**, 409–413 (1941)

Vermes, I., Molnar, D., Dull, G., Telegdy, G.: Functional interrelationship between cervical superior ganglia, pineal gland and hypothalamo-pituitary-adrenal system in rats. Acta physiol. Acad. Sci. hung. **45**, 53–62 (1974)

Vidmar, B.: The development *in vitro* of the embryonic pineal body of the fowl. J. Embryol. exp. Morph. **1**, 117–123 (1953)

Vigh, B., Vigh-Teichmann, I.: Vergleich der Ultrastruktur der Liquorkontaktneurone und Pinealozyten der Säugetiere. Verh. Anat. Ges. (Jena) **69**, 453–461 (1975)

Vigh, B., Vigh-Teichmann, I., Aros, B.: Comparative ultrastructure of cerebrospinal fluid-contacting neurons and pinealocytes. Cell Tiss. Res. **158**, 409–424 (1975)

Vigh-Teichmann, I., Vigh, B., Aros, B.: CSF contacting axons and synapses in the lumen of the pineal organ. Z. Zellforsch. **144**, 139–152 (1973)

Vitale-Calpe, A.: Contribution to the study of the epithalamus of *Bufo arenarum* (Hensel). I. Histological observations on the epiphysis cerebri. Acta anat. (Basel) **64**, 618–627 (1966)

Vivien, J.-H.: Ultrastructure des constituants de l'épiphyse de *Tropidonotus natrix* L. C.R. Acad. Sci. (Paris) **258**, 3370–3372 (1964a)

Vivien, J.-H.: Structure et ultrastructure de l'épiphyse d'un Chélonien *Pseudemys scripta elegans*. C.R. Acad. Sci. (Paris) **259**, 899–901 (1964b)

Vivien, J.-H.: Signes de stimulation des activités sécrétoires des pinéalocytes chez la Couleuvre *Tropidonotus natrix* L. traitée par des principes gonadotropes. C.R. Acad. Sci. (Paris) D **260**, 5370–5372 (1965)

Vivien, J.-H., Roels, B.: Ultrastructure de l'épiphyse des chéloniens. Présence d'un "paraboloide" et de structures de type photorécepteur dans l'épithelium sécrétoire de *Pseudemys scripta elegans*. C.R. Acad. Sci. (Paris) D **264**, 1743–1746 (1967)

Vivien, J.-H., Roels, B.: Ultrastructure synaptiques dans l'épiphyse des Chéloniens. Présence de rubans synaptiques au niveau des articulations entre cellules pseudosensorielles et terminaisons nerveuses dans l'épiphyse de *Pseudemys scripta elegans* et *Pseudemys picta*. C.R. Acad. Sci. (Paris) D **266**, 600–603 (1968)

Vivien-Roels, B.: Etude structurale et ultrastructurale de l'épiphyse d'un Reptile: *Pseudemys scripta elegans*. Z. Zellforsch. **94**, 352–390 (1969)

Vivien-Roels, B.: Ultrastructure, innervation et fonction de l'épiphyse chez les Chéloniens. Z. Zellforsch. **104**, 429–448 (1970)

Vivien-Roels, B.: L'épiphyse des Chéloniens. Etude embryologique, structurale, ultrastructurale; analyse qualitative et quantitative de la sérotonine dans des conditions normales et expérimentales. Thesis, Strasbourg 1976a

Vivien-Roels, B.: L'évolution de l'épiphyse des chéloniens: réduction de la fonction photoréceptrice et développement corrélatif de l'activité sécrétoire. Colloques internat. C.N.R.S., No. **266**, 263–278 (1976b)

Vivien-Roels, B., Arendt, J.: Variations circadiennes et circannuelles de la mélatonine épiphysaire chez *Testudo hermanni* G. (Reptile-Chélonien) dans des conditions naturelles d'éclairement et de température. Ann. Endocr. (Paris) **40**, 93–94 (1979)

Vivien-Roels, B., Humbert, W.: The lipopigments of the pineal gland of *Testudo hermanni* (Reptile, Chelonian): microprobe analysis and physiological significance. J. Ultrastruct. Res. **61**, 134–139 (1977)

Vivien-Roels, B., Petit, A.: Embryogenèse du complexe épiphysaire chez la Tortue mauresque *Testudo graeca* L. Ann. Embryol. Morph. **6**, 151–168 (1973)

Vivien-Roels, B., Petit, A.: Relative roles of light and temperature in the control of pineal serotonin (5HT) circadian and circannual variations in reptiles. Gen. comp. Endocr. **34**, Abstr. No. 37 (1978)

Vivien-Roels, B., Petit, A., Lichté, C.: Dosage spectrofluorimétrique de la sérotonine épiphysaire chez les reptiles. Etude des variations du taux de sérotonine en fonction de la période d'éclairement. C.R. Acad. Sci. (Paris) D **272**, 459–461 (1971)

Vodicnik, M.J., De Vlaming, V.L.: The effects of pinealectomy on pituitary prolactin levels in *Carassius auratus* exposed to various photoperiod-temperature regimes. Endocr. Res. Commun. **5**, 199–210 (1978)

Vogel, W.H., Orfei, V., Century, B.: Activities of enzymes involved in the formation and destruction of biogenic amines in various areas of human brain. J. Pharmacol. exp. Ther. **165**, 196–203 (1969)

Volkman, P.H., Heller, A.: Pineal N-acetyltransferase activity: Effect of sympathetic stimulation. Science **173**, 839–840 (1971)

Vollrath, L.: Synaptic ribbons of a mammalian pineal gland. Circadian changes. Z. Zellforsch. **145**, 171–183 (1973)

Vollrath, L.: Zur Funktion der "synaptic ribbons" der Säugerzirbeldrüse. Verh. anat. Ges. (Jena) **68**, 427–432 (1974)

Vollrath, L.: Die Struktur der Meerschweinchenzirbeldrüse unter dem Einfluß von Licht und Dunkelheit. Verh. anat. Ges. (Jena) **69**, 447–451 (1975)

Vollrath, L., Boeckmann, D.: Comparative anatomy of the rodent pineal complex. Gen. comp. Endocr. **34**, Abstract No. 39 (1978)

Vollrath, L., Howe, C.: Light and drug induced changes of epiphysial synaptic ribbons. Cell Tiss. Res. **165**, 383–390 (1976)

Vollrath, L., Huss, H.: The synaptic ribbons of the guinea-pig pineal gland under normal and experimental conditions. Z. Zellforsch. **139**, 417–429 (1973)

Vollrath, L., Schmidt, D.S.: Enzymhistochemische Untersuchungen an der Zirbeldrüse normaler und trächtiger Meerschweinchen. Histochemie **20**, 328–337 (1969)

Vollrath, L., Kantarjian, A., Howe, C.: Mammalian pineal gland: 7-day rhythmic activity? Experientia (Basel) **31**, 458–460 (1975)

Von Bartheld, F. Moll, J.: The vascular system of the mouse epiphysis with remarks on the comparative anatomy of the venous trunks in the epiphyseal area. Acta anat. (Basel) **22**, 227–235 (1954)

Von Eyl, O., Gusek, W.: Zur Frage von Altersveränderungen der Zirbeldrüse beim Menschen. Verh. dtsch. Ges. Path. **59**, 400–404 (1975)

Von Faber, H.: Vergleichende Untersuchungen über den Einfluß von Diäthylstilböstrol und der Kastration auf die innersekretorischen Organe des Hahnes. Z. mikr.-anat. Forsch. **61**, 1–36 (1954)

Von Haffner, K.: Über die progressive und regressive Entwicklung der Pinealblase (Parietalorgan) des Krallenfrosches (*Xenopus laevis* Daud.). Zool. Anz., Suppl. **15**, 93–100 (1951)

Von Haffner, K.: Die Pinealblase (Stirnorgan, Parietalauge) von *Xenopus laevis* Daud. und ihre Entwicklung, Verlagerung und Degeneration. Zool. Hb. Abt. Anat. u. Ontog. **71**, 375–412 (1952)

Von Haffner, K.; Untersuchungen über die Entwicklung des Parietalorgans und der Parietalnerven von *Lacerta vivipara* und das Problem der Organe der Parietalregion. Z. wiss. Zool. **157**, 1–34 (1953)

Von Studnitz, W.: Tryptophanhydroxylase im Corpus pineale beim Menschen. Experientia (Basel) **23**, 711 (1967)

Voss, H.: Beobachtung dreier selbständiger juxtapinealer Konkrementkörperchen an einem menschlichen Gehirn sowie topochemische Untersuchungen an ihren Kalkkonkrementen und an Kolloidkugeln im benachbarten Nervengewebe. Anat. Anz. **104**, 367–371 (1957)

Vriend, J., Reiter, R.J.: Free thyroxin index in normal, melatonin-treated and blind hamsters. Horm. Metab. Res. **9**, 231–234 (1977)

Vriend, J., Sackman, J.W., Reiter, R.J.: Effects of blinding, pinealectomy and superior cervical ganglionectomy on free thyroxine index of male golden hamsters. Acta endocr. (Kbh.) **86**, 758–762 (1977)

Vullings, H.G.B.: Influence of light and darkness on the hypothalamo-hypophysial system of *Rana esculenta* and the involvement of the pineal complex. Z. Zellforsch. **146**, 491–500 (1973)

Wagner, H.-J.: Darkness-induced reduction of the number of synaptic ribbons in fish retina. Nature (Lond.) New Biol. **246**, 53–55 (1973)

Wagner, H.-J.: Quantitative changes of synaptic ribbons in the cone pedicles of *Nannacara*: light dependent or governed by a circadian rhythm? In: Vision in fishes: New approaches in research. (M.A. Ali, ed.), pp. 679–686. New York: Plenum Press 1975a

Wagner, H.-J.: Der Einfluß von Licht und Tagesrhythmus auf die "Ribbon-Synapsen" in der Retina. Verh. anat. Ges. (Jena) **69**, 651–658 (1975b)

Wainwright, S.D.: Course of the increase in hydroxyindole-O-methyltransferase activity in the pineal gland of the chick embryo and young chick. J. Neurochem. **22**, 193–196 (1974)

Wainwright, S.D., Wainwright, L.K.: Regulation of the diurnal cycle in activity of serotonin acetyltransferase in the chick pineal gland. Canad. J. Biochem. **56**, 685–690 (1978)

Wake, K.: Acetylcholinesterase-containing nerve cells and their distribution in the pineal organ of the goldfish, *Carassius auratus*. Z. Zellforsch. **145**, 287–298 (1973)

Wake, K., Ueck, M., Oksche, A.: Acetylcholinesterase-containing nerve cells in the pineal complex and subcommissural area of the frogs, *Rana ridibunda* and *Rana esculenta*. Cell Tiss. Res. **154**, 423–424 (1974)

Walker, R.F., Wilhoft, D.C.: Differential effects of light on acetylserotonin methyltransferase (ASMT) activity in lizard brains. Amer. Zoologist **11**, 650 (1971)

Wallace, R.B., Altman, J., Das, G.D.: An autoradiographic and morphological investigation of the postnatal development of the pineal body. Amer. J. Anat. **126**, 175–183 (1969)

Wallen, E.P.: Photoperiod, the pineal gland, and reproductive function in the rat. Ph. D. Thesis. Kansas Univ. Diss./1–272, 1972

Wallen, E.P., Yochim, J.M.: Pineal hydroxyindole-O-methyltransferase (HIOMT) and reproductive cyclicity in the rat. Fed. Proc. **30**, 610 Abs. (1971)

Wallen, E.P., Yochim, J.M.: Pineal rhythmicity and its relation to reproduction in the rat. Biol. Reprod. **7**, 111 (1972)

Wallen, E.P., Yochim, J.M.: Rhythmic function of pineal hydroxyindole-O-methyl transferase during the estrous cycle: an analysis. Biol. Reprod. **10**, 461–466 (1974a)

Wallen, E.P., Yochim, J.M.: Pineal HIOMT activity in the rat: effect of ovariectomy and hormone replacement. Biol. Reprod. **10**, 474–479 (1974b)

Wallen, E.P., Yochim, J.M.: Photoperiodic regulation of the estrous cycle of the rat: role of the pineal gland. Biol. Reprod. **11**, 117–124 (1974c)

Waltimo, O., Talanti, S.: Histochemical localization of β-glucuronidase in the rat brain. Nature (Lond.) **205**, 499–500 (1965)

Walton, K.: Teratomas of pineal region and their relationship to pinealomas. J. Path. Bact. **61**, 11–21 (1949)

Warembourg, M.: Radioautographic study of the rat brain and pituitary after injection of ^3H dexamethasone. Cell Tiss. Res. **161**, 183–191 (1975)

Wartenberg, H.: Elektronenmikroskopische Untersuchungen an der Epiphysis cerebri der Katze. Anat. Anz., Suppl. ad **115**, 275–279 (1965)

Wartenberg, H.: Electron microscopic studies on nerve and glia processes and their relations to cells and vessels of the mammalian pineal organ. Anat. Rec. **154**, 439 (1966)

Wartenberg, H.: The mammalian pineal organ: Electron microscopic studies on the fine structure of pinealocytes, glial cells and on the perivascular compartment. Z. Zellforsch. **86**, 74–97 (1968)

Wartenberg, H.: Vergleichende Untersuchungen über die Feinstruktur des Pinealorgans der Säuger und anderer Wirbeltiere. Anat. Anz., Suppl. ad. **125**, 439–445 (1969a)

Wartenberg, H.: Vergleichende Untersuchungen über das Vorkommen von biogenen Aminen in den circumventrikulären Organen von Reptilien und Säugern. In: Zirkumventrikuläre Organe und Liquor (G. Sterba, ed.), pp. 161–165. Jena: Gustav Fischer 1969b

Wartenberg, H.: Pineal. In: The Encyclopedia of Microscopy and Microtechnique (P. Gray, ed.), pp. 440–445. New York: Van Nostrand Reinhold Company 1973

Wartenberg, H., Baumgarten, H.G.: Elektronenmikroskopische Untersuchungen zur Frage der photosensorischen und sekretorischen Funktion des Pinealorgans von *Lacerta viridis* und *L. muralis*. Z. Anat. Entwickl.-Gesch. **127**, 99–120 (1968)

Wartenberg, H., Baumgarten, H.G.: Untersuchungen zur fluoreszenz- und elektronenmikroskopischen Darstellung von 5-Hydroxytryptamin (5-HT) im Pineal-Organ von *Lacerta viridis* und *L. muralis*. Z. Anat. Entwickl.-Gesch. **128**, 185–210 (1969a)

Wartenberg, H., Baumgarten, H.G.: Über die elektronenmikroskopische Identifizierung von noradrenergen Nervenfasern durch 5-Hydroxydopamin und 5-Hydroxydopa im Pinealorgan der Eidechse (*Lacerta muralis*). Z. Zellforsch. **94**, 252–260 (1969b)

Wartenberg, H., Gusek, W.: Elektronenmikroskopische Untersuchungen über die Epiphysis cerebri des Kaninchens. Anat. Anz., Suppl. ad **113**, 173–181 (1964)

Wartenberg, H., Gusek, W.: Licht- und elektronenmikroskopische Beobachtungen über die Struktur der Epiphysis cerebri des Kaninchens. Progr. Brain Res. **10**, 296–316 (1965)

Wartenberg, H., Hadžiselimović, F., Seguchi, H.: Experimentelle Untersuchungen über die Passage der Blut-Gewebs-Schranke in den zirkumventrikulären Organen des Meerschweinchengehirns. Verh. anat. Ges. (Jena) **66**, 345–355 (1972)

Wartman, S.A., Branch, B.J., George, R., Newman Taylor, A.: Evidence for a cholinergic influence on pineal hydroxyindole-O-methyltransferase activity with changes in environmental lighting. Life Sci. **8**, (Part I), 1263–1270 (1969a)

Wartman, S.A., Newman Taylor, A., George, R.: Evidence for parasympathetic influence on pineal hydroxyindole-O-methyltransferase activity (HIOMT) with changes in environmental lighting. Fed. Proc. **28**, 443 Abs. (1969b)

Warwick, R., Williams, P.L.: Gray's Anatomy. 35th edit. Edinburgh: Longman 1973

Weber, G.: Tumoren der Glandula pinealis und ektopische Pinealozytome. Schweiz. Arch. Neurol. Neurochir. Psychiat. **90/91**, 473–509 (1963)

Weichselbaum, R., Patel, M., Das Gupta, T.K.: Influence of the pineal on wound healing. Nature (London) **254**, 349 (1975)

Weinberg, U., D'Eletto, R.D., Weitzman, E.D., Erlich, S., Hollander, C.S.: Circulating melatonin in man: Episodic secretion throughout the light-dark cycle. J. clin. Endocr. **48**, 114–118 (1979)

Weindl, A.: Verhalten der zirkumventrikulären Organe des Kaninchens nach intravenöser Trypanblau-Zufuhr. Naturwissenschaften **54**, 342 (1967)

Weindl, A., Schinko, I.: Scanning electron microscopy (SEM) of circumventricular organs. Scanning electron microscopy 1977, Vol. II, pp. 559–566. Proceeding of the Workshop on Biomedical Applications – SEM Studies of Circumventricular Organs. IIT Research Institute, Chicago, Ill., USA, March 1977

Weindl, A., Sofroniew, M.V.: Neurohormones and circumventricular organs. In: Brain-Endocrine interaction III. Neural hormones and reproduction (D.E. Scott, G.P. Kozlowski, A. Weindl, eds.), pp. 117–137. Basel: Karger 1978

Weiss, B.: Differences in the stimulatory effects of norepinephrine (NE) and sodium fluoride on adenyl cyclase (AC) of pineal gland and cerebellum. Fed. Proc. **27**, 752 Abs. (1968a)

Weiss, B.: Discussion of the formation, metabolism, and physiologic effects of melatonin. Advanc. Pharmacol. Chemother. **6**A, 152–155 (1968b)

Weiss, B.: Effects of environmental lighting and chronic denervation on the activation of adenyl cyclase of rat pineal gland by norepinephrine and sodium fluoride. J. Pharmacol. exp. Ther. **168**, 146–152 (1969)

Weiss, B.: Factors affecting adenyl cyclase activity and its sensitivity to biogenic amines. In: J.J. Blum (ed.), Biogenic amines as physiological regulators, pp. 35–73. Englewood Cliffs, N.J.: Prentice-Hall (1970)

Weiss, B.: On the regulation of adenyl cyclase activity in the rat pineal gland. Ann. N.Y. Acad. Sci. **185**, 507–519 (1971)

Weiss, B., Costa, E.: Adenyl cyclase activity in rat pineal gland: Effects of chronic denervation and norepinephrine. Science **156**, 1750–1752 (1967a)

Weiss, B., Costa, E.: Effect of denervation and environmental lighting on the norepinephrine reduced activation of adenyl cyclase of the rat pineal gland. Fed. Proc. **26**, 767 (1967b)

Weiss, B., Costa, E.: Regional and subcellular distribution of adenyl cyclase and 3′,5′-cyclic nucleotide phosphodiesterase in brain and pineal gland. Biochem. Pharmacol. **17**, 2107–2116 (1968a)

Weiss, B., Costa, E.: Selective stimulation of adenyl cyclase of rat pineal gland by pharmacologically active catecholamines. J. Pharmacol. exp. Ther. **161**, 310–319 (1968b)

Weiss, B., Crayton, J.: Gonadal hormones as regulators of pineal adenyl cyclase activity. Endocrinology **87**, 527–533 (1970a)

Weiss, B., Crayton, J.: Ovarian regulation of the norepinephrine-sensitive adenyl cyclase system of rat pineal gland. Fed. Proc. **29**, 615 (1970b)

Weiss, B., Crayton, J.W.: Neural and hormonal regulation of pineal adenyl cyclase activity. In: Greengard, P. and Costa, E. (eds.), Role of cyclic AMP in cell function, pp. 217–239. New York: Raven Press 1970c

Weiss, B., Strada, S.J.: Neuroendocrine control of the cyclic AMP system of brain and pineal gland. In: Advances in cyclic nucleotide research, Vol. 1, pp. 357–374. New York: Raven Press 1972

Weissbach, H., Axelrod, J.: The enzymatic biosynthesis of melatonin. Fed. Proc. **19**, 50 (1960)

Weissbach, H., Redfield, B.G., Axelrod, J.: Biosynthesis of melatonin: enzymatic conversion of serotonin to N-acetylserotonin. Biochim. biophys. Acta (Amst.) **43**, 352–353 (1960)

Weissbach, H., Redfield, B.G., Axelrod, J.: The enzymatic acetylation of serotonin and other naturally occurring amines. Biochim. biophys. Acta (Amst.) **54**, 190–192 (1961)

Weisz, J., O'Brien, L.V., Lloyd, T.: Methylation of estradiol-17-β by a partially purified preparation of bovine pineal hydroxyindole-O-methyltransferase. Endocrinology **102**, 330–333 (1978)

Wells, J.W.: Steroids in the pineal gland of the domestic fowl (*Gallus domesticus*): cholesterol and its biosynthesis. Comp. Biochem. Physiol. **40**B, 723–731 (1971)

Wells, S.A., Wurtman, R.J., Rabson, A.S.: Viral neoplastic transformation of hamster pineal cells in vitro: Retention of enzymatic function. Science **154**, 278–279 (1966)

Welser, J.R., Hinsman, E.J., Stromberg, M.W.: Fine structure of the canine pinealocytes. Amer. J. vet. Res. **29**, 587–599 (1968)

Welsh, M.G., Reiter, R.J.: The pineal gland of the gerbil, *Meriones unguiculatus*. I. An ultrastructural study. Cell Tiss. Res. **193**, 323–336 (1978)

Wendel, O.T., Waterbur, L.D., Pearce, L.A.: Increase in monoamine concentration in rat brain following melatonin administration. Experientia (Basel) **30**, 1167–1168 (1974)

Werner, R., Devečerski, V., Milin, J., Milin, R.: Morphodynamik der Schilddrüse der epiphysektomierten Ratten beim Kältestress. Verh. anat. Ges. (Jena) **66**, 223–233 (1971)

Wersäll, J., Bagger-Sjöbäck, D.: Morphology of the vestibular sense organ. In: H.H. Kornhuber, ed., Handbook of sensory physiology, Vol. VI, 1, pp. 123–170. Berlin-Heidelberg-New York: Springer 1974

Wersäll, J., Flock, A., Lundquist, P.G.: Structural basis for directional sensitivity in cochlear and vestibular sensory receptors. Cold Spr. Harb. Symp. quant. Biol. **30**, 115–132 (1965)

Wetterberg, L.: Melatonin in serum. Nature (Lond.) **269**, 646 (1977)

Wetterberg, L., Arendt, J.: Melatonin and menstrual cycle. Lancet **1975II**, 417

Wetterberg, L., Geller, E., Yuwiler, A.: Harderian gland: An extraretinal photoreceptor influencing the pineal gland in neonatal rats? Science **167**, 884–885 (1970a)

Wetterberg, L., Yuwiler, A., Ulrich, R., Geller, E., Wallace, R.: Harderian gland: Influence on pineal hydroxyindole-O-methyltransferase activity in neonatal rats. Science **170**, 194–196 (1970b)

Wetterberg, L., Eriksson, O., Friberg, Y., Vangbo, B.: A simplified radioimmunoassay for melatonin and its application to biological fluids. Preliminary observations on the half-life of plasma melatonin in man. Clin. chim. Acta **86**, 169–177 (1978)

Wetzig, H.: Die Entwicklung der Organe des Zwischenhirndaches (Epiphyse und Plexus chorioideus anterior) bei der Sturmmöve *Larus canus* L. Gegenbaurs morph. Jb. **101**, 406–431 (1961)

White, W.F., Hedlund, M.T., Weber, G.F., Rippel, R.H., Johnson, E.S., Wilber, J.F.: The pineal gland: a supplemental source of hypothalamic-releasing hormones. Endocrinology **94**, 1422–1426 (1974)

Wight, P.A.L.: Development of sympathetic innervation of the pineal of the domestic fowl. Acta morph. neerl.-scand. **9**, 47–56 (1971)

Wight, P.A.L., MacKenzie, G.M.: Dual innervation of the pineal of the fowl, *Gallus domesticus*. Nature (Lond.) **228**, 474–476 (1970)

Wight, P.A.L., MacKenzie, G.M.: The histochemistry of the pineal gland of the domestic fowl. J. Anat. (Lond.) **108**, 261–273 (1971)

Wiklund, L.: Development of serotonin-containing cells and the sympathetic innervation of the habenular region in the brain. A fluorescence histochemical study. Cell Tiss. Res. **155**, 231–243 (1974)

Wildi, R., Frauchiger, E.: Modifications histologiques de l'épiphyse humaine pendant l'enfance, l'âge adulte et le vieillissement. Progr. Brain Res. **10**, 218–233 (1965)

Wilkinson, M.: Growth and morphological differentiation of rat pineal cells in monolayer cultures. J. Endocr. **64**, 71 (1975)

Wilkinson, M.: The sensitivity of pineal gland β-receptors appears to be dependent upon calcium ions. Pflügers Arch. **373**, 209–210 (1978)

Wilkinson, M., Arendt, J.: Effects of oestrogen and progesterone on rat pineal N-acetyl transferase activity and melatonin production. Experientia (Basel) **34**, 667–669 (1978)

Wilkinson, M., Arendt, J., De Ziegler, D., Ruf, K.B.: Changes in sensitivity of pinealocyte β-receptors during precocious sexual maturation or after postnatal sterilisation with estrogen in the female rat. Ann. Biol. anim. **16**, 457–464 (1976)

Wilkinson, M., Arendt, J., Bradtke, J., De Ziegler, D.: Determination of a dark-induced increase of pineal N-acetyl transferase activity and simultaneous radioimmunoassay of melatonin in pineal, serum and pituitary tissue of the male rat. J. Endocr. **72**, 243–244 (1977)

Wilson, B.W., Lynch, H.J., Ozaki, Y.: 5-Methoxytryptophol in rat serum and pineal: Detection, quantitation, and evidence for daily rhythmicity. Life Sci. **23**, 1019–1024 (1978)

Wilson, J.F., Dodd, J.M.: The role of the pineal complex and lateral eyes in the colour change response of the dogfish, *Scylliorhinus canicula* L. J. Endocr. **58**, 591–598 (1973)

Winget, C.M., Warren, C.A., DeRoshia, C.W.: Interrelationships of the pineal gland, the diencephalon and the pituitary (*Gallus domesticus*). Amer. Zool. **7**, 732 (1967)

Winston, M., Johnson, E., Kelleher, J.K., Banerjee, S., Margulis, L.: Melatonin: cellular effects on live stentors correlated with the inhibition of colchicine-binding to microtubule protein. Cytobios **9**, 237–243 (1974)

Winters, K.E., Morrissey, J.J., Loos, P.J., Lovenberg, W.: Pineal protein phosphorylation during serotonin N-acetyltransferase induction. Proc. nat. Acad. Sci. (Wash.) **74**, 1928–1931 (1977)

Wislocki, G.B., Dempsey, E.W.: The chemical histology and cytology of the pineal body and neurohypophysis. Endocrinology **42**, 56–72 (1948)

Wislocki, G.B., Leduc, E.H.: Vital staining of the hematoencephalic barrier by silver nitrate and trypan blue, and cytological comparisons of the neurohypophysis, pineal body, area postrema, intercolumnar tubercle and supraoptic crest. J. comp. Neurol. **96**, 371–414 (1952)

Wolfe, D.E.: The epiphyseal cell: an electron-microscopic study of its intercellular relationships and intracellular morphology in the pineal body of the albino rat. Progr. Brain Res. **10**, 332–386 (1965)

Wolfe, D.E., Potter, L.T., Richardson, K.C., Axelrod, J.: Localizing tritiated norepinephrine in sympathetic axons by electron microscopic autoradiography. Science **138**, 440–442 (1962)

Wolstenholme, G.E.W., Knight, J.: The pineal gland. A Ciba Foundation Symposium. Edinburgh, London: Churchill Livingstone 1971

Wong, P.Y., Fritze, K.: Determination by neutron activation of copper, manganese, and zinc in the pineal body and other areas of brain tissue. J. Neurochem. **16**, 1231–1234 (1969)

Wong, R., Whiteside, C.B.C.: The effect of melatonin on the wheelrunning activity of rats deprived of food. J. Endocr. **40**, 383–384 (1968)

Wood, J.G.: Electron microscopy of pineal gland biogenic amines. J. Cell Biol. **55**, Abstr. No. 577 (1972)

Wood, J.G.: The effects of niamid and reserpine on the nerve endings of the pineal gland. Z. Zellforsch. **145**, 151–166 (1973)

Wood, J.G., Duncan, D.: The cytochemical localization of catecholamines and 5-hydroxytryptamine in the pineal gland. J. Cell Biol. **39**, 144a (1968)

Woolley, D.W., Van der Hoeven, T.: Prevention of a mental defect of phenylketonuria with serotonin congeners such as melatonin or hydroxytryptophan. Science **144**, 1593–1594 (1964)

Wragg, L.E.: Effects of pinealectomy in newborn female rats. Anat. Rec. **151**, 435–436 (1965)

Wragg, L.E.: Effects of pinealectomy in the newborn female rat. J. Anat. (Lond.) **120**, 391–402 (1967)

Wragg, L.E., Machado, C.R.S., Snyder, S.H., Axelrod, J.: Anterior chamber pineal transplants: Their metabolic activity and independence of environmental lighting. Life Sci. **6**, 31–38 (1967)

Wragg, L.E., Machado, C.R.S., Machado, A.B.M.: Serotonin rhythm independent of sympathetic nerves in the pineal body of the immature rat. Anat. Rec. **160**, 453–454 (1968)

Wright, N., Preslock, J.P.: Melatonin inhibition of pineal enzymes in *Coturnix* quail. Neuroendocrinology **19**, 177–184 (1975)

Wurtman, R.J., Anton-Tay, F.: The mammalian pineal as a neuroendocrine transducer. Recent Progr. Hormone Res. **25**, 493–522 (1969)

Wurtman, R.J., Axelrod, J.: Light and melatonin synthesis in the pineal Fed. Proc. **23**, 206 (1964a)

Wurtman, R.J., Axelrod, J.: Demonstration of hydroxyindole-O-methyl transferase, melatonin and serotonin in a metastatic parenchymatous pinealoma. Nature (Lond.) **204**, 1323–1324 (1964b)

Wurtman, R.J., Axelrod, J.: The pineal gland. Sci. Amer. **213**, 50–60 (1965)

Wurtman, R.J., Axelrod, J.: A 24-hour rhythm in the content of norepinephrine in the pineal and salivary glands of the rat. Life Sci. **5**, 655–669 (1966a)

Wurtman, R.J., Axelrod, J.: The effect of chlorpromazine and other drugs on the physiologic disposition of circulating melatonin. Nature (Lond.) **212**, 312 (1966b)

Wurtman, R.J., Kammer, H.: Melatonin synthesis by an ectopic pinealoma. New Engl. J. Med. **274**, 1233–1237 (1966)

Wurtman, R.J., Ozaki, Y.: Physiological control of melatonin synthesis and secretion: Mechanisms generating rhythms in melatonin, methoxytryptophol, and arginine vasotocin levels and effects on the pineal of endogenous catecholamines, the estrous cycle, and environmental lighting. J. Neural Transm., Suppl. **13**, 59–70 (1978)

Wurtman, R.J., Shein, H.M.: Lack of effect of increased pineal serotonin content on H^3-tryptophan uptake. J. Neural Transm. **36**, 177–181 (1975)

Wurtman, R.J., Altschule, M.D., Holmgren, U.: Effects of pinealectomy and of bovine pineal extract in rats. Amer. J. Physiol. **197**, 108–110 (1959)

Wurtman, R.J., Altschule, M.D., Greep, R.O., Falk, J.L., Grave, G.: The pineal gland and aldosterone. Amer. J. Physiol. **199**, 1109–1111 (1960)

Wurtman, R.J., Roth, W., Altschule, M.D., Wurtman, J.J.: Interactions of the pineal and exposure to continuous light on organ weights of female rats. Acta endocr. (Kbh.) **36**, 617–624 (1961)

Wurtman, R.J., Axelrod, J., Phillips, L.: Melatonin synthesis in the pineal gland: control by light. Science **142**, 1071–1073 (1963a)

Wurtman, R.J., Axelrod, J., Chu, E.W.: Melatonin, a pineal substance: effect on the rat ovary. Science **141**, 277–278 (1963b)

Wurtman, R.J., Axelrod, J., Fischer, J.E.: Melatonin synthesis in the pineal gland: effect of light mediated by the sympathetic nervous system. Science **143**, 1328–1330 (1964a)

Wurtman, R.J., Axelrod, J., Chu, E.W., Fischer, J.E.: Mediation of some effects of illumination on the rat estrous cycle by the sympathetic nervous system. Endocrinology **75**, 266–272 (1964b)

Wurtman, R.J., Axelrod, J., Barchas, J.D.: Age and enzyme activity in the human pineal. J. clin. Endocr. **24**, 299–301 (1964c)

Wurtman, R.J., Axelrod, J., Toch, R.: Demonstration of hydroxyindole-O-methyl transferase, melatonin and serotonin in a metastatic parenchymatous pinealoma. Nature (Lond.) **204**, 1323–1324 (1964d)

Wurtman, R.J., Axelrod, J., Potter, L.T.: The uptake of H^3-melatonin in endocrine and nervous tissues and the effects of constant light exposure. J. Pharmacol. exp. Ther. **143**, 314–318 (1964e)

Wurtman, R.J., Axelrod, J., Chu, E.W.: The relation between melatonin, a pineal substance, and the effects of light on the rat gonad. Ann. N.Y. Acad. Sci. **117**, 228–230 (1964f)

Wurtman, R.J., Axelrod, J., Snyder, S.H., Chu, E.W.: Changes of the enzymatic synthesis of melatonin in the pineal during the estrous cycle. Endocrinology **76**, 798–800 (1965a)

Wurtman, R.J., Axelrod, J., Toch, R.: Synthesis of melatonin by metastatic pinealoma tissue. Clin. Res. **13**, 250 (1965b)

Wurtman, R.J., Axelrod, J., Sedvall, G., Moore, R.Y.: Photic and neural control of the 24-hour norepinephrine rhythm in the rat pineal gland. J. Pharmacol. exp. Ther. **157**, 487–492 (1967a)

Wurtman, R.J., Axelrod, J., Chu, E.W., Heller, A., Moore, R.Y.: Median forebrain bundle lesions: blockade of effects of light on rat gonads and pineal. Endocrinology **81**, 509–514 (1967b)

Wurtman, R.J., Axelrod, J., Kelly, D.E.: The pineal. New York-London: Academic Press 1968a

Wurtman, R.J., Larin, F., Axelrod, J., Shein, H.M., Rosasco, K.: Formation of melatonin and 5-hydroxyindole acetic acid from ^{14}C-tryptophan by rat pineal gland in organ culture. Nature (Lond.) **217**, 953–954 (1968b)

Wurtman, R.J., Shein, H.M., Axelrod, J., Laren, F.: Incorporation of ^{14}C-tryptophan into ^{14}C-protein by cultured rat pineals: stimulation by l-norepinephrine. Proc. nat. Acad. Sci. (Wash.) **62**, 749–755 (1969)

Wurtman, R.J., Shein, H.M., Larin, F.: Mediation by β-adrenergic receptors of effect of norepinephrine on pineal synthesis of ^{14}C-serotonin and ^{14}C-melatonin. J. Neurochem. **18**, 1683–1687 (1971)

Wurzburger, R.J., Kawashima, K., Miller, R.L., Spector, S.: Determination of rat pineal gland melatonin content by a radioimmunoassay. Life Sci. **18**, 867–878 (1976)

Yamada, H., Ozawa, S., Kushima, S., Nakai, A.: Innervation of the pineal body and the subcommissural and supracommissural organs of the dog. Bull. Tokyo Med. Dent. Univ. **4**, 179–194 (1957)

Yamada, T.: The effect of electrical ablation of the nuclei habenulae, pineal body and subcommissural organ on endocrine function, with special reference to thyroid function. Endocrinology **69**, 706–711 (1961)

Yamashita, I., Hillman, J.C., Reiss, M.: Interrelation between human chorionic gonadotrophin inhibitory activity in plasma and urine and the rate of urinary gonadotrophin excretion. J. Endocr. **33**, 223–231 (1965)

Yamashita, K., Mieno, M., Shimizu, T., Yamashita, E.: Inhibition by melatonin of pituitary response to luteinizing hormone releasing hormone in vivo. J. Endocr. **76**, 487–491 (1978)

Yang, H.-Y.T., Goridis, C., Neff, N.H.: Properties of monoamine oxidase in sympathetic nerve and pineal gland. J. Neurochem. **19**, 1241–1250 (1972)

Yates, C.Y., Herbert, J.: Differential circadian rhythms in pineal and hypothalamic 5-HT induced by artificial photoperiods or melatonin. Nature (Lond.) **262**, 219–220 (1976)

Ying, S.-Y., Fiske, V.M.: Inhibition of ovulation by melatonin in the cyclic rat. Fed. Proc. **31**, 277 (1972)

Ying, S.-Y., Greep, R.O.: Inhibition of ovulation by melatonin in the cyclic rat. Endocrinology **92**, 333–335 (1973)

Yochim, J.M., Wallen, E.P.: HIOMT activity in the pineal gland of the female rat: effects of light. Biol. Reprod. **10**, 467–473 (1974a)

Yochim, J.M., Wallen, E.P.: Functions of the pineal gland of the rat during progestation: alterations in HIOMT rhythmicity. Biol. Reprod. **10**, 480–486 (1974b)

Yochim, J.M., Wallen, E.P.: Photoperiod, pineal HIOMT activity and reproductive function in the rat: a proposed model. Biol. Reprod. **11**, 125–132 (1974c)

Young, J.B., Landsberg, L.: Stimulation of the sympathetic nervous system during sucrose feeding. Nature (Lond.) **269**, 615–617 (1977)

Young, J.Z.: The photoreceptors of lampreys. II. The functions of the pineal complex. J. exp. Biol. Med. **12**, 254–270 (1935)

Yuwiler, A., Klein, D.C., Buda, M., Weller, J.L.: Adrenergic control of pineal N-acetyltransferase activity: developmental aspects. Amer. J. Physiol. **233**, E141–E146 (1977)

Zach, B.: Topographie und mikroskopisch-anatomischer Feinbau der Epiphysis cerebri von Hund und Katze. Zbl. Vet.-Med. **7**, 273–303 (1960)

Zadura, J., Roszkowski, J., Cakala, S.: Effect of pinealectomy on the weight and histological changes in the testes and adrenals of cockerels. Acta Physiol. pol. **20**, 117–124 (1969)

Zanoboni, A., Zanoboni-Muciaccia, W.: Experimental hypertension in pinealectomized rats. Life Sci. **6**, 2327–2331 (1967)

Zatz, M.: Effects of cholera toxin on supersensitive and subsensitive rat pineal glands: Regulation of sensitivity at multiple sites. Life Sci. **21**, 1267–1276 (1977)

Zatz, M.: Low concentrations of lithium inhibit the synthesis of cyclic AMP and cyclic GMP in the rat pineal gland. J. Neurochem. **32**, 1315–1321 (1979)

Zatz, M., Brownstein, M.J.: Central depressants rapidly reduce nocturnal serotonin N-acetyltransferase activity in the rat pineal gland. Brain Res. **160**, 381–385 (1979a)

Zatz, M., Brownstein, M.J.: Intraventricular carbachol mimics the effects of light on the circadian rhythm in the rat pineal gland. Science **203**, 358–361 (1979b)

Zatz, M., Romero, J.A.: Effects of calcium-free medium on the induction of serotonin N-acetyltransferase in the rat pineal. Biochem. Pharmacol. **27**, 2549–2553 (1978a)

Zatz, M., Weinstock, M.: Electric field stimulation releases norepinephrine and cyclic GMP from the rat pineal gland. Life Sci. **22**, 767–772 (1978b)

Zatz, M., Romero, J.A., Axelrod, J.: Diurnal variation in the requirement for RNA synthesis in the induction of pineal N-acetyltransferase. Biochem. Pharmacol. **25**, 903–906 (1976a)

Zatz, M., Kebabian, J.W., Romero, J.A., Lefkowitz, R.J., Axelrod, J.: Pineal beta adrenergic receptor: correlation of binding of ^{3}H-l-alprenolol with stimulation of adenylate cyclase. J. Pharmacol. exp. Ther. **196**, 714–722 (1976b)

Zboray, G.: The effect of continuous light or darkness on the subcommissural organ and glomerular zone of adrenals of the white rat. Acta biol. Acad. Sci. hung. **15**, 337–341 (1965)

Zieher, L.M., Pellegrino de Iraldi, A.: Central control of the noradrenaline content in the rat pineal and submaxillary glands. Life Sci. **5**, 155–161 (1966)

Zilles, K., Nickeleit, V.: Efferent connections from the brain to the frontal organ in *Rana temporaria* demonstrated by labeling with horseradish peroxidase. Cell Tiss. Res. **196**, 189–192 (1979)

Zimmerman, B.L., Tso, M.O.M.: Morphologic evidence of photoreceptor differentiation of pinealocytes in the neonatal rat. J. Cell Biol. **66**, 60–75 (1975)

Zimmerman, N.H.: Organization within the circadian system of the house sparrow. Hormonal coupling and the location of a circadian oscillator. PhD thesis. University of Texas at Austin, 1976

Zimmerman, N.H., Menaker, M.: Neural connections of sparrow pineal: Role in circadian control of activity. Science **190**, 477–479 (1975)

Zimmermann, P., Paul, E.: Reaktionsmuster verschiedener Mittel- und Zwischenhirnzentren von *Rana temporaria* L. nach Unterbrechung der Nervenbahnen des Pinealkomplexes. Z. Zellforsch. **128**, 512–537 (1972)

Zülch, K.J.: Biologie und Pathologie der Hirngeschwülste. In: Hdb. Neurochir., Bd. 3. Berlin-Heidelberg-New York: Spirnger 1956

Zurburg, W., Ebels, I.: Separation of pineal extracts by gelfiltration. I. Isolation from sheep pineals of a substance with special fluorescence characteristics. J. Neural Transm **35**, 117–124 (1974)

Zurburg, W., Ebels, I.: Separation of pineal extracts by gelfiltration. II. Idenfification and isolation of two indoles from sheep pineal glands. J. Neural Transm. **36**, 59–69 (1975)

Zweens, J.: Influence of the oestrous cycle and ovariectomy on the phospholipid content of the pineal gland in the rat. Nature (Lond.) **197**, 1114–1115 (1963)

Zweens, J.: The pineal lipid content in rat and the involvement of the epiphysis cerebri in the hypophysiogonadal interrelation. Ph. D. Dissertation, University Groningen 1964

Zweens, J.: Alterations of the pineal lipid content in the rat under hormonal influences. Progr. Brain Res. **10**, 540–551 (1965)

Zweig, M., Axelrod, J.: Relationship between catecholamines and serotonin in sympathetic nerves of the rat pineal gland. J. Neurobiol. **1**, 87–97 (1969)

Zweig, M.H., Snyder, S.H.: The development of serotonin and serotonin-related enzymes in the pineal gland of the rat. Commun. Behav. Biol. (A) **1**, 103–108 (1968)

Zweig, M., Snyder, S.H., Axelrod, J.: Evidence for a nonretinal pathway of light to pineal gland in newborn rats. Proc. nat. Acad. Sci. (Wash.) **56**, 515–520 (1966)

Appendix. Review Articles Not Mentioned in the Text

Axelrod, J.: Comparative biochemistry of the pineal gland. Amer. Zool. **10**, 259–267 (1970)

Axelrod, J.: The pineal gland. Endeavour XXIX. No. 108, 144–148 (1970)

Axelrod, J.: Neural control of indoleamine metabolism in the pineal. In: The pineal gland (G.E.W. Wolstenholme, J. Knight, eds.), pp. 35–52. Edinburgh-London: Churchill Livingstone 1971

Axelrod, J.: The pineal gland: A neurochemical transducer. Science **184**, 1341–1348 (1974)

Axelrod, J.: Relationship between catecholamines and other hormones. Recent Progr. Hormone Res. **31**, 1–35 (1975)

Axelrod, J., Wurtman, R.J.: The formation, metabolism and some actions of melatonin, a pineal gland substance. Res. Publ. Ass. nerv. ment. Dis. **43**, 200–211 (1966)

Axelrod, J., Wurtman, R.J.: The pineal gland: a biological clock. Probl. Act. Endocr. Nutr., Série **10**, 201–212 (1966)

Cardinali, D.P., Nagle, C.A., Rosner, J.M.: Pineal-gonadal relationships. Nature of the feedback mechanism at the level of the pineal gland. In: Neuroendocrine regulation of fertility. A potential new approach to human fertility regulation by interfering with neuroendocrine pathways. Proceedings of the International Symposium on Neuroendocrine regulation of fertility held at Simla, India, October 28–31, 1974, pp. 206–214. T.C. Anand Kumar (ed.). Basel: Karger 1976

Cohen, R.A., Wurtman, R.J., Axelrod, J., Snyder, S.H.: Some clinical, biochemical and physiological actions of the pineal gland. Ann. intern. Med. **61**, 1144–1161 (1964)

Geinitz, W.: Ergebnisse der Zirbeldrüsenforschung. Fortschr. Med. **81**, 357–362 (1963)

Gusek, W.: Alte und neue Kenntnisse zur Struktur und Funktion der Zirbeldrüse. Klin. Wschr. **45**, 1261–1262 (1967)

Horányi, B.: A corpus pineale szerkezete és müködisi jelentösége. Orv. Hetil. **107**, 1585–1997 (1966)

Ippen, H.: Das Zirbeldrüsenhormon Melatonin und die zentrale Pigmentsteuerung. Dtsch. med. Wschr. **86**, 307–314 (1961)

Kappers, J.A.: Epiphysis. In: Correlative anatomy of the nervous system (Crosby, E.C., Humphrey, T., Lauer, E.W., (eds,). New York: MacMillan Co. 1962

Kappers, J.A.: Innervation of the pineal organ: phylogenetic aspects and comparison of the neural control of the mammalian pineal with that of other neuroendocrine systems. Mem. Soc. Endocr. **19**, 27–48 (1971)

Karasek, M.: The role of the pineal body in mammals. Pol. Endocr. **22**, 315–327 (1971)

Kenny, G.C.T.: The epiphysis cerebri. In: The structure and function of nervous tissue (G.H. Bourne, ed.), Vol. VI, pp. 253–272. New York-London: Academic Press 1972

Krabbe, K.H.: La glande pinéale. Wld Neurol. **2**, 94–102 (1961)

Mastrangelo, D., Palma, A.: Moderne vedute sulla strutura e sulla funzione dell'epifisi (Rivista sintetica). Ric. Morf. **28/29**, 507–528 (1960–1963)

Mess, B., Trentini, G.P., Kovacs, I., De Gaetani, C.F.: Melatonin, cerebrospinal fluid, pineal gland interrelationships. Brain-Endocrine Interaction II. The ventricular system in neuroendocrine mechanisms. Proceeding of the second Intern. Symposium on Brain-Endocrine Function held in Shizuola, Jap. October 16–18, 1974. Ed. by K.M. Knigge, D.E. Scott, H., Kobayashi, S. Ishii. Basel: Karger 1975

Milcu, St.-M.: Epifiza, glandă endocrină. Bucharest: Ed. Acad. R.P.R. 1957

Minneman, K.P., Wurtman, R.J.: Effects of pineal compounds on mammals. Life Sci. **17**, 1189–1200 (1975)

Piechowiak, H.: Das Pinealorgan: Anatomie, Stoffwechselwege, Physiologie und Klinik. Dtsch. med. Wschr. **98**, 2088–2094 (1973)

Preslock, J.P.: Gonadal steroid regulation of pineal melatonin synthesis. Life Sci. **20**, 1299–1304 (1977)

Quay, W.B.: Endocrine effects of the mammalian pineal. Amer. Zoologist **10**, 237–246 (1970)

Quay, W.B.: The significance of the pineal. Mem. Soc. Endocr. **18**, 423–445 (1970)

Reiter, R.J., Sorrentino, S.: Reproductive effects of the mammalian pineal. Amer. Zoologist **10**, 247–258 (1970)

Relkin, R.: The pineal gland. New Engl. J. Med. **274**, 944–950 (1966)

Rodin, A.E.: The present status of the pineal gland. Alberta med. Bull. **26**, 2–7 (1961)

Rothschuh, K.E.: Descartes und die Zirbeldrüse. In: Wissenschaft, Wirtschaft u. Technik, K.H. Manegold, ed. Festschrift f. W. Treue zum 60. Geburtstag, pp. 438–447. München 1969

Thiéblot, L.: Physiology of the pineal body. Progr. Brain Res. **10**, 479–488 (1965)

Vaughan, G.M., Meyer, G.G., Reiter, R.J.: Evidence for a pineal-gonad relationship in the human. Progr. Reprod. Biol. **4**, 191–223 (1978)

Vollrath, L.: Über Bau und Funktion der Zirbeldrüse von Säugern. Materia Medica Nordmark **21**, 601–605 (1969)

Vollrath, L.: Die Zirbeldrüse – ein neuroendocriner Rezeptor und Effektor des Sympathikus. Hippokrates (Stuttg.) **45**, 236–238 (1974)

Vollrath, L.: Die Zirbeldrüsenfunktion bei Tier und Mensch unter dem Einfluß von Licht und Dunkelheit. Hippokrates (Stuttg.) **47**, 84–85 (1976)

Weindl, A.: Neuroendocrine aspects of circumventricular organs. In: Frontiers in neuroendocrinology, 1973 (W.F. Ganong and L. Martini, eds.), pp. 3–32. New York: Oxford University Press 1973)

Weindl, A., Schinko, I., Dommasch, D.: Circumventrikuläre Organe und Ventrikelsystem. In: Cerebrospinalflüssigkeit-CSF (Z.H.G. Meertens, ed.), pp. 66–78. Stuttgart: Georg Thieme (in press)

Wiederanders, R.E., Hirsch, S.M., Evans, G.E.: A theoretical schema of pineal function. Proc. North Dakota Acad. Sci. **21**, 171–179 (1967)

Wiener, H.: External chemical messengers. V. More functions of the pineal gland. N.Y. St. J. Med. **68**, 1019–1038 (1968)

Wurtman, R.J.: Effects of light and visual stimuli on endocrine function. In: Neuroendocrinology (L. Martini, W.F. Ganong, eds.), Vol. 2, pp. 19–59. New York-London: Academic Press 1967

Wurtman, R.J.: The pineal gland in relation to reproduction. Amer. J. Obstet. Gynec. **104**, 320–326 (1969)

Wurtman, R.J., Axelrod, J.: The physiologic effects of melatonin and the control of its biosynthesis. Probl. Act. Endocr. Nutr., Série **10**, 189–200 (1966)

Young, J.Z.: The pineal gland. Philosophy **48**, 70–74 (1973)

Zatz, M., Kebabian, J.W., O'Dea, R.F.: Regulation of beta-adrenergic function in rat pineal gland. Receptors and hormone action. Ed. by L. Birnbaumer, B.W. O'Malley, Vol. 3, pp. 195–219. New York: Academic Press 1978

Author Index

Page numbers in *italics* refer to the bibliography

Subject Index

Handbuch der mikroskopischen Anatomie des Menschen

Begründet von
W. v. Möllendorff
Fortgeführt von W. Bargmann
Herausgegeben von A. Oksche,
L. Vollrath

Band 1
Die lebendige Masse

Teil 1
Allgemeine mikroskopische Anatomie und Organisation der lebendigen Masse
Bearbeiter: G. Hertwig,
F. K. Studnička, E. Tschopp
Reprint der Erstauflage
Berlin 1929
1978. 453 zum Teil farbige Abbildungen. (4) XII, 626 Seiten
Gebunden DM 430,-
ISBN 3-540-07843-6

Teil 2
Wachstum und Vermehrung der lebendigen Masse
1929. 464 zum Teil farbige Abbildungen. X, 807 Seiten
DM 430,-
ISBN 3-540-01094-7

Teil 3
Chromosomes
in Mitosis and Interphase
By H. G. Schwarzacher
1976. 116 figures, 3 tables.
VIII, 182 pages
Cloth DM 136,-
ISBN 3-540-07456-2

Band 2
Die Gewebe

Teil 1
Epithel- und Drüsengewebe, Bindegewebe und blutbildende Gewebe, Blut
Bearbeiter: J. Brodersen,
A. Maximow, J. Schaffer
Reprint der Erstauflage
Berlin 1927
1978. 305 zum Teil farbige Abbildungen, 1 Tafel. (4) X, 703 Seiten
Gebunden DM 460,-
ISBN 3-540-07844-4

Teil 2
Stützgewebe, Knochengewebe. Skeletsystem
1930. 521 zum Teil farbige Abbildungen. VIII, 699 Seiten
DM 400,-
ISBN 3-540-01117-X

Teil 3
Gewebe und Systeme der Muskulatur
1931. 137 zum teil farbige Abbildungen. VI, 247 Seiten
DM 210,-
ISBN 3-540-01145-5

Teil 4
Gewebe und Systeme der Muskulatur
(Ergänzungsband zu Band 2/3)
1956. 40 Abbildungen.
VIII, 119 Seiten
DM 110,-
ISBN 3-540-02031-4
Einbanddecke DM 19,80
ISBN 3-540-02032-2

Teil 5
K.-H. Knese
Stützgewebe und Skelettsystem
1979. 299 Abbildungen in 677 Einzeldarstellungen,
24 Tabellen. XI, 938 Seiten
Gebunden DM 610,-
ISBN 3-540-08807-5

Band 3
Haut- und Sinnesorgane

Teil 1
Haut, Milchdrüse, Geruchsorgan, Geschmacksorgan, Gehörorgan
Bearbeiter: H. von Eggeling,
H. Hoepke, W. Volmer
Reprint der Erstauflage
Berlin 1927
1978. 321 zum Teil farbige Abbildungen. (4) VII, 505 Seiten
Gebunden DM 360,-
ISBN 3-540-07845-2

Teil 2
Auge
1936. 475 zum Teil farbige Abbildungen. VIII,782 Seiten
DM 360,-
ISBN 3-540-01226-5

Teil 3
Die Haut. Die Milchdrüse
(Ergänzung zu Band 3/1)
1957. 359 zum Teil farbige Abbildungen. VII, 524 Seiten
Gebunden DM 390,-
ISBN 3-540-02160-4

Teil 4
Das Auge und seine Hilfsorgane
(Ergänzung zu Band 3/2) 1964.
227 zum Teil farbige Abbildungen. XII, 662 Seiten
Gebunden DM 490,-
ISBN 3-540-03125-9

Band 4
Nervensystem

Teil 1
Nervengewebe. Das peripherische Nervensystem. Das Zentralnervensystem
Bearbeiter: M. Bielschowsky,
S. T. Bok, R. Greving, A. Jakob,
G. Mingazzini, P. Stöhr,
C. Vogt, O. Vogt
Reprint der Erstauflage
Berlin 1928
1978. 880 zum Teil farbige Abbildungen. (4) X, 1093 Seiten
Gebunden DM 540,-
ISBN 3-540-07846-0

Teil 2
Plexus und Meningen. Saccus vasculosus
1955. 176 zum Teil farbige Abbildungen. VI, 195 Seiten
DM 130,-
ISBN 3-540-01912-X
Einbanddecke DM 19,80
ISBN 3-540-01913-8

Teil 3
Sensible Ganglien
(Ergänzung zu Band 4/1)
1958. 298 zum Teil farbige Abbildungen. VIII, 485 Seiten
Gebunden DM 390,-
ISBN 3-540-02282-1

Teil 4
Das Neuron. Die Nervenzelle. Die Nervenfaser
(Ergänzung zu Band 4/1)
1959. 374 zum Teil farbige Abbildungen. XII, 763 Seiten
Gebunden DM 580,-
ISBN 3-540-02406-9

Teil 5
Mikroskopische Anatomie des vegetativen Nervensystems
(Ergänzung zu Band 4/1)
1957. 501 zum Teil farbige Abbildungen. XII, 678 Seiten
Gebunden DM 500,-ISBN 3-540-02161-2

Teil 7
Hypothalamus
(Ergänzung zu Band 4/1)
1962. 287 zum Teil farbige Abbildungen. XII, 525 Seiten
Gebunden DM 520,–
ISBN 3-540-02837-4

Teil 8
Das Kleinhirn
(Ergänzung zu Band 4/1)
1958. 197 zum Teil farbige Abbildungen. VIII, 323 Seiten
Gebunden DM 260,–
ISBN 3-540-02283-X

Teil 9
Allocortex
Bearbeitet von H. Stephan
1975. 465 zum Teil farbige Abbildungen. X, 998 Seiten
Gebunden DM 680,–
ISBN 3-540-07037-0

Teil 10
Neurologia I
1980. 278 zum Teil farbige Abbildungen in 607 Teilbildern. XIV, 840 Seiten
Gebunden DM 560,–
ISBN 3-540-09430-X

Band 5
Verdauungsapparat Atmungsapparat

Teil 1
Mundhöhle, Speicheldrüsen, Tonsillen, Rachen, Speiseröhre, Serosa
Bearbeiter: T. Hellman, S. Schumacher, E. Seifert, K. W. Zimmermann
Reprint der Erstauflage Berlin 1927
1978. 276 zum Teil farbige Abbildungen. (4) VII, 374 Seiten
Gebunden DM 420,–
ISBN 3-540-07847-9

Teil 2
Magen. Leber. Gallenwege
1932. 254 zum Teil farbige Abbildungen. X, 489 Seiten
DM 340,–
ISBN 3-540-01169-2

Teil 3
Zähne. Darm. Atmungsapparat
1936. 426 zum Teil farbige Abbildungen. XVI, 908 Seiten
Gebunden DM 480,–
ISBN 3-540-01227-3

Teil 4
Die Leber-Gallengangsysteme, Gallenblase und Galle
(Ergänzung zu Band 5/2)
Bearbeitet von J. Wallraff
1969. 183 zum Teil farbige Abbildungen. VII, 384 Seiten
Gebunden DM 290,–
ISBN 3-540-04530-9

Band 6
Blutgefäß- und Lymphgefäßapparat. Innersekretorische Drüsen

Teil 1
Blutgefäße und Herz, Lymphgefäße und lymphatische Organe, Milz
1930. 299 zum großen Teil farbige Abbildungen.
VIII, 584 Seiten
DM 340,–
ISBN 3-540-01118-8

Teil 2
Innersekretorische Drüsen I: Schilddrüse, Epithelkörperchen. Langerhanssche Inseln
1939. 152 zum Teil farbige Abbildungen. VIII, 306 Seiten
DM 190,–
ISBN 3-540-01269-9

Teil 4
Innersekretorische Drüsen III: Thymus, Paraganglien, Epiphyse. Lymphgefäßapparat
(Ergänzung zu Band 6/1)
1943. 236 zum Teil farbige Abbildungen. X, 535 Seiten
DM 360,–
ISBN 3-540-01294-X

Teil 5
Die Nebenniere. Neurosekretion
1954. 336 zum Teil farbige Abbildungen. XVI, 1199 Seiten
Gebunden DM 760,–
ISBN 3-540-01811-5

Teil 6
Die Milz
Bearbeiter: F. Tischendorf
1969. 325 zum Teil farbige Abbildungen. VIII, 968 Seiten
Gebunden DM 640,–
ISBN 3-540-04531-7

Band 7
Harn- und Geschlechtsapparat

Teil 1
Exkretionsapparat und weibliche Genitalorgane
1930. 422 zum großen Teil farbige Abbildungen.
VII, 574 Seiten
DM 360,–
ISBN 3-540-01119-6

Teil 2
Männliche Genitalorgane
1930. 245 zum Teil farbige Abbildungen. VIII, 399 Seiten
DM 340,–
ISBN 3-540-01120-X

Teil 3
Weibliche Genitalorgane. Das Ovarium
(Ergänzung zu Band 7/1)
1957. 120 zum Teil farbige Abbildungen. VI, 178 Seiten
DM 160,–
ISBN 3-540-02162-0
Einbanddecke DM 19,80
ISBN 3-540-02163-9

Teil 4
Tube, Vagina und äußere weibliche Genitalorgane
(Ergänzung zu Band 7/1)
1966. 121 zum Teil farbige Abbildungen. VI, 178 Seiten
Gebunden DM 295,–
ISBN 3-540-03546-X

Teil 5
W. Bargmann
Niere und ableitende Harnwege
1978. 181 zum Teil farbige Abbildungen in 255 Teilbildern, 12 Tabellen. VIII, 444 Seiten
Gebunden DM 320,–
ISBN 3-540-08568-8

Teil 6
G. Aumüller
Prostate Gland and Seminal Vesicles
1979. 142 figures (some in color) in 181 separate illustrations. X, 380 pages
Cloth DM 280,–
ISBN 3-540-09191-2

Springer-Verlag
Berlin
Heidelberg
New York

MIX
Papier aus verantwortungsvollen Quellen
Paper from responsible sources
FSC® C105338

If you have any concerns about our products,
you can contact us on
ProductSafety@springernature.com

In case Publisher is established outside the EU,
the EU authorized representative is:
Springer Nature Customer Service Center GmbH
Europaplatz 3, 69115 Heidelberg, Germany

Printed by Libri Plureos GmbH
in Hamburg, Germany